Acoustical Imaging
Volume 20

Acoustical Imaging

Recent Volumes in This Series:

Acoustical Imaging

Volume 20

Edited by

Yu Wei and Benli Gu

Southeast University
Nanjing, China

SPRINGER SCIENCE+BUSINESS MEDIA, LLC

The Library of Congress cataloged the first volume of this series as follows:

International Symposium on Acoustical Holography.

Acoustical holography; proceedings. v. 1–
New York, Plenum Press, 1967–

v. illus. (part col.), ports. 24 cm.

Editors: 1967–   . A. F. Metherell and L. Larmore (1967 with H. M. A. el-Sum)
Symposium for 1967–   held at the Douglas Advanced Research Laboratories, Huntington Beach, Calif.

1. Acoustic holography — Congresses — Collected works. I. Metherell. Alexander A., ed. II. Larmore, Lewis, ed. III. el-Sum, Hussein Mohammed Amin, ed. IV. Douglas Advanced Research Laboratories, v. Title.
QC244.5.I.5   69-12533

ISBN 978-1-4613-6286-9   ISBN 978-1-4615-2958-3 (eBook)
DOI 10.1007/978-1-4615-2958-3

Proceedings of the 20th International Symposium on Acoustical Imaging,
held September 12–14, 1992, at Southeast University, Nanjing, China

© 1993 Springer Science+Business Media New York
Originally published by Plenum Press, New York in 1993

PREFACE

The 105 theses contained in this book are selected from those whose authors were present at the 20th International Symposium on Acoustical Imaging, held at Southeast University, Nanjing, China, during September 12—14, 1992. It was the first time that the symposium had been held in China.

Our efforts to host the conference goes back to the 15th International Symposium on Acoustical Imaging held in Halifax, Canada, in 1986. We are glad that the 20th symposium has been successfully held at last. We are ardent for the symposium not only because we attach much importance to the field of acoustical imaging, but also because we admire the tradition of the serious academic exploration and friendly cooperation of the scholars attending the symposium.

The theses in this book are from 21 countries and those by Mr. G. Wade, Takuso Sato, J. F. Greenleaf, K. J. Langenberg, and Wencai Yang are the specially invited papers. These theses cover such important fields of acoustical imaging as follows:

1. Mathematics and physics of acoustical imaging;
2. Components and industry application;
3. Applications in medicine and biology;
4. Applications in nondestructive testing;
5. Applications in geophysics;
6. Underwater acoustical imaging.

All these theses reflect the latest progress in theory and technology. We are very grateful to all the authors who have provided these theses.

As compilers, we are grateful to the Honorary Chairman, Prof. G. Wade and Prof. Chongfu Ying. We'd like to thank the 56 members of the International Advisory Board for suggestions and help they gave to determine the specially invited speeches, to evaluate the theses and to organize the symposium. We are grateful to the colleagues in Southeast University for their work. We are especially indebted to Ms. Yuju Sun and the colleagues in the Department of Biomedical Engineering for their diligent work. It was their work that made the successful symposium and the compiling of this book possible.

We would also like to thank the National Natural Science Foundation of China, State Educational Commission of China, Chinese Geophysical Society and K. C. Wang Education Foundation of Hong Kong, which gave financial support to the symposium.

The 21th International Symposium on Acoustical Imaging will be held in Laguna Beach, California, U. S. A. under the Chairmanship of Prof. Jones.

Yu Wei
Benli Gu

Southeast University
Nanjing 210018, P. R. China

INTERNATIONAL ADVISORY BOARD

CONTENTS

MATHEMATICS AND PHSYSICS 1

APPLICATION IN MEDICINE AND BIOLOGY 1

APPLICATION IN MEDICINE AND BIOLOGY　2

SEISMIC IMAGING

NEW WAYS TO USE OLD IDEAS FOR
HIGH QUALITY ACOUSTIC IMAGES

Glen Wade

Department of Electrical and Computer Engineering
University of California
Santa Barbara, California 93106, USA

ABSTRACT

How to obtain images with sound has been the subject of much effort for many years. The modern research started with the work of the French in underwater search during World I. Shortly thereafter many valid ideas were advanced by workers in many nations on how to produce various kinds of acoustic images. The several generic approaches to acoustic imaging can be categorized into one or the other of three key modalities: pulse-echo imaging, intensity mapping, and phase-amplitude techniques. Each of these categories can be combined with the others to produce hybrid instruments, some of which are among the most powerful in the acoustic imaging domain.

INTRODUCTION

How to "see" with sound has intrigued engineers and scientists for years. Bats, whales and dolphins do it with ease. But humans have little natural ability for obtaining mental images with anything but light. However, the history of engineering and science is an impressive demonstration that technological solutions can compensate for the deficiencies of nature.

Modern research on acoustic imaging started with the work of the French in underwater search during World War I [L16]. Shortly thereafter many valid ideas were

Acoustical Imaging, Volume 20 Edited by Y. Wei
and B. Gu, Plenum Press, New York, 1993

advanced by workers in many nations on how to produce various kinds of acoustic images [P37], [S35a], [S35b], [W87]. Most of these early ideas could not be reduced to practical application because of inadequacies in the then existing technology. By now, however, because of technological advances and the emergence of clever new ways of applying the old ideas, acoustic images of unprecedented quality and usefulness are available for human exploitation.

The various generic approaches to acoustic imaging can be categorized into one or the other of three key modalities: intensity mapping, pulse-echo imaging, and phase-amplitude techniques, including holography and tomography [S84]. Each of these categories can be combined with the others to produce hybrid instruments, some of which are among the most powerful in the acoustic imaging domain. And for each there is a myriad of variants to the basic data acquisition and image reconstruction techniques that can be employed.

Although the above three modalities do indeed constitute the conceptual bases for the different types of acoustic imaging, by now systems exist that combine the approaches in such sophisticated ways that unambiguous categorization is difficult or impossible. Among the instruments so far produced are mechanically-scanning focused instruments, chirped pulse-echo instruments, and instruments involving holography, tomography, parametric excitation, phase conjugation, neural networks, random phase transduction, finite element methods, Doppler frequency shifting, pseudo inversion, Bragg diffraction and reflection, and a host of other approaches. The principles may be old but the ways in which they are employed are often very new.

Of what does imaging consist? There are two basic steps involved: the first is to gather the image data by means of detectors; the second, to manipulate these data with computers. The computers can be digital or analog, man-made or natural (that is, produced by nature). The second of the two steps is called image processing. It is all that is needed when we start with an image and wish to convert it into a better image. But most of the time we start with the object and wish to make an image of it. This is the usual case with ultrasonic images. Then, of course, we need both of the above steps: we need detection and computation.

Many of the most useful images produced with sound are two dimensional. However, sometimes the third dimension is desirable and can readily be obtained by holographic techniques. But even one-dimensional acoustical images are no rarity and are highly practical. An example of such an image is the A-scan of elementary pulse-echo sonar.

The three sections which follow deal with the three key modalities indicated above. The first of these sections treats the oldest of these modalities and is entitled "Pulsing Sound and Listening to Echoes." The other two modalities are treated respectively in sections entitled "Mapping the Intensity" and "Mapping the Phase and the Amplitude."

In the subsequent section, entitled "Employing Combinations of the Above," we discuss the technique of combining various modalities to produce hybrid imaging instruments, some of which are among the most powerful and interesting in the acoustic domain. The final section, entitled "Conclusion," contains a brief statement of what we might expect the future to hold based on what has already been accomplished.

PULSING SOUND AND LISTENING TO ECHOES

Can we form mental images by using our ears to acquire the data? Can we "see" with our ears? Would it be possible to go into a completely dark room and by means of clapping our hands and listening to the echoes obtain a mental image of the room and what's in it? We certainly could do the equivalent of this optically, for example with a pulsing flashlight. But it would clearly not be very practical to try to do this with sound rather than light, particularly if we wanted to obtain very much detail. Pulsing sound, I think we can agree, is not very good compared to light for obtaining mental images. Nevertheless sound is better than nothing.

For example, by clapping our hands and listening we can certainly tell whether we are in an anechoic chamber. In an ordinary room we hear reverberations of the sound from the walls, the ceiling, the floor, and the objects in the room. By listening to the echoes with a little skill we can say something about the size and the nature of the room. Reverberations are definitely important in the case of concert halls. Listening to high-quality music in a fine hall is distinctly more pleasurable than listening to it in an anachoic chamber.

A normal human ear can hear sound up to a frequency of about 15 KHz. This corresponds to 30,000 half cycles per second. Approximately 325 differences in volume are detectable by the ear. The data processing rate for the human audio system is therefore approximately $2 \times (\log_2 325) \times 30,000 = 0.5 \times 10^6 \text{ bits / sec}$. Human hearing is not fast at data processing when compared with the human visual system which can data process at a rate a couple of orders of magnitude faster. We therefore rely upon our eyes almost exclusively for producing mental images of the outside world.

Almost, but not quite. Sometimes sound does provide important data for human image processing, assuming a broad definition of the term. For example, imagine that we are offshore in a boat on a dark or foggy night. It is possible to clap our hands or shout sharply and then estimate the time before an echo returns from, say, a nearby cliff, in order to know the existence of the cliff and approximately where it is. Such an approach to data acquisition and processing is a simple version of the elaborate natural technology possessed by bats and porpoises.

A practical implementation of the procedure described has been useful to sailors since the days of the Phoenicians. Under conditions of fog or darkness when the shore or rocks were suspected to be nearby, it has long been the practice for men in boats to make sharp sounds of short duration and listen for echoes. Experienced fishermen are said to be able to detect a buoy in this way at several hundred meters.

I recently mentioned this to Bob Adler. He told me of a Vancouver ferry boat trip in the fog. He noticed the pilot paying more attention to the clearly-heard echoes from a sharp high-pitched whistle than from the radar screen in the pilot's cabin. The pilot produced pulses of sound and listened to their echoes in order to steer the boat through course-marking buoys that could be "seen" with sound, but not with light.

The "acoustic images" described in the preceding paragraphs are obviously one dimensional and are reminiscent of the A-scan of elementary pulse-echo sonar. We can easily make two dimensional images by laying a number of intensity modulated A-scan images of slightly displaced regions side by side.

These kinds of images are highly useful and much appreciated. They are employed in such diverse fields as nondestructive evaluation, medical diagnosis, oceanic search and exploration seismology. The advances in electronics and digital processing techniques over the last few decades have given rise to major improvements in the systems available and have permitted the development of new and significant scanning and processing methods.

One of the principal areas of use for these systems is in medical ultrasonic imaging [E74], [H79] where ultrasonic pulse-echo approaches are of substantial aid to clinical radiologists in complementing imagery from X-ray and nuclear cameras. Many of us have had the experience of seeing such acoustic images of our own internal organs when we ourselves have been the subjects of internal examinations in a hospital or clinic.

MAPPING THE INTENSITY

Intensity mapping is another commonly employed approach for "seeing" with our ears. When I was a boy I lived in the railroad town of Ogden, Utah. My home was on the slopes of a hill to the east of the town. Trains could be heard a long way off coming into town from the north. Many a night after bedtime I can recall being intrigued by the sounds of those trains. When I first would hear a particular train coming from far off the volume would be low, but after a short time it would increase substantially as the train neared the town. Some of the trains did not stop in town and did not seem to slow down very much but appeared to speed along at substantially undiminished velocity on tracks running north and south in the valley below my home. The tracks were situated generally to the west of the large buildings in the town thus putting the buildings between my home and the tracks. As a train would pass behind the buildings, the sound intensity I could hear would be diminished sharply, the sound having been cut off by the buildings.

Where streets intersected the town in such a way that clear avenues existed for the sound to proceed to my house, the intensity would suddenly increase. By the time the train had reached the other side of the town, the intensity again would be high. Without difficulty, I could obtain a crude one-dimensional north-south image of the density of buildings in the downtown region. It was easy for me as a youngster to mentally map the density profile of these buildings by recording in my mind the intensity I was hearing.

Historically, one of the first approaches to make use of intensity mapping was developed more than fifty years ago by Reimar Pohlman [P37]. It was based on a special technique Pohlman invented for making acoustic intensity visible with reflected light. This device, called the Pohlman cell, was the first practical high-resolution acoustical imaging tool that employed intensity mapping. It was described in detail in Volume 15 of this series on **Acoustical Imaging** [W87]. A modern example of intensity mapping that employs light-beam readout is the scanning laser acoustic microscope (SLAM), the subject of a number of papers in various volumes of this series and elsewhere over the last couple of decades [E74], [K72], [W89]. Another well-known, highly useful, modern example is also a microscope, i.e., the scanning acoustic microscope (SAM). This microscope also is the subject of numerous papers in this series and in many research journals [C92], [L74], [Q76].

MAPPING THE PHASE AND THE AMPLITUDE

The third approach is to map both the phase and amplitude. Phase-amplitude systems include holographic instruments for flaw detection and weld inspection. A hologram in its basic form is a record of a beam of coherent waves scattered from an object. The record consists of a pattern of fringes produced by the interference between the object beam and a coherent reference beam. The pattern itself contains information about both the amplitude and the phase of the waves scattered from the object.

After the advent of the laser in the early 1960's, a revolution in optical holography began. The successes achieved by it prompted many researchers to investigate the use of other forms of radiation to produce different kinds of holograms. Acoustic holography thus came into being [A65], [G65], [G75], [W75], [W89].

But before holography was understood or even named as such, a Russian scientist S.J. Sokolov invented a liquid-surface system for detecting flaws in metal test pieces. Sokolov's scheme is remarkably like modern liquid-surface holographic systems and was described in detail in Volume 15 of this series [W87]. As stated in that volume "conceivably...Sokolov can be regarded as having produced the first hologram, however crude, and doing so a number of years before Gabor's invention of holography. It is therefore possible to argue that, from a chronological standpoint, rudimentary acoustic holography preceded optical holography."

In other volumes of the series a number of other holographic systems have been described including point scanning systems [A65]. Other related systems in which phase and amplitude can be regarded as being mapped would include Sokolov's Debye-Sears diffraction system [S35b], Bragg diffraction systems [K66], and acoustical cameras employing scanning laser beams for reading out the phase and amplitude information associated with acoustical waves impinging upon solid surfaces. These latter systems can operate in real time and constitute what is sometimes referred to as "solid-surface" holography. Such a system, known as the scanning laser acoustic microscope (SLAM), has been profitably packaged by Sonoscan and was the first instrument commercially available for acoustic microscopy [K72].

EMPLOYING COMBINATIONS OF THE ABOVE

So far I've talked about relatively old ideas and how they are used in new systems. What about the new "ways" to use the old ideas? Pulse echo imaging, intensity mapping, and phase amplitude techniques do indeed constitute the conceptual bases for different types of acoustical imaging, but by now, systems exist that combine these approaches in such sophisticated ways that unambiguous categorization is difficult or impossible.

As previously stated, combining the approaches has brought into being mechanically-scanning focused systems, chirped pulse-echo systems, and systems involving holography, tomography, parametric excitation, phase conjugation, neural networks, random phase transduction, finite elements methods, Doppler frequency shifting, pseudo-inversion, Bragg diffraction and reflection, and a host of other systems. Thus by employing a combination of the three basic modalities, a would-be inventor can come up with imaging instruments that are difficult to conceive in any other way. Some of these hybrid systems are very powerful indeed.

An example of such a combination is the holographic version of SLAM, whose commercial name is HoloSLAM [W89]. This instrument is simply a scanning laser acoustic microscope in which phase information is recorded as well as amplitude information. Thus a holographic image is available rather than just an intensity image. One can take an additional step and employ tomographic principles also. When this is done we have the scanning tomographic acoustic microscope (STAM), an instrument previously treated in this series of volumes and a good example of a new way to employ old ideas in order to produce more powerful acoustical imaging [L85], [W87].

CONCLUSION

In spite of the progress described above, there remain many difficult questions to be resolved before acoustic imaging can be applied to many remaining practical problems. From what has already been accomplished, it can reasonably be expected that modified versions of the old systems will be put together in new and effective ways. We can expect also that new systems, still in the research stage, will employ various of the old ideas and thereby be enabled to emerge into important and viable use.

REFERENCES

A65   E.E. Aldridge, A.B. Clare, and D.A. Shepherd, "Ultrasonic Holography in Non-destructive Testing," _Acoustical Holography_, Vol. 3, Chap. 7, A.F. Metherell, Ed., New York: Plenum Press, 1971.

C 92   N. Chubachi, H. Kanai, T. Sannomiyac, and T. Wakahara, "Acoustic Microscope for Measuring Acoustic Properties by Micro-Defocusing Method," _Acoustical Imaging_, Vol. 19, pp. 685-689, H. Ermert and H.P. Harjes, Eds., New York: Plenum Press, 1992.

E74   K.R. Erickson, F.J. Fry, and L.P. Jones, "Ultrasound in Medicine—A Review," _IEEE Trans. on Sonics and Ultrasonics_, Vol. SU-21, No. 3, pp. 144-170, July 1974.

G65   P. Greguss, "Ultrasonic Holograms," _Res. Film_, Vol. 5:4, pp. 330-337, 1965.

G75   J.F. Greenleaf, S.A. Johnson, W.F. Samayoa, and F.A. Duck, "Algebraic Reconstruction of Spatial Distributions of Acoustic Velocities in Tissue from their Time-of-flight Profiles," _Acoustical Holography_, Vol. 6, pp. 71-90, N. Booth, Ed., New York: Plenum Press, 1975.

H79   J.F. Havlice and J.C. Taenzer, "Medical Ultrasonic Imaging: An Overview of Principles and Instrumentation," _Proc. IEEE_, Vol. 67, pp. 620-641, April 1979.

K66   A. Korpel, "Visualization of the Cross-Section of a Sound Beam by Bragg-Diffraction of Light," _Appl. Phys. Lett._, Vol. 9, p. 425, 1966.

K72   R.K. Kessler, P.R. Palermo, and A. Korpel, "Practical High-Resolution Acoustic Microscopy," _Acoustical Holography_, Vol. 4, pp. 51-71, G. Wade, Ed., New York: Plenum Press, 1972.

L16   P. Langevin and M.C. Chilowski, "Procédés et Appareils Pour la Production de Signaux Sous-marins Diregés et Pour la Localisation à Distance d'Obstacles Sous-marins," French patent 502913, 1916.

L74   R.A. Lemons and C.F. Quate, "Acoustic Microscope-scanning Version," _Appl. Phys. Lett._, Vol. 24, No. 4, pp. 163-165, February 1974.

L85   Z. Lin, H. Lee, and G. Wade, "Scanning Tomographic Acoustic Microscope: A Review," Vol. SU-32, pp. 168-188, March 1985.

P37   R. Pohlman, "Über Die Richtende Wirkung des Schallfeldes auf Suspensionen Nicht Kugelformiger Teilchen," _Zeitschrift für Physik_, Vol. 107, pp. 497-507, 1937.

Q76   C.F. Quate, "Imaging Using Lenses," _Acoustic Imaging_, pp. 241-303, G. Wade, Ed., New York: Plenum Press, 1976.

S35a   S.J. Sokolov, "Ultrasonic Oscillations and Their Applications," _Tech. Phys. USSR._, Vol. 2, p. 522, 1935.

S35b   S.J. Sokolov, "Über Die Praktishche Ausnutzung der Beugnung des Lichtes an Ultraschällwellen," _Phys. Z._, Vol. 36, p. 142, 1935.

S49   S.J. Sokolov, "Ultrasonic Microscope," <u>Akademia Nauk USSR.</u>, Doklady (Tekhnicheskaya Fizika), Vol. 64, p. 333-335, 1949.

S84   C.F. Schueler, H. Lee, and G. Wade, "Fundamentals of Digital Ultrasonic Imaging," <u>IEEE Trans. on Sonics and Ultrasonics</u>, Vol. SU-31, pp. 195-217, 1984.

W75   G. Wade, "Acoustical Imaging with Holography and Lenses," <u>IEEE Trans. on Sonics and Ultrasonics</u>," SU-22, Vol. 6, 1975.

W87   G. Wade, "A History of Acoustical Imaging," <u>Acoustical Imaging</u>, Vol. 15, pp. 1-28, H.W. Jones, Ed., New York: Plenum Press, 1987.

W89   A.C. Wey and L.W. Kessler, "Holographic Scanning Laser Acoustic Microscopy and Applications," <u>Acoustical Imaging</u>, Vol. 18, pp. 247-253, H. Lee and G. Wade, Eds., New York: Plenum Press, 1989.

IMAGING OF ACOUSTICAL NONLINEAR PARAMETERS AND ITS MEDICAL AND INDUSTRIAL APPLICATIONS: A VIEWPOINT AS GENERALIZED PERCUSSION

Takuso Sato, K. Yamashita, H. Ninoyu, K. Jhang and Y. Kosugi

The Graduate School at Nagatsuta
Tokyo Institute of Technology
4259 Nagatsuta, Midori-ku, Yokohama-shi 227, Japan

1. Introduction

Generally speaking, acoustical nonlinear parameters may be defined as the degrees of changes of linear acoustical parameters, which are derived under the assumption of infinitesimally small amplitudes, for respective finite amplitude perturbations.

These nonlinear parameters are related, for instance, to the faint changes of tissue's conditions at early stage of diseases or the value of residual stress or experience of over elastic deformation in metals which are observed far before the generation of cracks or flows.

In this paper, first, the process of giving perturbation and detecting the response is considered as a generalization of doctor's percussion and possible schemes of these generalized percussions are discussed. Then details of two systems practically constructed along this line are shown. They are new ultrasonic imaging system for tissue characterization aiming advanced medical diagnoses and nondestructive testing system to characterize metal's condition.

2. Generalized Percussions

Doctor's percussion is shown schematically in Fig. 1. Doctor gives knocking by fingers and listen to the response by using a stethoscope.

A scheme of ultrasonic version of the percussion may be as is shown in Fig. 2.

Acoustical Imaging, Volume 20 Edited by Y. Wei
and B. Gu, Plenum Press, New York, 1993

Here, knocking process is replaced by proper perturbations and the listening process is done by the ultrasonic detection followed by proper image formation.

Contrary to the case of linear imaging system where infinitesimally small amplitude is assumed and only first order interaction between waves and medium or static properties of the object are observed, this new scheme can detect 2nd order nonlinear interaction or dynamic characteristics.

Many kinds of perturbations and corresponding processings can be considered.

Percussions, detections and observed quantities in our concretely constructed systems are summarized in Table 1.

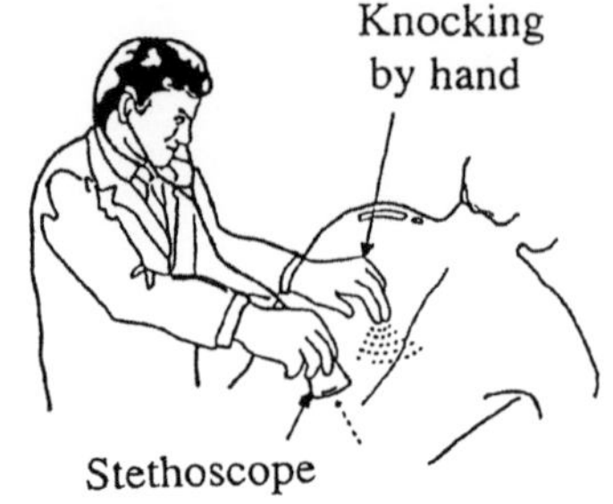

Fig. 1 Doctor's Percussion

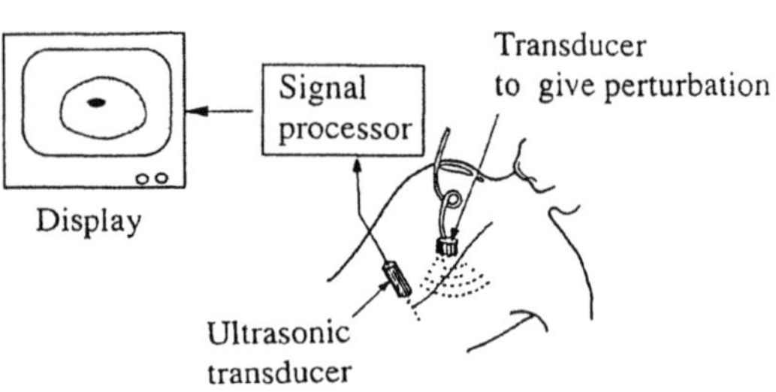

Fig. 2 Generalized Percussion Using Ultrasonic Waves

Table 1. Generalized Percussion Applied Measuring Systems

Subject	Percussion	Detection	Observed Quantity
Tissue Characterization (Human Body)	Static Pressure Δp_i	Change of Sound Velocity	Average of Nonlinear Parameter B/A or N
	Dynamic Pressure $\Delta p(t)$ ($\approx$1 atm)	Dynamic Change of Sound Velociy as Phase change (5 MHz, 1 deg)	Distribution of Nonlinear Parameter B/A or N
		Change of Reflected Wave Form	Movability M or Distortability D
	Low Frequency Vibration ($\approx$100 Hz)	Velocity and Distortion of Vibration ($\approx$500 cm/sec)	Hardness H, Nonlinearity N
Material Characterization (Metal)	Dynamic Stress $\Delta P(t)$ ($\approx$1 Mpa)	Dynamic Change of Sound Velocity as Phase Change (50 MHz, $\approx$1 deg)	Residure Stress, Limiting Stress for Elastic Deformation ($\approx$100 Mpa)
Gas Dynamics (Air)	Dynamic Pressure $\Delta p(t)$ ($\approx$0.5 atm)	Dynamic Change of Sound Velocity (40 KHz, $\approx$10 deg)	Velocity Distribution, Temperature Distribution

Another main special feature of the generalized percussion is the easiness of the image formation of the degree of interactions, because the region of perturbation can be localized and scanned as desired.

As further generalizations of percussions several other means can also be considered. For instance, temperature change by using focused high power ultrasonic waves and concentration of donors at the desired part in human body may be such a new direction.

Two practical systems constructed for tissue characterization aiming medical diagnoses and for material characterization aiming nondestructive metal's condition estimation are shown in the followings.

3. Simultaneous Imaging of Nonlinearity, Movability and Reflectivity of Soft Tissue

3.1 *System construction*

System construction is shown schematically in Fig. 3. Two coaxial transducers are used as in the previous system [1], [2]. The outer transducer (70 mm diameter) generates burst (2-3 waves) pumping waves of 350 KHz and gives about one atm of pressure variation at the observing region in the object (for instance breast or liver), while the inner transducer (30 mm diameter) is used for the generation of burst (about 5 waves) probing waves of 5 MHz.

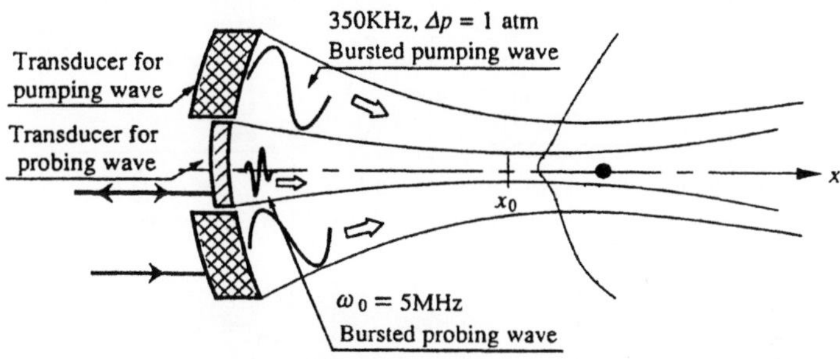

Fig. 3 Reflection-Type Ultrasonic Imaging System

The later waves are super imposed on the positive pressure part of the former waves at step I and on the negative pressure part at step II. The phase shift due to the pressure variation by the pumping waves reflected at the distance x from the transducer is estimated as the difference of the detected phases of the reflected waves in the above two steps.

It can be expressed as follows:

$$\Delta\varphi(x) = \Delta\varphi_+(x) - \Delta\varphi_-(x)$$

$$= \omega_0 \int_{x_0}^{x} N'(\xi)\left\{\Delta p_+(\xi) + \Delta p_-(\xi)\right\} d\xi$$

$$+ 2\cdot\frac{\omega_0}{c}M\left\{\Delta p_+(x) - \Delta p_-(x)\right\}$$

$$= \Delta\varphi_N(x) + \Delta\varphi_M(x)$$

where ρ_0 : density, c_0 : sound velocity under static pressure, ω_0 : angular frequency of the probing wave and N' is the equivalent phase shift parameter which is linearly related to the conventional nonlinear parameter B/A and M is the effective movability of the soft tissue. Formally N' and M are defined as follows

$$N' = - \left(\frac{\partial c^{-1}}{\partial p} \right)_s$$

and

$$M = - \left(\frac{\partial x}{\partial p} \right)$$

Flow chart of phase shift data processing to separate the nonlinearity term $\Delta\varphi_N$ and the movability term $\Delta\varphi_M$ to get N, M images as well as intensity image I is shown in Fig. 4 with examples of processed data.

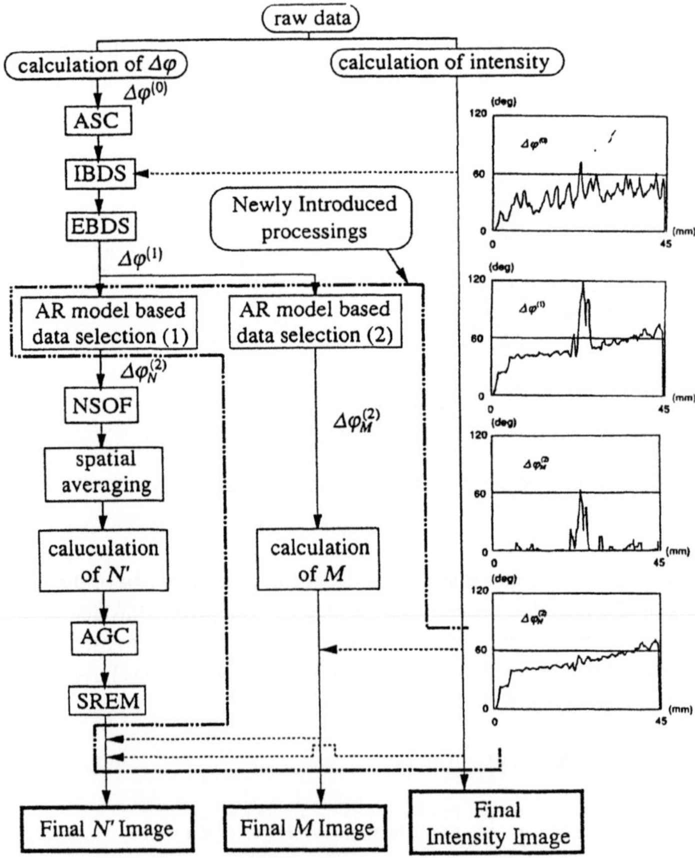

Fig. 4 Data Processings to get N', M and I Images

Special data processings IBDS (Intensity Based Data Selection), ASC(Adaptive Sensitivity Correction), EBDS (Energy Based Data Selection), NSOF (Non-Symmetric Order Statistics Filtering) and SREM(Sequential Region Extension Method) have been explained at the previous symposium[2].

Concretely, from the preprocessed phase data $\Delta\varphi_N^{(2)}(x_{i-m})$, ... $\Delta\varphi_N^{(2)}(x_{i-1})$ the gradient a_i and bias b_i are derived by mean of 1st order AR model and $\Delta\varphi^{(2)}_{est}(x_i)$ is obtained, then the upper and lower bounds $\Delta\varphi_{UB}$, $\Delta\varphi_{LB}$ are estimated.

If the data $\Delta\varphi^{(1)}(x_i)$ is between the bounds $\Delta\varphi^{(1)}(x_i)$ is used as $\Delta\varphi^{(2)}(x_i)$, otherwise $\Delta\varphi^{(2)}_{est}(x_i)$ is used. The derivative of thus obtained $\Delta\varphi_N^{(2)}$ gives N' image.

On the other hand, if $\Delta\varphi^{(1)}(x_i)$ is outside of the bounds the difference from the bounds, that is, $\Delta\varphi^{(1)}(x_i) - \Delta\varphi^{(2)}_{UB}(x_i)$ or $\Delta\varphi_{LB}(x_i) - \Delta\varphi^{(1)}(x_i)$ are considered due to the movement of the part of tissue or change of structure of the scatterers by the pumping waves and the equivalent movability image is derived using the definition of M.

3.2 *Results*

An example of images of a phantom with different N' parts is shown in Fig. 5. We can see the existence of the part with different N' most clearly in N' image in the case as expected. Second example is images of a tissue sample as is shown in Fig. 6. Artificially given knife cut is seen clearly in M image.

Thus, simultaneous display of these three kinds of images I.M.N' seems to be effective for more comprehensive diagnoses, since N' is related to the change at molecular level such as the change of ratio of free and bound waters, M is related to elastic properties such as the degree of fibrousation of tissues, while I is related to the change of density.

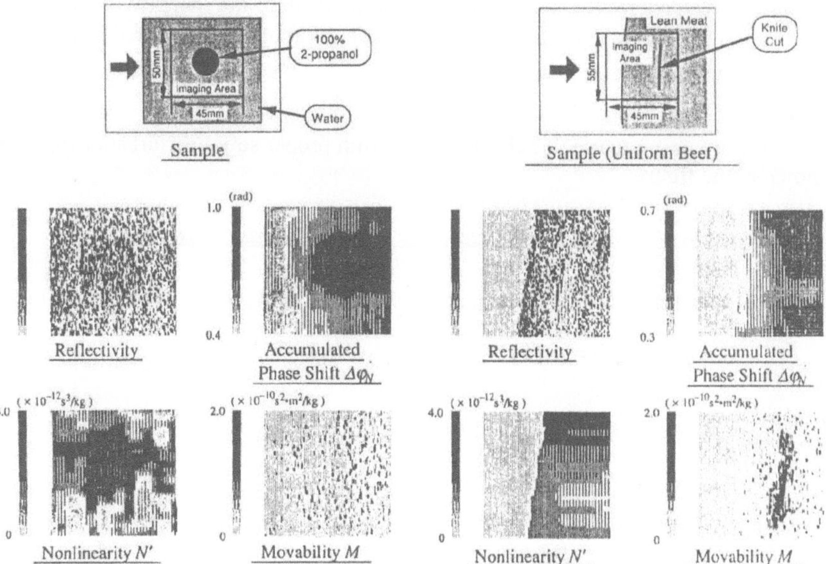

Fig. 5 Images of Phantom

Fig. 6 Images of Tissue Sample

4. Estimation of Stress Condition Inside of Metal using Nonlinear Acoustics

4.1 *Introduction*

Nondestructive estimation of the stress condition in metal is required in many field. The conditions may be expressed as the existence of residual stress from the manufacturing processes such as rolling and welding processes, or as the magnitude of the stress value to give the limit of the elastic deformation.

In this chapter, first the nonlinear dependency of sound velocity on stress in Al is observed and its expression by using high order elastic constants is given.

Then, a concrete measuring system which uses pumping waves to generate stress perturbation and ultrasonic probing waves to detect the phase change due to the pumping wave is constructed. Finally, it is applied for nondestructive detection of the state of metal such as the elastic deformation limit inside of Al alloy.

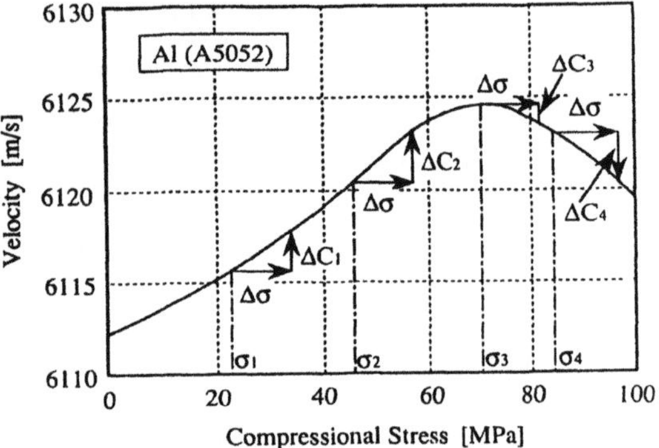

Fig. 7 Observed Result of Stress and Sound Speed Relation in Al

4.2 *Measurement of Nonlinear Dependence of Ultrasonic Sound Velocity on Stress*

A result of the sound velocity change measurement under applied stress is as shown in Fig. 7. We can see the monotone increase of ΔC with the increase of compressional stress up to the elastic deformation limit, which is estimated by other method as the value where the unrecoverable strain reaches 2.0×10^{-4}. After the limit value ΔC changes to decrease.

We are going to use this characteristics with proper stress perturbation $\Delta \sigma$ as is also shown in the figure.

4.3 *Expression of Nonlinear Dependence by using high order Elastic Constants*

Let us assume that the relation between stress T and strain S is given as follows

$$S_I = f_{IJ}T_J + g_{IJK}T_JT_K + h_{IJKL}T_JT_KT_L$$

where

$$\begin{array}{ccc} T_1 = T_{xx} & T_2 = T_{yy} & T_3 = T_{zz} \\ T_4 = T_{yz} & T_5 = T_{xz} & T_6 = T_{xy} \end{array}$$

and f, g and h are 1st, 2nd and 3rd order elastic constants, respectively.

Now the probing ultrasonic waves are passed through the material as is shown in Fig. 8.

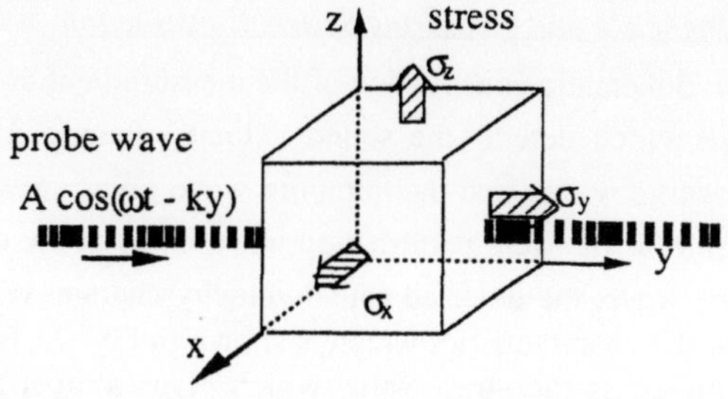

Fig. 8 Probing Ultrasonic
Wave and Stress

Then $S_2 = S_{yy}$ is derived as follows

$$S_2 = A\cos(\omega t - ky) \times [\ \{(f_{22} - \alpha f_{21}) + 2(g_{212} - \alpha g_{211})(\sigma_x + \sigma_z)$$
$$+ 3(h_{2112} - h_{2111})(\sigma_x + \sigma_z)^2\}$$
$$+ \{2(g_{222} - \alpha g_{212}) + 6(h_{2122} - \alpha h_{2112})\ (\sigma_x + \sigma_z)\}\sigma_y$$
$$+ 3(h_{2222} - \alpha h_{2122})\ \sigma_y^2\ ]$$

where

$$S_2 = S_{yy}, \quad \alpha = \frac{2f_{21}}{f_{22} + f_{21}}$$

On the other hand from the equation of motion we obtain

$$S_2 = \frac{k(x,y,z,t)^2}{\rho \omega^2}\ A(x,y,z)\cos\ \{\omega t - k(x,y,z,t)y\}$$

Thus from (2) and (5) we obtain

$$C(x,y,z,t) = \frac{\omega}{k(x,y,z,t)}$$

$$= \Big[\ \rho\{(f_{22} - \alpha f_{21}) + 2(g_{212} - \alpha g_{211})(\sigma_x + \sigma_z)$$
$$+ 3(h_{2112} - \alpha h_{2111})(\sigma_x + \sigma_z)^2$$
$$+ 2(g_{222} - \alpha g_{212})\sigma_y + 6(h_{2122} - \alpha h_{2112})(\sigma_x + \sigma_z)\sigma_y$$
$$+ 3(h_{2222} - \alpha h_{2122})\sigma_y^2\}\Big]^{-1/2}$$

or

$$C = \Big[\rho\frac{(1 + \nu)(1 - 2\nu)}{(1 - \nu)}\ [f_{22} + 2g_{222}\{\sigma_y - \nu(\sigma_x + \sigma_z)\}$$

$$+ 3h_{2222}\{\sigma_y - \nu(\sigma_x + \sigma_z)\}^2]\Big]^{-1/2}$$

where ν is Poisson ratio.

Schematic construction of the measurement system is shown in Fig. 9. The probe wave which detects the sound velocity change ΔC as its phase change is 50 MHz ultrasonic waves and the pumping wave gives stress perturbation $\Delta\sigma$ of Fig. 7. If the magnitude of stress perturbation wave is $\Delta\sigma$ at the crossing region of the wave and the probe wave, the detected sound velocity change ΔC is stress dependent as can be seen from the characteristics which is shown in Fig. 7. For example the elastic limit may be estimated as the stress value which gives almost zero sound velocity change or the boundary where the sign of ΔC is changed from plus to minus for the stress perturbation $\Delta\sigma$. (in the example at stress of about 70 MPa).

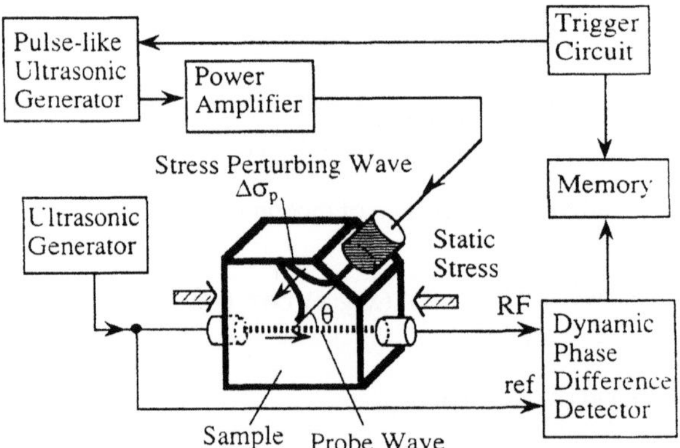

Fig. 9 Schematic Construction of Stress State Measuring System

Actually the constructed system can give about 0.3 MPa of stress perturbation with duration of 10μs at the depth of 20 mm from the surface and the dynamic phase of the ultrasonic wave can be detected with the accuracy of 0.05 deg.

The phase change $\Delta\phi$ is given by

$$\Delta\phi(t) \approx \rho\omega \frac{(1+\nu)(1-2\nu)}{(1-\nu)} \frac{C_{ref}^2}{C_{RF}} \int_{L-C_{RF}\cdot t}^{L} \Delta\sigma_p \, dy$$

$$\times (\cos\theta - \nu\sin\theta)[g_{222} + 3h_{2222}\{\sigma_y - \nu(\sigma_x + \sigma_z)\}]$$

where $\Delta\sigma_p$ is the stress perturbation, C_{ref} and C_{RF} are sound velocities without and with pumping waves, respectively.

4.5 *Characterization of the Metal's Condition*

The phase change $\Delta\phi$ is measured under the following three conditions.

 i) without stress ($\sigma=0$) : $\Delta\phi_0$

 ii) with any unknown stress ($\sigma=\sigma$) : $\Delta\phi_a$

iii) with the unknown stress and additional small stress ($\sigma=\sigma+\delta\sigma$) : $\Delta\phi_b$

Then the following value η is derived.

$$\eta\,(\sigma) = \frac{(\Delta\phi_b - \Delta\phi_a)}{\Delta\phi_0 \times \delta\sigma}$$

If we use the relation (6) it is reduced to

$$\eta\,(\sigma) = \frac{3h_{2222}}{g_{222}}$$

This is used as the characteristic value of the metal in this paper.

4.6 *Experimental Results and Discussions*

An example of observed maximum value of the phase change of the probing wave for the applied stress perturbation is shown in Fig. 10 (a).

On the other hand, Fig. 10 (b) shows the characteristic value η which are derived from the stress and sound velocity curve shown in Fig. 7.

From these results we may insist the following points.

i) The phase modulation $\Delta\phi$ shows change of its sign around the elastic deformation limit (in this case about 70 MPa). In other words nondestructive estimation of the elastic deformation at desired region given as the crossing region of probing and pumping waves inside metal is possible by the proposed method.

ii) The characteristic value η shows fairly large value in the region where plastic deformation is progressing (above 60 MPa in this case). In other words, the characteristic value η will be a proper measure to estimate the generation of plastic deformation.

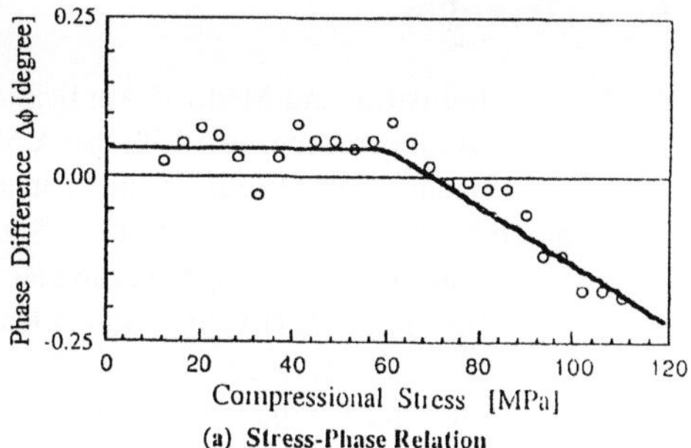

(a) Stress-Phase Relation

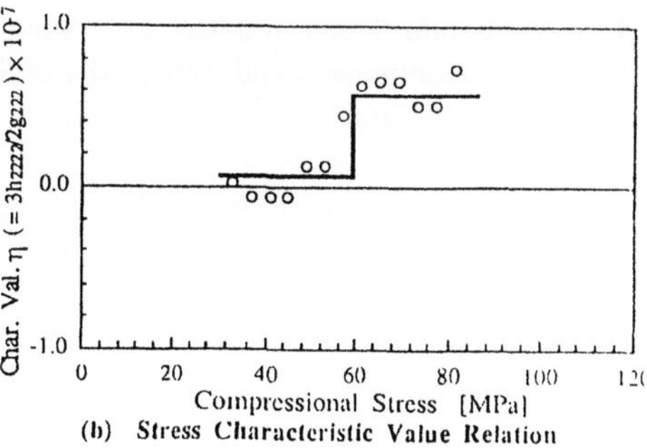

(b) Stress Characteristic Value Relation

Fig. 10 Observed Relation of Phase Shift and Characteristic Value for Stress

iii) The phase change $\Delta\phi$ and stress seems to have one to one correspondence in proper stress region (in the example from 50 MPa to 100 MPa). In other words, these is a possibility to estimate the stress value at any desired region in the metal (which is given as the crossing region of the probing and pumping waves) from the observed phase change.

In conclusion, these kinds of techniques seem to be very promising because of the possibility of getting new information about the object and easiness of its imaging.

5. Acknowledgements

The author would like to express his sincere thanks to the symposium chairman Professor Yu Wei, general secretary Professor Benli Gu and the member of the committee of this symposium for giving me the opportunity to present the works of our group at Tokyo Institute of Technology as an invited talk.
The content of chapter 3 is from reference 4 and that of chapter 4 is from reference 5, respectively.

6. References

1. T.Sato, "Industrial and Medical Application of Nonlinear Acoustics", Frontiers of Nonlinear Acoustics, 12th ISNA M.F.Hamilton Ed. Elsevier Applied Science, P96 (1990) (Invited paper).
2. T.Sato, E.Mori, et al. "A Few Effective Signal Processings for Reflection-Type Imaging of Nonlinear Parameter N of Soft Tissues" Acoustical Imaging, Vol.19, p363, H.Ermert and H.Harjes Ed. Plenum Press (1992).
3. W.Ma, T.Sato, et al. "Application of Nonlinear Dependence of Sound Velocity on Stress for Estimation of Elastic Deformation Limit Inside of Metal" Nondestructive Testing, Vol.41, No.7, p414 (1992) (in Japanese)
4. K.Yamashita, T.Sato and K.Jhang, "Simultaneous Imaging of Movability, Nonlinearity and Reflectivity of Soft Tissue" 17th International Symposium on Ultrasonic Imaging and Tissue Characterization, Arlington (June, 1992)
5. H.Ninoyu, T.Sato, K.Jhang and Y.Kosugi, "Nonlinear Elasto-Acoustics Applied Measuring System", 14th International Congress on Acoustics, Beijin (Sept., 1992)

PHONON IMAGING IN SINGLE AND MULTILAYERED

ANISOTROPIC SOLIDS

A.G. Every

Department of Physics
University of the Witwatersrand
P O WITS 2050, South Africa

INTRODUCTION

Recent developments in phonon imaging have increased its value as a tool for studying phonon reflection and transmission at surfaces and have extended its range into the ultrasonic frequency domain. In the process a number of interesting new phenomena have been observed, including reflection phonon imaging [1], refocusing and symmetry doubling in reflection [2], pseudo–surface wave structures in phonon imaging [3], total internal reflection of thermal phonons [4], phonon focusing of ultrasonic waves [5] and internal diffraction of ultrasound in crystals [6]. All of these investigations have been conducted on parallel plate samples and in most of them reflection or refraction at plane interfaces plays a central role.

In this paper we present a formulation of phonon focusing which is tailored specifically to plate geometries and which is readily generalised to multilayered solids. We establish a relation which determines the position dependence of the phonon intensity transmitted, via any mode sequence which may include back reflections, through a solid consisting of any number of parallel anisotropic layers. The layer thicknesses are assumed to be large compared with the dominant phonon wavelength and coherence effects that arise in superlattices [7] are not considered. An important attribute of phonon intensity patterns are the caustics that they contain. These are the result of bulk focusing, and are associated with lines of zero Gaussian curvature in the acoustic slowness surface. Phonon ray vectors are normal to the slowness surface and so the energy flux is proportional to the curvature of this surface. Variations in phonon intensity also arise from surface directivity effects, but not to the extent of producing singularities. The main emphasis of this paper is on bulk focusing, but in the example on Si we provide, we take account of selective mode conversion in reflection.

Our discussion will be couched in terms of ballistic phonon imaging with incoherent thermal phonons having frequencies in the region of 1 THz. Experiments of this type are usually carried out on crystalline samples at liquid helium temperatures (in order to obtain macroscopic phonon mean free paths) and employ laser or electron beam heating to generate the phonons and bolometric or tunnel junction techniques to detect them. Recently Every et al. [5] have shown that the same focusing effects can be observed at MHz ultrasonic frequencies using laser thermoacoustic generation and piezoelectric detection. In a modification of this technique using tone bursts and acoustic lenses to obtain improved spatial resolution, Weaver et al. [6] have observed diffraction patterns clothing phonon focusing caustics. The possibility of phonon imaging with ultrasonic waves had been earlier suggested on theoretical grounds by Novikov and Chernozatonskii [8].

THEORY: ANISOTROPIC PLATE

Figure 1 depicts the geometrical conditions pertaining to the transmission of ballistic phonons through an elastically anisotropic plate of thickness h. The phonons emanate from a small heated region of area a located on the lower surface, and after travelling at group velocity $\mathbf{V}$ through the sample, are detected at the point $\mathbf{R} = (X,Y,h)$ on the upper surface. Assuming perfectly diffuse radiation conditions, the incident phonon power at $\mathbf{R}$ is a sum of contributions [9,10]

$$P = \frac{a}{h^2} \frac{\pi}{120} \frac{k^4}{\hbar^3} \frac{T^4}{v^3} \frac{V\cos^4\theta}{} \frac{d\Omega_s}{d\Omega_v}, \tag{1}$$

for each point on the wave surface intersected by the source–detection direction. Because of the folding of the wave surface of an anisotropic solid, an individual acoustic branch may yield more than one ray pointing in the source–detection direction. In Eq. (1) k is Boltzmann's constant, $\hbar$ is Planck's constant, T is the source temperature, v is the phase velocity, $\cos^3\theta/h^2$ represents the solid angle per unit area subtended by an element of the detection surface at the source and there is a further factor of $\cos\theta$ for the projection of the source area in the propagation direction (Lambert's law).

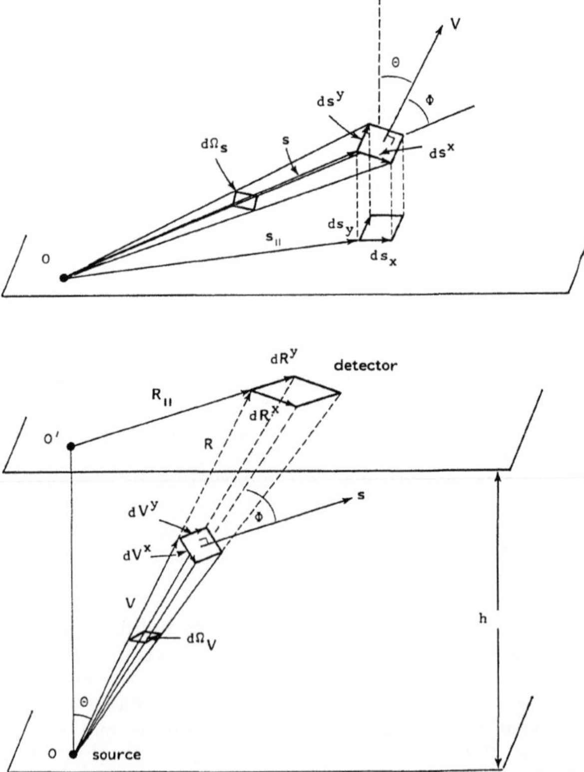

Fig. 1   Geometrical conditions pertaining to phonon transmission through an anisotropic plate.

The effects of phonon focusing are contained in the Maris phonon enhancement factor [11]

$$A = \frac{d\Omega_s}{d\Omega_v} , \qquad (2)$$

where $d\Omega_s$ is the solid angle element in slowness space that is mapped onto the solid angle element $d\Omega_v$ in group velocity space. The solid angle subtended by the element $ds^x \times ds^y$ of the slowness surface (see Fig. 1) is

$$d\Omega_s = |ds^x \times ds^y| \cos\phi/s^2 = ds_x ds_y \cos\phi/s^2\cos\theta , \qquad (3)$$

where ϕ is the angle between $\mathbf{V}$, which is normal to the slowness surface, and $\mathbf{s}$ and the factor $\cos\theta$ arises from the projection of $ds^x \times ds^y$ onto $ds_x ds_y$ in the $\mathbf{s}_{,,} = (s_x, s_y)$ plane (θ is the angle that $\mathbf{V}$ makes with the z–axis). Since $\mathbf{V}$ and $\mathbf{R}$ are parallel, the solid angle subtended at the origin by the element $d\mathbf{V}^x \times d\mathbf{V}^y$ of the wave surface is the same as that subtended by the physical surface element $d\mathbf{R}^x \times d\mathbf{R}^y$, i.e.

$$d\Omega_v = |d\mathbf{R}^x \times d\mathbf{R}^y| \cos\theta/R^2 = |d\mathbf{R}^x \times d\mathbf{R}^y| \cos^3\theta/h^2 . \qquad (4)$$

From Eqs. (1), (3) and (4) and the fact that $V\cos\phi = v = 1/s$ one obtains

$$P = \frac{a\pi}{120} \frac{k^4}{\hbar^3} \frac{T^4}{|J|} , \qquad (5)$$

where

$$J = \frac{|d\mathbf{R}^x \times d\mathbf{R}^y|}{ds_x ds_y} , \qquad (6)$$

is the Jacobian of the mapping $\mathbf{s}_{//} = (s_x, s_y) \longrightarrow \mathbf{R}_{//} = (X, Y)$.

Since $\mathbf{R} = (X, Y, h)$ is parallel to $\mathbf{V}$, which in turn is normal to the slowness surface, whose equation may be expressed in the form

$$s_z = s_z(s_x, s_y) , \qquad (7)$$

it follows that

$$\mathbf{R}_{//} = (X, Y) = -h \left[\frac{\partial s_z}{\partial s_x} , \frac{\partial s_z}{\partial s_y} \right] . \qquad (8)$$

For transmission in the reverse direction through the plate the same expression applies, but without the negative sign. From Eq. (8) it follows that the Jacobian J is given by

$$J = \frac{\partial(X,Y)}{\partial(s_x, s_y)} = \frac{\partial X}{\partial s_x} \frac{\partial Y}{\partial s_y} - \frac{\partial Y}{\partial s_x} \frac{\partial X}{\partial s_y}$$

$$= h^2 \left\{ \frac{\partial^2 s_z}{\partial s_x^2} \frac{\partial^2 s_z}{\partial s_y^2} - \left[\frac{\partial^2 s_z}{\partial s_x \partial s_y} \right]^2 \right\} . \qquad (9)$$

THEORY: MULTILAYERED SOLID

The above results are readily extended to transmission through a multilayered solid. Consider a path through a mutilayered sample consisting of path segments $i = 1, 2, \ldots$, each associated with a particular mode and located in a layer of thickness h_i, which we take to be positive whether propagation for that path segment is in the forward or the reverse direction. At each surface, on reflection or transmission there is conservation of $\mathbf{s}_{//}$, and the lateral displacements accompanying transit through each layer, whether in the forward or reverse direction, are additive. The resultant lateral displacement is therefore

$$\mathbf{R}_{//}^{T} = -h\left[\frac{\partial s_{z}^{T}}{\partial s_{x}}, \frac{\partial s_{z}^{T}}{\partial s_{y}}\right], \tag{10}$$

where

$$s_{z}^{T}(\mathbf{s}_{//}) = \sum_{i} \alpha_{i}.(\pm s_{z}^{i}(\mathbf{s}_{//})), \tag{11}$$

$h = \Sigma h_{i}$ and the $\alpha_{i} = h_{i}/h$ are weighting coefficients for the individual path segments. The positive sign is taken for transit through a layer in the forward direction and negative for transit in the reverse direction. The Jacobian for this path sequence is in turn given by

$$J = h^{2}.\left\{\frac{\partial^{2}s_{z}^{T}}{\partial s_{x}^{2}}\frac{\partial^{2}s_{z}^{T}}{\partial s_{y}^{2}} - \left[\frac{\partial^{2}s_{z}^{T}}{\partial s_{x}\partial s_{y}}\right]^{2}\right\}. \tag{12}$$

We see from Eqs. (10)–(12) that the focusing pattern for the path sequence is determined by an effective slowness surface whose equation (11) is the weighted mean of the equations of the slowness surfaces of the individual layers.

$\mathbf{R}^{T}$ and J are only defined in the domain of the $\mathbf{s}_{//}$ plane in which s_{z}^{i} is real for all path segments i. As $\mathbf{s}_{//}$ approaches the boundary of this domain the critical condition for one or possibly more of the path segments is approached and $\mathbf{R}_{//}$ for that segment diverges and hence so does $\mathbf{R}_{//}^{T}$. Beyond the boundary the wave associated with that path segment is evanescent and does not transmit energy through that layer.

ILLUSTRATIVE EXAMPLE

Figure 2 demonstrates how a phonon focusing pattern can be modified by reflection. It shows a set of Monte Carlo calculated multipass phonon focusing patterns for a <100>–oriented silicon single crystal obtained from a uniform distribution of approximately $2{\times}10^{4}$ values of $\mathbf{s}_{//}$, for various mode sequences involving longitudinal (L) and slow transverse (ST) path segments (there is negligible mode conversion between these modes and the fast transverse modes [3]). The L mode on its own shows no phonon focusing singularities. As the ST weighting of the mode sequence increases, it is evident how the focusing pattern evolves from one devoid of caustics to that of the ST mode on its own, which displays a prominent set of caustics.

The dashed lines in Figs. 2(b),(c) and (d) indicate the position of the scan line in the ultrasonic imaging experiments of Every et al. [5] on <100>–oriented Si. It is evident from these figures why their mST and L 4ST structures display cuspidal features, while their L 2ST structure does not. Figure 2(a) coresponds to the conditions for the L ST reflection phonon image of Si reported by Wichard and Dietsche [2]. The phonon intensity is very large but non singular along the $\{110\}$ planes. The caustics first appear at a value of $\alpha(ST)$ slightly larger than $1/2$. Such "precursors" to caustics are fairly common in phonon focusing patterns [3].

Further details on the subject matter of this paper are provided in Ref.[12]. W. Dietsche and R. Wichard are thanked for valuable suggestions. Financial support from the Foundation for Research Development is acknowledged.

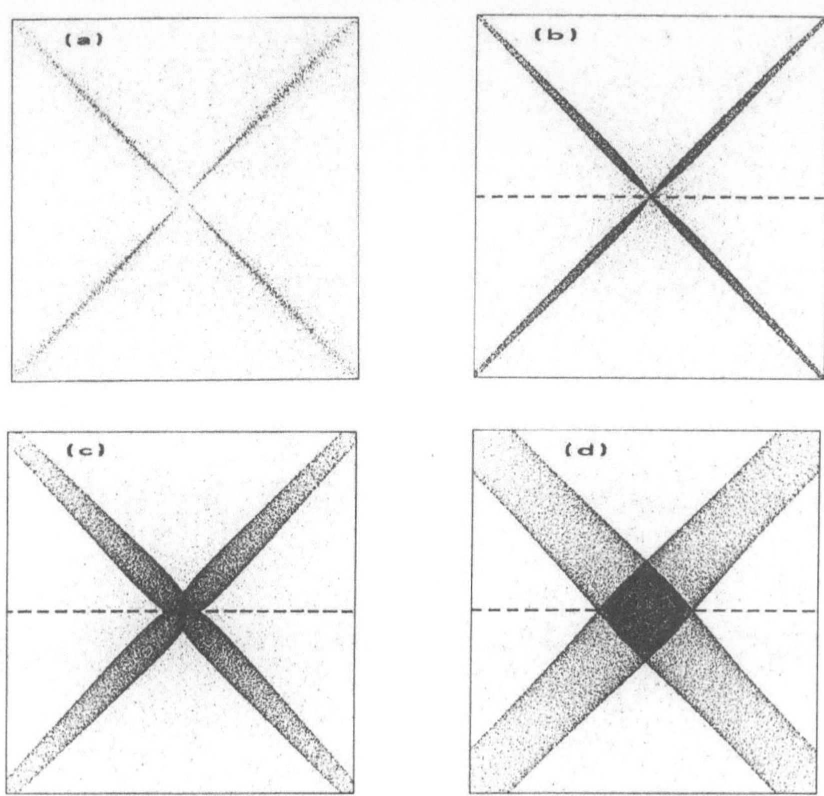

Fig. 2   Reflection phonon focusing patterns for a <100>—oriented silicon single crystal. (a) L ST, (b) L 2ST, (c) L 4ST and (d) ST.

REFERENCES

1. G A. Northrop and J.P. Wolfe, Phys. Rev. Lett. **52**, 2156 (1984).
2. R. Wichard and W. Dietsche, *19th Int. Conf. on Acoustical Imaging*, edited by H. Ermert and H. Harjes (Plenum, New York, 1991), in press.
3. A G. Every, G.L. Koos and J.P. Wolfe, Phys. Rev. B**29**, 2190 (1984).
4. C. Hoss, J.P. Wolfe and H. Kinder, Phys. Rev. Lett. **64**, 1134 (1990).
5. A.G. Every and Wolfgang Sachse, Phys. Rev. B**44**,(1991).
6. R.L. Weaver, M.R. Hauser and J.P. Wolfe, Phys. Rev. Lett. **68**, 2604 (1992).
7. S. Tamura, D.C Hurley and J.P. Wolfe, Phys. Rev. B**38**, 1427 (1988).
8. V.V. Novikov and L.A. Chernozatonskii, Sov. Phys. Acoust. **34**, 215 (1988).
9. O. Weis, Z. Angew. Phys. **26**, 325 (1969).
10. F. Rosch and O. Weis, Z. Phys. B**25**, 101 (1976); B**29**, 71 (1978).
11. H.J. Maris, J. Acoust. Soc. Am. **50**, 812 (1971).
12. A.G. Every, Phys. Rev. B**45**, 5270 (1992).

REFERENCES

WAVE FIELD SPLITTING, INVARIANT IMBEDDING, AND PHASE SPACE METHODS IN ACOUSTICAL IMAGING

Louis Fishman

Department of Mathematical and Computer Sciences
Colorado School of Mines, Golden, Colorado 80401 USA

INTRODUCTION

Due to the extremely large size of realistic three-dimensional ocean propagation problems, there is a desire to incorporate marching (one-way) methods into the solution algorithm for the elliptic, frequency-domain formulation. This has provided the impetus for the development of the "marching elliptic methods".[1] Such methods generally require (1) the backpropagation of a wave field component and (2) the knowledge of both the wave field and its normal derivative on an initial plane. The first requirement leads to an ill-posed problem, requiring a regularization which effectively filters out the evanescent portion of the spectrum[1] or a stability requirement which ultimately limits the spatial resolution.[1] For strongly backscattering environments, such filtering can result in spurious oscillations in the wave field, reflecting the important role played by the high-frequency portion of the spectrum under such conditions.[1] The second requirement is even more problematical. For the elliptic problem, both the wave field and its normal derivative cannot be independently specified on the initial plane.[1] Given the initial wave field, the determination of the corresponding normal derivative requires scattering data from the entire domain. Constructing this scattering information, however, effectively solves the original problem.

The above-mentioned problems notwithstanding, the original motivation is still sound. The goal only needs to be restated. It is now desirable to develop an algorithm for the elliptic propagation problem which can exploit marching methods as much as possible in a well-posed manner. The geometry of the problem provides the key insight. For the purpose of illustration, imagine a two-dimensional ocean waveguide with flat top and bottom surfaces divided into three distinct regions: left- and right-hand half-spaces separated by a transition region of arbitrary length. The two half-spaces are taken to be transversely inhomogeneous, while the transition region is both depth and range dependent. The vertical lines defining the transition region may or may not correspond to physical interfaces. If the vertical lines are imaginary, then the slab should be thought of as being smoothly imbedded in the larger medium.[1,2] Now, the ocean propagation problem is not in the form of the classical, well-posed elliptic boundary-value problem, where appropriate total wave field (or derivative) values are prescribed on the entire boundary.[1] In the ocean problem, boundary data is given on the top and bottom boundaries; however, no total wave field (or derivative) values are known on the vertical transition-region boundaries. Only sources in the left- and right-hand half-spaces are prescribed. The problem is a scattering problem. While radiation boundary conditions could be constructed beyond the sources in the two half-spaces to create a well-posed boundary-value problem (which could be treated by, for example, finite-element methods), this leads to an excessively large problem and obscures the physical picture.

Recognizing that typical ocean propagation problems are essentially scattering problems in terms of a transition region and transversely inhomogeneous half-spaces, wave field splitting, invariant imbedding, and phase space methods[1,2] reformulate the problem in terms of an operator scattering matrix characteristic of the transition region. The subsequent equations for the reflection and transmission operators are first-order in range, nonlinear (Riccati-like), and, in general, nonlocal. The system is well-posed, but stiff. The reflected and transmitted wave fields can be computed in a very efficient manner, while the wave field in the transition region can be computed by essentially a layer-stripping algorithm. Invariant imbedding methods focus directly on the reflection and transmission operators in the scattering matrix, resulting in a minimal and physically useful formalism and an efficient use of marching methods in the overall algorithm.

The reflection and transmission operator equations provide the framework for constructing inverse algorithms[1,2] based on, in principle, exact solution methods. This is directly applicable to the acoustic tomography problem[1] where the ocean features to be reconstructed are confined to the transition region and the sources and receivers are located in the half-spaces. Rotating the formulation by 90 degrees makes it appropriate for the seabed parameter reconstruction (reflection seismology) problem.[1] Algorithms based on this approach have already been developed and successfully implemented in the time-domain formulation of the acoustic problem in one spatial dimension.[2] The development of robust, numerical algorithms for multidimensional inversion based on exact solution methods is clearly a desirable goal since the computational algorithms currently being employed in these areas often involve either perturbative methods[1] (an initial linearization of the problem) or massive optimization (search) algorithms.[1]

THE DIRECT FORMULATION

For sound propagation in the ocean, the initial modeling is provided by the scalar Helmholtz equation,

$$\left(\nabla^2 + \bar{k}^2 K^2(\underline{x})\right)\phi(\underline{x}) = 0 \ , \tag{1}$$

where $K(\underline{x})$ is the refractive index field and $\bar{k}$ is a reference wave number. While the model described is two-dimensional, the analysis is n-dimensional. The geometry of the formulation suggests the following wave field decomposition. Split the wave field $\phi(\underline{x})$ into two components, $\phi^+(\underline{x})$ and $\phi^-(\underline{x})$, via the transformation

$$\begin{pmatrix} \phi(\underline{x}) \\ \partial_x \phi(\underline{x}) \end{pmatrix} = \begin{pmatrix} 1 & 1 \\ i\mathbf{B}_1 & i\mathbf{B}_2 \end{pmatrix} \begin{pmatrix} \phi^+(\underline{x}) \\ \phi^-(\underline{x}) \end{pmatrix} \ , \tag{2}$$

where $\mathbf{B}_1$ and $\mathbf{B}_2$ are the two operator solutions of the simple quadratic operator equation

$$\mathbf{B}^2 - \left(\bar{k}^2 K^2(\underline{x}) + \nabla_t^2\right) = 0 \ . \tag{3}$$

This results in the equivalent formulation[1,2]

$$\partial_x \begin{pmatrix} \phi^+(\underline{x}) \\[6pt] \phi^-(\underline{x}) \end{pmatrix} = \begin{pmatrix} \left(-\tfrac{i}{2}\partial_x \mathbf{B}_1^{-1} + 1\right) i\mathbf{B}_1 & \left(\tfrac{i}{2}\partial_x \mathbf{B}_2^{-1}\right) i\mathbf{B}_2 \\[12pt] \left(\tfrac{i}{2}\partial_x \mathbf{B}_1^{-1}\right) i\mathbf{B}_1 & \left(-\tfrac{i}{2}\partial_x \mathbf{B}_2^{-1} + 1\right) i\mathbf{B}_2 \end{pmatrix} \begin{pmatrix} \phi^+(\underline{x}) \\[6pt] \phi^-(\underline{x}) \end{pmatrix} , \qquad (4)$$

which holds in all three regions. Choosing $\mathbf{B}_1$ to correspond to the forward (outgoing) wave radiation condition and $\mathbf{B}_2$ to correspond to the backward wave radiation condition completes the identification. For the generally inhomogeneous transition region, $\phi^+(\underline{x})$ and $\phi^-(\underline{x})$ do not have a special physical interpretation. For the transversely inhomogeneous half-spaces, $K^2(\underline{x}) = K^2(\underline{x}_t)$ and (4) is diagonal. The diagonal system represents the exact decoupling (splitting) of the total wave field $\phi(\underline{x})$ into physically identifiable forward $(\phi^+(x,\underline{x}_t))$ and backward $(\phi^-(x,\underline{x}_t))$ wave field components in the transversely inhomogeneous environments. The forward evolution (one-way) equation,

$$\left(i/\bar{k}\right)\partial_x \phi^+(x,\underline{x}_t) + \left(K^2(\underline{x}_t) + (1/\bar{k}^2)\nabla_t^2\right)^{1/2}\phi^+(x,\underline{x}_t) = 0 , \qquad (5)$$

with $\mathbf{B} \equiv \left(K^2(\underline{x}_t) + (1/\bar{k}^2)\nabla_t^2\right)^{1/2}$ is the wave equation for propagation in a transversely inhomogeneous half-space supplemented with appropriate outgoing wave radiation and initial-value conditions.[1,2]

For wave propagation problems in the presence of two (generally different) transversely inhomogeneous half-spaces separated by a planar transition region of arbitrary length and inhomogeneity, (4) represents a type of two-point boundary-value problem. For the Helmholtz equation, designating the left and right boundaries of the transition region at $x = a$ and $x = b$, respectively, and generally locating a source in each half-space, the incident wave fields are connected to the scattered wave fields through the operator-valued scattering matrix $\underline{\underline{S}}(a,b)$,

$$\begin{pmatrix} \phi^+(b,\underline{x}_t) \\[6pt] \phi^-(a,\underline{x}_t) \end{pmatrix} = \underline{\underline{S}}(a,b) \cdot \begin{pmatrix} \phi^+(a,\underline{x}_t) \\[6pt] \phi^-(b,\underline{x}_t) \end{pmatrix} = \begin{pmatrix} \mathbf{T}^+(a,b) & \mathbf{R}^-(a,b) \\[6pt] \mathbf{R}^+(a,b) & \mathbf{T}^-(a,b) \end{pmatrix} \cdot \begin{pmatrix} \phi^+(a,\underline{x}_t) \\[6pt] \phi^-(b,\underline{x}_t) \end{pmatrix} . \quad (6)$$

The scattering matrix is defined in terms of the appropriate forward (right-traveling) and backward (left-traveling) reflection and transmission operators associated with the transition region. Invariant imbedding intuitively views the scattering matrix for a finite region as being composed of scattering matrices of a large number of contiguous subregions, and thus computes the effect of adjoining a very thin slab to the left-hand side of the transition region. This results in, for example,[1,2]

$$\left(i/\bar{k}\right)\partial_x \mathbf{R}^+(x,b) = \boldsymbol{\gamma}(x) + \boldsymbol{\delta}(x)\mathbf{R}^+(x,b) - \mathbf{R}^+(x,b)\boldsymbol{\alpha}(x) + \mathbf{R}^+(x,b)\boldsymbol{\beta}(x)\mathbf{R}^+(x,b) \quad (7)$$

and

$$\left(i/\bar{k}\right)\partial_x \mathbf{T}^+(x,b) = -\mathbf{T}^+(x,b)\boldsymbol{\alpha}(x) + \mathbf{T}^+(x,b)\boldsymbol{\beta}(x)\mathbf{R}^+(x,b) , \qquad (8)$$

with the initial conditions $\mathbf{R}^+(b,b) = 0$ and $\mathbf{T}^+(b,b) = \mathbf{1}$. In (7) and (8), $\boldsymbol{\alpha}(x)$, $\boldsymbol{\beta}(x)$, $\boldsymbol{\gamma}(x)$, and $\boldsymbol{\delta}(x)$ are related to the matrix operators in (4).[1]

In splitting the wave field in a manner which corresponds to the physical experiment and applying invariant imbedding to transform the scattering problem in (4) into an initial-value problem for the scattering operators, it is clear that an explicit representation of the square root Helmholtz operator $\mathbf{B}$ (and $\mathbf{B}^{-1}$, and derivative) and a one-way propagation theory are necessary. The formal one-way Helmholtz wave equation (5) can be recast and analyzed within the phase space framework as a Weyl pseudo-differential equation in the form[1,2]

$$(i/\bar{k})\partial_x\phi^+(x,\underline{x}_t) + (\bar{k}/2\pi)^{n-1}\int_{R^{2n-2}} d\underline{x}_t'd\underline{p}_t \,\Omega_{\mathbf{B}}\left(\underline{p}_t,(\underline{x}_t+\underline{x}_t')/2\right)$$

$$\cdot\exp\left(i\bar{k}\underline{p}_t\cdot(\underline{x}_t-\underline{x}_t')\right)\phi^+(x,\underline{x}_t') = 0 \ , \tag{9}$$

where $\Omega_{\mathbf{B}}(\underline{p},\underline{q})$ is the symbol for the Helmholtz operator $\mathbf{B} = \left(K^2(\underline{q})+(1/\bar{k}^2)\nabla_{\underline{q}}^2\right)^{1/2}$. In the Weyl pseudo-differential operator calculus, $\Omega_{\mathbf{B}}(\underline{p},\underline{q})$, the operator symbol, is defined through the Weyl composition equation

$$\Omega_{\mathbf{B}^2}(\underline{p},\underline{q}) = K^2(\underline{q}) - \underline{p}^2 \ = \ (\bar{k}/\pi)^{2n-2}\int_{R^{4n-4}} d\underline{t}d\underline{x}d\underline{y}d\underline{z}\,\Omega_{\mathbf{B}}(\underline{t}+\underline{p},\underline{x}+\underline{q})$$

$$\cdot\,\Omega_{\mathbf{B}}(\underline{y}+\underline{p},\underline{z}+\underline{q})\exp\left(2i\bar{k}(\underline{x}\cdot\underline{y}-\underline{t}\cdot\underline{z})\right) \ , \tag{10}$$

with $\Omega_{\mathbf{B}^2}(\underline{p},\underline{q})$ the symbol associated with the square of $\mathbf{B}$, $\mathbf{B}^2 = \left(K^2(\underline{q})+(1/\bar{k}^2)\nabla_{\underline{q}}^2\right)$.[1,2] Solution representations for pseudo-differential equations like (9) can be directly expressed in terms of infinite-dimensional functional, or path, integrals[1,2], following from the Markov, or semigroup, property of the propagator. This is detailed in [2]. The one-way marching algorithm is based on (1) the marching range step (following from the path integral), (2) a sophisticated symbol analysis (reflecting the detailed study of the (Helmholtz) Weyl composition equation (10)), and (3) Fourier component, or wave number, filtering in phase space (for increased efficiency, decreased computational time, and reduced error). The detailed numerical algorithm is discussed in [2]. Sufficiently accurate approximations of the square root ΨDO symbol over the relevant region of phase space result in very accurate numerical wave field calculations.[2]

The wave field splitting, invariant imbedding, and phase space methods can now be synthesized into an explicit solution method. As an example, for the limiting case of a transversely homogeneous environment, the equation for the reflection operator symbol takes the form

$$(i/\bar{k})\partial_x\Omega_{\mathbf{R}+} = \Omega_\gamma + (\Omega_\delta - \Omega_\alpha)\Omega_{\mathbf{R}+} + \Omega_\beta\Omega_{\mathbf{R}+}^2 \ , \quad \text{with } \Omega_{\mathbf{R}+}\Big|_{x=b} = 0 \ , \tag{11}$$

and where

$$\Omega_\gamma = (i/2\bar{k})\left(\frac{K(x)K'(x)}{K^2(x)-p^2}\right) \ , \quad \Omega_\beta = -\Omega_\gamma \ ,$$

and

$$\Omega_\delta - \Omega_\alpha = 2\left(K^2(x)-p^2\right)^{1/2} \tag{12}$$

In (12), an appropriate outgoing (forward) wave radiation condition is understood. Equation (11) is first-order in range, nonlinear (Riccati-like), and, in the general (transversely inhomogeneous) case, nonlocal. The system is well-posed, but stiff.[1] The stiffness arises from the local reflections and transmissions.

The algorithmic application of the solution method proceeds in the following manner. The basic idea is to compute the reflection and transmission operator symbols, and then, in the manner prescribed by the Weyl calculus, apply them to the appropriate incoming fields to produce the appropriate initial data for well-posed, one-way marching. Assuming a source in the left half-space only, the reflection and transmission equations (given symbolically in (7) and (8)) can be simultaneously marched from $x = b$ in incremental steps (Δx) to $x = a$, which then recovers the physical medium. Each incremental strip is transversely inhomogeneous in general. Starting from the physical problem at $x = a$, $\phi^-(a^-, z)$ is determined from the given $\phi^+(a^-, z)$, the determined $\Omega_{\mathbf{R}+}(a, b)$, and the Weyl symbol calculus. The continuous $\phi(a, z)$ and $\partial_x \phi(a, z)$ then follow from (2). $\phi^+(a^+, z)$ now follows from the inverse of (2).[1] The Weyl symbol calculus and the approximate constructions of $\Omega_{\mathbf{B}}$ [1,2] are used in the applications of (2) and its inverse. In the transversely inhomogeneous strip, $\phi^+(a^+, z)$ is now one-way marched to $x = (a + \Delta x)^-$, producing $\phi^+((a + \Delta x)^-, z)$. One now has the problem of an incoming wave field incident on a new slab with a known reflection symbol (operator). The above outlined procedure is then repeated, in effect, in a layer-stripping manner to $x = b$. Right- and left-moving wave fields are always propagated in their natural (well-posed) directions. The two-way nature of the elliptic problem in the transition region is accounted for by first marching the reflection symbol (operator) equation from $x = b$ to $x = a$, and then successively calculating the field by marching back from $x = a$ to $x = b$.

THE INVERSE FORMULATION

The reflection operator equation is the basis for constructing inverse algorithms[1,2] based on, in principle, exact solution methods. Algorithms based on this approach have been developed and successfully implemented in the time-domain formulation of the acoustic problem in one spatial dimension[2]. The reflection operator equation provides the basis for a layer-stripping approach which, in principle, is exact. The use of the invariant imbedding approach to incorporate "backscatter" in the direct, or forward, problem provides the opportunity to develop the analogous inversion algorithms for the fixed-frequency (elliptic) problem. There are significant differences between the elliptic problem and the hyperbolic formulation. In the time-domain problem, there is much more data available — the information from all frequencies is available, and further, is organized in the wave fronts. Following the wave fronts is the physical basis of the layer-stripping inverse algorithm. The medium is successively probed in range over time. This is not the case in the fixed-frequency problem. Here, the measurement of the reflection operator contains information from the entire transition region, not just the initial strip (at short times). This information must be inverted to reconstruct the square of the refractive index field $K^2(\underline{x})$.

The ideas and formalism involved can, however, be illustrated in a simple limiting case. Consider two transversely inhomogeneous half-spaces joined at the origin, a left half-space which is known and contains the source and a right half-space which is unknown. The reflection operator kernel for the interface scattering problem is assumed to be the measured data. The key idea is to first recover $\Omega_{\mathbf{B}_R}(\underline{p}, \underline{q})$ and then to reconstruct $K_R^2(\underline{q})$. In principle, the exact recovery of $K_R^2(\underline{q})$ then proceeds in the following steps.

1. The symbol for the reflection operator kernel $r(\underline{x}_t, \underline{x}'_t)$ is constructed from the measured data and the Weyl pseudo-differential operator calculus relation[1,2]

$$\Omega_{\mathbf{r}}(\underline{p}, \underline{q}) = \int_{R^{n-1}} d\underline{u} \exp(i\bar{k}\,\underline{p} \cdot \underline{u}) r\left(\underline{q} - \frac{\underline{u}}{2}, \underline{q} + \frac{\underline{u}}{2}\right) \ . \tag{13}$$

2. The symbol for the known left half-space $\Omega_{\mathbf{B}_L}(\underline{p}, \underline{q})$ is constructed from $K_L^2(\underline{q})$ by solving the (Helmholtz) Weyl composition equation (10).

3. The symbol (composition) relation for the interface scattering,[1,2]

$$\Omega_{\mathbf{B}_L}(\underline{p},\underline{q}) - \Omega_{\mathbf{B}_R}(\underline{p},\underline{q}) \;=\; (\bar{k}/\pi)^{2n-2} \int_{R^{4n-4}} d\underline{t}\, d\underline{x}\, d\underline{y}\, d\underline{z}$$

$$\cdot \Big(\Omega_{\mathbf{B}_L}(\underline{t}+\underline{p},\,\underline{x}+\underline{q}) + \Omega_{\mathbf{B}_R}(\underline{t}+\underline{p},\,\underline{x}+\underline{q}) \Big)$$

$$\cdot \Omega_{\mathbf{r}}(\underline{y}+\underline{p},\,\underline{z}+\underline{q}) \exp\Big(2i\bar{k}(\underline{x}\cdot\underline{y} - \underline{t}\cdot\underline{z}) \Big) \;,\qquad (14)$$

is solved for the unknown symbol $\Omega_{\mathbf{B}_R}(\underline{p},\underline{q})$. Equation (14) is just the starting equation for the initial half-space interface scattering problem in the invariant imbedding formulation in [1,2].

4. $K_R^2(\underline{q})$ is reconstructed from a direct computation of the (Helmholtz) Weyl composition equation

$$\Omega_{\mathbf{B}_R^2}(\underline{p},\underline{q}) = K_R^2(\underline{q}) - \underline{p}^2 \;=\; (\bar{k}/\pi)^{2n-2} \int_{R^{4n-4}} d\underline{t}\, d\underline{x}\, d\underline{y}\, d\underline{z}\, \Omega_{\mathbf{B}_R}(\underline{t}+\underline{p},\,\underline{x}+\underline{q})$$

$$\cdot \Omega_{\mathbf{B}_R}(\underline{y}+\underline{p},\,\underline{z}+\underline{q}) \exp\Big(2i\bar{k}(\underline{x}\cdot\underline{y} - \underline{t}\cdot\underline{z}) \Big) \;.\qquad (15)$$

The numerical calculations/inversions of the composition equations deserve further comment. Since singular operators are involved in the numerical reconstruction, the correct leading order terms accounting for the singular behavior are subtracted from the symbols following from the known asymptotics, and a finite basis set expansion is used to reduce the remaining problem to standard numerical linear algebra.[2]

There is a particularly attractive approximation to the above outlined scheme. The high-frequency symbol approximation[1,2] (a full-wave approximation distinct from geometrical acoustics) replaces (14) with

$$\Omega_{\mathbf{B}_L}(\underline{p},\underline{q}) - \Omega_{\mathbf{B}_R}(\underline{p},\underline{q}) \approx \Big(\Omega_{\mathbf{B}_L}(\underline{p},\underline{q}) + \Omega_{\mathbf{B}_R}(\underline{p},\underline{q}) \Big) \Omega_{\mathbf{r}}(\underline{p},\underline{q}) \qquad (16)$$

and (15) with

$$\Omega_{\mathbf{B}_R^2}(\underline{p},\underline{q}) = K_R^2(\underline{q}) - \underline{p}^2 \approx \Omega_{\mathbf{B}_R}^2(\underline{p},\underline{q}) \;,\qquad (17)$$

leading to the result

$$K_R^2(\underline{q}) \approx \Omega_{\mathbf{B}_L}^2(0,\underline{q}) \left(\frac{1 - \Omega_{\mathbf{r}}(0,\underline{q})}{1 + \Omega_{\mathbf{r}}(0,\underline{q})} \right)^2 \;.\qquad (18)$$

This approximation should be quite appropriate for many ocean/bottom environments and experimental parameters. As a brief illustration, the approximation in (17), $K^2(\underline{q}) \approx \Omega_{\mathbf{B}}^2(0,\underline{q})$, can be examined for the quadratic profile case for one transverse dimension, $K^2(q) = K_0^2 + \omega^2 q^2$. The exact Weyl symbol for this case is given by[1,2]

$$\Omega_{\mathbf{B}}(p,q) \;=\; -\Big(\exp(i\pi/4)\varepsilon^{1/2}/\pi^{1/2} \Big) \int_0^\infty dt\, \exp\big(i(Yt + X\tanh t) \big)$$

$$\cdot t^{-1/2}\mathrm{sech}t \left(iY + iX\,\mathrm{sech}^2 t - \tanh t \right) \qquad (19)$$

with $X = (1/\varepsilon)(\omega^2 q^2 - p^2)$, $Y = (1/\varepsilon)K_0^2$, and $\varepsilon = \omega/\bar{k}$ a dimensionless parameter that measures the medium variation on the wavelength scale. The accompanying figure illustrates the high-frequency reconstruction of the quadratic profile, $K^2(q) = 1 + q^2$, over a range of frequencies using the exact symbol given in (19) in (17) for $p = 0$.

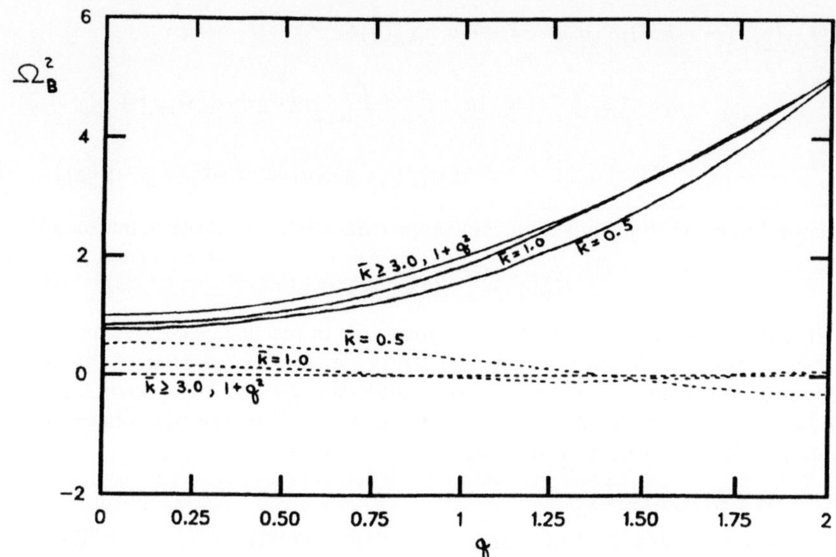

FIG. 1. $\Omega^2_{\mathbf{B}}(0,q)$ vs. q for the two-dimensional ($n = 2$) quadratic medium, $K^2(q) = 1 + q^2$, for $\bar{k} = 0.5$, 1.0, and ≥ 3.0. (——) $\mathrm{Re}\,\Omega^2_{\mathbf{B}}(0,q)$; (- - -) $\mathrm{Im}\,\Omega^2_{\mathbf{B}}(0,q)$.

Notice the excellent reconstructions as $\bar{k}$ increases (the dimensionless $\varepsilon = \omega/\bar{k} = 1/\bar{k}$ decreases). Moreover, the high-frequency approximate reconstruction for a real $K^2(\underline{q})$ has, in general, an error with both a real and an imaginary part. As the figure indicates, the deviation from zero imaginary part should serve, in some sense, as an indicator of the accuracy of the profile reconstruction.

There is an alternative formulation of the model inverse problem that, by choosing a different measured data set, results in a more compact analysis. Rather than measuring the reflection operator $\mathbf{r}$, the Dirichlet to Neumann map $\boldsymbol{\Lambda}$,[1]

$$\boldsymbol{\Lambda}\phi(0^-, \underline{x}_t) = \partial_\eta\,\phi(0^-, \underline{x}_t) = -\partial_x\,\phi(0^-, \underline{x}_t) \ , \tag{20}$$

where η is the outward normal at the interface, is taken as the measured data. Noting the definitions

$$\phi^-(0^-, \underline{x}_t) = \mathbf{r}\,\phi^+(0^-, \underline{x}_t) \quad \text{and} \tag{21}$$

$$\phi^\pm(x, \underline{x}_t) = \frac{1}{2}\left(\phi(x, \underline{x}_t) \mp (i/\bar{k})\mathbf{B}_L^{-1}\partial_x\phi(x, \underline{x}_t)\right) \tag{22}$$

in the left half-space, and comparing with (20) result in

$$\boldsymbol{\Lambda} = -i\bar{k}\,\mathbf{B}_L(1 + \mathbf{r})^{-1}(1 - \mathbf{r}) \ . \tag{23}$$

Expressing the reflection operator $\mathbf{r}$ in terms of $\mathbf{B}_L$ and $\mathbf{B}_R$ as indicated by (14) then yields the result

$$\boldsymbol{\Lambda} = -i\bar{k}\,\mathbf{B}_R \ . \tag{24}$$

Thus, once the Weyl operator symbol for $\boldsymbol{\Lambda}$ has been constructed, $K_R^2(\underline{q})$ follows from a direct computation of the composition equation

$$\Omega_{\Lambda^2}(\underline{p},\underline{q}) = -\bar{k}^2\,\Omega_{\mathrm{B}_R^2}(\underline{p},\underline{q}) \;=\; -\bar{k}^2\left(K_R^2(\underline{q}) - \underline{p}^2\right)$$

$$=\; (\bar{k}/\pi)^{2n-2}\int_{R^{4n-4}} d\underline{t}\,d\underline{x}\,d\underline{y}\,d\underline{z}\,\Omega_{\Lambda}(\underline{t}+\underline{p},\,\underline{x}+\underline{q})$$

$$\cdot\,\Omega_{\Lambda}(\underline{y}+\underline{p},\,\underline{z}+\underline{q})\exp\left(2i\bar{k}(\underline{x}\cdot\underline{y}-\underline{t}\cdot\underline{z})\right). \qquad (25)$$

Correspondingly, in the high-frequency approximation, the reconstruction takes the form

$$K_R^2(\underline{q}) \approx \left(-1/\bar{k}^2\right)\Omega_{\Lambda}^2(0,\underline{q})\;. \qquad (26)$$

The Dirichlet to Neumann map plays a pivotal role in recent inverse scattering results[1] and, even in this simple limiting case, efficiently captures the essence of the problem.

The analysis outlined in the contiguous half-space case can be extended, first to the single-layer transition region case, and then successively to the multi-layer transition region cases. The continuum limit must then be examined in detail.

ACKNOWLEDGMENTS

This work was supported by grants from ONR, AFOSR, NSF, and ASEE.

REFERENCES

1. L. Fishman, Wave field splitting, invariant imbedding, and phase space methods in direct and inverse scattering, *in*: "Proceedings of the Third IMACS International Conference on Computational Acoustics," D. Lee, A. Robinson, and R. Vichnevetsky, eds., North-Holland, Amsterdam (1992).

2. V.H. Weston, J.P. Corones, L. Fishman, and J.J. McCoy, "Wave Splitting with Applications to Wave Propagation and Inverse Scattering," SIAM, Philadelphia (1992).

ULTRASONIC TESTING FOR OXIDE HIGH TEMPERATURE SUPERCONDUCTORS

E. Biagi[1], E. Borchi[2], L. Masi[2], L. Masotti[1]

[1]Department of Electronic Engineering
[2]Department of Energetics
Via Santa Marta, 3
50139 Firenze

INTRODUCTION

The mechanism responsible for high temperature superconductivity in various oxide compounds is unknown, yet. Therefore any investigation on composition, structure, and phase transitions is of great value to characterize the physical properties of these systems. The apparent sensitivity of the propagation of an elastic wave to changes in the microstructural details of a material is well discussed.

In a previous work[1] the authors gave a preliminary material characterization, at room temperature, to evaluate the effect of the porosity and texture on the ultrasonic velocity. Various averaging methods were applied to estimate YBCO polycristal elastic constants from those of the corresponding mono-crystal. The obtained values were then compared with the experimental ones previously corrected for porosity.

In this paper longitudinal and shear sound velocity and attenuation are reported as a function of temperature and frequency for a single phase YBCO material, in order to characterize its acoustic behavior from 300 K to 50 K.

The velocity measurements carried out during cooling and heating have pointed out two distinct thermal hysteresis around 210 K and 120 K, for both longitudinal and shear waves .

Several experiments in the literature have shown that the attenuation of the wave is sensitive to inclusion, pores, grain boundary, interface boundary, dislocations. The relationship between changes in this microstructural features and changes in wave propagation properties has been semi-quantitatively established.

The attenuation was evaluated as the reduction of the global energy associated to the echo signal reflected from the back-wall of the sample in respect to a reference signal. The attenuation versus temperature monotone behavior presented no significative occurrence. An interesting correlation between velocity and attenuation was found when the attenuation is evaluated referring only to a limited zone around the maximum frequency of the echo signal. In this way, the influence of the measurement configuration errors on the spectral distribution shape was avoided.

The variation of the centroid of ultrasonic signal spectral distribution with temperature was also computed. The centroid behavior put in evidence the frequency selective scattering that the material induces on the ultrasonic wave. The shift observed in the centroid position values indicate that strong structural modifications take place inside the material during thermalcycling.

MATERIALS AND METHOD

Six polycristalline $YBa_2Cu_3O_{7-x}$ ($x < 0.1$) samples with density values varying in the range: 5.8 - 4.57 g/cm^3 were investigated. The specimens surfaces were polished to reduce roughness. All samples exhibited a partial axial texture along the forging axis. The texture degree decreases as porosity increases.

The thermocycling was performed in the temperature range 50-290 K. The temperature was varied with a mean rate of 1 K/min and was kept at each measurement point with a resolution higher than $1*10e^{-2}$.

The material resistivity data were acquired together with the ultrasonic signals during the temperature excursion through the transition point T_c. The T_c mean value for all the investigated samples was equal to 90 K.

The ultrasonic investigation was carried out by using a contact pulse echo technique (see fig 1). Two broad-band transducers for longitudinal and shear wave were used. The longitudinal transducer had nominal central frequency equal to 20 MHz, a 2.8 mm dia active surface, and a 4 μs delay line. The shear waves transducer, with a 5 MHz nominal central frequency, had 2.8 mm dia active surface and a 7 μs delay line. Honey was used as coupling with the specimen for both the transducers.

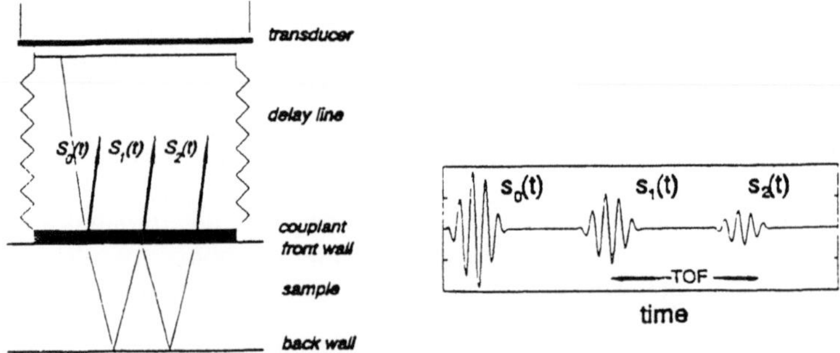

Figure 1. Transducer-specimen configuration for ultrasonic pulse echo methods.

The echo signals obtained by the ultrasonic inspection were digitalized with a 100 MHz sampling frequency and stored for subsequent processing.

Time of flight (TOF) measurements were obtained both by cross-correlation method, and phase detection techniques.

The cross correlation time of flight t_0 is the time shift τ for which the following expression reaches its maximum value.

$$\left| \lim_{T \to \infty} \frac{1}{T} \int_{-T}^{T} S_1(f) \, S_2(t+\tau) \, dt \right|$$

The used symbols are defined in figure 1: $S_1(t)$ and $S_2(t)$ are the first and second back-wall echoes.

The TOF is determined, using phase detection, by evaluating:

$$\phi_1(f) = \arctan\left[\frac{Im(S_1(f))}{Re(S_1(f))} \right]$$

$$\phi_2(f) = \arctan\left[\frac{Im(S_2(f))}{Re(S_2(f))} \right]$$

$$\Delta\phi = \phi_1 - \phi_2 \qquad \Rightarrow \qquad t_0 = \frac{L(\Delta\phi)}{2\pi f}$$

where $S_1(f)$ and $S_2(f)$ are the Fourier transform of first and second back-wall echoes. t_0 is the coefficient of the linear approximation of the phase difference $L(\Delta\phi)$.

A good agreement was found between the velocity data coming from the two evaluation techniques.

The TOF was measured with an 0.8% accuracy.

The velocity data discussed in the following paragraph as relative variation, in respect of the velocity at room temperature, were the mean values obtained on three different thermocyclings for three different position of the ultrasonic transducer over the sample.

In order to obtain a suitable evaluation of the attenuation it is important to refer to an appropriate signal reference. In the case of the above mentioned investigation techniques, the reference signal must have the same thermal history of the investigated sample to take into account the variations of the ultrasonic transducer response with temperature. The echo signal $S_0(f)$ coming from the delay line was assumed as the reference signal. No significative frequency content shift was observed in the $S_0(t)$ signal during thermocycling, but only strong amplitude variation.

The attenuation was computed in the frequency domain as the energy ratio between the spectral amplitude of $S_1(t)$ and $S_0(t)$. The energy was computed at -3 dB and -12 dB with respect to the maximum frequency amplitude of the first echo signal: $S_1(t)$. The irregular sample geometry and the inevitable variation of the transducer pressure over the sample induced scalloping and shape changes of the echo signal frequency content. These artifacts were avoided by evaluating the attenuation in the -3dB frequency range.

The frequency centroid position of the first echo signal, $S_1(t)$, was computed in a -3 dB maximum amplitude frequency range with the following expression:

$$f_c = \frac{\int_0^\infty S_1(f)\ f\ df}{\int_0^\infty S_1(f)\ df}$$

The accuracy of attenuation and centroid position measurement was equal to 5%.

RESULTS AND DISCUSSION

A strong measured data variance was found during the first few thermocyclings. This fact indicated that the samples were under a great deal of internal stress which gradually relaxes during each thermal cycle. The material passes from an intrinsic unstable state to a well annealed orthorhombic state.

For sake of brevity, only the results obtained for sample named Z are reported.

In figure.2 and 3 the temperature dependence of the sound velocity for longitudinal and shear waves is presented both for cooling and heating. The velocities increase with decreasing T. Two distinct hysteresis loops centered around 210 K and 120 K can be noted. The shear wave 120 K hysteresis is very weak (fig. 3). The main characteristics of the hystereses remain almost unchanged when changing the frequency of the ultrasonic waves. Qualitatively the 120 K hysteresis is in agreement with the experimental data obtained by several authors for YBCO compounds[2,3,4]. The 210 K hysteresis shows a peculiar behavior; the velocity variations during heating are larger than those obtained during cooling. These results are in opposition to those usually presented in the literature for the same material[2,3,4].

Sample Z. longitudinal velocity

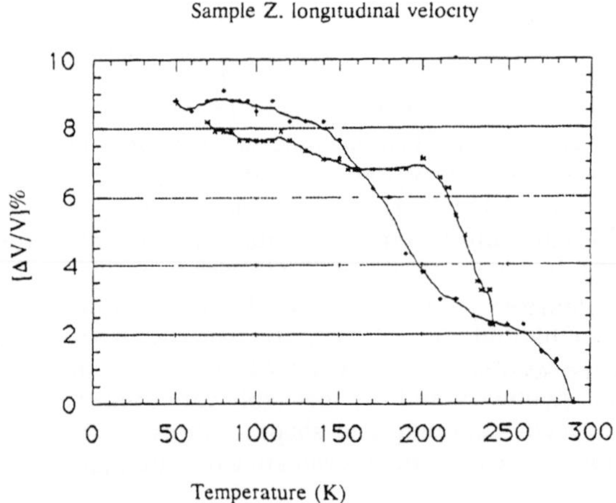

Figure 2. Longitudinal relative sound velocity: (ΔV/V)%, during cooling —•— and heating —▲—.

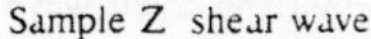

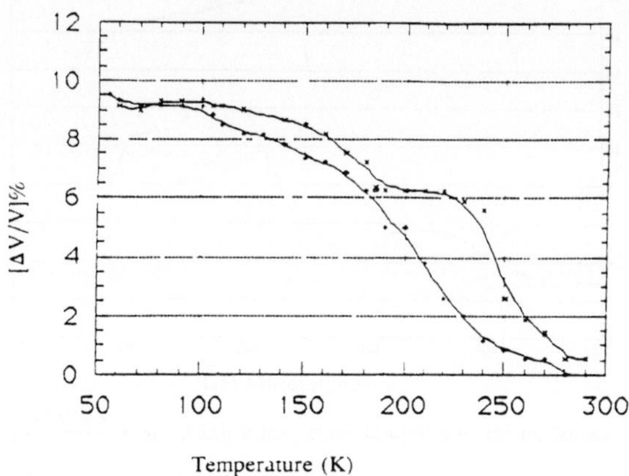

Figure 3. Shear relative sound velocity: $(\Delta V/V)\%$, during cooling —•— and heating —▴—.

The explanation of this disagreement is not possible according with the present knowledge of superconductive materials. Some microscopic mechanisms (e.g. the presence of a ferroelectric phase and structural phase transition) have been proposed to explain the hystereses[3,4,5].These hypotheses have been not completely supported by the experimental data[4]. On the other hand it is not possible to follow the interpretative models based on the material porosity because it was experimentally demonstrated that the hysteresis occurs also in single crystal ceramic material[6]. Results obtained in the present research work showed that a porosity density variation equal to 21% induces only a slight change in the hysteresis shape.

The attenuation data evaluated in the frequency range at -12 dB exhibits a strong data variance and no significant occurrence can be clearly noted except for a decreasing trend from 300 K down to 50 K.

The -3dB attenuation versus temperature for longitudinal and shear waves is reported in figure 4.

The longitudinal wave attenuation presents two high level regions which well correspond to the two hysteresis regions exploited by the velocity measurements. This behavior, similar to published results,[5,7] is associated by several authors to material relaxation phenomena[7].

In this temperature range, the energy attenuation of sound waves which propagate through a solid may be attributed to three different mechanisms: viscous friction, dislocation motion and scattering. It is very difficult to conclude, from the data obtained by simple experiments, which of these mechanisms play the major part in a given material at a particular temperature range. Moreover, different mechanisms appear to prevail for different ultrasonic frequency ranges.

The frequency centroid position versus temperature shows an interesting behavior that can be strictly correlated to the structural modification of the material. The results for shear and longitudinal waves are shown in figures 5 and 6 respectively.

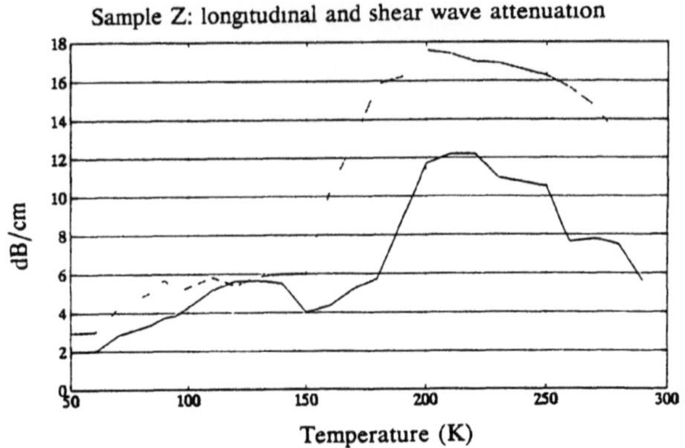

Figure 4. Attenuation for: logitudinal wave (solid line), shear wave (dotted line).

In the case of shear wave, the centroid value increases for a temperature region between 170 K and 120 K for all the investigated samples. Moreover a marked decrement in the centroid level can be noted around 110 K. This decrement is followed by a high amplitude variability in the temperature region around Tc and down to 50 K.

The selective scattering, depending on temperature, is evident in the reported centroid data for temperature lower than 120 K. The shift of the spectral amplitude distribution towards lower frequencies can be explained by the generation of internal structures that scatter the energy associated to the higher frequencies. To confirm this assumption it was verified that the spectral distribution of the signal backscattered by the internal structure of the material is characterized by a high frequency content (with a central frequency around 17 MHz).

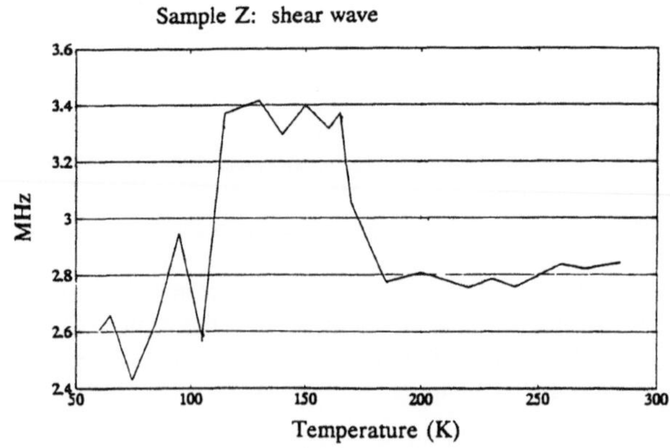

Figure 5. Frequency centroid position for shear wave

However, the present poor knowledge of these materials gives us little confidence to relate our data to specific phenomena.

It is only possible to affirm that strong modifications and rearrangements occur in the structural material organization.

The reported data for longitudinal waves behave similarly to those obtained for the shear waves. The strong decrement in the frequency centroid position occurs at 160 K. For the other investigated samples this discontinuity takes place inside the range: 140-160 K.

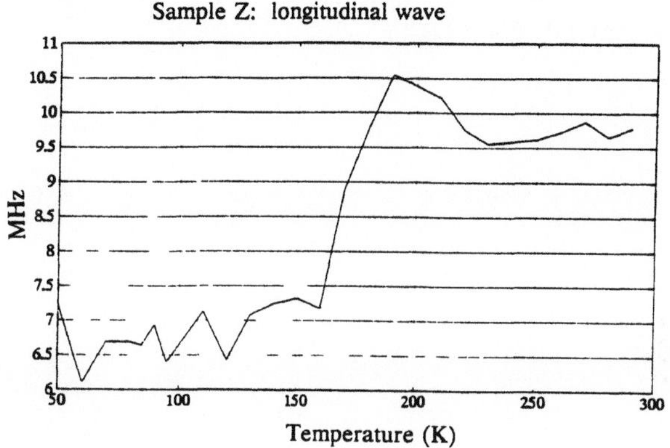

Figure 6. Frequency centroid position for longitudinal wave

CONCLUSIONS

No effect specifically correlated to the formation of the superconductive state has been observed. However, the evaluated ultrasonic parameters are powerful tools to follow the material structural modifications during thermocycling, and may provide interesting information on the physical properties of the new superconductive materials.

The experimental results have shown that several thermocyclings are necessary before to reach a stable annealed internal state.

REFERENCES

1. E.Biagi, E.Borchi, S.De Gennaro, and L.Masi,Elastic constant and moduli of polycrystallite $YBa_2Cu_3O_7$ and La_2CuO_4, J.Phys. D: Appl. Phys. 25 901-904 (1992).
2. H.M. Ledbetter, and S.A. Kim, Hysteretic phase transition in $YBa_2Cu_3O_7$ superconductors, Phys. Rev. B, Vol 38, no 16, 11857-11860 (1988).
3. V. Muller, C. Hucho, K. de Groot, D. Winau, D. Maurer, and K. H. Rieder,Evidence for ferroelectric phase transitions and exciton high-T_c superconductivity $YBa_2Cu_3O_7$ in from ultrasonic measurements, Sol.St.Comm., Vol.72, no.10, 997-1001 (1989).

4. L.N.Pal-Val, P.P.Pal-Val, V.D.Natsik, and V.I. Dotsenko,Comparative study of the low-temperature acoustic properties of CuO and YBCO ceramics, Sol. St. Communications. Vol.81,No9, 761-765,(1992).

5. L.G. Mamsurova, K.S. Pigalskiy, V.P. Sakun, and L.G. Scherbakova,First- and second-order phase transitions and oxygen rearrangement over the Cu1-04 chain in orthorhombic $YBa_2Cu_3O_{6-x}$at low temperature, Physica C, 167, 11-19, (1990).

6. T.J. Kim, j. Kowalewski, W. Assmus, and W. Grill, Propagation of ultrasonic waves and anomalies near T^c in single crystalline $YBa_2Cu_3O_{7-x}$high T_c superconductors, Z. Phys. B, Condensed Matter 78, 207-212, (1990).

7. T.J. Kim, B. Luthi, M. Schwarz, H. Kuhnberger, B. Wolf, G. Hampel, D. Nikl, and W. Grill, Valence fluctuation aspects of high temperature superconductors, J.Mag.Mag.Mat. 604-606 (1988).

INTERFEROMETRIC OBSERVATION OF ULTRASOUNDS
IN TRANSPARENT SOLIDS

X. Jia, G. Quentin, A. Boumiz and J. Beige

Groupe de Physique des Solides, Université Paris 7
Tour 23, 2 place Jussieu
75251 Paris Cedex 05
France

INTRODUCTION

The noncontact techniques based on the acousto-optic interactions have been widely used for studying the acoustic fields in the areas of medical diagnosis and nondestructive evaluation.[1,2] The effects of ultrasonic waves on the light transmission through transparent media, arise from refrative index variations induced by the pressure fluctuation of ultrasounds. These variations may be detected optically by deflection,[3] diffraction [4] or interference effects.[5] One of the most important application of these effects is the Schlieren technique which is employed for measuring acoustic beam profiles, essentially in the liquid medium.[6]

The optical method is also unviralled, beside piezoelectric transducers, in its ability to study acoustic waves inside solid materials, provided they are transparent optically. The most used technique is the photoelastic one which has been applied to the physical study of ultrasonic pulse scattering and the nondestructive testing.[7] The principle of this technique is to detect the polarization rotation of the light beam transmitting through the solid, as a result of the acoustooptic effect.

In the present paper, we describe an heterodyne interferometric method for detecting pulsed ultrasonic waves inside a transparent solid medium. This method, which was previously used for measuring the acoustic pressures in liquids and gases,[8] consists of an heterodyne interferometric probing of the refractive index variations associated with the ultrasonic waves. The heterodyne method combined with a broadband electronic precessing, allows us to detect the propagation of short pulsed elastic waves.

PRINCIPLE OF OPERATION

In this section, we begin with a brief description of the interferometric detection of refractive index varaiations induced by ultrasounds in a fluid medium, then describe the acoustooptic effects in an isotropic solid produced by elastic waves and show how longitudinal and shear waves are detected.

Heterodyne interferometric detection

The optical arrangement of the heterodyne interferometric detection is shown in Fig. 1.

Acoustical Imaging, Volume 20 Edited by Y. Wei
and B. Gu, Plenum Press, New York, 1993

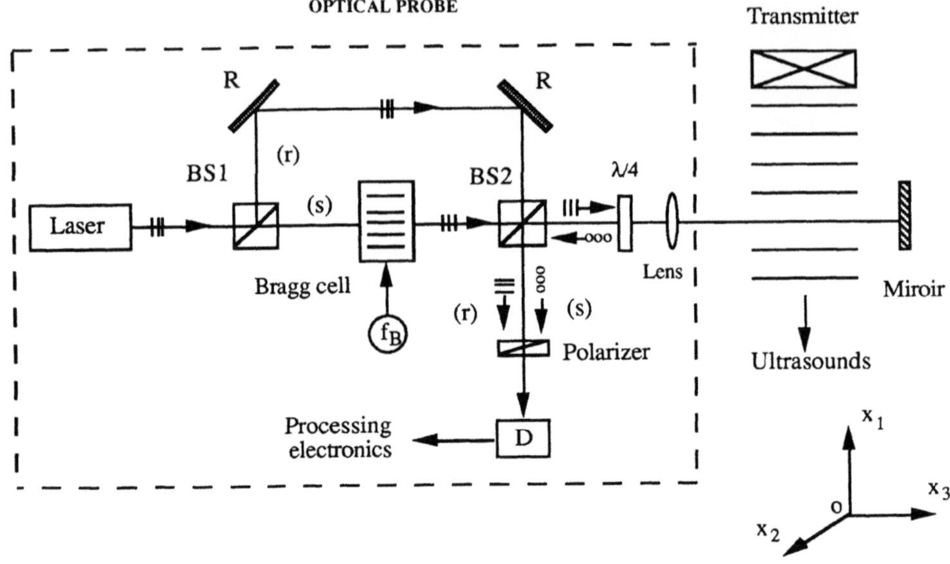

Figure 1 Optical arrangement of the heterodyne interferometric detection of ultrasonic waves.

The optical probe used in our experiments is a Mach-Zehnder type of heterodyne interferometer.[9] The horizontally polarized (ox_1) narrow laser beam is divided into the reference beam and probe one by a beam splitter (BS1). The reference beam (r) guided by optical reflectors is directed onto the photodiode. The probe beam shifted in frequency by Bragg cell (f_B = 70 MHz) crosses normally a transparent medium where the refractive index is changed by the acoustic wave, and is then reflected by a fixed mirror. This probe beam, vertically polarized (ox_2) after passing through the quarter wavelentgh plate, is reflected at a right angle by the polarized beam splitter (BS2) and confused with the reference beam. A polarizer oriented 45° between the reference and probe beams permits an efficient interference of the two beams. The electric fields of the reference and probe beams are given respectively by:

$$E_r = E_0 \exp j(2\pi f_L t + \phi_r)$$
$$E_s = E_0 \exp j(2\pi(f_L + f_B)t + \upsilon(t) + \phi_s)$$

where $\phi_r = (2\pi / \lambda_L) L_r$ and $\phi_s = (2\pi / \lambda_L) L_s$ (λ_L is the optical wavelength) are the phase constants corresponding to the optical paths L_r and L_s in the reference and the probe arms of the interferometer respectively, $\upsilon(t)$, defined as the Raman-Nath parameter, is the optical phase change due to the ultrasonic waves and proportional to the integrated pressure across (twice) the ultrasonic beam of diameter L. If the transmission medium is a fluid and the plane wave model is considered, one obtains:

$$\upsilon(t) = \frac{2\pi}{\lambda_L} (\Delta n) (2L) = \frac{2\pi\mu}{\lambda_L} p(t) (2L)$$

where μ is the photoelastic coefficient and $p(x, t)$ the pressure of either a continuous or pulsed acoustic wave.

The beating of the probe beam and the reference one on the photodiode results in a phase modulated current at frequency f_B :

$$i(t) = I_0 \cos(2\pi f_B t + \upsilon(t) + \phi_s - \phi_r)$$

With the aids of a broadband electronic processing,[9] the acoustic signal can be demodulated from the phase modulation photocurrent. The signal at the output is proportional to the Raman-Nath parameter and consequently to the acoustic pressure: $s(t) \sim \upsilon(t) \sim p(t)$.

Photoelastic effects

It is noted that the previously mentioned acoustooptic interaction in a fluid is a relative simple case, as the acoustooptic effect is characterized by only one photoelastic constant (μ) and the light conserves its polarization direction after transmitting through the ultrasonic field. In a solid, however, the situation becomes much more complicated. The optical anisotropy is induced when propagate the ultrasounds. The refractive index sphere of an initially isotropic solid is transformed into an elliposoid defined by three principle indices (n_1, n_2, n_3). In general, the solid becomes uniaxial if the elastic wave is longitudinal or biaxial if the wave is transversal.

It can be shown that,[10,11] for a <u>longitudinal wave</u> propagating along ox_1 (Fig. 1), the variations of the principle refractive indices are:

$$\Delta n_1 = -n^3/2 \ p_{11}S_{11}$$

$$\Delta n_2 = \Delta n_3 = -n^3/2 \ p_{12}S_{11}$$

where n is the refractive index of the isotropic solid, p_{ij} the photoelastic constant and S_{11} the longitudinal strain. It is noted that the three principle axes of refractive index are, in this case parallel, to ox_1, ox_2, ox_3 (Fig. 1). For a <u>shear wave</u> propagating along ox_1 and polarized in the plane x_2ox_3, the principle indices variations are:

$$\Delta n_1 = - (n^3 \ p_{66} \ S_{12}) \cos \alpha$$

$$\Delta n_2 = (n^3 \ p_{66} \ S_{12}) \cos \alpha$$

where is S_{12} is the shear strain, α the angle between the shear wave polarization and the axis ox_2.

Generally, the polarization of the light beam is rotated when passing through the solid if its polarization direction is not parallel to one of the principle index axes. This is what took place in our experiments. As being expected in this case, it is more difficult to express the Raman-Nath parameters υ in a solid medium than in a fluid medium. Even so, a detailed calculation[11] showed that longitudinal waves can be obtained by the optical configuration of Fig. 1, with an output signal proportional to the longitudinal strain S_{11}:

$$s(t) \sim n^3 (p_{11} + p_{12})S_{11}$$

As for shear waves, the shear strain S_{12} is found at the output of the probe:

$$s(t) \sim n^3 \ p_{66} \ S_{12} \cos \alpha.$$

if the optical arrangement of Fig. 1 is slightly modified by moving away the quarter wavelength plate.

EXPERIMENTAL RESULTS

The heterodyne interferometric method are employed for measuring various elastic waves propagating in transparent solids. The solid samples used in the experiments are made of fused quartz or plexiglas. The elastic waves are generated by piezoelectric transducers in contact with the solids. The excitation electric pulse has 150 V of amplitude and $0.1 \mu s$ of duration. The received signals from the optical probe are sampled and averaged by a digital oscilloscope, then recorded and plotted.

Longitidinal waves

Fig. 2a shows a typical strain waveform emitted by a 2 MHz-longitudinal transducer in a plexiglas sample. The first pulse corresponds to the direct wave arriving from the transducer and intersecting the optical probe beam (Fig. 1), the second one to the wave reflected at the interface plexiglas/air. The velocity of the longitudinal waves was measured to be 2720 m/s, in agreement with the value obtained by the acoustic echo method. As it appears clearly, the second pulse undergoes a phase shift of π after reflected at the interface. Such a π-phase shift is a well known wave phenomena which is explained by the fact that the acoustic impedance of plexiglas is greater than that of air. For further demonstrating this phenomena, a similar investigation was carried out at the interface plexiglas/duralumin realized by the adhesion. As shown in the figure 2b, no phase shift takes place, because the impedance of plexiglas is smaller than that of duralumin in this case. This result demonstrates that the present heterodyne interferometric method is applicable to short ultrasonic pulses measurement and allows to follow the propagation of ultrasonic waves inside the (transparent) transmitting medium.

Shear waves

The ultrasonic waves generated by a 2MHz-shear transducer were also investigated by the heterodyne interferometric method. The configuration of the optical probe used in this measurement is similar to the preceding one (Fig. 1) except that the quarter wavelength plate is moved away. As demonstrated in a previous work,[11] this optical arrangement favoures the detection of shear waves while the preceding one (Fig. 1) does longitudinal waves.

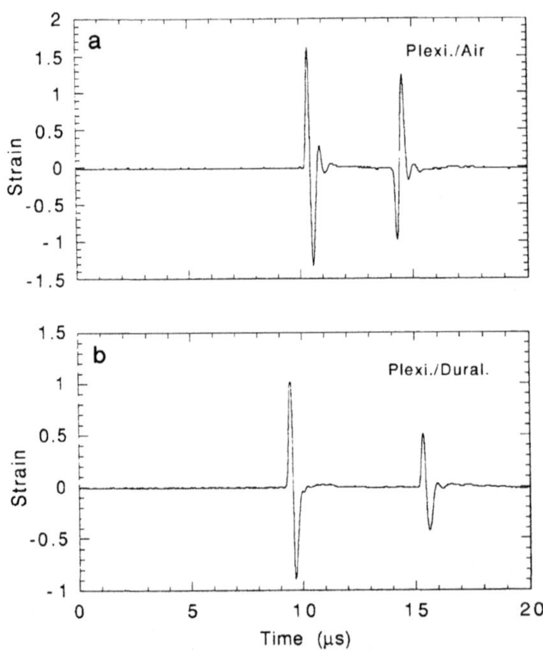

Figure 2 Optical heterodyne detection of the longitudinal waves and its reflections at the interfaces plexiglas/air (a) and plexiglas/duralumin (b).

Fig. 3 displays the optically detected waveforms of shear waves propagating in a fused quartz. Figures 3a, 3b and 3c correspond to the different situations in which the shear waves are polarized along ox_2, ox_3 and opposite ox_2, respectively (Fig. 1). The π-phase shift observed between the incident pulse and the reflected one in Fig. 3a and 3c is similar to the previous results for longitudinal waves. Moreover, the whole waveform (Fig. 3a) is observed to change phase when the wave is polarized in opposite direction (Fig. 3c). This π-phase change, as well as no signal observed in Fig. 3b, are explained by the factor $\cos \alpha$ included in the previously mentionned formula: $s(t) \sim n^3 p_{66} S_{12} \cos \alpha$. The shear wave velocity is measured as 3640 m/s close to the theoretical value in fused quartz.

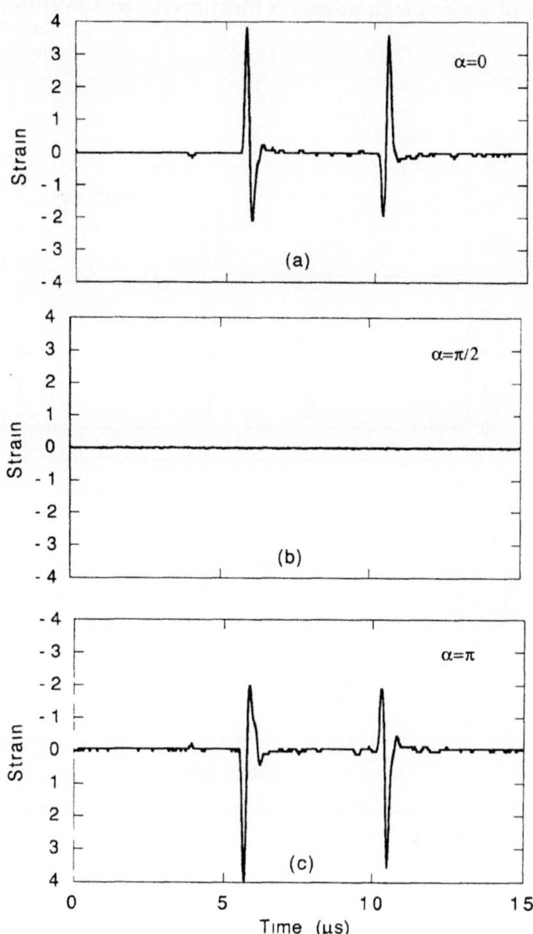

Figure 3 Optical heterodyne detection of shear waves polarized in different angles with respect to the axis: $\alpha=0$ (a), $\alpha=\pi/2$ (b), $\alpha=\pi$ (c).

Rayleigh waves

As another application, the heterodyne interferometric method was used for studying the propagation of surface acoustic waves (Rayleigh waves). These waves were generated by a wedge transducer coupled on fused quartz. Figures 4a and 4b illustrate the Rayleigh waves optically detected at distances of z=1.35 mm and 3.35 mm under the solid surface. The fact that the optical probe is able to follow the Rayleigh waves, point by point, inside the solid allows us to demonstrate the amplitude decay of the Rayleigh wave on leaving from the solid surface. The propagation velocity of the Rayleigh wave was measured to be 3265 m/s.

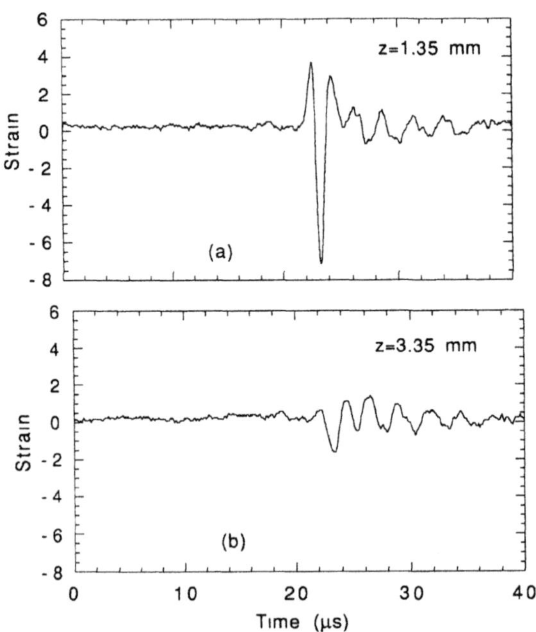

Figure 4 Longitudinal strain components of Rayleigh waves on fused quartz detected at distances of z=1.35 mm (a) and 3.35 mm (b) under the surface.

The preceding results (Fig. 4) were obtained with the same optical arrangement as that for measuring longitudinal waves (Fig. 1). As mentioned previously, this configuration favours the detection of longitudinal waves instead of shear waves. So the results shown in Fig. 4 correspond basically to the longitudinal strain components of Rayleigh waves. By opposition, the optical arrangement without the quarter wavelength plate in Fig. 1, is in favour of the shear wave measurement. Fig. 5 displays the waveform of Rayleigh wave detected by this optical configuration, which corresponds essentially to the shear strain component. We obseve that the waveforms of the two components are different. This can be explained by a phase difference of $\pi/2$ between the two components.[12] Indeed, the phase difference between the curves of Fig. 4a and Fig. 5, calculated from the Fourier transforms (Fig. 6), confirms qualitatively this $\pi/2$-phase shift. The discrepancy between the theory and experiment arises probably from the fact that the longitudinal and shear strain components are not seperated entirely with the preceding optical configurations.

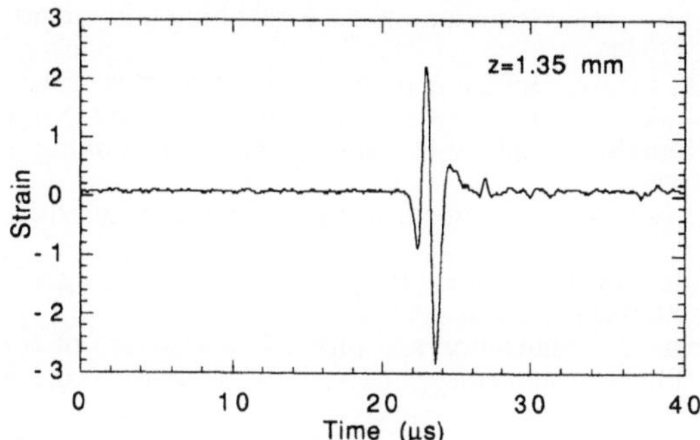

Figure 5 Shear strain component of the Rayleigh wave detected at a distance of z=1.35 mm under the solid surface.

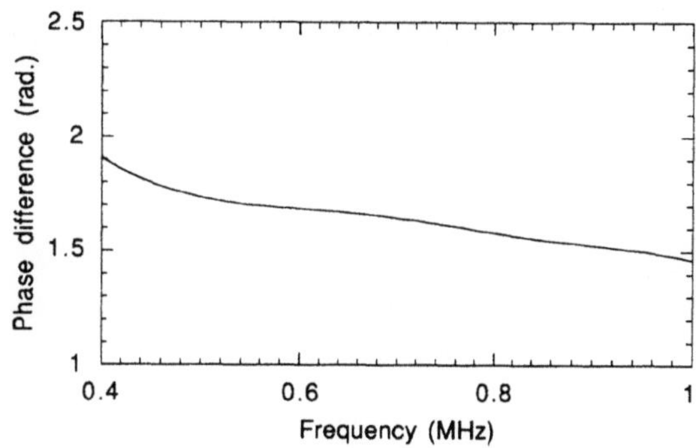

Figure. 6 Phase difference between the longitudinal and shear strain components of Rayleigh waves.

CONCLUSION

We have presented an heterodyne interferometric method for detecting ultrasonic pluses inside a transparent solid. The preliminary results obtained on the propagation of longitudinal, shear and Rayleigh waves show the high temporal and spatial resolutions of the present method. The phase changes of the bulk waves reflected at the interfaces, as well as the polarization of Rayleigh waves, have been demostrated experimentally.

Acknowledgement

The authors wish to thank Professor D. Royer for his interest and valuable discussions.

REFERENCES

1. W.A. Riley, Optics, in:"IEEE Guide for Medical Ultrasound Field Parameter Measurement," ANSI, New York (1990)
2. G. Hall, Ultrasonic wave visualization as a teaching aid in non-destructive testing, Ultrasonics March:57 (1974)
3. A.C. Tam and W.P. Leung, Optical generation and detection of acoustic pulse profiles in gases for novel ultrasonic absorption spectroscopy, Phy. Rev. Lett. 53:560 (1986)
4. J. Wolf, N.H. Neighbors and W.G. Mayer, Optical probing of ultrasonic pulses, Ultrasonics 27:150 (1989)
5. S. Nakai, Acoustic power measurement using an optical heterodyne method, Ultrasonics March:77 (1985)
6. G.S. Kino, Acoustic waves: Devices, Imaging, and Analog Signal Processing, Prentice-Hall, Inc., New Jersey (1987)
7. C.F. Ying, Photoelastic visualisation and theoretical analyses of scatterings of ultrasonic pulses in solids, in:"Physical acoustics," IXX, R.N. Thurston, ed., Academic Press, New York (1990)
8. X. Jia, G. Quentin, M. Lassoued and J. Berger, Quantitative measurement of ultrasonic pressures with an optical heterodyne interferometer, in Proc. IEEE Ultrason. Symp., p.1137 (1991)
9. D. Royer and E. Dieulesaint, Optical detection of sub-angström transient mechanical displacements, in Proc. IEEE Ultrason. Symp., p.527 (1986)
10. E. Dieulesaint, D. Royer and G. Sitbon, Acousto-optic position sensor, in Proc. IEEE Ultrason. Symp., p.480 (1980)
11. A. Boumiz, Rapport de stage DEA, Université Paris 7 (July 1992)
12. I.A. Viktorov, Rayleigh and Lamb waves, Plenum, New York (1967)

THE VISUALIZATION OF THE LOW FREQUENCY
ACOUSTICAL RADIATION FIELD *

Dejun Zhang, Xianhua Xia, Jin Yan
Jianzhen Cheng, Nianqiu Zhu, Qun Wang

Wuhan Institute of Physics
Chinese Academy of Sciences
Wuhan 430071, P. R. China

ABSTRACT

This paper researches the relationship between the acoustical source of low frequency and its radiation field by using Nearfield Acoustical Holography (NAH). Simulation gives the spatial distributions of the intensity vector of two pistons vibrating with various phase differences and in the plate in mode (0, 2). Experiment shows the surface pressure and intensity vector distribution of a loudspeaker. Finally, we discuss the unstability in NAH reconstruction.

INTRODUCTION

Nearfield Acoustical Holography (NAH) is a new acoustic imaging technique coming out of Holographic Imaging field and is the extension of the conventional acoustical holography. In recent years, the research on NAH is very active [1-6]. NAH records the hologram in the plane very close to the object plane (i. e. $d \ll \lambda$). It records not only the propagating wave components, but also the evanescent wave components. So the reconstructed images can have a resolution higher than $\lambda/2$. Choosing a Cartisian coordinate system, and letting the emitting plane be at $Z = Z_0$, receiving plane at $Z = Z_H (Z_H > Z_0 > 0$ and $d = Z_H - Z_0 \ll \lambda)$, we can reconstruct the pressure distribution at the emitting plane by

$$\psi(x, y, Z_0) = F^{-1}\{\overline{\psi}(k_x, k_y, Z_H) \overline{G}_D^{-1}(k_x, k_y, Z_H - Z_0)\} \tag{1}$$

---

* This project is supported by the National Natural Science Foundation of China and the State Key Laboratory of Acoustics.

where $\bar{\psi}(k_x, k_y, Z_H)$ and $\bar{G}_D(k_x, k_y, Z_H - Z_0)$ are the two dimensional Fourier transforms of $\psi(x, y, Z_H)$ and $G_D(x, y, Z_H - Z_0)$, respectively; F^{-1} represents two dimensional inverse Fourier transformation; G_D^{-1} the inverse of G_D (Green's function under Dirichled condition). Actually, Eq. (1) can be extended to the following case: to figure out the sound pressure field in an arbitrary plane $Z(Z > Z_0)$ from the known field $\psi(x, y, z_H)$ provided $d = Z_H - Z_0 \ll \lambda$. After the sound pressure distribution is figured out, the partial velocity $V = (V_x, V_y, V_z)$ can be found out by

$$V_a(x, y, z) = F^{-1}\{K_a\psi(k_x, k_y, Z)\}/\rho ck \tag{2}$$

where $a = x, y, z; k_z = (K^2 - K_x^2 - K_y^2)^{1/2}; \rho$ and c are the density and the sound velocity of the medium, respectively. Thus, we obtain the acoustical intensity distribution

$$I(x, y, z) = \mathrm{Re}[\psi \bar{v}^*]/2 \tag{3}$$

where $\mathrm{Re}[\ ]$ means to get the real part of the result, $*$ represents the conjugate of a complex.

NAH is used in the acoustical testing, especially in the low frequency testing. It is mainly used to analyse the relationship between the vibrating mode and the radiating field, for the source vibrating and radiating problems with irregular shapes and complex boundary conditions, which are very difficult or impossible to calculate theoretically.

I . SIMULANT VISUALIZATION OF ACOUSTICAL FIELD

Aiming to study the influences of the source vibration, especially the vibration phase, on the intensity vector, we simulate the following typical sound sources by using NAH method and get some interesting results.

(1) Two Piston Sources with Different Phases

Assume two pistons are at $Z = Z_0$ plane, their radii are both $a = 2.5$ cm, the distance between them is $D = 13$ cm, and they emit the sound wave of frequency $f = 340$ Hz to the $Z > Z_0$ space in the same pressure amplitudes. The wave length $\lambda = 1$ m in the air. We can calculate the pressure distribution at $Z = Z_H$ plane as known NAH data by using Kirchhoff integration. $d = Z_H - Z_0 = 5$ cm. From the known pressure distribution at $Z = Z_H$ plane, the pressure distribution at $Z = Z_0$ plane can be figured out by Eq. (1). Then the sound intensity at a plane, which contains the acoustic axis of the sources and is perpendicular to the hologram plane, can be woked out by Eq. (3). Figure 1 is the pressure distribution at $Z = Z_0$ plane reconstructed by using NAH method when two pistons vibrate at the same phase, and Figure 2 is the intensity vector. The energy flow emitted by the two pistons, after propagating a distance of 0.09λ, concentrate together in the direction of Z axis.

Figure 3 is the NAH energy flow distribution when the phase difference of the two vibrating pistons is π. We can see that the energy along Z axis is very small. The energy is mainly radiated in the directions, which have an angle $a = 35°$ with respect to the Z axis. Figure 4 shows the energy flow when the phase difference of the two sources is $\pi/2$.

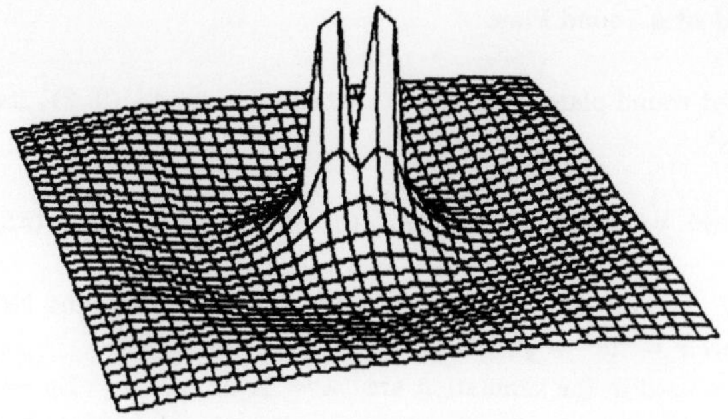

Figure 1. The pressure distribution at $Z = Z_0$ plane reconstructed by NAH method when two pistons vibrate at the same phase

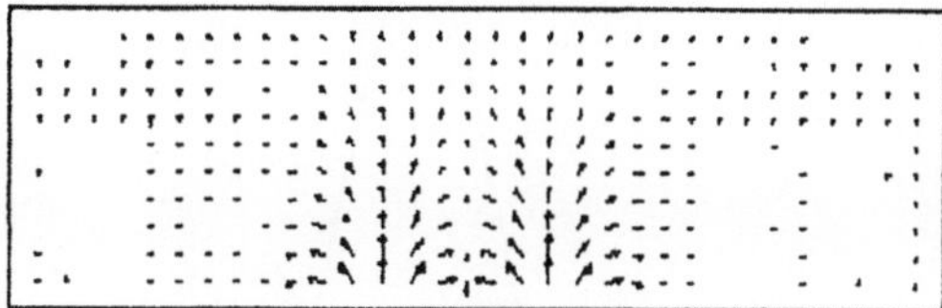

Figure 2. The intensity vector field

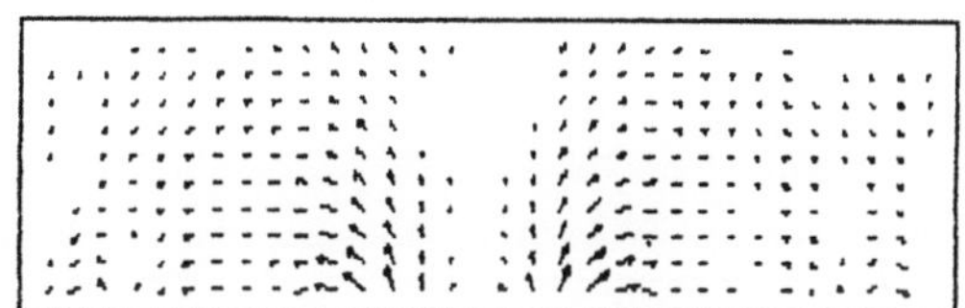

Figure 3. The NAH energy flow distribution when the phase difference of the two vibrating pistons is π

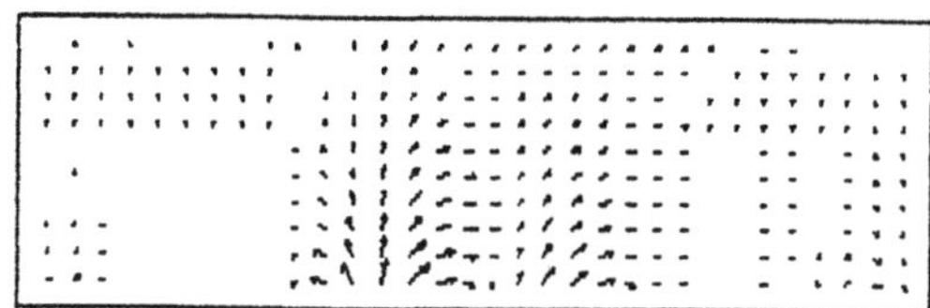

Figure 4. The energy flow when the phase difference of the two sources is $\pi/2$

(2) Radiation Field of a Sound Plate

When a baffled round plate with radius a vibrates in mode (0, 2), its surface normal velocity distribution is[8]

$$V(r) = J_0(\pi\beta_{02} \cdot r/a) - J_0(\pi\beta_{02}) I_0(\pi\beta_{02} \cdot r/a) / I_0(\pi\beta_{02}) \qquad (4)$$

where $\beta_{02} = 2.007$, $J_0(x)$ and $I_0(x)$ are zero-order Bessel function and Neumann function, respectively, and where $r = (x^2 + y^2)^{1/2}$.

The parameters used in the simulation are $\lambda = 1.02$ m, $d = Z_H - Z_0 = 0.02$ m, $a = 0.5$ m. After calculating the pressure distribution at $Z = Z_H$ plane as known NAH data, we reconstruct Figure 5a, the surface velocity distribution, and Figure 5b, the radiated energy flow of the plate. the circulating energy flow pattern can be found out from Figure 5b.

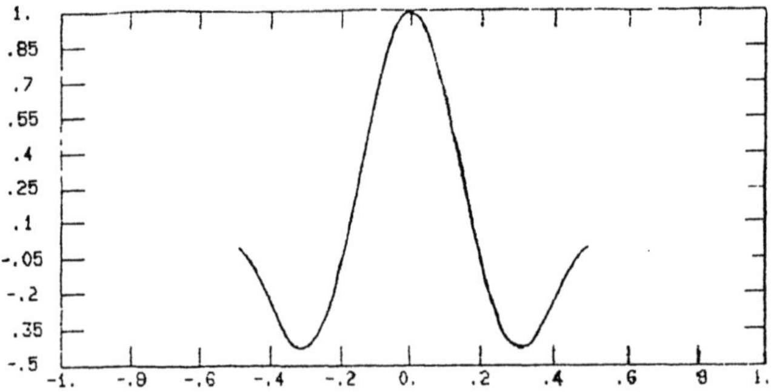

Figure 5a. The surface velocity distribution.

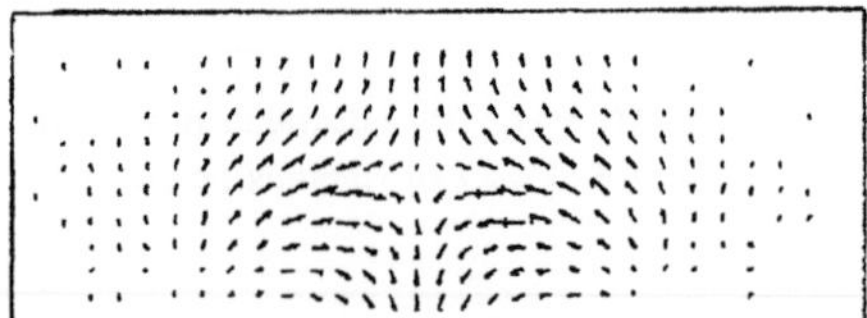

Figure 5b. The radiated energy flow of the plate

Figure 6a is the plate's surface velocity distribution with the center area shielded and Figure 6b is the plot of its projected vector intensity field in which the circulating energy flow disappears. Figure 7a is the plate's surface velocity distribution with the center area's vibrating phase inversed and Figure 7b is the plot of its projected vector intensity field from which we can see that the forward radiation has been intensified. This property can be the reason to process some sound sources with the phase compensation and center suppression methods to intensify their radiation efficiencies.

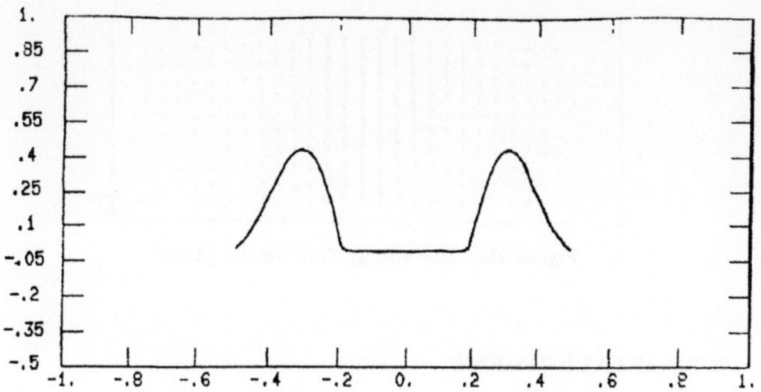

Figure 6a. The plate's velocity distribution

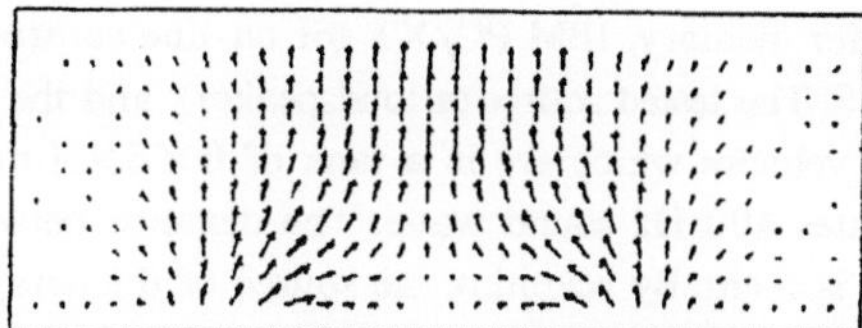

Figure 6b. The plot of the vector intensity field of the plate

II . UNSTABILITY IN RADIATION FIELD RECONSTRUCTION

The process of reconstruction of the source, as Eq. (1) shows, has an unstability. The small measurement errors, especially the phase errors will cause much greater errors in the reconstruction. This is a serious problem in NAH.

NAH data have many sources of errors. The environmental noise and the inconsistency of each element of the receiving array are the main factors. Simulation shows that while there are phase errors larger than 9° in NAH data , although the reconstructed mode of vibration is not influenced seriously, the direction of the energy flow is changed obviously. So, the amplitude and phase consistency correction of each receiving array element and a good environment are necessary when one does NAH experiment work.

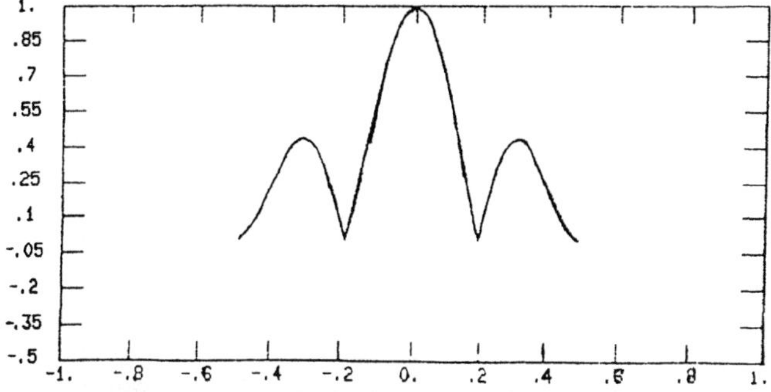

Figure 7a. The plate's surface velocity distribution with
the center area's vibrating phase inversed

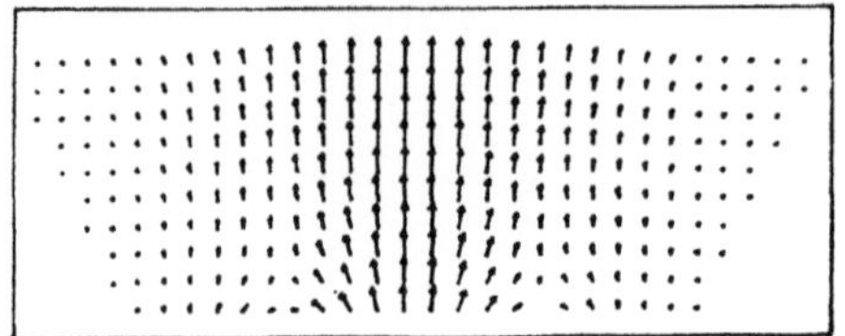

Figure 7b. The energy flow of the plate

Ⅲ. RESULTS OF NAH EXPERIMENT

In order to make an experiment we have set up a test system for NAH airborne sound radiation research. It contains: a 32 element linear array of microphones (element interval is 6. 5 cm, total length 2 m), 32 individual channel preamplifiers, a bandpass filter, a multiplexer, high speed A/D converter, buffer memory, IBM PC/XT for on-line control and preprocessing, and a network to a VAX 11/785. The tested source (a loudspeaker) and the array of microphones are installed on two scanning vehicles which are in a tank of $6\times5\times4$ m^3 and can scan and rotate in space. The source radiates 400 Hz sound wave. The distance between the source plane and the holographic plane, d, is 5 cm. By scanning the source or the array we can synthesize $32\times$ 32 square array hologram which are routed through a multiplexer to A/D and stored in a high speed buffer memory. After preprocessed by microcomputer, the data are transferred to VAX 11/785 for reconstruction.

Figure 8a is the reconstructed source surface pressure distribution when source is in the tank and the phases of the array elements are not corrected. Although there exists noise background, the pressure distribution is not influenced seriously. Figure 8b shows its plot of a projection of the vector intensity field. The energy flow is relatively confused because of the reverberation. Figure 9a is the reconstructed source surface pressure distribution when source is in the tank, which is established an anechoic chamber, and the microphones are calibrated rigidly. It

Figure 8a. The reconstructed source surface pressure distribution.

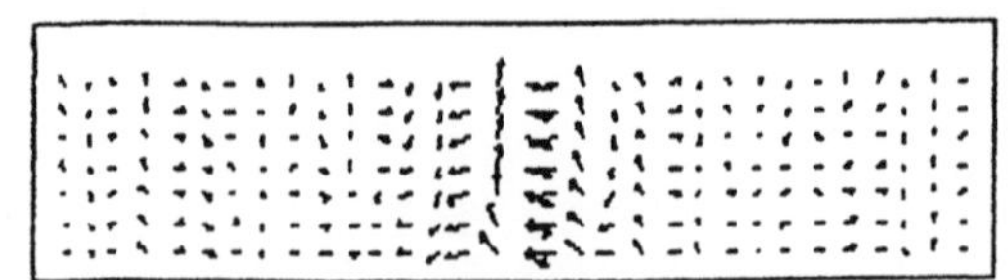

Figure 8b. The plot of a projection of the vector intensity field

shows that the background noise is reduced significantly. Figure 9b is the relative vector intensity field. We find that in the main radiation direction the sound energy flow concentrates and agrees with the theory.

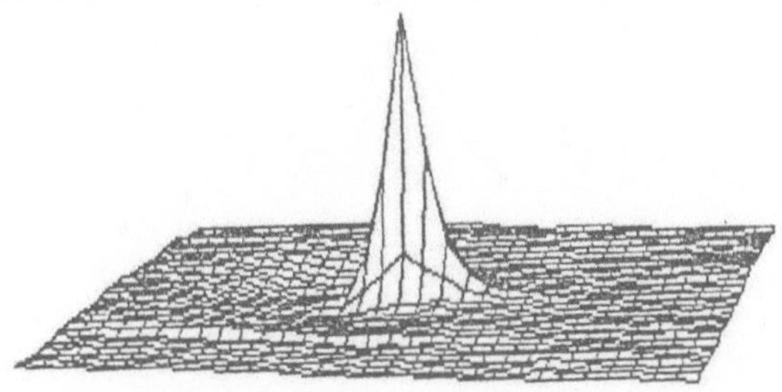

Figure 9a. The reconstructed source surface pressure distribution when source is in an anechoic tank

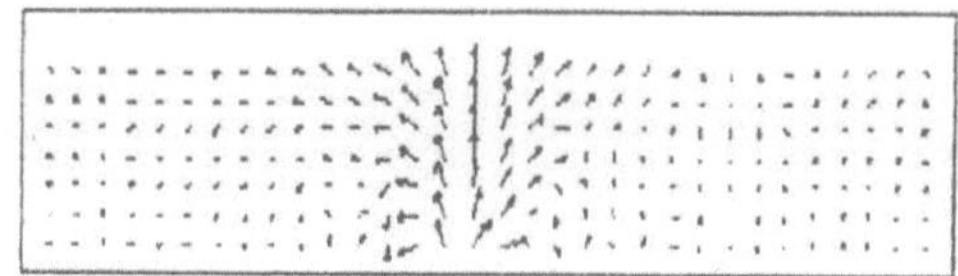

Figure 9b. The plot of a projection of the vector intensity field

REFERENCES

1. J. D. Maynard, "Near-field Acoustic Holography, Frontiers in Physical Acoustics", Italian Physical Society, (1986), p. 313.
2. E. C. Williams, J. D. Maynard, Holographic imaging without the wavelength resolution limit, *Phys. Rev. Lett.*, Vol. 45 (1980), p. 554.
3. J. D. Manard, et al, Nearfield acoustic holography: I . Theory of generalized holography and the development of NAH, *JASA*, Vol. 78 (1985), p. 1395.
4. W. A. Veronesi, et al, Nearfield acoustic holography: **I** . Holographic reconstruction algorithms and computer implementation, *JASA*, Vol. 81 (1987), p. 1307.
5. Thierry Loyau, et al, Broadband acoustic holography reconstruction from acoustic intensity measurements. I : Principle of the method, *JASA*, Vol. 81 (1987), p. 389.
6. Zhang Guxiang, et al, The prediction of sound field, in: "The Symposium of Chinese Underwater Acoustic Conference", 1990.
7. H. Fleischer, et al, Restoring an Acoustic Source from Pressure Data Using Wiener Filtering, *Acoustica*, Vol. 60 (1986), p. 172.
8. P. M. Morse, K. U. Ingard, "Theoretical Acoustics", McGraw-hill, (1986).

TRANSMISSION OF HOLOGRAPHIC IMAGING DATA OF MICRO- AND ACOUSTIC WAVES BY COMMUNICATION SATELLITE

Yoshinao Aoki Ryuichi Mitsuhashi and Shin Tanahashi

Department of Information Engineering, Hokkaido University
Sapporo 060, Japan

INTRODUCTION

In this paper, we propose a holographic image transmission method by communication satellite, where hologram data using long-wavelength waves such as microwaves and acoustic waves is transmitted and images are reconstructed from the received hologram data This technique is related to the technique of three-dimensional(3-D) image transmission by a communication satellite An experiment for hologram data transmission by a Japanese communication satellite (JCSAT-2) was carried out, where data of a Snow Search Radar System(SSRS)[1][2] was used for transmission experiment The SSRS is a 3-D imaging radar system used for searching of objects under accumulated snow We propose a system with a communication satellite which connects a searching field where the SSRS operates and a laboratory where we can process images with a high performance computer system The environmental conditions under which SSRS operations are carried out are not always good due to bad weather, various extraneous noises, and so on A method of hologram transmission has an advantage in that images can be transmitted even under such noisy and adverse conditions Experiments of transmission of SSRS data were performed to verify this advantage

PRINCIPLE OF SSRS

First we describe the principle of the holographic imaging system of SSRS for collecting data for transmitting images by a communication satellite Although holographic imaging systems using microwave and acoustical waves have good resolution with respect to the lateral direction compared with that of conventional radar and sonar Generally speaking however, resolution with respect to the range direction is poor compared with pulse echo imaging system One of the techniques for improving the resolution of range direction in holography is the use of wide-band signals, that is multi-frequency holography, where waves of stepping up frequencies are used to record holograms The information in the range direction is recorded as the phase rotation of the reflected signal from the targets like conventional hologram recording Consider the configuration as illustrated in Fig 1, where we neglect the effect of snow, and a target has a scattering distribution of $g(x_o, y_o, z_o)$ An antenna scans the 2-D hologram field and hologram data is collected in each frequency change The received signal $u(x, y, f)$ on the hologram plane, which is defined in terms of the lateral coordinates x and y, and frequency f, is given by

$$u(x, y, f) = \int_0^{+\infty} \int_{-\infty}^{+\infty} \int_{-\infty}^{+\infty} g(x_o, y_o, z_o)$$
$$\exp\left(-j2\pi f \frac{2r}{v}\right) dx_o dy_o dz_o \tag{1}$$

where v is the velocity of light, and r is the distance between target and the observation points We assume that r is within the distance where the following Fresnel approximation is valid,

$$r \simeq z_o + \frac{(x - x_o)^2 + (y - y_o)^2}{2z_o} \tag{2}$$

We express the frequency f by a frequency f_d deviated from a center frequency f_0 under the condition of $f_0 \gg f_d$, resulting in the following approximation,

$$-j2\pi f \frac{2r}{v} \simeq -j2\pi f \frac{2z_o}{v} - j2\pi f_0 \frac{(x - x_o)^2 + (y - y_o)^2}{v z_o} \tag{3}$$

Substituting (2),(3) into (1), resulting in the following approximation,

$$u(x, y, f) = \int_0^{+\infty} \exp\left(-j2\pi f \frac{2z_o}{v}\right) \int_{-\infty}^{+\infty} \int_{-\infty}^{+\infty} g(x_o, y_o, z_o)$$
$$\exp\left\{-j2\pi f_0 \frac{(x - x_o)^2 + (y - y_o)^2}{v z_o}\right\} dx_o dy_o dz_o \tag{4}$$

Replacing the inside of the integral of (4) by $h(x, y, z_0)$ as follows,

$$h(x, y, z_o) = \int_{-\infty}^{+\infty} \int_{-\infty}^{+\infty} g(x_o, y_o, z_o)$$
$$\exp\left\{-j2\pi f_0 \frac{(x - x_o)^2 + (y - y_o)^2}{v z_o}\right\} dx_o dy_o \tag{5}$$

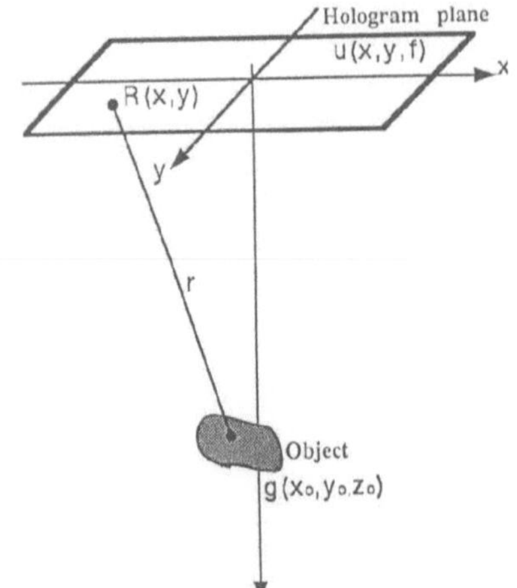

Fig 1 Coordinate system of multi frequency holography

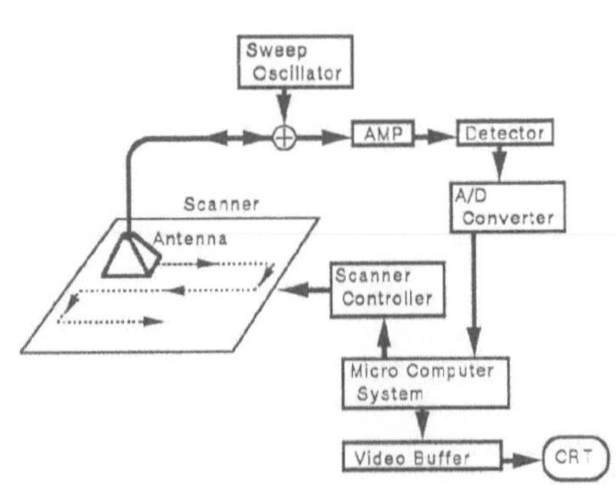

Fig 2 Block diagram of the snow search radar system

Using the relation of (5), we can rewrite (4) as follows,

$$u(x, y, f) = \int_0^{+\infty} h(x, y, z_o) \cdot \exp\left(-j2\pi f \frac{2z_o}{v}\right) dz_o \tag{6}$$

Here $u(x, y, f)$ represents a 1-D Fourier transform of $h(x, y, z_o)$ with respect to z_o. And $h(x, y, z_o)$ represents a 2-D Fresnel transform with respect to the lateral direction, which is independent of a frequency coordinate. Therefore, we find that the multi-frequency hologram $u(x, y, f)$ contains information on the lateral and range direction independently in the x,y sectional plane and frequency coordinate. We can obtain a reconstructed object image from a multi-frequency hologram in the following way. First, the multi-frequency hologram is compressed in the range direction by 1-D inverse Fourier transform with respect to the frequency. A reconstructed image is achieved from the compressed range data by an inverse Fresnel transform.

TRANSMISSION OF SSRS DATA BY COMMUNICATION SATELLITE

In order to transmit the collected data of SSRS located in the field to the signal processing system in the laboratory, we investigated the use a communication satellite. Communication by a satellite has certain advantages in that it can realize a high-speed and wide-area communication. This advantage guarantees a vast data transmission from wide searching fields using SSRS in practical use.

A block diagram of an experimental system of the SSRS is shown in Fig. 2. and SSRS collected data in Table 1. The hologram data is acquired at the sampling points of 2-D point array of 32×32 on the hologram plane by stepping up the frequency from 8GHz to 10GHz. The number of sampling points with respect to the frequency axis is 64. Fig. 3 is the photograph of the scanner part of the SSRS, where the horn antenna is scanned on the 2-D area to record holographic signals. A metallic messtin was used as targets buried in the snow in the configuration shown in Fig. 4. The recorded SSRS data is shown in Fig. 5. The size of one scanning data is about 130kbyte. Fig. 6 is a block diagram of the experimental communication system using a communication satellite. Experiments of transmitting data were carried out using Ku band frequency with JCSAT-2. The transmission speed was 64kbps.Therefore the time of transmis-

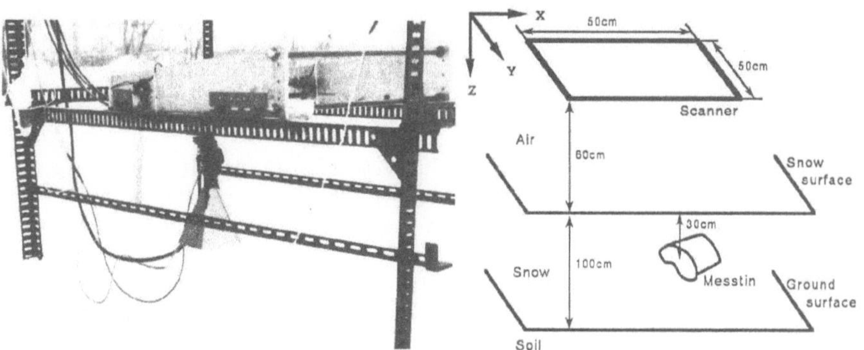

Fig.3 Experimental scanning antenna.   Fig.4 Geometry of experimental arrangement.

sion SSRS data was expected to be less than 20 seconds. However, since the receiving computer was one of rather low performance, the experiment was carried out in slow speed(about 400sec.). By replacing the computer with a high-speed one we can thereby realize high-speed transmission. Since the communication satellite used in this experiment is a commercially used one, data can be transmitted without error under normal use in good weather condition. We performed the experiments with artificially added noise.

Table 1 Parameter of experimentation.

Parameter	Value
Aperture Length	0.5×0.5m
Number of Sampling (Lateral Direction)	32×32
Frequency	$8.0 \sim 10.0$GHz
Number of Sampling (Frequency Direction)	64
Range Resolution	7.5cm
Lateral Resolution (z = 1.0m)	3.3cm

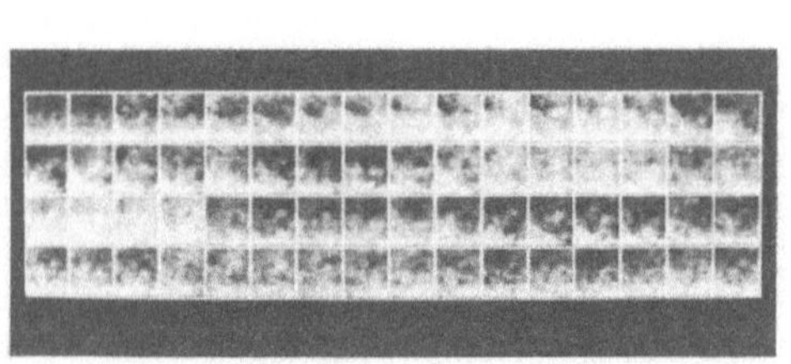

Fig.5 Recorded SSRS data.

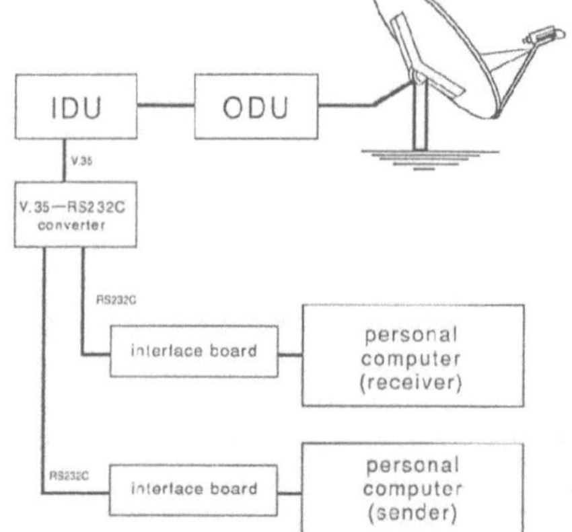

Fig.6 Block diagram of transmission of SSRS data.

ESTIMATION OF PERMITTIVITY

In order to reconstruct an image from collected SSRS data, it is necessary to know the relative permittivity of accumulated snow. The assumed relative permittivity at X-band microwave of dry snow depends on the accurate measurement of its density. In this section we propose a method of automatic estimation of the relative permittivity of snow utilizing certain characteristics of multi-frequency holography itself[3]. The penetration depth of the microwaves in accumulated snow is determined by the factors of wavelength of microwaves, water in snow, temperature, density of snow, and so on. Since the penetration length of microwaves in the experimental system here is estimated as the order of centimeters for wet snow, we chose dry snow in the snow season for performing experiments of SSRS. The wavelength in snow becomes shorter inverse-proportionally to the square root of the relative permittivity. If we assume the wavelength in snow is equal to that in air, we estimate the effect of snow by changing the distance according to the following distance parameters, z_o'(range direction),z_o''(lateral direction) given by

$$z_o' = \sqrt{\epsilon_r}\, z_o \tag{7}$$

$$z_o'' = \frac{z_o}{\sqrt{\epsilon_r}} \tag{8}$$

where ϵ_r is the relative permittivity of snow Substituting (7),(8) into (2) we obtain the following expression,

$$r_{\epsilon_r} = z_o' + \frac{(x - x_o)^2 + (y - y_o)^2}{2 z_o''} \tag{9}$$

Above equation means that we can not obtain focused reconstructed images without knowing the relative permittivity, that is the distance z' of (9). Conversely, when the reconstructed image is in focus, we can estimate the relative permittivity of snow by the following expression.

$$z_o'' = \frac{z_o'}{\epsilon_r} \tag{10}$$

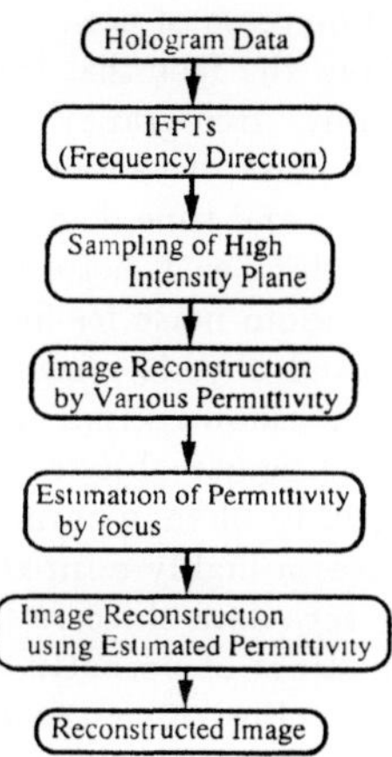

Fig.7 Algorithm of the reconstruction.

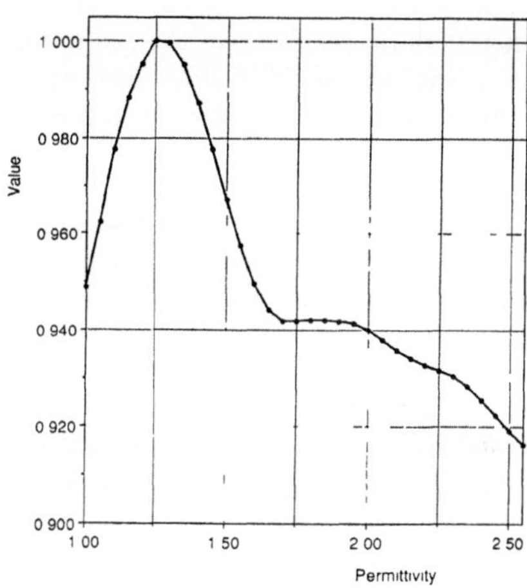

Fig.8 Estimation of focus for permittivity.

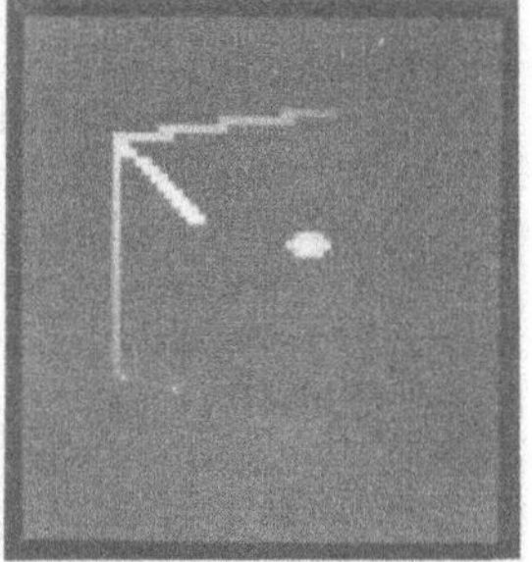

Fig.9 Reconstructed image.

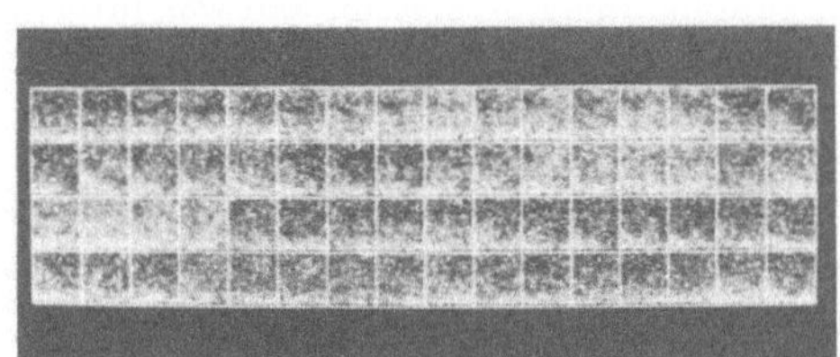

Fig.10 SSRS data with noise(30%).

The recorded SSRS data(Fig 5) is reconstructed by using the process as shown in Fig 7 The highest intensity data plane processed by the range compression is inversely Fresnel transformed with the variable of the relative permittivity The relative permittivity of snow can be estimated by analysis of reconstructed images We employed a weighting function in direct proportion to the spatial frequency of the images, for judgment of the relative permittivity Fig 8 is the result of analysis We estimated that the relative permittivity of the snow was 1 25(by Fig 8) Meanwhile the estimated relative permittivity by measured density of snow(0 1g/cm^3) was 1 2 ~ 1 3[3] This is in close agreement with previous techniques(by density of snow) Shown in Fig 9 is the reconstructed 3-D image using a direct search method[4] obtained by the proposed reconstruction method

EFFECT OF TRANSMISSION ERROR

The Ku band microwaves used in the transmission experiment are effected by the weather condition more strongly than in the case of low-frequency (for example C-band) microwaves In the experiment of satellite communication, we observed 17dB gain reduction during snow fall compared with fine weather condition This is one of the causes of generation of noise during the transmitting of SSRS data Moreover, the practical system of SSRS is supposed to be carried by a car, resulting in the requirement of a small sized system This means that we have to use an antenna of small aperture to transmit data, resulting in the reduction of gain and increase of noise

The advantage of SSRS data transmission comes from the fact that the SSRS depends upon the principle of a multi-frequency holography Holography has the good characteristic of recovering images from data defected by noise, because original object information is recorded as distributed information in space and time domain This fact is verified by the experiment of transmitting hologram data with noise added artificially Fig 11 are reconstructed images which added random noise for fig 5 data(ex fig 10) We could visualize adequately the reconstructed objects images even with some annexing noise But the reconstructed images (ex shown in fig 12) to which were added random noise were too poor to sufficiently recognize objects

The method of automatic estimation of the permittivity of snow form SSRS data is affected by added random noise Effect of noise causes a highly estimated value of permittivity Fig 13 is the result of noise data added to zero data Incorrect estimation of focussing due to noise leads to incorrect estimated values of permittivity of snow The result of SSRS data added noise shown in fig 14 Fig 15 is the ratio of the estimation value for the mean value This method could achieve correct permittivity data if noise is limited to within a few percent against true data

CONCLUSION

In this paper we proposed the transmission of holographic imaging data of micro- and acoustic waves by a communication satellite and SSRS, that is, a kind of microwave holographic imaging system, and was investigated by transmitting data of SSRS by a communication satellite From theoretical and experimental results, it is shown that hologram transmission has the advantage of noise-suppressing transmission and images are adequately reconstructed even during a long-distance transmission by a communication satellite Transmitting hologram data by a communication satellite has the potentiality of transmitting 3-D images of large capacity and can be applied to the transmission of 3-D images by micro- and acoustical waves Transmitting 3-D micro- and acoustical images of high quality is the subject of future study

ACKNOWLEDGEMENT

The authors thank the Collaboration Society for Integrated Satellites Communication, Japan Communications Satellite Company,Inc. and Sapporo Electronics Foundation for support in conducting the experiment. They also thank Dr.Y.Sakamoto of Hokkaido university for his helpful discussion in this research.

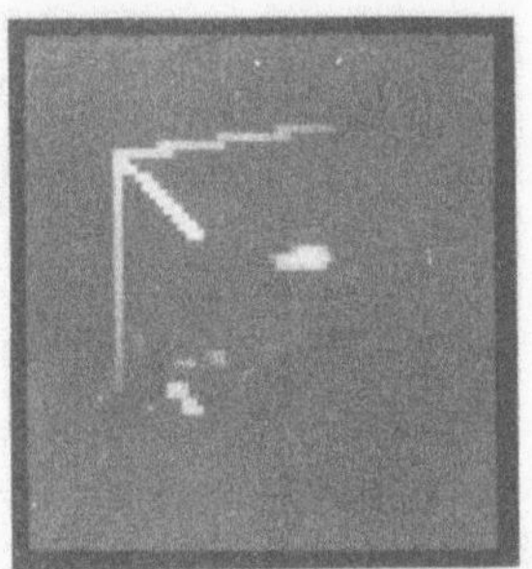

(a)noise/data = 10%

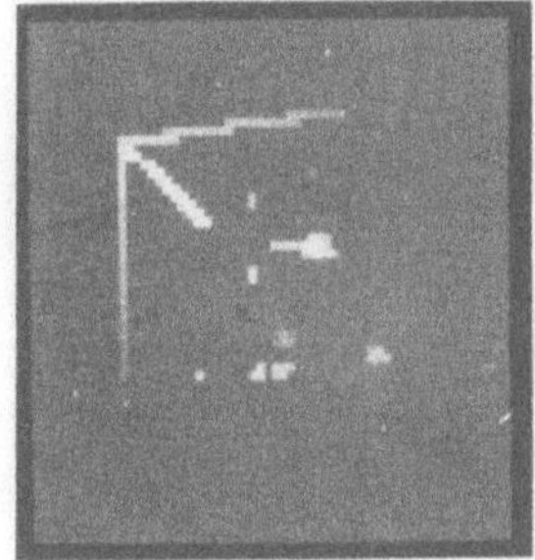

(b)noise/data = 30%

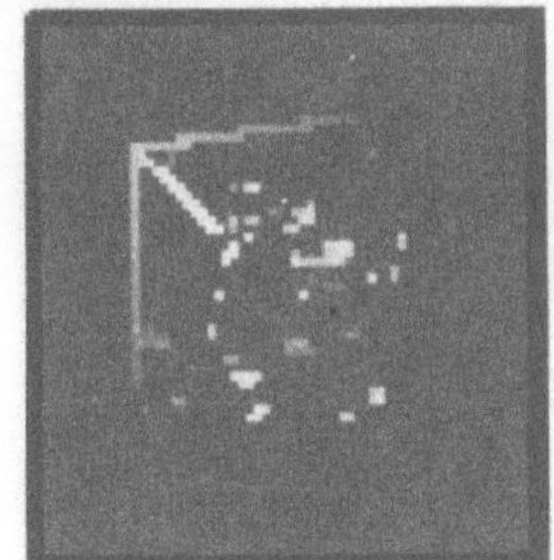

(c)noise/data = 50%

Fig.11 Reconstructed image using noisy data.

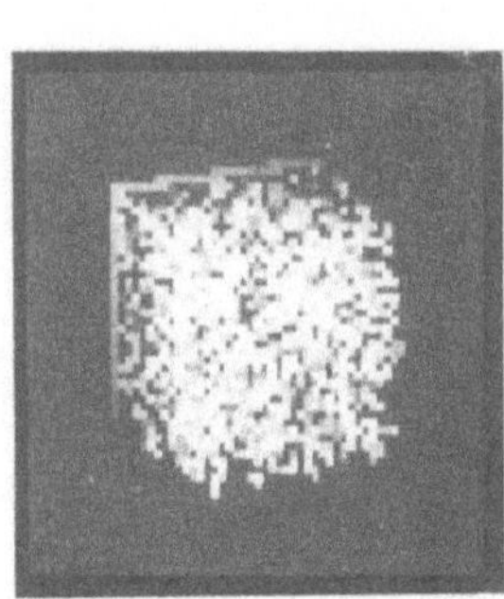

Fig.12 Reconstructed image + noise 10%.

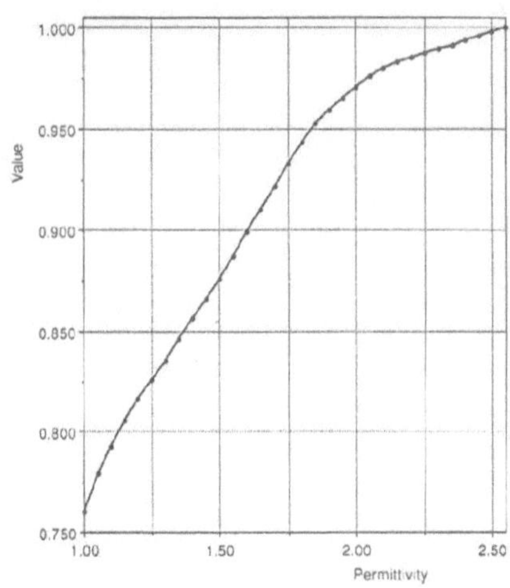

Fig.13 Estimation of permittivity for noise data(10%).

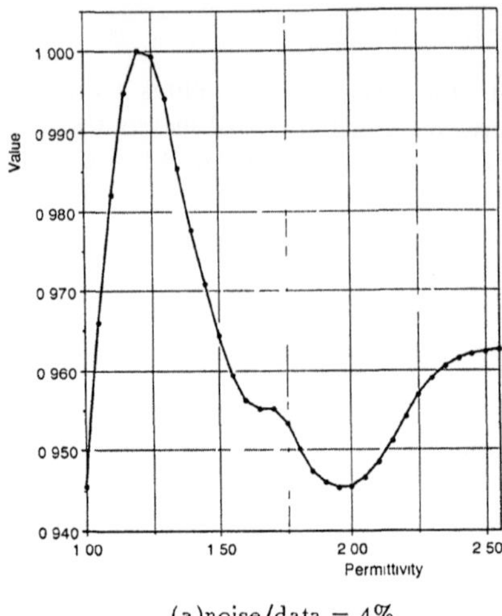

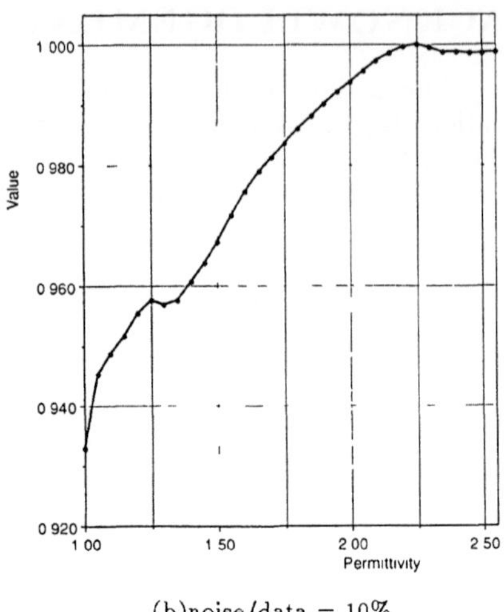

(a)noise/data = 4% (b)noise/data = 10%

Fig.14 Estimation of permittivity for noise added data.

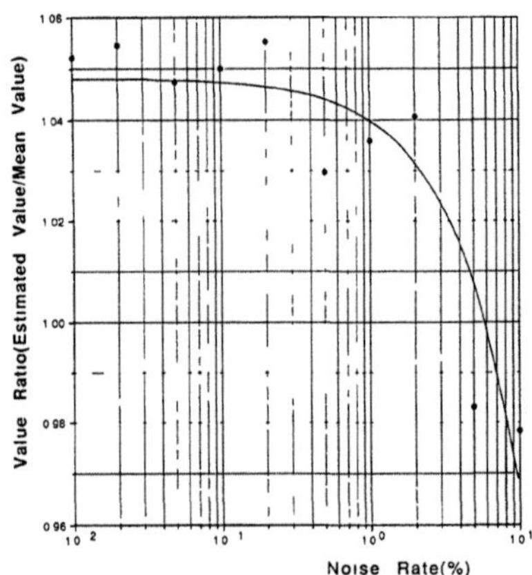

Fig.15 Value for mean ratio.

REFERENCES

[1] Y. Sakamoto, K. Tajiri, T. Sawai. and Y. Aoki, "Three-dimensional Imaging of Objects in Accumulated Snow Using Multifrequency Holography",IEEE Trans. Geosci. Remote Sensing, 26,4,pp.430-436(July 1988)

[2] Y. Aoki, Y. Takahashi, Y. Sakamoto, and M. Ikegami, " Technique of Volume Images of Buried Objects in Piled Snow by Acoustical and DisplayMicrowave Holographic Radar",Acoustical Imaging, vol.17,pp.285-294 (Plenum Pub Co. 1989)

[3]R. Mitsuhashi, Y. Sakamoto, and Y Aoki."A Method of Image Reconstruction for Snow Search Radar Considering the Propagation of Multi-Frequency Holograms in Snow", IEICE, Tech. Rep, A·P91-73(1991)

[4] Cumming W.A.,"The Dielectric Properties of Ice and Snow at 3.2 Centimeters", Jour.of Applied Phys.,23-7,pp.768-773(1952)

[5] T. Sato, R. Mitsuhashi, and Y Aoki."Development of 3D-Data Display System Based on Voxel Space",4th-SSICCA,pp.207-210(1990)

A 3-D OBJECT CLASSIFICATION METHOD COMBINING ACOUSTICAL IMAGING WITH PROBABILITY COMPETITION NEURAL NETWORKS

Sumio Watanabe and Masahide Yoneyama

Research and Development Center, RICOH Co., Ltd.
16-1, Shin-ei-cho, Kohoku-ku, Yokohama, 223 Japan

1. INTRODUCTION

It has been shown that acoustical imaging is useful for 3-D object recognition[1][2][3]. Different from the video-based systems, acoustical imaging enables us to obtain 3-D information directly from objects, even for metal or glass objects without regard to the environmental optical conditions. The low image resolution, which is a defect of acoustic images, is overcome by applying neural networks to recognition process. By these reasons, the neuro-ultrasonic 3-D visual sensor is now used for an automatic classification system in a lens production line[4].

Although neural networks automatically realize statistical inference even for unknown probability distributions, they still have some problems. (1) Conventional neural networks can classify several objects, but cannot identify objects. For example, they cannot answer 'unknown' for unlearned objects. (2) It is difficult to interpret what neural networks obtained by learning. By this reason, it is also difficult to improve neural networks when the ordinary learning mode is finished.

In this paper, a new 3-D object recognition system is proposed, combining acoustical imaging with a probability competition neural network (PCNN). First, we explain the 3-D acoustical imaging by combining acoustical holography with time flight ranging, and then propose 3-D image recognition method using a PCNN, which overcomes the above problems, and at last report experimental results. By the novel system, 3-D ultrasonic images are not only classified based on a posterior probability, but also identified by estimating the familiarity of the input samples.

2. 3-D ACOUSTICAL IMAGING

Fig.1 illustrates the relative positioning of a transmitter, a receiver array, and an object to be recognized. The transmitter illuminates the object with an incident wave, emitted at $t = 0$. The location of the transmitter is $\mathbf{r}_0 = (x_0, y_0, z_0)$. The sound pressure of the incident wave $P_{in}(\mathbf{r}, t)$ at location $\mathbf{r}$ and time t is then given by the following equation when $|\mathbf{r} - \mathbf{r}_0|$ is long compared to the wavelength.

$$
\begin{aligned}
P_{in}(\mathbf{r}, t) &= \Theta(ct - |\mathbf{r} - \mathbf{r}_0|)\exp(j\mathbf{k}_{in} \cdot (\mathbf{r} - \mathbf{r}_0) - j\omega t) \\
\mathbf{k}_{in} &= (k\sin\theta_0, 0, -k\cos\theta_0)
\end{aligned}
\tag{1}
$$

Acoustical Imaging, Volume 20 Edited by Y Wei
and B Gu, Plenum Press, New York, 1993

where c is the velocity of the sound, w is the angular frequency, k is the wave number vector, θ_0 is the illuminating angle, and $\Theta(x)$ is a step function whose value is 0 if $x < 0$, or 1 if otherwise. It is assumed that the reflection coefficient of the object is $\xi(x', y')$, and that its surface function is $z' = \zeta(x', y')$. The length $L(\mathbf{r}, \mathbf{r}')$ from the transmitter $\mathbf{r}_0$ to the receiver $\mathbf{r} = (x, y, z)$ through the path containing the surface point $\mathbf{r}' = (x', y', \zeta(x', y'))$ of the object is defined by $L(\mathbf{r}, \mathbf{r}') = |\mathbf{r}' - \mathbf{r}_0| + |\mathbf{r}' - \mathbf{r}|$. The complex sound pressure of the scattered wave at the receiver $\mathbf{r}$ is given by eq.(2) using the Kirchhoff approximation [4].

$$P(\mathbf{r}, t) = \frac{j \exp(jkr)}{4\pi r} \int_{-\infty}^{+\infty} dx' \int_{-\infty}^{+\infty} dy' \exp(j\mathbf{V} \cdot \mathbf{r}')\xi(x', y')\Theta(ct - L(\mathbf{r}, \mathbf{r}')) \quad (2)$$

$$\text{where} \qquad \mathbf{r} = (x, y, z), \qquad \mathbf{r}' = (x', y', \zeta(z', y')), \qquad \mathbf{V} = (V_x, V_y, V_z),$$

$$V_x = -k(\frac{x}{r} - \sin\theta_0), \quad V_y = -k(\frac{y}{r}), \quad V_z = -k(\frac{z}{r} + \cos\theta_0), \quad F(\mathbf{r}) = \frac{|\mathbf{V}|^2}{V_z}.$$

In the experiment, the height of the receiver array is set to z. The center of the receiver array is $(z \tan\theta_0, 0, z)$. Then, focusing on the origin causes $|\mathbf{r}'|$ to be small enough that $L(\mathbf{r}, \mathbf{r}')$ can be approximated by

$$L'(\mathbf{r}, \mathbf{r}') = r_0 + r - (2\cos\theta_0) \cdot \zeta(x', y'), \quad \text{where} \quad r = |\mathbf{r}|, r_0 = |\mathbf{r}_0| \quad (3)$$

Applying eq.(3) to eq.(2), and replacing t by the height z', using $z' = (r + r_0 - ct)/(2\cos\theta_0)$, then,

$$\begin{aligned}
P(\mathbf{r}, (r + r_0 - 2z'\cos\theta_0)/c) &= \frac{j\exp(jkr)}{4\pi r} \int_{-\infty}^{+\infty} dx' \int_{-\infty}^{+\infty} dy' \exp(j\mathbf{V} \cdot \mathbf{r}') \\
&\quad \cdot \xi(x', y')\Theta((2\cos\theta_0)(\zeta(x', y') - z')) \quad (4)
\end{aligned}$$

The inverse Fourier transform of the variables $(kx/r, ky/r)$ results in

$$\begin{aligned}
&\xi(x', y')\Theta((2\cos\theta_0)(\zeta(x', y') - z')) \exp(-2jk\cos\theta_0\zeta(x', y') + jkx'\sin\theta_0) \\
&= \frac{(kz)^2}{\pi} \int_{-\infty}^{+\infty} dx \int_{-\infty}^{+\infty} dy \exp(j\frac{k}{r}(xx' + yy')) \frac{P(\mathbf{r}, (r + r_0 - 2z'\cos\theta_0)/c)}{jF(\mathbf{r})r^3 \exp(jkr)} \quad (5)
\end{aligned}$$

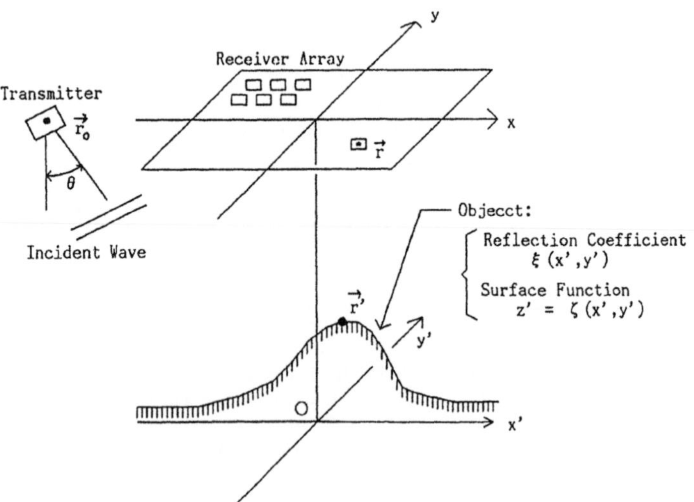

Fig.1 Arrangement of a Transmitter, a Receiver Array, and an Object.

Let $A(x', y', z')$ be the absolute value of the right-hand side of eq.(5), then

$$A(x', y', z') = \begin{cases} \xi(x', y') & \text{(if} \quad \zeta(x', y') \geq z') \\ 0 & \text{(otherwise)} \end{cases} \tag{6}$$

If the pressures $P(\mathbf{r}, (r + r_0 - 2z' \cos \theta_0)/c)$ are measured and substituted into eq.(5) and then evaluated according to eq.(6), the 3-D image of the object can be reconstructed.

3. A PROBABILITY COMPETITION NEURAL NETWORK

In this section, the learning structure of neural networks is explained, and a probability competition neural network (PCNN) is proposed (see [5], [6]).

Let $x \in R^M$ and $y \in R^N$ be an input and an output vector. A probability density function $P(w; x, y)$, where w is a parameter, is called an artificial neural network. For example, a function approximation neural network (FANN) is given by

$$P(w; x, y) = \frac{R(x)}{(2\pi\sigma)^{N/2}} \exp(-\frac{\|y - \varphi(w; x)\|^2}{2\sigma^2}) \tag{7}$$

where $R(x)$ is a density function on the input space, σ is a deviation of outputs, $\varphi(w; x)$ is an approximating function which is realized by a multi-layered perceptron (MLP), a radial basis function (RBF), and so on. Note that FANN cannot answer 'unknown' for unlearned objects because it does not esimate $R(x)$.

When x is input into a neural network, the output of the neural network is a random sample taken from the conditional probability density function, $P(w; y|x) = P(w; x, y)/P(w; x)$. The output of FANN is an expectation,

$$\varphi(w; x) = \int yP(w; y|x)dy = \frac{\int yP(w; x, y)dy}{\int P(w; x, y)dy}. \tag{8}$$

Note that $P(w; x) = \int P(w; x, y)dy$. The output function $\varphi(w; x)$ is calculated by a MLP in Fig.2.

The learning rule of neural networks is given by the maximum likelihood method. When teaching samples $\{(x_i, y_i); i = 1, 2, ..., S\}$ are given, the weight parameter w is optimized by

$$\frac{dw}{dt} = \sum_{i=1}^{S} \frac{\partial}{\partial w} \log P(w; x_i, y_i) \tag{9}$$

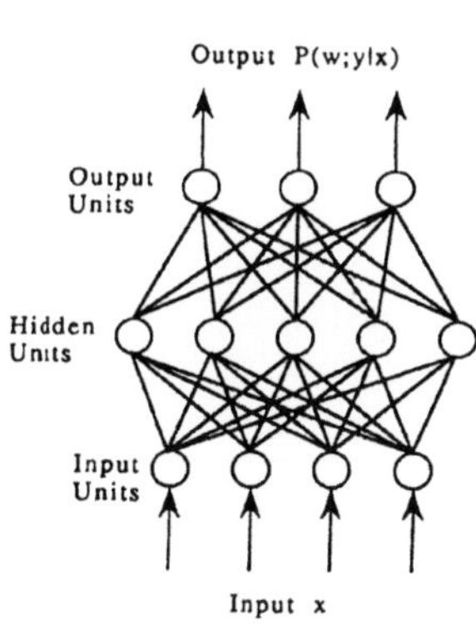

Fig.2 A Multi-Layered
Perceptron (MLP).

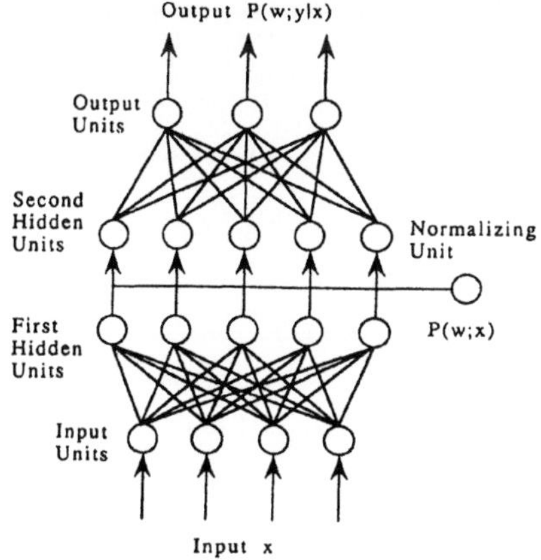

Fig.3 A Probability Competition
Neural Network (PCNN).

Note that the error backpropagation learning rule and the learning rule of the Boltzmann machine are derived from this unified learning rule.

A probability competition neural network (PCNN) is defined by the following probability density function.

$$P(w; x, y) = \frac{1}{Z(\theta)} \sum_{i=1}^{H} \exp(\theta_h) R(\xi_h, \rho_h; x) S(\eta_h, \sigma_h; y), \tag{10}$$

where $R(\xi, \sigma; x) = (1/\sigma^M) R((x - \xi)/\sigma)$, $S(\eta, \rho; y) = (1/\rho^N) S((y - \eta)/\rho)$, and $Z(\theta) = \sum_{h=1}^{H} \exp(\theta_h)$. For $R(x)$ and $S(x)$, a Gaussian function $(1/(2\pi)^{M/2})\exp(-\|x\|^2/2)$, a Laplace function $(1/2^M)\exp(-\|x\|^2)$, and other functions are used.

The output value of PCNN for a given input x is a sample taken from the conditional probability density function,

$$P(w; y|x) = \sum_{i=1}^{H} \alpha_h(x) S(\eta_h, \rho_h; y), \quad \alpha_h(x) = \frac{\exp(\theta_h) R(\xi_h, \sigma_h; x)}{\sum_h \exp(\theta_h) R(\xi_h, \sigma_h; x)} \tag{11}$$

The sample from this function is given by PCNN in Fig.3. The output of the first hidden unit is $o_h^{(1)}(x) = \exp(\theta_h) R(\xi_h, \sigma_h; x)$. The output of the normalizing unit is the sum the first hidden units. $o(x) = \sum_h o_h^{(1)}(x)$. Then we have $\alpha_h(x) = o_h^{(1)}(x)/o(x)$. The output vector of the second hidden units are set $(0, 0, 0, 1, 0, 0)$ (only h-th unit is 1.) with probability $\alpha_h(x)$. The output of PCNN is given by a sample from $S(\eta_h, \rho_h; y)$.

Remark that the output $o(x)$ of the normalizing unit is exactly equal to $P(w; x)$. Since $o(x)(= P(w; x))$ shows a familiarity of the input x, PCNN can answer 'unknown' when the normalizing unit outputs a small value.

The learning rule is given by the maximum likelihood method, eq.(9). After learning, the obtained parameters $w = \{\theta_h, \xi_h, \sigma_h, \eta_h, \rho_h; h = 1, 2, ..., H\}$ becomes to be the maximum likelihood estimator. It is known that w goes to the best parameter faster than any other estimators, when the number of the learning samples goes to infinite. (The maximum likelihood estimator is the BAN estimator.)

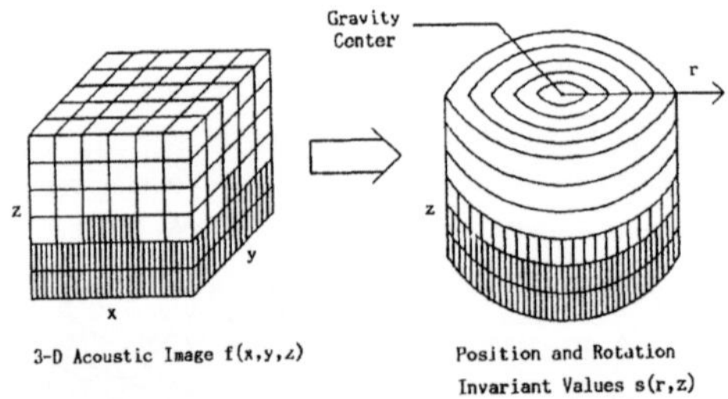

Fig.4 Calculation of the Feature Values.

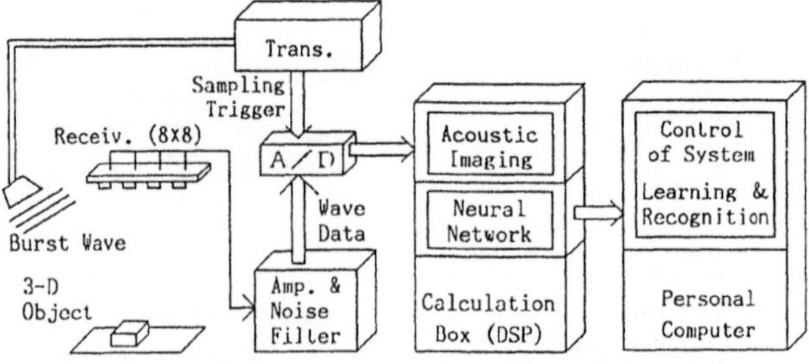

Fig.5 A System Block Diagram.

4. RECOGNITION OF 3-D ULTRASONIC IMAGES

For practical use of the object recognition system, for example, recognition of objects on the conveyer belt, it is very important to identify objects without regard to their positions and rotations. To apply neural networks to such a system, learning patterns of all locations and rotations are needed.

By this reason, to reduce the number of learning samples, and the dimension of input vectors, we use position and rotation invariant feature values. Let $f(x, y, z)$ be a 3-D image of the object. We define $s(r, z)$,

$$s(r, z) = \int_{D(r)} f(x, y, z)dxdy, \quad D(r) = \{(x, y); r^2 \le x^2 + y^2 < (r + a)^2\} \quad (12)$$

Fig.4 shows the method how to calculate this value. Note that, although $s(r, z)$ is theoretically invariant under parallel shifts or rotations, it distributes in experiments, because of the small number of image pixels, direction dependence of ultrasonic illumination, and environmental noises. The PCNN learns such a distribution by statistically optimal learning.

5. SYSTEM STRUCTURE

Fig.5 shows the system block diagram of the 3-D object classification system. For practicability of the system, a very small number of receivers (8 × 8) is used. Fig.6 is a photograph of the prototype neuro-ultrasonic 3-D visual sensor, which consists of a receiver array, a calculation box, and a personal computer. The recognition procedure is as follows.

(1) 40KHz ultrasonic burst waves (5 cycles of sine waves) are illuminated onto the object, and the sound pressures of scattered waves are measured by 8 by 8 receivers.

(2) The complex pressure, $P(\mathbf{r}, t)$, is calculated by the inner product of measured waves and referential sine or cosine waves.

(3) A 3-D image $f(x, y, z)$ is calculated from $P(\mathbf{r}', t)$ by acoustical imaging.

(4) Object categories are classified by a neural network.

In the experiments, we compared performance of PCNN with MLP.

6. EXPERIMENTAL RESULTS

(1) Experiments.

For experiments, 30 objects in Fig.7 are used. Objects in Fig.8 are used for testing neural networks. Three objects in Fig.8 are contained in Fig.7, and other three objects are not. A reconstructed images for a spanner in Fig.7 by acoustical imaging is shown in Fig.9. 12 cross sections of 3-D image is ordered from the heigher to the lower. A clear image was obtained compared with a small number of receivers.

Fig.10 shows examples of the position and rotation invariant feature values for a spanner, pliers, an object 'A', and 'B'. They are represented on a (r,z) plane.

Let L be the distance between the center of the object and the origin, and θ be the rotation angle of the object (Fig. 11). The following two groups were collected.

[Group.1] 30 objects in Fig.7 were placed at the origin $(L = 0)$, and rotated by $\theta = 0°$ and $45°$ degree. 10 samples were collected for each object and rotation angle.

[Group.2] 30 objects in Fig.7 were placed at $L = 20mm$, and rotated by $\theta = 0°$,5 ,10 ,15 ,...,45° degree. 10 samples were collected for each object and angle.

[Ex.1] The classification rates by PCNN were compared with those by MLP. Both neural networks were taught using Group.1, and tested using Group.2. The experimental results are shown in Fig.12. It is clear that both neural networks can classify objects with almost same performance. PCNN needs more hidden units than MLP.

[Ex.2] PCNN which were taught using group.1 was tested using objects in Fig.8. The outputs of the normalizing unit, which is theoretically equal to the familiarity of the input, is shown in Fig.13. They were small for unknown objects and large for known objects, which shows that PCNN can answer 'unknown' for unknown objects.

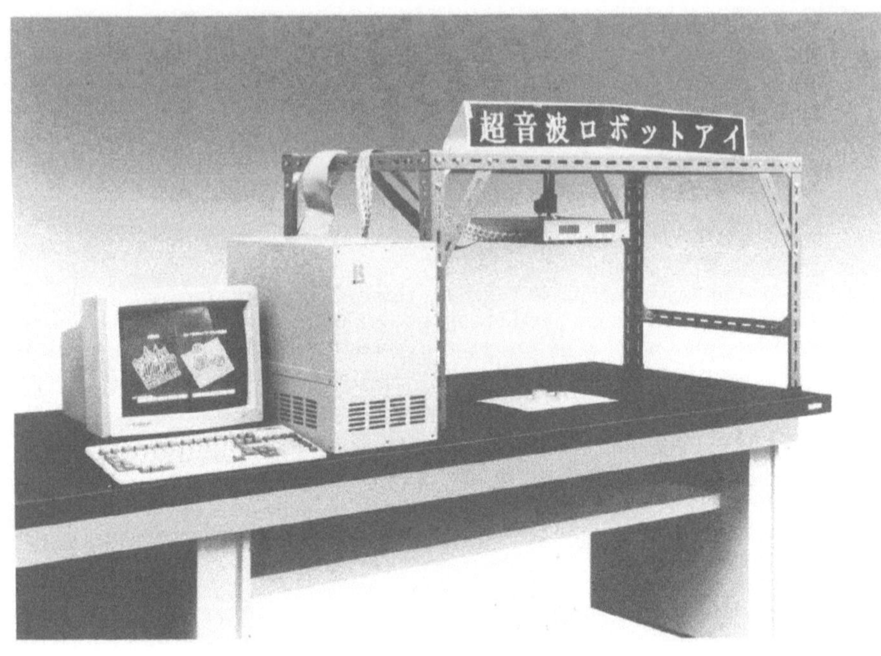

Fig.6 A Prototype Neuro-Ultrasonic 3-D Sensor.
An improved type is used in a lens production line.

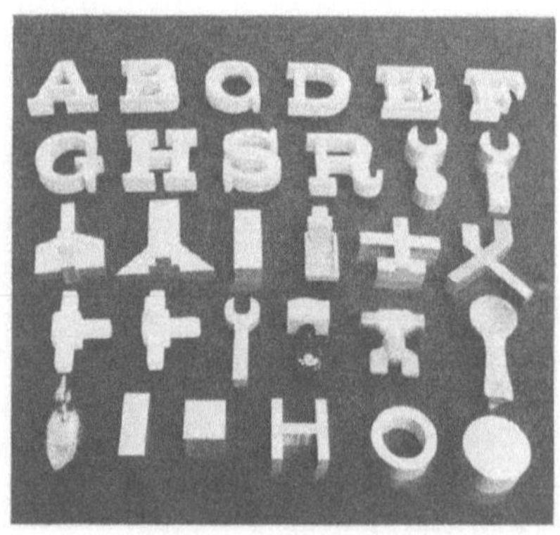

Fig.7 30 Objects Used in Experiments.

Fig.8 Known Objects
and Unknown Objects.

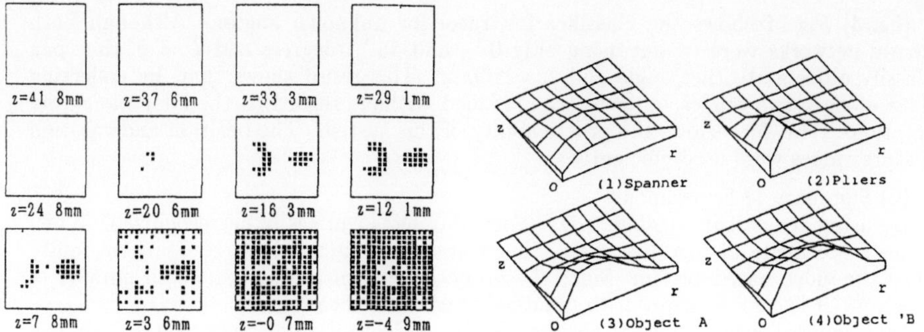

Fig 9 A 3-D Acoustical Image of A Spanner

Fig 10 Examples of Position and Rotation Invariant Feature Values

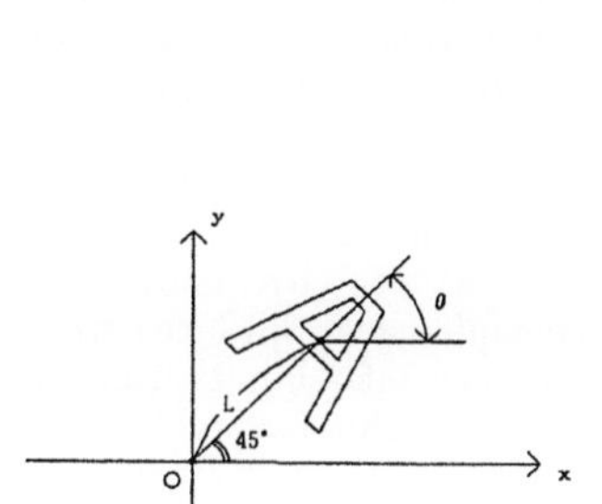

Fig 11 A Location and an Angle of the Object

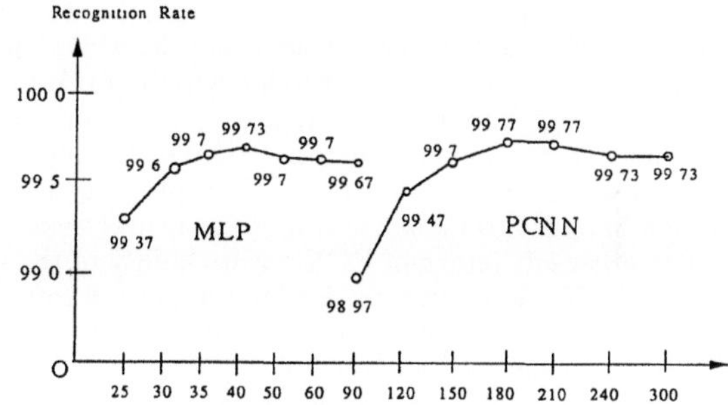

Fig 12 Recognition Rates by MLP and PCNN

Problem   Classification of 30 Objects
Input Data   Position and Rotation Invariant Values
Learning Data   Center=Origin 0 45 deg rotated
Tested Data   Center=20mm 0 5 10   45 deg rotated

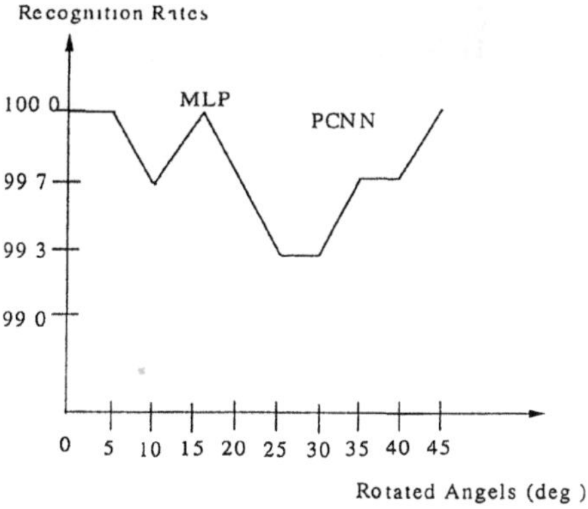

Fig 14 Recognition Rates and Rotated Angles

Learned   0 45 deg   Center=Origin

Tested 0 5 10   45 deg
Center=20mm from origin

Objects	Outputs of Normalizing Unit (Averaged 10 Samples log)	
Learned	Cube Block Spanner	5 5 12 3 7 2
Unknown	Pyramid Sphere Cylinder	136 3 72 1 56 5

Fig 13 PCNN can answer "Unknown

[Ex.3] Fig.14 shows the classification rates for unknown angles. Although both neural networks were taught using only 0° and 45° degrees and $L = 0$, they can classify objects at other angle and $L = 20mm$. This result shows that, by collecting data of only two angles, objects are classified without regard to their locations and rotations. This fact shows the practicability of this system. The position and rotation feature values were used efficiently.

(2) Summary of Experiments

By using acoustical imaging, 3-D shapes of objects can easily be recognized. When objects are made of metal or glass, this system has much better recognition ability than the video-based system. Since the proposed system uses only a small number of receivers, it is easy to apply it to practical object classification.

We proposed a novel neural network model, PCNN. Fig.11 shows that PCNN can classify objects with at least same performance as MLP. However, PCNN needs more hidden layer units than MLP, because MLP only classifies patterns but PCNN estimates the distributions of patterns.

Both neural networks infer based on the conditional density function $P(w; y|x)$. However, it should be emphasized that, when $P(w; x) \approx 0$ (the input is almost unknown), then $P(w; y|x) \approx 0/0$ (the output is not defined). That is to say, neural networks' outputs are with no assurance for unknown inputs. By this reason, it is very important to judge whether the inputs are known or unknown. Although MLP and the Boltzmann machine can not answer the familiarity of the inputs, PCNN can answer it, resulting in the assurance of the outputs. This property is important also because, when the system happens to be out of order, the neural network can notice the irregularity.

The other advantage of PCNN is its ability of the inverse inference. Since the structure of PCNN is symmetric for inputs and outputs, PCNN can infer inputs from the outputs by using the conditional probability $P(w; x|y)$ ([5],[6]). This fact means that PCNN can illustrate several input patterns for each category, and that PCNN can answer what is obtained by learning. This ability shows that PCNN can learn the concepts by "human-PCNN communication", even after the ordinary learning session.

7. CONCLUSION

In this paper, we proposed a new 3-D object classification system combining acoustical imaging with a probability competition neural network. By using acoustical imaging, 3-D information can easily be measured even for metal or glass objects. By using a PCNN. they are not only classified but also identified.

REFERENCES

[1] S.Watanabe and M.Yoneyama, "Ultrasonic Robot Eyes Using Neural Networks," IEEE Trans. on UFFC, Vol.37, pp.141-147, 1990.

[2] S.Watanabe and M.Yoneyama, "An Ultrasonic 3-D Visual Sensor Using Neural Networks," IEEE Trans. on Robotics and Automation, Vol.6, No.2, pp.240-249, 1992.

[3] S.Watanabe, et. al., 'An Application of Neural Networks to an Ultrasonic 3-D Visual Sensor," Proc. of IJCNN, (Singapole), pp.1397-1402, 1991.

[4] A.Ishimaru, "Wave Propagation in Random Media, Vol.2," Academic Press, New York, 1978.

[5] S.Watanabe and K.Fukumizu, "The Unified Neural Network Theory and Its Application to New Models," Proc. of the 2nd Int. Conf. on Fuzzy Logic and Neural Networks, pp.725-728, 1992.

[6] S.Watanabe and K.Fukumizu, "The Unified Neural network Theory and Proposal of New Models," to appear in Proc. of IJCNN (Beijing), 1992.

[7] S.Watanabe and M.Yoneyama, "An Ultrasonic Visual Sensor using a neural network and Its Application for Automatic Object Recognition," Proc. of IEEE Ultrasonics Symp., pp.781-784, 1991.

[8] S.Watanabe and M.Yoneyama, "An Ultrasonic Robot Eye Using Neural Networks," Acoustical Imaging, Vol.18, edited by H.Lee and G.Wade, pp.83-95, 1990.

SIGNAL WAVELET ANALYSIS IN ULTRASONIC IMAGE *

Jinqiu Mo,[1] Xiaojun Zhou,[1] Yaodong Cheng[1]
Jin Qi,[1] and Youlin Cheng[2]

[1]Mechanical Engineering Department
Zhejiang University
Hangzhou 310037

[2]Center of Nondestructive Evaluation
Chengdu Aircraft Corporation
Chengdu 610092

P. R. China

INTRODUCTION

The detection and analysis of the singularity of a signal is extended into the field of acoustic signal analysis. In processing of sonic signals with abrupt changes, e. g. , of ultrasonic reflected pulse signal the usages of signal processing methods of pure time or frequency domain are limited because of the singularity of these kinds of signals. In this paper, Wavelet Transform (WT), a time-frequency domain analysis method, is employed as a basic tool for A-mode ultrasonic signals feature extraction. And an implemental algorithm for generating of B-mode ultrasonic image based on zero-crossing detection of odd WT and quantification of even WT is also presented. In the end of the paper, a practical experiment example is given.

CONTINUOUS WAVELET TRANSFORM

Wavelet analysis is a new mathematical theory which has been developed in the past few

* This work is supported by Natural Science Foundation of Zhejiang Province ang Chengdu Aircraft Industrial Corporation, China.

years, and its successful application in engineering shows that it has great potentialities in signal and image processing. [1,2,3,4,5]

A wavelet is a function $\varphi(t) \in L^2(R)$, whose Fourier transform satisfies the following admissible condition:

$$\int_{-\infty}^{+\infty} \frac{|\hat{\varphi}(\omega)|^2}{|\omega|} < \infty. \tag{2.1}$$

Let us denote by ψ_s the dilation of φ by a scale factor s :

$$\psi_s(t) = \frac{1}{s}\,\varphi\left(\frac{t}{s}\right). \tag{2.2}$$

The WT of function $f(t) \in L^2(R^2)$ at scale s and position t is given by the convolution product:

$$W_s f(t) = f \otimes \psi_s(t), \tag{2.3}$$

We call a WT an even WT if φ is an even function and an odd WT if φ is an odd one.

Condition (2.1) implies that $\hat{\varphi}(0) = 0$, and thus WT is a multiresolution filtering operation with adjustable band-width when taking different scale s. So WT provides us a tool to analyze the local characteristics of a given signal in time-frequency domain.

ULTRASONIC REFLECTED PULSE MODEL

In the method presented in this paper, an ultrasonic reflected pulse signal $f(t)$ is split into two parts: slowly-varying part $f_0(t)$ and instantly-varying part $p(t)$

$$f(t) = f_0(t) + p(t) . \tag{3.1}$$

Here slowly-varying means loose stationary. We assume

(1) f_0 has finite energy to make sure that f_0 has Fourier Transform (FT) .

$$\hat{F}_0(t) = \int_{-\infty}^{+\infty} e^{-i2\pi\omega t} f_0(t)\, dt . \tag{3.2}$$

(2) f_0 has a zero mean, $E\left[f_0\right] = 0$, it is acceptable in general.

(3) f_0 is a band limited signal with frequency band-width B, which satisfies the relationship equation with the upper band frequency f_{max} of $p(t)$

$$f_{max} \gg B . \tag{3.3}$$

For the convenience of analysis, the instantly-varying part $p(t)$ in an ultrasonic reflected pulse signal $f(t)$ can be simply viewed as a pulse $\delta(t)$ or has triangular waveform $\Delta_{K,T}(t)$, they are:

$$\delta(t) = \begin{cases} \infty, & t = 0, \\ 0, & \text{otherwise} \end{cases} \quad \text{and} \quad \int_{-\infty}^{+\infty} \delta(t) = K \quad , \tag{3.4}$$

and

$$\Delta_{K,T}(t) = \begin{cases} K(t+T), & -T < t < 0 \\ -K(t+T), & 0 \leqslant t < T \\ 0, & \text{otherwise} \end{cases}, \tag{3.5}$$

where $K > 0$ is a constant.

A-MODE ULTRASONIC SIGNAL DETECTION USING WT

Choosing a real symmetric or antisymmetric function as basic wavelet φ and taking WT in the two sides of Equation (3.1), we get

$$W_s f(t) = f_0 \otimes \psi_s(t) + p \otimes \psi_s(t) = f_0 \otimes \psi_s(-t) + p \otimes \psi_s(-t). \tag{4.1}$$

Taking FT of $W_s f$, we have

$$\hat{W}_s f(\omega) = \hat{f}_0(\omega) \hat{\varphi}(-s\omega) + \hat{p}(\omega) \hat{\varphi}(-s\omega). \tag{4.2}$$

Assume φ has a pass-band $[-b, -a] \cup [a, b]$, and the frequency band of f_0 is $[-B, B]$, take

$$s < \frac{a}{B}, \tag{4.3}$$

that is equal to

$$[-B, B] \cap \left(\left[\frac{-b}{s}, \frac{-a}{s} \right] \cup \left[\frac{a}{s}, \frac{b}{s} \right] \right) = \varnothing, \tag{4.4}$$

then Equation (4.2) can be approximately written as

$$W_s f(t) \approx p \otimes \psi_s(-t) = W_s p(t), \ \left(s < \frac{a}{B} \right). \tag{4.5}$$

This means that after the WT operation, $f_0(t)$ is filtered out while $p(t)$ passes.

Equation (4.5) shows that the abruptly changing point of $p(t)$ can be detected by its WT. According to the local continuity and the uniqueness of $W_s p(t)$, if we set s small enough, then the abruptly changing point of $p(t)$ corresponds to a zero-crossing of $W_s p(t)$ when an odd WT is taken, and corresponds to a maximum point of $|W_s p(t)|$ analogously when an even WT is taken.[2,4]

For an ultrasonic reflected pulse signal composed of a slowly-varying wave f_0 and a pulse type wave p, if $s < a/B$, we have the following result:[4]

Take s to satisfy Equation (4.3) and suppose the basic wavelet to be odd. If t_0 is the point of the zero crossing of $W_s p$ with a sign-change $(-, +)$, it is a sharp point of $f(t)$ where peak goes upward and sign-change $(+, -)$ corresponds to a valley point. If the basic wavelet is even and t_0 is the maximum point of $|W_s p(t)|$, then it is a abruptly changing point of f while sign $+$ means the abrupt change is a passive peak, and sign $-$ the opposite.

WT BASED B-MODE ULTRASONIC IMAGING

In a practical ultrasonic system, the signals we get are blurred or distorted by the noise and the resolution limits of the detecting system. Using wavelet analysis method in ultrasonic image processing, we can reduce these blurring and distortion effectively.

As discussed above, the slowly-varying part in the ultrasonic reflected pulse signal is picked out when we take s to satisfy Equation (4. 3). The white noise, which has constant component along the frequency domain and zero mean in the time domain, will not affect the abrupt detection by WT, especially when using a zero-crossing method based on odd WT.

So we can form an implemental algorithm to generate the B-mode ultrasonic image from A-mode ultrasonic pulse signal peak/abrupt-point detecting and processing based on WT. That is, find the abrupt points of A-scan ultrasonic reflected pulse signal using the zero-crossing detection method based on odd WT and then get the amplitude of the abrupt wave using a modulus maxima detection method based on even WT. Thus, a B-mode ultrasonic image is basically constructed.

EXPERIMENTAL EXAMPLE

In the experiment, a computer controlled 3-directional scanner (Figure 1 (a)) was used. The ultrasonic detector was set in a pulse-echo mode and matched with a 2. 5 MHz, 2 mm diameter of ultrasound beam, 60 mm focal length transducer. The scan step size for data collection was 0. 2 mm. Figure 1 (b) gives the diagram of the experiment.

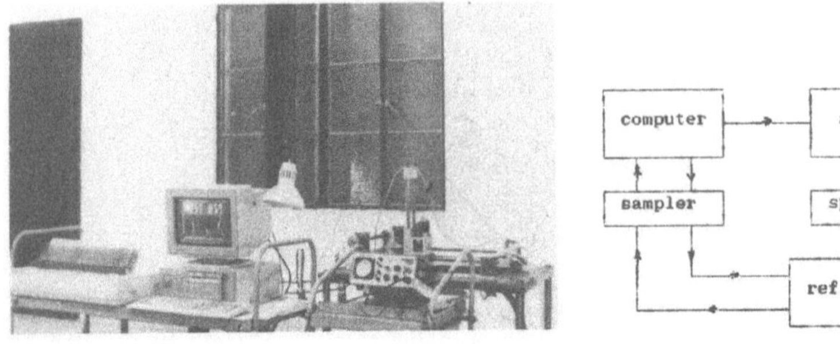

(a) The imaging system used
in the experiment

(b) The principle and procedure
of the experiment

Figure 1.   Experiment system

Figure 3 shows the experimental example of the application of our implementation. The experiment was taken on an Al specimen with a sealed oil hole. The scanning section drawing of it is shown in Figure 2. Figure 3 (a) is a B-mode image produced directly by the A-scan signal. Figure 3 (b) is a B-mode image generated with our implementation presented above.

Comparing Figure 3 (a) with 3 (b), it is very apparent that the quality of the image is greatly improved. In Figure 3 (a), the dark image located at the center produced by the upper-inner half surface of the oil hole (S in Figure 2) is severely stained with the image produced by the up surface of the specimen. And its shape is confusively upside down with the shape of S. Moreover, the small cake like dark image located at the middle down position of Figure 3 (a), which was generated by the second reflected sonic pulse by S, is unpleasantly big. In Figure 3 (b), however, the position of the oil hole can be easily located. The "dark cake" becomes a tiny piece of line and is no more so impressive.

Figure 4 was got when taking $s = 4$ while Figure 3 $s = 2$. More details of the image are presented in a smaller scale case. Some algorithm for removing the blurring and distortion brought in not only by noise and the resolution limits of imaging system, but also by certain physical effect, e. g. secondary reflection effect, should also be considered.

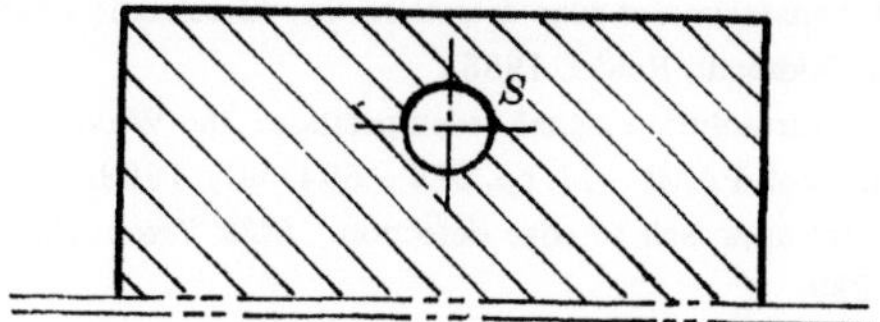

Figure 2.   The scanning sectional drawing of the specimen

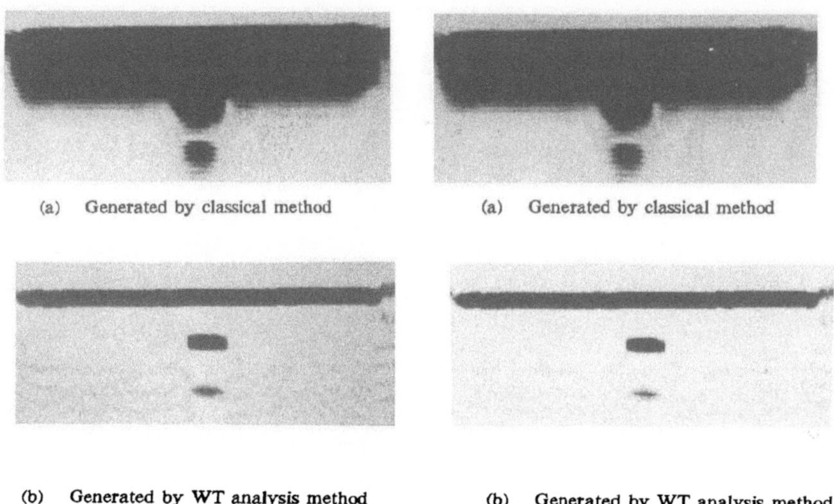

(a)   Generated by classical method

(a)   Generated by classical method

(b)   Generated by WT analysis method

(b)   Generated by WT analysis method

Figure 3.   The B-mode images ($s = 2$)

Figure 4.   The B-mode images ($s = 4$)

CONCLUSION

Wavelet analysis is an effective tool for the transient procedure analysis. We employ WT to construct a time-frequency domain characteristic recognition method and take WT as a link between B-mode image construction and the features extracted from the ultrasonic A-scan signals. The theory analysis and experimental example given show that WT can play an important part in the field of sonic signal and acoustic image processing. The result of the experiment shows that the implemental algorithm given in this paper is effective and efficient.

The problem we want to point out is that to solve different problem in applications, different kinds of wavelet should be chosen.

ACKNOWLEDGMENT

Thanks to Dr. Guosheng Hu for his former cooperational work which is helpful even to our present work.

REFERENCES

1. A. Grossmann, Wavelet transform and edge detection, in: "Stochastic Process in Physics and Engineering", M. Hazewinkel, ed. Dodrecht, Reidel, 1986.
2. S. Mallet, A theory for multiresolution signal decomposition: the Wavelet Representation, *IEEE Trans. Pattern anal. Machine Inteli.*, Vol. PAMI—11, pp. 674—694, July 1989.
3. J. Canny, A computational approach to edge detection, *IEEE Trans. Pattern Anal. Machine Inteli.*, Vol. 315, pp. 69—87, sept. 1989.
4. Xiao jun Zhou, Guosheng Hu, Zhongfang Tong, Yaodong Cheng, and Yumin Sun, Detection of monopulse-type signal using wavelet analysis, *J. Zhejiang Univ.*, Supplement, Mar. 1991.
5. Xiaojun Zhou, Guosheng Hu, Yaodong Cheng, Zhongfang Tong, and Junqiu Mo, Detection of step-type signal using wavelet analysis, *J. Zhejiang Univ.*, Vol. 26, Mar 1992.

ACOUSTICAL IMAGING USING CORRELATION TECHNIQUES: EXPERIMENTAL RESULTS

Mohamed A. Benkhelifa, Marcel Gindre, Jean-Yves Le Huérou and Wladimir Urbach

Laboratoire d'Imagerie Paramétrique URA CNRS n°1458
15 Rue de l'école de médecine
75006 Paris - France

INTRODUCTION

The implementation of matched filtering in acoustical imaging systems via correlation between the received signal and the delayed emitted signal allows a significant increase in the signal to noise ratio (SNR) of the echogram[1,2]. In correlation based systems, instead of a single periodic excitation function, a coded sequence is applied to the transducer, and the returning signals, corrupted by noise, are then correlated with a delayed replica of the driving sequence.

It has been shown that in non stationary media, circular correlations with maximal length, or m-sequences, lead to the best results[3]. Therefore in the following we will present experimental results obtained with m-sequences exclusively.

The maximal length sequence[4] is a sequence of binary digits which displays certain properties of randomness approximating those of white noise and is often referred as a pseudo random binary sequence or pseudo noise (P.N.). In fact, the sequence is perfectly deterministic and well defined mathematically.

An m-sequence is a two level signal of $N = 2^n - 1$ elements where n is an integer. Its generation is very simple and well suited for shift register implementation[5]. The corresponding periodic autocorrelation function, $C_{ee}(k)$, is

$$C_{ee}(k) = \sum_{j=1}^{N} e_i\, e_j \quad \text{with } k \geq 0 \text{ and } i = (j+k) \text{ modulo } N \tag{1}$$

where e_i is the ith bit of the sequence

EXPERIMENTAL SET UP

The digital correlation system assembled in our laboratory is shown in figure 1.

For maximum transmitted energy a phase shift keying (PSK) modulation was chosen. The binary code modulates a carrier at the central frequency of the transducer, $f_0 = 7.5$MHz. For good resolution the duration, τ, of each bit of the coding sequence should be as a short as possible, since after demodulation and correlation the signal output will be a triangular shaped waveform with base width 2τ. With the 2 MHz transducer's bandwidth (at - 3 dB) $\tau_{min} = 0.5$ µs. For practical reasons the experimental value of τ is actually 0.53 µs.

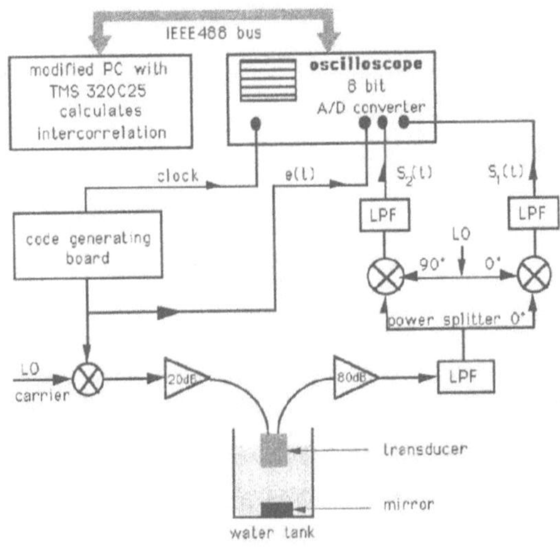

Figure 1 . Experimental set up used in correlation echography.
LPF is a low pass filter, LO : local oscillator signal, e(t), S_1(t) and S_2(t) are respectively the emitted code and the two received signals applied to the correlator.

To perform continuous emission, i.e. periodic correlation, we used a concave annular array. The concave transducer improves spatial resolution, and the eventual use of electronically focused annular phase arrays should allow us to vary the transducer's focal length around the geometrical preselected length of 10 cm. In the experiments reported here, the central part of the transducer (diameter ≈ 1.4 cm) is used as an emitter, whereas the nearest of six equi-area annuli is operated in receiver mode. In this configuration the lateral resolution is rather poor (figure 2). Nevertheless, it still allows an interesting comparison between the classical echography and correlation especially in echo A-mode, where the lateral resolution is unimportant. In the receiving unit the local carrier is derived from the same source as the transmitted carrier so that synchronous detection can be achieved. The return echoes are split into two channels and demodulated by the same

functions which are used for initial modulation but are separed from one another by a 90°
phase shift. The outputs, $S_1(t)$ and $S_2(t)$, from the two channels are correlated with delayed
replicas of the transmitted sequence, $e(t)$. The correlated results are then squared and
added, and the final result is displayed on the screen.

In our experimental set up a Lecroy oscilloscope is used as the data acquisition unit.
The demodulated 1023 bit signal is sampled and digitalized by an 8 bit A/D converter and
stored in the scope memory. The data acquired are then transferred to the microcomputer
memory where they are correlated with stored reference signals.

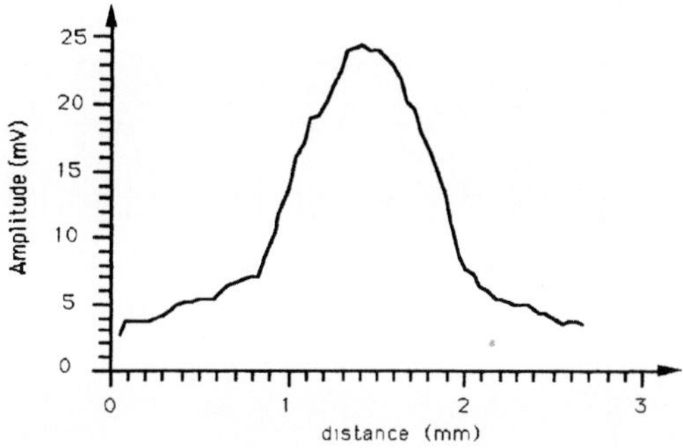

Figure 2 . Acoustical field in focal zone of the transducer

The beam width is 0.5mm à 3dB.

EXPERIMENTAL RESULTS

A-Type Images

Observations were first made using a block of stainless steel as a mirror.

In the classical pulse method the emitted signal was a 150 volt pulse.

For coded signal, the amplitude of the emitted signal was 1 volt. The code generated
at the frequency $f_c = \dfrac{1}{\tau} = 1.875$ MHz, which corresponds approximatly to the transducer
bandwidth at -3dB, was sampled at the frequency $f_s = k\, f_c$ where k is an integer. We
choose k = 4 i.e. $f_s = 7.5$ MHz, which is the central frequency of the transducer.

The classical echo-A type signal is shown in figure 3a. The SNR $\approx$ 30 dB is due to
the relatively low sensitivity of our transducer (no TGC was used).

The results obtained with the m-sequence is shown in figure 3b. The observed SNR
is nearly equal to 50 dB, which is 10 dB lower than the theoretically expected value. For a
code length of $N = 2^n - 1 = 1023$ bits the expected SNR value is 60 dB. This discrepancy
between theoretical and experimental values is mainly due to the limited bandwidth B of
the transducer. Indeed, if $E(f)$ is the spectral density of the emitted signal, the signal energy,
W, is simply

$$W = \int E(f)\, df \qquad (2)$$

and the energy transmitted to the medium by a transducer modelled by a Gaussian filter is

$$W_T = \int E(f) \exp \left\{ -\left[\frac{f - f_0}{B}\right]^2 \right\} df \qquad (3)$$

where B is the transducer bandwidth (at - 4.3dB).

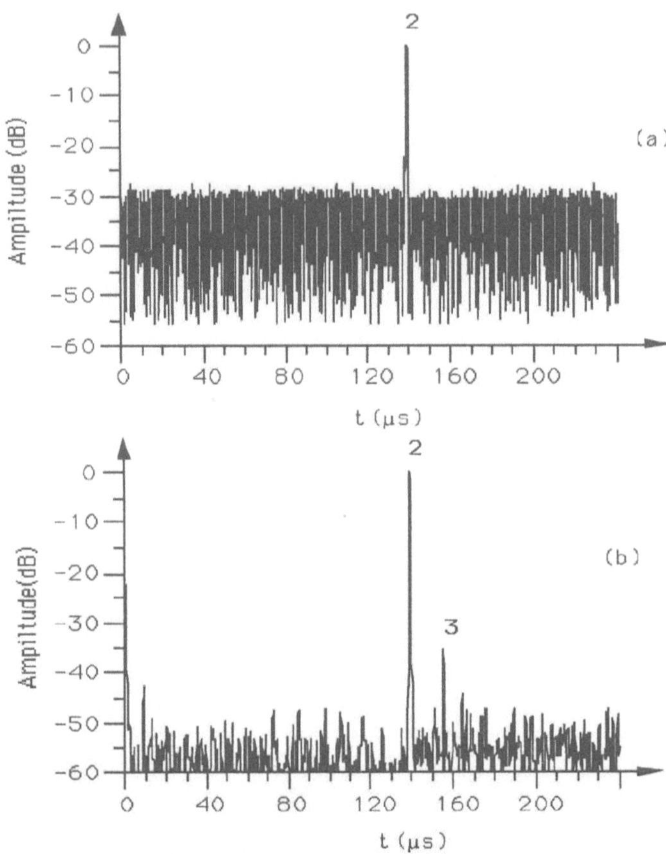

Figure 3 . A-type echogram obtained after a reflection on a metallic mirror submerged in a water tank

(a) with a classical pulse echography system (b) with the correlation system.

The label 2 corresponds to the front plane of the mirror and the label 3 to the back.

Thus the attenuation of the correlation peak is $10 \log_{10}\left[\frac{W_T}{W}\right]$, whereas the noise, mainly due to the electronics remains constant. With $f_0 = 7.5$ MHz and $B = 2$ MHz one obtains 6 dB which accounts for most of the 10 dB difference.

In figure 4 we present the results obtained when a small cylindrical tip 13 mm high and 0.2 mm diameter is placed in the front of the mirror as shown in figure 4a. The echo from the tip is designated by "1", the echo from the front plane of the mirror by "2" and its rear plane by "3". It is clear that the best result is obtained with m-sequences.

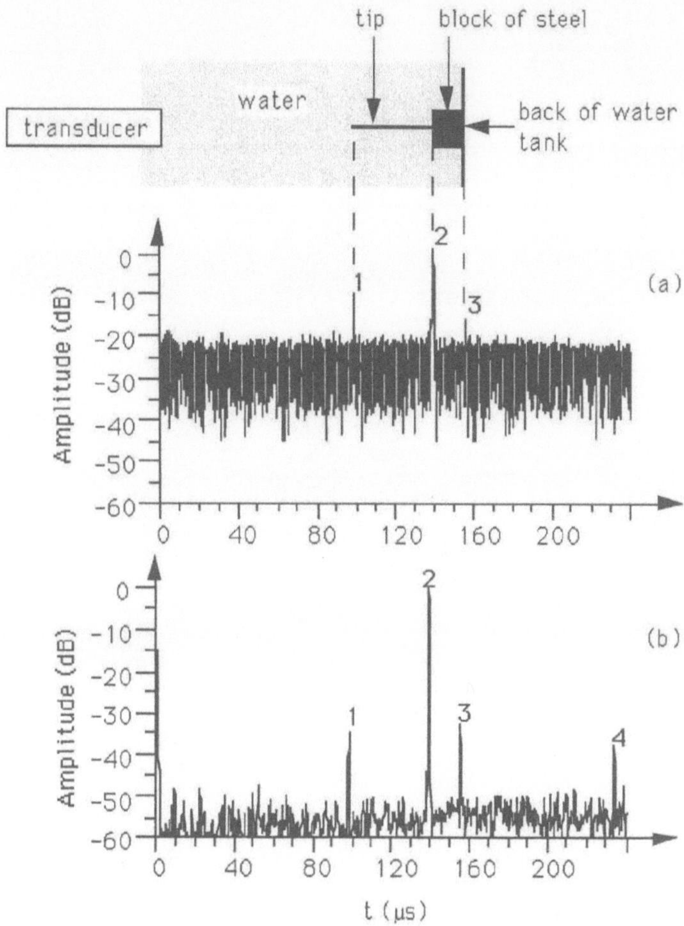

Figure 4 . Echo A type images obtained with a small cylindrical tip on a block of steel
with pulsed (a) and coded (b) signal

B-Type Images - Preliminary Results

A hole 3 mm in diameter and 4 mm in depth was drilled in the front plane of a
metallic block. With the same very broad ultrasonic beam we reconstructed from two
hundred A-type scans preliminary B-type images (see figure 5).

When the transducer is operating in the classical pulse mode, it yields the image
presented in fig 5a, where the bottom of the hole (lower line) is clearly visible. The
absence of a sharp discontinuity on the upper horizontal line of the echogram indicates that
the beam width is on the order of the diameter of the hole.

When m-sequences are used the walls of the drilled hole are clearly visible. At this
stage we do not have a clear explanation of the effect, but we suspect that this could be due
to mode conversion in the sample.

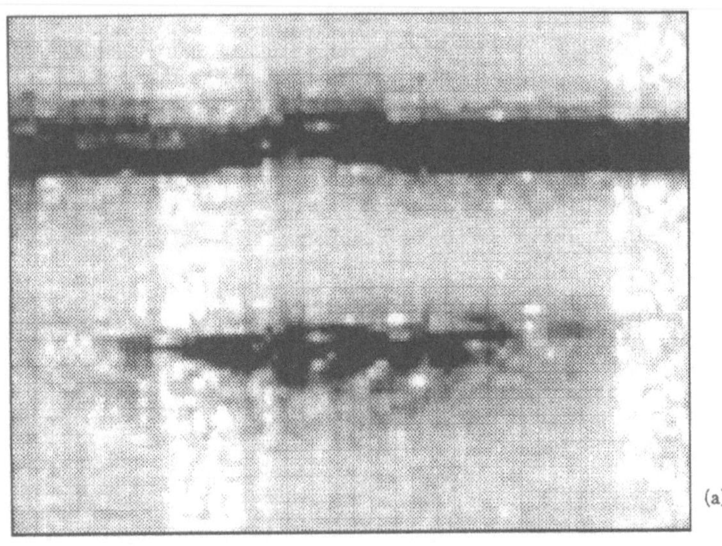

(a)

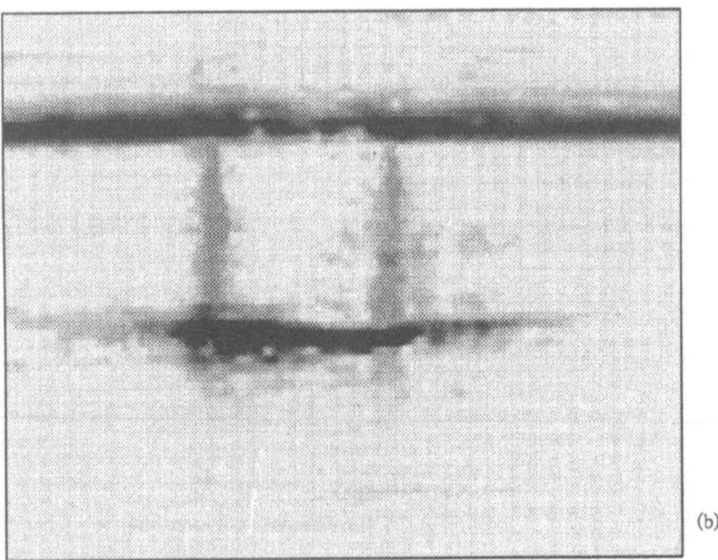

(b)

Figure 5 Echo B images of a small hole drilled in a block of steel (a) with classical pulse system and (b) with correlation system

CONCLUSION

The results presented in this article were obtained by acircular correlation of m-sequence signals. Since in this case the coded signal is emitted continuously, the system needs at least two transducers operating in emission and reception mode, respectively. In our opinion the multielement transducer such as the one that we used in this study is no longer a technical problem. Furthermore it has its own advantages since it could provide focus tracking and electronic apodization of the ultrasonic beam. In spite of several A-type echograms available in the literature[6] there is, to our knowledge, only two papers where B-type images are presented[7,8]. We hope that these our results will incerase interest in coded signals and adapted filtering techniques in echography.

Obtaining images of systems containing objects with less dramatic changes in acoustical impedance will be the next area of inquiry.

Acknowledgements

This project was supported by ANVAR grant n°A87069QAC, ARC grant n°6174 and with funds from INSERM n°88-90-10. It is a pleasure to thank Pr P. Kahn for a critical reading of the manuscript.

REFERENCES

1. B. B. Lee and E. S. Furgason, The use of correlation systems for real time ultrasonic imaging, Proc. of IEEE Ultrasonic Symposium, p 782-787, (1985).

2. M. Gindre, C. Lebeault, F. Rieuneau, W. Urbach and J. Perrin, Ultrasonic imaging using correlation techniques : Computer simulations and experimental results, in Acoustical Imaging, Vol. 17, Ed. Plenum Press (1988).

3. M. A. Benkhelifa, M. Gindre, J. Y. Le Huérou and W. Urbach, Echography with continuous ultrasonic waves, in Acoustical Imaging, Vol. 19, Ed. Plenum Press (1992).

4. J. Max, "Méthodes et techniques de traitement du signal et applications aux mesures physiques", Vol. 1, Ed Masson, Paris France (1987).

5. W. W. Peterson and E. J. Weldon, "Error correcting codes", Ed. MIT Press, Cambridge MA, (1972).

6. see for example :

    B. B. Lee and E. S. Furgason, An evaluation of NDE correlation flaw detection system, IEEE trans. on Sonics and Ultrasonics, Vol. SU 29, n°6, p 359-369, (1982).

    B. B. Lee and E. S. Furgason, High speed digital Golay code flaw detection system, Ultrasonics, Vol. 21, p 153-161, (1983).

    G. Hayward and Y. Gorfu, A digital hardware correlation system for fast ultrasonic data acquisition in peak power limited applications, IEEE trans. on UFFC, Vol. 35, n°6, p 800-808, (1988).

7. G. Hayward and Y. Gorfu, Low power ultrasonic imaging using digital correlation, Ultrasonics, Vol. 27, p 288-296, (1989).

8. M. O'Bonnell, Coded excitation system for improving penetrartion of real time phased array imaging systems, IEEE trans. on UFFC, Vol. 39, p 341-351, (1992).

MAXIMUM LIKELIHOOD ESTIMATION OF OBJECT PARAMETERS IN ULTRASOUND REFLECTIVE COMPUTERIZED TOMOGRAPHY

X. L. Wang and C. Q. Lan

Wuhan Institute of Physics
The Chinese Academy of Sciences
Wuhan 430071, P. R. China

INTRODUCTION

The applications of reconstruction from projections are in a diversity of disciplines, especially in X-ray medical radiology. Recently, ultrasound tomography techniques have been developed rapidly in a number of new areas, such as nondestructive evaluation, geophysical tomography and seismic tomography. However, the situation often arises in these areas where the number of views, the angular coverage and the number of samples within a view are severely restricted, and even the available data are corrupted by noise due to physical, geometrical, or economic constraints in the data acquisition. Thus, it is hard to reliably produce an accurate, high-resolution imagery of the cross section under these conditions. When the projection measurements are limited in number of view angle, or have high noise levels, only the information that is related to the object location, size and shape, can be extracted within the cross section. Moreover, in a number of applications, only the object-related information about the cross section is required.

Rossi and Willsky[1,2] first employed a maximum likelihood estimation (MLE) approach by which a few object parameters describing characteristics of the object in the cross section are estimated directly from noisy X-ray tomographic data.

The reflection mode tomography model using circular array proposed by Norton[3] is a new imaging approach which can provide quantitative images having both high lateral resolution and axial resolution. Therefore, it is valuable both in theory and application to do research work on that model. The estimation problem in reflection tomography has a number of potential applications in medical and industrial inspections. The examples include detecting / estimating organs and tumors in medical ultrasound tomography and detecting /characterizing cracks and flaws in industrial ultrasound tomography.

In this paper , the problem of locating a single object is considered in reflection tomography with the effects of the curvature of wave-fronts and the system temperal response being taking into account. Then , this procedure is extended to the problem of single object parameter estimation in reflection tomography. Simulation results show that this algorithm can both obtain good estimates of object parameters and lower the computation expenses.

BACKGROUND

Let $f(x)$ represent the reflectivity distribution function on a two-dimensional cross section of reconstructed object. The transducer is located at a point (X_{10}, X_{20}) on the circumference of radius R , at an angle φ measured from the positive x_1 axis , as shown in Fig. 1.

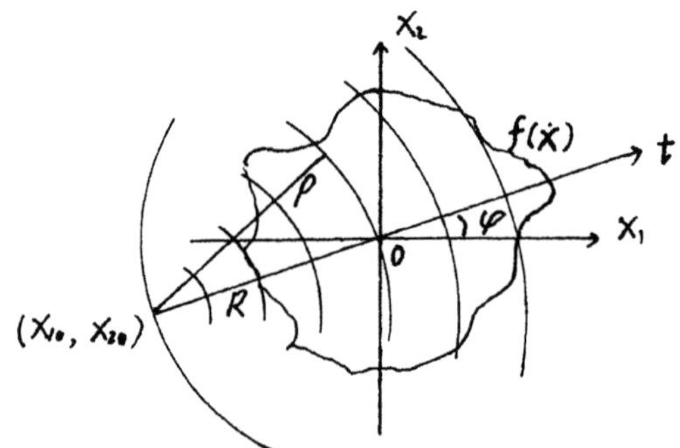

Fig. 1.   Scheme of ultrasound reflection mode CT
with single transmitting /receiving element

Consider below an idealized medium with the following properties: (1) the velocity of sound is constant; (2) the medium is weakly reflecting; (3) absorption is uniform over the region of interest. Suppose an infinitesimally narrow pulse of sound is emitted by the source , at the instant of time τ , the backscattered sound received by the transducer can be regarded as line integral of the reflectivity function $f(x)$ over circular arc $s(\rho,\varphi)$ indicated in Fig. 1,

$$g(\rho - R,\varphi) = \int_{s(\rho,\varphi)} f(x)\mathrm{d}s \tag{1}$$

where $\rho = \tau c/2$ and c is the speed of sound. Reflectivity distribution $f(x)$ can be denoted by the superposition of N objects,

$$f(x) = \sum_{k=1}^{N} r_k f_{0k}(x - x_{ck} ; \gamma_k) \tag{2}$$

Here the kth object is located at point x_{ck} and has reflectivity factor $r_k (f_{0k}(0 ; \gamma_k) = 1)$; γ_k is a finite dimensional vector of parameters characterizing the geometry parameters of kth object. For the problem of object parameters estimation in imaging applications ,[2] the function $f_{0k}(x ; \gamma_0)$, which characters a circularly-symmetric object located at the origin with norminal parameter vector γ_0 is usually known based on *a priori* knoweledge. It may , for example , be constant-valued in most cases.

In this paper, we consider the special case where only a single object ($N = 1$) exists and the object under consideration is simply parameterized with three important features of object geometry-size A , eccentricity λ and orientation θ . Here A is the radius of the ellipse, defined as the geometrical mean of its major and minor semi-axes λ is its axis ratio; and θ is its orientation, measured by the angle between its major axis and the x_1 coordinate axis. Thus, $r \cdot f_0(x - X_c ; A, \lambda, \theta)$ is a function that is zero everywhere except on an ellipse centered at point X_c , where it has the value r . Then the reflectivity distribution on the cross section may be represented as

$$f(x) = r \cdot f_0(x - X_c; A, \lambda, \theta) \tag{3}$$

MLE OF OBJECT LOCATION

Suppose $r \cdot f_0(x; y)$ is known, and consider the problem of MLE of object location. Let the noise-corrupted echo data be given by

$$y_s(\rho - R, \varphi) = g_s(\rho - R, \varphi; X_c) + n(\rho - R, \varphi) \tag{4}$$

where $n(\rho - R, \varphi)$ is a zero-mean white Gaussian noise process with spectral level $N_0 / 2$.

According to nonlinear parameter MLE method, the ML estimate $\hat{X}_c$ of the central location of the object is obtained by minimizing the squared error between the observed signal $y_s(\rho - R, \varphi)$ and the echo data $g_s(\rho - R, \varphi; x_c)$ of the function f assumed centered at the test location x_c integrated over ρ and summed over the view angles; i.e. the ML estimate of the central location minimizes the function

$$\varepsilon(x_c) = \frac{1}{N_0} \sum_\varphi \int_0^\infty |y_s(\rho - R, \varphi) - g_s(\rho - R, \varphi; x_c)|^2 d\rho \tag{5}$$

Expand the squared part in the integrand of (5) and make use of the fact that observed signal $y_s(\rho - R, \varphi)$ is independent of x_c , $\varepsilon(x_c)$ then can be replaced with the negative if the log likelihood function[4]

$$L(x_c) = \frac{2}{N_0} \sum_\varphi \int_0^\infty y_s(\rho - R, \varphi) \cdot g_s(\rho - R, \varphi; x_c) d\rho$$

$$- \frac{1}{N_0} \sum_\varphi \int_0^\infty |y_s(\rho - R, \varphi; x_c)|^2 d\rho \tag{6}$$

The ML estimate $\hat{X}_c$ is obtained by maximizing the log likelihood function within the cross section.

Compared with the computation of $L(x_c)$ in X-CT, the computation of $L(x_c)$ in ultarsound reflective computerized tomography (URCT) has the following differences:

(1) Rossi and Willsky proved that the second term is independent of the location x_c in X-CT(it is easy to be proved, so the proof is omitted).

(2) The first term in the case of that X-ray could be generated via a convoluted backprojection algorithm (CBP) which could greatly reduce the computation expenses; while in URCT $g_s(\rho - R, \varphi; x_c)$ it must be computed for a (possibly large) number of parameter values x_c and used as trial values in the first term.

(3) The imaging algorithms in X-CT only require a 180° range of view for the wavefronts being straight lines; while in URCT, if we measure line-integral data also over a 180° range of view, the estimate $\hat{X}_c$ will have errors to some extent due to the effect of curvature of wavefronts.

Thus, when we estimate the object location in URCT by MLE, the two terms of (6) must be computed for a number of parameter values x_c. Moreover, the range of view should be as close to 360° as possible in order to eliminate the estimate errors caused by curvature of wavefronts. So the computation of (6) is considerably large.

If we could progressively eliminate the effect of curvature of wavefronts on $y_s(\rho - R, \varphi)$ measured as the family of line integrals taken along the circular-arc wavefronts, and obtain the approximation $y^{(k)}(\rho - R, \varphi)$ (k is the iteration number) of $g_1(\rho - R, \varphi)$ formed as line integrals taken along straight lines, then the CBP algorithm can be utilized for the computation of $L^{(k)}(x_c)$, and only 180° view angle range is required, i. e. ,

$$
\begin{aligned}
L^{(k)}(x_c) &= \sum_{\varphi} \int_0^{\infty} y^{(k)}(\rho - R, \varphi) \cdot g_1(\rho - R, \varphi; x_c) \mathrm{d}\rho \\
&= \sum_{\varphi} \int_0^{\infty} y^{(k)}(\rho - R, \varphi) \cdot g_1(\rho - R - x_c \cdot \varphi, \varphi) \mathrm{d}\rho \\
&= \sum_{\varphi} B_{\varphi}(h_{\varphi} * y^{(k)})(x_c)
\end{aligned}
\tag{7}
$$

Here $x_c = (x_{1c}, x_{2c})$, $\varphi = (\cos\varphi, \sin\varphi)'$, B_{φ} is the backprojection operator, and h_{φ} is convolution kernel where $h_{\varphi} = g_{10}(R - \rho, \varphi)$. Note that $h_{\varphi} = g_{10}(R - \rho, \varphi) * h(R - \rho)$ as the time domain distortion of the imaging system defined by $h(\rho - R)$ summarizing the transducer characteristics, etc., is incorporated into measurement process.

According to the conclusion[5] that $y_s(\rho - R, \varphi)$ obtained along circular-arc wave-fronts equals $y_1(\rho - R, \varphi)$ plus an error term $e(\rho - R, \varphi)$, i. e. ,

$$
y_s(\rho - R, \varphi) = y_1(\rho - R, \varphi) + e(\rho - R, \varphi)
\tag{8}
$$

By employing the following iterative algorithm, from observed signal $y_s(\rho - R, \varphi)$ we can obtain the modification $y^{(k)}(\rho - R, \varphi)$ ($k = 1, 2, \cdots$) that gradually approach $y_1(\rho - R, \varphi)$ and the ML estimate $\hat{X}_c$ of the central location of the object in the case of curvature of wavefronts:

(1) Let $e^{(0)}(\rho - R, \varphi)$ of (8) be zero, then

$$
y^{(0)}(\rho - R, \varphi) = y_s(\rho - R, \varphi)
\tag{9}
$$

(2) Substituting (9) into (7), we obtain $L^{(0)}(x_c)$ and $\hat{x}_c^{(0)}$. Then we get the initial cross-sectional reflectivity distribution

$$
f^{(0)}(x) = r \cdot f_0(x - \hat{X}_c^{(0)}; \gamma)
\tag{10}
$$

where we have made use of the known object reflectivity function $r \cdot f_0(x; \gamma)$.

(3) By modifying $y_s(\rho - R, \varphi)$, we have

$$
y^{(1)}(\rho - R, \varphi) = y_s(\rho - R, \varphi) - (p_s(\rho - R, \varphi; \hat{X}_c^{(0)}) - p_1(\rho - R, \varphi; \hat{X}_c^{(0)}))
\tag{11}
$$

where

$$p_s(\rho - R,\varphi;\hat{X}_c^{(0)}) = \int_S f^{(0)}(x)\mathrm{d}s \tag{12a}$$

$$p_1(\rho - R,\varphi;\hat{X}_c^{(0)}) = \int_L f^{(0)}(x)\mathrm{d}s \tag{12b}$$

(4) Substituting $y^{(1)}(\rho - R,\varphi)$ into (8), $L^{(1)}(x_c)$ can be obtained. Then we get ML estimate $\hat{X}_c^{(1)}$ in the first step of the iteration, which should be closer to the object's central location X_c and $\hat{X}_c^{(0)}$. Thus after one step of the modification we obtain the cross-sectional reflectivity distribution

$$f^{(1)}(x) = r \cdot f_0(x - \hat{X}_c^{(1)};\gamma) \tag{13}$$

(5) we obtain p_s and p_1 by use of (13) with $f^{(1)}(x)$. If p_s and p_1 are substituted into (11) we have $y^{(2)}(\rho - R,\varphi)$.

The iteration is continued, and we gain the approximation $y^{(k)}(\rho - R,\varphi)$ of $y_1(\rho - R,\varphi)$ and the ML estimate $\hat{X}_c^{(k)}(k = 1,2,\cdots)$. When the convergence condition is satisfied, the iteration stops.

In order to illustrate and evaluate the ML procedure in URCT, a computer simulation study was performed for a single disk shaped object with radius $A = 0.3$ of which the reflectivity distribution is constant. When the object centers at $X_c = (0,0)$ and SNR $=$ 0dB where the measurement signal-to-noise ratio in dB is defined as follows:

$$\mathrm{SNR} = 10\ \log_{10} \frac{\text{signal variance}}{\text{noise variance}} \tag{14}$$

The estimation results generated with formula (6) and the CBP algorithm are shown in Fig. 2(a) and Fig. 2(b). In this section, the true objects are the circles shown with the solid line and the estimates are shown with dashed lines. It can be seen from Fig. 2(a) and Fig. 2(b) that the ML estimate $\hat{X}_c$ obtained by performing the CBP operation and that obtained using formula (6) are approximately equal and both of them are close to the object actual location X_c as the object location is far away from the transducer. When the object centers at $X_c = (0,-0.65)$ and SNR $=$ 0dB, the estimate results by the two methods are shown in Fig. 2(c) and Fig. 2(d). Comparing Fig. 2(c) with Fig. 2(d), we can find that the estimate obtained with CBP algorithm is remarkably in error as the object location nears the transducer, and the estimation result obtained with iterative algorithm proposed in this paper after five times of iteration is shown in Fig. 2(e). Comparing Fig. 2(c) with Fig. 2(e), we can see that the estimation error has been greatly reduced, and the estimation results approaches the result obtained directly with formula (6).

MLE OF OBJECT LOCATION AND SHAPE

Rossi, et al.[4] have proposed an effective procedure for the simultaneous estimation of object location, size, orientation and eccentricity, in which the four-parameter problem is considered as four separate subproblems with one parameter unknown at a time. The steps of this procedure are as follows:

(1) Initially assume that the object is a large circular object with the size of $A^{(0)}$ using *a priori* information, and estimate object location $\hat{X}_c^{(0)}$.

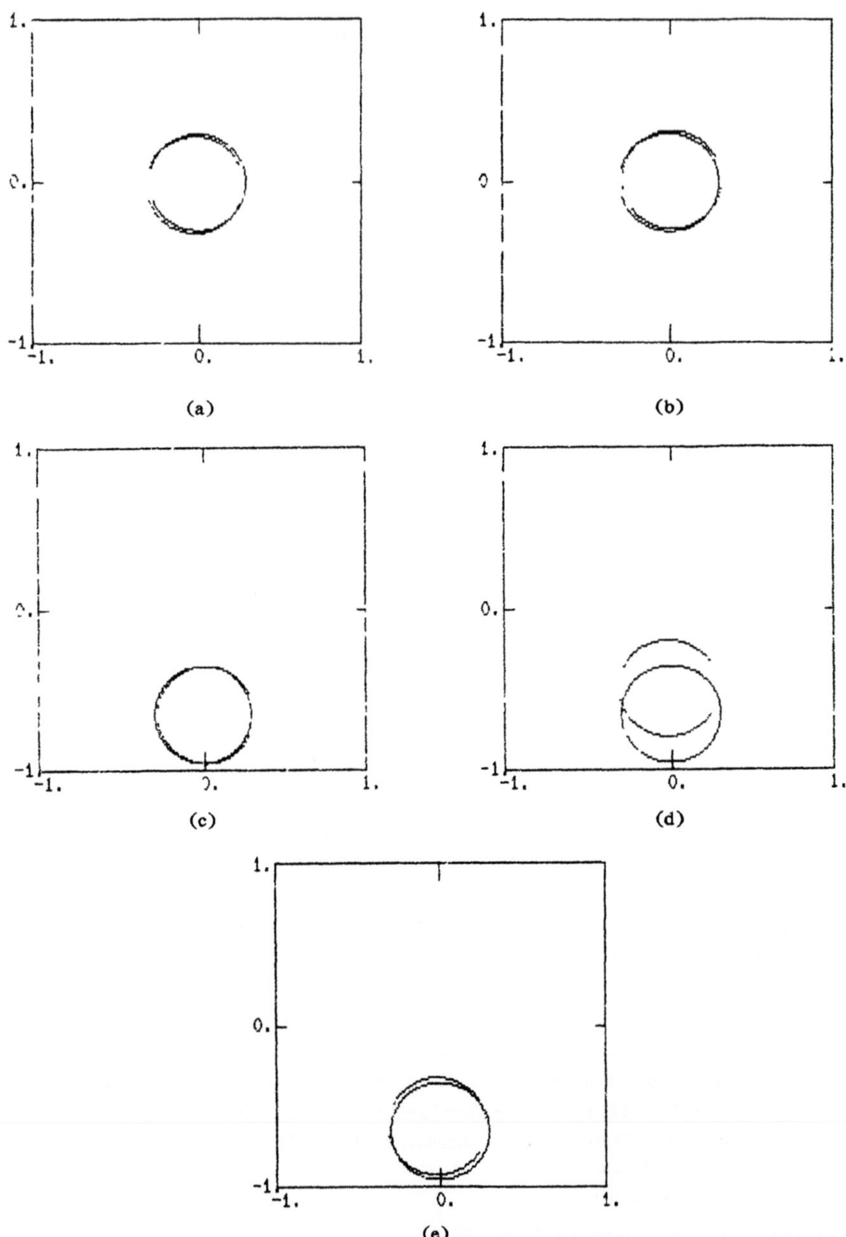

Fig. 2.   Object location estimation results at 0dB for circular object with $A = 0.3$. The solid lines represent the actual object, and the dashed lines represent the estimates of the object:

(a) Estimate generated using formula (6); $R = 3$; $X_o = (0,0)$, $\hat{X}_o = (-0.016,-0.016)$.
(b) Estimate generated using the CBP algorithm; $R = 3$; $X_o = (0,0)$, $\hat{X}_o = (0.016, 0.016)$.
(c) Estimate generated using formula (6); $R = 1.03$; $X_o = (0,-0.65)$, $\hat{X}_o = (-0.016,-0.651)$.
(d) Estimate generated using the CBP algorithm; $R = 1.03$; $X_o = (0,-0.65)$, $\hat{X}_o = (-0.016,-0.492)$.
(e) Estimate generated using iterative algorithm; $R = 1.03$; $X_o = (0,-0.65)$, $\hat{X}_o = (-0.016,-0.619)$.

(2) Giving the initial estimated location $\hat{X}_c^{(0)}$,estimate object size $\hat{A}^{(0)}$ still assuming that the object is circular.

(3) Giving the estimates of object location $\hat{X}_c^{(0)}$ and size $\hat{A}^{(0)}$,estimate object orientation assuming a nominal value $\lambda^{(0)}$ of eccentricity.

(4) Giving $\hat{X}_c^{(0)}$, $\hat{A}^{(0)}$ and $\hat{\theta}^{(0)}$,estimate eccentricity $\hat{\lambda}^{(0)}$.

(5) Update iteratively the estimates of location $\hat{X}_c^{(k)}$, size $\hat{A}^{(k)}$, orientation $\hat{\theta}^{(k)}$ and eccentricity $\hat{\lambda}^{(k)}$ ($k = 0,1,\cdots$) with the latest estimates of the remaining parameters.

Taking the effect of the curvature of wave-fronts on URCT into account,we modified the above produre；i. e. as the estimates $\hat{X}_c,\hat{A},\hat{\theta}$ and $\hat{\lambda}$ converge each time in iterations,we refashion the measured data y_e with the latest estimates of the object parameters.

The examples of single object parameters estimation algorithms are shown in Fig. 3(a) and Fig. 3(b). The geometry a in Fig. 3 is a true elliptical object with $X_c = (0. 0,-0. 6)$ and $\gamma = (0. 367,6. 0,0. 0°)$. The geometry b in Fig. 3(b) is the estimated object with $\hat{X}_c = (0. 016,-0. 492)$ and $\hat{\gamma} = (0. 29,9. 0,-1. 510°)$ generated by directly making use of the algorithm in X-CT. It is obvious that the estimations of X_c and γ are poor. The geometry b in Fig. 3(b) is the estimated object with $\hat{X}_c = (-0. 016,-0. 586)$ and $\hat{\gamma} = (0. 36,7. 74,-0. 503°)$ generated by use of the iterative algorithm after five times iteration in which the measured data are modified. It is clear that the estimation result illustrated in Fig. 3(a) is closer to the actual object location and shape parameters than that shown in Fig. 3(b).

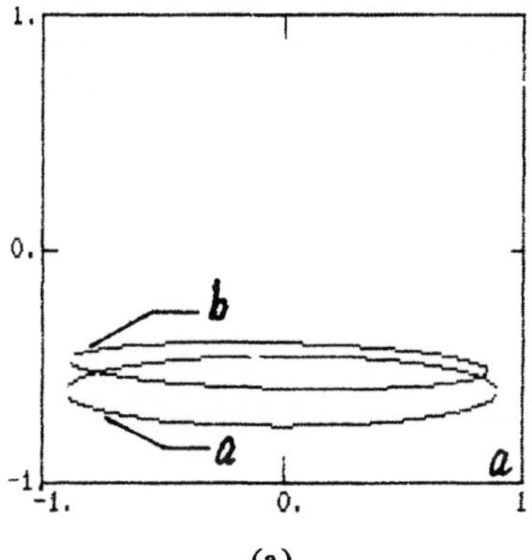
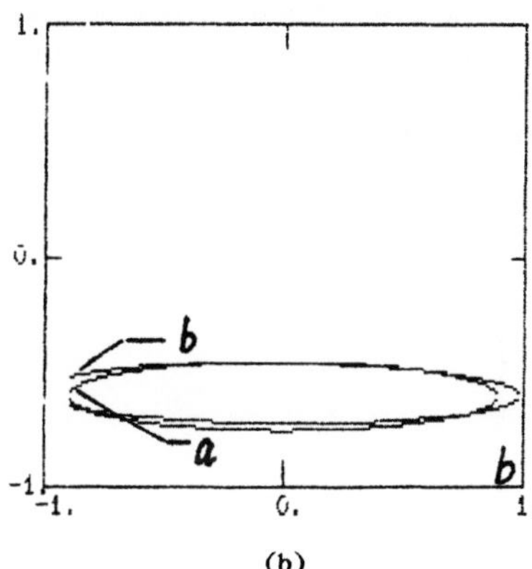

(a)   (b)

Fig. 3. Estimation of an elliptical object with SNR $=0$dB and $R =1. 5$. Geometry a reprerents the actual object($X_c = (0,-0. 6)$, $\gamma = (0. 367,6. 0°)$) and geometry b corresponds to the estimated object；
(a) Estimate generated by the algorithm in X-CT：$\hat{X}_c = (-0. 016,-0. 492),\hat{\gamma} = (0. 29,9. 0,-1. 51°)$.
(b) Estimate generated by the iterative algorithm：$\hat{X}_c = (-0. 016,-0. 587),\hat{\gamma} = (0. 36,7. 74,-0. 503°)$.

CONCLUSIONS

The method of MLE of object parameters which was first employed in X-ray tomography is introduced into ultrasound reflective computerized tomography . In this paper,when the object is far away from the ultrasound transducer,the MLE algorithm in X-CT given by Rossi,et al. can be directly used in URCT since the wave-fronts are approximately planewaves. However, when the object is in close distance from the transducer the estimation results are remarkably in error due to the effect of the curvature of wave-fronts. An iterative algorithm coping with this problem is then presented. The simulation results for a disk shaped object having constant reflectivity distribution show that the performance of this iterative algorithm is close to that by directly making use of the formula of log likelihood function while the former's computation complexity is greatly lower than the latter's. Moreover, combining this iterative approach with the algorithm for single object parameters estimation in X-CT, we can effectively estimate location and shape parameters of the object from the echo data.

The MLE procedure developed in this paper would be valuable in ultrasound tomography. Of course, a full set of view angles over the range $[0, \pi]$ are difficult to be achieved in practice, and maybe it is not realistic to assume only one object exists within the cross-section. The approach for the estimation of multiple objects will be presented in another paper.

REFERENCES

1. D. J. Rossi and A. S. Willsky, Reconstruction from projections based on detection and estimation of objects, *IEEE Trans. Acoust. , Speech Signal Process*, Vol. ASSP—32, 1984, pp886—906.
2. D. J. Rossi, A. S. Willsky and D. M. Spielman, Object shape estimation from tomographic measurements——A performance analysis, *Signal Processing*, Vol. 18, No. 1, 1989, pp63—87.
3. S. J. Norton and M. Linzer, Ultrasonic reflectivity tomography: Reconstruction with circular transducer arrays, *Ultrasonic Imaging*, Vol. 1, 1979, pp154—184.
4. H. L. Van Trees, "Detection, Estimation, and Modulation Theory, Part I", Wiley, New York, 1968.
5. K. K. Xu, C. Q. Lan and Y. H. Chen, An iterative filtered backprojection algorithm for ultrasound reflection mode tomography, in: "Proc. of the19th International Symposium on Acoustical Imaging", Bochum, Germany, Feb. 2—6, 1991.

ITERATIVE METHOD APPLIED TO ULTRASOUND REFLECTION MODE
TOMOGRAPHY WITH CIRCULAR AND PLANAR TRANSDUCER ARRAYS

C.Q. Lan, Y.H. Chen, K.K. Xu, L. Li, W. Xiong and S.W. Tong

Wuhan Institute of Physics
The Chinese Academy of Sciences
Wuhan 430071 P.R.China

INTRODUCTION

Since ultrasound reflection mode tomographic imaging method was proposed by S.J.Norton et al. [1,2], many scientists have been interested in this
method because of its simple measuring equipment required and   capability
of achieving high resolution in both range and azimuth directions simultaneously. It is difficult to obtain exact reconstruction of a two–dimensional
reflection mode tomography. Generally, the filtered backprojection method
used in X–ray tomography is employed. However, the distortion of   the reconstructed images will happen because of wavefront curvature when the
far field condition is not satisfied. T.K.Truong et.al.[3] proposed an extended
method named "An extended version of the tomographic transform convolution for the wavefront curvature" under paraxial approximation condition.
Unfortunately, it is difficult to calculate the second modified term in
Eq.(19) of the paper.[3]

In this paper, ultrasonic reflection tomographic reconstruction is discussed in the case where the transmitter and the receiver are separate elements which lie at different points on an array. An iterative formula is
derived based on Fourier Slice Theorem under the paraxial approximation
conditions. By using this method, the Fourier transform of the received
returning echoes can be modified iteratively. In this way, it can be expected that the distortion of the reconstructed image caused by the curvature of
integral lines would be eliminated.

ITERATIVE FORMULA

1. Single pair of separate transmit and receive elements in a circular array

Fig.1 shows the geometry of the imaging system where the transmitter
and receiver are separate elements which lie at different points on the circumference of an array. The Cartesian coordinate system XOY is fixed with

respect to the object imaged. f(x,y) denotes a cross–sectional function of acoustic reflectivity F(x,y,z) of the object. An omnidirectional source ultra-sonic transducer is located at point A(T) along a circular arc enclosing the object and a receiver is located at point B(R) along the same arc whose radius is R and origin is O. Thus, when a infinitesimal short ultrasound pulse is emitted at A, data are continuously recorded at B as a function of time. Outputs $P(\rho_1+\rho_2,\varphi)$ of the receiver R are line integrals of the reflectivity f(x,y) over an entire family of ellipses S whose foci are A and B. The lenghth of the semi–major axis of this ellipse is tc/2, where constant c is the

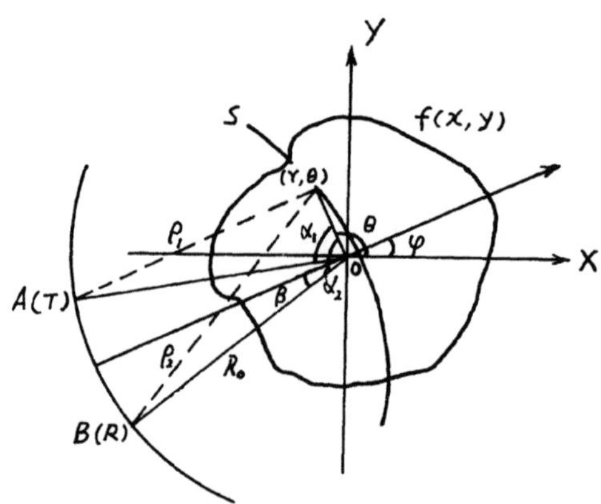

Fig.1 The geometry of the imaging system in which single pair of separate transmit
and receive elements are arranged on a circular arc enclosing the object imaged.

sound propagating velocity in the area of interest and $t=t_1+t_2$ the round trip delay. The outputs of the receiver R, called   projections, can be expressed[2]

$$P(\rho_1+\rho_2,\varphi)=\int_s f(x,y)ds \tag{1}$$

where $\rho_1+\rho_2=(t_1+t_2)c$, φ is the incident angle measured from the positive x-axis to the line that bisects angle $\angle AOB$. In the polar coordinate system, Eq.(1) can be   rewritten as

$$P(\rho_1+\rho_2,\varphi)=\int_0^\infty rdr \int_0^{2\pi} f(r,\theta)\,\delta\,[\bar{t}-(\rho_1+\rho_2)]d\theta \tag{2}$$

where

$$\bar{t}=\bar{t_1}+\bar{t_2}=c(t_1+t_2)$$

$$\bar{t}_{1,2}=\sqrt{R_0^2+r^2-2R_0 r\cos\alpha_{1,2}}$$

α_1, α_2 are angles between r and OA, OB, respectively. Setting $\rho=\rho_1+\rho_2$ and taking one–dimensional Fourier transform of projection $p(\rho,\varphi)$:

$$S(\omega,\varphi)=\int_{-\infty}^\infty p(\rho,\varphi)\exp(-j\omega\rho)d\rho \tag{3}$$

Substituting (2) into (3), when the distance between the object imaged and the transducer is longer than the size of the object   but   not   very   much, that is, the paraxial approximation is satisfied, we can obtain approximately [4]

$$S(\omega,\varphi)=\exp(-j\omega 2R_o)[F(u,v)-j\omega F^{(1)}(u,v)-\omega^2 F^{(2)}(u,v)/2+\ldots] \qquad (4)$$

where

$$u=2\omega\cos\beta\cos\varphi, \quad v=2\omega\cos\beta\sin\varphi$$

$$F(u,v)=\int_{-\infty}^{\infty}\int_{-\infty}^{\infty}f(x,y)\exp[-j(ux+vy)]dxdy$$

$$F^{(1)}(u,v)=\int_{-\infty}^{\infty}\int_{-\infty}^{\infty}f(x,y)G(x,y)\exp[-j(ux+vy)]dxdy \qquad (5)$$

$$F^{(2)}(u,v)=\int_{-\infty}^{\infty}\int_{-\infty}^{\infty}f(x,y)G^2(x,y)\exp[-j(ux+vy)]dxdy$$

$$G(x,y)=[\sin^2\beta\,(x\cos\varphi+y\sin\varphi)^2+\cos^2\beta\,(x\sin\varphi-y\cos\varphi)^2]/R_o \qquad (6)$$

When the distance between the object imaged and the transducer is much longer than the size of the object, that is, the far field condition is satisfied, $F^{(1)}(u,v)$, $F^{(2)}(u,v)$,... in Eq.(4) can be neglected. In this case, Eq.(4) becomes

$$S(\omega,\varphi)=\exp(-j\omega 2R_o)F(u,v) \qquad (7)$$

Eq.(7) shows, for the far field situation, when the separated transmit and receive elements form a fixed angle 2β to point O (Fig.1) and the angle between line bisecting the angle 2β and the positive x-axis is φ, the one-dimensional Fourier transform of "projection" received at point B gives the spectrum values of the two-dimensional Fourier transform of the object function $f(x,y)$ along the line $u=2\omega\cos\beta\cos\varphi$ and $v=2\omega\cos\beta\sin\varphi$. This is the famous Fourier Slice Theorem in X-ray tomography. When the transmit and receive elements T and R combine together, that is $\beta=0$, Eq. (7) becomes the formula of the Fourier Slice Theorem for ultrasonic reflective tomography with the same element serving both as transmitter and receiver. We can see from Eq.(7) that the length of this spectrum line is $\cos\beta$ times that in the case of $\beta=0$.

If keeping the separation between the transmitting and receiving elements constant (i.e. $\angle AOB = 2\beta$), making this pair of elements T and R scan along the circumference of a circle and changing the incident angle from 0° to 180°, the two dimensional Fourier transform spectrum line will be rotated from 0° to 180° around point O in (U,V) plane. Then the covered region in the spatial frequency domain is a disk whose radius is $2\omega\cos\beta$. In this fashion, we can reconstruct the unknown reflectivity function $f(x,y)$ employing the Direct Fourier Transform algorithm or the Filtered-Backprojection algorithm in the X-ray tomography.

2. Single pair of separate transmit and receive elements in a plane array

For some special situations, transmitting and receiving transducers must be arranged in a plane as shown in Fig.2. In the same way, we can obtain

$$S(\omega,\varphi)=\exp[-j\omega(R_o+R_1)]\times[F(u,v)-j\omega F^{(1)}(u,v)-\omega^2 F^{(2)}(u,v)/2+\ldots] \qquad (8)$$

where

$$F(u,v)=\int^{\infty}_{-\infty}\int^{\infty}_{-\infty}f(x,y)\exp[-j(ux+vy)]dxdy$$

$$F^{(1)}(u,v)=\int^{\infty}_{-\infty}\int^{\infty}_{-\infty}f(x,y)G_1(x,y)\exp[-j(ux+vy)]dxdy \qquad (9)$$

$$F^{(2)}(u,v)=\int^{\infty}_{-\infty}\int^{\infty}_{-\infty}f(x,y)G_1^2(x,y)\exp[-j(ux+vy)]dxdy$$

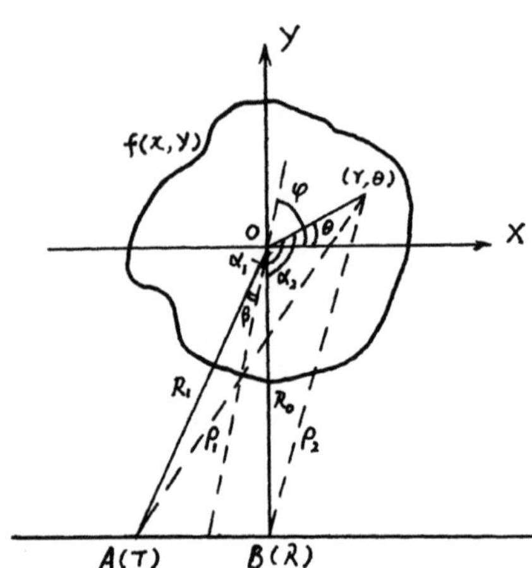

Fig.2 The geometry of the imaging system in which single pair of separate transmit end receive elements are aranged in a plane.

and it also can be shown[4]:

$$G_1(x,y)=x^2(1/R_1+1/R_0)/2-\sin2\beta\,[((x^2-y^2)\sin2\varphi)/2-xy\cos2\varphi]/R_1 \qquad (10)$$

When $R_0 \gg r$, i.e. the far field condition is satisfied, we can obtain

$$S(\omega,\varphi)=\exp[-j\omega(R_0+R_1)]F(u,v) \qquad (11)$$

Let

$$S(\omega,\varphi)=\begin{cases} \exp(-j\omega 2R_0)S'(\omega,\varphi) & \text{T/R on a circular arc} \\[6pt] \exp[-j\omega(R_0+R_1)]S'(\omega,\varphi) & \text{T/R in a plane} \end{cases} \qquad (12)$$

Substituting Eq.(12) into Eq.(4) or Eq.(8), we can obtain a unified relation for these two kinds of transducer arrays.

$$S'(\omega,\varphi)=F(u,v)-j\omega F^{(1)}(u,v)-\omega^2 F^{(2)}(u,v)/2+... \qquad (13)$$

Eq.(13) shows that the one-dimensional Fourier transform $S'(\omega,\varphi)$ of the projection $P(\rho_1+\rho_2,\varphi)$ can be expressed in terms of the two-dimensional Fourier transform values of the unknown reflectivity $f(x,y)$ of the object being imaged and the terms $G(x,y)f(x,y)$, $G^2(x,y)f(x,y)$, ..., at spatial frequencies of $u=2\omega\cos\beta\cos\varphi$ and $v=2\omega\cos\beta\sin\varphi$. If the radius R is much longer than the size of the object, Eq.(13) returns to the formula of the

Fourier Slice Theorem in X-ray tomography. At present, in reflection mode ultrasonic tomographic imaging, the filtered-backprojection algorithm based on the Fourier Slice Theorem is widely used. However, when the far field condition is not holds, $F^{(1)}(u,v)$, $F^{(2)}(u,v)$, etc. are unnegligible, this algorithm induces distortion because the integral line is curvature. We are unable to obtain $F(u,v)$ directly from Eq.(13) when terms $F^{(1)}(u,v)$, $F^{(2)}(u,v)$, etc. can not be neglected. In order to solve this problem, we propose an iterative formula

Interpolation

$$F_{(n+1)}(u,v)=S'(\omega,\varphi)+j\omega\mathcal{F}\{G(x,y)f_{(n)}(x,y)\}+\omega^2\mathcal{F}\{G^2(x,y)f_{(n)}(x,y)\}/2+\dots \quad (14)$$

Two dimensional inverse Fourier transform

where $\mathcal{F}$ expresses two-dimensional Fourier transform. An improved image can be expected by performing the iterative steps according to Eq.(14).

COMPUTER SIMULATION

In order to demonstrate how many iterations to be necessary to obtain the suitable image of an object, a computer simulation is carried out. As shown in Fig.3, we take the reflectivity distribution as

$$f(x,y)= \begin{cases} 0.1 & 0.1R_o<\sqrt{x^2+y^2}<0.3R_o \\ 0 & \text{elsewhere} \end{cases} \quad (15)$$

where R_o is the radius of an interested range to be imaged. For simplicity, we only make some simulations for the case in which a transducer acts as both transmitter and receiver.

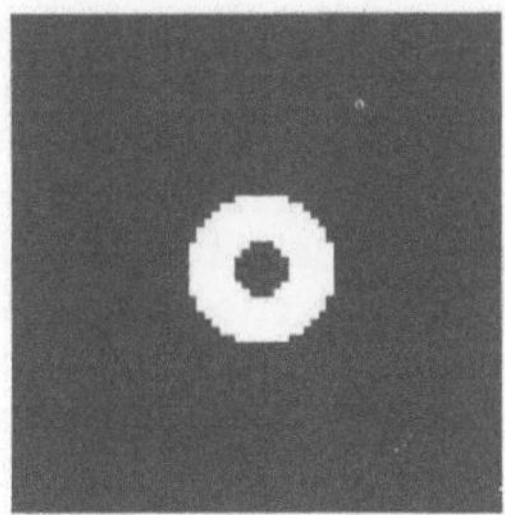

Fig.3   The distribution of the reflectivity function f(x,y) of an object to be imaged.

1.   An acoustic transducer is scaned along a circular arc with radius $R=1.25R_o$ as shown in Fig.1. The transducer emits an infinitesimal short acoustic pulse that propagats into the medium in a fan-beam pattern and then receives the echoes backscattered from the object. Every received echo is called a projection. 30 projections are obtained when the transducer is rotated around the object from $\varphi=0°$ to $180°$ and there are 64 sampling points

for each projection. Fig.4a shows the reconstruction image of the reflectivity distribution f(x,y) of an object shown in Fig.3 by using the algorithm in X-ray computer tomography. The distortion is arised because of the effect of wavefront curvature.

In order to improve the quality of reconstructed image, we use the iterative method according to the formula Eq.(14). Fig.4b and 4c are reconstructed images for one and ten loops of iteration, respectively, based on Fig.4a. As shown, Fig.4b and 4c are superior to Fig.4a.

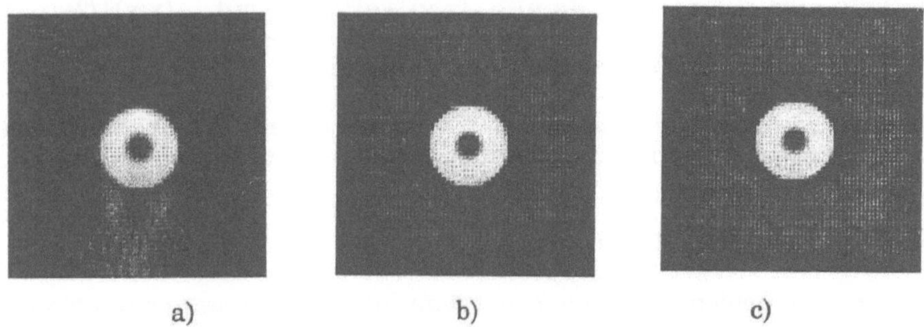

a)   b)   c)

Fig.4 Reconstructions of the Fig.3 on 30 projections from 0° to 180°.
a) Use the algorithm in X-ray tomography.
b,c) Use the iterative method proposed in this paper for one and ten loops of iteration respectively.

2. In the practical situation, when the transmitter and receiver can only be scanned along the line $y=-y_0$ in Fig.2, only a part of projections can be obtained because the variable range of the incident angle φ is limited. The reconstructed images will be distorted more seriously caused by both the integral line curvature and incomplete projections. In order to improve the quality of reconstruction images, an iterative method combined the iterative formula Eq.(14) in this paper and with iterative reconstruction-reprojection method in X-ray tomography proposed by M.Nassi [5] can be used. The iterative procedure is as follows. The reconstruction image $f_{(0)}(x,y)$ is obtained by employing filtered backprojection algorithm in X-ray tomography from incomplete projections. Using a priori knowledge about the object outline to correct $f_{(0)}(x,y)$ and reproject $f_{(0)}(x,y)$ to produce the missing projections in the region not illuminated by the incident wave. Calculate the one-dimensional Fourier transform of the all projections to obtain $S'(\omega,\varphi)$ from Eq.(12). Make two-dimensional Fourier transform of $f_{(0)}(x,y)G_1(x,y)$ and $f_{(0)}(x,y)G_1^2(x,y)$, respectively. Then, the modified spectrum $F_{(1)}(u,v)$ can be achieved from Eq.(14), and the improved value $f_{(1)}(x,y)$ of the ordinal reconstructed image $f_{(0)}(x,y)$ is obtained. Repeat iterative procedure above and stop when the satisfied reconstructed image $f_{(n+1)}(x,y)$ of f(x,y) is achieved.

In the simulations next, the transmit and receive elements T and R combine together in Fig.2, and an ultrasound transducer is scanned along a straight line $y=-y_0$ from $(-x_0,-y_0)$ to $(x_0,-y_0)$ where $x_0=\sqrt{3}y_0$. In this case, the incident angle φ is from 30° to 150° and 20 projections are obtained. There are 64 sampling data for each projection. Fig.5a shows the reconstruction image of the reflectivity distribution f(x,y) of an object shown in Fig.3 by only employing the filtered back–projection algorithm in X–ray tomography. The distortion is arised seriousely because of the effect of wavefront curvature and limited illumination angle. In order to improve the quality of reconstruction image, an iterative technique described above is used to eliminate imaging errors step by step. Fig.5b and 5c show the reconstructed images using the iterative technique for one and ten loops of iteration, respectively. By comparing the reconstruction results of Fig.5a with those of Fig.5b and Fig.5c, we can see obvious improvement due to the iterative revision.

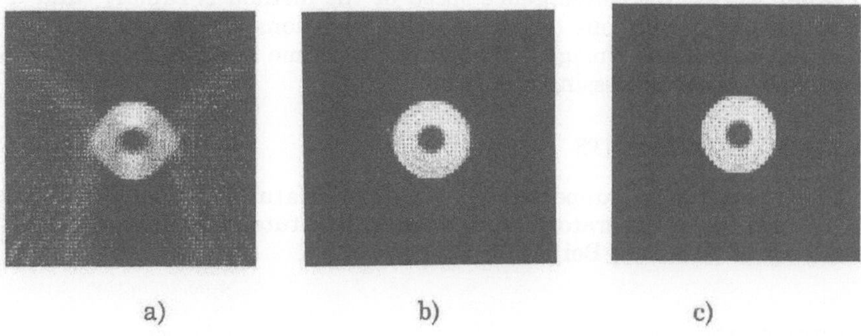

a)   b)   c)

Fig.5 Reconstruction of Fig.3 on 20 projections for scanning along a straight line.
a) Use the algorithm in X–ray tomography.
b,c) Use the iterative procedure discribed in this section for one and ten loops of iteration respectively.

CONCLUSION AND DISCUSSION

In this paper, we have developed a uniform iterative formula for ultrasonic reflection mode tomographic imaging in the different receive modes under paraxial approximation. The simulations show that an improved reconstructed image can be obtaind by eliminating iteratively the effect caused by the curvature of integral lines.

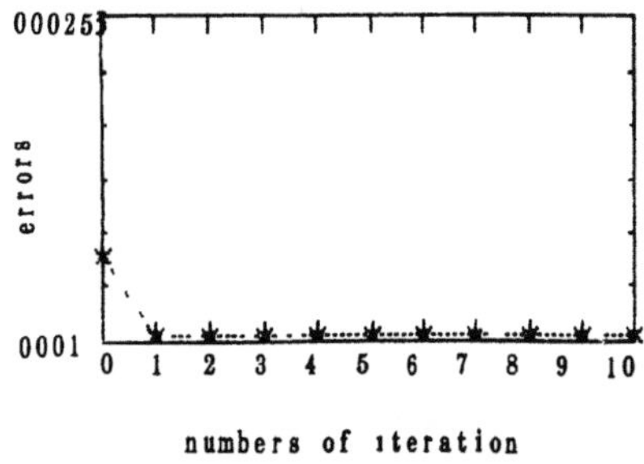

Fig.6 Iterative convergence in Fig.4

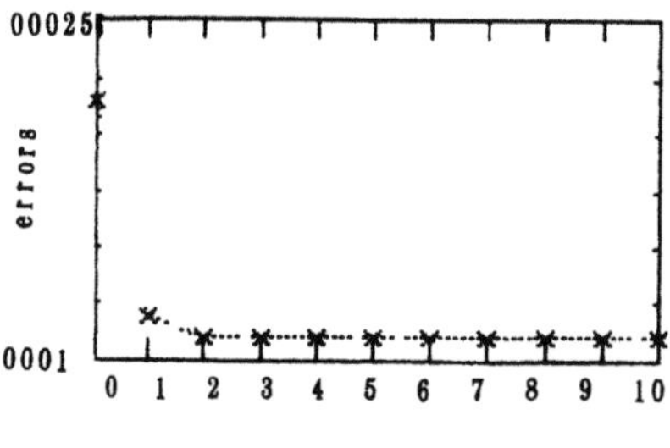

Fig. 7 Iterative convergence in Fig.5

When a transducer scanes along a straight line or a single pair of separate transmit and receive elements in a plane array, the obtained projections are generally incomplete. The reconstruction images using filtered backprojection algorithm in X-ray tomography are very poor. A revision technique combined an itrative method proposed in this paper with an iterative reconstruction-reprojection method in X-ray tomography can be used to eliminate the errors of reconstructed images. The computer simulation results have demonstrated that this iterative method is effective for improving the reconstruction image of an object.

The iterative convergence lines in Fig.4, and Fig.5 are shown in Fig.6 and Fig.7, respectively. We can see that the reconstruction errors without iteration are serious and the errors in Fig.7 is more serious than that in Fig.6. However, the errors have reached very small after only one loop of iteration. In other words, the convergence speed of the method is rapidly and even if $\varphi=30°$ to $150°$, only one or two loops of iterations are necessary to obtain a good reconstruction image. Therefore, the time required for data acquirement and signal processing is short.

ACKNOWLEDGEMENTS

This research is supported by National Natural Science Fundation of China and State Laboratory of Acoustics, Institute of Acoustics, The Chinese Academy of Sciences, Beijing, P.R.China.

REFERENCES

1. S.J.Norton and M.Linzer, " Ultrasonic reflectivity tomography : reconstruction with circular transducer array ", Ultrasonic Imaging 1, pp. 154-184, (1979).
2. S.J.Norton, " Reconstruction of a reflectivity field from line integrals over circular paths," JASA, Vol.67(3), Mar. 1980, pp.853-863.
3. T.K.Truong and I.S.Reed, " An Extended Version of the Tomographic Transform Convolution for the Wavefront Curvature ", ACTA ELECTRONICA SINICA, CHINA, Vol.15, No.3, pp. 11-14, May 1987.
4. C.Q.Lan and Y.H.Chen, " The Fourier slice theorem for ultrasonic reflectivity tomography and the correction to the effect caused by curvature of integral lines, " CHINESE JOURNAL OF ACOUSTICS, Vol.12, No.1, Jan. 1993, pp. 1-12.
5. M.Nassi, W.R.Brody, B.P.Medoff and A.Macovski, " Iterative reconstruction-reprojection: an algorithm for limited data cardiac-computed tomography ", IEEE Trans. on Biomedical Engineering, Vol.29, No.5, May 1982, pp. 333-340.

MOTION EFFECTS AND COMPENSATION IN SYNTHETIC APERTURE ACOUSTICAL IMAGING

Xianhua Xia, Dejun Zhang

Wuhan Institute of Physics
The Chinese Academy of Sciences
Wuhan 430071, P. R. China

ABSTRACT

The analysis of the motion effects of the imaged object on imaging is important for scanning array synthetic aperture. In this paper the motion effects on imaging are discussed when the imaged object moves in the direction parallel to or perpendicular to the scanning direction. The discussion of the motion compensation is also given in the paper. The computer simulations are made to verify the theoretical analyses.

I . INTRODUCTION

Linear receiving array scanning could be used to get holograms in acoustical holography (according to the existing theory, source scanning is equivalent to receiving array scanning[1]). As the scanning needs time to obtain a hologram and during this period the relative motion between the imaging system and the object will affect the quality of the image, the investigation of the motion effects and the compensation would be of great help to synthetic aperture acoustical imaging[2,3]. In this paper the problem has been thoroughly discussed and some results of simulative experiments are given.

II . MOTION EFFECTS

Usually the scanning time is not very long so we assume that the relative motion is a uniform rectilinear one. In this paper we provide that: The scanning linear receiving array is placed parallel to X axis, the number of its elements is $N = 64$ and the interval between the adjacent elements is $d_x = 12\text{mm}$, The sampling interval along the scanning direction which is supposed to be in Y direction is $d_y = d_x$ and the number of the sampling points is N . The wavelength

$\lambda = 5$mm. The sound velocity $C = 1500$m/sec. The distance between the object plane and the receiving plane, Z_0, is 10m. The emitting source locates at $(x_t, y_t, 0)$ at time t, the coordinates of one of the receiving elements are $(x, y_r(t), 0)$ where $y_r(t) = y_{r0} + v_s t$ and v_s the scanning velocity of receiving array. The aperture function is $\sigma(x, y)$, where $y = v_s t$. Let $O(x_0, y_0)$ be the distribution of the plane object field on the plane $z = -z_0$ at $t = 0$. We use complex notation for acoustical signal. The carrier of the acoustical wave, $\exp\{-j2\pi ft\}$, and the amplitude factor of the spherical wave are omitted because of their unimportance. What is to be considered is the phase factor of the signal. As a reference for the following discussion, Figure 1 is the image of a motionless object in the shape of letter "T" imaged by the above system and this object will be used in this paper as the test object for all simulative experiments.

Now we are ready to analyze the motion effects. We first consider a simple situation where a point object, located at $\bar{r}_0 = (x_0, y_0, -z_0)$ at time $t = 0$, moves with velocity $\bar{v}_0 = (v_x, v_y, v_z)$, and at time t it moves to the position $\bar{r}_0(t) = \bar{r}_0 + \bar{v}_0 t$. $T(t)$ and $R(t)$ are the distances from the object position $\bar{r}_0(t)$ to the emitting source and the receiving element, respectively. The holographic signal comeing from the emitting source via the point object and received by the receiving element at $(x, y_r(t), 0)$ is

$$u(x, t) = \exp\{jk[T(t - \Delta t) + R(t - \Delta t)]\} \tag{1}$$

where $k = 2\pi/\lambda$, $\Delta t = R(t - \Delta t)/C$. let $l_i = v_i t$, $i = x, y, z$, then

$$\Delta t = \{[x - x_0(t) + l_x]^2 + [y_r(t) - y_0(t) + l_y]^2 + [z_0 + l_z]^2\}^{1/2}/C$$

$$= \{l_x^2 + l_y^2 + l_z^2\}^{1/2}/v_0 \tag{2}$$

Now we consider three situations: the imaged object moves in X, Y or Z direction.

(1). Object Moving in X Direction

Let $l_y = l_z = v_y = v_z = 0$. From Eq. (2) we obtain

$$l_x = \left\{ v_x^2[x - x_0(t)] + v_x\sqrt{v_x^2[x_0(t) - x]^2 + [C^2 - v_x^2]R^2(t)} \right\} \Big/ (C^2 - v_x^2)$$

$$\approx v_x R(t) \Big/ \sqrt{C^2 - v_x^2} \tag{3}$$

the above approximation is valid when $v_x/C \ll 1$. Hence

$$R(t - \Delta t) = \{[x - x_0(t) + l_x]^2 + [y_r(t) - y_0]^2 + z_0^2\}^{1/2}$$

$$\approx a_x R(t) + v_x[x - x_0(t)]/C \tag{4}$$

where $a_x = C/(C^2 - v_x^2)^{1/2}$. Similar to the above deduction, we can get

$$T(t - \Delta t) \approx a_x T(t) + v_x[x_t - x_0(t)]/C \tag{5}$$

Substituting Eqs. (4) and (5) into Eq. (1), considering $t = y/v_s$, then expanding the distance terms $R(t)$ and $T(t)$ in a binomial series and retaining only the first two terms, we obtain the hologram

$$u(x, y) = A_1 \exp\{jk[a_z x^2 + (a_z + 2a_z b_z^2)y^2 - 2x(a_z x_0 + w_{z1})$$

$$- 2y(a_z y_0 - y_{r0} - 2a_z b_z x_0 + w_{y1}) - 2a_z b_z xy]/2z_0\} \tag{6}$$

where $b_z = v_z/v_s$, $w_{z1} = -v_z z_0/C$, $w_{y1} = 2b_z v_z z_0/C + a_z b_z x_t + y_{r0} - a_z y_{r0}$. A_1 is a complex independent of x, y. If we integrate the above hologram data and the function

$$h(x_i, y_i, x, y) = \exp\{-jk(x^2 + y^2 - 2xx_i - 2yy_i + 2y_{r0}y)/2z_0\} \tag{7}$$

we obtain the reconstructed image of the point object on the plane $z_i = -z_0$

$$S_z(x_i, y_i) = \iint \sigma(x, y) \exp\{jk[(a_z - 1)x^2 + (a_z + 2a_z b_z^2 - 1)y^2$$

$$+ 2(x_i - a_z x_0 - w_{z1})x + 2(y_i - a_z y_0 + 2a_z b_z x_0 - w_{y1})y$$

$$- 2a_z b_z xy]/2z_0\}dxdy \tag{8}$$

Take the orthonormal transformation V

$$V = \begin{bmatrix} v_{11} & v_{12} \\ -v_{12} & v_{11} \end{bmatrix} \qquad v_{11} = \frac{a_z}{2 + 2q_1}, \qquad v_{12} = \frac{a_z}{2 + 2q_2} \tag{9}$$

$$q_1 = a_z + a_z b_z^2 - 1 + a_z b_z \sqrt{1 + b_z^2}$$

$$q_2 = a_z + a_z b_z^2 - 1 - a_z b_z \sqrt{1 + b_z^2} \tag{10}$$

to transform the position coordinates

$$\begin{bmatrix} x \\ y \end{bmatrix} = V \begin{bmatrix} x_1 \\ y_1 \end{bmatrix}, \qquad \begin{bmatrix} x_i - w_{z1} \\ y_i - w_{y1} \end{bmatrix} = V \begin{bmatrix} x_{i1} \\ y_{i1} \end{bmatrix} \tag{11}$$

and

$$\begin{bmatrix} x_0 \\ y_0 \end{bmatrix} = D \begin{bmatrix} x_{01} \\ y_{01} \end{bmatrix} \tag{12}$$

where

$$D = a_z^{-2} \begin{bmatrix} a_z & 0 \\ 2b_z & a_z \end{bmatrix} V \tag{13}$$

then Eq. (8) can be expressed as

$$S_z(x_i, y_i) = A \iint \sigma(x_1, y_1) V^T \exp\{jk[q_1 x_1^2 + q_2 x_2^2$$

$$+ 2(x_{i1} - x_{01})x_1 + 2(y_{i1} - y_{01})y_1]/2z_0\}dx_1dy_1 \tag{14}$$

where T denotes the transpose.

From the above analysis we can directly deduce the image distribution of the object $O(x_0, y_0)$

$$S_{x1}(x_i, y_i) = \iint O[(x_{01}, y_{01}) D^T] S_x(x_i, y_i) \, dx_{01} dy_{01} \tag{15}$$

From Eqs. (14) and (15) one can see that the effect of the motion will distort the image of the object. To further discuss the effect we separately consider each factor in Eq. (15).

(a) To find out the image position we first omit the effect of the finite aperture and the focussing factor. In Eq. (14) we let

$$\sigma(x_1, y_1) V^T \exp\{jk(q_1 x_1^2 + q_2 x_2^2)/2z_0\} = 1 \tag{16}$$

then

$$S_{x1}(x_i, y_i) = A_2 O[(x_i - w_{x1}, y_i - w_{y1} + 2b_x x_i/a_x)/a_x] \tag{17}$$

As $a_x \gg 1$, the image is amplified by factor a_x. There is a shift w_{x1} in X direction. The shift in Y direction, $w_{y1} - 2b_x x_i/a_x$ is dependent not only on x_i and y_{r0} but also on the X coordinate of the object position, which will cause the severe distortion of the image.

(b) According to Eq. (14) the motion of the object leads to defocussing and the focussing distances along x_1 and y_1 axes, instead of z_0, are

$$z_{x1} = z_0/(1 + q_1), \quad z_{y1} = z_0/(1 + q_2) \tag{18}$$

(c) The finite receiving aperture (represented by $\sigma(x, y)$) still causes diffractive blurring on the image. The focussed image distribution of the point object is the Fourier transformation of σ.

(2) Object Moving In Y Direction

Letting $l_x = l_z = v_x = v_z = 0$, solving Eq. (2) and making the similar approximation we obtain

$$l_y = v_y R(t) \Big/ \sqrt{C^2 - v_y^2}$$

$$R(t - \Delta t) \approx a_y R(t) + v_y[y - y_{r0} - y_0(t)]/C \tag{19}$$

$$T(t - \Delta t) \approx a_y T(t) + v_y[y_i - y_0(t)]/C$$

where $a_y = C \Big/ \sqrt{C_2 - v_y^2}$. The hologram data of a point object

$$u(x, y) = \exp\{ja_y k[((x_i - x_0)^2 + (y_i - y_0 - b_y y)^2 + z_0^2)^{1/2}$$

$$+ ((x - x_0)^2 + (y + y_{r0} - y_0 - b_y y)^2 + z_0^2)^{1/2}]$$

$$+ jkv_y[y + y_{r0} + y_i - 2y_0 - 2b_y y]/C\} \tag{20}$$

where $b_y = v_y/v_s$. Expanding the root factor of Eq. (20) in a binomial series and ignoring the third and higher terms, multiplying the result with Eq. (7) and then integrating the product we obtain, on the plane $z_i = -z_0$, the image of the object

$$S_y(x_i, y_i) = A_3 \iint \sigma(x, y) \exp \{jk[(a_y - 1)x^2 + (a_y - 1 + 2a_yb_y^2 - 2a_yb_y)y^2$$

$$+ 2(x_i - a_yx_0)x + 2(y_i - ma_yy_0 - w_{y2})y]/2z_0\}dxdy \qquad (21)$$

where $m = 1 - 2b_y$ and $w_{y2} = a_yb_y(y_{r0} + y_i) + (1 - a_y)y_{r0} - v_yz_0(1 - 2b_y)/C$. Extending the result to an object of the distribution $O(x_0, y_0)$, we obtain the image of the object

$$S_{y1}(x_i, y_i) = \iint O(x_0, y_0) S_y(x_i, y_i) dx_0dy_0 \qquad (22)$$

(a) Similar to the previous derivation, we omit the effect of the finite aperture and the focussing factor by letting

$$\sigma(x, y) \exp \{jk[(a_y - 1)x^2 + (a_y - 1 + 2a_yb_y^2 - 2a_yb_y)y^2]/2z_0\} = 1$$

then

$$S_{y1}(x_i, y_i) = A_3O[x_i/a_y, (y_i - w_{y2})/ma_y] \qquad (23)$$

One can see the image is a_y times the imaged object in X direction, a_ym times the object and shifted by an amount W_{y2} in Y direction, It is worth noticing that when $b_y = 0.5$ ($m = 0$) the image on the plane $z_i = -z_0$ contracts to the straight line $y_i = w_{y2}$ despite of the original location of the object. So such a situation ought to be avoided in acoustical imaging.

(b) Similar to the case discussed before, the effect of the focussing factor

$$\exp\{jk[(a_y - 1)x^2 + (a_y - 1 + 2a_yb_y^2 - 2a_yb_y)y^2]/2z_0\}$$

is that the focussing distance changes from z_0 to $z_{ix} = z_0/a_y$ in X direction and $z_{iy} = z_0/a_y(1 + 2b_y^2 - 2b_y)$ in Y direction. The effect of the finite aperture $\sigma(x, y)$ on imaging is still to blur the image because of the diffraction.

(3) Object Moving in Z Direction

Letting $l_x = l_y = v_x = v_y = 0$, $a_z = C/\sqrt{C^2 - v_z^2}$, $b_z = v_z/v_s$, $y_{r0} = x_i = y_i = 0$ and $z_0 \gg v_z\Delta T$ (ΔT is the scanning time needed to get the hologram), solving Eq. (2) and making the similar approximation we obtain the hologram data of a point object

$$u(x, y) = A_4 \exp \{jk(x^2 + y^2 - 2xx_0 - 2yy_0)/2z_0 + j2kb_zy - j2kb_zyv_z/C\} \qquad (24)$$

and the reconstructed image on the plane $z_i = -z_0$

$$S_z(x_i, y_i) = \iint u(x, y)\sigma(x, y)h(x, y, x_i, y_i)dxdy$$

$$= A_4 \iint \sigma(x, y) \exp \{jk[(a_z - 1)(x^2 + y^2) + 2(x_i - a_zx_0)x$$

$$+ 2(y_i - a_zy_0 - w_{y3})y]/2z_0\}dxdy \qquad (25)$$

where $w_{y3} = 2b_z v_z z_0/C - 2b_z z_0$. the image of the object $O(x_0, y_0)$ is

$$S_{z1}(x_i, y_i) = \iint O(x_0, y_0) S_z(x_i, y_i) \, dx_0 dy_0 \tag{26}$$

From Eq. (25) one can see that the focussing distance becomes $z_{ix} = z_{iy} = z_0/\alpha_z$, and the image is amplified by the factor α_z and shifted by an amount w_{y3}. The influence of the object motion in Z direction on imaging is less than in the other two directions.

Ⅲ. MOTION COMPENSATION

From the above analyses we know that besides the distortion of the shape and position of the image the motion of the imaged object will cause defocussing. In computer reconstruction of the hologram the defocusssing can be easily compensated provided that the information of the motion is known. Figures 2, 3 and 4 show the images of the system after the defocussing compensation when the motion velocities of the object, (v_x, v_y, v_z), are known. Figure 2 is the image of the object moving in X direction with $\alpha_x = 0.02$ and $b_x = 0.2$. Figure 3 is the image of the object moving in Y direction with $\alpha_y = 0.02$ and $b_y = 0.2$ in Figure 3 (a), $\alpha_y = 0.05$ and $b_y = 0.5$ in Figure 3 (b). Figure 4 is the image of the object moving in Z direction with $\alpha_z = 0.02$ and $b_z = 0.2$. Figure 5 has the same imaging condition as Figure 3 (a) but without compensation.

Figure 1. The image of the motionless object

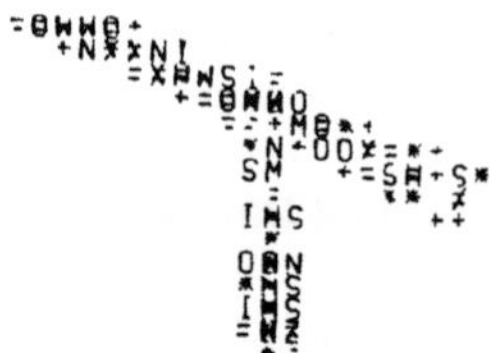

Figure 2. The image of the object moving in X direction, $\alpha_z = 0.02$, $b_z = 0.2$

(a) $\alpha_y = 0.02$, $b_y = 0.2$

(b) $\alpha_y = 0.05$, $b_y = 0.5$

Figure 3. The image of the object moving in Y direction

We should keep in mind that the compensation in the above examples is the defoucussing distortion and it has nothing to do with the distortions of the shape and the position. When the emitting source is a point one, the image distortion of the shape and the position can be compensated to a certain extent from the known values of b_x, b_y, b_z, x_l, y_l, y_{r0} and z_0. But the effect of the compensation is conditioned by concrete situations.

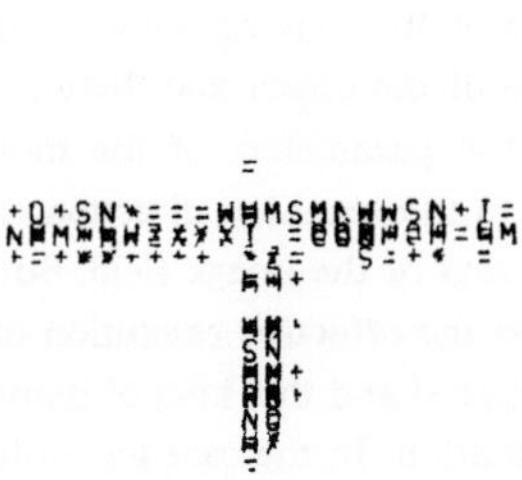

Figure 4.   The image of the object moving in Z direction, $a_z = 0.02$, $b_z = 0.2$

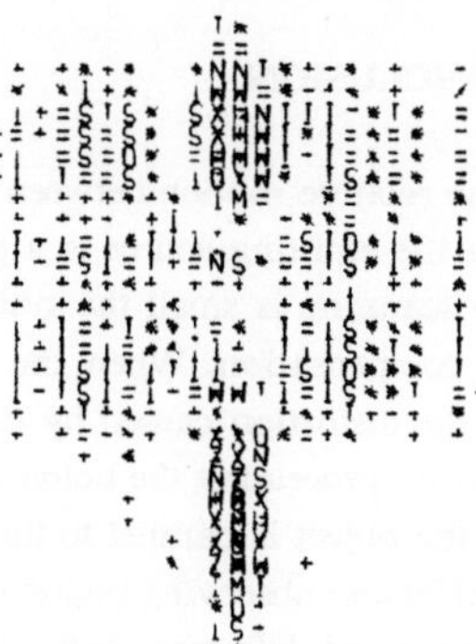

Figure 5.   The image of the object moving in Y direction without defocussing, $a_y = 0.02$, $b_y = 0.2$

Figure 6.   The image of the object moving in X direction with two point emitters, $a_x = 0.02$, $b_x = 0.2$

For the case where the object motion is in X or Z direction, the distortionless image can be obtained by compensating. When the object motion is in Y direction the distortion compensation can also be made, but the effect of the compensation depends on the absolute value of m . For $|m| > 1$ ($b_y < 0$ or $b_y > 1$), the degradation of the image can be counteracted by compensating but the maximum lateral resoluble distance decreases. For $|m| < 1$ ($0 < b_y < 1$) the quality of the image is damaged by the waste of the resolution of the system. For $b_y = 0.5$ the system can not image the moving object at all.

The above discussion is on the basis of single point source emitting. If the source is not a point one the situation becomes more complex. The effects of any kind of the compensation are not satisfactory enough because the directivity pattern of the acoustical field of the emitting source makes the amplitude and the phase vary from point to point on the object plane. So generally speaking, for an imaging system with multi-source emitting, the satisfactory effect of the motion compensation will be very difficult or impossible to obtain. Figure 6 is an example for this case, where the emitting source consists of two point transducers separated by a distance $(x_{t1} = y_{t1} = 0, x_{t2} = 10\lambda z_0/Nd_x, y_{t2} = 0)$ and the object moves in X direction with $a_x = 0.02$ and $b_x = 0.2$. It gives us a good illustration for the effect of multi-source emitting on the object imaging.

V. CONCLUSION

The relative motion between the imaged object and the imaging system is very common. In the case the emitting source is a point one, when the ratio of the moving velocity of the object and the scanning is small the system can obtain the image of the object and there is no need of motion compensation. When the ratio is not small and the parameters of the movement are known the distortion caused by the motion perpendicular to the scanning direction can be compensated by processing the hologram and rearranging the data of the image field. But if the motion of the object is parallel to the scanning direction either the effective resolution of the image or the effective observing region of the system will be damaged and this kind of damage can not be compensated. We must pay attention to this kind of situation. In the case the emitting source is a complex one, generally the motion compensation is limited or can not be made because of the dependent relation between the distortion and the position of the source.

REFERENCES

1. B. P. Hilderbrand, et al, "An Introduction to Acoustical Holography", Plenum Press, New York (1972).
2. R. K. Muller, Acoustical holography, *Proc. IEEE*, 59:1319 (1971).
3. L. Schlussler, et al, Motion limitation of an acoustical holographic system utilizing a scanning linear arrays, *Opt. Eng.*, 16(5): 426 (1977).

INVERSION OF ACOUSTIC VELOCITY WITH
MOMENT MULTI-GRID ALGORITHM*

Benli Gu, Dongyun Deng

Department of Biomedical Engineering
Southeast University
Nanjing 210018, P. R. China

INTRODUCTION

The inverse scatterring imaging has a background of wide application in medicine, nonde-structive testing, seismic imaging, etc. The iterative inverse algorithm, with the moment method to discretize the integral equations, was developed by S. A. Johnson and others in [1—3]. In our work of numerical simulation, we find that this kind of algorithm possesses three attractive characteristics: it has less limitation to the acoustic velocity of the imaging area, less sensitivity to the data of limited angle and it is capable of inversing the velocity and attenuation simultane-ously. The shortcomings of this method are that the computational requirement increase rapidly with the increase of the number of grid cells, and the stability become worse as well. Therefore, it is valuable to decrease the computational requirement and to improve its stability. In this pa-per we have done some work on this subject and achieved relatively satisfactory results.

MULTI-GRID METHOD

For general numerical methods, once the fields have been discretized, the after operations are all on the fixed grids, no longer referring to the form of grids. The fundamental principle of multi-grid methods is as follows: employ different scales of grids to discretize the interesting field, calculate and smooth the residual difference on small grids, rectify it on the large grids and adopt constraint operator and interpolation operator to realize the transformation of the val-ue in the nodes between different scale grids. This kind of method has been extensively applied to the forward problems of elliptic partial differential equations[4]. In this paper, we use it to the

---

* This work was supported by the National Natural Science Foundation of China.

inverse scattered tomography, which belongs to inverse problem. As far as we know, it is the first time the method has been used to such an inverse problem.

Let's take bi-grid method as an example to demonstrate the procedure. Let A represent the imaging area, which is discretized to A_h with certain method, where h is the grid parameter, we name it small grid, and then we formulate the corresponding discrete equations defined on A_h. At the same time another large grid is employed to discretize the same area into A_H, where H is the corresponding large grid parameter. As on A_h, we can also formulate the discrete equations defined on A_H. Although more accurate numerical solution will be achieved by solving the discrete equations on the small grid A_h, it requires much computation time and its stability is poor. The key that the multi-grid method has high efficiency is that the forward calculation of the procedure is on small grid A_h, so that the accuracy is insured, whereas we change the inverse process from A_h to A_H, because the ill-posed inversion calculation is on the large grid, the computational requirement is decreased and the stability is improved. Multi-grid method possesses some lure features, that is, only $O(N)$ operations are required and the convergence rate is independent of the grid scale. As a result, the total computational requirement of iterative procedure is much less than that both the forward and the inverse calculations are on small grid, whereas the total accuracy is not lost and the stability has been improved as well.

Using linear operator L_h to represent the discrete equations defined on A_h

$$L_h U_h = f_h, \qquad U_h \in G(A_h) \tag{1}$$

where U_h is the objective distribution and f_h the field data, $G(A_h)$ represents the grid functional space on the small grid. Correspondingly, for the large grid, the discrete equations defined on A_H are as follows:

$$L_H U_H = f_H, \qquad U_H \in G(A_H) \tag{2}$$

The operators connecting the functions between the small and the large grids are denoted by

$$I_h^H : G(A_h) \rightarrow G(A_H)$$

$$I_H^h : G(A_H) \rightarrow G(A_h)$$

where I_h^H is named constraint operator, I_H^h is the interpolation operator.

Let d_h, V_h and d_H, V_H represent the residual difference of the field data and the error of objective distribution encountered on solving Eqations (1) and (2), respectively. In order to solve Equation (1) with numerical method, the iterative procedure of error rectification in Kth iteration can be described as

$$U_h^{K+1} = U_h^K + V_h^K \tag{3}$$

$$L_h V_h^K = d_h^K \tag{4}$$

At first K equals zero, U_h is the given initial approximate value. Transfer Equation (4) from small grids to large grids, so as to calculate it on large grids

$$\tilde{L}_H V_H^K = d_H^K$$

where $\tilde{L}_H = I_h^H L_h$, $V_H^K = I_h^H V_h^K$, $d_H^K = I_h^H d_h^K$. However, $\tilde{L}_H$ may be very complicated, generally $\tilde{L}_H$ can be replaced approximately by L_H.

$$L_H V_H^K = d_H^K \tag{5}$$

Through simple calculation, we get

$$V_H^K = L_H^{-1} d_H^K = L_H^{-1} I_h^H d_h^K \tag{6}$$

Employ interpolation operator to transfer V_H^K to small grids

$$\hat{V}_h^K = I_H^h V_H^K \tag{7}$$

Finally, regard $\hat{V}_h^K$ as a rectified value of U_h^K defined on small grids, a new and more accurate approximate value can be achieved

$$U_h^{K+1} = U_h^K + \hat{V}_h^K \tag{8}$$

Thus, a recurrence to rectify the node value has been fulfilled. But if we only do this rectifying procedure, the results would not be accurate enough, even could not converge yet. The equations of residual difference defined on large grids only embody the information of the low spatial frequency of residual error of objective distribution on small grids. If the residual error calculated on small grids is transferred to the large grids with constraint operator only, an inaccurate result would be got because of not effectively constraining the component of high spatial frequency. Therefore, it is necessary to smooth the component of high frequency before the operation of rectification operator of the large grids. Likewise, interpolation of Equation (7) will increase the component of high frequency of rectified value defined on the small grids. So it is also necessary to smooth after the operation of rectification operator of large grids. This performance is usually completed with relax iteration, here another feature of relax iteration is employed, that is attenuating the high frequency component.

Multi-grid method is based on bi-grid, it is the link of bi-grid, the only difference is that the calculation order of the multi-grid method must be considered.

ALGORITHM DERIVATION

First, we derive the rectification operator of residual error, the fundamental formula of common iterative inversion of moment method are acoustic wave equations, whose integral form in frequency domain are

$$U(\bar{r}) - U_i(\bar{r}) = \int O(\bar{r}')U(\bar{r}')g(\bar{r}', \bar{r})d\bar{r}' \tag{9}$$

$$U_s(\bar{r}) = \int O(\bar{r}')U(\bar{r}')g(\bar{r}', \bar{r})d\bar{r}' \tag{10}$$

where U, U_i, U_s are the total field, illuminative field and the scattering field, respectively. g is two dimensional Green function, $O(\bar{r})$ is the anomalous coefficient of objective function.

$$g(\bar{r}', \bar{r}) = -\frac{i}{4}H_0^{(1)}(k_0|\bar{r} - \bar{r}'|) \tag{11}$$

$$O(\bar{r}) = 1 - C_0^2/C^2(\bar{r}) \tag{12}$$

where $C(\bar{r})$ and C_0 are the velocity distribution and background velocity, respectively; $H_0^{(1)}$ is the Hankel function and k_0 is the wave number in background medium. Equation (9) is used for calculating inner total field $U(\bar{r})$, and Equation (10) is used for calculating the scattering field in the point measured.

Assume the increment of $O(\bar{r})$, $U(\bar{r})$, $U_s(\bar{r})$ to be $\delta O(\bar{r})$, $\delta U(\bar{r})$, $\delta U_s(\bar{r})$, the function of rectification operator of residual error is to solve $\delta O(\bar{r})$ from $\delta U_s(\bar{r})$, from Equation (10) we get

$$\delta U_s(\bar{r}) = \int \delta O(\bar{r}')U(\bar{r}')g(\bar{r},\bar{r}')d\bar{r}' + \int O(\bar{r}')\delta U(\bar{r}')g(\bar{r},\bar{r}')d\bar{r}' \qquad (13)$$

Equation (13) is the accurate rectification equation, but it is too complicated. Considering that the variation of $O(\bar{r})$ is small per every iteration, we could make linearization, that is, neglect the variation of $U(\bar{r})$ in Equation (13), therefore Equation (13) becomes

$$\delta U_s(\bar{r}) = \int \delta O(\bar{r}')U(\bar{r}')g(\bar{r},\bar{r}')d\bar{r}' \qquad (14)$$

Thus, the rectification operator of residual error is as the same as the discrete operator of integral equation. But $U(\bar{r})$ varied actually, if $U(\bar{r})$ is fixed on the chosen initial value all the time, the results may be very bad or the iteration dispersed. Thus we adjust the total field $U(\bar{r})$ by Equation (9) after solving Equation (14) each time, that is, $U(\bar{r})$ is adjusted via the rectified $O(\bar{r})$.

From above analysis and combining the characteristics of moment method, we select the parameters in the multi-grid methods as follows.

1. Division of the Field

In this paper, differing from the usual way such as standard-coarse-grids, half-coarse-grids and red-black-coarse-grids[5], the fashion shown in Figure 1 is employed.

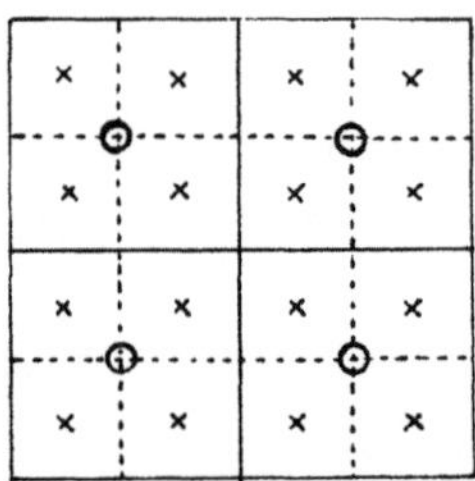

Figure 1. Division of the field. $\times$ and $\bigcirc$ are located in the center of small and large grids, respectively

2. Constraint Operator

The node value of large grids equals the average of the four adjacent node values of small grids.

3. Rectification Operator of Residual Error on Large Grids

We use Equation (14) to rectify $O(\bar{r})$ and Equation (9) to adjust the total field $U(\bar{r})$.

4. Interpolation Operator

If the small grid is located on the frame of the domain, let the node value equal the node value of nearest large grid, otherwise, let it equal the weight mean of node values of the four adjacent nodes of large grids, the weight is inversely proportional to the distance.

5. Smoothing Operator

The common method, which uses relaxation iteration method in some iteration steps, takes considerable computation time and may disperse. In this paper, we use the Fourier transform method, at first transform the object into domain of spatial frequency, then filter out the component of high frequency, finally transform it back to spatial domain. As FFT technique is employed, the total operation is very fast and has a clear physical meaning, the algorithm convergence is insured, and the degree of smoothness can be controlled easily.

The computer realization can be summarized as follows:

① Employ common iterative inversion of moment method, solve the operational equations on large grids, then interpolate the result into small grids, thus we get the initial value of $O(\bar{r})$ and $U(\bar{r})$, respectively; meanwhile, give the accuracy δ.

② Smooth the $O(\bar{r})$ resulted from above step by direct and inverse FFT with filter.

③ Calculate the residual difference $\delta U_s(\bar{r})$ on small grids and smooth it with FFT.

④ Transfer the residual difference to large grids with Constraint operator.

⑤ Solve the rectification equations on large grids.

⑥ Interpolate the rectified value of $O(\bar{r})$ into small grids and smooth it with FFT. Thus we get the rectified value $\delta O^K(\bar{r})$ on small grids.

⑦ Calculate the new value of object

$$O^{K+1}(\bar{r}) = O^K(\bar{r}) + \delta O^K(\bar{r})$$

⑧ Calculate the new total field $U(\bar{r})$ with Equation (9).

⑨ Calculate the mean square error, if it is greater than δ then go to step ②, otherwise go to step ⑩.

⑩ Quit the iteration, print the results or show the image.

NUMERICAL RESULTS

As an example, we employed the imaging system of cross-hole geometry, which was used in geophysical application frequently. As illustrated in Figure 2, the source moved in a series of discrete points in the range $- 2.5\lambda$ to 2.5λ, where λ represents the acoustical wavelength on the background, the interval between adjacent points was 0.5λ and at each point the source emitted once. In the right borehole, the receivers were located also from $- 2.5\lambda$ to 2.5λ, but the interval was 0.25λ. A square area $2\lambda \times 2\lambda$ located in the middle of two boreholes was regarded as the imaging area, where the velocity was unknown, and this area was divided into $16 \times 16 = 256$ grids with each grid $0.125\lambda \times 0.125\lambda$, it was the small grids of bi-grid method, the large grid was chosen as $0.25\lambda \times 0.25\lambda$.

The model adopted was two $0.5\lambda \times 0.5\lambda$ square domains embedded in the homogeneous background with anomalous coefficient $O(\bar{r}) = 0.3$. Both the numerical results obtained by multi-grid method and common iterative inversion of moment method were illustrated for comparison. It can be seen from Figure 3 that the two algorithms have equal accuracy, however, the

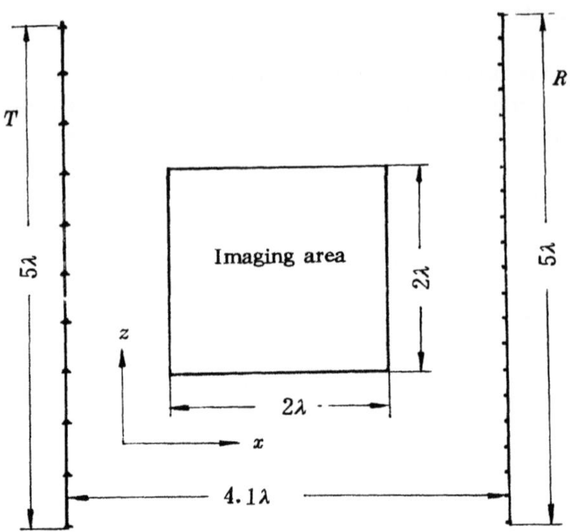

Figure 2.   The imaging system of cross-hole geometry

multi-grid method requires only 5 iterations to go to the stable value.

On the personal computer of Boat-386 with 80287 assistant unit, the common iterative inversion of moment method consumed 16 minutes and 35 seconds, while the multi-grid algorithm only consumed 1 minute and 54 seconds for the same problem. In addition, in the latter, the γ factor for overcoming the ill-posed problem was less than that of the former (decreased 1—2 quantitative order). It was shown the stability of the multi-grid method was stronger than that of common iterative inversion of moment method. We also did some calculations for simulating different situation, such as the $O(\bar{r})$ varied from 0.1 to 0.3 including the imaginary part to represent the attenuation of objectives, and reconstruction from scattering field with 30% noise and so on. The conclusion was the same as that mentioned above.

CONCLUSION

For the purpose of decreasing the computation time of the common iterative inversion of moment method and improving its stability, a new algorithm, namely moment multi- grid method is put forward. The numerical simulations with bi-grid show that the algorithm proposed requires only 1/8 of computation time compared with the conventional method, without any loss of accuracy and without adding any limited condition. The stability of this new algorithm is stronger than that of the old one as well. It demonstrates that it is practicable and effective to introduce the multi-grid method to reconstruction of tomography. Of course, the number of the grid layers can't be more because of the characteristics of the moment method itself. The further work is to choose more suitable parameters and to achieve credible results for more complicated model. We are filled with hope to this.

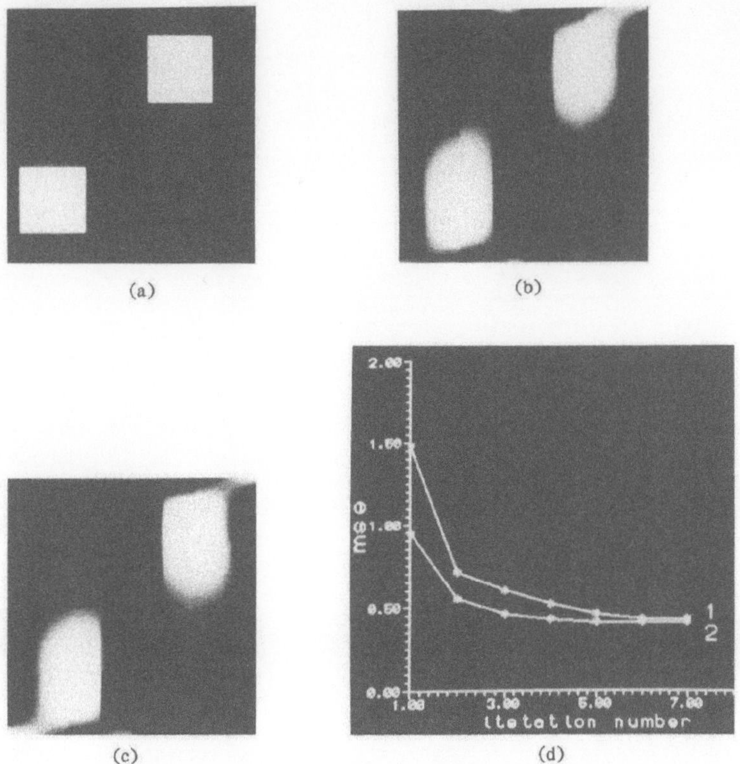

(a)   (b)

(c)   (d)

Figure 3. The model and the results of iterative inversion. (a) The model, (b) The results of 7 iterations with common iterative inversion with moment method, (c) The results of 7 iterations with moment multigrid algorithm, (d) The curves of mean square error of the two methods (curve 1 is of (b), curve 2 is of (c)).

REFERENCES

1. J. H. Richond, Scattering by a dielectric cylinder of arbitrary cross-sectional shape, *IEEE Trans. on AP*, Vol. 1. P331—341, 1965.

2. S. A Johnson and M. L Tracy, Inverse scattering solution by a Sinc basis, multiple source, moment method——part I：theory, *Ultrasonic Imaging*, Vol. 5. P. 361—375, 1983.

3. Y. M. Wang and W. C. Chew, Reconstruction of two-dimensional refractive index distribution using the Born iterative and distorted Born iterative method, in："Acoustical Imaging", Vol. 18, Plenum Press, U. S. A. 1990.

4. W. Hackbusch, "Multi-Grid Methods and Applications", Springeverlag, Berlin, 1985.

5. Changfa Xu, "The Numerical Method for Solving Partial Difference Equations", The Press of Huazhong University, China.

A MODIFIED ALGORITHM FOR ULTRASOUND

HOLOGRAPHIC Q IMAGING

Qin Zhengdi

Computer Technology Laboratory
Technical Research Centre of Finland
90571 Oulu, Finland

INTRODUCTION

It is essential to use linear approaches for non-linear problems in many scientific fields. In the back-propagation theory of ultrasound holography, the imaging process is generally accomplished by twice of Fourier transform[1]. When the propagation field and the data acquisition geometry are linear, the reconstruction is elementary and the fast algorithm can be applied in digital imaging techniques. If the geometry is non-linear (e.g., circular or spherical sensor array), the process is much more complicated and needs much more computation power.

In our earlier study, a circular array ultrasound holographic imaging method was found[2]. In the method, the linear array process was applied directly to reconstruct a circular array image. The geometry differences were compensated for in the k-space. The important step to derive the reconstruction process is based on a circular-rectangle geometrical transform. We gave the name "Q imaging" to this method since the letter "Q" symbolises the analytical process. The circular line in the letter "Q" represents a circular array and the bar at the bottom of it represents a radius of the circle. In the analysis, the circular area is cut by the radius and stretched (transformed) into a rectangle. The image reconstruction process is performed in the rectangular area using the back-propagation principle.

In our earlier analysis, the transformed rectangular area was supposed to be homogenous with a uniform propagation velocity which is the same as that in the circular area before the transform. The phase difference of the holograms between the two geometric systems is compensated for in the frequency domain of the linear array reconstruction process for the rectangular image. In order to get the best focus of the image, an additional compensation factor ([2], Equation 11) was found (rather than proved) to give a further compensation due to the spatial expansion phenomena.

In order to give a better explanation of the spatial expansion phenomena, we found in our later study that the transformed rectangular area should be treated as a non-linear field with a gradient propagation velocity[5]. In this paper, a modified algorithm is given for the "Q imaging" principle in which the compensation is based on the non-linear velocity field of the transformed rectangular area.

GEOMETRICAL TRANSFORM

Figure 1-a describes a cylindrical wave propagation field. In which the propagation media is supposed to be homogenous with constant propagation velocity. A line wave source is radiating a cylindrical wave from the origin of the circle to the infinite. If the object and the transducer array are cylindrical and with the same origin of the circle, the system is a two-dimensional structure. All the parameters in the dimension perpendicular to the circle (or parallel to the axial origin) are constant.

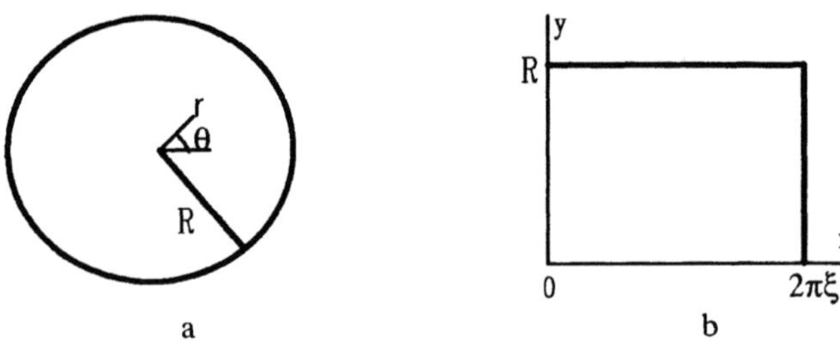

Fig. 1 Geometrical Transform

Let's use a radius R of the circle to cut the circular area and stretch (transform) it into a rectangular area (Figure 1-b). The transform functions from the polar coordinates into the rectangular coordinates are:

$$y = r \qquad (r > 0)$$
$$x = \xi\theta \qquad (2\pi > \theta \geq 0, \ \xi > 0) \tag{1}$$

where r, θ are the polar coordinates and x, y are the rectangular coordinates; ξ is a spatial expansion coefficient with the length dimension. The meaning of ξ is that if $\xi = R$, the length of the circumference with radius R will keep the same when it is transformed into the rectangular area. In our study, we generally let the spatial expansion coefficient be the radius of the transducer array [2].

After the transform, the circular area becomes a rectangular area. A circumference becomes a line segment parallel to the X-axis. A radius becomes a line segment parallel to the Y axis. The line wave source in the circular geometry becomes a plane wave source from the plane at y = 0 and travels parallel to the y direction. So a cylindrical wave in the circular geometry becomes to a plane wave in the rectangular geometry. The change of the geometry also changes the homogeneity of the propagation field. The propagation velocity in the transformed rectangular area is no longer constant. According to the transform functions (1), the wavelength λ' in the rectangular area can be derived as:

$$\lambda' = \xi\lambda/y \qquad (y > 0) \tag{2}$$

where λ is the constant wavelength in the original circular area.

It is a non-linear wave propagation field with the gradient propagation velocity along the y direction. The change of the geometry makes the wavelength on the circumference r = ξ keeps the same in both the geometries (λ' = λ; r = x). But the wavelength on r > x has been compressed (λ' < λ; r > ξ) and the wavelength on r < ξ has been expanded (λ' > λ; r < ξ). Since the wavelength is the function of y and the travelling direction is parallel to the y, the wave front is a plane wave.

BACK-PROPAGATION FOR IMAGE RECONSTRUCTION

In Fig. 1-b, the back propagation function between the line $y = R_1$ and $y = R_2$ can be derived by infinitely dividing the area along Y direction:

$$H_1 = \lim_{\Delta y > 0} \prod_{R_1}^{R_2} H_{\Delta y}$$

$$= \exp\{ 2\pi j \int_{R_1}^{R_2} \sqrt{(1/\lambda')^2 - (f_x)^2}\, dy \} \quad (\ |f_x \lambda'| \leq 1\ ) \tag{3}$$

where $H_{\Delta y}$ is the back-propagation transfer function for a infinite small distance Δy in which the wavelength can be treated as constant; f_x is the abscissa of the spectrum of the hologram.

The next step is to check the differences between Fig.1-a and Fig. 1-b. Since Fig. 1-b is a transformed geometry from Fig. 1-a, are the two system equivalent? The answer should be found by examining the phase difference between the holograms on the receiving apertures in both systems sine the holographic imaging is the phase sensitive method.

let's consider two points, $P_1(R_1, \theta_1)$ and $P_2(R_2, \theta_2)$ in the original circular field. The phase loss of the propagation from P_1 to P_2 is:

$$\phi = \frac{2\pi}{\lambda}[R_1^2 + R_2^2 - R_1 R_2 \cos(\theta_1 - \theta_2)] \tag{4}$$

After the change of the geometry, the two points become $P'_1(\xi\theta_1, R_1)$ and $P'_2(\xi\theta_2, R_2)$ in the rectangle coordinate. The phase loss from P'_1 to P'_2 is given by:

$$\phi' = 2\pi \int_{P'_1}^{P'_2} \frac{1}{\lambda'}\, dl \tag{5}$$

where dl is the line differential along the curve of possible travelling route from P'_1 to P'_2.

To a fixed propagation direction, the phase loss along the path can be decided by the so-called trace-back method using the refraction principle. In an other word, the phase loss in (5) can be described as wavefront-angle-dependent. Once the phase is decided in a certain geometry like Fig. 1-b, a clear map can be drown for the propagation path. The phase loss between P_1 to P_2. in (4) can be also rewrite as wavefront-angle-dependent based on the geometry transform relation (1). Eventually, the phase difference between Fig. 1-a and Fig. 1-b is derived as:

$$\phi' - \phi \;=\; \Phi(\alpha) \tag{6}$$

Where α is the direction of the propagation or wavefront-angle.

It is well known that the spectrum (f_x) of the hologram has the following relation with the wavefront-angle (α) in the spatial domain [3]:

$$f_x \;=\; \sin(\alpha)/\lambda \tag{7}$$

So, the circular array imaging system (Fig. 1-a) can be treated as a distorted linear array system with non-linear propagation media. The phase distortion of the hologram can be described as the so-called wavefront-angle-dependent and be compensated for by a simple multiplication factor in the frequency domain [4]. With the phase compensation, the linear array imaging process is directly applied to reconstruct the circular array image:

$$\mathbf{u}_l \;=\; \mathbf{F}^{-1}\{\; \mathbf{H}_l \cdot \exp[-j\Phi(f_x)] \cdot \mathbf{F}\{\mathbf{u}_h\}\} \tag{8}$$

where $\mathbf{u}_h$ is the hologram on the circular array; $\mathbf{u}_l$ is the image of the object circumference; $\mathbf{F}$ and $\mathbf{F}^{-1}$ are the linear Fourier transform and inverse Fourier transform respectively.

In the digital imaging techniques, a limited number, say, N elements are uniformly distributed on the hologram circumference. After the discrete Fourier transform, the spectrum is an N-element vector. The calculation for the phase difference are performed for each element of the spectrum. The compensation factor is also an N-element vector. When a circular array geometry has been fixed, the calculation needed only once and the compensation factor is stored and be used for later reconstruction.

On the other hand, the non-linear back propagation function $\mathbf{H}_l$ is also an N-element vector and the geometry dependent. The compensation vector and $\mathbf{H}_l$ can be merged into a single vector and be stored as a look-up table to form a special vector for the back propagation function of the circular system. From practical sense, the reconstruction time needed for a circular array image is exact the same as for a normal linear array image.

REFERENCES

[1] J. W. Goodman, "Introduction to Fourier optics," New York: McGraw-Hill, 1968.

[2] Z. D. Qin, et al, "Circular-array ultrasound holography imaging using the linear-array approach," *IEEE Trans. UFFCl. 36*, No. 5, pp. 485-493, Sep. 1989.

[3] Z. D. Qin and J. Ylitalo, "Ultrasound holographic Q imaging", IEEE 1991 Ultrasonics Symposium, Proceedings p1235-1238, Dec. 1991.

[4] Z. D. Qin, et al, "Frequency domain compensation for inhomogeneous layer in ultrasound holography", *IEEE Trans. UFFC,* Vol.36, no. 1, pp. 73-79, Jan. 1989.

A COMPARISON OF INTERMEDIATE, RYTOV AND BORN TRANSFORMATIONS IN ACOUSTIC TOMOGRAPHY

Zhen-Qiu Lu[1], Xiao-Yan Han[2], Yan-Yun Zhang[1]
and Guo-Ping Chen[1]

1: Department of Physics, Nankai University, Tianjin 300071, P.R.C
2: Inst. of Automatic Control, Academia Sinica, Beijing 100008, P.R.C

INTRODUCTION

Acoustic transmission diffraction tomography, which is a technique for reconstructing one or several parameters of an object to be imaged from the observed scattered acoustic field outside of the object, is discussed in this paper. Because the wave length of the incident wave is on the order of 1mm, diffraction effects of the insonifying wave must be taken into account. Conventionally, the first-order Born and Rytov approximation inversion algorithms have been used on the assumption that the parameters of the object to be imaged deviate slightly from the same parameters of the surrounding medium, i.e., inhomogeneities are weakly scattering[1-6]. For an object characterized by a wave number, the method of the first-order Born perturbation approximation is only valid when the product of the size of the object and the deviation in the circular wave number is small. The method of the first-order Rytov approximation is only valid when the following three conditions are satisfied:

(a) The relative wave number deviation is small.

(b) The product of the size of the object and the derivation of the relative wave number deviation along the direction perpendicular to the incident wave (hereinbelow, referred to as the effect of inhomogeneity) is small.

(c) On wave number discontinuities, including the boundary of the object, junction lines must be 'steep' enough (hereinbelow, referred to as the effect of discontinuity).

The first condition above mentioned was noticed but the other two have been neglected before[7-8]. Due to the effect of discontinuity, the distortion is considerable large on the boundary of the object when the first-order Rytov approximation is used to reconstruct a single uniform cylinder surrounded by a homogeneous medium.

The aim of this paper is to investigate advantages and disadvantages of the first-order Born, Rytov and intermediate transformations for reconstruction in detail and to use the intermediate transformation, which is a combination of Born and Rytov transformations in a sense, for reconstruction algorithms in order to overcome difficulties caused by the Born and Rytov transformations.

GENERALIZED SCATTERED WAVES AND RECONSTRUCTION ALGORITHMS

Diffraction tomography has been formulated as an approximate inversion of the inhomogeneous Helmholtz equation

$$\nabla^2 \psi(\mathbf{x}) + k_0^2 \psi(\mathbf{x}) = -k_0^2 f(\mathbf{x}) \psi(\mathbf{x}) \tag{1}$$

where

$$f(\mathbf{x}) = \left(\frac{k(\mathbf{x})}{k_0}\right)^2 - 1 \qquad k(\mathbf{x}) = \begin{cases} k_0 & \mathbf{x} \in D^* \\ k_i(\mathbf{x}) & (\mathbf{x}) \in D \end{cases} \tag{2}$$

in which $k(\mathbf{x})$ is the circular wave number, D and D^* are domains occupied by an object to be imaged and its surrounding medium, respectively.

We introduce a generalized transformation[4,8,9]

$$\chi = \psi_0 g(\alpha) \qquad \alpha = \frac{\psi}{\psi_0} \tag{3}$$

where $\psi_0 = Ae^{-i\omega t}$ is a single-frequency incident wave, which satisfies the homogeneous wave equation

$$\nabla^2 \psi_0 + k_0^2 \psi_0 = 0 \tag{4}$$

and $g(\cdot)$ is a twice differential function with $g(1) = 0$. $\chi(\mathbf{x})$ is called a generalized scattered wave[9]. After the generalized transformation (3), Eq.(1) becomes

$$\nabla^2 \chi + k_0^2 \chi = -k_0^2 f \psi_0 \frac{dg}{d\alpha} \alpha + \psi_0 \frac{d^2 g}{d\alpha^2} (\nabla \alpha)^2 \tag{5}$$

If $g(\alpha) = \alpha - 1$, we obtain the Born transformation

$$\chi = \psi - \psi_0 = \psi_s, \qquad \nabla^2 \psi_s + k_0^2 \psi_s = -k_0 f \psi \tag{6}$$

If $g(\alpha) = \ln \alpha$, we obtain the Rytov transformation

$$\chi = \psi_0 \ln \frac{\psi}{\psi_0} \qquad \nabla^2 \chi + k_0^2 \chi = -k_0^2 f \psi_0 - \psi_0 (\nabla \ln \alpha)^2 \tag{7}$$

If $g(\alpha) = m(\alpha^{\frac{1}{m}} - 1)$, we obtain the intermediate transformation

$$\chi = \psi_0 m\left[\left(\frac{\psi}{\psi_0}\right)^{\frac{1}{m}} - 1\right], \nabla^2 \chi + k_0^2 \chi = -k_0^2 f \psi_0 \alpha^{\frac{1}{m}} - \psi_0 (1 - \frac{1}{m}) \alpha^{\frac{1}{m}} (\nabla \ln \alpha)^2 \tag{8}$$

When m=1, Eq.(8) is reduced to the Born transformation and when $m \to \infty$, it approaches the Rytov transformation.

In the following, we give the reconstruction formulae approximately according to the equation

$$\nabla^2 \chi + k_0^2 \chi = -k_0^2 f \psi_0 \tag{9}$$

This is a first-order perturbation approximation of (5), (6), (7) and (8).

Taking the Sommerfeld radiation conditions into account,Eq.(9) becomes the integral equation[6,9]

$$\chi(\mathbf{x}) = \int G(|\mathbf{x} - \mathbf{x}\prime|)[-k_0^2 f(\mathbf{x}\prime)\psi(\mathbf{x}\prime)]d^n\mathbf{x} \tag{10}$$

where

$$G(|\mathbf{x}|) = \frac{k_0^{\frac{n}{2}-1}}{4(2\pi|\mathbf{x}|)^{\frac{n}{2}-1}i^{n-1}}H_{-(\frac{n}{2}-1)}^{(1)}(k_0|\mathbf{x}|) \tag{11}$$

is the n-D free-space Green function[10], $H_{-(\frac{n}{2}-1)}^{(1)}$ is the Hankel function of the first kind of $-(\frac{n}{2}-1)$ order. When $n = 2$,Eq.(11) reduces to

$$G(|\mathbf{x}|) = H_0^{(1)}(k_0|\mathbf{x}|)/4i \tag{12}$$

Which is a divergent circular cylindrical wave.

According to (10), we obtain the following 2-D reconstruction formulae in the frequency domain[6]

$$f(x\prime, y\prime) = \frac{1}{\pi A k_{0i}^2} \int\int \nu_1 S(x_r, \nu_y, \theta)e^{-i2\pi\nu_1 x_r + i2\pi(ux\prime+vy\prime)}dudv \tag{13}$$

and

$$f(x\prime, y\prime) = \frac{1}{\pi A \bar{k}_{0i}} \int\int |\nu_y| S(x_r, \nu_y, \theta)e^{i2\pi(-\nu_1 x_r + ux\prime + vy\prime)}d\nu_y d\theta =$$

$$= \frac{1}{\pi A \bar{k}_{0i}} \int\int |\nu_y| rect(\frac{\nu_y}{2\bar{k}_0})S(x_r, \nu_y, \theta)e^{i2\pi(-\nu_1 x_r + ux\prime + vy\prime)}d\nu_y d\theta \tag{14}$$

where $\qquad \bar{k}_0 = \frac{k_0}{2\pi}$

$$\begin{cases} u = (\sqrt{\bar{k}_0^2 - \nu_y^2} - \bar{k}_0)\cos\theta + \nu_y \sin\theta \\ v = \nu_y \cos\theta - (\sqrt{\bar{k}_0^2 - \nu_y^2} - \bar{k}_0)\sin\theta \qquad |\nu_y| \le \bar{k}_0 \end{cases} \tag{15}$$

$$\nu_1 = \sqrt{\bar{k}_0^2 - \nu_y^2} \qquad |\nu_y| \le \bar{k}_0 \tag{16}$$

$$S(x_r, \nu_y, \theta) = \int_{-\infty}^{\infty} \chi(x_r, y, \theta)e^{-i2\pi\nu_y y}dy \tag{17}$$

rect $(\cdot)$ is the rectangular function, θ the rotation angle of the object and x_r the position where a receiver array is located (see Fig.1).

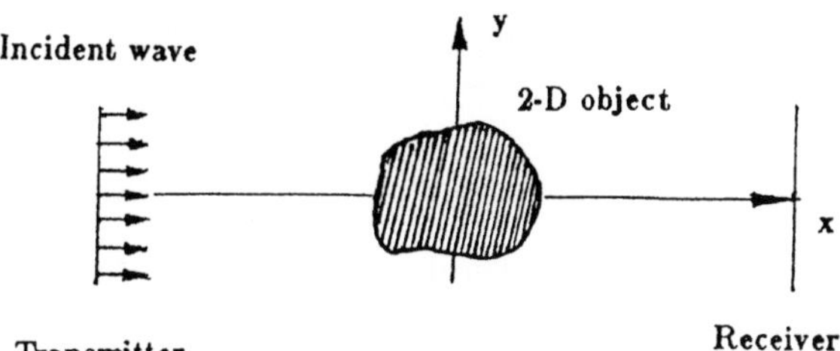

Figure 1. Geometry for 2-D structure reconstruction.

ADVANTAGES AND DISADVANTAGES OF THE FIRST-ORDER BORN AND RYTOV PERTURBATION APPROXIMATION ALGORITHMS

Eq.(5) can be rewritten as

$$\nabla^2\chi + k_0^2\chi = -k_0^2\psi_0 f_c \tag{18}$$

where

$$f_c = f\frac{dg}{d\alpha}\alpha - \frac{1}{k_0^2}\frac{d^2g}{d\alpha^2}(\nabla\alpha)^2 = f + D \tag{19}$$

and

$$D = f(\frac{dg}{d\alpha}\alpha - 1) - \frac{1}{k_0^2}\frac{d^2g}{d\alpha^2}(\nabla\alpha)^2 \tag{20}$$

D is considered as a distortion term under the first-order approximation. The first-order perturbation approximation is only valid when the distortion term is small

$$|D| << |f| \tag{21}$$

This reduces to

$$|D_B| = |(\alpha - 1)f| << |f| \tag{22}$$

$$|D_R| = |\frac{1}{k_0}\nabla\ln\alpha|^2 << |f| \tag{23}$$

and

$$|D_I| = |(\alpha^{\frac{1}{m}} - 1)f - (1 - \frac{1}{m})\alpha^{\frac{1}{m}}(\frac{\nabla\ln\alpha}{k_0})^2| << |f| \tag{24}$$

for the first-order Born, Rytov and intermediate approximations respectively.

In the following, we discuss advantages and disadvantages of the first-order Born and Rytov approximations. We assume that there is a discontinuity line inside the object. For simplicity, we choose the geometrical optics approximation (see Fig.2)

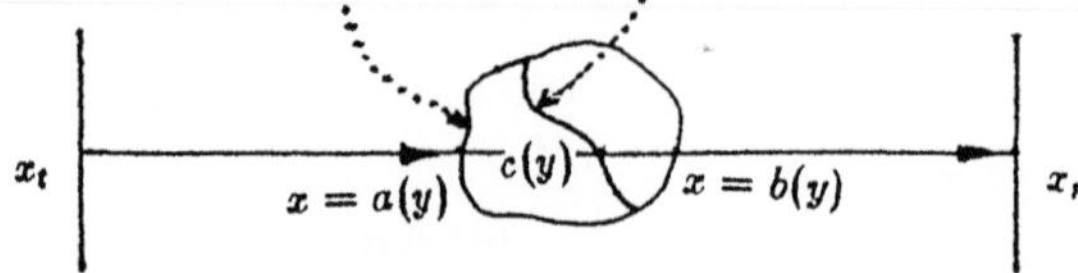

Figure 2. Geometrical optics approximation.

$$\psi = Aexp\{i\int_{x_t}^{x}k(x,y)dx\} =$$

$$= A \begin{cases} exp\{i(\int_{x_i}^{a(y)} k_0 dx + \int_{a(y)}^{x} k_i dx)\} & a \le x < c \\ exp\{i(\int_{x(t)}^{a(y)} k_0 dx + \int_{a(y)}^{c(y)} k_i dx + \int_{c(y)}^{x} k_i dx)\} & c \le x < b \\ exp\{i(\int_{x_i}^{a(y)} k_0 dx + \int_{a(y)}^{c(y)} k_i dx + \int_{c(y)}^{b(y)} k_i dx + \int_{b(y)}^{x} k_0 dx)\} & x \ge b \end{cases} \qquad (25)$$

Thus we have

$$\alpha = \frac{\psi}{\psi_0} = \begin{cases} exp\{i\int_{a(y)}^{x}(k_i - k_0)dx\} & a \le x < c \\ exp\{i[\int_{a(y)}^{c(y)}(k_i - k_0)dx + \int_{c(y)}^{x}(k_i - k_0)dx]\} & c \le x < b \\ exp\{i[\int_{a(y)}^{c(y)}(k_i - k_0)dx + \int_{c(y)}^{b(y)}(k_i - k_0)dx]\} & x \ge b \end{cases} \qquad (26)$$

Under the first-order Rytov approximation

$$|D_R| = |\frac{1}{k_0}\nabla \ln \alpha|^2 =$$

$$= \begin{cases} |(\frac{k_i-k_0}{k_0})^2 + [\int_{a(y)}^{x} \frac{\partial}{\partial y}(\frac{k_i-k_0}{k_0})dx - \frac{k_i(a(y),y)-k_0}{k_0}\frac{da(y)}{dy}]^2| \\ \qquad\qquad\qquad a \le x < c \\ |(\frac{k_i-k_0}{k_0})^2 + [\int_{a(y)}^{x} \frac{\partial}{\partial y}(\frac{k_i-k_0}{k_0})dx - \frac{k_i(a(y),y)-k_0}{k_0}\frac{da(y)}{dy} - \frac{k_i(c(y)+0,y)-k_i(c(y)-0,y)}{k_0}\frac{dc(y)}{dy}]|^2 \\ \qquad\qquad\qquad c \le x < b \\ |\int_{a(y)}^{b(y)} \frac{\partial}{\partial y}(\frac{k_i-k_0}{k_0})dx - \frac{k_i(a(y),y)-k_0}{k_0}\frac{da(y)}{dy} + \frac{k_i(b(y),y)-k_0}{k_0}\frac{db(y)}{dy} - \frac{k_i(c(y)+0,y)-k_i(c(y)-0,y)}{k_0}\frac{dc(y)}{dy}|^2 \\ \qquad\qquad\qquad x \ge b \end{cases}$$

$$(27)$$

where $k_i(c(y) + 0, y)[k_i(c(y) - 0, y)]$ is the limit of $k_i(x, y)$ when x approaches $c(y)$ from the right [left] side of $c(y)$.

Obviously, the distortion in the Rytov's method results from three factors, they are

(a) The relative wave number deviation.

(b) The product of the size of the scatterer and the derivative of the relative wave number deviation along the direction perpendicular to the incident wave.

(c) The product of the relative wave number deviation and the slope of the discontinuity line with the y-axis.

For an homogeneous cylinder surrounded by an homogeneous medium, the model approximation error using the Rytov transformation comes from (a) and (c). Because the boundary of the circular cylinder is $x^2 + y^2 = R^2$, where R is the radius of the cylinder, we have

$$a(y) = -\sqrt{R^2 - y^2} \qquad\qquad b(y) = +\sqrt{R^2 - y^2} \qquad (28)$$

Therefore the error coming from (c) is

$$-\frac{k_i(a(y), y) - k_0}{k_0}\frac{da(y)}{dy} + \frac{k_i(b(y), y) - k_0}{k_0}\frac{db(y)}{dy} = -2\frac{k_i - k_0}{k_0}\frac{y}{\sqrt{R^2 - y^2}} \qquad (29)$$

When y approaches to R,Eq.(29) becomes very large even if the relative wave number deviation is very small. So, the first-order approximation method under the Rytov transformation is valid for large-sized objects with small deviations in relative wave number only if the effects of inhomogeneity and discontinuity could be ignored.

On the other hand, the first-order approximation under the Born transformation is valid on condition that

$$|\alpha - 1| = |exp\{i \int_{a(y)}^{b(y)}(k_i - k_0)dx\} - 1| << 1 \qquad (30)$$

i.e., the product of the size of the object and the deviation in the circular wave number is small. The first-order Born approximation method has no effects of inhomogeneity and discontinuity.

INTERMEDIATE TRANSFORMATION

Under the first-order intermediate approximation,

$$D_I = D_B(m) - (1 - \frac{1}{m})\alpha^{\frac{1}{m}} D_R \tag{31}$$

where

$$D_B(m) = (\alpha^{\frac{1}{m}} - 1)f \tag{32}$$

is a distortion term in the first-order Born approximation method for an object whose circular wave number deviation from the one of its surrounding medium is less than the deviation for the scatterer by a factor of m. We have

$$|D_I| \leq |D_B(m)| + (1 - \frac{1}{m})|\alpha^{\frac{1}{m}}||D_R| \tag{33}$$

in particular, for the phase-type object, i.e., k_i is real, we have

$$|D_I| \leq |D_B(m)| + (1 - \frac{1}{m})|D_R| \tag{34}$$

In the sense of Eq.(31), D_I is a combination of $D_B(m)$ and D_R. When m increases, $|D_B(m)|$ decreases and $(1 - \frac{1}{m})|D_R|$ increases. When $m = 1$, $D_B(m) = D_B$ and the second term on the right side of Eq.(31) disappears. When $m \rightarrow \infty$, $D_B(m)$ approaches zero and $(1 - \frac{1}{m})D_R$ approaches D_R. So reconstruction using the intermediate transformation can enlarge the region of validity in the first-order perturbation approximation because it can not only increase valid sizes of objects to be imaged, but also reduces the effects of inhomogeneity and discontinuity. For example, if $m = 2$, the valid size of the object can increase by a factor of 2 and the effects of inhomogeneity and discontinuity can decrease by a factor of 2.

COMPUTER SIMULATIONS

We conducted computer simulations on a single homogeneous circular cylinder with its wave number real (i. e.,an object of pure phase type).

In the following, comparisons of tomograms obtained by the three methods are given to demonstrate the above effects.

For a single circular homogeneous cylinder, we have

$$k(x,y) = \begin{cases} k_0 d & x^2 + y^2 \leq R^2 \\ k_0 & \text{elsewhere} \end{cases} \tag{35}$$

and

$$f(x,y) = \begin{cases} d^2 - 1 & x^2 + y^2 \leq R^2 \\ 0 & \text{elsewhere} \end{cases} \tag{36}$$

where d is a real constant, $k_0 = \frac{2\pi}{\lambda}$, $\lambda = 0.5mm$.

Simulations were performed for an object of phase type and $d = 0.92$ and $R = 3\lambda$, using the first-order Born, Rytov and intermediate $(m = 2, 30)$ approximations(see Fig.3).

These examples show advantages and disadvantages of the first-order Born and Rytov approximations, and also show advantages of the intermediate transformation over the Born and Rytov transformations. Using the intermediate m=2 transform,the distortion near the boundary of the cylinder due to discontinuity is less than the distortion in the Rytov approximation by a factor of 2 and the distortion due to the size of the cylinder is nearly as the same as the distortion for a cylinder of 1.5λ radius in the Born transformation. Obviously, when $m >> 2Rk_0|d - 1| \approx 3$,for example, $m = 30$, the error in the first-order Born approximation disappears,and the first-order intermediate approximation is reduced to the first-order Rytov approximation.

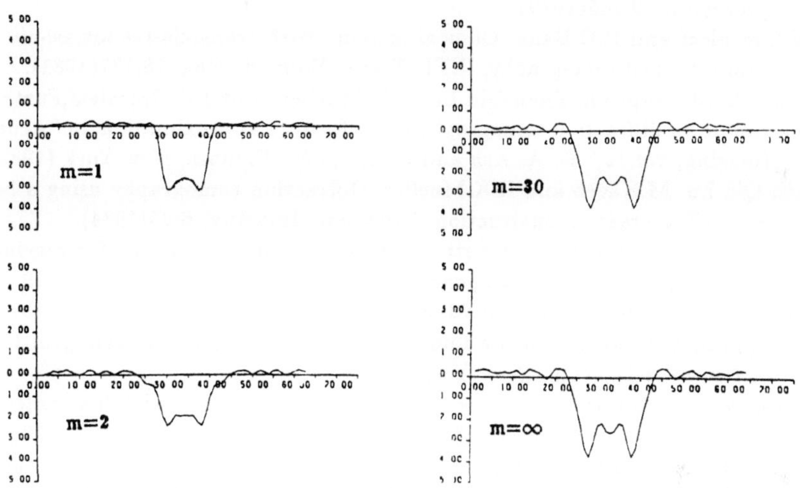

Figure 3. Reconstructed profiles of a circular cylindrical object of $3\ \lambda$ radius and $d = 0.92$ using the intermediate transformations $m = 1(Born), 2, 30, \infty(\text{Rytov})$.

CONCLUSIONS

We have discussed advantages and disadvantages of the first-order Born, Rytov and intermediate approximations. The distortion in the first-order Rytov approximation results from three factors, which are related to the relative wave number, inhomogeneity inside the scatterer and the size of the scatterer, and discontinuity inside and on the boundary and slopes of discontinuity lines. Many of them were ignored in the previous work because examples taken before were too simple to show those effects. On the other hand, the distortion in the first-order Born approximation results only from the deviation in circular wave number and the size of the scatterer, but it is very sensitive to the size. The distortion in the first-order intermediate approximation is a combination of the Born's and Rytov's and reconstruction using appropriate intermediate transformation can enlarge the region of validity in the first-order perturbation approximation because it can not only increase valid sizes of objects to be imaged but also reduce the effects of inhomogeneity and discontinuity.

ACKNOWLEDGMENT

This work was supported in part by the National Natural Science foundation of China.

REFERENCES

1. R.K.Mueller, M.Kaveh and G.Wade, Reconstructive tomography and applications to ultrasonics, Proc. IEEE, 67:567(1979).
2. M.Kaveh, R.K.Mueller and R.D.Inversion, Ultrasonic tomography based on perturbation solution of the wave equation, Computer graphics and image processing, 9:105(1979).
3. J.F.Greenleaf and R.C.Bahn, Clinical imaging with transmissive ultrasonic computerized tomography, IEEE Trans. Biomed. Eng.,28:177(1981).
4. M.Kaveh, M.Soumekh, Zhen-Qiu Lu, R.K.Mueller, and J.F.Greenleaf,Further results on diffraction tomography using Rytov's approximation, in:"Acoustical Imaging, vol.12," E. A. Ash and K. Hill, eds. Plenum, New York (1982).
5. Zhen-Qiu Lu, M.Kaveh and R.K.Mueller, Diffraction tomography using a beam wave: Z-average reconstruction, Ultrasonic Imaging, 6:95(1984).
6. Zhen-Qiu Lu, Multidimensional structure diffraction tomography for varying object orientation through generalized waves, Inverse Problems,1:339 (1985).
7. M.Kaveh, M.Soumekh, and R.K.Mueller, A comparison of Born and Rytov approximations in acoustic tomography, in :"Acoustical Imaging,vol.11," J. Powers, ed.,plenum,New York (1982).
8. M.Slaney, A.C.Kak and L.E.Larsen, Limitations of imaging with first-order diffraction tomography, IEEE Trans Microwave Theory Tech,32: 860(1984).
9. Zhen-Qiu Lu, JKM perturbation theory, relaxation perturbation theory and their applications to inverse scattering: theory and reconstruction algorithms, IEEE Trans. Ultrason. Ferroelectrics, Freq. Contr., 33:722(1986).
10. I. M. Gel'fand and G. E. Shilov, "Generalized Functions Vol.1," Academic, New York (1953).

EXACT RECONSTRUCTION OF A SCATTERER AND COMPLETENESS RELATIONS FOR THE SCATTERING SOLUTIONS

Zhen-Qiu Lu

Department of Physics, Nankai University , Tianjin 300071, P.R.C.

INTRODUCTION

Over the last two decades, there has been a considerable amount of work on the inverse scattering problems, especially for quantum and acoustic waves[1-10]. But up to now, it is still very difficult to reconstruct a higher dimensional scatterer from the observed scattered field outside the scatterer. The difficulties in reconstructing a higher dimensional scatterer lie in diffraction effects. What is the way out? One notes that the one dimensional inverse scattering problems have been worked out well[1-5]. The key ideas in solving the one dimensional inverse scattering problems are as follows:

(1) Introduce a field transform to simplify the relation between the underlying scatterer (or potential) and the direct scattered field.

(2) Use the completeness relation for this direct scattering solution system.

So the completeness relation plays a central role in the exact reconstruction of a scatterer. But up to now, only the completeness relation for the Schrödinger equation on the half line was given correctly [1-2,4]. Almost all papers and books, for example, Refs.1,5 and 10, introduced wrong 'completeness formulas' for the Schrödinger equation on the full line, and even used them to treat inverse problems. Ref.11 gave a correct form of completeness relation for the Jost solutions, but did not gave mathematically rigorous proof and put strong constraints on the underlying potential. Also there were the same problems in three dimensions as in one dimension about completeness relations. In order to solve the higher dimensional inverse scattering problems, especially for acoustic waves of practical use, we must make detailed researches into the completeness of direct scattering solution systems. First, we discuss the completeness relations for the solution systems of the Schrödinger equation on the full line in detail. And then discuss the relationship between the exact reconstruction of a scatterer, or the uniqueness of the inverse scattering problem and the completeness of scattering solution, and from this point of view, it is very easy to see the wrongness of so called 'completeness formulas,' at a glance, without resort to careful calculations. Finally, point out why such wrong completeness formulas were used so long by so many authors.

Acoustical Imaging, Volume 20 Edited by Y. Wei
and B. Gu, Plenum Press, New York, 1993

SCATTERING SOLUTIONS; SCATTERING AND JOIST MATRICES AND THE NATURE OF THE SPECTRUM OF THE SCHRÖDINGER OPERATOR

First let us introduce several different solutions of the Schrödinger equation on the full line

$$Lz = -\frac{d^2z}{dx^2} + V(x) = \lambda z \qquad -\infty < x < \infty \tag{1}$$

that satisfies different boundary conditions. Here we suppose that $V(x)$, the potential or scatterer is a real function of x, continuous on $-\infty < x < \infty$ and obeys the constraint

$$\int_{-\infty}^{\infty} dx(1 + |x|)|V(x)| < \infty \tag{2}$$

The scattering solutions which will be used later are as follows:

(1) The Jost solutions which satisfy the boundary conditions

$$f_1 \sim e^{ikx} \qquad x \sim +\infty \ , \qquad f_2 \sim e^{-ikx} \qquad x \sim -\infty \tag{3}$$

where $\lambda = k^2$ and $'\sim'$ denotes asymptotic approximation. For real k we have

$$f_p(x, -k) = f_p^*(x, k) \qquad p = 1, 2 \tag{4}$$

where '$*$' denotes complex conjugation. We shall combine these two functions into a two component column vector denoted by f.

(2) The physical solutions, which satisfy the boundary conditions

$$\psi_1(x, k) \sim \begin{cases} s_{11}(k)e^{ikx} & x \sim +\infty \\ e^{ikx} + s_{12}(k)e^{-ikx} & x \sim -\infty \end{cases} \tag{5a}$$

$$\psi_2(x, k) \sim \begin{cases} e^{-ikx} + s_{21}(k)e^{ikx} & x \sim +\infty \\ s_{22}(k)e^{-ikx} & x \sim -\infty \end{cases} \tag{5b}$$

in which the matrix

$$S(k) = \begin{pmatrix} s_{11} & s_{12} \\ s_{21} & s_{22} \end{pmatrix} \tag{6}$$

is called the S-matrix of the equation (1) and has the following properties: the symmetry and the unitary

$$s_{11}(k) = s_{22}(k) = T(k) \ , \qquad S(k)S^{\dagger}(k) = I \tag{7}$$

and for real $k, S(-k) = S^*(k)$, where '$\dagger$' denotes Hermitian conjugation and I a 2×2 unit matrix. $T(k)$ is called the transmission coefficient, $s_{12} = R_l$ the reflection coefficients to the right and to the left respectively. The column vector of the physical solutions will be denoted by ψ.

(3) The regular solutions, which satisfy the boundary conditions

$$g(x, k) = \begin{pmatrix} g_1(x, k) \\ g_2(x, k) \end{pmatrix} = \begin{pmatrix} 1 \\ 1 \end{pmatrix}, \quad \frac{dg}{dx} = \begin{pmatrix} \frac{dg_1}{dx} \\ \frac{dg_2}{dx} \end{pmatrix} = ik \begin{pmatrix} 1 \\ -1 \end{pmatrix} \qquad x = 0 \tag{8}$$

The regular solutions g_1 and g_2 are integral functions of k for each fixed x.

(4) The regular solutions, which satisfy the real boundary conditions

$$\phi(x, k) = \begin{pmatrix} \phi_1(x, k) \\ \phi_2(x, k) \end{pmatrix} = \begin{pmatrix} 1 \\ 0 \end{pmatrix}, \quad \frac{d\phi}{dx} = \begin{pmatrix} 0 \\ -1 \end{pmatrix} \qquad x = 0 \tag{9}$$

The regular solutions ϕ_1 and ϕ_2 are real-valued and also integral functions of k for each fixed x.

There are some equations among these scattering solutions. They are

$$\psi = Tf = \left(\begin{array}{ccc} R_t(k)f_2(x,k) & + & f_2(x,-k) \\ R(k)f_1(x,k) & + & f_1(x,-k) \end{array} \right) \tag{10}$$

$$g = JTf = J_0 f = \alpha\phi \tag{11}$$

in which α is a 2×2 constant matrix

$$\alpha = \left(\begin{array}{cc} 1 & -ki \\ 1 & ki \end{array} \right)$$

$J_0 = TJ$, the normalized Jost matrix which possesses the following properties:

$$\det J = \frac{1}{T} \ , \quad J(k)S(k) = \sigma J(-k)\sigma \ , \quad \sigma JJ^\dagger = (JJ^\dagger)^*\sigma \tag{12}$$

and for real $k, J(-k) = J^*(k)$, where σ is a 2×2 matrix

$$\sigma = \left(\begin{array}{cc} 0 & 1 \\ 1 & 0 \end{array} \right)$$

and 'det' means determinant.

Under the constraint Eq.(2),the nature of the spectrum of the operator L is as follows:There is a continuous spectrum in $\lambda > 0$,with a point-spectrum in $\lambda < 0$,which consists of a finite number of $\lambda_n = -\eta_n^2, n = 1(1)M$. The $k_n = i\eta_n$ are simple poles of the transmission coefficient T(k) and

$$\frac{d}{dk}\frac{1}{T(k)}|_{k=i\eta_n} = -iN_{1n}N_{2n} \ , \qquad N_{pn}^2 = \int_{-\infty}^{\infty} dx f_p^2(x,k) \qquad p = 1,2 \tag{13}$$

where N_{pn} are the norms of the real-valued eigenfunctions $f_p(x,k_n)$ corresponding to the eigenvalues $\lambda = \lambda_n$. Most of the mentioned results in this section can be found in many papers and books,for example,in Ref.5.

COMPLETENESS RELATION FOR THE REGULAR SOLUTION ϕ

According to Eq.(2), we have $|x||V(x)| \leq C_1|x|^{-1-\epsilon}$ for large x where C_1 and ϵ are positive constants. Hence we have $V(x) \geq -C_1|x|^{-2-\epsilon} \geq -C_2 x^2$ for x large enough such that $|x|^{4+\epsilon} \geq \frac{C_1}{C_2}$ where C_2 is any positive constant. Therefore the operator L is in the limit-point case at $+\infty$ and $-\infty$ [12-18].

Thus we have the following completeness relation for the regular solutions ϕ[12]

$$\int_{-\infty}^{\infty} \tilde{\phi}(x,k)d\rho_\phi(\lambda)\phi(y,k) = \delta(x-y) \tag{14}$$

where ' $\sim$ ' denotes matrix transpose, $\rho_\phi(\lambda)$ is symmetric spectral matrix corresponding to the regular solutions ϕ,and $\delta(\cdot)$ the Dirac-δ function.

$$\rho_\phi(\lambda) = \left(\begin{array}{cc} \rho_{11}(\lambda) & \rho_{12}(\lambda) \\ \rho_{21}(\lambda) & \rho_{22}(\lambda) \end{array} \right) \tag{15}$$

Now let us derive the spectral matrix. According to Ref.12, we have

$$\rho_{11}(\lambda) = \lim_{\epsilon \to 0+} \frac{1}{\pi} \int_0^\lambda du \, \Im \frac{1}{m_\infty(u + i\epsilon) - m_{-\infty}(u + \epsilon)} \tag{16a}$$

$$\rho_{12}(\lambda) = \rho_{21}(\lambda) = \lim_{\epsilon \to 0+} \frac{1}{\pi} \int_0^\lambda du \, \Im \frac{m_{-\infty}(u + i\epsilon)}{m_\infty(u + i\epsilon) - m_{-\infty}(u + i\epsilon)} \tag{16b}$$

$$\rho_{22}(\lambda) = \lim_{\epsilon \to 0+} \frac{1}{\pi} \int_0^\lambda du \, \Im \frac{m_\infty(u + i\epsilon) m_{-\infty}(u + i\epsilon)}{m_\infty(u + i\epsilon) - m_{-\infty}(u + i\epsilon)} \tag{16c}$$

where $\epsilon \to 0+$ means that ϵ tends to zero in the positive direction,' $\Im$' denotes taking the imaginary part of a complex number,

$$m_\infty(\lambda) = \lim_{x \to +\infty} -\frac{\phi_1(x, k)}{\phi_2(x, k)} \ , \qquad m_{-\infty}(\lambda) = \lim_{x \to -\infty} -\frac{\phi_1(x, k)}{\phi_2(x, k)} \tag{17}$$

using Eqs.(11),(10) and (3),we obtain

$$g \sim J \begin{pmatrix} T e^{ikx} \\ R e^{ikx} \ + \ e^{-ikx} \end{pmatrix} \qquad\qquad x \sim +\infty \tag{18a}$$

$$g \sim J \begin{pmatrix} R_l e^{-ikx} \ + \ e^{ikx} \\ T e^{-ikx} \end{pmatrix} \qquad\qquad x \sim -\infty \tag{18b}$$

Let

$$0 < \arg \lambda = \arg k^2 < \pi \tag{19}$$

for $\lambda = u + i\epsilon, \epsilon > 0$,and hence we have

$$\Im k = \Im \sqrt{\lambda} > 0, \quad \lim_{x \to +\infty} e^{ikx} = 0, \quad \lim_{x \to -\infty} e^{-ikx} = 0 \tag{20}$$

Thus substituting for ϕ_1 and ϕ_2 in Eqs.(17) from Eqs.(11) and then using asymptotic representations (18a) and (18b) results in

$$m_\infty(\lambda) = \frac{j_{12} + j_{22}}{j_{12} - j_{22}} ki \ , \qquad m_{-\infty}(\lambda) = \frac{j_{11} + j_{21}}{j_{11} - j_{21}} ki \tag{21}$$

where j_{pq} are elements of the Jost matrix.

For $\lambda > 0$ which constitutes the continuous spectrum of the operator L [1-3,5], we have

$$\frac{d\rho_{11}}{d\lambda} = \frac{1}{\pi} \lim_{\epsilon \to 0+} \Im \frac{1}{m_\infty(\lambda + i\epsilon) - m_{-\infty}(\lambda + i\epsilon)} = \frac{1}{2\pi k} \Re[T(j_{11} - j_{21})(j_{22} - j_{12})] \tag{22a}$$

where '$\Re$' means taking the real part of a complex number. Similarly,for real k we have

$$\frac{d\rho_{12}}{d\lambda} = \frac{d\rho_{21}}{d\lambda} = -\frac{1}{2\pi} \Re[iT(j_{11}j_{12} - j_{21}j_{22})] \tag{22b}$$

and

$$\frac{d\rho_{22}}{d\lambda} = \frac{k}{2\pi} \Re[T(j_{11} + j_{21})(j_{12} + j_{22})] \tag{22c}$$

Hence for $\lambda > 0$,we obtain

$$\frac{d\rho_\phi}{d\lambda} = \frac{1}{2\pi k} \Re \begin{pmatrix} [T(j_{11} - j_{21})(j_{22} - j_{12})] & -k[iT(j_{11}j_{12} - j_{21}j_{22})] \\ -k[iT(j_{11}j_{12} - j_{21}j_{22})] & k^2[T(j_{11} + j_{21})(j_{22} + j_{12})] \end{pmatrix} = \frac{1}{2\pi k} \Re \theta \tag{23}$$

where the matrix is denoted by θ.

As mentioned in section 2, for $\lambda < 0$ there exists a finite number (M) of $\lambda = \lambda_n = -\eta_n^2$ which constitutes the point-spectrum of the operator L and $k = i\eta_n$ are simple poles of the transmission coefficient T(k). In this case, ρ is a step matrix function of λ having a jump at each eigenvalue λ_n and otherwise constant[12-18] and

$$\frac{d\rho}{d\lambda} = \sum_{n=1}^{M} (\rho|_{\lambda=\lambda_n+0} - \rho|_{\lambda=\lambda_n-0})\delta(\lambda - \lambda_n) \qquad \lambda < 0 \qquad (24)$$

holds according to the derivative of generalized functions[14].

According to Eqs.(16) or by calculating the residue of the Green function $G(x, y, \lambda)$ at $\lambda = \lambda_n$, we obtain the jump of the spectral matrix at the eigenvalue λ_n

$$\Delta\rho|_{\lambda=\lambda_n} = \rho|_{\lambda=\lambda_n+0} - \rho|_{\lambda=\lambda_n-0} = \frac{1}{2N_{1n}N_{2n}}\tilde{\alpha}(J_0\sigma\tilde{J}_0)^{-1}\alpha|_{\lambda=\lambda_n} \qquad (25)$$

Thus the completeness relation for the regular solutions of the Schrödinger equation can be written as

$$\int_{-\infty}^{\infty} \tilde{\phi}(x, k)d\rho_\phi(\lambda)\phi(y, k) = \delta(x - y) \qquad (26)$$

where

$$\frac{d\rho_\phi}{d\lambda} = \sum_{n=1}^{M} \Delta\rho|_{\lambda=\lambda_n}\delta(\lambda - \lambda_n) \qquad \lambda < 0 \qquad (27)$$

Thus we complete the proof of the completeness relation for the regular solutions ϕ with full mathematical rigor.

COMPLETENESS RELATION FOR THE JOST SOLUTIONS

In the following, we shall prove that the completeness formula for the Jost solutions is

$$\frac{1}{2\pi}\int_{-\infty}^{\infty} dk T(k)f_1(x, k)f_2(y, k) + \sum_{n=1}^{M}\frac{1}{N_{1n}N_{2n}}f_1(x, k_n)f_2(y, k_n) = \delta(x - y) \qquad (28a)$$

It can be rewritten as

$$\frac{1}{8\pi}\int_0^\infty \frac{d\lambda}{k}[f^\dagger(x, k)f(y, k) + \tilde{f}(x, k)f^*(y, k) +$$

$$+\tilde{f}(x, k)\begin{pmatrix} R(k) & 0 \\ 0 & R_1(k) \end{pmatrix}f(y, k) + f^\dagger(x, k)\begin{pmatrix} R^*(k) & 0 \\ 0 & R^*(k) \end{pmatrix}f^*(y, k)] +$$

$$+\frac{1}{2}\sum_{n=1}^{M}\frac{1}{N_{1n}N_{2n}}\tilde{f}(x, k_n)\sigma f(y, k_n) = \delta(x - y) \qquad (28b)$$

From Eqs.(8), we obtain $g(k) = \sigma g^*(k)$ for real k. And then, from this equation, Eqs.(11) and (12), we obtain $\tilde{f} = f^\dagger J_0^\dagger \sigma \tilde{J}_0^{-1}$ for real k. Substituting it into Eq.(28b) results in

$$\int_{-\infty}^{\infty} f^\dagger(x, k)d\rho_f f(y, k) = \delta(x - y) \qquad (28c)$$

where

$$\frac{d\rho_f}{d\lambda} = \begin{cases} \frac{1}{8\pi k}[2I + J_0^\dagger \sigma \tilde{J}_0^{-1}\begin{pmatrix} R & 0 \\ 0 & R_l \end{pmatrix} + \begin{pmatrix} R^* & 0 \\ 0 & R_l^* \end{pmatrix}(J_0^*)^{-1}\sigma J_0] & \lambda > 0 \\ \sum_{n=1}^{M} \frac{1}{2N_{1n}N_{2n}}\sigma\delta(\lambda - \lambda_n) & \lambda < 0 \end{cases} \tag{29}$$

From Eqs.(11),we have

$$f = J_0^{-1}\alpha\phi \;, \qquad\qquad f^\dagger = \tilde{\phi}\alpha^\dagger(J_0^\dagger)^{-1} \tag{30}$$

because ϕ is real-valued. Substituting it into Eq.(28a) or Eq.(28b) results in

$$\int_{-\infty}^{\infty} \tilde{\phi}d\rho\phi = \delta(x - y) \tag{31}$$

where

$$\frac{d\rho}{d\lambda} = \frac{1}{8\pi k}\left(2\frac{d\rho_1}{d\lambda} + \frac{d\rho_2}{d\lambda} + \frac{d\rho_2^*}{d\lambda}\right) \tag{32a}$$

$$\frac{d\rho_1}{d\lambda} = \frac{d\hat{\rho}_1}{d\lambda} = \alpha^\dagger(J_0 J_0^\dagger)^{-1}\alpha \tag{32b}$$

$$\frac{d\rho_2}{d\lambda} = \frac{d\tilde{\rho}_2}{d\lambda} = \alpha^\dagger\sigma\tilde{J}_0^{-1}\begin{pmatrix} R & 0 \\ 0 & R_l \end{pmatrix}J_0^{-1}\alpha = \tilde{\alpha}\tilde{J}_0^{-1}\begin{pmatrix} R & 0 \\ 0 & R_l \end{pmatrix}J_0^{-1}\alpha \tag{32c}$$

for $\lambda > 0$ and

$$\frac{d\rho}{d\lambda} = \sum_{n=1}^{M} \frac{1}{2N_{1n}N_{2n}}\tilde{\alpha}(J_0\sigma\tilde{J}_0)^{-1}\alpha\delta(\lambda - \lambda_n) \qquad \lambda < 0 \tag{32d}$$

Obviously for $\lambda < 0$,we have $\frac{d\rho}{d\lambda} = \frac{d\rho_f}{d\lambda}$. And to prove Eqs.(28),it is therefore sufficient to prove the matrix equation

$$\frac{d\rho_1}{d\lambda} + \frac{d\rho_2}{d\lambda} = 2\theta \qquad \lambda > 0 \tag{33}$$

According to Eqs.(12),we have $|j_{12}|^2 + |j_{22}|^2 = |j_{12}|^2 + |j_{11}|^2$ and hence we have

$$\frac{d\rho_1}{d\lambda} = \begin{pmatrix} |j_{11} - j_{21}|^2 + |j_{22} - j_{12}|^2 & ki(j_{11}^*j_{21} + j_{12}^*j_{22} - j_{21}^*j_{11} - j_{22}^*j_{12} \\ ki(j_{11}^*j_{21} + j_{12}^*j_{22} - j_{21}^*j_{11} - j_{22}^*j_{12}) & k^2(|j_{11} + j_{12}|^2 + |j_{22} + j_{12}|^2) \end{pmatrix} \tag{34a}$$

and

$$\frac{d\rho_2}{d\lambda} = \begin{pmatrix} R(j_{12} - j_{22})^2 + R_l(j_{11} - j_{21})^2 & ki[R(j_{12}^2 - j_{22}^2) + R_l(j_{11}^2 - j_{21}^2)] \\ ki[R(j_{12}^2 - j_{22}^2) + R_l(j_{11}^2 - j_{21}^2)] & -k^2[R(j_{12} + j_{22})^2 + R_l(j_{11} + j_{21})^2] \end{pmatrix} \tag{34b}$$

It follows from Eq.(16b) that

$$s_{pp}j_{pp} + s_{qp}j_{pq} = j_{qq}^* \;, \qquad\qquad s_{pp}j_{qp} + s_{pq}j_{pp} = j_{qp}^* \tag{35}$$

for $p \neq q$. Using them,we obtain

$$2T(j_{11} - j_{21})(j_{22} - j_{12}) - R(j_{12} - j_{22})^2 - R_l(j_{11} - j_{21})^2 = |j_{11} - j_{21}|^2 + |j_{12} - j_{22}|^2 \tag{36a}$$

$$2T(j_{11} + j_{21})(j_{22} + j_{12}) + R(j_{12} + j_{22})^2 + R_l(j_{11} + j_{21})^2 = |j_{11} + j_{21}|^2 + |j_{12} + j_{22}|^2 \tag{36b}$$

and

$$2T(j_{21}j_{22} - j_{11}j_{12}) - R(j_{12}^2 - j_{22}^2) - R_l(j_{11}^2 - j_{21}^2) = j_{11}^*j_{21} + j_{12}^*j_{22} - j_{12}^*j_{11} - j_{22}^*j_{12} \tag{36c}$$

Thus the proof of Eq.(33) is completed and hence Eqs.(28) holds.

Substituting Eq.(10) into Eq.(28a) and using $\frac{1}{T(k_n)} = 0$ results in

$$\frac{1}{2\pi}\int_{-\infty}^{\infty} dk[f_1(x,k)f_1^*(y,k)+R(k)f_1(x,k)f_1(y,k)]+\sum_{n=1}^{M}\frac{1}{N_{1n}^2}f_1(x,k_n)f_1(y,k_n) = \delta(x-y)$$

$$(28d)$$

COMPLETENESS RELATIONS FOR THE REGULAR SOLUTIONS g AND FOR THE PHYSICAL SOLUTIONS

It follows from Eqs.(28c) and (11) that

$$\int_{-\infty}^{\infty} g^\dagger d\rho_g g = \delta(x-y) \tag{37}$$

where

$$\frac{d\rho_g}{d\lambda} = \begin{cases} \frac{1}{8\pi k}[2(J_0 J_0^\dagger)^{-1} + \sigma \tilde{J}_0^{-1}\begin{pmatrix} R & 0 \\ 0 & R_l \end{pmatrix} J_0^{-1} + (J_0^\dagger)^{-1}\begin{pmatrix} R^* & 0 \\ 0 & R_l^* \end{pmatrix}(J_0^*)^{-1}\sigma] & \lambda > 0 \\ \sum_{n=1}^{M}\frac{1}{2N_{1n}N_{2n}}\tilde{J}_0^{-1}\sigma J_0^{-1}\delta(\lambda - \lambda_n) & \lambda < 0 \end{cases}$$

$$(38)$$

The matrix ρ_g is the spectral matrix corresponding to the regular solution g.

Inserting Eq.(10) into Eq.(28c) results in

$$\int_{-\infty}^{\infty} \psi^\dagger \frac{1}{|T|^2}d\rho_f\psi = \int_0^{\infty} \psi d\rho_\psi\psi + \sum_{n=1}^{M}\frac{1}{2N_{1n}N_{2n}}f^\dagger(x,k_n)f(y,k_n) = \delta(x-y) \tag{39}$$

where

$$\frac{d\rho_\psi}{d\lambda} = \frac{1}{8\pi k|T|^2}[2I + J_0^\dagger \sigma \tilde{J}_0^{-1}\begin{pmatrix} R & 0 \\ 0 & R_l \end{pmatrix} + \begin{pmatrix} R^* & 0 \\ 0 & R_l^* \end{pmatrix}(J_0^*)^{-1}\sigma J_0] \qquad \lambda > 0 \tag{40}$$

COMPLETE INFORMATION FOR EXACT RECONSTRUCTION AND COMPLETENESS RELATIONS FOR SCATTERING SOLUTIONS

The inverse problem of scattering theory is to determine the underlying potential or scatterer, by the direct scattering data, for example, so called 'scattering matrix.' In the case of the Schrödinger equation, the reflection coefficient R and the bound states determined the whole of the scattering matrix[1-3,5]. Do the reflection coefficient and the bound states determine the potential? The answer is no. In Ref.15, the correct derivation of the Marchenko equation on the full line based on the completeness relation for the Jost solutions,i.e., Eq.(28d) showed that the norming constants of the Jost solutions corresponding to the bound states $\lambda = \lambda_n < 0$ are indispensable to exactly reconstruct the potential. Why are the norming constants so important? The answer is obvious:the completeness relations for the scattering solutions,the transforms of which determine the potential,such as the Jost solutions f and the regular solutions g, need the norms. That is to say, only the completeness relation can provide the complete information for exact reconstruction,and any relation formula that does not provide the complete information is no completeness relation. From this point of view, we can also come to the conclusion that formulas,such as

$$\frac{1}{4\pi}\int_{-\infty}^{\infty}dk|T|^2[f_1^*(x,k)f_1(y,k)+f_2^*(x,k)f_2(y,k)]+\sum_{n=1}^{M}\frac{1}{N_{1n}^2}f_1(x,k_n)f_1(y,k_n)=\delta(x-y)$$

(41)

in Refs. 1 and 10, and

$$\frac{1}{2\pi}\int_{-\infty}^{\infty}dk\psi_1^*(x,k)\psi_1(y,k)+\sum_{n=1}^{M}\frac{1}{N_1^2}f_1(x,k)f_1(y,k)=\delta(x-y)$$

(42)

where $\psi_1^*\psi_1=|T|^2f_1^*f_1$,and Eq.(4.4) in Ref.5 [1], none of them are completeness relations because the reflection coefficient is not incorporated in these formulas,which are formed by the scattering solutions determining the underlying potential. In comparison with Eqs.(28d),(38) and(39), we see that only in the case of reflectionlessness i.e., $R=R_l=0$ then $|T|=1$, can Eqs.(41),(42) and Eq.(4.4) in Ref.5 hold.

DISCUSSIONS

One can not refrain from asking why so many papers and books made mistakes so long,at least more than one decade in the case on the full line. According to me,there are three main reasons. First,those papers and books only introduced those wrong formulas,but did not used them to derive equations such as the Marchenko equation; second,even though some used them to deduce other equations ,no one gave any example to illustrate them because calculation involved, especially for not very simple one,for instance,with reflection,is very difficult and tedious; third,they did not think over the relationship between the completeness of scattering solutions and the complete information. One may also ask where such mistakes came from. For example,on page 333,Ref.1 said: "The general theorem for the self-adjoint extension of $\frac{-d^2}{dx^2+V(x)}$ in $L^2(R)$ can be obtained in a standard way(Deif and Trubowitz, 1979)." and on page 16,Ref.10 said:"Here we state the spectral theorem $\cdots$. The proof is standard(see e.g. Coddington-Levinson) and is omitted." But in fact, Ref.13 did not give the formula,which was claimed to be given by Ref.10.

Another example,Ref.5 stated that it(referring to the completeness formula on the full line)was well known and gave Ref.3 as its reference. But Ref.3 did not give such a formula either.

The similar problem in three dimensions occurred more than two decades ago [16,6-7]. If the potential $V(x,y,z)$ in three dimensions does not depend on two spatial coordinates y and z,the scattering problem in three dimensions reduces to the one in one dimension. Therefore,the 'completeness formula' in three dimensions[6,26],the counterpart of Eq.(42) in one dimension, is incorrect either. We mention this in passing and shall discuss in detail in a sequel of this paper.

Considering the importance of completeness relations for inverse scattering problems,we shall make further researches into them. Next,we shall give an example of the completeness relations for the scattering solution systems to the Schrödinger equation on the full line and show the difference between Eq.(29) and its counterpart in Ref.5. Also we shall give and prove completeness relations for scattering solution systems of the acoustical wave equation with rotation parameter as a compensation,which are

---

[1]In order to conveniently compare these formulas with their corresponding formulas in this paper,we use our symbols in stead of theirs.

very important for solving intermediate and strong scattering problems.

ACKNOWLEDGMENT

This work was supported in part by the National Natural Science Foundation of China.

REFERENCES

1. K.Chadan and P.C.Sabatier,"Inverse Problems in Quantum Scattering Theory,"
    2nd ed.,Springer,New York (1989).
2. V.A.Marchenko,"Sturm-Liouville Operators And Applications," Birkauser Verg,
    Basel (1986).
3. L.D.Faddeev,Properties of the S-matrix of the one dimensional Schrödiger
    equation,Trudy Mat.Inst.Steklov 73:314(1964).
4. L.D.Faddeev,The inverse problem in the quantum theory of scattering,J.Math.
    Phys.,4:72(1963).
5. R.G.Newton,Inverse scattering 1. one dimension,J.Math.Phys.,21:93(1980).
6. R.G.Newton,Inverse scattering 2. three dimensions,J.math.Phys..21:1698(1980).
7. R.G.Newton,Inverse scattering 3. three dimensions,continued,J.Math.Phys.,
    22:2192(1981).
8. R.K.Mueller, M.Kaveh and G.Wade, Reconstructive tomography and applications
    to ultrasonics, Proc. IEEE, 67:567(1979).
9. D.Colton and P.Monk,A new method for solving the inverse scattering problem
    for acoustic waves in an inhomogeneous media:1 and 2 Inverse Problems,
    5:1013 and 6:695(1990).
10. D.Deift and E.Trubowitz,Inverse scattering on the line,Comm.Pure.Appl. Math..
    32:121(1979).
11. G.Eilenberger,"Solitons," Springer,Berlin (1981).
12. E.C.Titchmarch,"Eigenfunction Expansions,Parts 1," 2nd ed.,Clarendon Press,
    Oxford (1962).
13. E.A.Coddington and N.Levinson,"Theory of Ordinary Differential Equations,"
    McGraw-Hill,New York (1955).
14. Zhen-Qiu Lu,"Equations of Classical and Modern Mathmatical Physics,"
    Shanghai Scientific and Technical Publishers,Shanghai (1991).
15. Zhen-Qiu Lu,A new derivation of the Marchenko equation on the full line based
    on the completeness relation for the Jost solution system, (1991).
    (unpublished).
16. R.G.Newton,"Scattering Theory of Waves and Particles," Spring-Verlag,
    New York (1982).

ITERATIVE METHOD FOR ACOUSTICAL WAVE INVERSION WITH SPARSE DATA[1]

Longji Tang[2]

Hunan Computing Center
Changsha, Hunan 410012
P.R.China

INTRODUCTION

This paper studies wave–theoretical and computational method for acoustical wave non–linear inversion with sparse data, for example multi–source cross–hole imaging in geophysical prospecting. Synthetic researches show [3] that for weak scatter problems Born approximations are valid, diffraction tomography performs slightly better than a single iteration of non–linear inversion. However, if the background velocities increase systematically with depth, diffraction tomography is ineffective whereas non–linear inversion yields useful images. For more complicated medium, iterative non–linear inversion method will be a prospective method [1].

To solve the acoustical wave non–linear inversion, linearized inverse procedures based on asymptotic ray theory have been developed by many authors (Cohen and Bleisein [2], ···), the iterative optimization methods [1] based on the least–squares criterion in functional spaces for time–domain inversion have been established by Tarantola [1] and Tarantola's results have been transformed into the frequency–domain acoustical wave inversion [3]. This paper presents a new frequency–domain method that can be applied to acoustical wave inversion with sparse data and in media of arbitrary a prior complexity. In this paper, first we reduce the acoustical wave inversion into a functional minimum problem. Secondly, we derive the gradient of the functional by using Lagrangian multiplier method [4] different from [1]. Finally, we give a computing scheme of preconditioned gradient method.

[1] This research was supported by the Chinese National Science Foundation under grant 49290200.

[2] Present addess: Department of Mathematics, The Pennsylvania State University, University Park, PA 16802, USA.

ACOUSTICAL WAVE INVERSE PROBLEM

We consider the following acoustical wave equation:

$$\triangle u + \frac{\omega^2}{c^2(x)} = F(x,x_s;\omega), \qquad x \in \Omega \tag{1.1}$$

$$u(x,x_s;\omega) = 0 \qquad\qquad x \in S \tag{1.2}$$

where $\Omega = R^+ \times R$ or $R^+ \times R \times R$, S denotes the surface of the medium Ω, $c(x)$ is wave velocity, ω is frequency.

The acoustical wave inverse problem is how to determine the wave velocity $c(x)$ from sparse measured data:

$$u^*(x_r,x_s;\omega_i) \tag{1.3}$$

$$i = 1,2,\cdots,k, \quad r = 1,2,\cdots,m, \quad s = 1,2,\cdots,n$$

The typical example of the acoustical wave inverse problem is the multi–source cross–hole acoustical imaging. In cross–hole imaging, x_r, $r = 1,2,\cdots$,m indicate the coordinates of foci, and x_s, $s = 1,2,\cdots$,n denote the coordinates of receivers.

It is well known that above–mentioned inverse problem is a strong non–linear problem, because acoustical compress u depends on wave velocity $c(x)$ non–linearly, i.e., u is non–linear functional with respect to $c(x)$: $u = u(x,x_s;\omega_i,c(x))$. Moreover, the inverse problem is also unstable. Therefore the inverse problem can be reduced to the following regularization[5] constrained non–linear functional minimum problem:

$$J(c(x)) \triangleq \frac{1}{2} \sum_s \sum_i \sum_r \left| u(x_r,x_s;\ \omega_i;\ c(x)) - u^*(x_r,x_s;\ \omega_i) \right|^2$$

$$+ \frac{\alpha}{2} \| c - c_{prior} \|_Q^2 = \min_{c(x) \in D} \tag{2.1}$$

subject to

$$e(c(x),u) \overset{\triangle}{=} \triangle u + \frac{\omega_i^2}{c^2(x)} \cdot u - F(x,x_s;\omega_i) = 0 \tag{2.2}$$

$$i = 1,2,\cdots,k$$

where $\| \cdot \|_Q^2 = (Q \cdot , \cdot)_{L^2(\Omega)}$ is weight norm in $L^2(\Omega)$ space, $Q = Q(x) > 0$, $\alpha > 0$ is regularizing parameter. And

$$D = \{ c(x) \in L^\infty(\Omega):\ \underline{c} \leqslant c(x) \leqslant \bar{c} \} \bigcap E \tag{2.3}$$

in which E is a closed, bounded, convex subset of space X which is a Hilbert space such that the embedding of X into L^∞ is compact.

LAGRANGIAN MULTIPLIER METHOD AND SOLUTION OF THE OPTIMIZATION PROBLEM

In fact, (2.1) and (2.2) can be regarded as a non–linear programming model with constrain. Thus by adopting Lagrangian multiplier method for solving non–linear programming problem in functional space, the constrained minimum problem (2.1) and (2.2) can be converted into an unconstrained functional minimum problem:

$$J_v(c(x)) \overset{\triangle}{=} J(c(x)) + \sum_s \sum_i (v,e(c,u))_{L^2(\Omega)} = \min_{c(x) \in D} \tag{3.1}$$

where $v = v(x,x_s;\omega_i;c)$ is multiplier function and

$$(v, e(c,u))_{L^2(\Omega)} = \int_\Omega v(\triangle u + \frac{\omega_i^2}{c^2(x)} u - F)dx$$

In order to solve the minimum problem (3.1) by using gradient methods, we will first give the gradient of functional (3.1) for $c(x)$ as follows.

Suppose δc is a small variation of $c(x)$, $\delta u = u(c+\delta c) - u(c)$, $v \in H_0^1(\Omega)$ and $u(x,x_s;\omega_i;c)$ is continuous with respect to wave velocity $c(x)$. From δ-function property and expansion:

$$\frac{1}{(1 + \frac{\delta c}{c})^2} = 1 - \frac{2\delta c}{c^3} + \frac{3(\delta c)^2}{c^4} - \cdots \tag{3.2}$$

we can obtain the first-order variation of the functional $J_v(c(x))$:

$$\delta J_v = \sum_i \sum_s \int_\Omega \{\triangle v + \frac{\omega_i^2}{c^2} v + \sum_r \delta(x - x_r)[u(x_r,x_s;\omega_i;c) - u^*(x_r,x_s;\omega_i)]\}\delta u\, dx$$

$$+ \int_\Omega - \frac{2}{c^3(x)} \sum_i \sum_s \omega_i^2 u(x,x_s;x_i;c)v(x,x_s;\omega_i;c)\delta c\, dx$$

$$+ \alpha \int_\Omega Q(c - c_{prior})\delta c\, dx \tag{3.3}$$

In (3.3), let multiplier function v satisfy the following wave equation:

$$\triangle v + \frac{\omega_i^2}{c^2(x)} v = - \sum_r \delta(x - x_r)[u(x_r,x_s;\omega_i;c) - u^*(x_r,x_s;\omega_i)], \quad x \in \Omega \tag{3.4}$$

$$v(x,x_s;\omega_i) = 0, \qquad x \in S \tag{3.5}$$

$$i = 1,2,\cdots,k, \quad s = 1,2,\cdots,n$$

Hence, we have

$$\delta J_v = \int_\Omega [- \frac{2}{c^3(x)} \sum_i \sum_s \omega_i^2 u(x,x_s;\omega_i;c)v(x,x_s;\omega_i;c) + \alpha Q(c - c_{prior})]\delta c\, dx \tag{3.6}$$

By definition of the functional derivative of J_v with respect to the function $c(x)$, thus have the gradient formula of the functional J_v:

$$\frac{\partial J_v}{\partial c}(x) = - \frac{2}{c^3(x)} \sum_i \sum_s \omega_i^2 u(x,x_s;\omega_i;c(x))v(x,x_s;\omega_i;c(x))$$

$$+ \alpha Q(c(x) - c_{prior}(x)) \tag{3.7}$$

Remark: For multi-point-source problem, i.e., $F(x,x_s;\omega) = - \delta(x - x_s)$, $s = 1,2,\cdots,n$, we have

$$\frac{\partial J_v}{\partial c}(x) = \frac{2}{c^3(x)} \sum_i \sum_s \omega_i^2 G(x,x_s;\omega_i;c) \sum_r G(x,x_s;\omega_i;c)[G(x_r,x_s;\omega_i;c)$$

$$- u^*(x_r,x_s;\omega_i)] + \alpha Q(c(x) - c_{prior}(x)) \tag{3.8}$$

where $G(x,x';\omega_i;c)$ is Green function which satisfies the following wave equation:

$$\triangle G + \frac{\omega_i^2}{c^2(x)} G = - \delta(x - x'), \quad x \in \Omega \tag{3.9}$$

$$G(x,x';\omega_i) = 0 \qquad x \in S \tag{3.10}$$

It is convenient to use a preconditioned gradient algorithm [1] to solve the functional minimum problem (3.1). This gives

$$c_{N+1}(x) = c_N(x) - \mu_N H_N \frac{\partial J_{\bullet}}{\partial c_N(x)}(x) \tag{3.11}$$

Where H_N is an arbitrary positive definite operator, called the "preconditioner" and which is suitably chosen to accelerate the convergence.

CONCLUSIONS

This paper mainly discusses a new computational method for solving acoustical wave non—linear inversion and establishes a preconditioned gradient scheme by using Lagrangian multiplier method. It is evident that our method is a combination of mathematical programming and regularization [5] . Since the main part of the functional $\quad J(c(x))--\hat{J} = \frac{1}{2} \sum_i \sum_t \sum_r \left| u(x_r,x_s; \omega_i; c) - u^{*}(x_r,x_s; \omega_i) \right|^2$ is non—convex, optimization problem (3.1) only has local minimum solution, and regularizer $\frac{\alpha}{2} \| c - c_{prior} \|_Q^2$ in (2.1) will improve the non—convex property of the functional $\hat{J}$. Also, a proper choice of preconditioner H_N in (3.11) will also strengthen computational stability and can speed up the iterative convergence.

REFERENCES

1. A.Tarantola, Inverse Problem Theory, Amsterdam—Oxford—New York—Tokyo, (1987)
2. Bleistein, N.. Cohen,J.K. and Hagin, F.G., Two and one—half dimensional inversion with an arbitrary reference, Geophysics, Vol.52, 26—36,(1987)
3. R.G. Partt and M.H. Worthington, Inverse theory applied to multi—source cross—hole tomography, Geophysical Prospecting, Vol 38, 287—310,(1990)
4. Tang Longji and Li Youming, Lagrangian multiplier method for solving acoustical wave imaging, in preparation,(1992)
5. A.N. Tikhonov and A.V. Arsenin, Solutions of Ill—Posed Problems, Wiley, New York,(1977)

TCCR NUMERICAL METHOD FOR SOLVING TWO-DIMENSIONAL INVERSE SCATTERING PROBLEM

Ganquan Xie, Longji Tang, Wei Liu and Wen Li

Hunan Computer Technology Research Institute
Hunan Computing Center
Changsha, Hunan 410012, China

ABSTRACT

This paper[1] is devoted to the study of the two-dimensional inverse scattering problem from numerical point of view. [1] has converted the problem into solving an interesting Radon's integral geometry equation of the first kind. We present the TCCR method based on the ideas of [2],[3] for solving the integral geometry equation, and give several satisfactory numerical results. It is found that the method is numerically stable and convergent fast.

1. INVERSE SCATTERING PROBLEM AND ITS NONLINEAR INTEGRAL EQUATION

The object of this report is to describe some computational results on a two-dimensional inverse scattering problem. The problem has been reduced to a nonlinear integral geometry equation by [1]. The nonlinear integral equation is ill-posed: the solution of the equation depends discontinuously on the data. Since the data in practical circumstances are measured quantities, measurement errors can be magnified by the solution process and can produce large instabilities in the computed solution. To copy with the instabilities, we establish a TCCR (Time Convolution Characteristic Regularization) algorithm based on the ideas of [2],[3] for solving the nonlinear integral equation.

The two-dimensional scattering problem is the following. Let u(x,y,t) be a solution of

$$\frac{\partial^2 u}{\partial t^2} - \left(\frac{\partial^2 u}{\partial x^2} + \frac{\partial^2 u}{\partial y^2}\right) + q(x,y)u = 0, \qquad (x,y)\in R^+ \times R,\ t > 0, \qquad (1.1)$$

---

[1]This research was supported by the Chinese National Science Foundation under grant 1860551 and the Joint Earthquake Science Foundation under grant 86022.

with initial condition

$$u(x,y,t) = 0, \qquad (x,y) \in R^+ \times R, \ t \leqslant 0, \tag{1.2}$$

and boundary condition

$$\frac{\partial u}{\partial x}(0,y,t) = \delta(y,t), \qquad y \in R, \ t > 0. \tag{1.3}$$

where $q(x,y) \in C(R^+ \times R) \cap L(R^+ \times R)$ and $q(x,y) \geqslant 0, \ (x,y) \in R^+ \times R$.

Our object is to determine $q(x,y)$ by giving measure data on surface boundary of $x = 0$:

$$u(0,y,t) = f(y,t) = -\frac{1}{\pi}\frac{H(t-y)}{\sqrt{t^2-y^2}} + f_*(y,t), \tag{1.4}$$

This is the so-called inverse scattering problem of 2-D.

In order to solve above-mentioned inverse scattering problem, [1] has proved the following result.

Theorem 1 [1]. Suppose that $q(x,y) \in C(R^+ \times R)$, then 2-D inverse scattering potential problem will be reduced to the following nonlinear integral equation

$$\frac{1}{\pi^2} \iint\limits_{D[y,t]} q(\xi,\eta) \int_{\sqrt{\xi^2+(\eta-y)^2}}^{t-\sqrt{\xi^2+\eta^2}} \frac{d\tau}{\sqrt{((t-\tau)^2-\xi^2-\eta^2)}\sqrt{(\tau^2-\xi^2-(\eta-y)^2)}}\, d\xi d\eta$$

$$= \frac{1}{\pi} \iint\limits_{D[y,t]} q(\xi,\eta) \int_{\sqrt{\xi^2+(\eta-y)^2}}^{t-\sqrt{\xi^2+\eta^2}} \frac{v(\xi,\eta,t-\tau)}{\sqrt{(\tau^2-\xi^2-(\eta-y)^2)}}\, d\tau d\xi d\eta + f_*(y,t). \tag{1.5}$$

where $D[y,t]$ denotes half disk of ellipse:

$$t \geqslant \sqrt{(\xi^2+\eta^2)} + \sqrt{(\xi^2+(\eta-y)^2)} \tag{1.6}$$

Here the proof of the Theorem 1 is omitted.

(1.5) can be reduced to the following form by means of elliptic integral:

$$\frac{2}{\pi^2} \iint\limits_{D[y,t]} \frac{K(k)}{\sqrt{(t^2-(\sqrt{(\xi^2+\eta^2)}-\sqrt{(\xi^2+(\eta-y)^2)})^2)}}\, q(\xi,\eta)d\xi d\eta$$

$$= \frac{1}{\pi} \iint\limits_{D[y,t]} q(\xi,\eta) \int_{\sqrt{\xi^2+(\eta-y)^2}}^{t-\sqrt{\xi^2+\eta^2}} \frac{v(\xi,\eta,t-\tau)}{\sqrt{(\tau^2-\xi^2-(\eta-y)^2)}}\, d\tau d\xi d\eta + f_*(y,t), \tag{1.7}$$

where $k = k(\xi,\eta;y,t)$ is the modulus of elliptic integral of the first kind and

$$k = \frac{\sqrt{(t^2-(\sqrt{(\xi^2+\eta^2)}+\sqrt{(\xi^2+(\eta-y)^2)})^2)}}{\sqrt{(t^2-(\sqrt{(\xi^2+\eta^2)}-\sqrt{(\xi^2+(\eta-y)^2)})^2)}}$$

Moreover, $v(x,y,t)$ satisfies initial-boundary value problem [1]

$$\frac{\partial^2 v}{\partial t^2} - \left(\frac{\partial^2 v}{\partial x^2} + \frac{\partial^2 v}{\partial y^2}\right) + q(x,y)v = \frac{1}{\pi}q(x,y)\frac{H(t-\sqrt{(x^2+y^2)})}{\sqrt{(t^2-x^2-y^2)}},$$

$$(x,y) \in R^+ \times R, \ t > 0 \tag{1.8}$$

$$v(x,y,t) = 0, \qquad (x,y) \in R^+ \times R, \ t \leqslant 0, \tag{1.9}$$

$$\frac{\partial v}{\partial x}(0,y,t) = 0 \qquad y \in R, \ t > 0, \tag{1.10}$$

and

$$f_*(y,t) = v(0,y,t) \tag{1.11}$$

2. TCCR ALGORITHM

For convenience, we write (1.7) as form of nonlinear operator equation

$$G(q;y,t) = Aq - B(q) - j_s(y,t) = 0 \tag{2.1}$$

where A is linear operator and

$$Ah = \frac{2}{\pi^2} \iint_{D(y,t)} \frac{K(k)}{\sqrt{(t^2 - (\sqrt{\xi^2 + \eta^2} - \sqrt{\xi^2 + (\eta - y)^2}))^2}} h(\xi,\eta)d\xi d\eta \tag{2.2}$$

and B is a nonlinear operator and

$$B(q)h = \frac{1}{\pi} \iint_{D(y,t)} h(\xi,\eta) \int_{\sqrt{\xi^2 + (\eta - y)^2}}^{t - \sqrt{\xi^2 + \eta^2}} \frac{v(q;\ \xi,\eta,t - \tau)}{\sqrt{(\tau^2 - \xi^2 - (\eta - y)^2)}} d\tau d\xi d\eta \tag{2.3}$$

By [1] We have

$$G(q;y,t) = v(q;0,y,t) - j_s(y,t) \tag{2.4}$$

In order to establish our algorithm, we derive Frechet derivative operator now. For this purpose, an auxiliary problem similar to [2] is introduced

$$\frac{\partial^2 W}{\partial t^2} - (\frac{\partial^2 W}{\partial x^2} + \frac{\partial^2 W}{\partial y^2}) + q(x,y)W = 0, \qquad (x,y) \in R^+ \times R, \ t > 0 \tag{2.5}$$

$$W(x,y,t) = 0, \qquad (x,y) \in R^+ \times R, \ t \leqslant 0. \tag{2.6}$$

$$\frac{\partial W}{\partial x}(0,y,t) = \delta(y - \eta,t), \qquad y \in R, \ t > 0, \tag{2.7}$$

Theorem 2. Let q(x,y) belong to $C(R^+ \times R)$ and $q(x,y) \geqslant 0$. h(x,y) be increment of q(x,y), then Frechet derivative operator of nonlinear operator G(q;y,t) can be expressed as the following form

$$G'_q(q;y,t)h(\xi,\eta)$$

$$= \frac{2}{\pi^2} \iint_{D(y,t)} \frac{K(k)}{\sqrt{(t^2 - (\sqrt{\xi^2 + \eta^2} - \sqrt{\xi^2 + (\eta - y)^2}))^2}} h(\xi,\eta)d\xi d\eta$$

$$- \frac{1}{\pi} \iint_{D(y,t)} h(\xi,\eta) \int_{\sqrt{\xi^2 + (\eta - y)^2}}^{t - \sqrt{\xi^2 + \eta^2}} \frac{v(q;\ \xi,\eta,t - \tau)}{\sqrt{(\tau^2 - \xi^2 - (\eta - y)^2)}} d\tau d\xi d\eta$$

$$- \frac{1}{\pi} \iint_{D(y,t)} h(\xi,\eta) \int_{\sqrt{\xi^2 + (\eta - y)^2}}^{t - \sqrt{\xi^2 + \eta^2}} \frac{W_s(q;\ \xi,\eta,\tau;\ y)}{\sqrt{((t - \tau)^2 - \xi^2 - \eta^2)}} d\tau d\xi d\eta$$

$$+ \iint_{D(y,t)} h(\xi,\eta) \int_{\sqrt{\xi^2 + (\eta - y)^2}}^{t - \sqrt{\xi^2 + \eta^2}} W_s(q;\xi,\eta,\tau;y)v(q;\xi,\eta,t - \tau)d\tau d\xi d\eta$$

$$\tag{2.8}$$

where $W_s(q;x,y,t;\eta)$ is the part of scattering of $W(q;x,y,t;\eta)$.

As follows, we let

$$G'_q(q;y,t)h = Ah - B(q)h - C(q)h + D(q)h \tag{2.9}$$

where operators A and B are the same as A and B in (2.2) and (2.3), respectively. But operators C(q) and D(q) are also linear operators, and

$$C(q)h = \frac{1}{\pi} \iint_{D(y,t)} h(\xi,\eta) \int_{\sqrt{\xi^2 + (\eta - y)^2}}^{t - \sqrt{\xi^2 + \eta^2}} \frac{W_s(q;\ \xi,\eta,\tau;\ y)}{\sqrt{((t - \tau)^2 - \xi^2 - \eta^2)}} d\tau d\xi d\eta \tag{2.10}$$

$$D(q)h = \iint_{D(y,t)} h(\xi,\eta) \int_{\sqrt{\xi^2 + (\eta - y)^2}}^{t - \sqrt{\xi^2 + \eta^2}} W_s(q;\xi,\eta,\tau;y)v(q;\xi,\eta,t - \tau)d\tau d\xi d\eta \tag{2.11}$$

Now, we establish TCCR iteration for solving nonlinear integral equation (2.1)

$$(A - B(q_N) - C(q_N) + D(q_N) + \alpha I)h_N = f_s(y,t) - v_N(0,y,t) \tag{2.12}$$

$$q_{N+1} = q_N + \omega_N h_N, \qquad N = 0,1,2,\cdots, \quad \alpha > 0 \tag{2.13}$$

where $q_0(x,y)$ is initial value and $h_N(0,y) = 0$.

It has been shown that iterative scheme (2.12)–(2.13) is not only very stable but also convergent fast. Whereas the computational work of F–derivative operator is very large. If computing net has p dissection points in the direction of y, then we must solve p initial–boundary value problems at each stage of iteration. To over-come the difficulty, a modified TCCR iteration is presented.

$$(A - B(q_{N_1}) - C(q_{N_2}) + D(q_{N_3}) + \alpha I)h_N = f_s(y,t) - v_N(0,y,t) \tag{2.14}$$

$$q_{N+1} = q_N + \omega_N h_N, \qquad N = 0,1,2,\cdots, \ 0 \leqslant N_1,N_2,N_3 \leqslant N, \ \alpha > 0 \tag{2.15}$$

where $h_N(0,y) = 0$, i.e. $q(0,y)$ is a given function. ω_N is a relaxation factor.

We will establish the numerical scheme of (2.14)–(2.15) in § 3 and will give several numerical results in § 4.

3. THE NUMERICAL SCHEME

First, we suppose
$$q(x,y) = \begin{cases} q(x,y), & (x,y) \in \Omega \\ q^*(x,y), & (x,y) \in R^+ \times R \setminus \Omega \end{cases}$$
where Ω is computed region and it is bounded $q(x,y)$ is the sought scattering poten-tial function and $q^*(x,y)$ is known function, such as $q^*(x,y) = 0$.

Secondly, we introduce a mesh in (x,y,t) space, $(x,y) \in \Omega$, $t > 0$ with $x = N_1 h_1$, $y = N_2 h_2$, $t = M\tau$ and take $h_1 = h_2 = \tau$ in this paper. The computational method on wave field $v(x,y,t)$ and $w_s(x,y,t)$ is based on a modified Von Neumann's al-ternating direction implicit finite difference scheme. It is stable without condition as $4\bar\eta - 1 > 0$.   And its coefficient matrix can maintain the form of tridiagonal matrix. Therefore the discrete algebraic equation can be solved by using double sweep methods.

Finally let center our attention on establishing discrete iteration scheme of (2.14)–(2.15). Assume that mesh has n inner points in Ω and express them as $p(x_j,y_j)$, $j = 1,2,\cdots,n$ and let $r(0,y_i)$, $i = 1,2,\cdots,m$ denote mesh nodes in y–axis. Then we make n ellipses (see fig.1) with focal points $(0,0)$ and $(0,y_i)$, $i = 1,2,\cdots,m$, respectively.

$$D[y_i,t_j]: \ t_j \geqslant \sqrt{(\xi^2 + \eta^2)} + \sqrt{(\xi^2 + (y_i - \eta)^2)} \tag{3.1}$$

where $i = 1,2,\cdots,m$, $j = 1,2,\cdots,M$ and $M \times m = n$. Moreover, the n ellipses must cover Ω, i.e.

$$\Omega \subseteq \bigcup_{i,j=1}^{n} D[y_i,t_j] \tag{3.2}$$

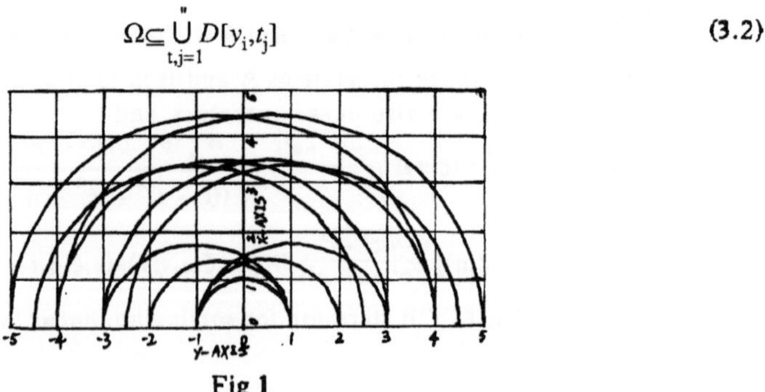

Fig 1

In each ellipse $D[y_i, t_j]$, we compute all integral of (2.8) by adopting numerical integration method (for example, two-dimensional Simpson's method), (2.14)–(2.15) are discretized as the following discrete scheme:

$$(\bar{A} - \bar{B}_{N_1} - \bar{C}_{N_2} + \bar{D}_{N_3} + \alpha I)\bar{h}_N = F_N \tag{3.3}$$

$$\bar{q}_{N+1} = \bar{q}_N + \omega_N \bar{h}_N, \qquad N = 0,1,2,\cdots, \ \alpha > 0, \ 0 \leqslant N_1, N_2, N_3 \leqslant N, \tag{3.4}$$

I is unit matrix, $\bar{A}, \bar{B}_{N_1}, \bar{C}_{N_2}$ and $\bar{D}_{N_3}$ are $N \times N$ order matrix.

The iterative scheme (2.14)–(2.15) or (3.3)–(3.4) has great flexibility. As N_1, N_2, N_3 are taken different values, we can obtain several useful computational schemes:

(1) partial simple Newton–type iterative scheme

In (3.3), taking $N_1 = N$, $N_2 = N_3 = 0$, we have

$$(\bar{A} - \bar{B}_N - \bar{C}_0 + \bar{D}_0 + \alpha I)\bar{h}_N = F_N \tag{3.5}$$

$$\bar{q}_{N+1} = \bar{q}_N + \omega_N \bar{h}_N, \qquad N = 0,1,2,\cdots, \ \alpha > 0 \tag{3.6}$$

Clearly, the initial–boundary value problem (1.8)–(1.10) is only solved once each step of iteration.

(2) simple Newton–type iterative scheme

In (3.3), getting $N1 = N2 = N3 = 0$, then

$$(\bar{A} - \bar{B}_0 - \bar{C}_0 + \bar{D}_0 + \alpha I)\bar{h}_N = F_N \tag{3.7}$$

$$\bar{q}_{N+1} = \bar{q}_N + \omega_N \bar{h}_N, \qquad N = 0,1,2,\cdots, \ \alpha > 0 \tag{3.8}$$

Computational work of the scheme (3.7)–(3.8) is less than that of the scheme (3.5)–(3.6).

(3) In particular, iterative schem (3 3)–(3.4) can be reduced to simpler form as starting value $q_0 = 0$.

$$(\bar{A} + \alpha I)\bar{h}_N = F_N \tag{3.9}$$

$$\bar{q}_{N+1} = \bar{q}_N + \omega_N \bar{h}_N, \qquad N = 0,1,2,\cdots, \ \alpha > 0 \tag{3.10}$$

Obviously, computational work can be reduced to minimum degree by using iteration scheme (3.9)–(3.10). Computing simulations in next section will show that satisfactory results can be obtained either by using the iteration (3.5)–(3.6) or by adopting (3.9)–(3.10).

§ 4. NUMERICAL RESULT AND CONCLUSION

Our first experiment was aimed at retrieving a given function

$$q(x,y) = \begin{cases} 1, & (x,y) \in \Omega_0 \\ 0, & (x,y) \in \Omega \setminus \Omega_0 \end{cases}$$

by obtaining $v(0,y,t) = f_s(y,t)$ by computation. In Fig.2 and Fig.3 we give the results of the first and seventh iterations using the computed $f_s(y,t)$ and the correct $q(x,y)$ as the first guess $q_0(x,y) = 0$, $(x,y) \in \Omega$ by adopting the scheme (3.5)–(3.6). Higher iterations remain the same and differ by a few percent from $q(x,y)$. The region $\Omega \setminus \Omega_0$ is not shown in this figure. The computed iterates oscillate about the true value of $q(x,y)$, 0 in this range with a maximum excursion of 0.05. In the following Fig.2––7:

———— exact. –––––– guess. $\cdots\cdots\cdots$ iteration.

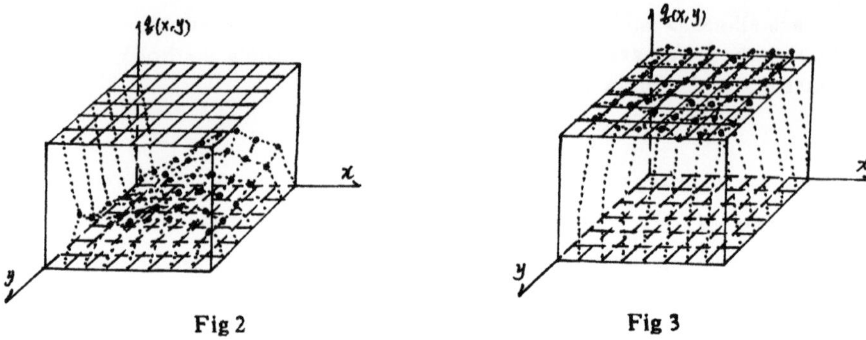

Fig 2   Fig 3

The next experiment we tried was more interesting. We kept the same potential function q(x.y) as before. but added 5% random noise to $f_s(x,y)$. The results of this calculation are shown in fig.4 and fig.5, where for clarity only the first and seventh iterates are exhibited. Higher iterates were computed but they differed insignificantly from the seventh. Again the region $\Omega \setminus \Omega_0$ is omitted. The oscillatory errors are larger than the first case.

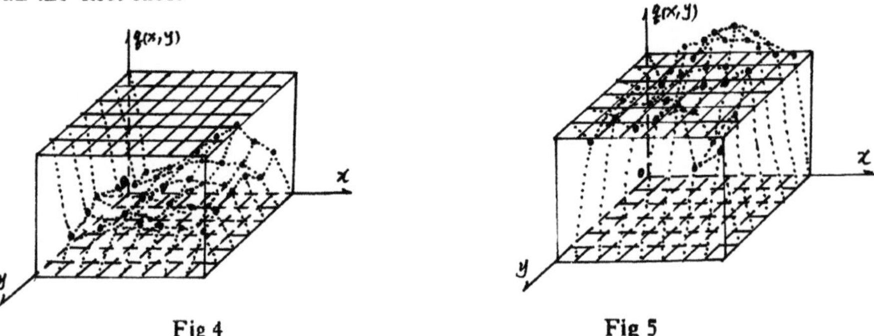

Fig 4   Fig 5

In our final experiment we took known funtion

$$q(x,y) = \begin{cases} \sqrt{r^2 - x^2 - y^2}, & x^2 + y^2 \leqslant r^2 \\ 0, & x^2 + y^2 \geqslant r^2 \end{cases}$$

Next we recomputed the "measured data" $f_s(y,t) = v(0,y,t)$. Then we began our iteration (3.9)–(3.10) with $q_0(x,y) = 0$, $(x,y) \in \Omega$. The first and eight iterations are plotted in Fig.6 and Fig.7. Other situations are similar to the first example.

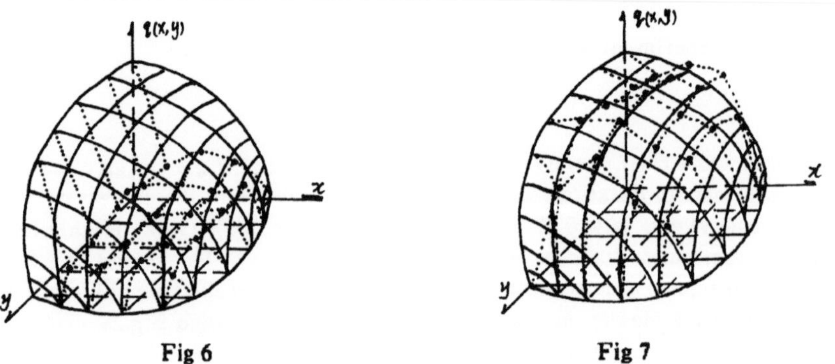

Fig 6   Fig 7

The numerical results comfirm that TCCR method can reconstruct potentials reasonably accurately. The method is relatively costly because the iterations are complete sweeps of the (x,y,t)—space. Here we needed 2000K of memory. Striking an average, each iteration required 120 sec. for Ω with 100 inner nodes on a M—150 computer.

Acknowledgments: The authors would like to thank Professor P.D.Lax and Professor Feng Kang for their encouragement and guidance. And we gratefully acknowledge the support of the Joint Seismological Science Foundation and Hunan Province Scientific and Technological commission.

REFERENCES

1. Ganquan Xie, Jianhua Li. Nonliear Integral Equation of Inverse Scattering Problem of Wave Equation and Iteration, in preparation, JCM.
2. Ganquan Xie, Jianhua Li, A New Characteristic Iterative Method for Solving Scattering Potential Inverse Problem of 3—D Wave Equation, Scientia Sinica, No.4,(1988).
3. Tiknonov,A.N. and Arsein,V.Y., On the Solution of Ill—Posed Problem, Johu Wiley and Sons. New York,(1977).
4. Longji Tang, Ganquan Xie, Wei Liu, Wen Li, Numerical Solution for Solving the Nonlinear Integral Equation of One—Dimensional Inverse Scattering Problem, in preparation.

A TIKHONOV REGULARIZATION METHOD

FOR IMAGE RECONSTRUCTION

Chengbin Peng, William L. Rodi and M. Nafi Toksöz

Earth Resources Laboratory
Department of Earth, Atmospheric and Planetary Sciences
Massachusetts Institute of Technology
Cambridge, MA 02139, U.S.A.

INTRODUCTION

Many problems of image reconstruction from projection data have the following mathematical form

$$A\, m = d \tag{1}$$

where m is the unknown model, d is the observed data and A is a known operator. If A is independent of the model m, the tomographic problem is linear; otherwise it is nonlinear. In this paper, we restrict ourselves to the linear case.

A is usually an integral operator with a kernel which is not a function but a distribution. For example, in the straight ray tomography the data is related to the model by

$$\int_V \delta(x \cos \theta + y \sin \theta - t)\, m(x,y)\, dx dy = d(t,\theta) \tag{2}$$

where (t,θ) is a pair of parameters describing one particular ray path, and $V \subset \mathbf{R}^2$ is the region where $m(x,y)$ is defined. ($\mathbf{R}$ denotes the set of real numbers.)

Radon (1913) showed that (1) has a well-defined, unique solution when error-free, infinite data from all view angles are available. Otherwise, the problem is ill-posed in the sense of Hadamard (1923), i.e., a solution does not exist if $d \notin \text{Ran}(A)$, the range of the operator A; it is not unique when $\text{Null} < (A) \neq \{0\}$, where $\text{Null}(A)$ is the null space of A; and it is unstable when the inverse of A (when it exists) is not continuous. For ill-posed problems, some method of regularization or smoothing is necessary to obtain well-defined, stable solutions.

Smoothing and regularization have played a role in many of the conventional reconstruction algorithms for projection data. For example, some non-iterative methods employ bandlimited windowing or interpolation (Rowland, 1979), while in the case of

Acoustical Imaging, Volume 20 Edited by Y. Wei
and B. Gu, Plenum Press, New York, 1993

iterative methods the image is usually smoothed after each iteration with a nine point moving average filter so as to remove the "salt and pepper" effect (Herman, 1981; Natterer, 1986). Such approaches are ad hoc in that they do not address how the windowing and smoothing operators they use affect the reconstruction, particularly in the presence of noise.

In this paper, we propose a tomographic reconstruction algorithm based on Tikhonov regularization with a gradient differential operator. This algorithm defines a solution to (1) to be that model m which is the smoothest among all solutions that yield a given fit to the data d. In addition, our algorithm computes the uncertainty of the solution in the sense of Backus and Gilbert (1968, 1970), which is of particular importance for experiments with irregular geometry and limited view angles. Numerical tests with both the medical and crosshole seismic geometries are implemented and the results are compared with those obtained by the Algebraic Reconstruction Technique (ART).

FORMULATION

Mathematical Preliminaries

In this section, we present a brief summary of the theory of Tikhonov regularization with differential operators. This theory is discussed extensively in the literature (see, for example, Tikhonov and Arsenin, 1977; Locker and Prenter, 1980; Groetsch, 1984; Bertero, 1986; Rodi, 1989; Yanovskaya and Ditmar, 1990; Wahba, 1990).

Let $m \in \mathcal{M}$ and $d \in \mathcal{D}$, where both $\mathcal{M}$ and $\mathcal{D}$ are Hilbert spaces. We let $(\cdot, \cdot)$ and $||\cdot||$, respectively, denote the inner product and norm on each space. Let $A: \mathcal{M} \rightarrow \mathcal{D}$ be a bounded everywhere defined linear operator. The adjoint of A is then also bounded and everywhere defined linear operator $A^*: \mathcal{M} \rightarrow \mathcal{D}$ satisfying

$$(Am, d) = (m, A^*d), \qquad \text{for all } m \in \mathcal{M},\ \lceil \in \mathcal{D}. \tag{3}$$

Let $H: \mathcal{M} \rightarrow \mathcal{H}$ be a closed, densely defined linear operator into a Hilbert space $\mathcal{H}$ with domain $\text{Dom}(H) \subset \mathcal{M}$ and range $\text{Ran}(H) \subset \mathcal{H}$. The adjoint operator $H^*: \mathcal{H} \rightarrow \mathcal{M}$ is then closed and densely defined such that

$$(Hm, h) = (m, H^*h), \qquad \text{for all } m \in \text{Dom}(H),\ h \in \text{Dom}(H^*). \tag{4}$$

In solving equation (1) by the method of regularization, we introduce the functional

$$\Phi(m) = ||Am - d||^2 + \alpha||Hm||^2 \tag{5}$$

defined on $\text{Dom}(H)$. α is a nonnegative real number called the regularization parameter. The regularized solution of (1) is an elment $\hat{m} \in \text{Dom}(H)$ that minimizes the quadratic functional Φ, i.e.

$$\Phi(\hat{m}) = \inf_{m \in \text{Dom}(H)} \Phi(m). \tag{6}$$

A unique solution to (6) exists if the following conditions are satisfied (Locker and Prenter, 1980):

(1) The intersection of the null spaces of A and H is null, i.e.,

$$\text{Null}(A) \cap \text{Null}(H) = 0$$

(2) The range of operator H, Ran (H), is closed

(3) There exists $\beta > 0$ such that

$$||Am|| \geq \beta||m||, \qquad \text{for all } m \in \text{Null}(H) \tag{7}$$

Given (1)–(3), the solution satisfies the so called regularized normal equation

$$A^*A\hat{m} + \alpha H^*H\hat{m} = A^*d \tag{8}$$

and as $\alpha \to 0$ this solution converges to the least square solution of equation (1)

When H is a differential operator, the adjoint operator H^* $\mathcal{H} \to \mathcal{M}$ depends on the boundary conditions imposed on m at ∂V In this paper, we let H be the gradient operator, $H = \nabla$, with Dirichlet boundary conditions, $m = 0$ on ∂V This implies that H^* is a divergence operator, $H^* = -\nabla$ and solution to equation (8) can be obtained via the Green's function of the operator $H^*H = -\nabla^2$

Tikhonov Regularization With a Gradient Operator

The problem we consider is to image an object which can be modeled by a real function m of two dimensional position $\vec{r} \equiv (x, y)$ Outside a given area V we assume m has a known constant value m_0 The objective is then to determine the unknown function m $V \to \mathbf{R}$ We presume the existence of projection data between various sources and receivers located in the model region V or its boundary

The model space is assumed to be $\mathcal{M} \subset L_2(V)$, a subspace of functions that are square integrable in V The data space is $\mathcal{D} = \mathbf{R}^M$ where M is the number of measurements (ray paths) We choose $H = \nabla$, the gradient operator The operator A $\mathcal{M} \to \mathcal{D}$, in this case, is defined as

$$(Am)_i = \int_V a_i(\vec{r})m(\vec{r})\,dV(\vec{r})$$

where $a_i(\vec{r}) = \delta(x\cos\theta_i + y\sin\theta_i - t_i)$ is a distribution along the i th ray path This choice of operators A and H satisfies the three conditions given by Locker and Prenter (1980), which follows from the fact that $\text{Null}(H)$ is the one dimensional space of functions which are constant over V

Given that A is linear we will assume with no loss of generality that $m_0 = 0$ This may always be achieved by redefining m as $m - m_0$ and d as $d - Am_0$ Once this is done we have $m \equiv 0$ on the boundary of V and $H^* = -\nabla$

For this particular setting, the regularized normal equation (8) reduces to

$$A^TE^{-1}Am - \alpha\nabla^2m = A^TE^{-1}d \tag{9}$$

where A^T is the transpose of A given by

$$(A^Tc)(\vec{r}) = \sum_{i=1} c_ia_i(\vec{r})$$

for $c \in \mathbf{R}^M$ A^T maps an element in $\mathbf{R}^M$ to a distribution over V E is the covariance matrix of the observation d and is assumed to be a strictly positive matrix

Let $g(\vec{r}, \vec{r}')$ be the Green's function of the 2 D Laplacian operator on V with vanishing value on the boundary

$$\nabla^2 g(\vec{r}, \vec{r}') \;=\; -\delta(\vec{r} - \vec{r}') \tag{10}$$
$$g(\vec{r}, \vec{r}') \;=\; 0, \quad \vec{r} \in \partial V \tag{11}$$

The solution of (9) is (Peng et al., 1992)

$$\widehat{m} = GA^T(AGA^T + \alpha E)^{-1}d,\tag{12}$$

where the transformation G is defined by

$$(Ga)(\vec{r}) = \int_V g(\vec{r}, \vec{r}')a(\vec{r}')\,dV(\vec{r}).$$

For a practical problem, ∂V, the boundary of V, may be very irregular and the computation of the Green's function $g(\vec{r}, \vec{r}')$ will in general require numerical evaluation. An implicit assumption of all tomographic algorithms is that no contribution to the data comes from the region outside V, i.e., where $m \equiv m_0$, which is zero for the problem at hand. In this paper, we relax this assumption by replacing V with the smallest circular disk which encompasses V. This allows us to use an analytical expression for $g(\vec{r}, \vec{r}')$. The details of this approach are given in Peng et al. (1992).

Smoothness Parameter α

The choice of smoothness parameter α for general regularization problems has been addressed in the literature (e.g., Groetsch, 1984; Bertero, 1986; Wabba, 1990). Here, we derive a constraint on α suitable for the particular problem in this paper.

The choice of the smoothing parameter α is related to a tolerance allowed in the misfit of the data by the reconstructed model $\widehat{m}$, such that

$$||d - A\widehat{m}|| \leq \delta.\tag{13}$$

From (12) we have

$$A\widehat{m} = AGA^T(AGA^T + \alpha E)^{-1}d$$

and thus, denoting that $AGA^T = S$,

$$d - A\widehat{m} = \alpha E(S + \alpha E)^{-1}d.$$

Substituting into (13) obtains

$$||\alpha E(S + \alpha E)^{-1}d|| \leq \delta.\tag{14}$$

or

$$\alpha \leq \frac{\delta}{||E(S + \alpha E)^{-1}d||}\tag{15}$$

Assuming $E = \sigma^2 I$ and using the eigenvalue decomposition of matrix S, we get

$$
\begin{aligned}
\alpha \;\leq\;& \frac{\delta}{||\sigma^2 U(\Lambda + \alpha\sigma^2 I)^{-1}U^T d||}\\[6pt]
=\;& \frac{\delta}{||\sigma^2 \sum_{k=1}^{M} u_{ik}\frac{1}{\lambda_k+\alpha\sigma^2}u_{jk}d_j||}\\[6pt]
\leq\;& \frac{\delta}{\frac{\sigma^2}{\lambda_{max}+\alpha\sigma^2}||\sum_{k=1}^{M} u_{ik}u_{jk}d_j||}\\[6pt]
=\;& \frac{\lambda_{max} + \alpha\sigma^2}{\sigma^2}\,\frac{\delta}{||d||}
\end{aligned}
$$

where I is the identity matrix, U is an orthogonal matrix with each column consisting of an eigenvector of S, and Λ is a diagonal matrix containing the eigenvalues of S. λ_{max} is the largest eigenvalue of S.

Denoting $\beta = \delta/||d||$, the percentage of error of data misfit, we get

$$\alpha \leq \frac{\lambda_{max}\beta}{\sigma^2(1-\beta)} \tag{16}$$

A less rigorous lower bound for α can be written as

$$\alpha \geq \frac{\lambda_{min}\beta}{\sigma^2(1-\beta)} \tag{17}$$

for the case of worst data misfit (i.e, wheb equality holds in (13)), where λ_{min} is the smallest eigenvalue of S.

Resolution Operator

The reconstruction $\hat{m}$ can be related to the known true model m as

$$\hat{m}(\vec{r}) = A^\dagger A m + A^\dagger e \tag{18}$$

where e is the observational error and

$$A^\dagger = GA^T(AGA^T + \alpha E)^{-1}.$$

We will refer to $A^\dagger$ as the "Tikhonov inverse" of A.

The operator

$$\Omega = A^\dagger A, \tag{19}$$

is known as the resolution operator. Three measures are used to characterize the distorting effects of Ω, they are

- The amplitude bias (zero moment, a scalar):

$$\rho(\vec{r}) = \int_V \Omega(\vec{r}, \vec{r}')d\vec{r}'$$

- The mislocation (first moment, a vector):

$$\vec{\Delta}(\vec{r}) = \frac{\int_V \Omega(\vec{r}, \vec{r}')(\vec{r}' - \vec{r})d\vec{r}'}{\int_V \Omega(\vec{r}, \vec{r}')d\vec{r}'}$$

- The spatial smearing (second moment, a tensor):

$$\Lambda(\vec{r}) = \frac{\int_V \Omega(\vec{r}, \vec{r}')(\vec{r}' - \vec{r} - \vec{\Delta}(\vec{r}))(\vec{r}' - \vec{r} - \vec{\Delta}(\vec{r}))^T d\vec{r}'}{\int_V \Omega(\vec{r}, \vec{r}')d\vec{r}'}$$

$\rho(\vec{r})$ has information about whether the reconstruction gives the true amplitude of the exact input model at a given position $\vec{r}$, $\vec{\Delta}(\vec{r})$ tells us whether a given feature of the reconstructed image is at the same location as the true model, and $\Lambda(\vec{r})$ shows how the true model has been spatially smeared in various directions. The reconstructed model is approximately related to the input model by

$$\hat{m}(\vec{r}; \vec{r}_0, a) \sim \rho(\vec{r}_0)m(\vec{r}; \vec{r}_0 + \vec{\Delta}(\vec{r}_0), a) * \text{Gau}(\vec{r}; \vec{r}_0 + \vec{\Delta}(\vec{r}_0), \Lambda(\vec{r}_0)) \tag{20}$$

for a true model of a circular disk with a small radius a centered at $\vec{r}_0$, where $\text{Gau}(\vec{r}; \vec{r}_0, \Lambda(\vec{r}_0))$ represents a Gaussian distribution with mean $\vec{r}_0$ and covariance matrix Λ, and $*$ denotes the spatial convolution.

Analytical expressions for $\rho(\vec{r})$, $\vec{\Delta}(\vec{r})$ and $\Lambda(\vec{r})$ can be found in Peng et al., (1992).

Variance of the Reconstructed Model

When the inverse operator $A^\dagger$ acts on the data which is contaminated by noise, the error in the data will enter into the reconstructed model and the distribution of the error in the model space is determined by the geometry of experiment, the trade-off parameter α and the statistical properties of noise (as characterized by the data covariance matrix E). From (18), the expectation of the reconstructed model will be related to the expectation of measured data by

$$E[m(\vec{r})] = A^\dagger(\vec{r}) \, E[d]$$

and the variance of $m(\vec{r})$ will be related to the data variance by

$$Var[m(\vec{r})] = (A^\dagger(\vec{r}))^T \, E \, A^\dagger(\vec{r})$$

For the case $E = \sigma^2 I$, the variance of model is simply

$$Var[m(\vec{r})] = \sigma^2 \sum_{i=1}^{M}(A_i^\dagger(\vec{r}))^2$$

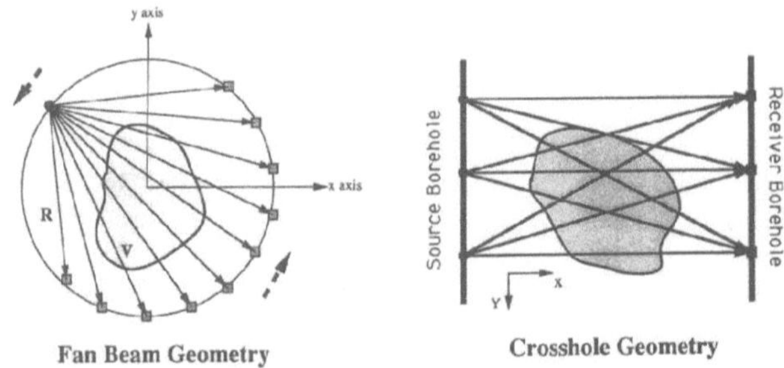

Fan Beam Geometry

Crosshole Geometry

Figure1. (Left) A fan beam geometry of data acquisition in a medical tomography: one source and an array of receivers are employed on a circle which encloses the object (shaded area), the source and detectors are rotated simultaneously to offer 360° coverage of view angles; (Right) A crosswell geometry of data acquisition in a geophysical application: Array of sources and receivers are deployed in two boreholes, the object to be imaged is in between.

COMPARISON WITH THE ALGEBRAIC RECONSTRUCTION TECHNIQUE (ART)

Numerical tests are implemented with two types of data acquisition geometry. In the first example, a medical application, a fan beam geometry is assumed with source and receivers on a circle with radius 1.5 m encompassing the imaging region. 36 sources are uniformly distributed around the circle, and 10 receivers which span the semicircle across the source (Herman, 1981). The model we choose to study is a homogeneous disk with $m = 0.1$ embedded in a homogeneous background of $m_0 = 0$. The anomalous disk has a diameter of 0.8 m and is centered inside the imaging region ($-1 \leq x \leq 1$ m, $-1 \leq y \leq 1$ m). In the second example, we limit the view angles in a crosswell geometry. 20 sources are located in a source well. For each source position, 20 receivers are located in a receiver well which is 2 km apart. The input model is a homogeneous disk with $m = 0.083$ s/km above the background slowness. The disk has a radius 0.8 km and is centered in the imaging region. Figure 1 shows both of these data acquisition geometries.

In each of these tests, reconstructions by our analytic smoothest model reconstruction algorithm and the the Algebraic Reconstruction Technique (ART) are presented. The version of ART we used here is described in Chapter 11 of the textbook by Herman (1981). The relaxation parameter is chosen to be $\lambda = 0.05$ and the smoothing weights are $w_1 = 1.0$, $w_2 = 0.5$ and $w_3 = 0.2$, where notations follow Herman (1981).

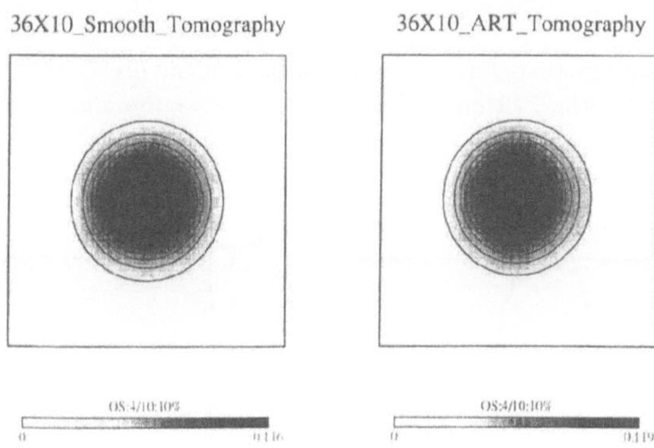

Figure 2. Reconstructions for a circular object with 36 sources and 10 receivers in a fan beam geometry: (Left) Smoothest model algorithm with $R = 1.5$, $\alpha = 0.1$ and $\sigma^2 = 1.0$; (Right) Algebraic reconstruction technique (ART).

Figure 2a shows the reconstruction by our algorithm, as implemented with analytic expression for the solution, with a fan beam data acquisition geometry. As a comparison, Figure 2b shows that by the algebraic reconstruction technique (ART). We see these two techniques give very similar results.

To test the performance of our reconstruction algorithm in the presence of noise, uncorrelated zero-mean uniformly distributed noises are added to the synthetic data with magnitude restricted to (-10%, 10%) of the maximum data residual (standard deviation is about 6 %). Figures 3a and 3b show the reconstructions from the smoothest model algorithm and ART method with a fan beam geometry respectively. The test shows that our algorithm works even better in the presence of noise than ART, which is evident in the plots of a cross section of the reconstructed image at $y = 0$, as shown in Figure 4.

Figure 5 shows the reconstructions of both the smoothest model algorithm and ART under a crosswell geometry. In this example, a Gaussian distributed random noise with zero mean and 6 ms standard deviation is added to the travel time data after the background contribution being subtracted. Again our algorithm shows better reconstruction in the presence of data noise.

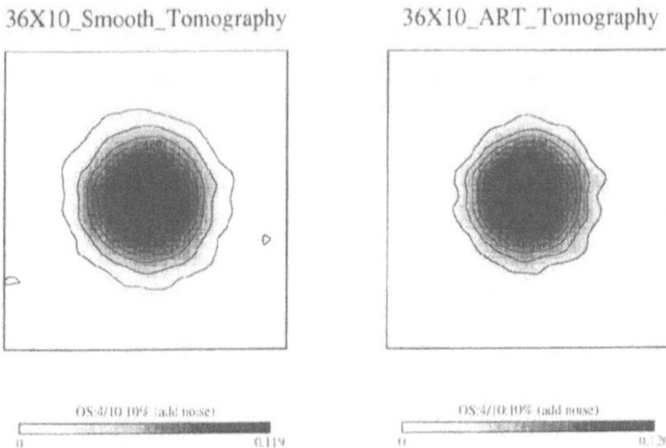

Figure 3. Reconstructions for the circular object in the presence of random noise: (Left) Smoothest model algorithm; (Right) Algebraic reconstruction algorithm (ART). The choice of parameters are the same as those in Figure 2.

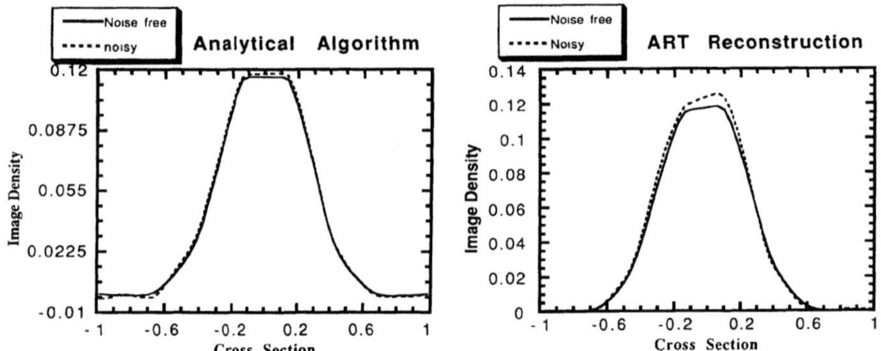

Figure 4 Cross section profiles along $y = 0$ in Figure 3: (Left) Smoothest model reconstructions with 10 % data noise (dash line) and noise free (solid line); (Right) ART reconstructions with 10 % data noise (dash line) and noise free (solid line).

As byproducts our algorithm offers the spatial resolution and variance of the reconstructed model. For a given choice of trade-off parameter α, we plot the amplitude bias $(\rho(\vec{r}))$, mislocation $(\vec{\Delta}(\vec{r}))$ and spatial smearing $(\sqrt{tr\Lambda(\vec{r})})$ as well as the variance as a function of position in the imaging area. Figures 6 shows those measures under a fan beam data acquisition geometry. We see the amplitude bias is small in the middle of the imaging area. Away from the center, however, the reconstruction overestimates the true model by as large as 30 %, while significant underestimation occurs near the boundary of the imaging area. The plot of spatial smearing shows smaller values at the center compared those near the edges. The mislocation vectors are near zero at the center region, but grow as one moves towards the edges. An object close to the edge would be located closer to the center in the reconstructed image. The reconstruction near the center also has smaller variance than at the edges, which means that the central portion of the image is less affected by the data noise. By changing the trade-off parameter α, we find that as α increases, the spatial smearing and mislocation grow while the model variance decreases, thus exhibiting the expected trade-off between estimation bias and variance (Backus and Gilbert, 1970).

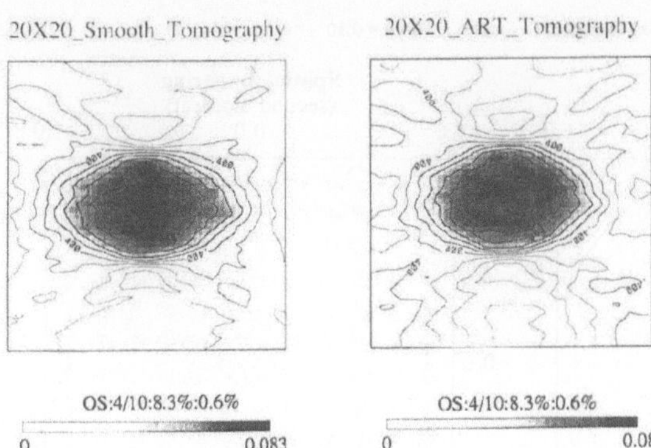

Figure 5. Reconstructions for the circular object with 20 sources and 20 receivers arranged as a crosswell geometry: (Left) Smoothest model Algorithm; (Right) ART. In this example, a random noise with zero mean and 6 ms standard derivation is added to the travel time data.

Figure 7 shows the spatial resolution and model variance of the analytic algorithm under a crosswell data acquisition geometry. Significant amplitude overestimation occurs at the central portion of the imaging area, while large underestimation occurs near the two vertical boundaries close to the source and receiver arrays. The spatial smearing is smaller and isotropic at the center portion compared to those near the edges. The mislocation vector points horizontally towards the center and the amplitude grows as one moves to the two edges coinciding with the sources and receivers. The reconstruction near the center suffers larger variance near the center than at the edges, which is different from that of medical tomography.

CONCLUSIONS

We have presented a new algorithm for image reconstruction from projections. This algorithm, based on Tikhonov regularization with a gradient differential operator, yields the smoothest reconstruction among all possible solutions that satisfy the measured data within a given accuracy, in addition to the resolution and variance of the reconstruction. For the case of linear ray tomography over a circular model region, we are able to derive an analytical inverse operator (referred to as the Tikhonov inverse) which is independent on reconstructed model. The dependence of the Tikhonov inverse on the data is only through its statistical properties, i.e., the covariance operator E.

Numerical tests show that the new algorithm compares favorably with the ART method and is more stable than ART in the presence of strong noise. Additionally, our algorithm has high execution speed compared to ART, and would be especially useful for repetitive experiments with fixed source-receiver geometry and noise variance in that the inverse operator can be precomputed and then applied to each new set of data.

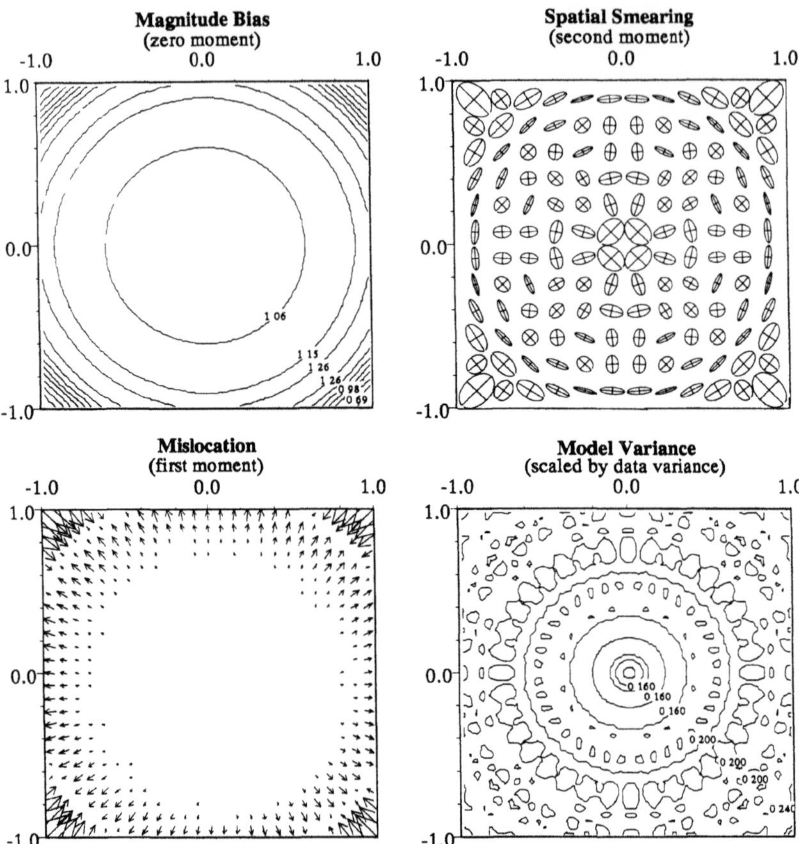

Figure 6. Resolution measures and variance of the smoothest model reconstruction with a fan beam geometry: Magnitude bias is the image of an unity model; Spatial smearing measures the sidelobe of the resolution operator; Mislocation indicates the relative position of the image and the true model; Model variance is related to the confidence of image.

This algorithm can also be applied to experiments with irregular geometry by extending to a circle encompassing the imaging region, without any artificial contribution to the data (after subtracting the background contribution). The size of the circle actually controls the smoothness of the reconstruction in the sense that it confines the total power of model perturbation that accounts for the observed data residual inside the circle, so the larger the circle is, the smoother will be the image for a given data misfit. Mathematically it plays its role in λ_{max}, the largest eigenvalue of the ray geometry matrix, which can be shown to be proportional to R^2 as $R \rightarrow \infty$.

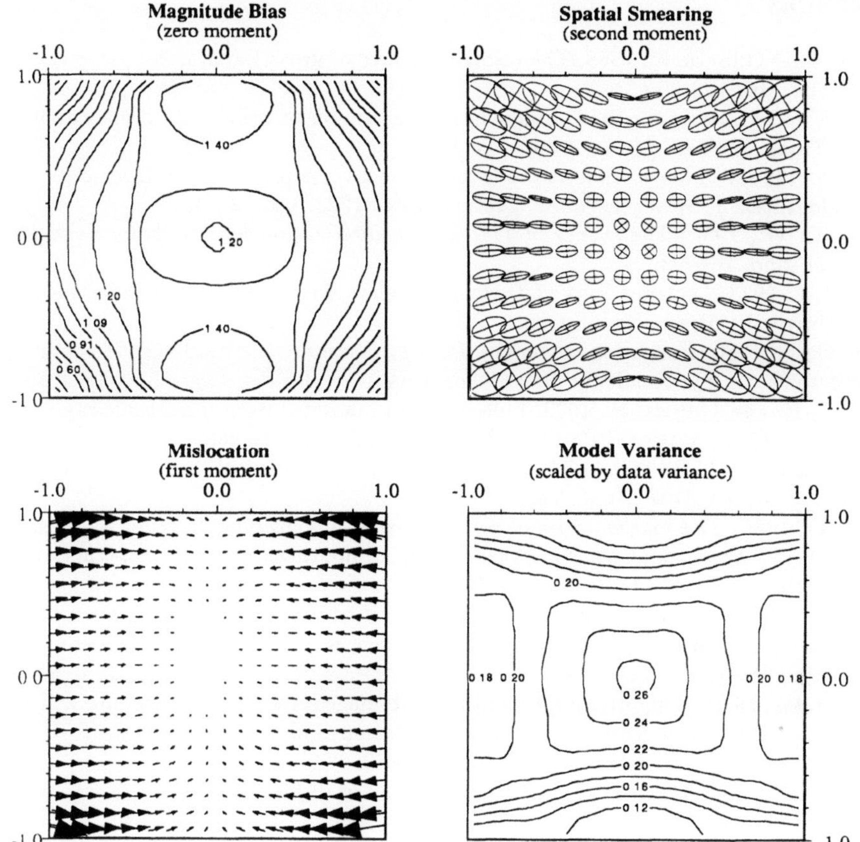

Figure 7. Resolution measures and variance of the smoothest model reconstruction with a crosswell geometry. The model variance has been smoothed for plotting.

ACKNOWLEDGEMENTS

This work was sponsored by the ERL/nCUBE Geophysical Center for Parallel Processing at the Earth Resources Laboratory, MIT. The first author holds an nCUBE fellowship. We wish to thank Joseph Matarese for many valuable discussion about nonlinear tomography and helpful suggestions for improving this paper.

REFERENCES

Backus, G. and Gilbert, F., 1968, The resolving power of gross Earth data: *Geophys. J. R. astr. Soc.*, **16**, 169-205.

Backus, G. and Gilbert, F., 1970, Uniqueness in inversion of inaccurate gross Earth data: *Phil. Trans. R. Soc. Lond.*, **A266**, 123-192.

Bertero, M., 1986, Regularization methods for linear inverse problems, in Inverse Problems, G. Talenti (Ed.), Springer Verlag, Lecture Notes 1225, pp. 52 - 112.

Groetsch, C. W., 1984, *The theory of Tikhonov regularization for Fredholm Equations of the first kind*, Pitman.

Hadamard, J., 1923, *Lectures on the Cauchy problem in linear partial differential equations*, Yale University Press, New Haven.

Herman, G. T., 1980, *Image reconstruction from projections*: the fundamentals of computerized tomography: Academic Press, New York.

Johnson, L. E. and Gilbert, F., 1972, Inversion and inference from teleseismic ray data: in Method in Computational Physics (B.A. Bolt, ed.), **12**, 231-266.

Locker, J. and P. M. Prenter, 1980, Regularization with differential operators I: general theory, *J. Math. Anal. Appl.*, **74**, 504 - 529.

Natterer, F., 1986, *The mathematics of computerized tomography*, J. Wiley & Sons.

Peng, C., Rodi, W. L. and M. N., Toksoz, 1992, Smoothest model reconstruction from projections, submitted to *Inverse Problem*.

Rodi, W. L., 1989, *Regularization and Backus-Gilbert estimation in nonlinear inverse problems: application to magnetotellurics and surface waves*, Ph.D. Thesis, Pennsylvania State University.

Rowland, S. W., 1979, Computer implementation of image reconstruction formulas, in *Image Reconstruction from Projections*: Implementation and Application (G. T. Herman, ed.): Springer-Verlag, Berlin and New York, 9-80.

Tikhonov, A. N. and Arsenin, V. Y., 1977, *Solutions of ill-posed problems*, V. H. Winston and Sons, Washington, D. C.

Wahba, G, 1990, *Spline Models for Observational Data*, Society for Industrial and Applied Mathematics, Philadelphia.

Yanovskaya, T. B. and Ditmar, P. G., 1990, Smoothness criteria in surface wave tomography: *Geophysical J. Int.*, **102**, 63-72.

IMAGING AND INVERSE SCATTERING IN NONDESTRUCTIVE EVALUATION WITH ACOUSTIC AND ELASTIC WAVES

K.J. Langenberg, K. Mayer, P. Fellinger, R. Marklein

Dept. Electrical Engineering
University of Kassel
3500 Kassel
Germany

INTRODUCTION

Ultrasonic nondestructive testing of solid materials requires the assessment of location, shape and size of defects. If appropriate scattering data are received within a synthetic aperture as function of time in — for instance — a pulse-echo mode, defect imaging in terms of B- or C-scan display of the data is available. In contrast to that, heuristic imaging algorithms like SAFT (Synthetic Aperture Focusing Technique) try to invert the scattering process in order to reconstruct the defect geometry via an image. As a matter of fact, such heuristic imaging concepts cannot answer — apart from performing a large number of test experiments — the question whether their image output is a quantitative reconstruction of the defect or not: An inverse scattering *theory* is required. One particularly successful approach is based on a linearization of inverse scattering in terms of either the Born or Kirchhoff approximation and results in a generalized diffraction tomography, which contains SAFT as a special approximation (Langenberg, 1987; Langenberg, 1989; Mayer et al., 1990). Unfortunately, the linearization can often not be tolerated, and, hence, apart from developing algorithms without relying on this approximation, numerical techniques to model ultrasonic acoustic and elastic wave scattering should be available for the computation of synthetic data to be utilized as input to imaging procedures. This interactive imaging approach — combining modeling and imaging — is called CAI for Computer Aided Inspection.

In the present paper we give a number of examples for CAI, mainly based on figures. An overview of the pertinent equations can be found in a recent publication (Langenberg et al., 1993). Our numerical modeling techniques are refered to as AFIT (Acoustic Finite Integration Technqiue) and EFIT (Elastodynamic Finite Integration Technique) (Marklein et al., 1993).

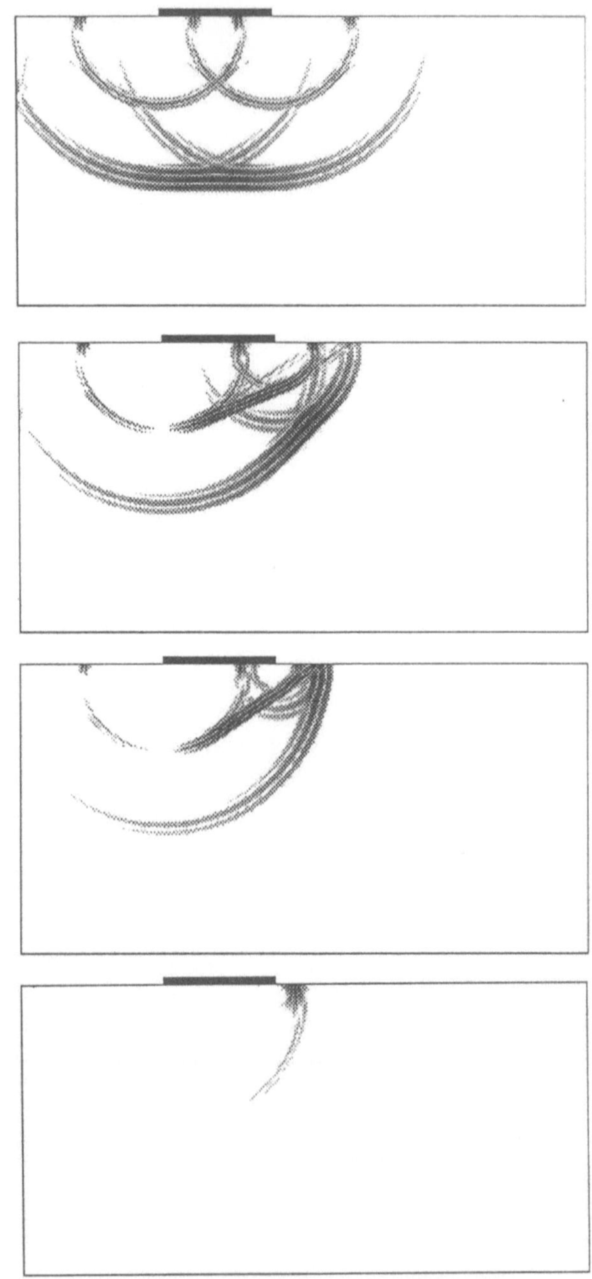

Figure 1. Various transducers modeled with the 2D EFIT code

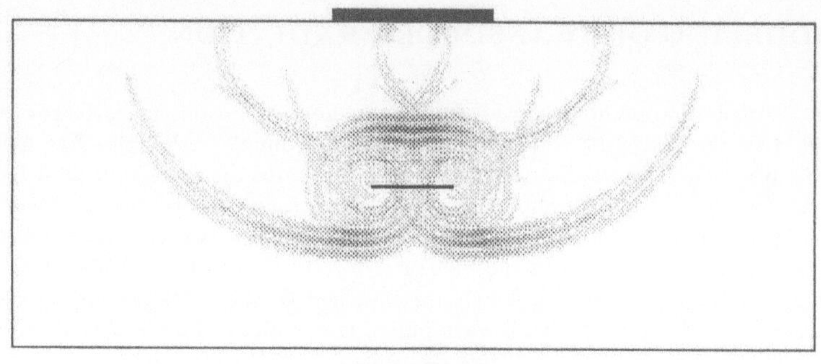

Figure 2. Pressure wave crack scattering, top: EFIT; bottom: AFIT.

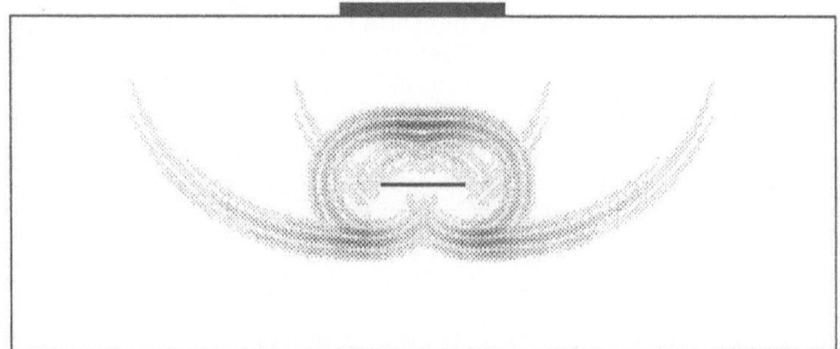

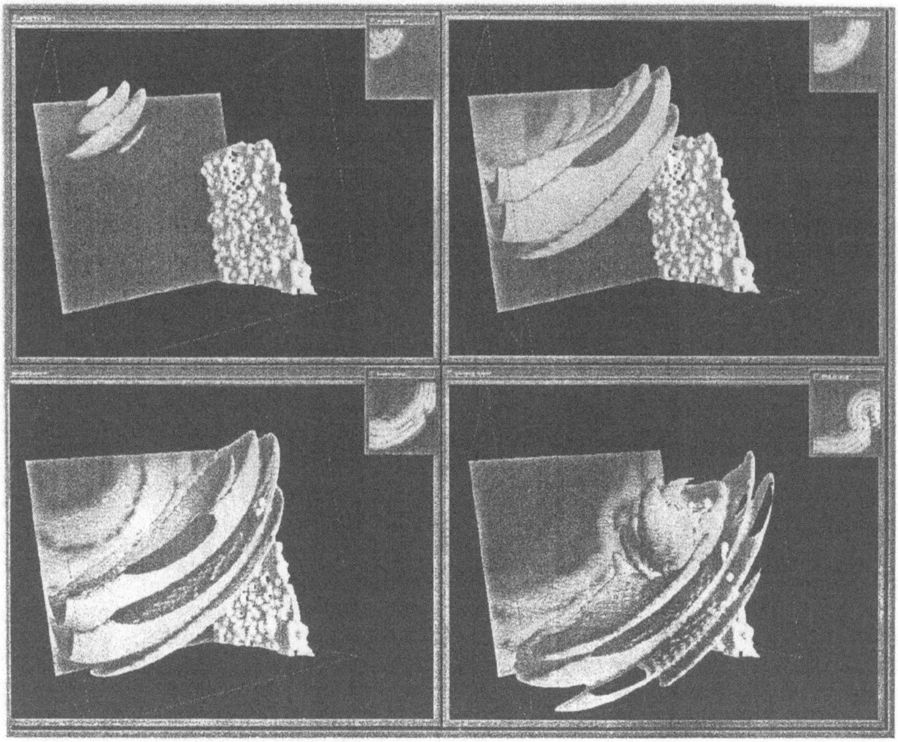

Figure 3. 3D AFIT model of rough crack scattering

EFIT MODELING OF TRANSDUCER RADIATION

Fig 1 displays time domain wavefront snapshots of various piezoelectric transducers as modeled with the 2D EFIT code A "steel test specimen" with stress free boundaries is excited by a prescribed stress distribution within the finite aperture indicated by the black bar The first model of fig 1 implies a homogeneous amplitude distribution resulting in the well known "geometric optical" pressure wavefront as well as two circular cylindrical wave fronts emanating from the edges of the aperture, for this kind of excitation shear waves are only present as edge contributions, which are attached to the corresponding pressure waves via head waves In addition, Rayleigh wave pulses travel along the stress free surface

For the next model, a linear time delay has been implemented in the excitation of the aperture resulting in a steering of the pressure wave "beam" This steering is automatically accompanied by the appearance of a "geometric optical" shear wave being inclined with regard to the pressure wave according to the law of refraction Due to the existence of a critical angle for the refracted pressure wave, a further increase of the time delay results in a 45° shear wave accompanied by the so called subsurface longitudinal wave (Langenberg et al , 1990) as superposition of the circular cylindrical wavefronts from the edges For an even further increase (not shown in fig 1) these two wavefronts get "out of phase", leaving us with a simple shear wave angle probe As soon as the critical angle for the shear wave is reached, the only remaining wave field consists of Rayleigh waves, as demonstrated in the last example of fig 1

AFIT AND EFIT MODELING OF CRACK SCATTERING

Let us consider crack scattering Fig 2 shows 2D EFIT (top) and 2D AFIT (bottom) wavefronts scattered by a surface parallel crack with stress free boundaries, as transducer the top model of fig 1 was utilized Notice, the purely acoustic AFIT code, which does not deliver mode converted shear waves, nevertheless reproduces the scattered pressure waves sufficiently well This supports the application of *scalar* imaging schemes to appropriately time gated wavefronts

Fig 3 gives an example for 3D AFIT modeling of pressure wave scattering by a twodimensional rough crack In principle, the resulting synthetic data could be used to demonstrate the quality of imaging algorithms, but very often, the appropriate experimental mode is pulse echo, and 3D pulse echo simulations are very time consuming Therefore, in the following, we restrict ourselves once more to two dimensions

SAFT IMAGING WITH EXPERIMENTAL AND MODELING DATA

AFIT Data

Fig 4 shows the geometry for a pulse echo experiment, where the synthetic aperture is appropriately indicated A 45° shear wave probe is *acoustically* simulated with AFIT, and the resulting wavefront snapshots as scattered by the rough crack are displayed in fig 4, open boundary conditions have been applied to the specimen's side and bottom "surfaces" The same simulation is repeated for every transducer location within the synthetic aperture, and the pertinent time domain signals are stored, resulting in the rf data field of fig 5, which shows the data as function of time and scan coordinate Applying the FT SAFT inverse scattering scheme (Mayer et al , 1990) to these synthetic data yields the images of fig 5, the bottom image being superimposed by the crack geometry Obviously, only distinct scattering centers on the rough defect can be imaged that way, i e using *this* experimental mode and *this* particular imaging technique

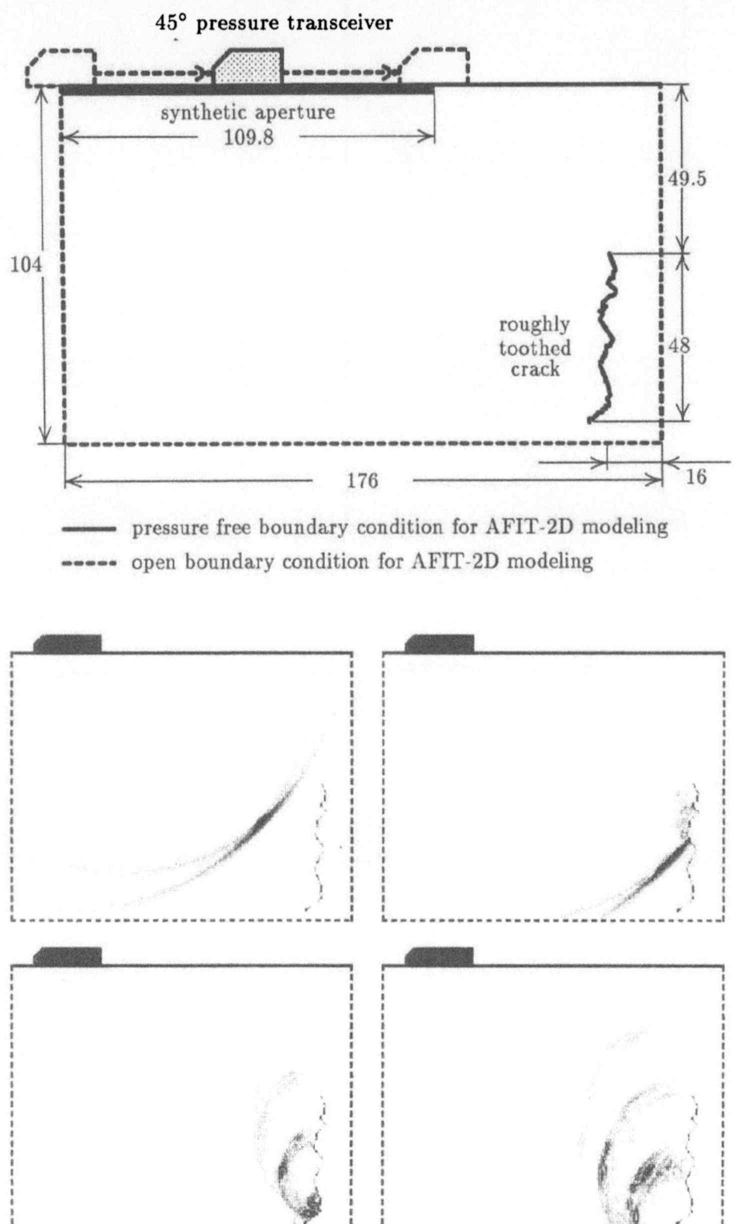

Figure 4. AFIT simulation of a pulse-echo experiment: geometry and wavefronts.

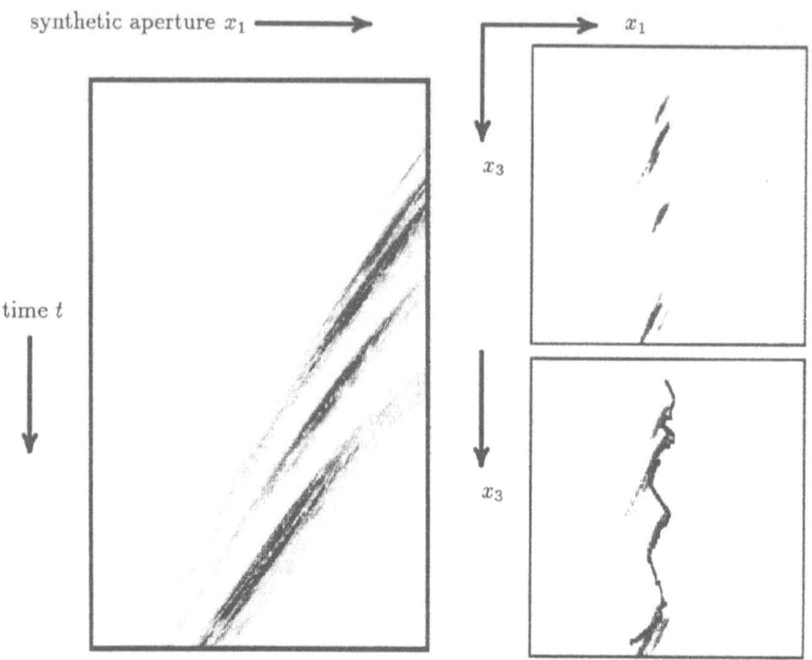

Figure 5. AFIT-simulation of a pulse-echo experiment: rf-data and FT-SAFT image (bottom: crack geometry superimposed).

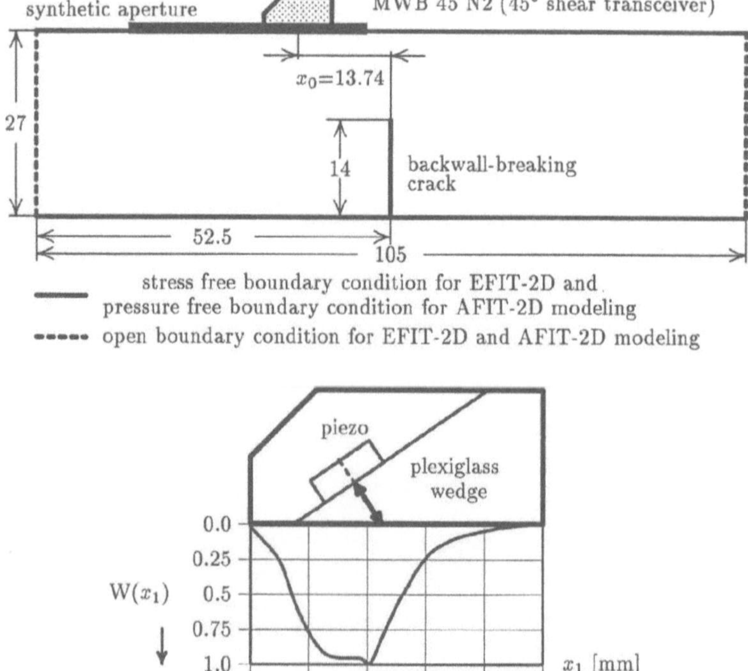

Figure 6. EFIT-simulation of a pulse-echo experiment: geometry and probe model.

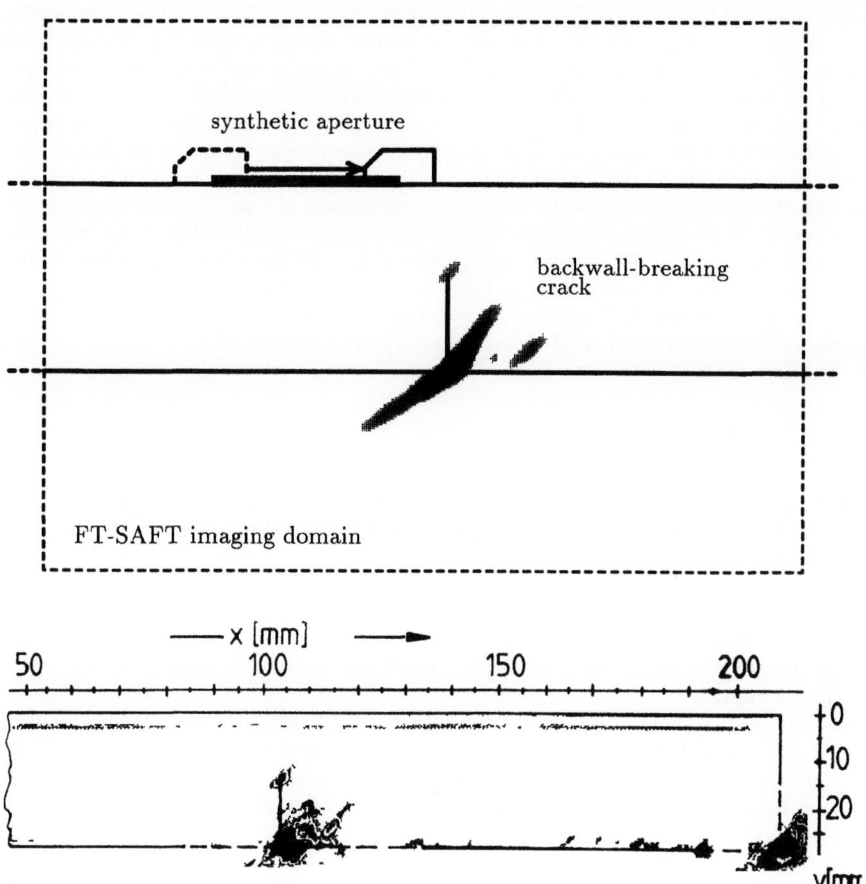

Figure 7 EFIT simulation of a pulse echo experiment FT SAFT image from synthetic data (top) and experimental data (bottom)

EFIT and Experimental Data

We will now check the FT-SAFT imaging algorithm against *elastodynamic* simulation
data as provided by the EFIT code. To be able to compare the results with measurements
made by the Fraunhofer Institute for Nondestructive Testing in Germany (Marklein et al.,
1993) we purposely choose the geometry of fig. 6: A backwall breaking fatigue crack is
illuminated by a 45° shear wave probe from various positions within a synthetic aperture. To
ensure our model to be as close to reality as possible, the displacement amplitude distribution
along the radiating aperture of the piezoelectric probe has been carefully measured, and the
result is given in fig. 6; this amplitude distribution enters the EFIT code in the transmitting as
well as in the receiving mode. The comparison of single simulated and experimental A-scans
turns out to be excellent; the physical interpretation of the various observed A-scan pulses
is easily possible on the basis of the elastodynamic wavefront evaluation in the time domain
— for instance, mode conversion and surface wave effects on the crack faces are apparent
—, which gives rise to a precise understanding of the resulting rf-data field (Marklein et al.,
1993). Therefore, the output of FT-SAFT either for synthetic or for simulated data is no
longer surprising: Fig. 7 reveals the image to be composed of a weak indication of the crack
tip compared to the strong crack root image, which is formed out of the corner reflection of
the shear wave. In addition, ghost images are produced by the resonances of the crack, i.e.
the Rayleigh surface waves traveling up and down from root to tip.

As demonstrated by this example, we believe that a combination of modeling and
imaging will be the future of nondestructive testing in terms of CAI.

REFERENCES

Langenberg, K.J., 1987, Applied inverse problems of acoustic, electromagnetic and elastic
waves, *in*: "Basic Methods of Tomography and Inverse Problems," P. Sabatier, ed.,
Adam Hilger, Bristol.

Langenberg, K.J., 1989, Introduction to the special issue on inverse problems, *Wave Motion*
11:99.

Mayer, K., Marklein, R., Langenberg, K.J., and Kreutter, T.,1990, Three-dimensional imaging system based on Fourier transform synthetic aperture focusing technique, *Ultrasonics* 28:241.

Langenberg, K.J., Fellinger, P., Marklein, R., Zanger, P., Mayer, K., and Kreutter, T., 1993,
Inverse methods and imaging, *in*: "Evaluation of Materials and Structures by Quantitative Ultrasonics," J.D. Achenbach, ed., Springer-Verlag, Vienna.

Marklein, R., Fellinger, P., Langenberg, K.J., and Walte, F., 1993, The ultrasonic modeling
codes AFIT and EFIT as quantitative NDE tools, *in*: "Review of Progress in Quantitative Nondestructive Evaluation," D.O. Thompson, D.E. Chimenti, eds., Plenum Press,
New York.

Langenberg, K.J., Fellinger, P., and Marklein, P., 1990, On the nature of the so-called subsurface longitudinal wave and/or the surface longitudinal "creeping" wave, *Res. Non-
destr. Eval.* 2:59.

MULTISCALE ANALYSIS OF ULTRASONIC IMAGE[*]

Xiaojun Zhou,[1] Yueming Sun,[1] Guosheng Hu[2]
Zhongfang Tong,[1] and Xingli Wang[3]

[1]Mechanical Engineering Department
[2]Mathematic Department
Zhejiang University
Hangzhou 310027

[3]Nondestructive Test Division
Hangzhou Boiler Works
Hangzhou 310010

P. R. China

INTRODUCTION

Different kinds of ultrasonic image generating and processing methods in nondestructive interrogation of materials have been developed and some interesting subjects therefore are brought about, e. g. in an ultrasonic C-scan image case:

A) To evaluate the image, we want to reduce the blurring and distortion introduced by the noise and the resolution limits of the imaging system.

B) To recognize different kinds of faults, we need get not only the features from the C-scan image grey-grade or color representation information directly, but also the ones from some kinds of treatments of the image data.

C) In computer application occasions, e. g. in an automatic NDE system, we need to represent the whole image or some interesting details in a computer-understood and compressed way.

The conventional methods of solving these problems are to treat them respectively. In this paper, however, we propose an unique method of generating and processing the C-scan image from rf A-scan signals based on wavelet transform analysis. Firstly, we use one dimensional wavelet transform (1D-WT) as a bandpass filter whose band width is adjustable for processing rf A-scan signals to get local features and form original C-scan image (so called in this paper).

---

* The work is supported by the Natural Science Foundation of Zhejiang Province, China.

And then a further multiscale analysis algorithm based on two dimensional wavelet transform (2D-WT) for processing the original C-scan image to extract globe features is presented. Thus an implemental algorithm of pseudo-3 dimensional wavelet analysis is constructed. At the end of the paper, some simulation and experimental results are given.

1D-WT AND MULTISCALE ANALYSIS OF SONIC SIGNAL

Wavelet analysis is a newly developed time-frequency (scale) domain analysis method. The formalism of the continuous wavelet transform was first introduced by J. Morlet, A. Grossmann et al. [1,2,3] Recently, S. Mallat presented the concept of multiscale (multiresolution) analysis and the unique frame of construction of orthonormal wavelet bases. [4,5] The former is used as the theory foundation of our analysis method presented in this paper, and the latter is employed as the main tool of the realization of our implementation.

Complex valued function $\psi(x)$ is called a basic wavelet if and only if its Fourier transform $\hat{\psi}(\omega)$ satisfies the following admissible condition:

$$\int_{-\infty}^{+\infty} \frac{|\hat{\psi}(\omega)|^2}{|\omega|} d\omega = C_\psi < +\infty. \tag{1}$$

To denote by ψ_s the dilation of ψ by a scale factor s :

$$\psi_s(x) = \frac{1}{s}\psi\left(\frac{x}{s}\right). \tag{2}$$

for the finite-energy function $f(x) \in L^2(R)$, its wavelet transform is defined as

$$Wf(s, x) = f * \psi_s(x), \tag{3}$$

where $*$ is an convolution operator.

The invertibility of the wavelet transform ensures us to reestablish $f(x)$ from $Wf(s, x)$: [2]

$$f(x) = \frac{1}{C_\psi} \int_0^{+\infty} Wf(s, u) \bar{\psi}_s(u - x) du \frac{ds}{s}, \tag{4}$$

where $\bar{\psi}_s(x)$ denotes the complex conjugate of $\psi_s(x)$.

Remark: condition (1) and the definition Equation (3) imply that ψ_s acts as a filter whose band width is adjustable and the wavelet transform is a multiscale filtering operation (more strict mathematic expression can be found in Mallat[4,5]), Factor s makes the pass-band width variable.

Compared with other time-frequency domain analysis methods, e. g. short time Fourier transform, wavelet transform provides us a tool with which we can analyze thoroughly the local characteristics of a given function (signal).

Like the applications of wavelet transform in other fields, [3,4,5,8,9] the detection of the singularity of the signals is essential in our processing of the ultrasonic signals. The following theorem allows us to analyse the singularity of $f(x)$ form $Wf(s, x)$ and the proof can be found in M. Holshneider and P. Tchamitchian. [6]

Theorem: To function $f(x) \in L^2(R)$, let $[a, b]$ be an interval of R, and $0 < \alpha < 1$. For any $\varepsilon > 0$, $f(x)$ is uniformly Lipschitz α over $[a + \varepsilon, b + \varepsilon]$, if and only if for any $\varepsilon > 0$ there exists a constant A_ε such that for $x \in [a + \varepsilon, b + \varepsilon]$ and $s > 0$,

$$|Wf(s, x)| \leqslant A_s s^a. \tag{5}$$

To use the 1D-WT for the local multiscale time-frequency domain feature extracting, the modulus maxim[4,5] and some other methods[8,9] are employed. Certain constructions of eigenvectors are established by the foremost multiscale analysis across some intervals of the rf A-scan signals which allows us to get the objective pattern recognized. For the details about the pattern recognition method we used, it is out of the subject of this paper and will be discussed in our cooperators' another paper in preparing.

2D-WT AND SONIC IMAGE FEATURE EXTRACTION

The definition of 2D-WT can be simply taken as the extension of 1D-WT. That is, for two-variable function $f(x, y)$, define its 2D-WT as:

$$Wf(r, s_1, s_2, x, y) = f * \psi_{r s_1 s_2}(x, y). \tag{6}$$

Where $\psi_{s_1 s_2}$ denotes the dilation of the wavelet function $\psi(x, y)$:

$$\psi_{s_1 s_2}(x, y) = \frac{1}{\sqrt{s_1 s_2}} \psi\left(\frac{x}{s_1}, \frac{y}{s_2}\right), \tag{7}$$

whose Fourier transform satisfies the admissible condition

$$\int_{-\infty}^{+\infty} \frac{|\hat{\psi}(\omega_x, \omega_y)|^2}{|\omega_x \omega_y|} d\omega_x d\omega_y < \infty, \tag{8}$$

and r is a rotation factor.

It is easy to see that this definition is suitable for the occasions where the two variables x and y have different physical meanings. In the other hand, however, it is not convenient and will occupy too big a memory and storage resource in computer digital approach or application cases. So in this paper we use its another form of definition.

Let us take two functions $\psi^1(x, y)$ and $\psi^2(x, y)$, which vanish rapidly and whose double integrations are both zero, and denote $\psi_s^1(x, y)$ and $\psi_s^2(x, y)$, respectively, the dilations of ψ^1 and ψ^2 by the same scale factor s:

$$\psi_s^1(x, y) = \frac{1}{s^2} \psi^1\left(\frac{x}{s}, \frac{y}{s}\right), \tag{9a}$$

$$\psi_s^2(x, y) = \frac{1}{s^2} \psi^2\left(\frac{x}{s}, \frac{y}{s}\right). \tag{9b}$$

If there are two constants $B \geqslant A > 0$ that the Fourier transforms of ψ^1 and ψ^2 satisfy:

$$A \leqslant \int_0^\infty (|\hat{\psi}^1(s\omega_x, s\omega_y)|^2 + |\hat{\psi}^2(s\omega_x, s\omega_y)|^2) \frac{ds}{s} \leqslant B, \tag{10}$$

then for any function $f(x, y) \in L^2(R^2)$, its two components form wavelet transforms with respect to $\psi^1(x, y)$ and $\psi^2(x, y)$ can be given as:

$$W^1 f(s, x, y) = f * \psi_s^1(x, y), \tag{11a}$$

$$W^2 f(s, x, y) = f * \psi_s^2(x, y). \tag{11b}$$

The theorem given below indicates that the 2D-WT (given by Equation (11)) retains all the information of the ordinary signal $f(x, y)$.

Theorem: To dual wavelet function $\psi^1(x, y)$ and $\psi^2(x, y)$ of ψ^1 and ψ^2, if they satisfy the condition:

$$\int_0^\infty \left[\hat{\psi}^1(s\omega_x, s\omega_y) \, \hat{\bar{\psi}}^1(s\omega_x, s\omega_y) + \hat{\psi}^2(s\omega_x, s\omega_y) \, \hat{\bar{\psi}}^2(s\omega_x, s\omega_y) \right] \frac{ds}{s} = 1 \tag{12}$$

then $f(x, y)$ can be reconstructed as:

$$f(x, y) = \int_0^s \int_{-\infty}^{+\infty} \frac{1}{s^2} \left[\overline{\psi^1\left(\frac{u-x}{s}, \frac{v-y}{s} \right)} W^1 f(s, u, v) \right.$$

$$\left. + \overline{\psi^2\left(\frac{u-x}{s}, \frac{v-y}{s} \right)} W^2 f(s, u, v) \right] \frac{ds}{s} du dv. \tag{13}$$

Proof: Take $\psi^1(x, y)$ and $\psi^2(x, y)$ and let their Fourier transforms be:

$$\hat{\psi}^1(s\omega_x, s\omega_y) = \frac{\hat{\psi}^1(s\omega_x, s\omega_y)}{\int_0^\infty \left[|\hat{\psi}^1(s\omega_x, s\omega_y)|^2 + |\hat{\psi}^2(s\omega_x, s\omega_y)|^2 \right] \frac{ds}{s}}, \tag{14a}$$

$$\hat{\psi}^2(s\omega_x, s\omega_y) = \frac{\hat{\psi}^2(s\omega_x, s\omega_y)}{\int_0^\infty \left[|\hat{\psi}^1(s\omega_x, s\omega_y)|^2 + |\hat{\psi}^2(s\omega_x, s\omega_y)|^2 \right] \frac{ds}{s}}, \tag{14b}$$

It is easy to prove that the condition (12) is satisfied. The Fourier transforms of $W^1 f$ and $W^2 f$ with respect to x and y can be written as:

$$\hat{W}^1 f(s\omega_x, s\omega_y) = \hat{\psi}^1(s\omega_x, s\omega_y) \hat{f}(s\omega_x, s\omega_y), \tag{15a}$$

$$\hat{W}^2 f(s\omega_x, s\omega_y) = \hat{\psi}^2(s\omega_x, s\omega_y) \hat{f}(s\omega_x, s\omega_y). \tag{15b}$$

According to the Plancheral's Formula, the right hand side of Equation (13) can then be written as:

$$\frac{1}{4\pi^2} \int_0^s \int_{-\infty}^{+\infty} \int_{-\infty}^{+\infty} \left[\hat{\psi}^1(s\omega_x, s\omega_y) \, \overline{\hat{\psi}^1(s\omega_x, s\omega_y)} + \hat{\psi}^2(s\omega_x, s\omega_y) \, \overline{\hat{\psi}^2(s\omega_x, s\omega_y)} \right]$$

$$* \exp[i(x\omega_x + y\omega_y)] \hat{f}(\omega_x, \omega_y) \frac{ds}{s} d\omega_x d\omega_y$$

$$= \frac{1}{4\pi^2} \int_{-\infty}^{+\infty} \int_{-\infty}^{+\infty} \exp[i(x\omega_x, y\omega_y)] \hat{f}(\omega_x, \omega_y) d\omega_x d\omega_y \tag{16}$$

Since this is the Fourier transform of function of $f(x, y)$, we have established Equation (13).

In our application of 2D-WT to ultrasonic NDE image processing, the partial derivatives along x and y of a two dimensional smoothing function $\theta(x, y)$ are taken as two wavelets:

$$\psi^1(x, y) = \frac{\partial \theta(x, y)}{\partial x}, \tag{17a}$$

$$\psi^2(x, y) = \frac{\partial \theta(x, y)}{\partial y}, \tag{17b}$$

where $\theta(x, y)$ can be any function whose double integration is nonzero and which converges to zero at infinity. [4, 5]

In this case, the wavelet transform is equivalent to those multiscale transforms which have been developed in computer vision, especially in image processing, [7] but were studied before the development of the wavelet formalism.

To an ultrasonic image, say, a C-scan image, the information of the characteristics of the flaws initially detected is all expressed together in the image space in a color or grey-grade representation way. We can take ψ^1 and ψ^2 (with Equation (17)) as a pair of orthonormal multiscale filters that different kinds of time -frequency features can be extracted for some typical purpose of detection.

ALGORITHM OF IMPLEMENTATION

To use the multiscale analysis method based on wavelet analysis for ultrasonic C-scan image generating and processing, we implement a pseudo-3 dimensional wavelet analysis method consisting of:

(1) on-line rf A-scan signals gathering and digitizing during raster-type scanning of the specimen;

(2) on-line axially multiscale analysis of the rf A-scan signals using 1D-WT to get the local time-frequency features of the signals;

(3) on-line pattern recognizing of the local features according to the knowledge of the possible faults of the specimen to generate the so called original C-scan image and storing the image data;

(4) applying a further multiscale analysis based on 2D-WT to the original C-scan image data for some practical usage such as default edge extraction, typical flaw image enhancement, or blurring reduction and so on.

Although people could extend the wavelet transform into three dimensional form in theory, the huge demand of computer memory and storage resource for three dimensional data blocks makes it difficult to form a relevant implementation. In another respect worthy of remark, researchers have developed some kinds of implemental algorithms using different pseudo-3D filters for the purpose of C-scan image generating and processing. [9, 10, 11, 12] but most of them require off-line analysis thoroughly which is not convenient and strictly limits the detection area because of the great demand of computer memory or storage resource to store the original signal data. The rapid decay speed of the wavelets enables us to implement an on-line generating and a later processing of C-scan image presented in this paper.

SIMULATION AND EXPERIMENT

The multiscale analysis algorithm of the implementation based on pseudo-3D WT for the default edge extraction, typical flaw image enhancement, and blurring reduction is programmed using Borland C^{++} (V2. 0) language in an ALR PC/386 (IBM PC/386 compatible) microcomputer. The rf A-scan signal sampling rate is 20 MHz and the central frequency of the ultrasonic pulse signals is 2. 5 MHz.

Figure 1 (b) and 1 (c) show the two components of 2D-WT at scale 2 of the simulation data of Figure 1 (a) which is taken as an original C-scan image of 180×160 pixels. Figure 1 (d) and 1 (e) show the modulus and phase of the 2D multiscale analysis results, respectively. The last result of the simulation test in Figure 1 (f) shows that the multiscale detector has better effects in horizontal and vertical directions along the filtering directions where the 2D-WT was taken.

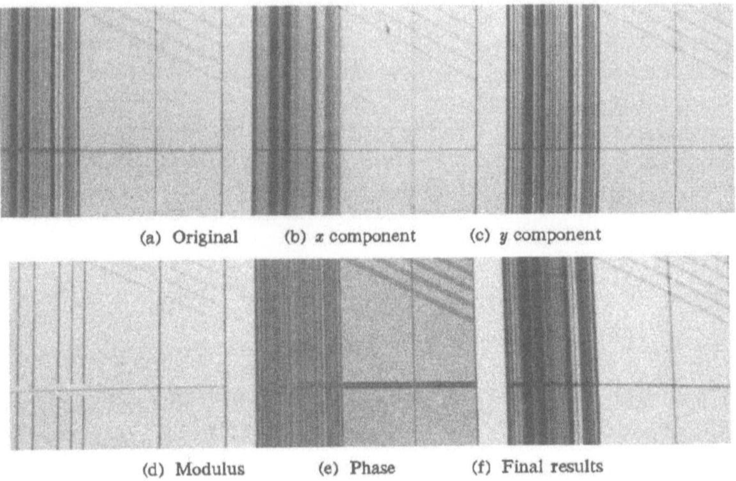

Figure 1.   Simulation Analysis

The specimen detected in the first experiment is a piece of carbon fiber enforced composites material of scan size of 36 mm $\times$ 32 mm. The original C-scan image generated by an on-line filter of rf A-scan signals is shown in Figure 2 (a). The result in Figure 2 (f) shows that the enhancement effect is apparent.

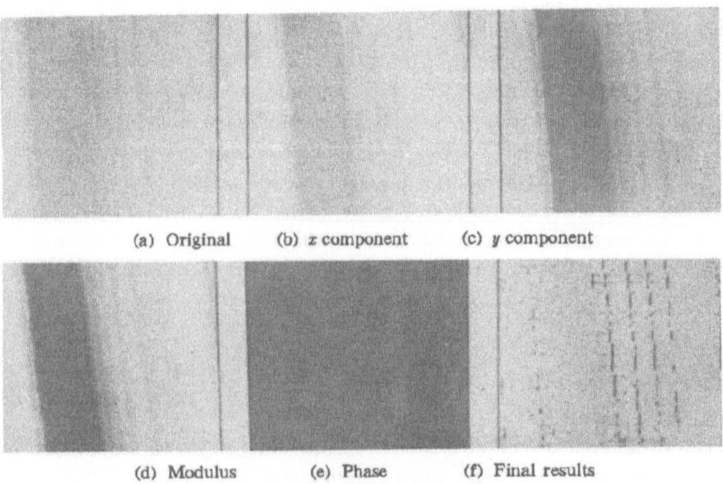

Figure 2.   The first experiment

Figure 3 presents the second experiment taken to a T-type specimen of two welded steel plates. The holes driven in one plate was covered and welded over by the other one. The scan sizes are 10 mm $\times$ 120 mm. Figure 3 (a) is the scan result of the holes and Figure 3 (f) gives their edges detected by the multiscale analysis.

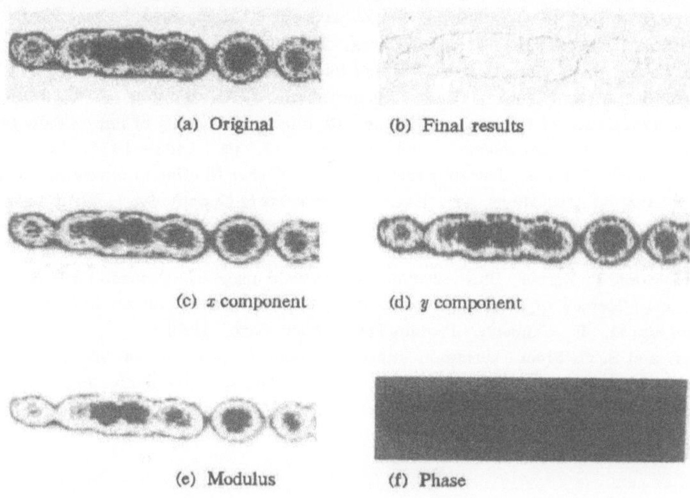

(a) Original

(b) Final results

(c) x component

(d) y component

(e) Modulus

(f) Phase

Figure 3.　The second experiment

CONCLUSION

The multiscale analysis based on wavelet transform gives us an effective and efficient tool to perform an pseudo-3D C-scan image generating and processing implemental operation. It improves the resolution in depth and meet our need for the reduction of blurring and distortions introduced by the noise. More advantages of this implementation are also apparent in its applications in defective feature extraction, e. g. for typical default image enhancing and flaw edge detecting, and in default pattern recognition.

Although an on-line local feature extracting algorithm based on 1D-WT for the generating of original C-scan image is proposed in this paper, we would point out that it still limits the scanning speed in some high-speed detection case.

REFERENCES

1. Y. Meyer, in: "Ondelettes", Hermann, Paris, 1990.
2. A. Grossmann and J. Morlet, Decomposition of hardy functions into square integrable wavelet of constant shape, *SIAM J. Math.* vol. 15, pp. 723—736, 1984.
3. A. Grossmann, Wavelet transform and edge detection, in: "Stochastic Process in Physics and Engineering," M. Hazewinkel, Ed. Dodrecht, Reidel, 1986.
4. S. Mallat, A theory for multiresolution signal decomposition: the wavelet representation, *IEEE Trans. Pattern anal. Machine Intell.*, vol. PAMI—11, pp. 674—693, July 1989.
5. S. Mallat, Multiresolution approximation and wavelet orthonormal bases of L2, *Trans. of the American Mathematical Society.* vol. 315, pp. 69—87, Sept. 1989.

6. M. Holschneider and P. tchamitchian, Regularite local de la fonction nondefferentiable de Riemann, in: "Les Ondelettes en 1989, Lecture Note in Mathematics", P. G. Lemarie, Ed. New York: Springer Verlog, 1989.

7. J. Canny, A computational approach to edge detection, *IEEE Trans. Pattern Analysis and Machine Intelligence*, vol. 8, pp. 679—698, 1986.

8. Xiaojun Zhou, Guosheng Hu, Zhongfang Tong, Yiaodong Cheng, and Yueming Sun, Detection of mono-pulse-type signal using wavelet analysis, *J. Zhejiang Univ.*, supplement, Mar. 1991.

9. Xiaojun Zhou, Guosheng Hu, Yiaodong Cheng, Zhongfang Tong, and Jinqiu Mo, Detection of step-type signal using wavelet analysis, *J. Zhejiang Univ.*, vol. 26, Mar. 1992.

10. B. T. Smith, J. S. Heyman, A. M. buoncristiani, E. D. Blodgett, J. G. Miller, and S. M. Freeman, Correlation of the Deply technique with ultrasonic imaging of impact damage in graphite-epoxy composites, *Materials Evaluation*, vol. 47, No. 12, pp. 1408—1416, Dec. 1989.

11. P. Karpur, and B. G. Frock, Two dimensional pseudo-Wiener filtering in ultrasonic imaging for nondestructive evaluation applications, in: "Review of Progress in Quantitative Nondestructive Evaluation", vol. 8A, pp. 743—750, Ed. D. O. Thompson and D. E. Chimenti, Plenum press, New York, 1990.

12. B. G. Frock, and P. Karpur, Blur reduction in ultrasonic image using pseudo three-dimensional Wiener filtering, in: "Review of Progress in Quantitative Nondestructive Evaluation", vol. 9, Ed. D. O. Thompson and D. E. Chimenti, Plenum Press, New York, 1990.

13. P. Karpur, and B. G. Frock, Ultrasonic image restoration by pseudo three-dimensional power spectrum equalization, in: "Review of Progress in Quantitative Nondestructive Evaluation", vol. 9, Ed. D. O. Thompson and D. E. Chimenti, Plenum Press, New York, 1990.

COMPUTER-CONTROLLED ACOUSTIC TRANSDUCTION
FOR REAL-TIME THREE-DIMENSIONAL IMAGING

Jing-ao Sun and Anni Cai

Department of Radio Engineering
Beijing University of Posts & Telecommunications
Beijing 100088, China

ABSTRACT

This paper describes a new acoustic imaging technique which involves the use of a novel device, computer-controlled acoustic transducer (ComCAT). When a ComCAT system operating, both acoustic beam focusing and scanning are programmatically controlled by means of a computer-generated Fresnel-zone pattern. The zone pattern can be shifted horizontally and vertically, therefore both two- and three-dimensional images can be obtained.

INTRODUCTION

Linear-phased array transducers have been widely used in acoustic imaging for medical diagnosis. This technique requires a large number of small railroad-tie-shaped transducer elements to be arranged next to each other to form a line array. When the number of elements is large, it requires very complex circuitry and the systems are expensive. The image quality of linear phased-array systems is often disappointing due, in part, to the existence of large sidelobes in beam pattern, especially when the beam is deviated from the central axis of the array [1]. In addition, the sensitivity of a linear-phased array system is limited by the factor that the array is only capable of providing focusing in one lateral dimension. In order to find a better technique, many researchers have explored the possibilities using annular array transducers [2-4].

However, the main problem that exists in annular array systems is that the transducers can only be scanned manually or mechanically. In order to replace these clumsy scanning methods, a potentially attractive technique involving the use of an opto-acoustic transducer (OAT) was suggested [2]. In this technique a light pattern of a Fresnel zone plate is used to induce acoustic beam focusing and steering with the transducer stationary. Unfortunately, at the frequencies interested for medical imaging, the sensitivity of the OAT is much less than practically needed, although the idea is excellent.

In this paper, we present a new technique which may effectively replace the manual or mechanical scanning mechanisms with electrical beam scanning for the annular array systems.

SYSTEM DESCRIPTION

A simplified block diagram of our proposed system is shown in Fig.1. The key element involved is a ComCAT. The ComCAT consists of a two-dimensional electrical switch array in connection with a two-dimensional shift register. These operational elements are fabricated on a piezoelectric plate. The acoustic beam focusing is induced by a computer-generated Fresnel-zone plate pattern. Beam scanning is performed by shifting the electrical zone-plate pattern in the shift register. Advances in integrated circuit technology has provided a solid basis for developing the ComCAT.

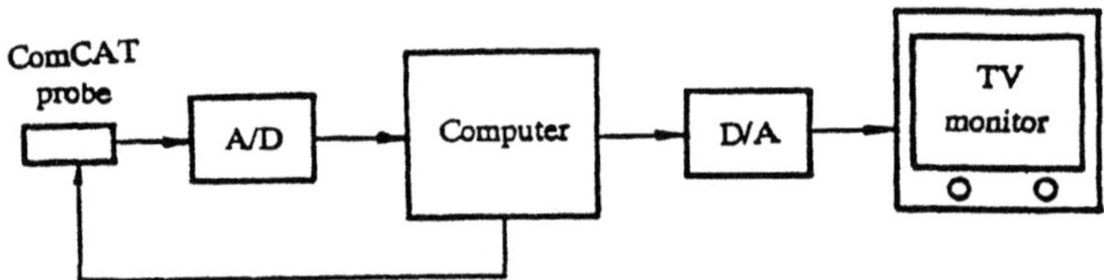

Fig.1 A simplified diagram of the proposed system.

As it can be seen from Fig.1, such a system becomes the sonic equivalent of a computer-controlled scanning electron microscope. It is inherently simple both in construction and in operation because it does not require any mechanical means for scanning the sound beam; it has no acoustic lenses; nor does it have a high-density array of transducers.

BASIC CONFIGURATION OF THE ComCAT

A schematic illustration of a side view of the ComCAT is shown in Fig.2. A piezoelectric plate is employed for acoustic-wave generation and, also, as a substrate. One side of the plate is covered with a metal layer and a two-dimensional switching array is fabricated on the other side of the plate. One of the two terminals of each switch is connected to the plate; the other, to a common lead. This lead and the metal layer serve as the two electrodes needed to furnish AC excitation to the ComCAT to provide insonification with switch K in position 1. During signal reception, switch K is turned to position 2. The common lead serves as a signal output terminal and the metal cover as the common ground. Each switch is separately controlled by a computer-generated signal via a two-dimensional shift register which is fabricated on top of the switch array. One row of memory cells in the shift register is represented by

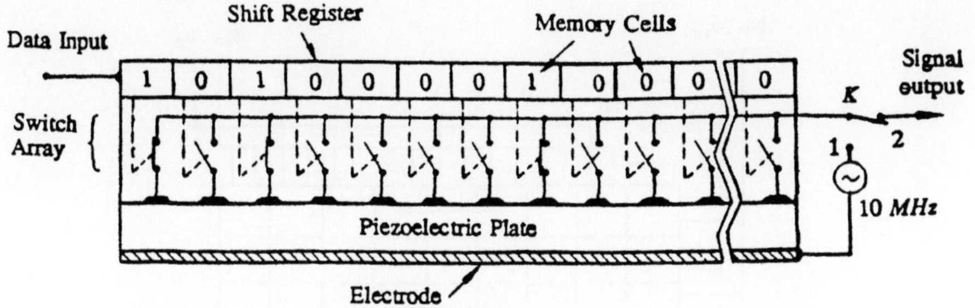

Fig.2 Schematic diagram showing a single
row of elements of a ComCAT.

the small boxes at the top of Fig.2. The cells and the
switches have a one-to-one mapping with respect to each
other.

The input signal, generated by a computer, is
corresponding to a Fresnel-zone plate pattern. As such, it
controls the generation of a focused, scanned acoustic beam
so as to provide the radiation needed for imaging the tissue
structure under observation.

PRINCIPLES OF OPERATION OF ComCAT

As shown by Fig.2, a memory cell and a switch connected
together with the corresponding small area of the
piezoelectric plate, form an operating element of the ComCAT.
The presence of a 1 at a particular cell closes the
corresponding switch, and a 0 at a cell means that the switch
there is open.

When all the shift register cells are properly loaded,
the desired pattern for controlling the ultrasound pulse
transmission and echo signal reception is completed. The
pattern is then shifted to the right, one cell at a time. At
the beginning of each step, the switch K turns to position 1.
A 10 MHz signal is applied across the piezoelectric plate via
the closed switches of the array on the small areas which are
in contact with the switches and an ultrasound pulse is
transmitted into the object being observed. Then K is
switched to position 2 and the ComCAT is ready for echo
reception. The induced electrical signals in the
piezoelectric plate by the echoes impinging on the areas
connected to the closed switches are summed at the common
lead. After the last echo has returned, the pattern is
shifted once more to the right and the above procedure is
repeated.

Fig.3 illustrates how a computer-generated zone pattern
is loaded and shifted. During operation, the pattern loading
of all the rows is accomplished via a single data-input line.
This is accomplished by using a vertical shift register (VSR)
as shown at the left of the figure. The elements of the VSR
are shift register cells only and are not attached to any of
the switches of the array that is connected to the
piezoelectric plate.

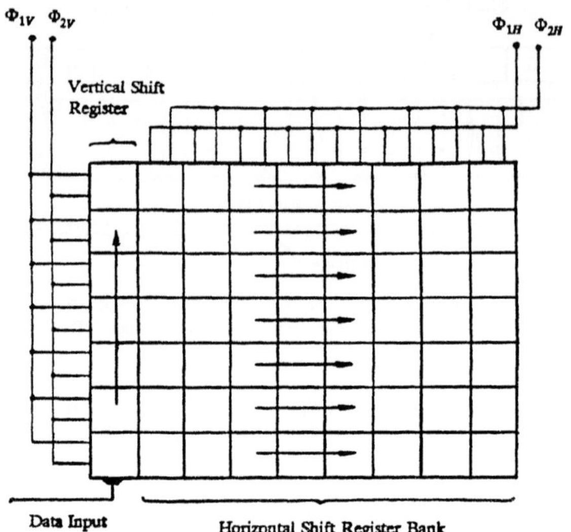

Fig.3 A schematic illustration of the two-dimensional shift register.

First, the data from a computer for each column are read into the VSR by a series of clock signals labeled Φ_{1V} and Φ_{2V}. After one column of data has been loaded into the VSR, it is transferred into the leftmost column of the horizontal shift register (HSR). Repeated cycles of the Φ_{1H} and Φ_{2H} signals move the pattern, column by column, in the direction of the arrows. As each new column of data is loaded, the pattern already in the HSR bank is shifted one column to the right.

At each step of the pattern movement, a pulse of sound waves with a synthesized wave front of Fresnel-zone plate pattern is launched and converges to a point directly in front of the pattern center during propagation. At the time of detection, switch K (referred to Fig.2) is turned to position 2. The echo pulses reflected back from the focal point of the ComCAT impinge the piezoelectric plate and induce signal currents on all ComCAT elements. If the computer-generated Fresnel zone pattern is designed such that the positive zones are represented by data 1s and negative zones, by 0s, then the switches of the array corresponding to the positive zones are closed. Since all the echoes reflected from the focal point can be regarded as being in phase when arriving at the positive zones [3], the induced current signals, after passing through the switches, are added in phase on the common lead of the switches and are output from terminal 2 of switch K. Thus, the common lead is time-sequenced to furnish AC excitation to the piezoelectric plate during transmission, and to deliver imaging information from the ComCAT when the echo signals are received.

As the pattern center is shifted from the left end to the right, one scan is accomplished and a two-dimensional B-scan image can be obtained. At the case when three-dimensional images are needed, two-dimensional beam scanning

is necessary. To do so, one just has to perform the above
scan procedure successively with the pattern center one row
lower in the HSR for each scan.

It may be worth mentioning that a ComCAT can be designed
to have its areas corresponding to positive and negative
Fresnel zones both active (called two-phase operation) [3].
In this case, the sensitivity of the ComCAT will be doubled.

DATA READ-IN SPEED

To check if the ComCAT design is suitable for real-time
imaging, an important thing is to find out if the required
data read-in speed of the ComCAT is adequate. As we have seen
from the ComCAT operation, the HSR is operated at a low data
shifting rate, while VSR at a high rate. So the speed of the
VSR is the focus of our concern.

Referring to Fig.3, let N_c be the number of memory cells
in each row and N_r be the total number of rows of the HSR
bank, respectively. When starting a scan, $N_c N_r$ bits data are
needed to establish one frame of the pattern. As described
previously, the pattern takes N_c steps when moving from left
to right to complete one scan. To support the pattern
shifting, N_r bits (one column) of data needs to be loaded via
the VSR each time the pattern moves. Thus, the data loaded
via the VSR during the time of one scan is $N_c N_r$ bits. In the
case of two-dimensional imaging, one scan corresponds to one
image frame. Let the required imaging rate of the system for
a particular application be R_f frames per second. The data
read-in speed R_r of the VSR is

$$R_r = R_f [N_c N_r + N_c N_r] = 2 R_f N_c N_r$$

In medical diagnosis, the highest required image rate is
about 40 frames per second [5]. If the desired lateral
resolution is 128 for a particular application, the number of
memory cells in each row (N_c) should at least be 128. Let the
row number N_r also be 128 to form a square active aperture of
the ComCAT. In this case, the R_r will be 1.31 X 10^6
bits/second. In practice, a shift register can be built   to
have a data shift speed less than 0.1 *μs/bit* without
difficulty.

For three-dimensional real-time imaging, higher data
read-in speed is required. Another loading scheme of the
ComCAT has been devised to meet the needs [3].

OPERATING VOLTAGE OF THE ComCAT

When ComCAT operating, the AC voltage applied to the
piezoelectric plate to provide insonification is controlled
by the switch array, requiring that the semiconductor switch
must be able to withstand this voltage.

Both theoretical calculation and experiment measurements
show that the peak AC operating voltage of ComCAT needed in
an ophthalmic inspection system is lower than 8 volts [3].

YSTEM PERFORMANCES

Since the ComCAT is capable of focusing an acoustic beam
n two lateral dimensions, like a spherical lens, the
sensitivity of the ComCAT system may be significantly better
than that of a linear-phased array system, which focuses the
sound beam in one lateral dimension, like a cylindrical lens.
This is the reason why the AC excitation for sound radiation
in an ophthalmic inspection system could be as low as 8
volts.

A detail theoretical analysis has been performed to
explore the focusing properties of the ComCAT [3,6]. The
results show that a system using the ComCAT will give much
more uniform lateral resolution over the entire view field,
more uniform sensitivity and lower sidelobe levels,
especially when the beam is deflected, than systems using
linear-phased arrays. Thus, better images can be expected.

CONCLUSION

A new acoustic imaging method involving a novel device,
ComCAT, for applications in medical diagnosis has been
discussed. The ComCAT system is inherently simple both in
construction and in operation, and is capable of electrically
performing lateral beam scanning in either one or two
dimensions. The sensitivity and resolution of the system may
be significantly better than the systems using linear-phased
array.

REFERENCES

1. A. Macovski, "Ultrasonic Imaging Using Arrays," *Proc.
   IEEE*, 67:484-495 (1979).
2. K. Wang and G. Wade, "A Scanning Focused Beam System for
   Real-Time Diagnostic Imaging," *Acoustic Holography*, N.
   Booth Ed., Plenum Press, New York, 6:213-228 (1974).
3. J. A. Sun, *Computer-Controlled Acoustic Transduction for
   Real-Time Three-Dimensional Imaging*, Ph.D. Dissertation,
   UCSB, (1989).
4. D. Dietz, S. Parks and M. Linzer, "Expanding-Aperture
   Annular Array," *Ultrasonic Imaging*, 1:56-75 (1979).
5. H. E. Karrer, "Phased Array Acoustic Imaging Systems," in
   Physics and Engineering of Medical Imaging, R. Guzzardi,
   Ed., 319-339 (1987).
6. J. A. Sun and G. Wade, "The Focusing Performance of Zone-
   Plate Transducers," *Acoustic Imaging*, H. Shimizu *et al.*,
   Ed., Plenum Press, New York, 17:551-559 (1989).

A REGULARIZATION RECONSTRUCTION ALGORITHM
OF ULTRASONIC TOMOGRAPHY FOR 3-D STRESSES
IN SLIGHTLY ORTHOTROPIC MATERIALS

Jin Shaowu, Xu Keke, Lan Congqing

Wuhan Institute of Physics
The Chinese Academy of Sciences
P.O. Box 71010, Wuhan 430071, P.R. China

1. INTRODUCTION

The non–destructive measurement of stresses in solids is an important problem in engineering. Present techniques have only successed in special case. Ultrasonic tomography is considered as a promising technique to solve this problem. However, few literatures have disscused this problem[1]–[2]. It is known that the presence of large stresses in solids causes small changes in sonic velocity(the acoustoelastic effect). This raises the possibility of using ultrasonics as a stress measurement techique. By propagating the waves at different angles of incidence, it is possible to use tomographic method to reconstruct stress tensors from the traveltimes of sound waves. For slightly orthotropic materials, the reconstruction of stress can be formulated as the solution of the system of linear equations if straight raypaths of sound wave are assumed. This equation is usually ill–posed, which leads to significant stability problems when using matrix inversion methods. Iterative reconstruction techniques such as ART and SIRT are effective methods, but they introduce artifacts. If we select the total strain energy or shear strain energy as stabilizing function, we can get the regularization solution of the equation system.

2. RECONSTRUCTION ALGORITHM

Let a coordinate system be orientated in the axes of symmetry of a slightly orthotropic stressed material. If a sound wave propagates in a cross–section parallel to one of the coordinate planes, its velocity is given by[3]:

$$v = v_0(1 + a\,\sigma_a + b\,\sigma_b + c\,\sigma_c + d\,\sigma_d + e) \tag{1}$$

Acoustical Imaging, Volume 20 Edited by Y. Wei
and B. Gu, Plenum Press, New York, 1993

Where, v_0 is the wave velocity of the material in its unstressed state; a,b,c, d,e are coefficients related to the sound wave mode, the angle of incidence and material constants; σ_a, σ_b, σ_c, σ_d are the components of stress tensor $\sigma_{ij}(i,j=1, 2,3)$, depending on the direction of the cross-section. It should be noted that (1) is based on the assumption that the square of the direction cosine of the incidence angle is much larger than any of the strains. If a sound wave traverses the object along the path l, the travel time t can be written:

$$t= \int (1/v)dl = \int (1/v_0)(1-a\sigma_a-b\sigma_b-c\sigma_c-d\sigma_d-e)dl \qquad (2)$$

This integral can be discretized to:

$$\sum(a\sigma_{aj}+b\sigma_{bj}+c\sigma_{cj}+d\sigma_{dj})l_{ij}=(1-e)L_i-v_0t_i \qquad (3)$$

$$j\in N_i , \qquad i=1,2,..,M$$

Where, l_{ij} is the path length of the ray i through the pixel j, N_i is the set of all pixels which intersect the ray, L_i is the whole length of raypath, M is the total number of rays. For simplicity, all stress components σ_{aj} to σ_{dj} in pixel are denoted by symbol σ_j, the productions of l_{ij} and a, b, c, d by a_{ij}. It is convenient to treat σ_j and a_{ij} as vetors. The equation (3) can be rewritten as follows:

$$\sum a_{ij}\sigma_j=p_i , \quad i=i,..,M , \quad j\in N_i \qquad (4)$$

where $p_i=(1-e)L_i-v_0t_i$. From (4), we can use traditional ART method to reconstruct stresses:

$$\sigma^0=0$$

$$\sigma_j^{n+1}= \sigma_j^n+ \lambda a_{ij}(p_i-\sum a_{ij}\sigma_j^n)/\sum a_{ij}^2 j\in N_i, \ j\in N_i \qquad (5)$$

where the relaxation factor λ satisfies $0<\lambda<2$.

Equation (4) contains the length of the raypath L_i which must be measured. The measurement of the raypath can be avoided if we utilize two waves with different modes propagating in the same path simultaneously. (3) can be rewritten:

$$1/Li=(1-e-\sum(a\sigma_{aj}+b\sigma_{bj}+c\sigma_{cj}+d\sigma_{dj})(l_{ij}/L_i))/(v_0t_i), \ j\in N_i$$

$1/L_i$ is the same for the two kinds of waves, so we have:

$$(1-e^I-\sum(a^I\sigma_{aj}+b^I\sigma_{bj}+c^I\sigma_{cj}+d^I\sigma_{dj})(l_{ij}/L_i))(v_0^{II}t_i^{II})$$

$$=(1-e^{II}-\sum(a^{II}\sigma_{aj}+b^{II}\sigma_{bj}+c^{II}\sigma_{cj}+d^{II}\sigma_{dj})(l_{ij}/L_i))(v_0^I t_i^I), \ j\in N_i$$

where, l_{ij}/L_i is the relative path length, which can be determined by using geometry; I,II are the superscripts referring to the two waves. This equation can be rewritten in the form:

$$\sum (a^{I} - \alpha_i a^{II})(l_{ij}/l_i)\, \sigma_j = (1-e^{I}) - \alpha_i(1-e^{II}), \quad j \in N_i \qquad (6)$$

where, $\alpha_i = (v_0^I t_i^I)/(v_0^{II} t_i^{II})$. This equation is simular to (4), therefore the same re-construction can be used.

Although ART method is convenient algorithm in practical application, it usually introduces artifacts, especially when the measurements contain the errors. In fact, both (4) and (6) can be written in the matrix equation:

$$AX=B \qquad (7)$$

where X stands for the stresses. Because of the errors of measurements, we can not assume that the right side of (7) is given with absolute accuracy. We may suppose only that we get an element B_δ satisfying the inequality

$$\| B - B_\delta \| < \delta$$

where the number δ is determined by the accuracy of the instruments. Here the norm of the space is dictated by the organization of the system of measurements. As a rule, this is either the norm in the space C (we hnow an estimate of maximal error of the measurements) or the norm in L_2 (we know the mean square error). In this case, we can not get the exact solution of the equation (4). If we select the total strain energy or shear strain energy as a stabilizing function, we get an approximate solution (regularization solution in the sense of Thihnov).

For a stressed object ,the strain energy, per unit volume, is

$$U_c = (\sigma_{11}^2 + \sigma_{22}^2 + \sigma_{33}^2 + 2(\nu+1)(\sigma_{23}^2 + \sigma_{13}^2 + \sigma_{12}^2)$$
$$-2\nu(\sigma_{22}\sigma_{33} + \sigma_{11}\sigma_{33} + \sigma_{11}\sigma_{22}))/2E \qquad (8)$$

and the shear strain (specific) energy, per unit volume, is

$$U_s = (\sigma_{11}^2 + \sigma_{22}^2 + \sigma_{33}^2 + 3(\sigma_{23}^2 + \sigma_{13}^2 + \sigma_{12}^2)$$
$$-(\sigma_{22}\sigma_{33} + \sigma_{11}\sigma_{33} + \sigma_{11}\sigma_{22}))(1+\nu)/3E \qquad (9)$$

Either (8) or (9) can be written as the following form

$$U = \sigma^{T} D \sigma$$

where, $\sigma^{T} = (\sigma_{11}, \sigma_{22}, \sigma_{33}, \sigma_{12}, \sigma_{23}, \sigma_{13})$, D is 6×6 real symmtry positive definite matrix. Therefore, the total strain energy or the total share strain energy in the object to be reconstructed may be written as follows:

$$\Phi(X) = X^{T} C X$$

where C is matrix determined by D.
Selecting Φ as stabilizing function and minimizing the following function

$$\Psi(X, \alpha) = (AX-B)^{T}(AX-B) + \alpha X^{T} C X$$

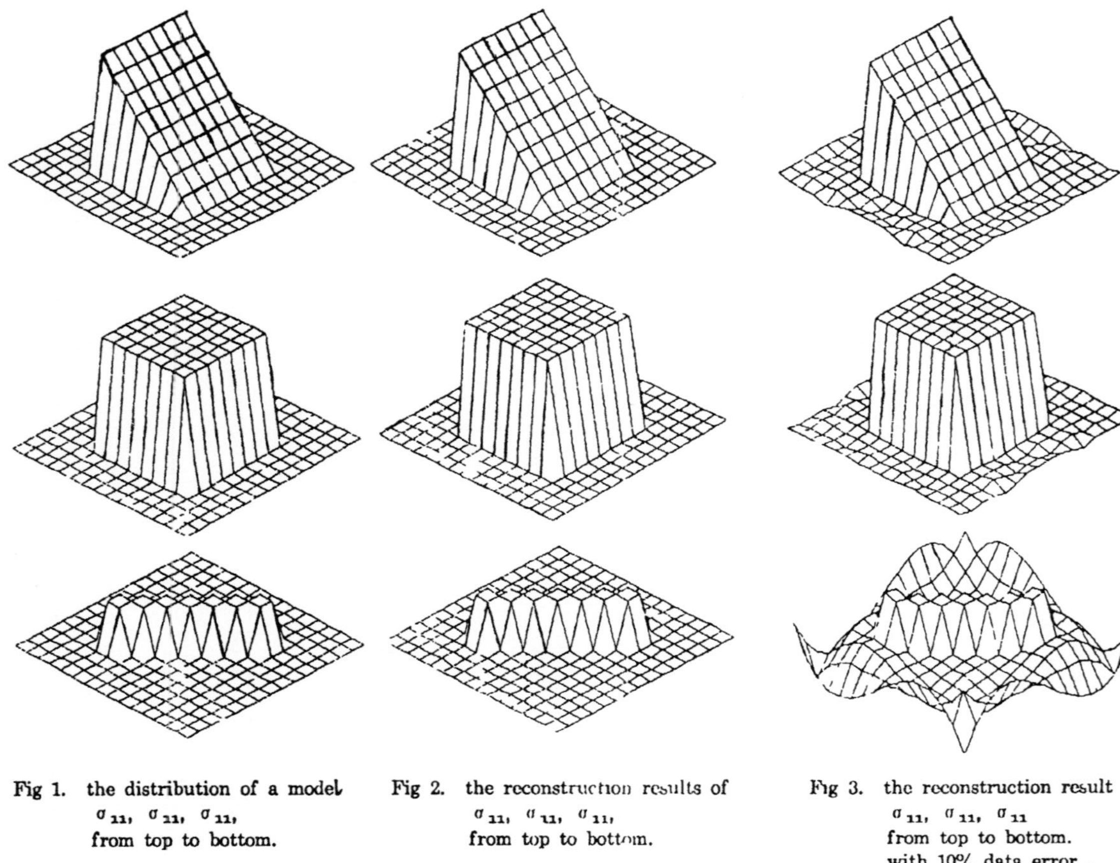

Fig 1. the distribution of a model σ_{11}, σ_{11}, σ_{11}, from top to bottom.

Fig 2. the reconstruction results of σ_{11}, σ_{11}, σ_{11}, from top to bottom.

Fig 3. the reconstruction result of σ_{11}, σ_{11}, σ_{11} from top to bottom. with 10% data error .

we can get an approximate solution[4]:

$$(A^T A + \alpha C)X = A^T B \tag{10}$$

i.e.

$$X = (A^T A + \alpha C)^{-1} A^T B \tag{11}$$

It is usually troublesome to determine the regularization parameter α. To avoid seleting α strictly, we can use the following extrapolation method:

To select the regularization paramter $\alpha_1 > \alpha_2 > ... > \alpha_n > 0$, solve the equation (10), such that

$$\| (A^T A + \alpha C)X_{i+1} - A^T B \| < \| (A^T A + \alpha C)X_i - A^T B \| , \quad i=1,..., n-1$$

the final solution can be written:

$$X = \sum \beta_i X_i , \quad \beta_i = \prod_{i \neq j} \frac{\alpha_i}{\alpha_i - \alpha_j}$$

3. SIMULATION

To demonstrate the performance of the alogrithm, a numerical model is designed. Figures 1 represents the stress distrubutions of a slightly orthotropic aluminum plate (10cm$\times$10cm) in plane stress state. Figure 2 shows the result of reconstruction without date errors. From Figure 2, we can see that the result of reconstruction agrees with the model. Figure 3 shows the result of reconstruction with 10% data errors, which is also proximate to the model. It is verified that the algorithm is stable and effective.

4. SUMMARY AND CONCLUSION

A reconstruction algorithm for stresses in solids has been presented in this paper. Using this method, we can determine three-dimensional stresses in a slightly orthotropic object by measuring the traveltimes of sound waves propagating in the cross-sections parallel to the symmetry planes of material. A numerical model is designed to test the effectiveness of our algorithm, the results of the simulation are good. Further theoretical and experimental work will be carried out in the future.

REFERENCES

[1]   Hildbrand, B. P., et al., EPRI (1977) Np-338.
[2]   Hildebrand, B.P, Mat. Eval. 39(4),383-390
[3]   Wu,K. C., The stress-acoustic relationship for off-axis ultrasonic wave in solids, to be published.

[4] M. M. LAVRENT'EV et al, ill-posed problems of mathematical physics and analysis. Translations of Mathematical Monographs, Volume 64.   American Mathematical Socity.

RECONSTRUCTING TEMPERATURE AND VELOCITY
FLOW FIELDS OF FIRE INDUCED SMOKE

G. Höfelmann

Department of Communication Engineering, Duisburg University
Bismarckstraße 90, D-4100 Duisburg 1, Germany

INTRODUCTION

In automatic fire detection it is of great importance to know the propagation phenomenon of smoke in the initial state of a fire. This information can be used to determine the optimal positions of fire detectors. Theoretical models have been made some years ago[1] but have only been partly proved by experiment. For verification, measuring of the spatial and temporal distribution of the extinction coefficient, temperature and velocity flow field is required. Due to its non-invasive nature, computer tomography (CT) is a suitable means.

The extinction coefficient has recently been reconstructed in a vertical plane by infrared computer tomography our laboratory[2] and it was found to agree with the theoretical model. Temperature and velocity flow fields can be reconstructed in principle by ultrasonic computer tomography from a set of propagation time measurements.

In the case under consideration, ultrasonics means ultrasonics in a gas phase within a frequency range of $20KHz < f < 40KHz$. Therefore the spatial resolution is reduced compared to applications in water. Serious problems in the tomographic data acquisition of a thermally driven smoke arise from stochastic fluctuations in the temperature distribution which are fast compared to the data acquisition time. The temperature distribution may change significantly between the first and the last time of flight measurement due to these fluctuations. To reduce the influence of this effect only a small number of time of flight measurements is taken (199) and all receivers inside the transmitter aperture work in parallel.

According to Fermat's principle, in an inhomogeneous medium acoustic waves propagate along curved rays which are not known a priori. The velocity flow field of the medium also causes non straight lined rays. The curved rays and the times of flight are computed with a ray linking procedure. In the first part of this paper the procedure is introduced and simulation results are shown.

An appropriate method to reconstruct temperature and velocity flow fields from a

Acoustical Imaging, Volume 20 Edited by Y. Wei
and B. Gu, Plenum Press, New York, 1993

small number of data is to combine the 'algebraic reconstruction technique' (ART) and the vector ART. A measurement result acquired with the developed experimental set-up is presented. Simulation results of a temperature and a velocity flow field are shown.

RAY TRACING IN A STATIONARY MEDIUM

The ray tracing methods described here are applied in computer tomographic imaging, hence only two dimensional problems (x,y-plane) are being examined. The propagation of a ray in an inhomogeneous medium is described by the Fermat principle, which states that the energy propagates along that path g_k which makes its time of flight t_k a minimum

$$t_k = \int_{g_k} \frac{ds}{v} = \frac{1}{n_0 \cdot v_0} \cdot \int_{g_k} n\,ds = \frac{\sqrt{T_0}}{v_0} \cdot \int_{g_k} \frac{ds}{\sqrt{T}} \rightarrow \min \tag{1}$$

where $v_0 = 331.6\,\text{m/s}$, $n_0 = 1$, $ds = (dx^2 + dy^2)^{1/2}$ is a path element, n is the refractive index, $T_0 = 293.1°\text{K}$ and v, the velocity of sound. The variational integral eq.(1) is solved by the Euler equations and yields a scalar non-linear differential equation, where $y(x)$ describes the desired ray path

$$\frac{d^2 y(x)}{dx^2} = \frac{1}{n(x,y)} \cdot \left[\frac{\partial n}{\partial y}(x,y) - \frac{\partial n}{\partial x}(x,y) \cdot \frac{dy(x)}{dx} \right] \cdot \left[1 + \left(\frac{dy(x)}{dx} \right)^2 \right] \tag{2}$$

Starting from the well-known Snell's law of refraction, the ray paths may be described by a system of three nonlinear differential equations[3]

$$\frac{dx}{ds}(s) = \cos(\alpha(s)), \qquad\qquad \frac{dy}{ds}(s) = \sin(\alpha(s))$$

$$\frac{d\alpha}{ds}(s) = \frac{1}{n(x(s),y(s))} \cdot \left[\cos(\alpha(s)) \cdot \frac{\partial n(x(s),y(s))}{\partial y} - \sin(\alpha(s)) \cdot \frac{\partial n(x(s),y(s))}{\partial x} \right] \tag{3}$$

where s represents the arc length. Both ray tracing equations are suitable for calculating propagation paths.

RAY LINKING IN A STATIONARY MEDIUM

The numerical solution of the ray tracing equations requires initial-values. The common problem in computer tomography is to find the path linking a receiver with a transmitter. This is a boundary-value problem, called ray linking. Nonlinear boundary-value problems can be transformed into initial value problems and solved by iterative procedures like the shooting method or the Newton-Raphson method. The iteration is stopped when the endpoint of the ray lies in a given circle around the receiver position.

Another iterative method is to estimate the initial-values from the boundary-values by means of eq.(2) and the assumptions that the ray has a parabolic shape and is only slightly bended[4]. This method results in the best initial guess of the first launch angle for the iteration process. The method is briefly discussed.

The transmitter is positioned at $(x_{Tr}, y_{Tr}) = (0,0)$ and the receiver at $(x_R, y_R) = (x_R, 0)$, which can always be achieved by a coordinate transform. A quantity $m(x)$ is defined as $m(x) = dy(x)/dx$. Now eq.(2) is written in a general form $d^2y(x)/dx^2 = F(x,y(x),m(x))$. Suppose there is an estimator $y^*(x)$ which converges to $y(x)$ with

$$\frac{d^2 y^*(x)}{dx^2} = F(x,y^*(x),m^*(x)) =: f(x) \tag{4}$$

where the second derivative is only a function of x. Eq.(4) is integrated twice. Now eq.(2) is inserted and the quadratic therm $m^{*2}(x)$ is neglected. Then the refractive index and its partial derivatives are expanded in Taylor series in the y direction. With the given boundary-values and the abbreviations according to eq.(5) in which $y^*(x)$ is assumed to be a parabola,

$$N_x(x,y) := \frac{1}{n(x,y)} \cdot \frac{\partial n(x,y)}{\partial x} \ , \qquad N_y(x,y) := \frac{1}{n(x,y)} \cdot \frac{\partial n(x,y)}{\partial y} \qquad \text{and}$$

$$I(x) := (x_R - x) \cdot [\, x \cdot (x_R - x) \cdot N_y^2(x,0) + (x_R - 2x) \cdot N_x(x,0)\,] \tag{5}$$

it follows the estimate of the initial value, or the launch angle $\dfrac{dy^{*(1)}}{dx}(x_{tr}=0)$ for the first iteration in the calculation of the ray path:

$$\frac{dy^{*(1)}}{dx}(x_{tr}=0) = \frac{1}{x_R} \int_0^{x_R} (x_R - x) \cdot N_y(x,0)\, dx \cdot \left[1 - \frac{1}{x_R^2} \int_0^{x_R} I(x)\, dx \right]^{-1} \tag{6}.$$

For smooth progress of a refractive index distribution eq.(6) yields a very good estimation of the launch angle for the first iteration step. The ray linking problem is calculated by first estimating the launch angle according to eq.(6). and then solving eq.(3) by means of the shooting method. Eq.(3) is preferred because it converges even if dy/dx approaches infinity.

RAY LINKING IN A NONSTATIONARY MEDIUM

If the medium is nonstationary, its movement is described by the velocity flow field $\vec{v}_{fl}$. The influence of $\vec{v}_{fl}$ is taken into account by a time variant coordinate transform. An inertial system (x,y) is given and a coordinate system (x',y') is defined that moves at the flow velocity of the medium in the position vector $P(t)$ at the head of the wave front along the ray g_k, fig.1. For any arbitrary ray point $P(t)$ we have

$$\vec{r'}(t) = \vec{r}(t) - \vec{r}_L(t); \qquad \vec{r}(t) = (x(t),y(t))$$
$$\vec{v'}(t) = \frac{d\vec{r'}(t)}{dt} = \vec{v}(t) - \vec{v}_L(t) \tag{7}.$$
$$t' = t$$

Since $\vec{v}_{fl}(P(t)) = \vec{v}_{fl}(x(t),y(t)) = \vec{v}_L(x(t),y(t))$, it follows that

$$\vec{r_L}(t) = \int_0^t \vec{v_L}(x(\tau),y(\tau))\,d\tau = \int_0^t \vec{v_{fl}}\,(P(\tau))\,d\tau \tag{8}.$$

Displacement vector $\vec{r_L}(t)$ is calculated from the velocity flow field $\vec{v_{fl}}(P(t))$ although $\vec{v_{fl}}(P(t))$ is not known a priori because $P(t)$ depends on $\vec{v_{fl}}$.

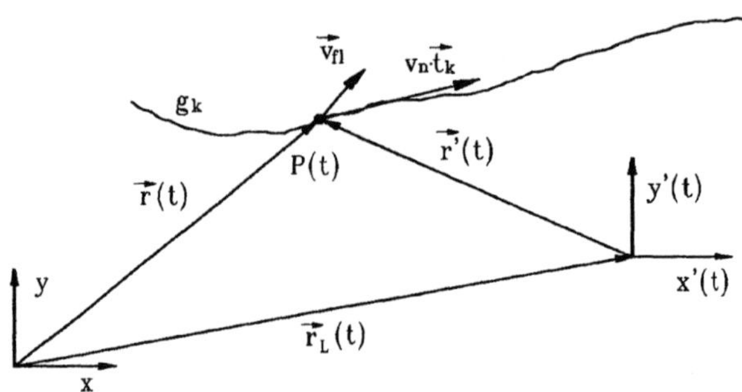

Figure 1. Coordinate transform relating the inertial system (x,y) and the moving system (x',y').

The sonic velocity v in a nonstationary medium is given by $v \cong v_n + \vec{v_{fl}} \cdot \vec{t_k}$, where $v_n >> |\vec{v_{fl}}|$ is that part of the velocity which only depends on n and $\vec{t_k}$ is the tangent vector of the k-th ray. The time of flight in the nonstationary medium is

$$t_k = \int_{g_k} \frac{ds}{v} \cong \int_{g_k} \frac{ds}{v_n + \vec{v_{fl}} \cdot \vec{t_k}} \cong \int_{g_k} \frac{ds}{v_n} - \int_{g_k} \frac{\vec{v_{fl}} \cdot \vec{t_k}}{v_n^2}\,ds \cong \int_{g_k} \frac{ds}{v_n} \tag{9}.$$

Since the physical quantities are indepentend of the chosen coordinate system it follows

$$t_k \cong \int_{g_k} \frac{ds}{v_n} = \frac{1}{n_0 v_0} \cdot \int_{g_k} n(\vec{r}) \cdot v(\vec{r}) \cdot dt = \frac{1}{n_0 v_0} \cdot \int_{g'_k} n(\vec{r'} + \vec{r}) \cdot v'(\vec{r'}) \, dt' \tag{10}.$$

Minimizing the right integral yields the same differential equations as minimizing eq.(1). Therefore it makes no difference if one solves the ray tracing equation for a stationary medium in the inertial system or for a nonstationary medium in the (x',y') system.

The coordinate transform is implemented in the Runge-Kutta algorithm which solves the ray linking equations (eq.(3)). The independent variable in the difference equations of this algorithm is updated stepwise to consider the displacement vector $\vec{r_L}(t_i)$ due to the flow field in every single step $s(t_i)$ along the ray ($i=1,(1),I\leq256$).

SIMULATION RESULTS OF THE RAY LINKING METHOD

Some simulation results achieved with the introduced ray linking method are shown here. First the velocity flow field was defined to be zero and the refractive index was a 'Maxwell's fish-eye'. Since the 'fish-eye' ray paths are described by a family of circles, the performance of the algorithm derives from the ability of a ray to form a circle after one revolution (fig.2). Fig.3 shows the linked ray in the refractive index field designed as a scattering lense. The refractive index decreases from $n = 1$ in the middle to $n = 0.8$. The circles are lines of constant refractive index, the velocity flow field is defined to be zero. The plotted area is normalized to a unit square. Fig.4 depicts the calculated ray, linking the transmitter Tr and the receiver R in an inhomogeneous flow field $\vec{v}_{fl} = (0, v_0 \cdot (x+1)/2)$. The refractive index is defined to be 1 at all points. The linked ray for the superposition of the refractive index defined for fig.3 and the flow field from fig.4 is presented in fig.5. To compute the ray path 10 iterations are required.

This ray linking algorithm is used in the simulations of the computer tomographic reconstruction to compute the ray paths and the time of flight data.

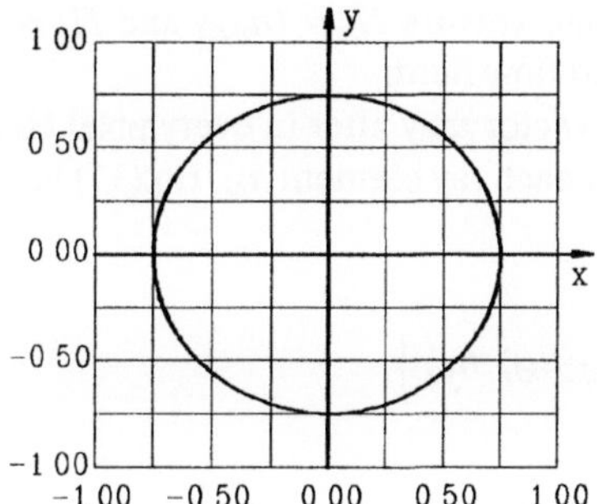

Figure 2. Traced ray in a 'fish-eye' shaped refractive index field.

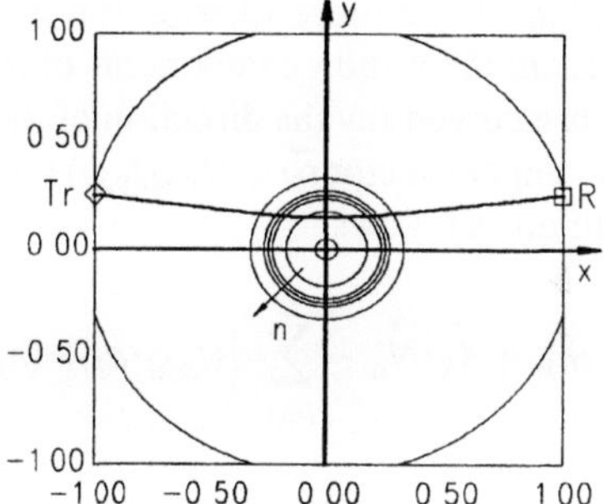

Figure 3: Ray linking in a diverging lense like refractive index field.

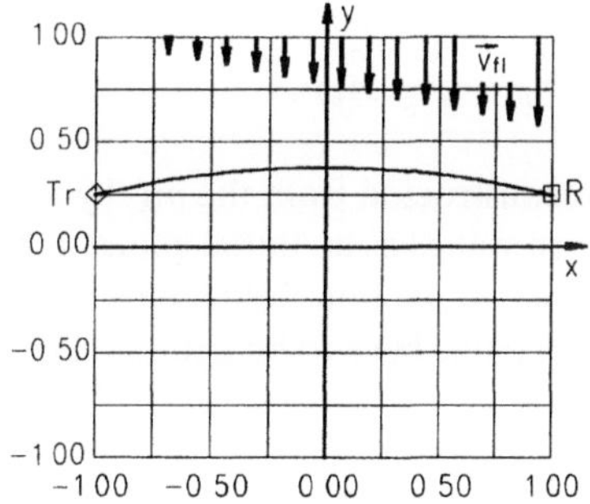

Figure 4. Ray linking in an inhomogeneous velocity flow field

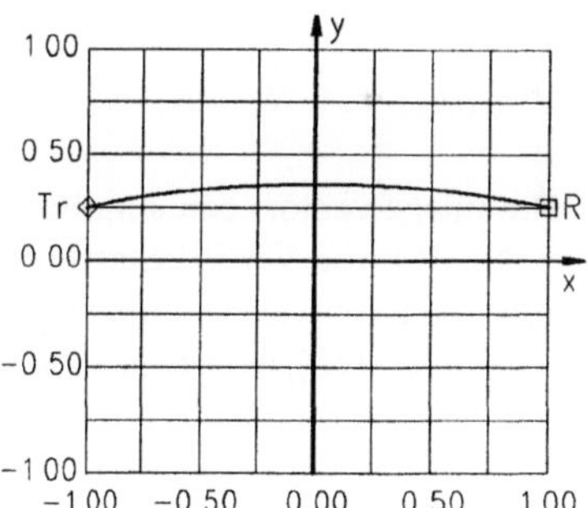

Figure 5. Ralinking in a refracting and nonstationary medium.

COMPUTER TOMOGRAPHIC RECONSTRUCTION

Computer tomography is an image processing method for aquisition and reconstruction of spatially inhomogeneously distributed physical parameters in a cross sectional plane of a 3-dimensional object, e.g. a cloud of smoke. The values of the parameters of the object cross section are called the object functions. The object functions

$\vec{v_{fl}}(\vec{r})$ and $n(\vec{r})$ are reconstructed from a set of time of flight (TOF) data t_k, measured along the ray paths g_k, $k = 1,(1),\ K$, with K being the total number of rays. Appropriate reconstruction methods for a small number of measurement data are the algebraic reconstruction technique and the vector algebraic reconstruction technique.

For the computational implementation a square grid of $64 \times 64 = L$ pixels is superimposed to the cross-sectional plane and the physical quantities are sampled on this grid. If g_k is a straight line ray, it follows the discrete form of eq.(9)

$$m_k = \vec{A}_k \cdot \vec{N}_n - (\vec{A}_k \cdot \vec{N}_x, \vec{A}_k \cdot \vec{N}_y) \cdot \vec{t}_k \tag{11}.$$

m_k represents the calculated or measured time of flight along the k-th ray. The row vector $\vec{A}_k = (a_{kl})$ represents the discrete k-th ray path, where a_{kl} is the length of the k-th ray in the l-th pixel , $l = 1,(1),L$. The image vector (column vector) $\vec{N}_n = (n_{n,l})$ represents the sampled temperature distribution and the image vectors $\vec{N}_x = (n_{x,k})$ and $\vec{N}_y = (n_{y,k})$ represent the x and y components of the sampled flow field.

If g_k is a curved ray the direction of the tangent vector may alter in every pixel. Hence a lokal tangent vector $\vec{t}_{kl} = (t_{x,kl}, t_{y,kl})$ is defined for each ray element a_{kl}. Eq.(11) has to be modified as follows:

$$m_k = \vec{A}_k \cdot \vec{N}_n - \sum_{l=1}^{L} \left\{ (t_{x,kl} \cdot a_{kl}) \cdot (n_{x,l}) + (t_{y,kl} \cdot a_{kl}) \cdot n_{y,l} \right\} \tag{12}.$$

$t_{x,kl} \cdot a_{kl}$ is the x component of a ray element a_{kl} and $t_{y,kl} \cdot a_{kl}$ is its y component. By defining the vectors $\vec{A}_{x,k} = (a_{x,kl}) = (t_{x,kl} \cdot a_{kl})$ and $\vec{A}_{y,k} = (a_{y,kl}) = (t_{y,kl} \cdot a_{kl})$ it follows eq.(13) and therefrom the matrix form eq.(14) which includes all K rays.

$$m_k = \vec{A}_k \cdot \vec{N}_n - \vec{A}_{x,k} \cdot \vec{N}_x + \vec{A}_{y,k} \cdot \vec{N}_y \tag{13}$$

$$\overset{\leftrightarrow}{M} = (m_k) = \overset{\leftrightarrow}{A} \cdot \vec{N}_n - \overset{\leftrightarrow}{A}_x \cdot \vec{N}_x + \overset{\leftrightarrow}{A}_y \cdot \vec{N}_y = \overset{\leftrightarrow}{A} \cdot \vec{N}_n - (\overset{\leftrightarrow}{A}_x, \overset{\leftrightarrow}{A}_y) \cdot \overset{\leftrightarrow}{N} \tag{14}.$$

The image vectors can simultaneously be reconstructed from the m_k by a combination of ART and VART. Starting with the mean values as an initial guess in the first iteration step the image vectors of the temperature and the flow field are reconstructed by eq.(15) to eq.(18), where any two successive ray paths have to be anti-parallel.

$$\vec{N}_n^{(1)} = <\vec{N}_n> ; \quad (\vec{N}_x^{(1)}, \vec{N}_y^{(1)}) = (<\vec{N}_x>, <\vec{N}_y>) \tag{15}$$

$$d_k^{(i)} := \overset{\leftrightarrow}{A} \cdot \vec{N}_n - \vec{A}_{x,k} \cdot \vec{N}_x^{(i)} + \vec{A}_{y,k} \cdot \vec{N}_y^{(i)} - m_k \tag{16}$$

$$\overset{\leftrightarrow}{N}^{(i+1)} = \overset{\leftrightarrow}{N}^{(i)} - \frac{(\vec{A}_{x,k}^{T}, \vec{A}_{y,k}^{T})}{|\vec{A}_k| \cdot |\vec{A}_k^{T}|} \cdot d_k^{(i)} \cdot \frac{1}{2} \tag{17}$$

$$\vec{N}_n^{(i+1)} = \vec{N}_n^{(i)} - \frac{\vec{A}_k^{T}}{|\vec{A}_k| \cdot |\vec{A}_k^{T}|} \cdot d_k^{(i)} \cdot \frac{1}{2} \cdot (-1)^i \tag{18}$$

The superscript *(i)* indicates the *i*-th iteration, the algorithm stops after 5 to 7 iterations. Since the ray paths and the object functions are not known for a set of measured TOF data, the reconstruction starts with the assumption that straight-lined rays connect the transmitter/receiver pairs. Now two types of reconstruction procedures can be created. The first one continues with straight-lined ray paths in every iteration. The second one calculates curved rays in the *i*-th iteration on the basis of the image vectors from the *(i-1)*-th iteration and takes these ray paths to update the image vectors for the *i*-th iteration and so on. The computational time of the second method is considerably higher but the computational results are only slightly better. Therefore the results presented in the following are calculated by means of straight lines. The time of flight data are calculated by the introduced ray linking algorithm.

EXPERIMENTAL SET-UP AND MEASUREMENT RESULT

The experimental set-up consists of two main parts: a parallel computer with 11 transputers where the reconstruction algorithm, eq. (15) to (18), is implemented plus the measuring device. The cross sectional horizontal measurement plane is surrounded by a rectangular metallic frame of 2x2m on which 44 ultrasonic transducers are mounted.

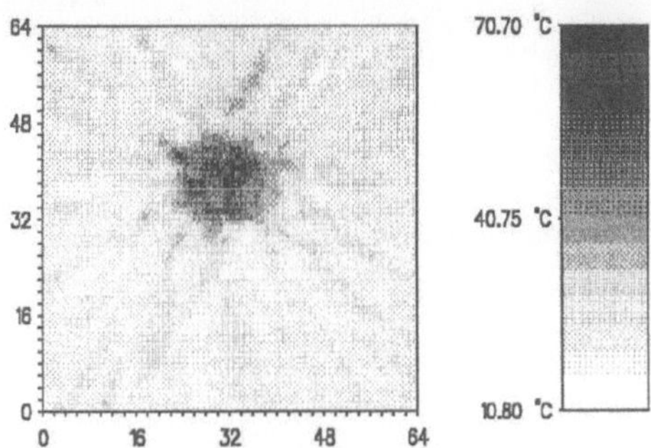

Figure 6. Measured temperature distribution above a hotplate. 4096 pixels are reconstructed from 199 time of flight data.

Below the measurement plane is a hotplate with a few pieces of wood to produce heat and smoke. It takes 2.0 seconds to measure 199 TOF data and to reconstruct one image of 64x64 pixels of the temperature distribution. Due to the small number of measured data it was not possible to reconstruct the velocity flow field. Fig.6 shows a measured and reconstructed image of the temperature distribution. An additional thermometer was installed above the hotplate and the measured temperature was 79°C.

COMPUTATIONAL RESULTS

For simulation, analytically given images have been generated to provide an approximate representation of the temperature and velocity flow field in an enclosure in a vertical plane above the hotplate. A generated temperature distribution is depicted in fig.7 and its reconstruction in fig.8. In fig.9 the generated velocity flow field is shown, the reconstructed image follows in fig.10. Both reconstructed images have been computed simultaneously according to eq.(15) to (18) from the same set of 1200 TOF data. The reconstructed images agree with the originals, only the magnitude of the reconstructed flow field differs clearly. But one has to keep in mind that 3x64x64 pixels have been reconstructed from only 1200 data.

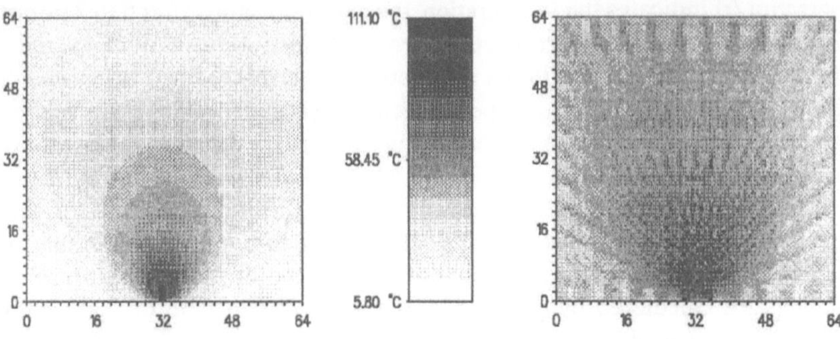

Figure 7. Generated temperature distribution.

Figure 8. Reconstructed temperature distribution.

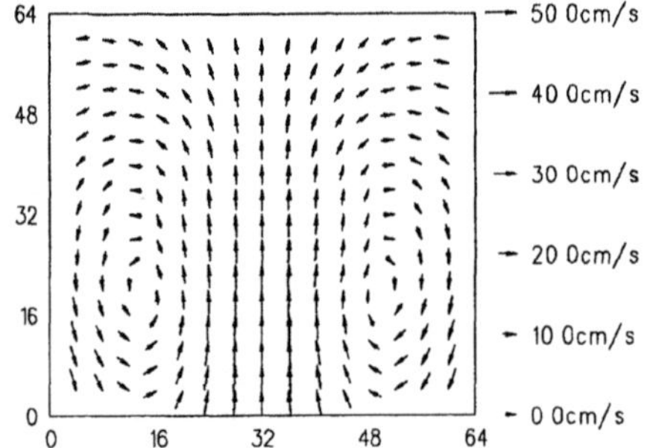

Figure 9. Generated velocity flow field, (1cm = 0.01m)

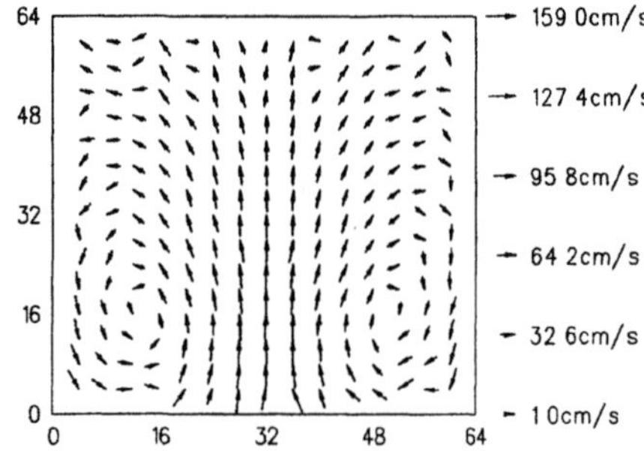

Figure 10. Reconstructed velocity flow field.

SUMMARY

A ray linking procedure was developed that traces rays in a nonstationay medium even if the velocity flow is in the same order of magnidute as the sonic velocity. The measurement result of a spatial temperature distribution in a cross sectional plane of an aerosol cloud was depicted. Simulation results of simultaneously calculated temperature distributions and flow fields have been shown.

REFERENCES

1. H.R.Baum, R.G.Rehm and G.W.Mulholland: Prediction of heat and smoke movement in enclosure fires. *Fire Safety J., Vol.6*, pp.193-201 (1983).
2. P.Beckord and G.Höfelmann: On-line measurement of a buoyant plume produced by a smoldering fire in an enclosure. *J. of Aerosol Science. 21/1* (1990).
3. A.C.Kak and A.H.Andersen: Digital ray tracing in two-dimensional refractive fields. *JASA 72 5*, pp. 1593-1606, Nov (1982).
4. B.Siemund, W.Wegmann, G.Hofelmann and P.Beckord: Iteratives Verfahren zur Schallwegberechnung und Anfangswertbestimmung. *Acustica, Vol.67* (1989).
5. S.A.Johnson, J.F.Greenleaf and M.Tanaka: Reconstructing three-dimensional temperature and fluid velocity vector fields from acoustic transmission measurements. *ISA Transactions Vol.16, 3* (1977).

ARTIFICIAL INTELLIGENCE AS AN APPROACH
TO IMPROVE ULTRASONIC LOG SCANNING

Wei Han, Rolf Birkeland

The Norwegian Institute of Wood Technology
N-0314 Oslo 3, Norway

ABSTRACT. Scanning of logs for optimum material utilization becomes more and more important in wood industry Ultrasound tomography seems to be a good solution to log scanning because of its capability of scrutinizing a log both externally and internally or the external geometry and the internal defects However, the substantial computations and the complex tomogram translation make the industrial application difficult This paper presents principles of applying artificial intelligence (AI) to ultrasonic scanning for the purpose of improving scanning efficiency The presentation involves the integration of AI in signal acquisition and treatment, and defect computation The AI functions involved include learning, knowledge based reasoning, fuzziness handling and neural network application An intelligent ultrasonic system is introduced to illustrate the principles discussed The results given by the illustrating system indicate that the combination of ultrasound and AI opens up a promising future for a practical log scanning solution

KEYWORDS. Ultrasound, AI log scanning

INTRODUCTION

Scanning of logs is a process of allocating, identifying and dimensioning defects in logs to provide information for on-line nondestructive evaluation (NDE) of the logs Ultrasonic method used to scan logs generally has the advantages of being cheap to implement, safe to human health and easy to understand

However there are several problems with ultrasonic scanning of logs They are from three aspects, i e wood, ultrasound and scanning speed Wood is inhomogeneous and heterogeneous in structure, and has varieties of defects Wood has a property of high acoustic attenuation Besides, shape of logs can be very irregular to cause scanning difficulties The problems with ultrasound lie in the need for coupling, and in low signal level and complex responses on logs The requirement on high scanning speed in on-line operations often contradictes the requirement on scanning resolution, which causes extra difficulty in the system implementation

To deal with these problems, techniques of artificial intelligence (AI) were studied. Generally, AI functions include:

a. learning i.e. the ability of the system to accumulate experience and to organize this knowledge so that it can be a basis for improved and faster reasoning;
b. knowledge-based reasoning, based on the use of available expertise, earlier experience and commonly accepted rules - very often taking a form of expert systems;
c. the ability to handle imprecision, incompleteness and randomness of information - by fuzzy set theory and neural networks;
d. the ability to adapt to new situations through task shifting and strategy modification;
e. distributed information-processing.

AI were felt to have the potentials of improving scanning efficiency and enabling practically good scanning accuracy.

AN ULTRASONIC LOG SCANNING SYSTEM

In ultrasonic non-destructive evaluation (NDE), the three most common techniques are through-transmission, pulse-echoing and mode conversion. To utilize the maximum potential of ultrasound in examining log quality, the combination of the three techniques was used (Han & Birkeland 1992). The pulse-echoing technique was used to obtain the log profile in a cross-section of interest, while the through-transmission technique to detect internal defects. The mode conversion, or the 'grain-tracing' technique connected with log scanning, was taken as short cut to find knots that extend to the log surface. A log scanning system was implemented as shown in Figure 1. To scan a log, the transducers move back and forth lengthwise over it while it is rotating in coupling water.

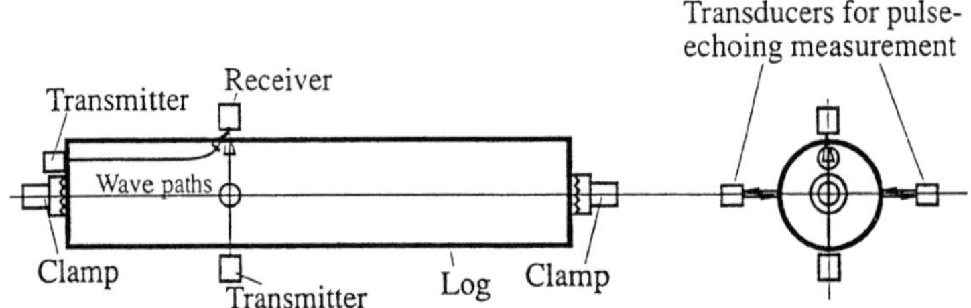

Figure 1. An arrangement for ultrasonic scanning of logs (Han & Birkeland 1992)

KNOWLEDGE AND DATA STRUCTURE

On its sources, knowledge that may be useful for scanning wood can be roughly divided into two categories:

a. Generic knowledge: this category includes knowledge from specialized publications, textbook, and research reports and knowledge from experts of related problems.
b. Real-time knowledge: this refers to knowledge that is learnt by the scanning system from on-line scanning operations. Real-time knowledge concerning wood includes that obtained from the normal operations of scanning.

This classification takes into account both expert system function and learning function of AI. Generic knowledge may be used as reference to check real-time knowledge or to support reasoning, when real time knowledge is absent or incomplete.

Knowledge can also be classified according to the properties of wood. It concerns:

a. knowledge about silvicultural and structural properties of wood and wood defects;

b   knowledge about physical reactions of wood and wood defects on ultrasound,

c   knowledge about end uses of the wood material, including quality criteria and grading
rules

The knowledge categories a and b include both the generic and the real-time knowledge
while category c includes only the generic type

HOW TO LEARN

Learning is a process of obtaining knowledge of or skill in a given subject by study,
instruction, practice or experience The purpose of learning is to achieve a potential that makes
a given performance possible There is a general principle in learning It is that knowledge
obtained from learning is always about a part and should be representative for a wider
population Concerning the performance of wood scanning, the subject is about wood quality -
conditions of sound wood and defects Therefore, a learning function in a wood scanning
system is defined to be obtaining knowledge about a given volume of testpieces from training
on or experiencing a part of the volume The purpose of this is to efficiently and effectively
examine the rest of the volume This definition embodies two possible learning approaches, i e
the approach through training and the approach through experiencing

The learning approaches can be further classified into the following

a) training - learning from samples

off-line training acquiring knowledge, prior to normal scanning operations, from
deliberately selected samples which should be very representative,

on-line training acquiring knowledge from samples that are on-line selected according
to pre-defined models

b) experiencing - learning from normal scanning operations - on-line only

model based   accumulating experience through parameter modification on pre-defined
knowledge structure,

self organizing accumulating experience iteratively by the reasoning mechanism itself

For wood scanning, there is a practical principle for training (Han 1991) The idea is that a
learning procedure is taken to a part of each logpiece in order to measure the properties that
have significant similarity or regular relations with the rest of the logpiece A new course of
learning on each log provides a profit of piecewise adjustment, or that piece-to piece variation
is taken into consideration Many properties in an individual woodpiece can approximately
taken as being uniform over its length, such as moisture content, solid wood density, ring width
and even knot growing pattern

The common properties that can easily be accessed are those about sound wood The
defects, though having varieties of forms, often show some definite relations to the sound
wood Those relations can be used to form knowledge models Studies in non-destructive
evaluation (NDE) of wood tend to present the relations

Neural networks are efficient approaches to implement learning functions in wood scanning
Neural networks mimic the way the human brain is handling knowledge The learning
capability of the networks is closely related to the associativeness and the distributiveness of
the networks memory Some networks, such as the so-called Kohonen network which is based
on the competitive filter associative memory (Caudill & Butler 1990), have self-organizing
memory architecture which enables the system to learn by itself or automatically Such neural
networks can obtain knowledge both through training and through experiencing

SIGNAL TREATMENT ASSISTED BY LEARNING AND KNOWLEDGE

The main sensing method of the above-given system is through-transmission, i e to transmit
waves across the log diameter To penetrate logs of 20 cm in diameter, which are commonly
seen in Norwegian sawmills, waves of as low as 100-300 kHz in frequency have to be used

Waves as such tend to be disturbed severely by the test objects and a lot of noise is produced in the received waveforms. The main sources of noise are wave leakage through coupling water, scattering by wood structure and water disturbance. They tend to generate extra peaks which may have magnitude levels similar to or higher than those of peaks of interest. Extra knowledge has to be applied tell useful peaks from them.

Use of the knowledge about the wave velocity through sound wood and defects, knowledge about sizes of largest possible defects and log diameter measure may lead to prediction of a range where the useful peaks may highly possibly appear. Most of the noise was proved to be out side such a range. For each new diameter measure, a new range is to be predicted, which is a concept of active framing. The knowledge pieces can be from a knowledge base or from on-line learning procedures. Piecewise on-line learning is preferred for a better recognition. It is because the effect of piece-to-piece variation can be eliminated in the recognition process.

HOW TO CONDUCT REASONING

To reason is to think and to draw conclusions. In conventional way of reasoning, judgement is taken on yes-no information and conclusion is also formulated in yes-no forms. Mathematically, 1 is used to represent 'true' and 0 is used to represent 'false'. Accordingly the conventional computers were built.. Such a way of reasoning has to base itself on completely and precisely formulated information. It does not work, whereas, when the problem can not be precisely formulated. To cope with this, the fuzzy sets theory was developed (Zadah 1965). Information pieces are formulated with degrees of belief, or truth values taking (0, 1), inclusive. A property is described by a set of truth values, a fuzzy set or a membership function, corresponding to all possibilities of the property. Reasoning on so formulated information is based on the operations of related fuzzy sets. The fuzzy sets theory, unfortunately, is not yet a well-developed tool in realizing intelligence. To implement sophisticated systems, it can not easily and effectively be used. Neural networks are believed to be better tools in this aspect.

The information involved in reasoning processes can be presented by a series of numerical rules and data. To use the rules and data, a reasoning mechanism has to be implemented. It is the so-called concept, rule-based system. An expert system, which possesses a knowledge base and an inference engine as its main components, is such a system. A well designed expert system can conduct reasoning on incomplete and imprecise information in real time. This is the main feature of the second-generation expert systems.

The information can also be presented in the form of neural network. If so, the reasoning mechanism is the network itself.

Reasoning tasks in wood scanning are mainly connected with detecting and identifying wood defects.

IDENTIFICATION OF DEFECTS BY A RULE-BASED SYSTEM

Identification wood defects can be performed by a procedure similar to that policemen identify criminals. The procedure is illustrated in Figure 2. Features of different defects are formulated in wave patterns. Because of the complexity of wood defects, it is complex and difficult to construct a unique wave pattern for each defect type with on-line acquired signals. A practical way we found was to classify defects into groups and construct a unique pattern for each group. To identify a defect, two steps of reasoning are needed. The first step is taken to identify the group a wood defect belongs to. A pattern recognition procedure is imposed on each acquired waveform. Then, the second step is taken to tell of which particular type the defect is. Exclusion of the other identities in the group is accomplished by using extra rules and data. The reasoning mechanism of this identification approach is a hierarchy as shown in Figure 3.

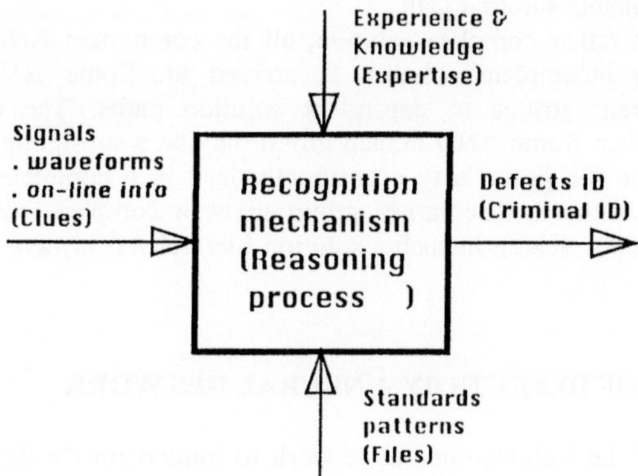

Figure 2. A procedure of identifying wood defects vs. that of policemen identifying a criminal.

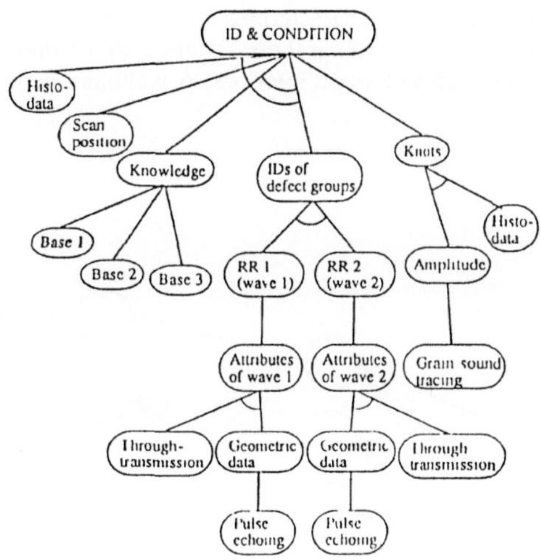

Figure 3. A rule-based reasoning hierarchy for log scanning (Han & Birkeland 1992).

This is a rule-based system. Identification of defects is based on both the on-line acquired information and off-line given knowledge. The hierarchy takes a form of an OR/AND tree. In such a tree, the root is formed by the descriptions of the log interiors and leaves are information sources, i.e. the different sensing arrangements and knowledge bases. The connections between the three tasks and information sources are OR- and AND-branches, which are formed by different intermediate states of information.

An OR-branch leads to an independent solution path. An AND-branch shows a dependent-solution path, or incomplete solution path.

In order to reach a rather complete solution, all the component AND-branches must join force. Otherwise, an incomplete solution is arrived at. Some AND-branches may be components in different groups of dependent solution paths. The different groups are alternatives to each other. Some AND-branch group may be a sub-group of another one. This relation means that the sub-group may occasionally lead to a complete solution. If not, the larger group should be used. The larger group gives a complete solution by employing additional AND-branches. Search in such a solution hierarchy is always guided by IF-THEN-ELSE rules.

IDENTIFICATION OF DEFECTS BY A NEURAL NETWORK

We use the idea of the Kohonen neural network to implement the defect-detection system. Kohonen neural network is a network using competitive filter associative memory. It consists of three layers, an input layer, an intermediate competitive layer and an output layer. Each node, or neurode in terms of neural network theory, in the input layer corresponds to an input component or variable, while each neurode in the output layer presents a pattern or class. The intermediate layer stores the information each training procedure or experienced sample provides and plays the role of a reasoning mechanism.

Well known, wood defects can be characterized by acoustic velocity, acoustic attenuation coefficient, and frequency response. Taking these as inputs of a Kohonen network, the outputs of the network should be identities of wood interiors. A Kohonen network as such is shown in Figure 4.

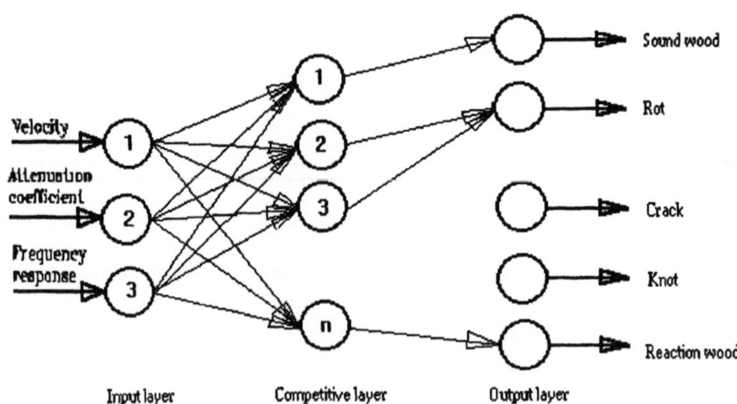

Figure 4. A Kohonen neural network used for ultrasonic identification of lumber defects.

The intermediate layer stores information about different wood interiors through weight vectors of its neurodes. The weight vectors are determined in a training procedure. When an input set is fed to the net during the normal operation, the level of correlation between the input set and each weight vector is computed. The neurode that provides the highest correlation wins the competition and triggers the defect class it represents. The output layer presents the result.

This network has an important property concerning relations between its effectiveness, efficiency and size. The more samples are used in training, the more neurodes are constructed in the intermediate layer and therefore the larger is the size of the network. As a result, the network becomes more effective in making a right decision but spends more time in doing so. Therefore, it is important that the samples used in training the network be as representative and few as possible.

To release the strictness of being representative, the network should be added with a feedback loop running from the output side to the input side. This feedback loop allows the net to accumulate its experience and improve its performance. Then the competitive layer has to be modified or expanded.

Simulation of the Neural Network

The rule-based system as given in the previous section was simulated and presented in Tables 3 and 4 in the authors' another paper (Han & Birkeland 1992). Each of the tables corresponded to a different cross-section of a common logpiece. They demonstrated a fuzzy set approach for ultrasonic identification of log defects.

A	B	C	D	E	F	G	H	I	J	K	L	M	N	O
Training sets		Length	Weight vectors		Testing sets		Length	Normalization		Dot products			Results	Results
v0(j)	a0(j)	L(v0 a0)	v0 j	a (j)	v	a(i) (inputs)	L(v a)	v (i)	a (i)	(v0 1) a0 (1))	(v0 (2) a0 (2))	(v0 3 a0 3)	fr neu a network	fr X ray tomogr
1 0267	1 7767	2 05197	0 5003	0 8658	0 99	1 72	1 984565	0 49885	0 86669	0 999999	0 997775	0 984403		
0 8500	1 7350	1 93203	0 4400	0 8980	1 09	1 72	2 036296	0 53529	0 84467	0 999165	0 994003	0 99 999		
1 5050	1 7900	2 33862	0 6435	0 7654	1 25	1 88	2 257632	0 55368	0 83273	0 998029	0 991402	0 993696		
					0 89	1 71	1 927745	0 46168	0 88705	0 999028	0 999704	0 976065	t knot	knot
					0 86	1 84	2 031059	0 42342	0 90593	0 996239	0 999832	0 965901	t knot	knot
					0 89	1 73	1 945508	0 45746	0 88923	0 998808	0 999808	0 975022	t knot	
					0 98	1 95	2 182407	0 44905	0 89351	0 998302	0 999948	0 972881	t knot	knot
					0 98	1 79	2 040711	0 48022	0 87715	0 999734	0 998971	0 980421		
Sound wood samples														
0 93	76				0 93	1 46	1 731040	0 53725	0 84342	0 999067	0 993776	0 991308		
0 93	1 62				1 24	1 98	2 336236	0 53077	0 84752	0 999369	0 994601	0 990270		
1 16	1 95				0 84	1 66	1 860430	0 45151	0 89227	0 998459	0 999917	0 973515	t knot	
1 0267	1 7767 (average)				0 79	2 10	2 243680	0 35210	0 93596	0 986555	0 995421	0 942986	t knot	
Tilting kno samples					1 21	1,91	2 261017	0 53516	0 84475	0 999 71	0 994049	0 990979		knot
0 83	1 68				1 21	1 91	2 261017	0 53516	0 84475	0 999171	0 994049	0 990979		
0 87	1 79				1 29	1 52	1 993615	0 64707	0 76243	0 983889	0 969360	0 999989	h knot	kno'
0 8500	1 7350 (average)				0 97	1 48	1 769548	0 54816	0 83637	0 998422	0 992245	0 992933		
Horizonta knot samples					0 97	1 55	1 828497	0 53049	0 84769	0 999381	0 994635	0 990224		
1 04	1 92				1 07	1 47	1 818186	0 58850	0 80850	0 994470	0 984960	0 997557	h knot	knot
1 47	1 66				1 17	1 45	1 863169	0 62796	0 77824	0 988019	0 975153	0 999796	h knot	
1 5050	1 7900 (average)				1 11	1,75	2 072342	0 53563	0 84446	0 999149	0 993989	0 991053		

<table>
<tr><td>Data of training set
Knowledge base / intermediate layer</td><td>Data of testing sets
Singal parameters / input layer</td><td>Reasoning procedure
The competitive layer</td><td>Output
layer</td></tr>
</table>

Figure 5. Simulation of a Kohonen neural network for identification of wood defects

In the present paper, we will present a simulation of the neural network approach which was mentioned above. The simulation is based on the same raw data as used in that paper. The procedure of the simulation is shown in Figure 5. In this example, only ultrasonic velocity and attenuation coefficient of the most significant peak in each acquired 'waveform are taken as inputs. To train the network, three sets are used. They correspond to sound wood, tilting knot and horizontal knot, respectively. The values of a set are derived from averaging over a couple of samples of a same wood interior type. The training samples are chosen from the table 3. The testing samples are from the table 4 which is connected to another log cross-section.

The simulation shows that neural network is much simpler to implement and performs better than the fuzzy set method, 95% vs. 83% on average in identity hit ratio.

SUMMARY

Ultrasonic scanning is a process of information acquisition and analysis. The efficiency of processing and the effectiveness of using information are very important. Techniques of artificial intelligence provides a neat way to improve this. These techniques involve representation of knowledge and data, implementation of reasoning mechanisms and use of the mechanisms. They may be introduced to signal treatment, defect detection and identification. The highly recommended approach for AI implementation is the use of neural networks. Neural networks, taking the Kohonen network as an example, have the advantage in self information-organizing, automatic learning, easy implementation and simple reasoning computation, when compared with rule-based systems.

REFERENCES

1. Caudill, M. & C. Butler. 1990. Naturally Intelligent Systems. A Bradford Book, the MIT Press.
2. Han, W. 1991. An intelligent ultrasonic system for log scanning. Ph. D. dissertation submitted to the Norwegian Institute of Technology, Trondheim, Norway.
3. Han W. & R. Birkeland. 1992. Ultrasonic scanning of logs. Industrial Metrology 2:253-281.
4. Yoshiaki, Y. & J. I. Tsujii. 1982. Artificial Intelligence: Concepts, Techniques and Applications. John Wiley & Sons, Chichester.
5. Zadah, L. A. 1965. Fuzzy sets. Information and Control 8:338-353.

STUDIES ON THE SYSTEM OF ULTRASONIC TEST OF CONCRETE

Lu Jiecheng, Zhuang Zhenquan and Dai Yingxia

Department of Electronic Engineering
University of Science and Technology of China
Hefei, Anhui, People's Republic of China

INTRODUCTION

Based upon the correlation between interior flaw and parameters such as ultrasonic velocity, the first ultrasonic amplitude and the received ultrasonic frequency, a microcomputer testing system has been developed, which comprises of three elementary parts, namely, data sampling, signal processing and flaw judging. It can be applied in nondestructive testing of the internal flaws of the structure concrete. This paper concentrates on the introduction of the fast calculating algorithms of audiotime and the first ultrasonic amplitude, the spectral analyzing method of the received waveform and the two flaw judging procedures on the basis of fuzzy mathematical theory and production rule respectively. Finally, an example of detection is provided. All computing data and results of testing can be output in the form of graphic or literal through the CRT or printer.

DETECTING PRINCIPLE

The detecting system consists of two parts: hardware and software. The hardware, with the CPU as its nucleus, includes data sampling and other I/O devices, supporting circuitry of amplifying, shaping, delaying trigger, and testing equipments like ultrasonic pulser and transducers. The software, in modular structure, is composed of one main module and the submodules under its control, which are respectively capable of: data sampling, signal processing, correlation parameter extracting, flaw synthetical judging, CRT displaying and printer output of testing result. All data transfer among these modules through disc files. The operational block diagram of this system is shown in figure 1.

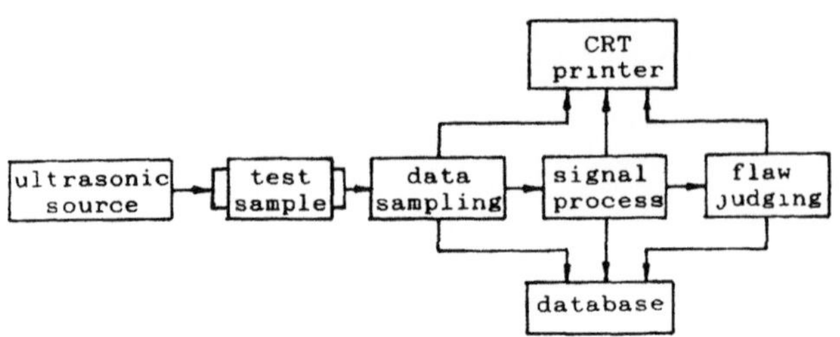

Fig. 1   The operational block diagram
of the ultrasonic testing system

The Received Wave and Its Sampling

The emitter of the ultrasonic transducer converts signals of electronic pulses into ultrasound, which passes through the tested medium before being received by the receiving transducer, amplified, filtered and then transferred to the input of the A/D convertor. At the same time, the emitting pulses are shaped, delayed and passed on to the input terminal of the A/D synchronization signal in order to sample the ultrasonic received waves in the preceding periods which include the 1st. There are three types of data sampling: individual and continuous sampling, repetitive scanning, with eight sampling frequencies for select, the highest being 10MHz.

Signal Processing Module

Signal processing module executes all kinds of calculation and analyses of the sampled data or those stored in disc files which include pre—processing of sampled data, detecting of the ultrasonic transit time and the first received amplitude, computing of the maximum, minimum and average of the signals, and analyzing of frequency spectra, power spectra and corrrelation, etc. Theoretical analysis and experiments prove that ultrasonic velocity[1-2], the first wave amplitude, main frequency of the received wave and waveform are important parameters reflecting interior flaws, therefore this module is dedicated mainly to the programs for detection of ultrasonic velocity, the first wave amplitude and analyses of spectra. The data from the calculation of these programs will be input to the flaw judging module so that reasonable judgments on the interior flaw of the concrete can be reached.

Flaw Judging Module

This module implements synthetical judging on the main ultrasonic parameters and determines the location and range of the interior flaw in the tested structural concrete. To increase flexibility and generality, two judging methods have been employed in this module: the Fuzzy judging based on Fuzzy math theory and intelligentized judging based on production rule. It must be pointed out that, whichever method is adopted, flaws can only be located preliminarily, to exactly determine the range, elaborate detecting must be done near the doubtful tested points.

SIGNAL PROCESSING AND FLAW JUDGING METHOD

The signals input into computer memory are digital—coded sequences of the received ultrasonic signals already sampled and quantified, which may either be those from solid points, or others travelling through flaws, but anyway are mixed with random noises. How to recognize the useful signal, i.e how to extract the characteristic message about flaws from the coded sequence, is the key to flaw judging. The main task of signal processing is to precisely calculate the ultrasonic parameters (or correlation parameters) related with flaw, including ultrasonic transit velocity (audiotime), the first ultrasonic amplitude and other parameters of waveform spectral analysis. The function of the judging program is generally to evaluate the above parameters according to certain criteria so that exact location of the interior flaw can be realized.

Detecting Method of Ultrasonic Parameters

By microcomputer, many methods can be utilized to detect sonic velocity and the first wave amplitude. What adopted here are Uni—and Bidirectional recursive comparing algorithms and direct reading of the audiotime and the first ultrasonic amplitude from the sampling wave through cursor manipulation on the screen. Through FFT., it is convenient to obtain frequency, power and correlation spectra and parameters such as frequency of the received wave.

Recursive Comparing Algorithms of Audiotime and First Wave Amplitude   This is a fast automatic method of calculating the audiotime (sonic velocity) and the first ultrasonic amplitude, which is divided into uni—and bidirectional recursive methods. The former, starting with any point between A and B in fig. 2(b), repeatedly compares the sampling codes rightward till piont D, thereby detects the initial point B and peak C of the 1st ultrasonic wave one after another, and subsequently the audiotime, the velocity, and the 1st wave amplitude can be obtained; While the latter, begins with anywhere between A and C, first detects the peak of 1st wave, from which continues recursive comparing leftward till the initial point is obtained. The flowchart of the recursive comparing methods is depicted in fig. 3.

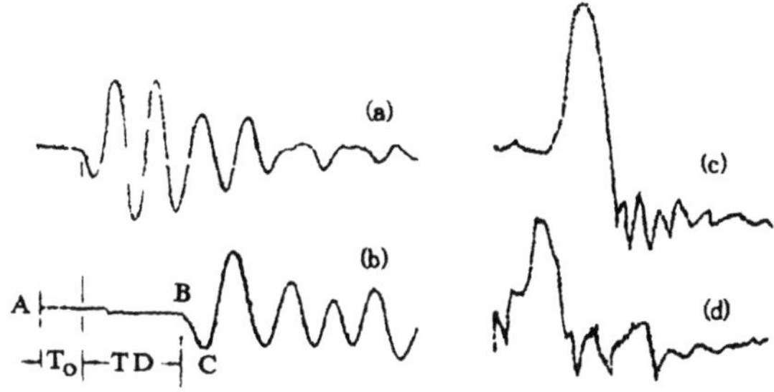

Fig. 2   Emitting receiving wave and their amp. spectra
  (a) Emitting signal   (b) Received signal
  (c) Amp. spectra of the emitting signal
  (d) Amp. spectra of the received signal

Read Audiotime, 1st Wave Amplitude and Signal Period Using Cursor
Two cursors t_1 and t_2, set in the software, can be moved left and rightward along the sampling waveform on the screen through keyboard operation; Also, five counters are placed on the screen to count and display the respective position of the sampled point where the cursor is, the corresponding quantified eletronic level and the time difference between the two cursors, thus by placing one of the cursors at the 1st sampled point, the other at the initial point (peak) of the 1st ultrasonic wave, the transit time can be shown immediately on the screen; If t_1 and t_2 are placed at two peaks (or troughs) of the waveform, respectively, the signal period can be read directly, and frequency can be calculated through exp. (1)

$$1/f = r \times (t_2 - t_1) \times T_s \qquad (1)$$

Ts in the exp. (1) is the sampling period; r is a proportional constant, its value is determined by the position of the two cursors. $r = 1$ if t_1 and t_2 are at two ajacent peaks (trough), respectively; $r = 2$ if $(t_2 - t_1)$ is half the signal period and so on. Obviously, the time delayed by the testing circuitry is canceled because of the substraction of two times, therefore precise frequency can be computed through exp. (1).

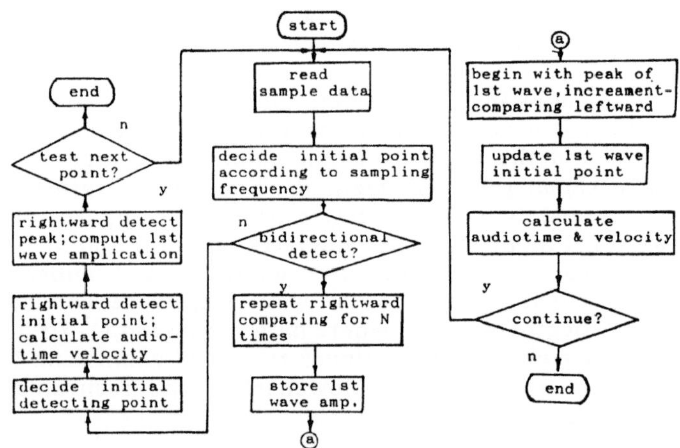

Fig. 3　Flowchart of the detecting of audiotime
(sonic velocity) and 1st wave amplitude

Spectral Analysis and Detecting of Main Signal Frequency[3]　In order to investigate from all respects the frequency variation and energy distribution of the received wave of the ultrasonic passing through different media, two spectral analyzing programs have been designed, one for spectral analysis obtaining amplitude spectra, phase spectra and main freq.; the other calculating the estimated value of power spectrum through the correlation method, obtaining correlation function, power density function and main signal frequency. The main frequencies produced by the two programs are completely identical. The freq. analyzing steps are as follows:

Sample with a time interval of Δ, the limited length discrete signals are shown in exp. (2)

$$x(n\Delta), \ n = 0, 1, \cdots, N_0 - 1 \tag{2}$$

Analyze the $N = 2^k$ terms taken from exp. (2),which are described as

$$\widetilde{x}_n = x(n\Delta), \ n = 0, 1, \cdots, N - 1 \tag{3}$$

Window process exp. (3), producing

$$x_n = \widetilde{h}_n \widetilde{x}_n, \ n = 0, 1, \ldots, N - 1 \tag{4}$$

Calculate frequency spectrum using FFT technique

$$X_m = \sum_{n=0}^{N-1} x_n e^{-i2\pi mn/N}$$
$$= U_m + iV_m \qquad (0 \leqslant m \leqslant N - 1) \tag{5}$$

Smooth the spectrum of amplitude

$$\widetilde{A}_m = P_m \times A_m = \sum_{j=-2}^{2} P_j A_{m-j} \tag{6}$$

Compute the main frequency f_M of the amplitude spectrum

$$\text{Let} \qquad f_M = \frac{mM}{N\Delta} \ (0 \leqslant mM \leqslant \frac{N}{2}) \tag{7}$$

where mM satisfies

$$\widetilde{A}_{mM} = \max\{\widetilde{A}_m, m = 0, 1, \ldots, N/2\} \tag{8}$$

mM can be calculated easily by computer, therefore f_M is available; Power spectrum can be obtained through similar procedure. Finally,all curves of the spectral analyses can be plotted on the screen or by printer, with which the composite frequencies of the received ultrasonic wave can be analyzed. The amplitude spectra corresponding to fig. 2(a) and 2 (b) are illustrated in fig. 2(c) and 2(d),respectively.

Flaw Judging Method

The principal methods used in current computerized concrete judging, as reported in relevant literature, are freq. analysis, sonic velocity—amp. detecting,probability and statistical method, synthetical evaluating on the basis of fuzzy mathematical theory, etc. However, intelligentized judging utilizing expert system is not yet introduced in any article as far as we know.

Fuzzy Synthetical Judging[4] The method based on the concept of grade of membership of fuzzy mathematics, uses sonic velocity u_1 , first wave amp. u_2, received — wave frequency u_3 and waveform correlation coefficient u_4 (maybe other correlation parameters too) to constitute factor domain U

$$U = \{ u_1, u_2, u_3, u_4 \} \tag{9}$$

Upon U a fuzzy subset $\underset{\sim}{A}$ is defined, whose vector form is expressed as

$$A = (\mu_{\underset{\sim}{A}}(u_1), \mu_{\underset{\sim}{A}}(u_2), \mu_{\underset{\sim}{A}}(u_3), \mu_{\underset{\sim}{A}}(u_4)) \tag{10}$$

Where $\mu_A(u_i)$ represents the grade of membership of element u_i in fuzzy subset $\underset{\sim}{A}$, and reflects the relative contribution of each ultrasonic parameter to the flaw evaluation.

There are only two kinds of outcomes of the flaw judging: normal v_1 and abnormal v_2, which form the comment domain

$$V = \{ v_1, v_2 \} \tag{11}$$

Upon V a fuzzy subset $\underset{\sim}{B}$ is defined, whose vector form is expressed as

$$\underset{\sim}{B} = (u_{\underset{\sim}{B}}(v_1), u_{\underset{\sim}{B}}(v_2)) \tag{12}$$

$u_{\underset{\sim}{B}}(v_1)$ and $u_{\underset{\sim}{B}}(v_2)$ in exp. (12) are results of evaluation, representing the grade of membership of normality and abnormality, respectively. Subsequently, utilizing factor domain U and comment domain V, a fuzzy relation $\underset{\sim}{R}$ is established, as shown in exp. (13)

$$\underset{\sim}{R} = \begin{array}{c} \\ \mu_1 \\ \mu_2 \\ \mu_3 \\ \mu_4 \end{array} \begin{array}{cc} v_1 & v_2 \\ \begin{bmatrix} r_{11} & r_{12} \\ r_{21} & r_{22} \\ r_{31} & r_{32} \\ r_{41} & r_{42} \end{bmatrix} \end{array} \tag{13}$$

The matrix element r_{ij} indicates the grade of membership of the quality of the tested concrete point in v_j, when considering factor u_i only, this is the result of single parameter evaluating; Whereas $u_{\underset{\sim}{B}}(v_i)$ in exp. (12) is the result of multiparameter — evaluating.

From the four sonic parameters of the tested point, the average and variance of each parameter, we can obtain matrix $\underset{\sim}{R}$. For $\underset{\sim}{A}$, we choose $\underset{\sim}{A} = (0.4, 0.3, 0.2, 0.1)$ (the value may differ slightly with different people's preference). Assuming $\underset{\sim}{A}$ and $\underset{\sim}{R}$ are known, then $\underset{\sim}{B}$ can be computed through exp. (14)

$$\underset{\sim}{B} = \underset{\sim}{A} \cdot \underset{\sim}{R} \tag{14}$$

With the evaluation of each tested point obtained through the above procedure, we can understand whether there is any flaw within the concrete in the tested region, or if any range and degree are affected by flaws. We have referred to reference 4 for details.

Judging Method Based on Rules[5] This is an expert system method on the basis of rule — expression, using production rule to make synthetical judgment about interior flaw. In production system described by rules, all detecting information and expertise are expressed in the form of "If...Then...", and stored in rule base. An "If...Then..." rule may either be a " Premise ... Conclusion " or a "Situation...operation"pair, where "premise" and "situation"are premise assertions and situation fact respectively, indicating conditions; while the "conclusion" may either be the ultimate target of the problem solving or some intermediate conclusions; "operation" is the certain act to control the system.

The basic architecture of the rule judging system,as illustrated in fig. 4, is made up of three elementary sections: rule base, general data base, and controlling strateries. In the rule base,there are statements in the form of "If...Then...",for example

> If the velocity of the sonic wave received from the tested point is less than the difference between its average and two times its standard diviation
>
> Then there may be flaw within the tested point

General data base is sometimes called context,current data base or temporary memory, which is used to store the condition (fact or assertion) at the lefthand of each production exp. The role of controlling strategies is to explain the application of rules.

For the convenience of establishing and applying rules,rules in the program are divided into two categories:signal processing rules and flaw judging rules. The signal processing rules are set up for the production of intermediate facts, that is, they supply the necessary evidence for flaw — judging; The flaw judging rules are for the direct judging of the interior flaw location, determining whether there is flaw in the tested point by matching the condition part of the rule with the correlation parameters. If the answer is affirmative,then a "+" is labeled at the corresponding point in the stereoscopic graph of the tested structural member, otherwise "—" is assigned.

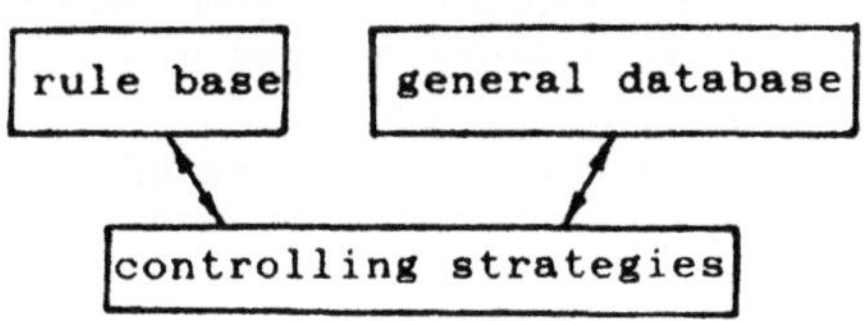

Fig. 4 Basic architecture
of the rule judging system

THE RESULT OF DETECTION

The results of detection produced through this system's testing and investigation on all kinds of concrete test samples are consistent with those by ultrasonic detector. Take a cubic test sample with a side length of 20m for example,respectively test the solid sample and the one with a crack on the surface that extands inward. The freq. of the transducer is 100KHz; Use butter as coupling agent,choose 5MHz for the sampling

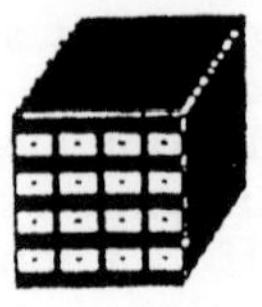

Fig. 5　solid sample

Fig. 6　Flaw detecting using fuzzy method

Fig. 7　Flaw detecting using rule method

freq. Shown in fig. 5 is the result of the penetrating testing on the solid sample, with all sixteen tested points labeled with " —"s, indicating no flaw in the test sample. The detecting result of the clefted one, using fuzzy evaluation method, is demonstrated in fig. 6, the white area showing the extension of the flaw. Fig. 7 illustrate the outcome of rule detection.

It is to be pointed out that structure concrete or cement products is elastic — viscid and plastic complicated coagulation of many materials, in which sonic wave will be substantially scattered and attenuated when transitting. Also, the choice of transducers, the coupling condition and the testing distance will all affect the detecting outcome.

CONCLUSION

• Interior flaw of structure concrete can be synthetically detected through ultrasonic velocity, 1st ultrasonic amplitude, and spectral analyzing parameters of the received waveform.

• Detecting of sonic velocity and 1st wave amplitude through recursive comparing method is more precise and immediate than that by sliding the vernier of a ultrasonic detector.

• Fuzzy evaluating and rule judging are efficient methods in interior flaw synthetical judging.

REFERENCES

1. Chung , H . W . and Law , K . S . Concrete International : Desiqna Construction, 5—10(1983), 42:49
2. Lin Weizheng, Nondestructive Testing, China, 6(1981), 29:31
3. Cheng Qiansheng , " Mathematical Theory of Digital Processing of Signal",China,167:173 (1979)
4. Lu Mingliang, Nondestructive testing, China, 13—11 (1991),301:303
5. Fu Jingsun , Cai Zixin , Xu Guangyou , " Artificial Intelligence and Its Application", China,277:290 (1987)

A MANUAL COLOR SCAN SYSTEM BASED ON
THE CAMERA-SEE POSITION SENSING

Jiaou Li, Jimao Chen

Beijing Aeronautical Manufacturing Technology Research Institute
P. O. B 863—11, Beijing 100024
P. R. China

INTRODUCTION

The defects of metal or non-metal materials formed during manufacturing process will severely damage the capacity of materials. NDT & NDE methods are engaged to meet the needs of controlling the quality of those materials. Ultrasonic C-scan is the main method for this purpose. As most of the automatic mechanical C-scan systems are designed for some kinds of definitely shaped structures, it is difficult to inspect complicated shaped structures or assembled structure on site. Therefore it is necessary to develop a manual C-scan system.

The manual C-scan systems engaged in industrial applications fall into two categories. One is based on mechanical position sensing, the other is on the sonic position sensing. The repeating position-sensing precision of those two systems can not meet the needs of the structure on-field inspection satisfactorily.

Mechanical Position-Sensing

The position-sensing of mechanical scanners is based on the connecting rods mechanism which determines the probe's position by obtaining the lengths of rods and the angles between two rods, and converting them into latitude and longitude coordinates.

Because of the complexity of its mechanism and low repeating precision, this mechanical manual scanner isn't good enough to be applied to the on-field NDT inspection.

Sonic Position-Sensing

The principle of sonic position-sensing assembly is shown in Figure 1. P is the source of sound. Two receivers are mounted on A and B, respectively. The digital processor can measure the flight time t_1 and t_2 from P to A and B, respectively, and convert them into distance R_1 and R_2.

Taking A and B as the centers of circle 1 and circle 2, and using R_1 and R_2 as their radii, the intersection (P) of the two circles is the center of the probe.

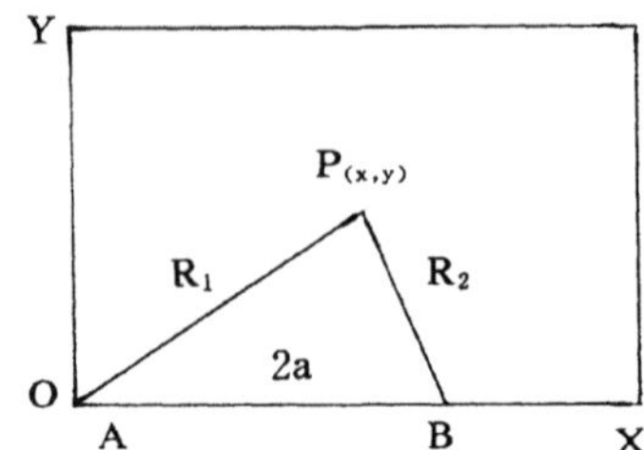

Figure 1.   Principle of sonic position sensing

The circle's equation:

$$\begin{cases} (x - 2a)^2 + y^2 = R_2^2 \\ x^2 + y_2 = R_1^2 \end{cases} \qquad \Rightarrow Y \geqslant 0 \qquad \begin{cases} x = \dfrac{R_1^2 - R_2^2 + 4a^2}{4a} \\ y = \sqrt{R_1^2 - x^2} \end{cases}$$

Compared with mechanical scanners, the sonic position assembly is light, small and easy to handle, but its noise has a harmful effect on human nerve system. The receiver of sound is exposed to environment with sonic disturbance.

NDT research in BAMTRI is leading the manual C-scan development in P. R. China. Since 1987, some systems based on mechanical position sensing, sonic position sensing and mouse position sensing have been developed, but we found that their repeating position precision cannot meet the needs of composite material's on site image inspection in air and airspace industries. Recently BAMTRI developed a new kind of manual C-scan system, which is based on camera see position sensing. We think that is the best way to solve complicated shaped structure and on-field inspection.

SYSTEM CONFIGURATION

To form an ultrasonic C-scan image, we must obtain two important data:
(1) type of anomaly in the detected material;
(2) positioning of anomaly in the surface of the detected material.

The characteristics of anomaly are obtained from ultrasonic echo wave through feature abstraction by computer. As shown in Figure 2, focus probe A receives the echo waves from the inner structure of materials. The ultrasonic detector divides the echo wave into front wave, anomaly wave, and back wave in radiate frequency form. The signal digitalizer can convert the analogy signal into digital signal, and then put them into computer. The computer classify the anomaly by analyzing the feature characteristics, such as U_f (front wave), U_{lm} (anomaly wave), Δt (interval), f (number of anomaly wave) and U_b (maximum valve of back wave).

A feature array of the anomaly is formed, $\overline{P} = (U_f, U_{lm}, t, f, U_b)$, to assist to evaluate the characteristics of anomalies. A computer pattern recognition program analyses the feature array $\overline{P}$ and gives the type of anomaly it represents.

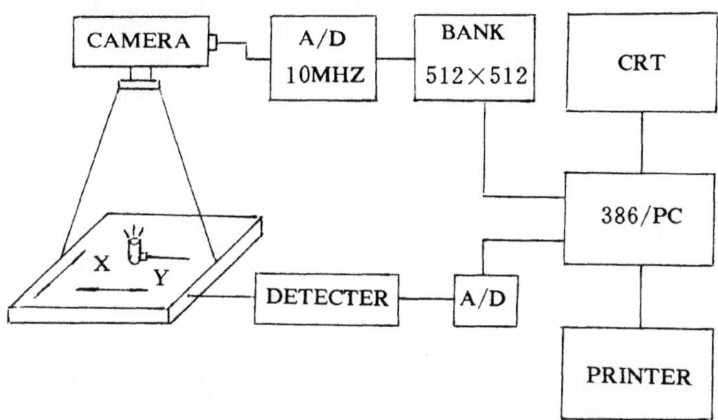

Figure 2.   System configuration of SSDC-I manual C-scan system based on camera see position sensing

A novel approach based on camera see technique is engaged in SSDC-I for probe's position sensing. As shown in Figure 2, a CCD camera see the inspection field with a speed of 50 fields per second. A bright point mounted on the top of the probe can be discerned according to its grey level, which is different from other objects around it. The image signals from camera are sent to image digitalizer through a 10MHz A/D converter to form a digitalized two dimensional image, and stored in RAM bank on the graphic card. A particular computer program can trace the probe's position by searching and calculating a 512×512 matrix according to its characteristic on grey level and feed the coordinates back to main processor of the computer. Combining the position of the probe with the type of anomaly, which is represented in color, to form a two dimensional color C-scan image.

Camera See Position Seeing

As shown in Figure 5, we devise such a probe with a small lamp mounted on its top, which represents the position of the detected point. The camera, which is about 700 mm above the inspection area, views the probe with a speed of 50 fields per second to form a series of two dimensional images (50 images per second). The analogy image can be converted into digital image of 512×512 pixels matrix. The computer can trace the position of the probe by discerning the bright point from other objects around it in grey level. And finally we can get its coordinates in Cartesian coordinate system Oxy.

As shown in Figure 3, to contract the data to be processed, we trace the probe A in a scope of 50×50 pixels matrix. With a limited scanning speed of manual handling, we can assume that the next traced point A_{i+1} be located in 50×50 pixels matrix with the center of A_i, which is the point now being traced.

Let H_{min} represent the valve in grey level, and H_g the grey level at point g.

If $H_g > H_{min}$, then $(X_g, Y_g) \in A_{i+1}$; if $H_g < H_{min}$, then $(X_g, Y_g) \notin A_{i+1}$.

If we find N point which belongs to A_{i+1}, the coordinates of A_{i+1} are:

$$A_{i+1} \in (X_{i+1}, Y_{i+1}) \qquad \begin{cases} x_{i+1} = \left(\sum_{j=1}^{N} X_j \right) \Big/ N \\ y_{i+1} = \left(\sum_{j=1}^{N} Y_j \right) \Big/ N \end{cases}$$

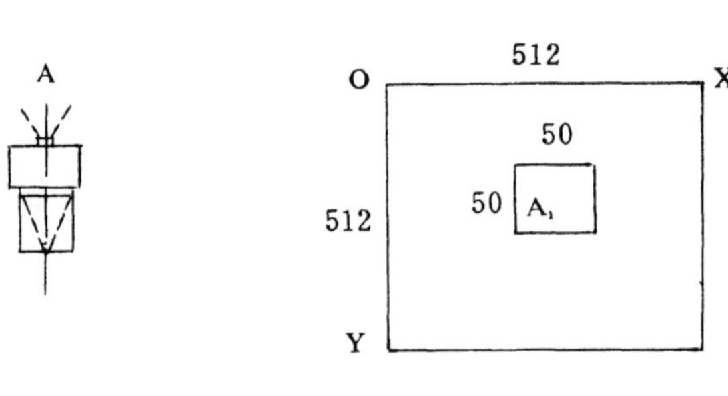

Position Sensor   Tracing Mode

Figure 3. Tracing principle of position sensing

To locate the initial point $A(x, y)$, we must search the whole scope of 512×512 pixels matrix, until we find it. From point A, the position sensing unit trace point $A(x, y)$ and find the group of points which belong to A_{i+1} in grey scale. The computing processor calculates the geometric center of those mass points using the equation shown above.

High Speed Date Acquisition

The 10MHz ultrasonic probe emits ultrasonic pulse with a width of 50ns and receives the echo wave from the detected structure, which must be amplified and be displayed in radiant frequency. The echo wave contains a large amount of information about the anomalies in the materials. In our camera-see system, we use a echo wave digitalization unit to convert high frequency signal into digital signal and put them in RAM.

For echo waves with a width of 50ns, and amplitude of $+2.5V$, it can be acquired with a speed of 10 points for each wavelength. That is one point every 5ns. A superspeed A/D converter is needed for this high frequency acquisition which will be too expensive. A step-increase acquisition technique is engaged in SSDC-I to solve the problem. As shown in Figure 4, we obtain only part of echo wave points and obtain the whole wave through n times acquisitions. This technique will lower the requirement for A/D converter.

The acquisition width T and step delay Δt are controlled by a logical control units. The repeating scanning times n should meet

$T = n\Delta t$, n is determined by the detected material.

The acquisition data are stored in array B, Figure 5 shows the arrangement of the data in Array B. The thicker the detected material is the less the acquisition width T and repeating times should be.

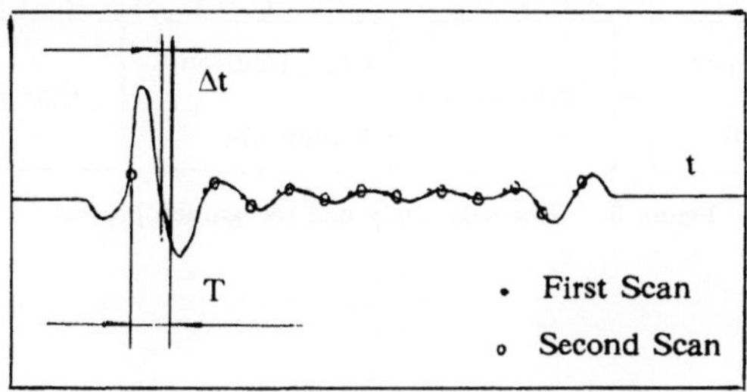

Figure 4. A step-increase acquisition method

Computer Pattern Recognition in Composite Materials Evaluation

A general pattern recognition process can be described as in Figure 6. Analytical process founded the bases of classification, that is, to erect classification functions. The recognition process is to apply classify functions to materials evaluation. First we must erect a feature vector $\overline{P}$ which is connected with the method of inspection and materials to be detected. For composite materials (CFRP), $\overline{P}$ consists of U_f, U_{lm}, U_b, f, and t, where U_f represents the amplitude of front wave, U_{lm} the amplitude of anomaly wave, U_b the amplitude of back wave, f the counted numbers of anomaly wave above a supposed valve. Δt the interval between front wave and back wave.

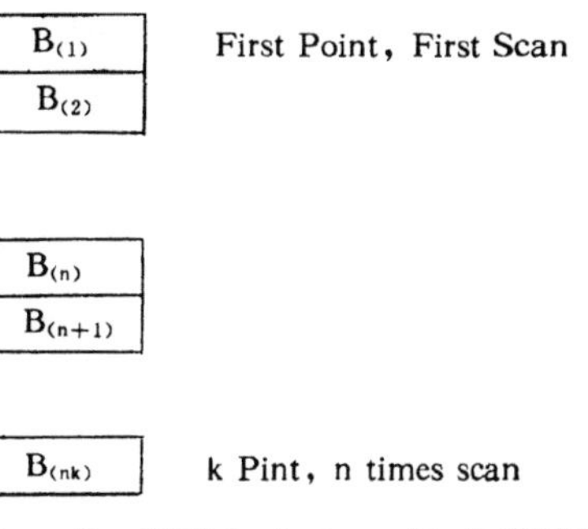

Figure 5. Digital echo wave data in RAM

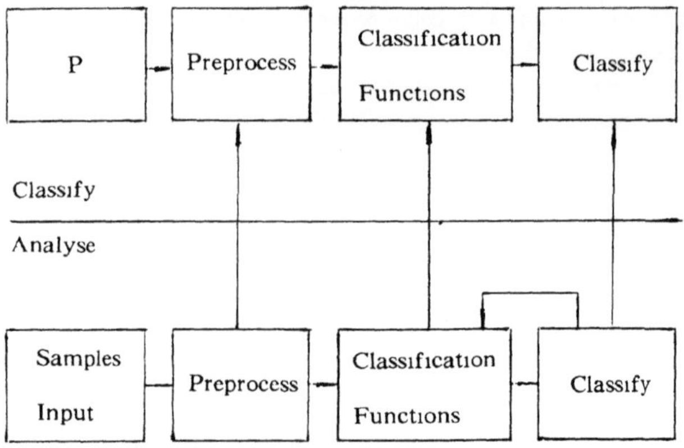

Figure 6.   Anomaly study and recognition process

Vector $\overline{P}$ composed of (U_f, U_{lm}, U_b, f) can reflect ordinary defects in composites such as delamination, porosity, void, inclusion, rich or poor in resin. Figure 7 shows the features of anomaly in echo waves.

In pattern recognition, we engage a novel approach based on the fuzzy theory to determine what extent P belongs to a particular type.

$F_i(p)$ is a family of classification functions , which serves as a tool to determine how much a vector P belongs to an anomaly type i .

If there are n types of anomalies, and $\overline{B} = (F_1, F_2, ... , F_n)$, then

$$\overline{P} \cdot A = \overline{B} \qquad A = \begin{Bmatrix} a_{11} & a_{12} & \cdots & a_{1n} \\ a_{21} & a_{22} & \cdots & a_{2n} \\ \cdots & \cdots & \cdots & \cdots \\ a_{51} & a_{52} & \cdots & a_{5n} \end{Bmatrix}$$

where A is a parameters matrix, which comes from repeating learning of typical anomalies and can be changed with the increase of collection of experimental recording. $\overline{P}$ belongs to the maximum of $\overline{B}$.

EXPERIMENT

By using camera-see position sensing unit, the scanning area can change with scanning resolution.

Let r represent scanning resolution, and s represent scanning area.

If $s = 1024$ mm $\times 1024$ mm, then $r = 2$ mm $\times 2$ mm;

if $s = 512$ mm $\times 512$ mm, then $r = 1$ mm $\times 1$ mm .

we can get required scanning area and scanning resolution by adjusting the distance between the camera and the test sample.

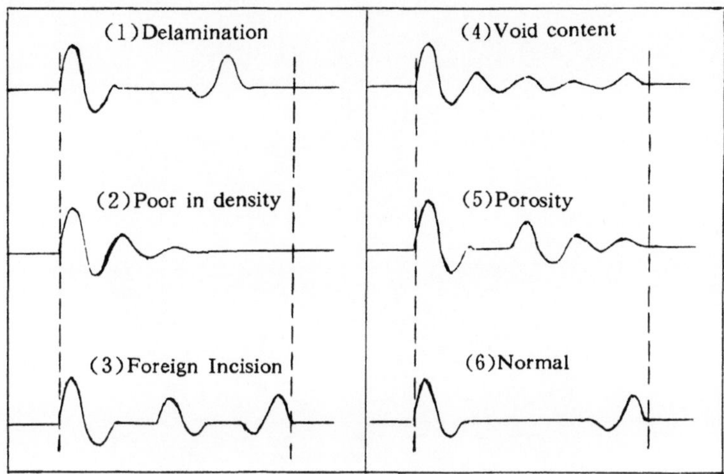

Figure 7. Typical defects in composite (CFRP)

Sample 1 Inspection

Sample Description:

This is a 245 mm $\times$ 245 mm CFRP laminates. The ply orientation angle is $\pm 30°$. The ply group is of 16 laminates. There are five manmade delaminations, one of them is similar to five pointed-star, imbedded in the middle lamina of the slat.

Parameters set: $A_1 = 145$, $A_2 = 145$, $A_3 = 143$, $B_N = 120$, $H = 1000$ mm.

The ratio of real area to recording area: $245/95 = 2.58 : 1$.

CONCLUSION

In the SSDC-I manual Color-scan system, several techniques are developed such as camera-see position sensing, high speed data acquisition and fuzzy computer pattern recognition with following features:

(1) Manual C-scan inspection for complicated shaped structure and on field inspection.

(2) Changing the scanning area and scanning resolution by adjusting the distance between the camera and the test sample. In an area of 512 mm $\times$ 512 mm, the scan resolution $\leqslant 2$ mm.

(3) It is easy to handle because of no mechanical connection between the probe and the camera.

(4) Fuzzy computer pattern recognition, color image display and recording

(5) Menu-handling of software.

Figure 8.   Image of sample 1

REFERENCES

1.   A. Yates, Corrosion-damage images in real time from manual scanning, *Materials Evaluation*, Vol. 47, August 1989.

2.   F. H. Chang, Prototype of an in-service inspection system for composites, *Materials Evaluation*, Vol. 43, April 1985.

3.   J. B. Nestleroth, Physically basic ultrasonic feature mapping for anomaly classification in composites, *Materials Evaluation*, Vol. 43, April 1985.

4.   James H. Brahney, Inspection and repair of composite aerospace structures, *Materials Evaluation*, Vol. 44, December 1986.

PULSECOMPRESSION IN PULSE-ECHO-MODE WITH SPLITTED CHIRP SIGNALS

M. Pollakowski[1], H. Ermert[1], and L. von Bernus[2]

[1] Ruhr Universität Bochum, Institut für Hochfrequenztechnik
Universitätsstr. 150, D-4630 Bochum, Germany
[2] Siemens AG, Hammerbacherstr. 12+14, D-8520 Erlangen, Germany

ABSTRACT

A frequently encountered measurement configuration for ultrasonic inspection of steel components consists of a single element transducer insonifiing a specimen of limited depth in pulse echo mode.

In this case, the gain in signal-to-noise ratio achievable by pulsecompression technique is limited by the fact, that the duration of the transmitted chirp signals must not be longer than twice the time-of-flight from the tramitting element to the imaging area. This is why the transducer is unable to transmit and receive simultaneously.

A new class of splitted chirp signals allows to overcome this limitation. A long chirp signal is divided into several parts, each of twice the length of the time-of-flight. These parts are transmitted with time delays determined by the depth of the imaging area.

The received echo signals can be pulse-compressed by a single filtering operation. Therefore a considerable gain in data aquisition speed, compared to averaging methods, can be achieved.

INTRODUCTION

Pulse compression is a technique to improve the signal-to-noise ratio[1-3]. Instead of short pulses, signals with a large time-bandwidth-product (e.g. chirp-signals[4], pseudo-random-signals[5]) are applied. As these exhibit a single-peaked autocorrelation-function, they can be compressed to short pulses by appropriate correlation- or pulse-compression-filters.

Most frequently chirp-signals (linear frequency modulated signals) are used for pulse-compression[6]. The theoretical achievable enhancement of the signal-to-noise-ratio $SNRE_{th}$ can be calculated using

$$SNRE_{th} = \sqrt{B \cdot T}$$

(with: B = bandwidth, T = duration of the chirp signal)

Acoustical Imaging, Volume 20 Edited by Y Wei
and B. Gu, Plenum Press, New York, 1993

During ultrasonic inspections, usually the same transducer is used for transmission and reception of the acoustig signals (pulse-echo-mode operation[7]). This configuration opposes severe restrictions on both the bandwidth and the duration of the applied signals, and therefore also the achievable gain in SNR is limited.

SIGNAL BANDWIDTH LIMITATIONS

Ultrasonic transducers inherently show bandpass-like behaviour. Therefore it makes no sense to apply chirp signals with frequency-components far beyond the spectral window determined by the transducer.

Detailed analysis of this problem has shown, that the optimum gain in SNR can be achieved, if the bandwidth of the chirp-signal is 1.14 times the 6-dB bandwidth of the transducer[8]. Due to a spectral mismatch of the rectangular-shaped chirp-signal-spectrum and the bell-shaped transducer-spectrum, the achievable gain in SNR then is given by:

$$SNRE \; = \; 0.77 \; \cdot \; \sqrt{B_T \cdot T}$$

(with: B_T = 6-dB bandwidth of the transducer)

SIGNAL DURATION LIMITATIONS

Figure 1 shows a typical NDE situation with a contact-technique transducer scanning an aperture a of a specimen of thickness d:

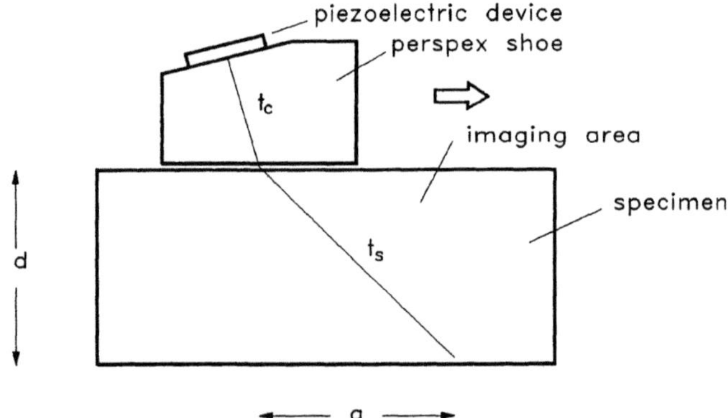

Figure 1. Geometry of a typical NDE situation

The transducer consists of a piezoelectric device, mounted on a perspex shoe. An acoustic signal first travels through the perspex-shoe and reaches the surface of the specimen after a time t_c. Then it enters the specimen and a first echo signal from the interesting imaging area would reach the piezoelectric device at the earliest after $2 \cdot t_c$. As during this first period no information about the the specimen can be collected, this whole time-slot ('transmit-time-slot') can be used for the transmit signal.

With this upper limit of $2 \cdot t_c$ for the duration of the chirp signal, an upper limit of the achievable SNR-enhancement $SNRE_{max}$ can be written as:

$$SNRE_{max} \; = \; 0.77 \; \cdot \; \sqrt{B_T \cdot 2 \cdot t_c}$$

SNR ENHANCEMENT BY AVERAGING

The most straightforward way to further enhance the SNR seems to be the application of signal-averaging. If one assumes the noise to be white and uncorrelated, the averaging of N measurements yields an $SNRE_{av}$ of:

$$SNRE_{av} = \sqrt{N} \cdot 0.77 \cdot \sqrt{B_T \cdot 2 \cdot t_c}$$

(with: N = number of measurements) as the signal increases by a factor of N, while the noise increases by a factor of $\sqrt{N}$. Yet this averaging strategy has the disadvantage of slowing down the measurement process.

SNR ENHANCEMENT BY SIGNAL SPLITTING

This disadvantage can be avoided, if one unique feature of the frequency modulated signals is utilized: if one splits a chirp signal in several parts, these parts occupy different frequency-ranges in the frequency domain. Therefore a filter can be designed, that accomplishes the appropriate delays for the different parts of the chirp signal and allows pulse compression within one filtering operation. The time-delay-vs-frequency function of such a filter has to be non-continuous.

With this kind of 'splitted-chirp-signals', N subsequent transmit-time-slots can be filled with parts of one long chirp signal. Each part having the duration of $2t_c$, the achievable $SNRE_{sp}$ then is:

$$SNRE_{sp} = 0.77 \cdot \sqrt{B_T \cdot N \cdot 2 \cdot t_c}$$

THE DURATION OF ONE MEASUREMENT CYCLE

A complete measurement cycle consists of one transmit-time-slot and one receive-time-slot, as it is shown in figure 2:

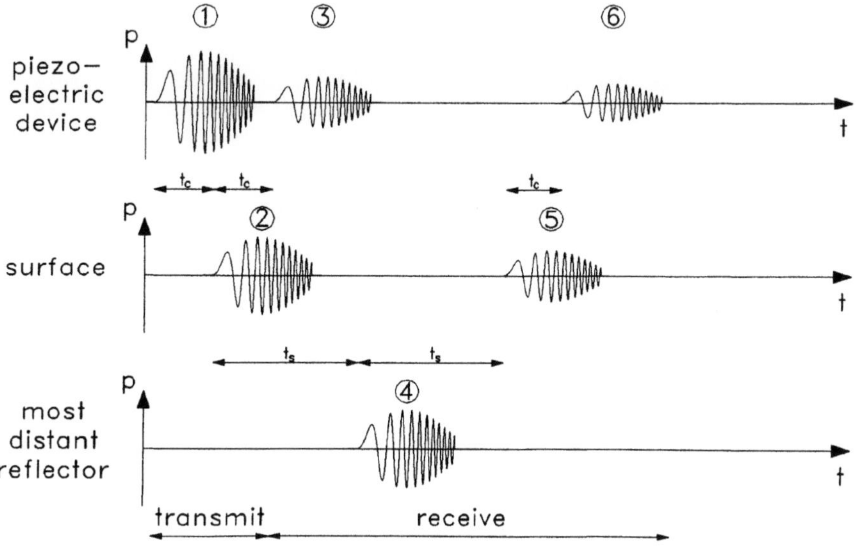

Figure 2. Timing relations of one complete measurement cycle

Signal 1 is the excitation waveform. It reaches the surface after t_c (signal 2) and the surface-echo reaches the piezoelectric device after $2 \cdot t_c$ (signal 3). The time-delay from the surface to the most distant reflector (at the border of the imaging area) is t_s (signal 4). This latest echo needs another t_s to travel back to the surface (signal 5) and another t_c to reach the piezoelectric device (signal 6). The transmit-time-slot is finished as soon as the complete echo has been received.

Therefore the duration of the receive-time-slot is $2t_c$, the duration of the transmit-time-slot is $2t_s + 2t_c$ and the duration of the complete measurement cycle is $2t_s + 4t_c$.

MEASUREMENT EXAMPLE

If we consider a system with a scan-buffer of 4096 points and a sampling rate of 20 MHz, scans of a duration of 204,8 μsec can be aquired. A transducer with a 20 mm perspex-shoe has a time-of-flight t_c of 8 μsec. The time-of-flight t_s within a 10mm steel block is 17 μsec for longitudinal waves ($c_l = 5920\frac{m}{s}$) and 31 μsec for shear waves ($c_s = 3255\frac{m}{s}$).

Therefore a chirp signal of a total duration of 48 μsec, splitted into 3 parts could be applied. This yields a measurement cycle of $\frac{204,8}{3}$ μsec $= 68,26$ μsec and a t_s of 18,13 μsec. The imaging depth then is restricted to 107 mm (imaging with longitudinal waves) or 59 mm (imaging with shear waves).

Using a transducer with a bandwidth B_T of 2.8 MHz (6-dB frequencies at $f_l = 1.6$ MHz and $f_u = 4.4$ MHz), the following SNRE figures can be expected:

$$SNRE_{max} = 0.77 \cdot \sqrt{2.8MHz \cdot 16\mu sec} = 5.25 \hat{=} 14dB$$
$$SNRE_{sp} = \sqrt{3} \cdot SNRE_{max} = 8.93 \hat{=} 19dB$$

(with: $SNRE_{max}$ = SNRE achievable with conventional pulse compression technique, $SNRE_{sp}$ = SNRE achievable with splitted-chirp pulse compression technique).

TRANSMIT SIGNALS

Figure 3 compares the conventional chirp signal $x(t)$ (left) to the splitted-chirp signal $x_{sp}(t)$ (right):

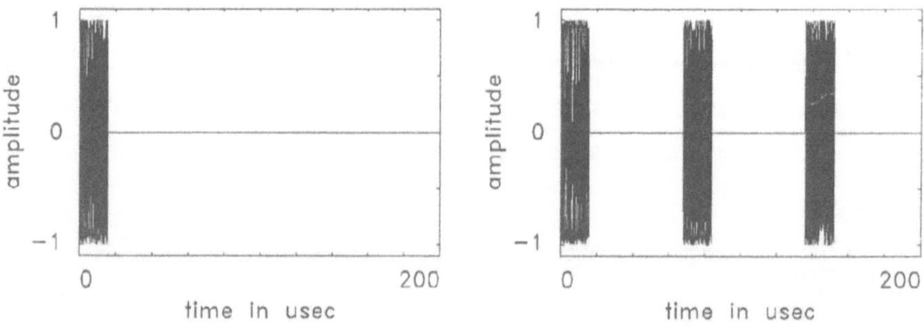

Figure 3. Comparison of conventional chirp-signal (left) and splitted chirp-signal (right)

The conventional chirp signal $x(t)$ of duration $2t_c$ and bandwidth $B = f_u - f_l$ can be described by the formula

$$x(t) = rect(\tfrac{2t-2t_c}{2t_c}) \, cos(2\pi(f_l t + \tfrac{B}{4t_c}t^2))$$

$$rect(\tfrac{t}{T}) = \begin{cases} 1 & : & |t| < T \\ 0 & : & |t| \geq T \end{cases}$$

The splitted chirp signal $x_{sp}(t)$ can be considered as a sum of three delayed chirp signals $x_1(t)$, $x_2(t)$, and $x_3(t)$:

$$x_{sp}(t) = x_1(t) + x_2(t - 2t_s - 4t_c) + x_3(t - 4t_s - 8t_c)$$

$$x_1(t) = rect(\tfrac{2t-2t_c}{2t_c}) \, cos(2\pi(f_l t + \tfrac{B/3}{4t_c}t^2))$$

$$x_2(t) = rect(\tfrac{2t-2t_c}{2t_c}) \, cos(2\pi((f_l + \tfrac{B}{3})t + \tfrac{B/3}{4t_c}t^2))$$

$$x_3(t) = rect(\tfrac{2t-2t_c}{2t_c}) \, cos(2\pi((f_l + 2\tfrac{B}{3})t + \tfrac{B/3}{4t_c}t^2))$$

Figure 4 shows the spectra (magnitude) of the parts of the splitted chirp-signal $x_{sp}(t)$ and demonstrates, that they occupy different frequency-ranges:

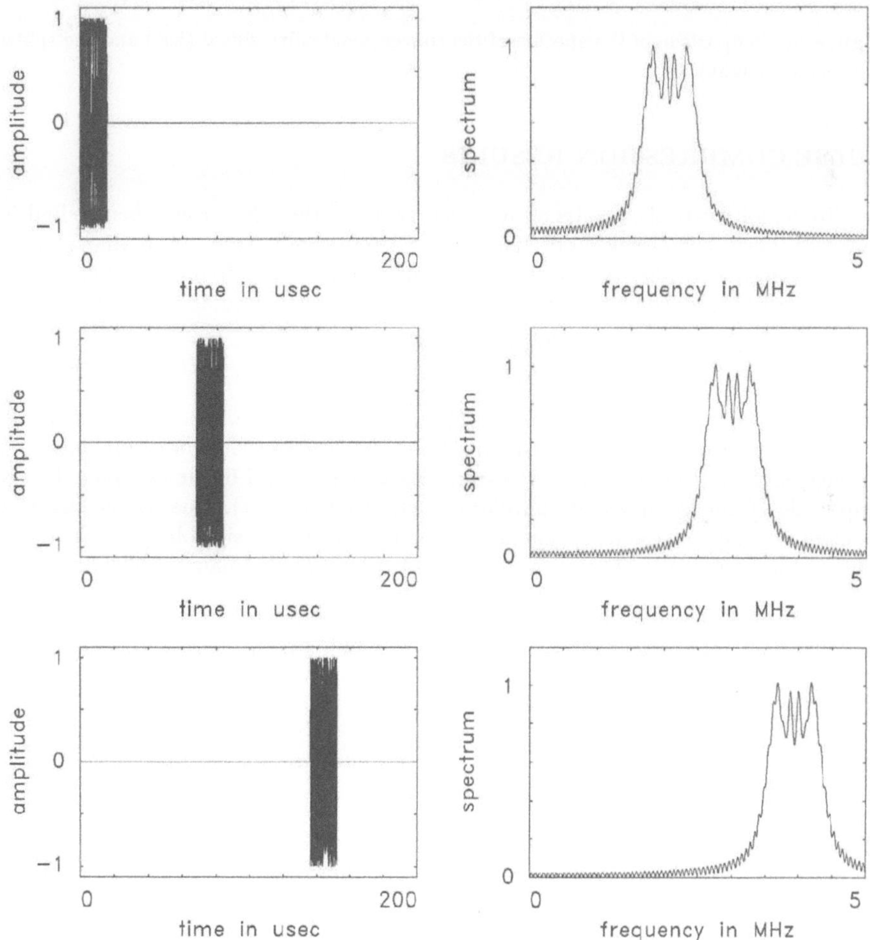

Figure 4. The three parts of the splitted chirp-signal (left) and their spectra (right)

Comparing the spectrum of the conventional chirp-signal (figure 5, left) and the spectrum of the splitted chirp-signal (figure 5, right), one can see the gain in amplitude, but also some disturbances at the frequencies, where the different parts of the splitted signal overlap:

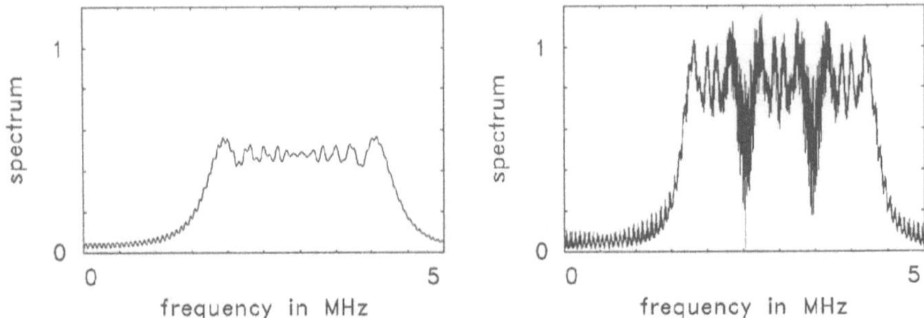

Figure 5. Comparison of the spectra of the conventional chirp-signal (left) and the splitted chirp-signal (right)

PULSE COMPRESSION RESULTS

To demonstrate the results of pulse-compression, the echos from a hemicylindrical backwall have been measured. They have been compressed by a digital pulse-compression-filter (phase-conjugate filter) operating in the frequency domain. The filter transfer functions $H(\omega)$ and $H_{sp}(\omega)$ have been calculated according to

$$H(\omega) = exp(arg(X^*(\omega)))$$
$$H_{sp}(\omega) = exp(arg(X_{sp}^*(\omega)))$$

(with: $X(\omega) = \mathcal{F}\{x(t)\}$ and $X_{sp}(\omega) = \mathcal{F}\{x_{sp}(t)\}$ the spectra of the transmit signals)

Figure 6 shows the normalized output signals of the pulse-compression filter. The amplitude of the compressed splitted-chirp-signal (right) is 1.68 times larger than the amplitude of the compressed conventional chirp-signal (left). Due to the disturbed spectrum of the splitted-chirp-signal additional compression-sidelobes appear at both sides of the main pulse. Their amplitude is 24 dB below the main amplitude.

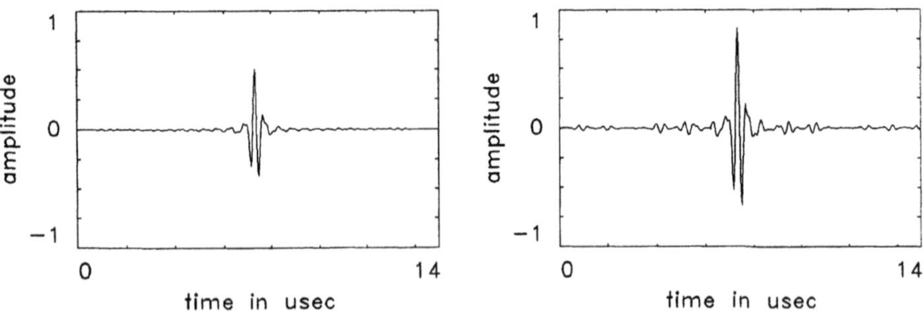

Figure 6. Comparison of the compressed echos of the conventional chirp-signal (left) and the splitted chirp-signal (right)

CONCLUSIONS

It has been shown that for a given test-geometry and a given transducer, an upper limit of SNR enhancement achievable by conventional pulse-compression-techniques can be calculated.

The application of splitted chirp-signals yields an further SNR enhancement without increasing the measurement time, as it makes full use of the available scan-buffer-length. By splitting a long chirp-signal into shorter parts and transmitting this parts in subsequent time-slots, the time-bandwidth-product of the transmitted signal can be increased. Pulse compression can be achieved with one single filtering operation, as it is the case with conventional chirp-signals.

Measurements show that the achievable gain in SNR (1.68) is slightly below the calculated value (1.72). Additional compression-sidelobes with an amplitude 24 dB below the main lobe occure due to disturbances in the spectrum of the splitted chirp-signal.

REFERENCES

1. Barton, David K.: 'Radars Volume 3: Pulse Compression', Artech House, Dedham (MA), 1975
2. Cook, Charles E.; Bernfeld, Marvin: 'Radar Signals', Academic Press, New York, 1967
3. Skolnik, Merrill I.: 'Introduction to Radar Systems' McGraw-Hill, Aukland, 1962
4. Cole, R. C.: 'Properties of swept FM waveforms in medical ultrasound imaging', IEEE 1991 Ultrasonics Symposion, pp. 1243-1248
5. Hayward, Gordon; Gorfu, Yonael: 'A digital hardware correlation system for fast ultrasonic data acquisition in peak power limited applications', *IEEE Trans. on Ultrasonics, Ferroel. and Frequ. Contr.*, Vol. 35, No. 6, November 1989 pp. 800-808
6. Patrick, N.; Sawatari, Takeo; Zilinskas, Gene: 'Signal Processing in Acoustical Imaging', *Proceedings of the IEEE*, Vol. 67, No. 4, April 1979, pp. 496-510
7. Krautkrämer, Josef; Krautkrämer, Herbert: 'Werkstoffprüfung mit Ultraschall', Springer Verlag, Berlin, 1986[5] 'Ultrasonic testing of materials', Springer Verlag, New York
8. Pollakowski, M.; Ermert, H.; von Bernus, L.: 'The optimum bandwidth of chirp signals in ultrasonic applications', *Ultrasonics* (submitted for publication)

ADVANCES IN STUDY OF LARGE-SCALE MICROSTRUCTURE OF REAL CRYSTALS USING METHODS OF ACOUSTIC MICROSCOPY

Vadim Levin

Center of Acoustic Microscopy
Russian Academy of Sciences
Institute of Chemical Physics
Kosygin Str.4, 117334 Moscow

INTRODUCTION

AM visualization of bulk structures is hard problem even for isotropic specimens. But the surprising thing is that the problem is simplified for anisotropic bodies. When a convergent beam comes into anisotropic medium, energy of the beam is redistributed in the directions. There appears a set of directions of preferential transfer of acoustic energy. Due to this phenomenon the incident focused beam can be transformed interface in one or several collimated beam penetrating deeply inside the specimen. Previously [1] we have proposed to use feature of acoustic propagation in anisotropic media for AM visualization. Purposes of this paper - 1) theoretical analysis of convergent acoustic beam propagation in anisotropic specimens and of output signal formation for these objects; 2) development of methods of AM visualization of anisotropic specimens; 3) application of these methods for studying large-scale microstructure of some kinds of crystals.

THEORETICAL FOUNDATIONS OF THE METHOD

Assume a convergent acoustic beam, produced by a acoustic lens in coupling liquid is incident on a plane surface of an anisotropic specimen. The beam partially penetrates into the specimen inside which at the depth d there is a plane inhomogeneity reflected sound (Fig.1). Acoustic signal is pulse with the carrier frequency ω nd the envelope $\beta(t)$. The field in a liquid may be represented as a set of plane waves within wave vectors $\{k_x, k_y, k_z = (k^2 - k_x^2 - k_y^2)^{1/2}\}$, where $k = \omega/c$, c - sound velocity in the liquid. In any plane z the alternative pressure $P(\vec{r}, t)$ can be expressed via the space spectrum $A(k_x, k_y)$ of the beam in the focal plane $z = \xi$ of the acoustic lens:

$$P(x,y,z,t) = \iint dk_x dk_y A(k_x, k_y) e^{ik_x x + ik_y y + ik_z(z-\xi)} B[t - \tau - \frac{z-\xi}{c_z}] e^{-i\omega t},$$

(1)

where $c_z = c\, k_z/k$, τ - the time of signal propagation up to the focal plane, $(z-\xi)/c_z$ - the time of propagation from the focal plane to up to the plane of observation. At the interface $z=0$ each plane wave (k_x,k_y) is partially reflected and partially

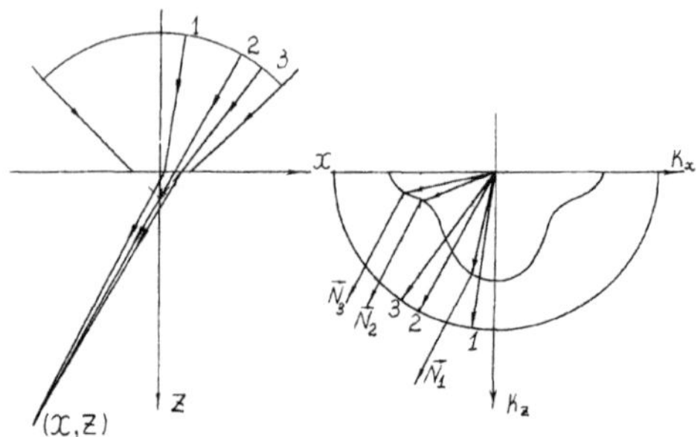

Fig. 1. Propagation of a convergent beam in anisotopic medium.

penetrates into the specimen as elastic waves of different polarizations α. Reflection and transmission are characterized by coefficients $R(k_x,k_y)$ and $T(k_x,k_y)$. The field of the displacement vector $\vec{u}(\vec{r},t)$ inside the specimen may be written in the form:

$$U(x,y,z,t)= \sum_\alpha \iiint dk_x dk_y \vec{e}_\alpha(k_x,k_y) T_\alpha(k_x,k_y) e^{-i\xi(k^2-k_x^2-k_y^2)^{1/2}} *$$

$$e^{ik_x x+ik_y y+ik_z^\alpha z} B[t-\tau-\xi/c_z-z/c_{gz}^\alpha]e^{-i\omega t}. \tag{2}$$

Here $\vec{e}_\alpha(k_x,k_y)$, $k_z^\alpha(k_x,k_y)$ - the polarization vector and z-component of the wave vector $\vec{k}_\alpha$ determined from Christoffel-Green equations, $c_{gz}^\alpha(k_x,k_y)= \partial\omega_\alpha/\partial\vec{k}$ - the vector of ray velocity. The integrand of (2) equation involves the fast - oscillating factor $exp(i\Phi(k_x,k_y))$ with the phase $\Phi(k_x,k_y)=k_x x + k_y y+zk_z^\alpha(k_x,k_y)-\xi(k^2-k_x^2-k_y^2)^{1/2}$. Value of the integral (2) will be determined mainly by integration over domains near stationary points of the phase $\Phi(k_x,k_y)$ for which the equations (3) hold:

$$\partial\Phi/\partial k_x = x + \xi\, k_x/k_z + z\,\partial k_z^\alpha/\partial k_x = 0$$

$$\tag{3}$$

$$\partial\Phi/\partial k_y = x + \xi\, k_y/k_z + z\,\partial k_z^\alpha/\partial k_y = 0.$$

Acoustic anisotropy of a medium is defined by the shape of the wave vector surface (or slowness surface). Vector $\vec{N} = \{-\partial k_z^\alpha/\partial k_x; -\partial k_z^\alpha/\partial k_y; 1\}$ gives at once the direction of external normal to the slowness surface at the point with a wave vector $\vec{k}_\alpha=\{k_x,k_y,k_z^\alpha\}$ and the direction of the ray velocity for the same vector $\vec{k}_\alpha$.

When the beam is focused onto the face of the object ($\xi=0$)

the equations (3) reduce to

$$x/z = -\partial k_z^\alpha/\partial k_x = N_x/N_z; \qquad y/z = -\partial k_z^\alpha/\partial k_y = N_y/N_z. \tag{4}$$

The equations hold only for those points of slowness surface in which the normal is parallel to the direction from the beam focus to the observation point. Near the stationary points z-component of wave vector $\vec{k}_\alpha$ can be expanded in a power series of $(k_x-k_{x_0})$ and $(k_y-k_{y_0})$:

$$k_z^\alpha(k,k) = k_z^\alpha + N_x(k_x-k_{x_0}) + N_y(k_y-k_{y_0}) + a_{2x}(k_x-k_{x_0})^2 + a_{2y}(k_y-k_{y_0})^2$$

$$+ a_{3x}(k_x-k_{x_0})^3 + \dots \tag{5}$$

Value of the field at the observation point can be evaluated using standard procedure of asymptotic evaluation of integrals. This point is ordinary when the coefficient of quadratic terms in the expansion (5) are nonzero: $a_{2x} \neq 0$ and $a_{2y} \neq 0$. In this case $\vec{U}(x,y,z,t)$ is divergent wave:

$$\vec{U}(x,y,z,t) \approx$$

$$kT_\alpha(k_x,k_y)\vec{e}_\alpha(k_x,k_y)A_\alpha(k_x,k_y)\frac{\exp\{ik_x x + ik_y y + ik_z^\alpha z - i\omega t\}}{r}$$

$$B(t-\tau-r/c_g^\alpha), \tag{6}$$

the amplitude of which reduces in inverse proportion to the beam focus-to-observation point distance r. Here c_g^α - the value of ray velocity in this direction, k- a coefficient involving an angle dependence on position of the observation point.

When $a_{2x} = a_{2y} = 0$ the wave vectors surface in the vicinity of the point (k_x,k_y,k_z^α) is nearly flat. The wave amplitude of the field decreases in the direction of the observation point much more slowly: $u \sim 1/r^{2/3}$ [1]. By experiment it appears as if a quasicollimated beam had propagated along this direction. There are many other types of topological features of the slowness surfaces that cause the particular character of propagation of an incident convergent beam in anisotropic medium. They must be topics of future studies.

There is a simple ray interpretation of the results. Focused acoustic beam in liquid can be represented as a set of rays each of that propagates in the direction of its own wave vector $\vec{k}$. The ray correspond to the vector $\vec{k} = \{k_x,k_y,k_z\}$ crosses the interface at the point $\{\xi k_x/k_z, \xi k_y/k_z, 0\}$ shifted in respect to the beam axis. At the interface the ray can break down into some refracted rays with different polarizations. Directions of these rays coincides with directions of external normals to different cavities of the slowness surface of a crystal at the points $\{k_x,k_y,k_z^\alpha\}$ with the carried with the ray is specified by the wave vector $\vec{k}_\alpha$ and the time of arrival of the ray at the observation point is defined by the value c_g of ray velocity.

If there is a group of incident rays whose wave vectors

after refraction belong to a flattened off part of a slowness surface, all these rays should be parallel with the external normal to this flat region. The rays form a collimated beam carrying the significant part of energy of the transmitted beam. Change of the phase $\Delta\varphi$ on a path l: $\Delta\varphi=k_n l$, is constant along each of the rays because of the projection k_n of a wave vector $\vec{k}$ on the normal $\vec{n}$ to the slowness surface at the point $\vec{k}$ is invariable for the flat region on the slowness surface. It follows, that a distribution of phase of oscillations over the beam cross-section doesn't change while propagating the beam.

At a depth d inside of the specimen the transmitted radiation is reflected from a plane inhomogeneity. Reflection, accompanied by mode conversion ($\alpha \rightarrow \beta$) is characterized by reflection coefficients $R_{\alpha\beta}(k_x,k_y)$. Reflected radiation goes back to the front surface of the specimen and partly passes through the interface with transmission coefficients $\tilde{T}_\beta(k_x,k_y)$.

Reflected field in a coupling liquid is

$$Pr(r,t)=$$

$$\iint dk_x dk_y R(k_x,k_y) A(k_x,k_y) e^{ik_x x+ik_x y-i(k^2-k_x^2-k_y^2)^{1/2}(z+\xi)-i\omega t} *$$

$$B[t-\tau-(\xi+z)/c_z]+ \sum_\alpha \iint dk_x dk_y T_\alpha(k_x,k_y) R_{\alpha\beta}(k_x,k_y) \tilde{T}_\beta(k_x,k_y) A(k_x,k_y) *$$

$$e^{ik_x x+ik_x y+i[\tilde{k}_z^\beta+k_z^\alpha]d-i\sqrt{k^2-k_x^2-k_y^2}(z+\xi)-i\omega t} *$$

$$B[t-\tau-(z+\xi)/c_z-d/\tilde{c}_{gz}^\beta-d/c_{gz}^\alpha],\tag{7}$$

where $\tilde{c}_z^\beta$ and $\tilde{k}_z^\beta$ - ray velocity and z component of a wave vector for waves travelling inside the specimen towards its surface. Particularly, equation (7) defines the space spectrum of the reflected beam in the focal plane of the lens. According to (3) the output signal $V(t)$ of acoustic microscope can be expressed by the convolution integral of the space spectra of reflected field and the field produced by the lens, both in the focal plane of the lens:

$$V(t)=\iint dk_x dk_y P(k_x,k_y) R(-k_x,-k_y) e^{-i2\xi(k^2-k_x^2-k_y^2)^{1/2}} B(t+2\xi/c_z-2\tau) *$$

$$e^{-i\omega t} + \iint dk_x dk_y P(k_x,k_y) \sum_{\alpha\beta} \tilde{T}_\alpha(k_x,k_y) R_{\alpha\beta}(k_x,k_y) T_\beta(k_x,k_y) *$$

$$e^{i\{-2\xi(k^2-k_x^2-k_y^2)^{1/2}+[k_z^\beta+k_z^\alpha]d\}} B(t+2\xi/c_z-2\tau-d/\tilde{c}_{gz}^\beta-d/c_{gz}^\alpha) e^{-i\omega t}.\tag{8}$$

Here $P(k_x,k_y)=A(k_x,k_y) A(-k_x,-k_y)$ - the aperture function of microscope. Signal contains 2 components, that correspond to acoustic pulses reflected from the face of a specimen and from internal inhomogeneity. The integrand in (8) involves a fast-oscillating factor $\exp[i\Phi_v(k_x,k_y)]$, $\Phi_v(k_x,k_y) = -2\xi(k^2-k_x^2-k_y^2)^{1/2}+ d [\tilde{k}_z^\beta(k_x,k_y) + k_z^\alpha(k_x,k_y)]$. As before the value of the integral (8) is defined by integration over the vicinity of stationary points Φ_v. Their positions are defined by equations

$$\partial\Phi_v/\partial k_x = 2\xi k_x/k_z + d[\partial\tilde{k}_{\beta z}/\partial k_x + \partial k_{\alpha z}/\partial k_x] = 0$$

$$\partial\Phi_v/\partial k_y = 2\xi k_y/k_z + d[\partial\tilde{k}_{\beta z}/\partial k_y + \partial k_{\alpha z}/\partial k_y] = 0.$$

(9)

These conditions are satisfied only for those rays in the incident beam that leave the specimen (after double refraction at the front surface and reflection at internal surface) at points disposed symmetrically about the beam axis. Real paths of the ray inside the specimen are of no significance, particularly. They can escape from the incidence plane (Fig.2).

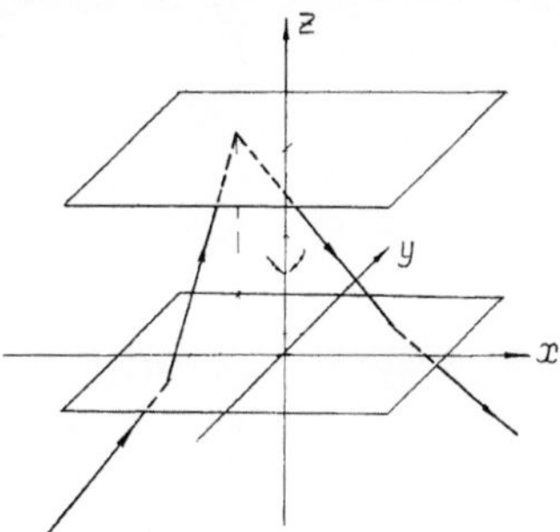

Fig.2. Scheme of output signal formation in anisotropic objects.

When conditions (9) are satisfied only for a finite number of wave vectors, the value of integral (8) is small. Signal from internal obstacle is weak against the background of the signal from the face of the specimen. Visualization of internal structures situated at a depth is possible when conditions (9) are satisfied for some cones of directions within the angle spectrum of the incident beam. In this case the incident beam forms the collimated beam inside the specimen which remains collimated after reflection at internal inhomogeneity.

METHODS OF OBSERVATION OF INTERNAL STRUCTURES

In a course of an experiment it is possible to change 2 parameters of microscope - 1) distance between a specimen and the acoustic lens and 2) position and width of electronic gate that defines beginning and duration of the work of the lens as a receiver. Control over these parameters is taken as a basis of methods of AM visualization of anisotropic objects.

The method of gate position control. The method is based on division in time signals produced by echo-pulses from the front surface and from internal structures. It can be used for study materials with the topological features of the slowness surface that provide the formation of the collimated beam directed normally to the specimen surface. The incident beam is focused on the specimen face. In the specimen a thin collimated beam appears. Its cross-section sizes of which are compared with a wave length. After reflection it goes back along the same direction. Signals reflected from different obstacles are shifted in time. Changing gate position we can form the output signal caused by pulses reflected at different depth. It makes AM visualization and reconstruction of 3-d structure of an object available.

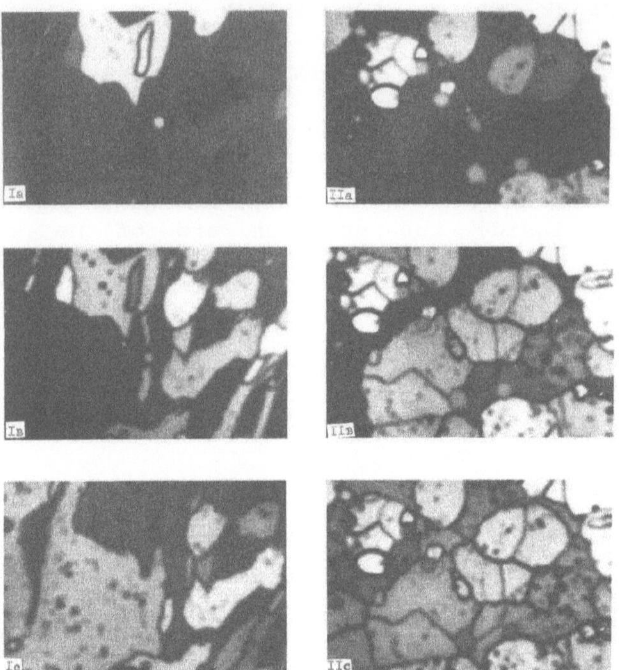

Fig. 3. Acoustic images of integral defect structure of crystal
pyrographite samples. I-specimen with small block structure,
II-specimen with large block structure.

Advantages of the method are evident. Limiting depth, up to
what the visualization can be realized is given by ultrasonic
attenuation and diffraction losses and does not depend on the
working distance of the lens. The focusing of beam at the
object surface gives maximum resolution deeply inside the
specimen. The method gives dramatic results for crystal
pyrographite (Fig. 3 and [1]). But the method can be used for
studying high anisotropic layered materials (like graphite)
only. The other defect of the method is existence of artifacts
caused by repeated re-reflections inside the specimen.

The method of combined control. There is wider field of
application for the other method in which a time position of
the electronic gate and a position of the lens with respect to
the object change simultaneously. It can be used for studying
structure of highly anisotropic materials for which there are
inclined flat domains of the slowness surfaceoriented
symmetrically about the axis of the incident beam.

After refraction the incident converge beam forms several
collimated beams symmetric about the beam axis. The beams are
formed, usually, by peripheral part of incident rays.
Conditions (9) can be not valid for these beams at a given
position of the lens. But shifting the lens towards the object
we can achieve fulfillment of conditions (9) for all rays
involved in collimated beam. If it is known orientation of a
flat region of the slowness surface (given by the angle Θ_g
between normals to the specimen surface and to the slowness
surface in this region) and it is known a mean angle Θ of in
incidence of rays which forms the collimated beam after

refraction, the necessary magnitude of the lens shifting is expressed by the depth of position of internal reflecting surface d:

$$\xi = d \ tg\Theta_g/tg\Theta \ . \tag{10}$$

Artifacts due to re-reflection are absent for this method but the limiting depth depends on the working distance of the lens 1: $d_{lim} = 1 \ tg\Theta/tg\Theta_g$.

By this method we studied the defect structure of Bi-2212 phase of high temperature superconductive crystal Bi-Sr-Ca-Cu-O for which the slowness surface contains flattened out areas[2].

Quantitative measurements. When the lens is in optimal position (10) the output signal V(t) gives possibility to measure more parameters - time interval Δt between ultrasound pulses reflected from internal and front surfaces.

According to (8) Δt is expressed by the depth of inhomogeneity position d and magnitude of ray velocity Cg in the direction of propagation of beam and by the angle Θ_g defining this direction:

$$\Delta t = d/(c_g cos\Theta_g) \tag{11}$$

The relations (9,11) connect two measurable parameters $\xi, \Delta t$ to four values, characterized as internal structure (depth of inhomogeneity position), as anisotropy and elastic properties of the specimen (ray velocity C_g and angles Θ, Θ_g). When two of them are known local magnitude of two other can be obtained from (9) and (11).

APPLICATIONS OF THE METHOD FOR STUDYING
DEFECT STRUCTURE OF CRYSTALS

Applications of AM in materials science is defined by its potentials. Resolution of acoustic microscopes does not exceed $0,5\mu m$ and it can be used for studying large -scale elements of crystal microstructure: cracks and delamina- tions, inclusions and intergrowthes of phases, grains, twins, domains.

We studied defect structure of some layered crystals (pyrographite, HTS cuprates of Bi and rare earths, hexagonal boron nitride α-BN) using proposed methods.Their elastic anisotropy caused by a difference between the interatomic interaction in the plane of an atomic layer and the interlayer interaction. Nature of atomic structure gives the special character of imperfections in layered crystals - most of them are extended plane defects, oriented along atomic layers.

Dominant elements of defect structure of crystal pyrographite are delaminations. They are extensive in the based plane and thin (Fig.3). In Bi-2212 phase HTS crystal main elements of the structure are inclusions of the other non-superconductive 2201 phase in a form of bands (Fig.4). Using proposed methods we also observed non-plane defects in hexagonal boron nitride. The sample of hexagonal α-BN contains spherical inclusions formed by bent atomic sheets like a head of cabbage. In the acoustic images (Fig. 5) they look as systems of concentric rings.

All images were made with industrial acoustic microscope Elsam at different working frequencies: f=100Mhz (pyrographite), f=1.6GHz (Bi-2212 crystal), f=200 and 400 MHz (αBN). Field of view is 0.65*1 mm.

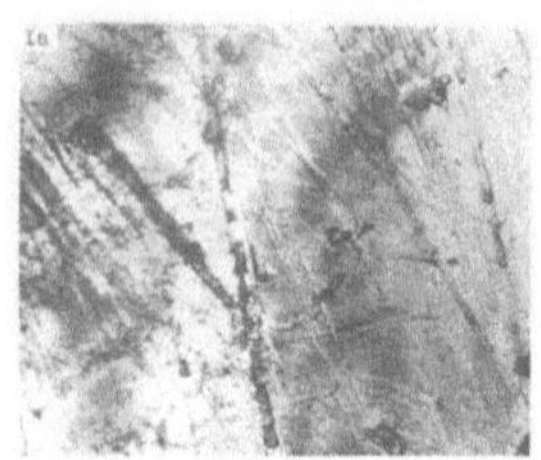

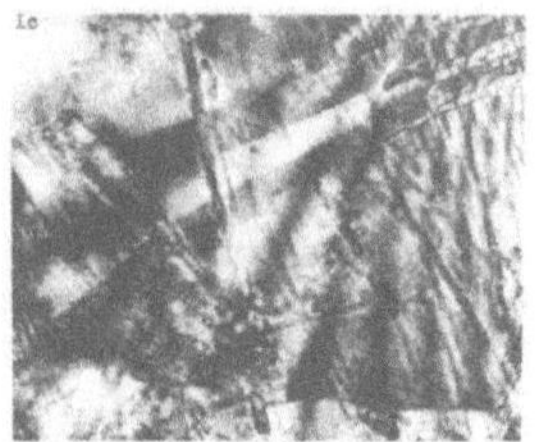

Fig. 4. Acoustic images of Bi-2212 phase Bi-Sr-Ca-Cu-O crystal.
I-image of the front surface ($\xi=0$), II and III - images of the
interior of the sample for different position of the gate
(shift of the lens $\xi=30$ mkm towards the sample surface).

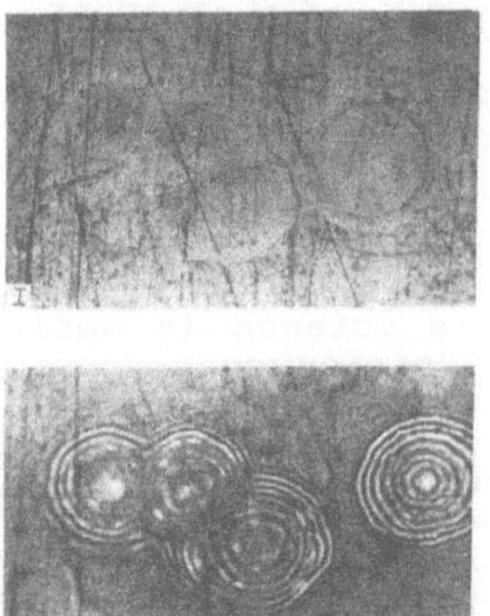
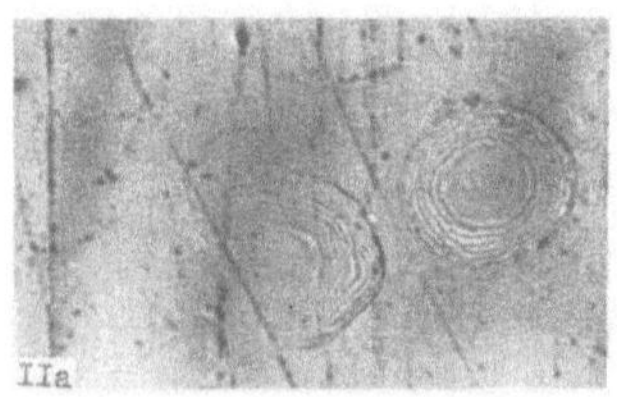
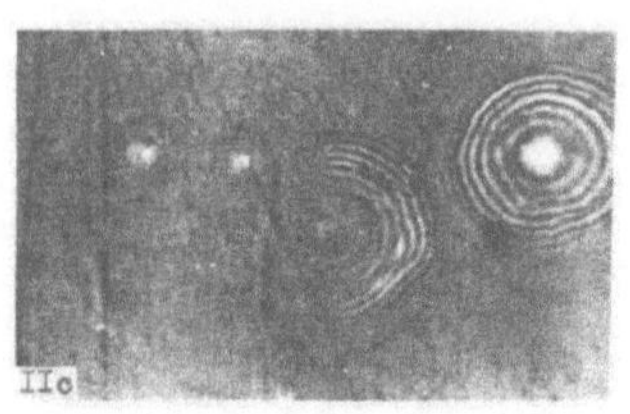

Fig. 5. Optical (I) and acoustic (II) images of α-BN sample.
IIa-IIc refer to different position of the gate.

REFERENCES

1. V.M.Levin,R.G.Maev et al,Study of structure and properties
of highly anisotropic materials by acoustic microscopical methods, in Acoustical Imaging, H.Ermert and H.-P. Harjes, ed.,
NY, London (1991).
2. L.A.Chernozatonskii,V.M.Levin et al, Phonon Focusing
influence on the formation of acoustic images in crystals, ibid.
3. C.F.Quate,A.Atalar,H.K.Wickramasinghe,Acoustic microscopy
with mechanical scanning-a review,Proc.IEEE. 67:1092(1979).

NEARFIELD SCANNING ACOUSTIC MICROSCOPY

A. Kulik, J. Attal[1], G. Gremaud

Ecole Polytechnique Federale de Lausanne,
Institut de genie Atomique, CH-1015 Lausanne, Switzerland

1) Permanent address: Laboratoire de Microacoustique
Universite Montpellier II, F-34095 Montpellier, France

INTRODUCTION

The resolution of a Scanning Acoustic Microscope is directly related to the used wavelength. The practical limit at room temperature is situated around 2 GHz. The main limitation comes from the excessive attenuation in coupling water, which is proportional to f^2, where f is the working frequency. Furthermore, scattering and attenuation in the material rarely allow one to obtain images from the depth of more a few wavelengths in the GHz frequency range. Technical difficulties related with high frequencies make the instruments bulky and expensive.

One of the possibilities to overcome these problems, is the use of a superresolution system. One of the first experiments was performed by Ash and Nichols[1], who used electromagnetic waves and obtained the spatial resolution of $\lambda/60$. Nearfield Scanning Optical Microscopes are actually commercially available. One of the first acoustic systems was designed by Durr et al [2] obtaining the resolution of $\lambda/4$. A so called "Pin" Scanning Acoustic Microscope was successfully built by Zieniuk[3] having the resolution of 10 µm at 35 MHz, working in transmission. Using micromachining, Khuri-Yakub et al [4] achieved a resolution of 1000 Å working at 135 MHz. Takata et al[5], combined a Scanning Tunneling Microscope (STM) and an acoustic system, using a vibrating tip at 70 kHz as the acoustic source. Here, due to extremely short distance between the tip and the surface, short range interactions were sufficiently strong to excite detectable acoustic signal in transmission. Uozumi et al [6] propagated pulsed ultrasound of 1.2 MHz in the STM tip. In the last two systems a spatial resolution of the order of 100 Å was obtained. The lowest frequency (32.7 kHz) set-up was built by Gunthner et al.[7] They used a vibrating tuning fork near the sample surface. The sharp corner of the tuning fork was acting as a tip and spatial resolution was better than 3 µm. Furthermore, the influence of the air pressure on coupling was studied.

From the above cited literature we can deduce that there is no clear relation between obtainable resolution and wavelength when working in near field.

EXPERIMENTAL

We used Mason's horn[8], working at its resonant frequency (near 1 MHz) to obtain superresolution images with a Continuous Wave Setup. A ceramic transducer (PZT) of 10 mm diameter was bonded to the Aluminium horn which was tapered exponentially down to 0.1 mm. The length of the horn was chosen to insure the resonant frequency of the system to lie inside the transducer bandwidth. After the first experiments we found that the resonant frequency and the amplitude of vibrations depend heavily on the depth of the immersed part of the horn. In order to avoid lateral radiation, we suppress the acoustic leaks using an expansed polyurethan sponge. The sponge surface was protected from water by silicon rubber. The transducer system was suspended in a node of vibrations and mounted on a XYZ motorized table. This computer driven table allows controlled approach of the sensor and raster scanning.[9] The overall response of the system - so called s_{11} reflection scattering parameter - was measured using a Network Analyzer working at the constant, resonant frequency of the sensor. We measured the s_{11} scattering parameter as a function of the spatial position of the transducer over the sample.

RESULTS

Typical behaviour of the resonant frequency and s_{11} amplitude as a function of the distance tip-sample is presented at Fig. 1. It is easy to locate the point of mechanical contact between the tip and sample. Amplitude s_{11} decays as a function of distance, while changes in resonant frequency are difficult to detect outside contact region.

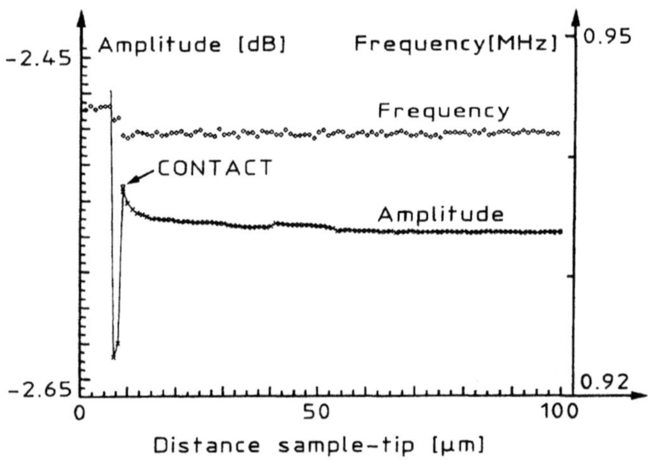

Fig. 1. Amplitude of the s_{11} of the horn and its resonant frequency as a function of the distance between the tip and sample. Sample material : glass, input power 20 dBm.

A test sample made from a printed circuit (PCB) board (glass fiber reinforced epoxy with a bonded thin copper layer). Different figures were etched in the copper layer using a standard PC photoresist technology. We approached the tip down to contact and moved up a few µm. At this stage CW images were taken [Fig. 2,3]. A clear contrast from the remaining Cu layer (white) was observed, as well as diagonal stripes corresponding to the shape of the glass fibers under the surface. The latter contrast is even better visible on Fig 3 - it has the form of stripes almost perpendicular to the copper stripes.

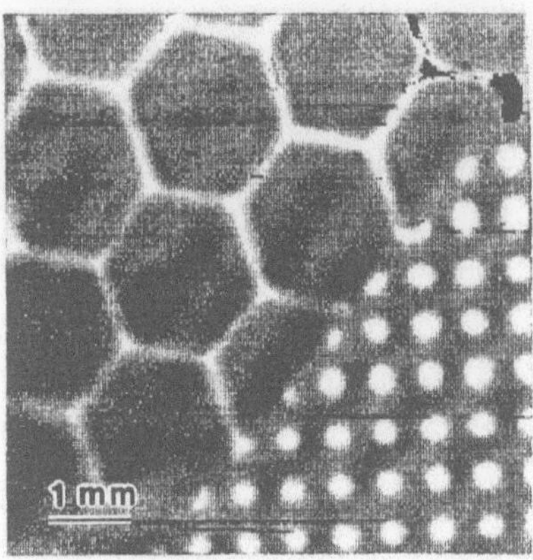

Fig. 2. CW superresolution image obtained using 1 MHz Mason's horn. sample: printed circuit board. Contrast inversion in upper right corner corresponds to the mechanical contact tip - sample.

Fig. 3. CW image of the different area of the same sample as Fig. 2. Note diagonal stripes, perpendicular to the copper stripes. They correspond to the structure of the glass fibers used in the composite. Contrast inversion is due to contact tip-sample.

From the line scans we could estimate the spatial resolution of the system to be 0.1 mm. This means that the resolution was equal to the diameter of the used horn. The maximum distance between the transducer and the sample where images were obtainable was of the order of 50 μm. Spatial resolution diminishes quickly with the increasing working distance.

CONCLUSIONS

CW acoustic superresolution images were obtained using a relatively simple set-up. The observed images present subsurface features, probably related to the local changes of the acoustic impedance of the underlying material. The apparent spatial resolution of the system was determined by the geometry of the used horn. The most difficult practical problem of the superresolution setups is the knowledge of the real working distance between the sensor and the sample surface. The use of an independent signal allowing distance measurement eg. tunneling[5] current in a TAM could be a right solution of the problem. A good revue of the available methods was presented by Wickramasinghe[10].

REFERENCES

1. Ash and G. Nichols, Superresolution aperture scanning microscope, *Nature* , 237:510 (1972)
2. W. Dürr, D.A. Sinclair and E.A. Ash, High resolution acoustic probe, *Electronic Letters*, 21:805 (1980)
3. J.K. Zieniuk and A. Latuszek, Non-conventional pin scanning ultrasonic microscopy, *in*: "Acoustical Imaging vol 17", H. Shimizu et al, Plenum Press, new York (1989)

4. B.T. Khuri-Yakub et al, Near field acoustic microscopy, *in*: "Scanning Microscopy Instrumentation", SPIE, San Diego (1992)

5. K. Takata et al, Tunneling acoustic microscope, *Appl. Phys. Lett.* 55(17):1718 (1989)

6. K. Uozumi and K. Yamamuro, A possible novel scanning ultrasonic tip microscope, *Jap. J. Appl. Phys.* 28(7):1297 (1989)

7. P. Günther et al, Scanning near-field acoustic microscopy, *Appl. Phys. B*, 48:89 (1989)

8. W.P. Mason and R.F. Wick, A barium titanate transducer capable of large motion at an ultrasonic frequency, *J. Ac. Soc. Am.* 23(2):209, (1951)

9. A. Kulik et al, Continuous Wave Reflection Scanning Acoustic Microscope, *in*: "Acoustical Imaging vol 17", H. Shimizu et al, Plenum Press, new York (1989)

10. H.K. Wickramasinghe, Scanning probe microscopy : current status and future trends, *J. Vac. Sc. Technol. A* 8(1):363 (1990)

ACOUSTIC MICROSCOPE LENS OPTIMIZATION
FOR SUBSURFACE IMAGING

Konstantin Maslov

Institute of Chemical Physics Russian Academy of Science
Kosygin st., 4, Moscow, 117334, Russia

INTRODUCTION

The acoustic microscope image of internal structure of solid sample is complicated by high spherical aberrations, that can limit a penetration depth of a microscope[1], and by formation of acoustic waves of different polarizations in the sample body and on the sample surface[2,3]. A short ultrasonic pulse passed through acoustic lens and reflected by uniform isotropic solid layer split in five pulses due to an ultrasonic wave reflection at surface, Raleigh wave formation on the sample surface[4], longitudinal and transverse wave reflection from back surface, and longitudinal to transverse wave conversation on back surface of the sample. In simple case of uniform solid layer the pulses is time resolved, but if investigated body contain discontinuities at different depth the image would complicate by interference of it. It is especially true for high frequency ultrasonic microscope that use a long lasting ultrasonic pulses[5].

The non spherical lens can be used[6] to reduce the aberration for given propagation mode. Unfortunately shape of this lens depended on both material sound velocity and depth of focusing. The special spherical coupling piece can be introduced in front of conventional acoustic lens[7] that radically reduce the aberration but it produced extra reflections diminishing a sensitivity of microscope.

A method of suppression of Raleigh wave[8] or specular reflection[9] by using an acoustic lens with satiable transducer to exclude some excitation angles is will known.

An analysis of usefulness of different propagation modes for flaw detection and a choice a proper aperture of the lens and transducer to rise acoustic microscope sensitivity and diminish an influence of mode interference on image formation is a scope of present paper.

ANGLE DEPENDENCE OF REFLECTION COEFFICIENT FROM
LIQUID FILED FLAT GAP IN ISOTROPIC SOLID

A flat gap of h_i width parallel to the surface of uniform isotropic body displaced at depth - h is used as a model of a flaw in solid. The gap may be empty or liquid filled. The ray tracing model[3,10] is used to calculate the resulting transducer voltage. For relatively long ultrasonic pulses when ultrasonic wave can be considered as monochromatic the complex reflection coefficients corresponded to chosen angle of incidence can be calculated as a combination of reflection and transmission coefficients for flat ultrasonic wave[11]. The viscous wave formation in liquid was neglected due to small wavelength of viscous wave.

The resulting reflection coefficients for three propagation modes: shear - K_s, longitudinal - K_l and converted from shear to longitudinal and vice versa on gap surface - K_c for thin gap when lateral shift of the ray within gap is negligible is respectively:

$$K_1 = W\, D_1 \left(B_{11} + D_1 W_1 N_1 V_1 \right) \qquad (1)$$

$$K_t = P\, D_t \left(B_{t1} + D_t P_1 N_1 V_1 \right) \qquad (2)$$

$$K_c = W\, D_t B_{1t1} + P\, D_1 B_{t11} + \left(W\, D_t P_1 D_{11} + W_1 D_{t1} P\, D_1 \right) N_1 V_1 \qquad (3)$$

with:

$$N_1 = \sum_{n=0}^{\infty} \left(V_1^2 \exp(-2jk_1 h_1 \cos(a_1)) \right)^n = \frac{V_1 \exp(-2jk_1 h_1 \cos(a_1))}{1 - V_1^2 \exp(-2jk_1 h_1 \cos(a_1))} \qquad (4)$$

where V and $B_t = B_1$ is a reflection coefficients for liquid-solid and solid-liquid interface, B_{tl} and B_{lt} is shear to longitudinal and longitudinal to shear wave transform coefficients, P, W, D_l, D_t is a transmission coefficients for liquid-solid interface, and solid-liquid interface for longitudinal and shear waves respectively[10]; a is an angle of refraction; k is wave vector, j - imaginary unity, i indicate liquid filled in a gap.

Figure 1 shows reflection coefficients for some liquid filled in 10^{-3} mm. gap at 1.1 mm. depth in silicon sample. The acoustic impedances of materials is equal 842, 1498, 2237, and 2423 g/cm^{2}sec, for decane, water, dibromethane, and glycerol respectively. The reflection coefficients have approximately the save shape for large variety of samples if it's ultrasound velocities satisfy the equation $c < c_s < c_l$.

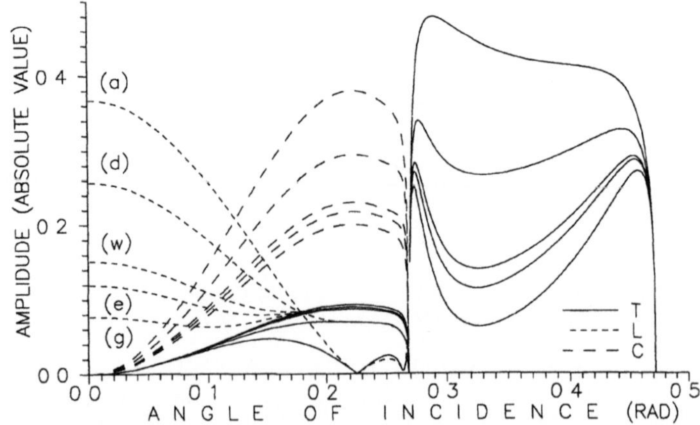

Figure 1. Reflection properties of a 10^{-3} mm. flat liquid filled gap displaced 1.1 mm. depth in silicon sample: no liquid (a), decane (d), water (w), dibromethane (e), and glycerol (g). Longitudinal wave - L, transverse wave- T, wave modes conversion on gap surface - C.

The reflection coefficient for longitudinal wave in solid is high for aperture angles smaller than approximately 1/2 of the critical angle. The reflection coefficient for waves with polarization transformed on back sample surface becomes high with the aperture angles larger than 1/2 of critical and rapidly drop near it. For angles larger than critical only shear wave reflection exists and also rich the same order value.

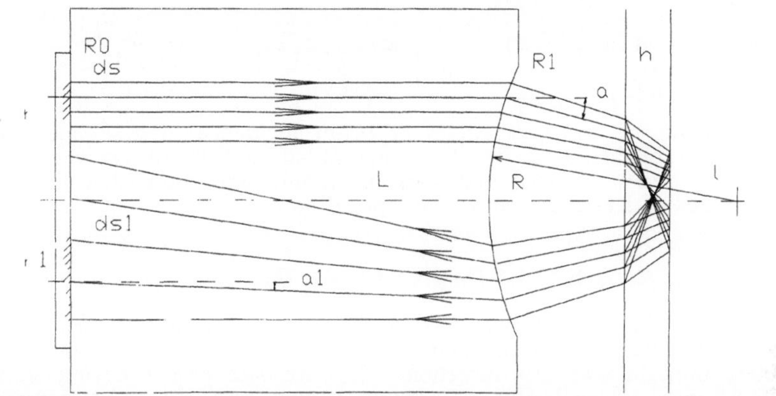

Figure 2. Lens and sample geometry using for received acoustic signal calculation.

THE TRANSDUCER VOLTAGE INDUCED BY THE ACOUSTIC REFLECTION FROM GAP

The geometry of acoustic lens system is presented on figure 2. A spherical acoustic lens of radius of curvature R and with aperture R_1 was made in silicon buffer rode of length L against electroacoustic transducer of R_0 radius.

To analyze the propagation of both pulsed and continuos wave in a complex case of a real lens and thick solid we use a simplified ray traced theory. The resulting transducer voltage cam be presented as a sum of includes of the rays passed through the lens surface and reached of transducer[10,13]. The phase shift of a ray from axial one can be calculated using geometrical ray path difference and phase shift due to complex nature of reflection coefficients. An amplitude of a ray can be estimated from energy conservation low, so cold energy tube method[10] that is not a good approximation but is valid due to large value of kR and specular character of reflection within critical angle aperture with reflection coefficients do not rapidly changing with incidence angle. The result can be expressed as:

$$U = \frac{1}{2\pi R_0^2} \int K(a) \exp(-jkx) \frac{\sqrt{2\pi r \, dr \cos(a_1)}}{\sqrt{2\pi r_1 \, dr_1}} \, 2\pi r_1 \, dr_1 \qquad (5)$$

with:

$$\sin(a) = r \left(\sqrt{(n_0 R)^2 - r^2} - \sqrt{L^2 - r^2} \right) / (n_0 R^2) \qquad (6)$$

$$a_1 = a_2 - a - \arcsin(n_0 \sin(a_2)) \qquad (7)$$

$$\sin(a_2) = \frac{r}{n_0 R} + 2\,\frac{l}{R}\,\sin(a) - \frac{h}{R}\,\cos(a)\,(\tan(a_t) + \tan(a_1)) \qquad (8)$$

$$r_1 = (\,R + L\,)\tan(a_1) + n_0 R\,\sin(a_2)\,/\,\cos(a_1) \qquad (9)$$

$$x = \frac{1}{n_0}\left(\frac{L - L\,(1 - \cos(a_2 - a))}{\cos(a_1)} + L + L - \sqrt{L^2 - r^2}\,\right) +$$

$$+ \frac{r - L\,\sin(a_2 - a)}{\sin(a)} + h\left(\frac{c/c_1 - c_1/c}{\cos(a_1)} + \frac{c/c_t - c_t/c}{\cos(a_t)} \right) \qquad (10)$$

where: a, a_2, a_1, a_1, a_t is angles of beam incidence to a sample, lens and transducer and refraction angles for longitudinal and transverse wave in solid sample; c_1, c_t and c - sound velocities in sample and immersion liquid; n_0 is refraction index of lens material; x - ray optical path

NUMERICAL AND EXPERIMENTAL RESULTS

We used an improved version of low frequency acoustic microscope made in our laboratory[14] with a scanning step of 0.01 mm. in a plane of a sample surface and 0.0002 mm. in lens axis direction. All ultrasonic probes having transducer diameter 6.5 mm., silicon buffer rode length 30 mm. and lens aperture diameter 6 mm. Four periods of carrier frequency of 25 Mhz were used for transducer excitation when different propagation modes must be distinguished and 50 period if mixture signal is used.

To measure an influence of gap width on transducer voltage a model system consisting of a 1.1 mm. thick glass plate with a convex lens made of the same material pressed to it's back surface was used. The gap was filled with low viscosity high acoustic impedance liquid - dibromethane. The gap width was controlled optically using interference ring picture in monochromatic light.

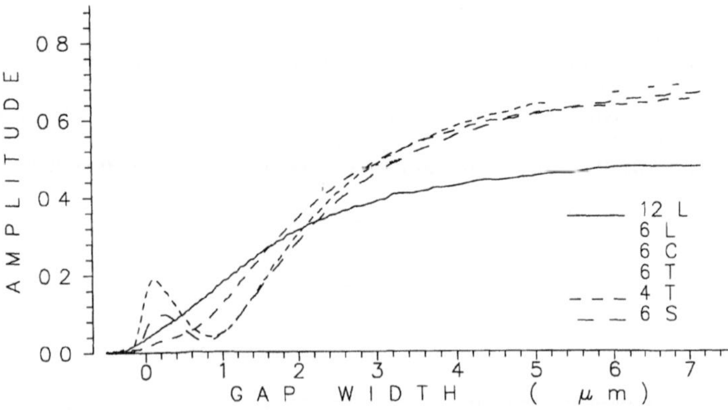

Figure 3. The acoustic microscope signal from dibromethane filled gap in silicon for longitudinal wave - L, transverse wave - T, converted wave - C and for lens with absorber of 5 mm. diameter in a central part of it - S. Numbers in legends mean lens radius of curvature.

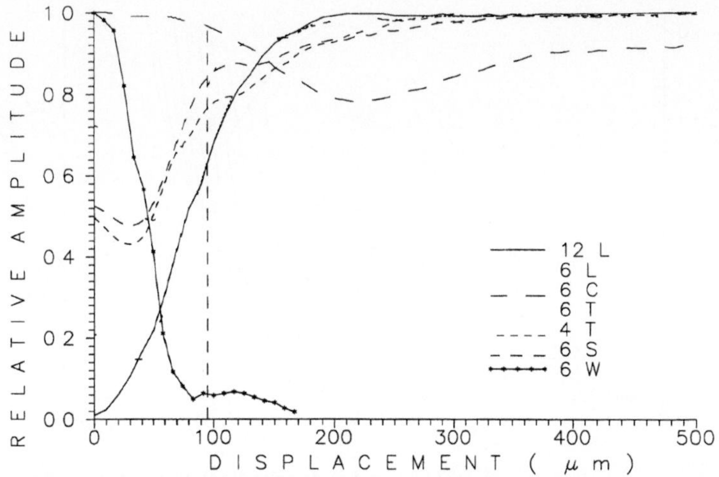

Figure 4. One dimensional image of 0.19 mm. diameter region of a contact of 1.25 mm. glass ball pressed to back surface 1.1 mm. glass plate for longitudinal - L, transverse - T and converted - C waves, and for lens with absorber - S. Numbers on legends mean lens curvature radius in mm., curve - W is an image of 0.04 mm. glass ball immersed in a water.

The distance between two first optical interference rings was about 1 mm. so gap boundaries can be considered as parallel. The result is presented on figure 3.

Signal amplitude is normalized on a signal from free back surface for the same lens and wave propagation mode. It can be seen, the sensitivity is high for acoustic modes having a large transverse component e.g. for large aperture lens and longitudinal wave, relatively small aperture lens and shear wave, and highest for mode conversion on a defect surface.

As a model for defect of limited dimensions a glass ball pressed to back side of 1.1 mm. glass plate was used. A diameter of contact region of two glass surfaces as measured optically was 0.19 mm., which for is equivalent to 0.19 mm. diameter disc introduced in glass at 1.1 mm. depth parallel to glass surface. The results is presented in figure 4.

In spite of one dimensional case, the highest signal achieved for longitudinal wave. The lateral resolution is approximately the same for longitudinal and shear waves and compatible with pressure distribution width in the acoustic lens focuses (see curve W on figure 4). The image of a contact region in mode converted wave contain large aberrations and has a weak contrast. The image radius is approximately equal to the distance at a back surface of a sample from lens axis to the place of stationary phase ray reflection.

To estimate the influence of aberration on a signal amplitude a ray optical path was calculated for particular case of 1.1 mm. silicon glass and lens of 6 mm. radius of curvature. The results is presented in figure 5 and figure 6 and measured as an geometrical path difference in water between ray corresponding the chosen angle of incidence to the sample front surface and axial beam. Dashed line indicate the angles where acoustic beam dos not reached a transducer.

For longitudinal wave (see figure 5) there is a large region of incidence angles within approximately 1/2 of a critical angle where ray path difference variation is an order of one micron. It allows to suppose a penetration depth of a pulse mode acoustic microscope of an order of 1 mm. for up to 1 Ghz ultrasound frequencies.

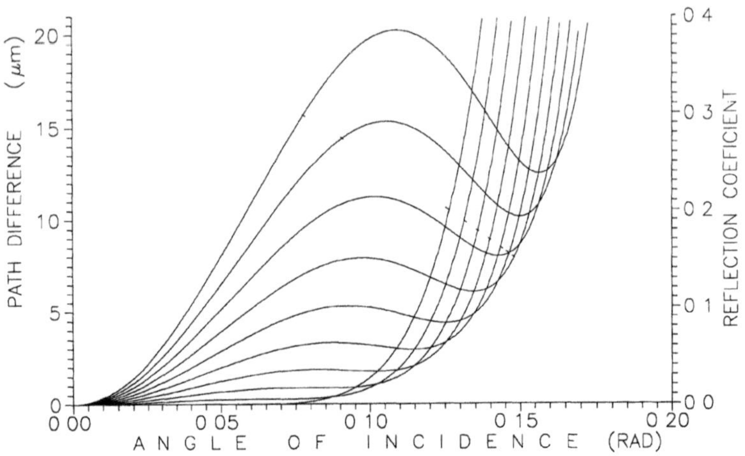

Figure 5. Acoustic lens ray path difference for longitudinal wave propagation in 1.1 mm. silicon layer for depth of focusing changed from 4.1 mm. to 5 mm. with 0.1 mm. step.

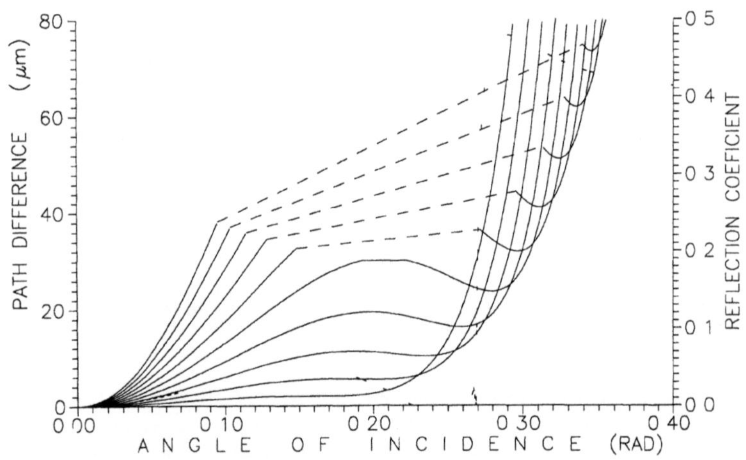

Figure 6. Acoustic lens ray path difference for transverse wave propagation in 1.1 mm. silicon layer for depth of focusing changed from 2.5 mm. to 3.4 mm. with 0.1 mm. step.

For transverse wave, in angle diapason where ultrasound wave reflection coefficient is high, the ray path change rapidly with aperture angle (see figure 6). There is only the narrow aperture angle diapason near the stationary phase angle where ray path does not essentially changed. It limit a penetration depth of acoustic microscope working with transverse wave by some tens of wavelength.

So, to optimize a lens for transverse wave under surface imaging it is possible to limit transducer area by narrow ring corresponding to stationary phase angle.

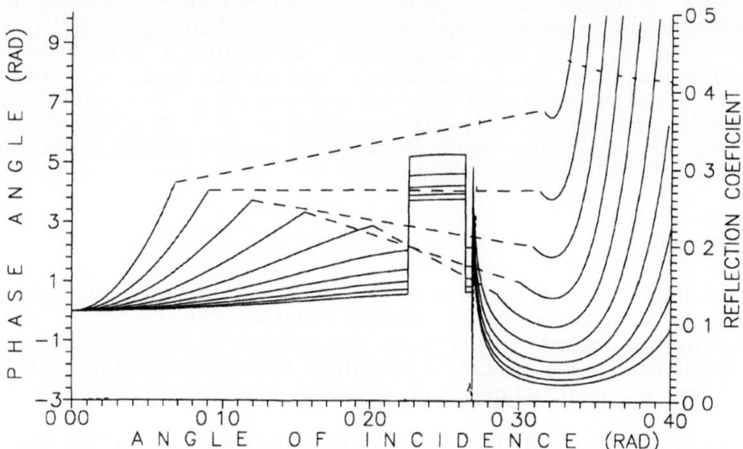

Figure 7. Acoustic lens ray phase shift for transverse wave in silicon layer. Layer sickness varied exponentially from 0 1 mm. (lower curve) to 2.56 mm. (upper curve). Depth of focusing corresponds to stationary phase incidence angle of 0.32 rad.

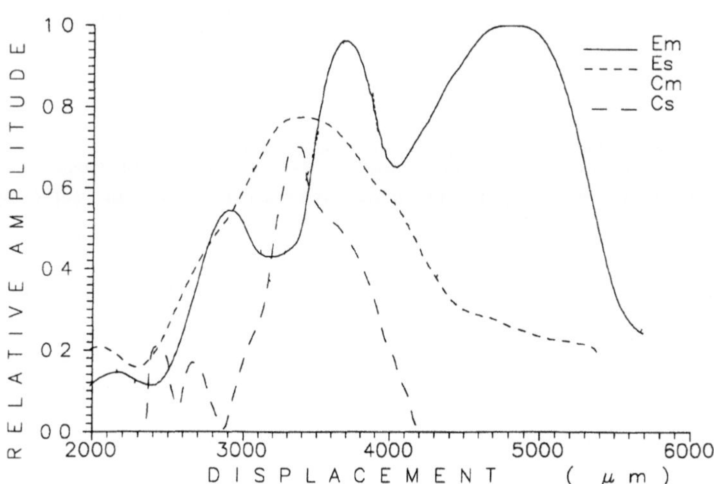

Figure 8. V(z) curves of 1.1 mm. thick silicon plate for long lasting pulse: calculated (C_m, C_s) and experimental (E_l, E_s) results for obvious lens and for lens with 5 mm. diameter absorber placed on the lens surface.

The aperture ring diameter depends on sample ultrasonic velocities but not from the depth of a reflected surface in a sample (figure 7). Thus the amplitude response of an acoustic lens can be improved and also front surface specular reflection and the other wave propagation modes can be suppressed.

The result of numerical calculation of transducer voltage in a form of acoustic signature[13] along with experimental results for flat silicon glass of 1.1 mm. thick and lens radius R of 6 mm. is presented on figure 8. The theoretical and experimental results is in a satisfactory agreement. When a round acoustic absorbing material occupying 3/4 of a lens area was placed in a central region of a lens a complex character of V(z) curve is simplified up to that for shear wave due to switching off the other wave propagation modes although the maximum amplitude of a signal dropped insignificantly.

CONCLUSIONS

In one dimensional case the most sensitive for thin liquid filed delamination detection is an acoustic lens of an aperture equal to critical angle for longitudinal wave when using a regime of wave mode conversation at the delamination surface.

For small defects the maximum amplitude response achieved for longitudinal wave and lens of approximately half a critical angle aperture with good lateral resolution.

The large aperture lens and shear wave propagation in sample can be used. The amplitude response of it can be improved by absorber plate introduced in a central part of the lens and by using a special transducer of a form of concentric narrow ring. The penetration depth of such a system was not as good as for small aperture lens bat it can be used for thin specimens or low defect depths due to strong suppression of surface reflection and better traveling time resolution.

REFERENCES

1. A. Atalar, Penetration depth of the scanning acoustic microscope, *IEEE Trans. Sonics Ultrason.*, 32:164 (1975).
2. N.J. Burton, NDE application o scanning acoustic microscopy, *IEE Proc.*, 134:283 (1987).
3. J.K. Wang, C.S. Tsai, Reelection acoustic microscopy for thick specimens, *J. Appl. Phys.*, 55:80 (1984).
4. M.R. Weaver, M.W. Daft, A.D. Briggs, A Quantitative acoustic microscope with multiple detection modes, *IEEE Trans. Ultrason., Ferroelec. Freq. Control.*, 36:554, (1989).
5. M. Hoppe, J. Bereiter-Hahn, Application of scanning acoustic microscopy - survey and new aspects, *IEEE Trans. Sonics Ultrason.*, 32:289 (1975).
6. J. Perdijon, Acoustical system for the focusing of a broad ultrasonic beam, *Material Eval.*, 48:497 (1990).
7. I.R. Smith, D.A. Sinclair, H.K. Wickramasihghe, NDE of solids with a mechanically B-scanned acoustic microscope, *in:* "Acoustic Imaging", Vol 12, Plenum, New York (1982).
8. M. Nicoonahad, P. Sivaprakasapillai, E.A. Ash, Raleigh wave suppression in reflection acoustic microscopy, *Electron. Lett.* 19:906 (1983).
9. A. Atalar, Improvement of anisotropy sensitivity in the scanning acoustic microscopy, *IEEE Trans. Ultrason. Ferroelec. Freq. Contr.*, 36:246 (1978).
10. C.F. Quate, Microwaves acoustic and scanning microscopy, *in:* "Proc. of Rank Prize Conference on Scanning Image Microscopy", Academic, New York (1980).
11. A. Atalar, An angular spectrum approach to contrast in reflection acoustic microscopy, *J. Appl. Phys.*, 49:5130 (1978).
12. L.M. Brekhovskikh, "Waves in Layered Media", Academic, New York (1980).
13. W. Parmon, H.L. Bertoni, Ray interpretation of the material signature in the acoustic microscopy, *Electr, Lett.*, 15:684, (1979).
14. K.I. Maslov, Acoustic scanning microscope for investigation of subsurface defects, *in:* "Acoustic Imaging", Vol 19, Plenum, New York (1992).

LOW-TEMPERATURE SCANNING ACOUSTIC MICROSCOPY IN RANGE OF 85-225K

R.G. Maev[1], L.A. Chernozatonskii[1], A.F. Denisov[1], G.S. Abilov[1], M.A. Buchny[1], V. Muller[2], C. Hucho[2], and D. Maurer[2]

[1]Center of Acoustic Microscopy, Institute of Chemical Physics Russian Academy of Sciences, Moscow, 117977, Kosygin Str. 4, Russia
[2]Freie Universitat Berlin, Farbereich Physic, Arnimallee 14, D-1000 Berlin 33, Germany

INTRODUCTION

Although acoustic microscopy is well-known to be an effective non-destructive control of local elastic and structural properties in the bulk of both opaque and transparent materials, the applicability of commercial scanning acoustic microscopes (SAM) is limited to temperatures near ambient, because SAM calls for the use of on immersion liquid. On the other hand, since the discovery of high-Tc superconductivity, there is a rapidly growing interest in non-destructive low-temperature examination of ceramics, crystals and devices made out of these materials. It is interesting to investigate other materials also in low-temperature region.

INVESTIGATIONS

We extended the applicability of SAM to temperatures as low as 85K using liquid propane as adequate immersion liquid. An adequate low-temperature chamber - Fig.1 (Chernozatonskii, Abilov et al.) as well as a speciallens-adapter had to be developed in order to overcome the difficulties arising from the large temperature gradient between the lens and its scanning unit. The images of the grating (Fig.2) were obtained at different and it could be demonstrated that the setup (ELSAM plus the cryostat) operates properly in the range of 85-225K. Furthermore a $SrTiO_3$ single crystal (used a well-known substrate for high-Tc films) was investigated in this range. No unusual structure was observed in sample at temperatures higher than 108K. However a subsurface structure appeared at about 108K and became more pronounced at lower temperatures. The subsurface structure looks like scratches which sometimes are parallel to each other. This structure is associated with the formation of domains due to a structural transition near 106K. After heating structure disappeared. We also observed the acoustic image in Bi-HTSC single crystals in the temperature range of 85-225K. The observed structure, which is indicative for a domain wall, cannot be observed if focusing on the surface with a delay time usually applied in SAM for obtaining both surface and subsurface images. We determined surface acoustic wave

Fig. 1

(SAW) velosity on the (a,b)plane of $Bi_2Ca_2Sr_2Cu_3O_x$ crystal at 100K from V(Z) curve - Fig.3.

In such crystals we also studied phonon focusing effect, which was predicted and calculated for acoustic image case by Chernozatonskii et al. (1988,1991) and at first was observed by Every et al.(1990) in transmission variant of laser induced acoustic pulse and by Levin et al (1991) in reflect variant of SAM. Due to this effect, subsurface structures may be observed not only focusing on the respective subsurface region but also focusing on the surface of the sample and changing the delay time of the receiver window in appropriate manner. In real space this means that an acoustic beam focused in "point" on the surface of a layered specimen, for example on (001) plane of $Bi_2Sr_2Ca_2Cu_3O_x$ crystal - Fig.4 were we see fast transversal mode computer image of point source, will propagate in the interior of the sample in a predominantly unfocused manner as shown in Fig.5. In this Figure we see the artificial cross on the opposite surface of the Bi-oxide crystal (300 m thick) after choosing gate position corresponds to "there-here" propagation of transversal acoustic waves (Z=0, T=100K). As indicated by the arrow in our imagings, a subsurface crack becomes visible not only by subsurface focusing. We mention that the observed microcracks at T=100-120K was not visible at room temperature. It originates from internal stress developing upon cooling the sample.

Although we have confined ourselves to a few examples only, we may state that a low-temperature acoustic microscope provides valuable information about crystal perfection in particular HTSC compounds which otherwise are hard to get.

200 MHz   88°K   1000x720 mcm
200 MHz   102 K   1000x720 mcm
200 MHz   205 K   1000x720 mcm

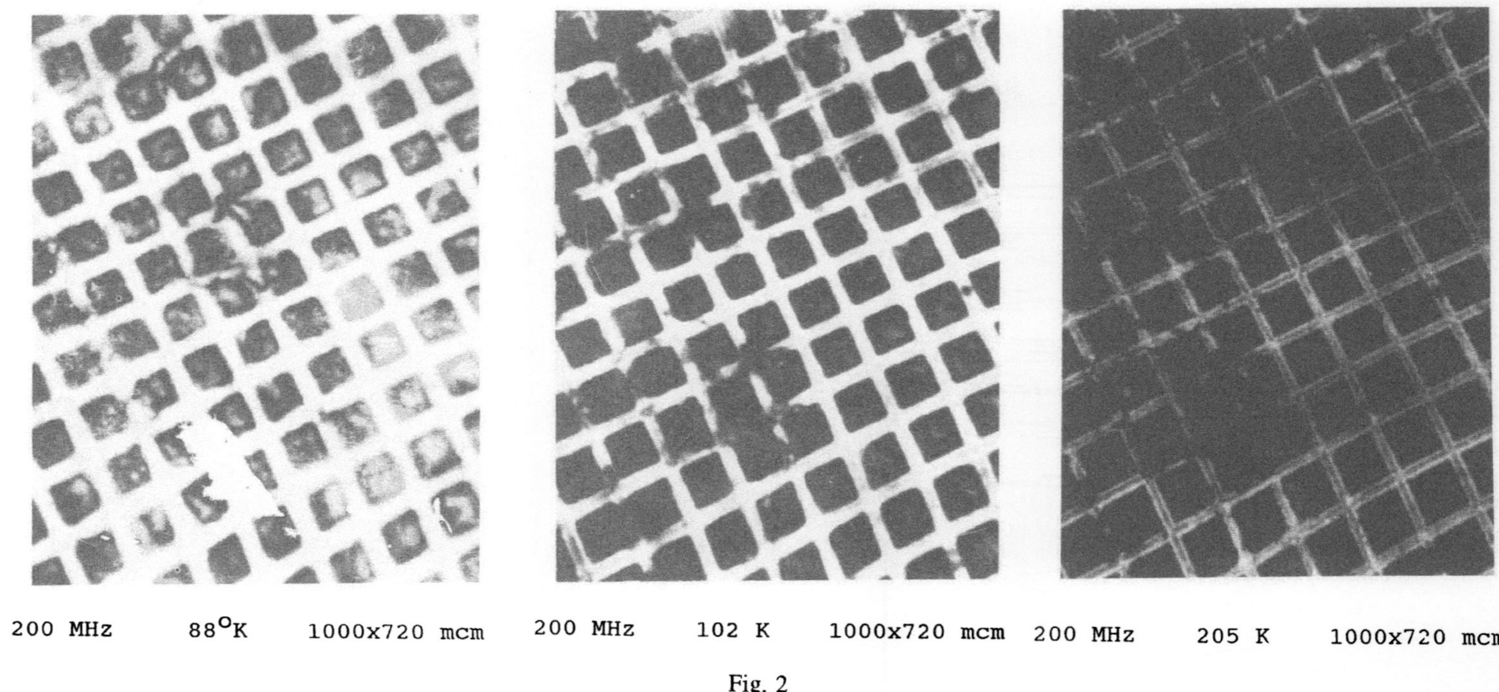

Fig. 2

Delta_Z   =   3.216 micron
Vlsaw   =   2643.9 m/sec
100
90
80
70
60
50
40
30
20
10
-7.68   0.0   7.68   15.36   23.04   30.72   38.40   45

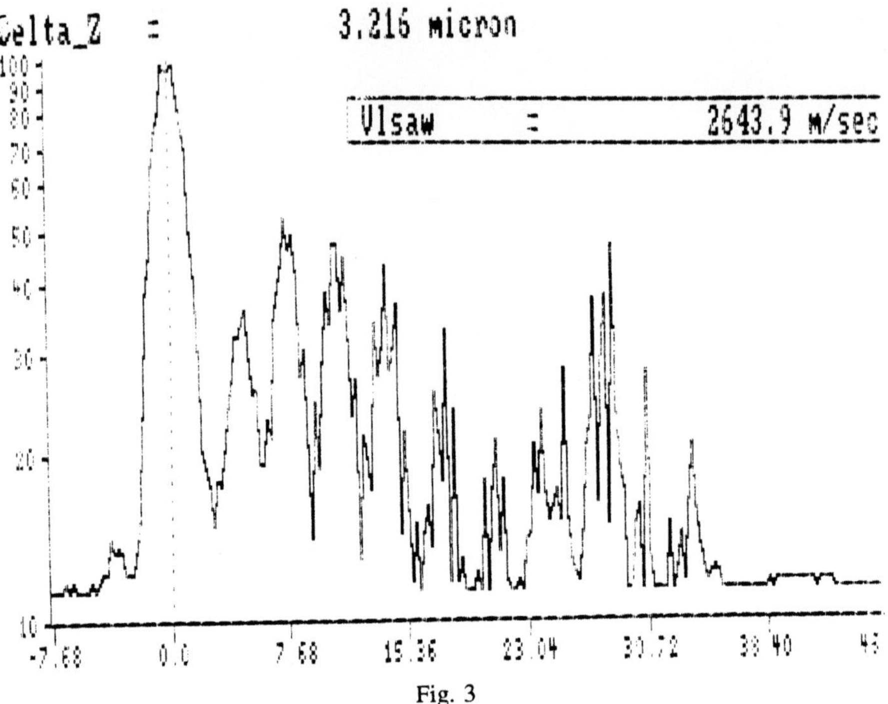

Fig. 3

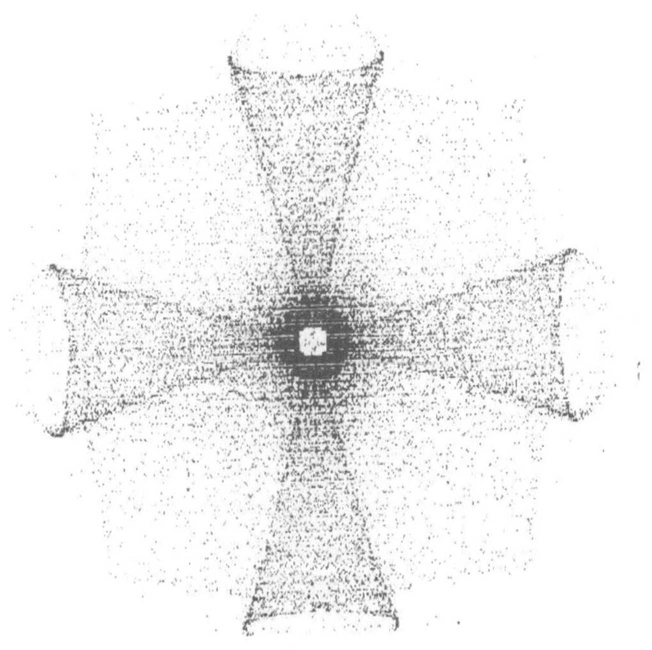

Fig. 4

Fig. 5   Sample: BiCaSrCuO Steshov's

ACKNOWLEDGEMENTS

We are very grateful to T.Schuring in providing us with a $SrTiO_3$ crystal and to S.M.Steshov in providing us with Bi-oxide crystal.

REFERENCES

Chernozatonskii L.A., Abilov G.S. et al., 1991, Low-temperature chamber for scanning acoustic microscope. Russian patent from 29 oct. 1991
Chernozatonskii L.A.& Novicov V.V., 1988, About possibility of ultrasound concentration observation. *Acoust.J.Soviet Phys.*, 343
Chernozatonskii L.A. et al., 1991, Phonon focusing influence on the formation of acoustic images in crystals. Acoustical Imaging 19, ed. H.Ermert & H.-P.Harjes,N.Y. London
Levin V.M. et al. 1991, Study of structure and properties of highly anisotropic materials. ibid.
Every A.G. et al., 1990, Phonon focusing and mode conversation effect in silicon at ultrasonic frequencies. Phys.Rev. Lett. 65, 1146

THEORY OF CONTINUOUS WAVE SCANNING ACOUSTIC MICROSCOPE

S. Sathish[1], G. Gremaud[2], A. Kulik[2]

[1]Analytical Services and Materials Inc, 07, Research Drive,
Hampton, VA 23665, USA
[2]Institute de genie Atomique, EPFL, CH- 1015, Lausanne,
Switzerland

INTRODUCTION

The very first reflection Scanning Acoustic Microscope (SAM) was built on the principles of continuous waves [1]. Though it produced good images it was abandoned because of sever degradation in images due to the standing waves [2]. The next and present generations of SAMs use the most popular pulse-echo technique for imaging as well as for quantitative measurements. It is capable of producing good images and the contrast has been explored in detail. Ray and wave theories have been worked out to explain the contrast [3,4,5,6]. Very good agreement between the theory and experiment have been obtained for the V(z) curves[5,7]. Robust software techniques have been developed to analyze the V(z) curves which can determine the Surface wave velocity and attenuation to an accuracy better than one part in thousand[8]. Although the pulse echo SAM is a powerful tool in materials characterization, a valuable information of phase is lost in this technique. Several alternatives have been proposed [9,10] to get obtain both amplitude and phase information. One such microscope has been Continuous Wave Reflection Scanning Acoustic Microscope (SAMCRUW)[11]. It has been used to obtain both amplitude and phase images. It can also provide quantitative information through unique CWV(z) as well as conventional V(z). It is quite versatile because it can be used as a CW SAM or even like a conventional pulse echo SAM with quite ease. Moreover it has been built basically with off the shelf commercial equipment. It has been used for characterization of different kinds of materials [12,13,14].

Although SAMCRUW has provided very useful information about elastic properties of materials all this has been derived with simple physical intuition and drawing parallels with the pulse echo SAM. CWV(z) is much more complex than its counter part. It is a mixture of standing waves in water between lens and specimen, and the surface waves propagating at the interface. In this sense this is a unique V(z). There has been no mathematical approach to obtain the CWV(z) curve. This paper is the first attempt to obtain a mathematical model for CWV(z). Our approach to calculate the CWV(z) is parallel to that of the theories of CW ultrasonic interferometer[15]. This is combined with the conventional V(Z) theory of SAM to obtain a complete description of CWV(z) curve.

In a SAMCRUW with a Network Analyzer the input reflection coefficient, (S_{11} parameter) which is defined as the ratio of the reflected energy to the input energy to the system is measured. To obtain CWV(z) S_{11} of the Transducer-Lens-Water-Sample combination is measured as a function of position of the sample. This is similar to a continuous wave ultrasonic interferometer which consists of a flat transducer and a flat reflector, a combination popularly used in measurement of ultrasonic velocity and attenuation in liquids. In an interferometer once the standing waves are set up between the transducer and the reflector, changes in the current flowing through the oscillator circuit which excites the piezoelectric transducer is measured as a function of the position of the reflector. Although the basic idea is to measure the energy variation it is acheived through a current measurement which is much more simple. A series of maxima and minima, with varying amplitude are observed. The distance between the consecutive minima or maxima corresponds to one wavelength of sound in the liquid. Variation in the amplitude of maxima or minima provides the attenuation in the liquid. If a Network analyzer had been used instead of measuring the current it would have measured S_{11} which is an another way to measure the energy variation. In a CWV(z) obtained from SAMCRUW a series of maxima and minima corresponding to one wavelength in water and an overriding variation due to the Rayleigh wave at the interface between water and specimen are observed. In this regard an interferometer could be considered as a special case of a SAMCRUW.

THEORY OF CONTINUOUS WAVE V(Z)

Let us consider a simple picture of the transducer-lens-water-sample combination at focus as shown in Fig.1. A continuous wave coherent radiation is applied to the transducer. This propagates down the lens rod and converges to a point on the specimen at focus in presence of water. After reflection it propagates back to the lens and reaches the transducer. The incident waves and the reflected waves which are propagating in opposite directions will set up a standing wave pattern as in a conventional CW ultrasonic interferometer. Let us consider a plane at a distance x from the transducer. The average particle displacement due to the wave propagating in the downward direction is

$$\xi_d = \xi_0 \; e^{i\omega t} \; e^{-ikx} \qquad\qquad (1)$$

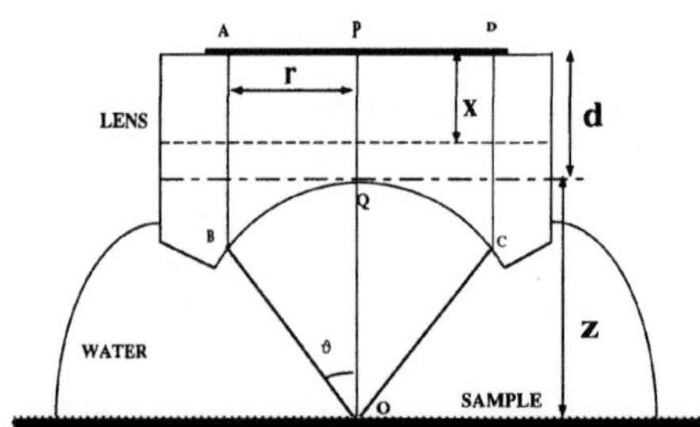

Fig.1. Acoustic lens indicating the distances used in the model

This wave propagates down the lens rod and meets the lens-water interface. To calculate the amplitude of the average particle displacement due to the reflected wave at this plane, we follow the conventional theory of SAM [6].Each wave propagates at an angle ϑ with respect to the normal to the specimen surface and passes through the lens with a pupil function $P_1(\vartheta)$. All these waves meet at the point O that is at a distance z from the lens. The specimen will reflect these waves with a reflection coefficient $R(\vartheta)$. These waves will cross the lens with a pupil function $P_2(\vartheta)$. Now we can combine all these wavelets to obtain the wave form and the average particle displacement on the plane by summing over the total area (A) of the transducer

$$\xi_r = \xi_0 \frac{e^{i\omega t}}{A} \int P_1(\vartheta)\, R(\vartheta)\, P_2(\vartheta)\, e^{-ik[d + 2z\cos\vartheta + (d-x)]}\, 2\pi r\, dr \tag{2}$$

$$R(\vartheta) = \frac{Z_L \cos^2(2\vartheta_{Sh}) + Z_S \sin^2(2\vartheta_{Sh}) - Z_w}{Z_L \cos^2(2\vartheta_{Sh}) + Z_S \sin^2(2\vartheta_{Sh}) + Z_w} \tag{3}$$

Where $Z_w = \dfrac{\rho_w V_w}{\cos\vartheta}$, $Z_L = \dfrac{\rho V_L}{\cos\vartheta_L}$ and $Z_S = \dfrac{\rho V_{Sh}}{\cos\vartheta_{Sh}}$

and $\sin\vartheta_L = \left[\dfrac{V_L}{V_w}\right]\sin\vartheta$, and $\sin\vartheta_{Sh} = \left[\dfrac{V_{Sh}}{V_w}\right]\sin\vartheta$

Here V_w is the velocity of sound in water and V_L and V_{Sh} are respectively longitudinal and shear wave velocity of sound in the specimen.

Total particle displacement on this plane becomes

$$\xi = \xi_d + \xi_r$$

$$\xi = \xi_0\, e^{i\omega t}\left[e^{-ikx} + \frac{e^{-2ikd}}{A}\, e^{ikx} \int P(\vartheta)\, R(\vartheta)\, e^{-2ikz\cos\vartheta}\, 2\pi r\, dr \right] \tag{4}$$

where $P(\vartheta) = P_1(\vartheta)\, P_2(\vartheta)$ is the composite pupil function

So, the mechanical impedance in the transducer (at $x = 0$) due to this particle displacement is given as

$$Z_{mech} = \left[\frac{P}{\dot{\xi}}\right]_{x=0} \tag{5}$$

Where Pressure, $\quad P = -\rho_0 V_l^2 \dfrac{\partial \xi}{\partial x} \tag{6}$

Using Eq(4) and Eq (5) in Eq (4) we obtain the variation of the mechanical impedance at the transducer due to the reflector at a distance z from the lens.

$$Z_{mech} = \rho_0 V_l \left[\frac{1 - \dfrac{e^{-2ikd}}{A}\displaystyle\int P(\vartheta)\, R(\vartheta)\, e^{-2ikz\cos\vartheta}\, 2\pi r\, dr}{1 + \dfrac{e^{-2ikd}}{A}\displaystyle\int P(\vartheta)\, R(\vartheta)\, e^{-2ikz\cos\vartheta}\, 2\pi r\, dr} \right] \tag{7}$$

As we can recognize each of the integrals represent the expression for conventional V(z). If instead of a lens with an opening angle of ϑ a flat transducer is used the expression reduces to the familiar expression for the CW ultrasonic interferometric equation. In an interferometer variation in Z_{mech} produces corresponding variations in the current through the oscillator circuit which is directly related to the energy changes.When a network analyzer is used to capture the variation of variation of S11, this is also directly related to the energy and hence Z_{mech}.

For an acoustic lens in fig 1, with, $r = f \, Sin \, \vartheta$, where f is the focal length, variation in Z_{mech} [Eq(7)] is defined as CWV(z), which is given as,

$$CWV(z) = (\rho_0 \, V_l) \frac{1 - 2\pi f \, \frac{e^{-2ikd}}{A} \int P(\vartheta) \, R(\vartheta) \, e^{-2ikz \cos \vartheta} \sin \vartheta \, \cos \vartheta \, d\vartheta}{1 + 2\pi f \, \frac{e^{-2ikd}}{A} \int P(\vartheta) \, R(\vartheta) \, e^{-2ikz \cos \vartheta} \sin \vartheta \, \cos \vartheta \, d\vartheta} \tag{8}$$

COMPARISON WITH EXPERIMENTAL RESULTS

CWV(Z) was obtained experimentally at a frequency of 237.325 MHz on an isotropic polycrystalline WC specimen with SAMCRUW [11] employing a spherical lens which has a half the opening angle of 60°. To compare this with the theoretical CWV(z)

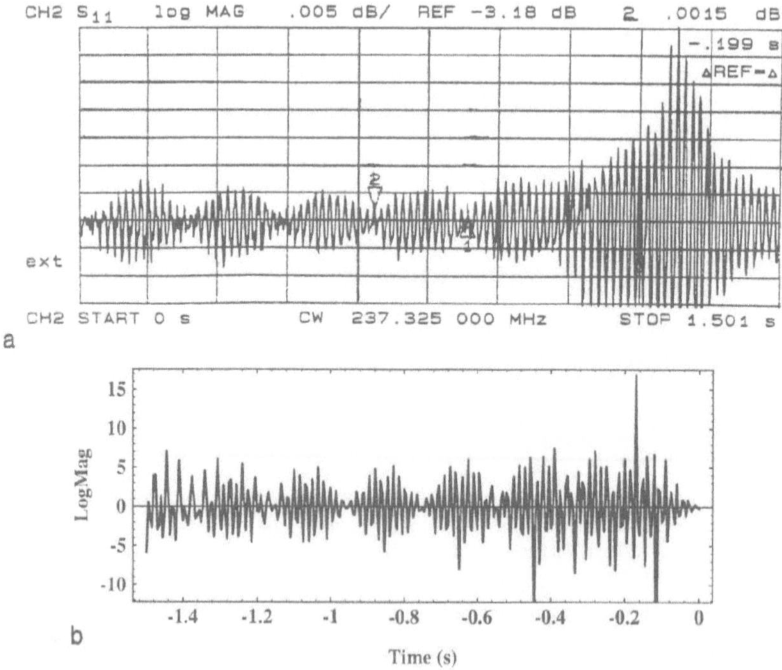

Fig.2.(a) Experimental CWV(Z) curve for an isotropic polycrystalline WC
    (b)Theoretically calculated CWV(Z) curve, Specimen WC, Frequency: 237.325 MHz

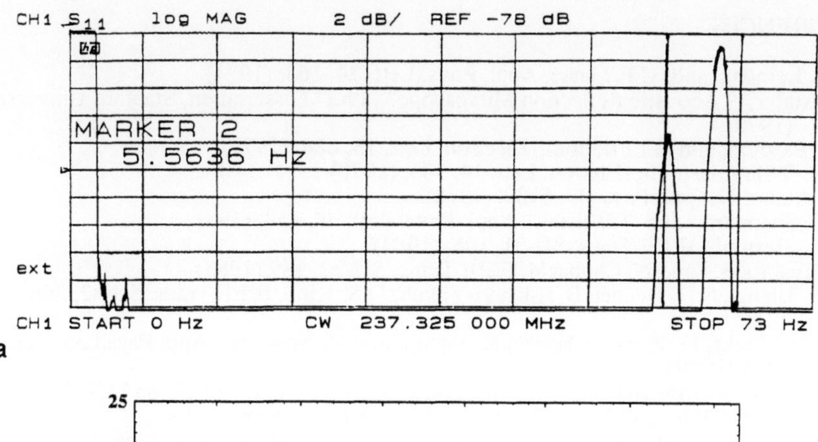

a

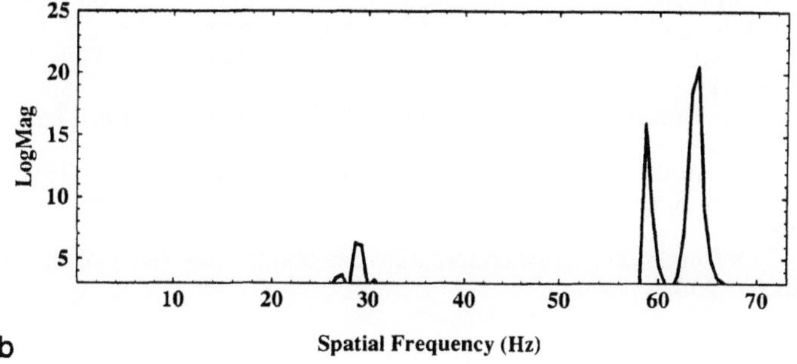

b

Fig. 3. (a) Spatial Frequency spectrum of experimental CWV(Z) [FFT of Fig.2 a]
(b) Spatial Frequency spectrum of the theoretical CWV(Z) [FFT of Fig.2b]

Eq (8) was numerically evaluated. A top hat pupil function $P(\vartheta)$ which is unity inside and zero outside the opening angle of the lens has been employed. As input parameters to calculate $R(\vartheta)$ of the specimen, longitudinal velocity 7.053 km/s and shear wave velocity as 4.155 km/s and density as 1500 kg/m3 [16] have been used. The experimental results are shown in Fig 2 along with the theoretically evaluated results. The two curves are qualitatively resemble each other. To compare spatial frequency response of the two curves in Fig 3 we show the FFTs obtained for the experimental curve from the Network Analyzer and the FFT of the theoretical curve. As it can be seen clearly the positions of the water and the Rayleigh peaks in FFTs are in good agreement. There is some discrepency in the absolute magnitudes of the curves. This should be attributed to the simplified form of the theory which does not take into account the detailed structure of the lens and employs a simple form of pupil function.

CONCLUSIONS

In this paper we have presented a simple theory of Continuous wave V(z) following the approach of theories of interferometer and conventional pulse echo SAM. The theoretical results are in qualitative agreement with the experiments.

REFERENCES

1.R.A Lemons,.and C.F. Quate, Appl. Phys. Lett. 24, 163, (1974)
2.A. Atalar. " Acoustic Reflection Microscope" , Ph.D.Dissertation, Stanford University,
    (1978)
3. W. Parmon, and H.L. Bertoni, Electron. Lett. 15, 684, (1979)
4.H.K. Wikramasinghe, Eletron. Lett. 14, 305, (1978)
5. A. Atalar. J. Appl. Phys. 49, 5130, (1978)
6. C.J. Sheppard, and T.Wilson,   Appl. Phys. Lett. 38, 858, (1981)
7. H.L. Bertoni, IEEE Trans. SU-31, 105, (1984)
8. J. Kushibiki, . and N. Chubachi, IEEE Trans. SU-32, 189,(1985)
9.K.K. Liang., S.D. Bennet, B.T. Khuri-Yakub, G.S. Kino, IEEE Trans. SU-32, 266,
    (1985)
10.S.W. Meeks, D. Peter, D. Horne, K. Young, and V. Novotny., Appl.Phys.Lett., 55,
    1835 (1989)
11. A Kulik, G Gremaud, and S. Sathish, Acoustical Imaging. Vol 17, pp 71, H.Shimizu,
    N.Chubachi, and J.I. Kushibiki ed, Plenum Press, New York.
12. S.Sathish, M. Mendik, A. Kulik, G. Gremaud, P. Wachter, Appl. Phys. Lett. 51,
    167, (1991)
13. M. Mendik, S. Sathish, A. Kulik, G. Gremaud, P. Wachter, J. Appl. Phys. 71,
    2830, (1992)
14. S. Sathish, A. Kulik, G. Gremaud, Solid. State. Commun., 77, 403, (1991)
15. R.T. Beyer, and S.V. Letcher, " Physical Acoustics" Academic Press, New York,
    (1969)
16. D. Mari, " Deformation a haute temperature des composites WC-Co", EPFL Doctoral
    Thesis No 938 , (1991)

STUDY OF LASER ANNEALED FERROMAGNETIC AMORPHOUS RIBBONS BY SCANNING ELECTRON-ACOUSTIC MICROSCOPY

M. Urchulutegui[1], J. Piqueras[1], I. Tanarro[2]
and C. Aroca[1]

[1]Dpto. Física de Materiales, Fac. Ciencias
 Físicas, Universidad Complutense de Madrid
 28040 Madrid, Spain
[2]Inst. Estructura de la Materia, C.S.I.C.
 Madrid, Spain

INTRODUCTION

Scanning Electron Acoustic Microscopy (SEAM) was developped in 1980 (1)(2) and has been mainly used in the last few years in the characterization of metals and semiconductors. SEAM has been also applied to the observation of different surface and subsurface features in ceramics (3) and domain walls in amorphous ferromagnetic alloys (4).

In SEAM a chopped electron beam generates an acoustic signal in the specimen surface due to the interaction between the solid and the primery electron beam. The signal is usually detected by a transductor attached to the bottom surface of the sample and used to form a scanned image in a scanning electron microscope (SEM).

In a previous work (4), SEAM has been used to investigate the domain structure of amorphous, iron and cobalt based, magnetic ribbons. The alloys have respectively high (27.10^{-6}) and low ($0.3.10^{-6}$) magnetostriction constants (λ_s). SEAM contrast was found to originate in the magnetic walls due to the associated mechanical stresses. Recently (5) a high power laser has been used as a heat source to anneal the same kind of amorphous ribbons inducing high anisotropies due to internal stresses. In the present work SEAM has been applied to the observation of the laser annealed ribbons and the effect of local laser annealing on the generated stress distribution and the related magnetic structure has been investigated.

EXPERIMENTAL METHOD

SEAM observations were performed in a Cambridge S4-10 scanning electron microscope. The experimental arrangement has been previously described (3) and basically consists of a chopping system attached to a function generator to create an intensity modulated beam (40-240 kHz), a specimen-transducer assembly and an amplifying system. The electron-acoustic signal is detected by a piezoelectric ceramic transducer. The amplification is carried out by a lock-in and a video amplifier at the reference frequency f or at 2f.

Measurements were performed in samples with high ($Fe_{78}B_{13}Si_9$) and low ($Co_{70}Fe_5Ni_2Mo_3B_5Si_{15}$) magnetostriction constants. They were locally irradiated for 0.5 s through a continous 5-W Ar laser focused by a suitable microscope objective to obtain a spot size of 10 μm. Different arrays were induced with different symmetries, one regular with distances between annealed spots 1.2x1.2 mm^2 (fourfold symmetry) and an irregular one with 0.6x2.4 mm^2 (twofold symmetry).

RESULTS AND DISCUSSION

The laser impact area includes a recrystallized region surrounded by an annealed zone. Both regions are observed in the electron-acoustic images of the two materials investigated.

We describe first the results concerning to the high magnetostrive samples in fourfold and twofold symmetry. In fourfold symmetry the laser spots are the origin of a complex structure as Fig. 1 shows. The electron-acoustic contrast consists of bright arms along radial directions connecting

single laser   impacts. In addition, in some regions of the
sample, black and white alternating fringes placed between two
adjacent laser impacts are observed in the SEAM image (Fig. 2b).
The laser spots correspond to the bright areas marked A as
comparison with the secondary electron image (Fig. 2a) shows.

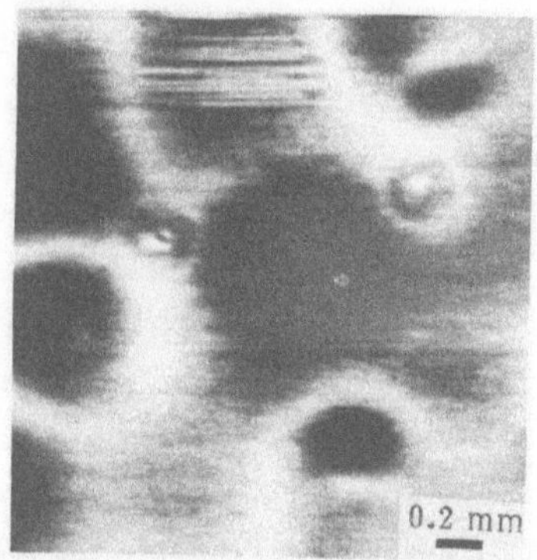

Figure 1. High magnetostrictive sample (fourfold symmetry).
Linear SEAM image at 27 keV and 49.31 kHz.

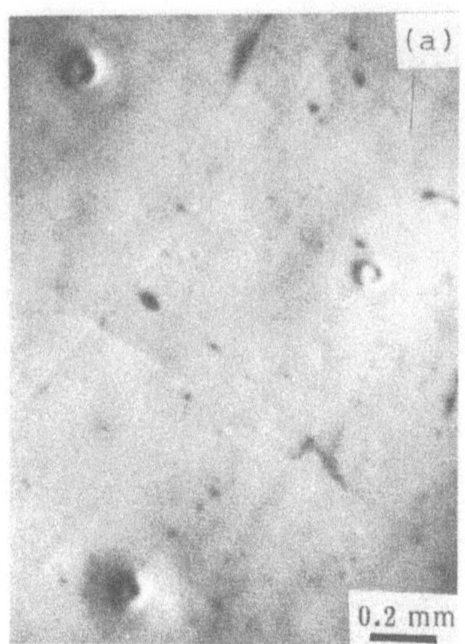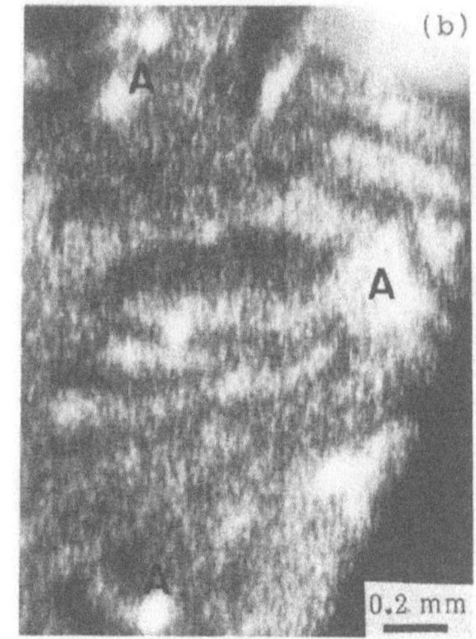

Figure 2. High magnetostrictive sample (fourfold symmetry)
(a) Secondary electron image, (b) linear SEAM image at 20
keV and 55.43 kHz.

These SEAM observations can be discussed in terms of anisotropies induced by local laser annealing and the stress model proposed in ref. 5. The impacted area includes a recrystallized region surrounded by an annealed zone whose volume decreases. As a consequence of the change in thermoelastic properties in the impacted areas electroacoustic contrast appears. In Fig. 1 the laser spot circles of about 10 μm diameter, are surrounded by larger circles corresponding to the annealed areas, while the structure of bright arms joining adjacent spots is possibly associated to the laser induced stresses described in ref. 5. On the other hand our SEAM images do not reveal the magnetic structure observed by Aroca et al. (5) in these samples by using Bitter techniques and consisting of domain walls in radial directions. This can be explained if the high stress gradient in the proximity of laser impacts masks the possible SEAM contrast due to a magnetoelastic effect. Fig. 3 shows the secondary electron (a) and electroacoustic (b) images of a twofold array of annealed areas. The SEAM contrast corresponds mainly to the annealed areas surrounding the laser impacts and is probably related to the induced stress distribution.

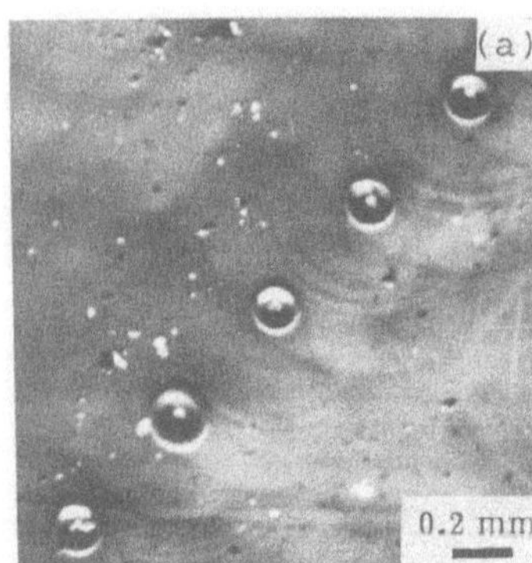

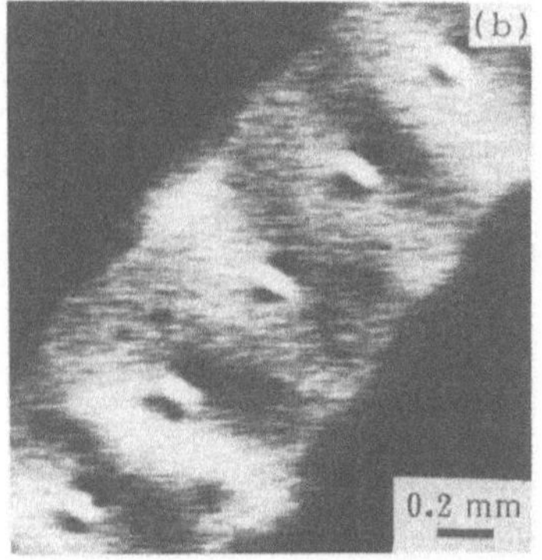

Figure 3. High magnetostrictive sample (twofold symmetry). (a) Secondary electron image, (b) linear SEAM image at 20 keV and 97.24 kHz.

In the fourfold symmetry array the minimum of the laser induced stress is found (5) in the regions equidistant from two impacts. In those regions the anisotropy would be close to that of the as-cast material. The SEAM contrast of such material has been found (4) to consist of dark-white alternating fringes associated to domain walls or to boundaries of closure structures. The fringes were found in regions with anisotropy normal to the surface. Since it has also been proved (5) that stripe domains exist in the regions between laser spots in high magnetostrictive samples, we suggest that the bright-dark areas in the SEAM image of Fig. 2b correspond to stripe domain walls and the associated stress field.

In the low magnetostrictive samples the magnetic structure previously described was not observed. As figure 4 shows, only the laser spot surrounding areas produce a marked SEAM contrast. The array of induced stress predicted by the theoretical model (5) has not been observed in this kind of amorphous alloys. Following Aroca et al. (5), an easy plane perpendicular to the surface of the sample will be induced and therefore stripe domains around the impact will appear. However, if the domain walls are surrounding the rows of laser impacts, il would be difficult to separate both domain walls and tensile stress contrast in the electron-acoustic images.

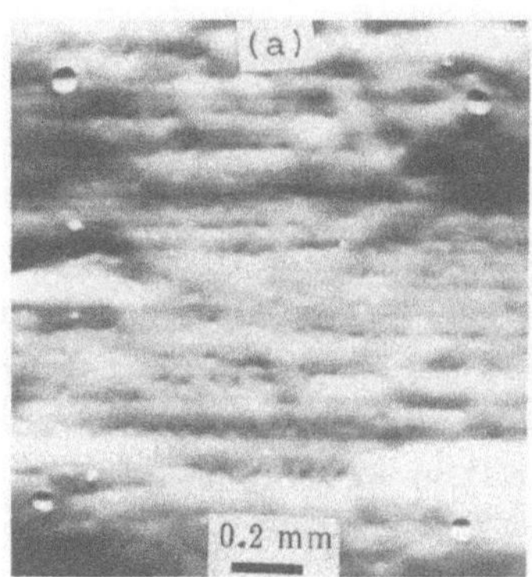
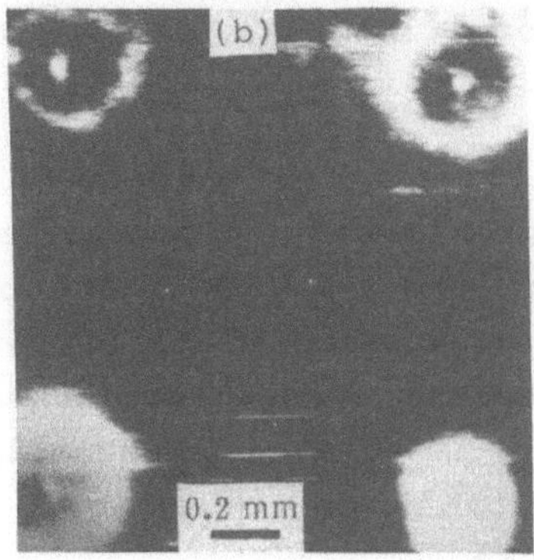

Figure 4. Low magnetostrictive sample (fourfold symmetry). (a) Secondary electron image, (b) linear SEAM image at 20 keV and 20.56 kHz.

For this reason, we cannot identify the dark and bright areas in electron-acoustic images as those in Fig. 4b. This contrast can be produced either by the stress distribution near the annealed areas or the circular domain walls around the laser spots. The fact that the stress distribution has not been

observed in the low magnetostrive samples would be a consequence
that the stress gradient is very small for low-λ_s values and
therefore the magnetoelastic effects in the electron-acoustic
signal would be negligible at a long distance from the laser
impact.

Finally, figure 5 proves the existence of a non-linear
electron-acoustic contribution to the signal that can be mainly
due to a mechanism related to a magnetic coupling (e.g. non-

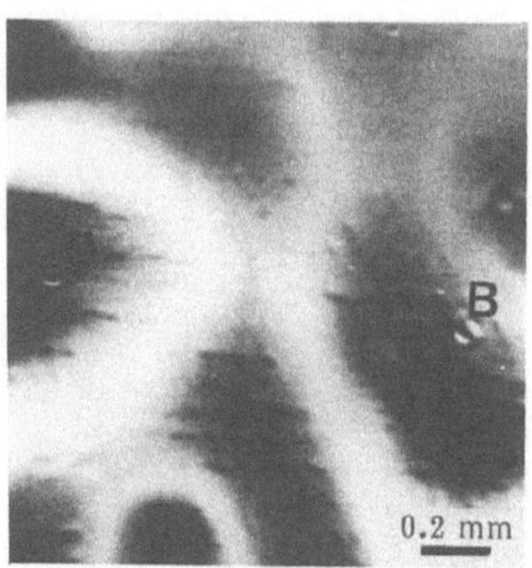

Figure 5. Low magnetostrictive sample (fourfold symmetry)
Nonlinear (2f) SEAM image at 20 keV and 104.08 kHz.

linear variation of magnetostriction constant with temperature or
the approximately quadratic dependence of strain with M(T). The
electron-acoustic contrast consists of dark areas around the
laser spots (marked B) surrounded by bright annular regions.
Nevertheless, such electron-acoustic contrast does not seem to be
directly associated to magnetic domains around the laser impacts
and probably corresponds to a more complex electron-acoustic
mechanism.

ACKNOWLEDGEMENTS

This work has been supported by DGICYT (Project PB90-1017)
and Volkswagen Foundation.

REFERENCES

1. E.Brandis and A.Rosencwaig. Appl. Phys. Lett. **37**, 98 (1980)
2. G.S.Gargill, Nature **286**, 691 (1980)

3. M.Urchulutegui, J.Piqueras and J.Llopis. J. Appl. Phys. $\underline{65}$(7), 2677 (1989)

4. M.Urchulutegui, J.Piqueras and C.Aroca. Appl. Phys. Lett. $\underline{59}$(8), 944 (1991)

5. C.Aroca, M.C.Sánchez, I.Tanaro, P.Sánchez, E.López and M.Vázquez. Phys. Rev.B, $\underline{42}$(13), 8086 (1990)

IMAGE PROCESSING FOR THE MEASUREMENT OF CRACK
DEPTH USING THE SCANNING ACOUSTIC MICROSCOPE

D.D. Bennink, D. Knauss, T. Zhai,
G.A.D. Briggs and J.W. Martin

Department of Materials
University of Oxford
Oxford OX1 3PH, UK

INTRODUCTION

The ability to measure the depth of surface breaking fatigue cracks, especially while they are still in the short crack stage, is of great importance for the study of crack growth behaviour.[1-3] Time-resolved scanning acoustic microscopy offers the possibility to accomplish such depth measurements in a nondestructive manner,[4-6] and thus the ability to actually monitor crack growth throughout the fatigue process. In time-resolved acoustic microscopy pulses of very short temporal extent are used to excite the SAM lens, so that echoes from the individual scattering sites within a specimen can be separately identified in the received signal.[7-9] The measurement of crack depth therefore relies on the behaviour of the crack tip as a scattering site, with the depth being determined by measuring the arrival time of the signals caused by diffraction of the acoustic waves at the tip. Although the crack tip diffraction signal is much weaker in strength than specular scattering, it has been successfully used to measure longer cracks (>1 mm) at lower frequencies (< 10 MHz), where the technique is referred to as Time-of-Flight Diffraction (TOFD).[10,11]

The combination of TOFD and time-resolved scanning acoustic microscopy presents difficulties due not only to the very weak strength of the crack tip signals, but also to the presence of a very strong specular reflection from the specimen surface. This strong surface reflection occurs because the same lens/transducer acts as both the transmitter and the receiver in the scanning acoustic microscope. This paper describes the image processing undertaken in order to increase the strength of the tip diffracted signals relative to both the other signals present and the background noise level. Further image processing, used to identify crack tip signals and to extract crack depth information, is also discussed. Results are presented for a 40 µm deep crack in Al-Li alloy 8090.[12]

TIME-RESOLVED MEASUREMENTS

A simplified block diagram for the time-resolved scanning acoustic microscope at Oxford University[13] is shown in Figure 1. A very short duration voltage spike is generated by a step-recovery diode and used to drive the SAM lens through the transmitting amplifier with the SPDT switch in the T position. After the excitation pulse has fired, the SPDT

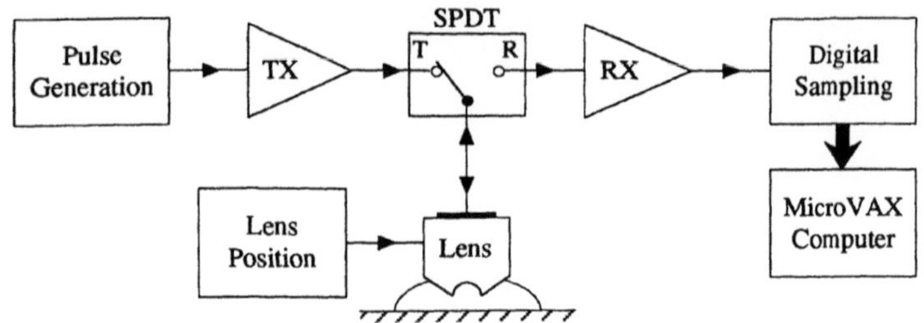

Figure 1. Simplified block diagram for the time-resolved acoustic microscope at Oxford University.

switch is moved to the R position so that any echo pulses returned from the specimen can be amplified for reception. The output of the receiving amplifier is digitally sampled at 512 equally spaced time locations within a selected time window, and this information is transferred to a MicroVAX II computer for image processing and storage. A complete image is generated by repeating this process after each movement of the lens as it is stepped through 512 equally spaced positions. Such an image thus consists of the signal amplitude, plotted on a 256 level grey scale for display, at 512 temporal versus 512 spatial locations, traditionally referred to as a B-scan image.

For the measurement of crack depth, the lens is scanned perpendicularly across the surface length/mouth of the crack. A sketch of this is given in Figure 2, together with a ray theory interpretation of the principle signals appearing in such a B-scan image.[12,14-16] The displacement of the lens from its initial position is labelled as x, d is the depth of the crack tip below the surface, Δx is the displacement of the crack tip from directly below the crack mouth, and Δz is the lens defocus value. Ray 1 is normal to the specimen surface and the ray path 1-1 (ray 1 from the lens to the surface, reflection from the surface, and ray 1 back to the lens) represents the strong specular reflection signal (the time-of-flight trace labelled SR in Figure 2b). Excitation of Rayleigh waves occurs for rays 2 and 3 which intersect the surface at the Rayleigh critical angle. Ray paths 2-3 and 3-2 thus represent the signal for Rayleigh wave transmission across the crack, or simply along the surface if the crack is not present (the trace RT in Figure 2b). Rayleigh wave reflection from the crack (the crossing traces RR in Figure 2b) also occurs for ray paths 2-2, the longer time for the situation shown, and 3-3, the shorter time. Ray 4 intersects the crack mouth and ray path 4-4 represents the signal for crack mouth diffraction (the hyperbolic time-of-flight trace MD in Figure 2b). Finally, ray 5 meets the surface at the correct angle so that when refracted into the specimen it intersects the crack tip. Ray path 5-5 thus represents the signal for crack tip diffraction (the quasi-hyperbolic trace TD in Figure 2b).

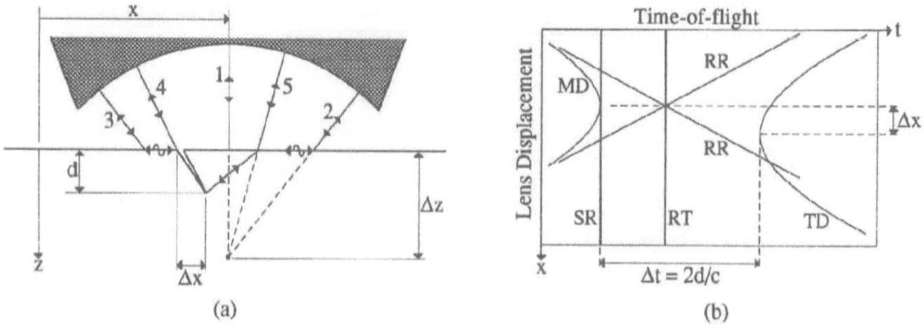

Figure 2. Ray theory interpretation of the time-resolved measurements: (a) ray diagram; (b) B-scan image.

The time-of-flight trace TD for tip diffraction is based on scattering from a single, well defined point. Similar diffraction arcs can also be expected to arise from other "point-like" scattering sites, such as small voids or kinks in the subsurface crack profile. For such point scattering sites the spatial and temporal location of the diffraction arc yields both the depth of the scattering site and its position along the scan line. For example, the vertex of trace TD in Figure 2b is displaced in x by the amount Δx from the Rayleigh wave reflection crossing point (which is assumed to originate from scattering at the crack mouth), and is displaced in t by the amount $\Delta t = 2d/c$ from the specular reflection. The constant c is the velocity for the wave mode corresponding to that segment of ray 5 which is within the specimen. Since solids support both longitudinal and transverse wave modes, there are actually two possibilities for ray 5. Therefore, although only one is shown in Figure 2b, there are also three possible tip diffraction arcs: both specimen ray segments for ray path 5-5 longitudinal, both segments transverse, or one segment longitudinal and one transverse.

IMAGE PROCESSING

The image processing to be presented here falls roughly into two categories: signal enhancement and signal identification. Signal enhancement refers to the processing steps used to increase the strength of the tip diffracted signals relative to both the other signals present and the background noise level, while signal identification refers to the processing steps used to actually identify crack tip signals and to extract depth information.

Signal Enhancement

Figure 3a shows the B-scan image, after filtering to remove a cable echo due to an impedance mismatch in the electronics, obtained from a scan across the midpoint of a surface breaking fatigue crack in Al-Li alloy 8090. The crack was 70 µm in surface length, and the total lens/scan displacement was 185 µm. The time window digitised was 100 ns in length and selected to include all the signals of interest (Figure 2b). For all images presented in Figure 3, time runs horizontally from left to right and lens displacement runs vertically from top to bottom, as in Figure 2b, with brightness representing signal amplitude. To remove the cable echo, a region of the original B-scan image is selected, usually at the top or bottom, in which each time line may be written as $v(t) = v_s(t) + v_e(t)$, where $v_e(t)$ is the temporally separated echo of the actual signal $v_s(t)$. The product filter

$$H(f) = 1 - \frac{\overline{V}_e(f)\overline{V}^*(f)}{\left|\overline{V}(f)\right|^2 + Q^2} \tag{1}$$

is then calculated and applied to each time line of the entire image, where $\overline{V}(f)$ and $\overline{V}_e(f)$ are average spectra over the image lines in the selected region for the full time line and for the temporally separated cable echo respectively, the asterisk denotes complex conjugation, and Q is a constant (set at 10% of the maximum value of $|\overline{V}(f)|$). The label CE in Figure 3a marks the location of the cable echo in the original, unfiltered image. The strong set of vertical black and white bars in the left half of Figure 3a contains both the specular reflection and the Rayleigh wave transmission signals, as labelled. The two signals are not separated because of both their finite temporal width, approximately 10 ns, and the low defocus value, -40 µm. Just discernible from the specular and Rayleigh transmission signals in the upper half of the image are the Rayleigh wave reflection signals. No tip diffracted signals are evident in this figure, and they only become evident in the B-scan image if the contrast setting is increased dramatically.

The crack tip signals are evident in Figure 3b, which shows the processed B-scan image after subtraction of the specular and Rayleigh transmission signals. Removal of the specular and Rayleigh transmission signals is accomplished for each time line of the image by subtracting a scaled version of a reference line, representing the signals to be removed,

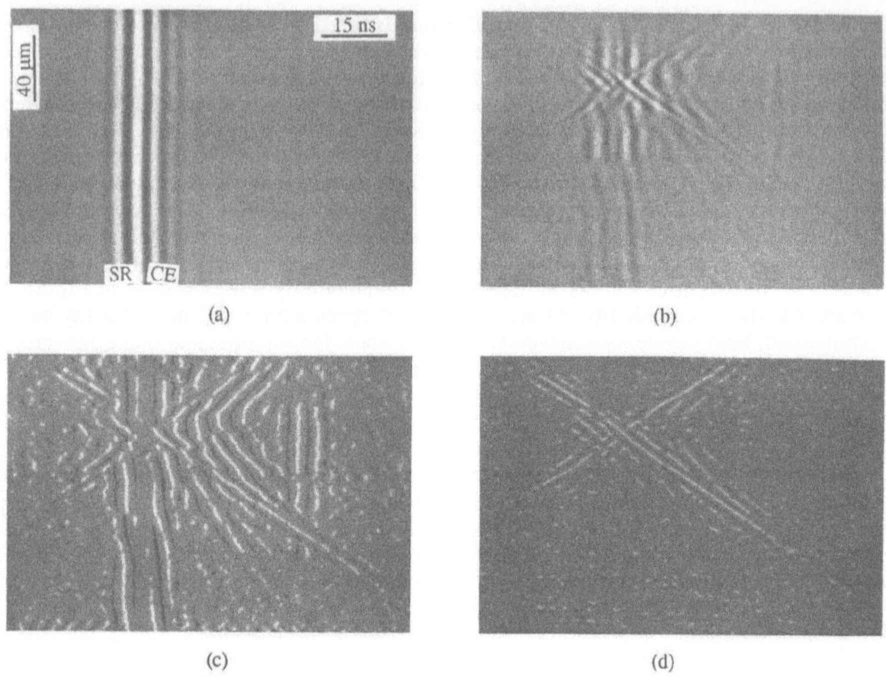

Figure 3. B-scan images generated from a scan across the midpoint of a surface breaking fatigue crack in Al-Li alloy 8090: (a) time-resolved data with cable echo filtered out; (b) specular reflection and Rayleigh wave transmission signals removed; (c) peak detection applied by rows to fully enhance crack tip signals; (d) peak detection applied by columns to enhance Rayleigh wave reflection signals for selection of reference at crossing point.

from a shifted version of the original line. The algorithm is designed to eliminate signals which do not change as the lens is moved. First, the original line is shifted in order to align it with the reference line, the shift amount being determined by the location of the maximum in the cross-correlation between the two lines. A quadratic interpolation is used to find the maximum location and the time shift is applied in the frequency domain, thus allowing for shift values which are not integer multiples of the sample spacing. Second, a scaled version of the reference line is subtracted from this shifted original line, the scale factor being determined by a least-squared-error fit between the two lines. This step produces the processed line. Finally, the reference line is updated as a moving average over a chosen number of the prior shifted original lines, thus allowing for some flexibility in the handling of slow variations in the signals to be removed. The processing is of course not perfect, but even though some of the cable echo, specular reflection, and Rayleigh wave transmission signals can still be seen in the processed image, the enhancement of the crack tip signals is significant.

Even further enhancement of the crack tip signals is evident in Figure 3c, which shows the processed image obtained by applying a peak detection algorithm by rows, i.e. local peaks or extrema are looked for in each individual time line, to the B-scan image of Figure 3b. The use of peak detection by rows produces an image somewhat more similar to Figure 2b, and the crack tip signals, finally put on an equal basis with the other signals present, stand out clearly. The algorithm works by first calculating a local least-squared-error quadratic fit about each point in the line, using a selected number of its neighbours on either side. The peak location, whether it be for a maximum or a minimum, is then determined for each local quadratic fit. If the peak is located within the interval between the point to which the fit applies and its nearest neighbour on either side, then a peak indication is given, either +1 for a positive peak (local maxima) or -1 for a negative peak (local minima). However, to avoid false peak detections, a local linear fit is also applied and peak indications are thrown out if the linear fit is sufficiently better than the local quadratic fit. Indications are also thrown out for positive minima and negative maxima, and if the range covered by the quadratic fit over the interval for which it was determined is sufficiently less than the average residue for the fit. All of these screening tests are used to decrease the sensitivity of the algorithm to noise and "sufficiently" generally means determine by trial and error. As can be seen from Figure 3c, the algorithm can work quite well, and the crack tip signals are now on an equal basis with the other signals present. Of course the noise is now also on an equal basis, but fortunately peak indications which are apparently due to noise seem to appear only in regions separate from those containing signal peak indications.

The peak detection algorithm when applied by columns, where local extrema are looked for in each individual spatial line (at fixed time), works very well for extracting the Rayleigh wave reflection signals, as shown in Figure 3d. The Rayleigh reflection signals are important because their crossing point provides a well defined temporal and spatial reference point, necessary for the application of TOFD. For this purpose a set of crossing lines can be overlaid on the B-scan image and moved about until a satisfactory reference point is found. The slope of the lines must of course be consistent with the velocity of Rayleigh waves (the longitudinal, transverse and Rayleigh wave velocities for this material are $c_L = 6.8$ μm/ns, $c_T = 3.6$ μm/ns and $c_R = 3.4$ μm/ns).[12]

Signal Identification

A simple pattern recognition algorithm is used to perform signal identification. The algorithm is based on adding together the values at those image locations which correspond to the ray theory time-of-flight trace (i.e. trace TD in Figure 2b). For convenience the process will hereafter be referred to as SART, for Sum-Along-Ray-Trace (not to be confused with SAFT, although the two are as similar in nature as they are in name). The SART image is therefore given by

$$g(x,z) = \int dx' \, f\{x', T(x'|x,z)\} \tag{2}$$

where $f(x,t)$ is the B-scan image to be processed and $T(x'|x,z)$ is the ray theory time-of-flight trace as a function of lens/scan position x' for a scattering site located at the point (x,z) within the specimen. In practice Eq. (2) is calculated as a sum and for each lens position the image values are taken for the discrete time location nearest to the ray theory time-of-flight value. The SART calculations can be done for any one of the three possible diffraction arcs: L, T, or M (longitudinal, transverse, or mixed mode). For example, Figure 4b show the T SART calculations for the B-scan image in Figure 4a. The image in Figure 4a is a windowed and filtered version of the B-scan image in Figure 3c. The windowing was done to exclude the spurious peak detections due to both noise and the remainder of the cable echo, and the filtering was essentially a box convolution in time used to widen the signal peak traces. In Figure 4b, depth z runs vertically from top to bottom, starting from $z = 20$ μm at the top, and displacement x runs horizontally from right to left.

As shown in Figure 4b, the SART calculations produce a large number of bright and dark patches of varying degree on a complicated, lower amplitude background. These patches represent point scattering sites for which the diffraction arcs more closely fit the signal peak traces. Applying peak detection both by rows and by columns to Figure 4b, and then taking the product of the two results, yields the local two-dimensional maxima and minima for the SART image. The more extreme maxima and minima obtained in this manner represent essentially the scattering site locations for which the diffraction arcs best overlap the positive and negative signal peak traces. However, it is important to remember

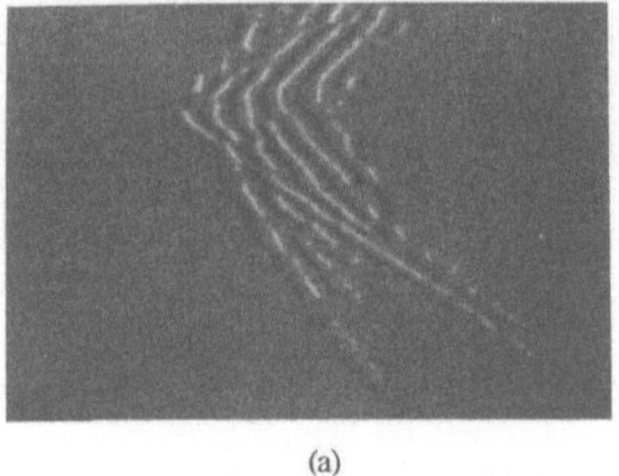
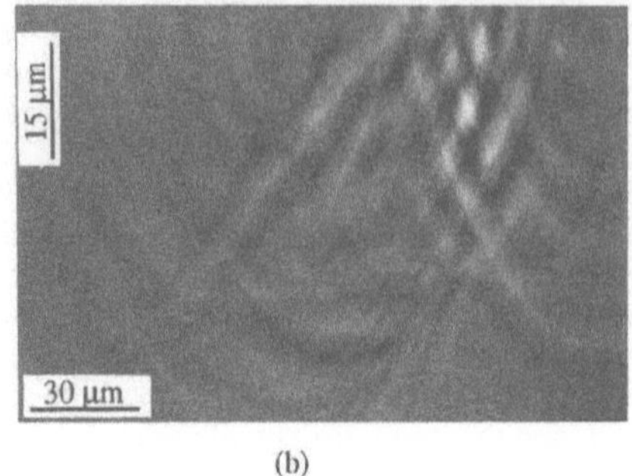

(a)   (b)

Figure 4. SART computations: (a) windowed and filtered version of the B-scan image in Figure 3c; (b) T SART results for the B-scan image in (a).

that the patterns produced by the experimental crack tip signals will deviate from the ideal ray theory predictions. The extent to which this deviation occurs depends on several factors, including the anisotropic and inhomogeneous nature of the specimen (for this work the material was assumed to be isotropic and homogeneous). Therefore, the identification of the best fit scattering sites is checked subjectively by eye by overlapping the ray theory time-of-flight trace for each individual indication on the original SART input image and deciding if the diffraction arc corresponds sensibly to the experimental data.

Figure 5a summarises the results obtained by applying the process described above to the L, T and M SART calculations. The scattering site indications for the three SART calculations, and for positive and negative extrema, which have survived the screening process are plotted as depth versus displacement (now relative to the Rayleigh wave reflection crossing point). The indications cluster about the average point of $x = -5.4$ μm and $z = 38.5$ μm with a standard deviation of $\Delta x = 4.4$ μm and $\Delta z = 5.8$ μm. Also plotted in Figure 5a is the basic geometry of the crack subsurface profile measured from Figure 5b (an acoustic micrograph at 550 MHz of a through section at the scan location). This crack geometry has been positioned in Figure 5a based on the assumption that the Rayleigh wave reflection appears to originate from the point halfway between the crack mouth and the long crack segment perpendicular to the surface (the wavelength for Rayleigh waves at the pulse centre frequency is 15 μm). The large number of scattering site indications which survive screening is not surprising considering the complex geometry of the crack near its tip. It is also not surprising that the details of this complex tip geometry are not readily discernible in the acoustic measurements, given that a transverse wave will travel a total of 18 μm in only 5 ns. Furthermore, the first crack segment, at approximately 45° to the surface, does not run deep enough to be detected, since any signals caused by it are buried in what still remains of the specular and Rayleigh transmission signals.

DISCUSSION

The results presented in this paper illustrate the kind of image processing necessary to successfully apply the combination of TOFD and time-resolved scanning acoustic microscopy to the depth measurement of surface breaking cracks. Due to the very weak strength of the crack tip signals, and to the confocal arrangement of the scanning acoustic microscope (same lens/transducer acting as both transmitter and receiver), image processing must first be undertaken to enhance the strength of the tip diffracted signals. The dominant specular reflection and Rayleigh transmission signals can be removed by taking advantage of the fact that they do not vary (or only very slowly) with lens position. Peak detection can then be used to fully bring out the crack tip signals. Once the crack tip

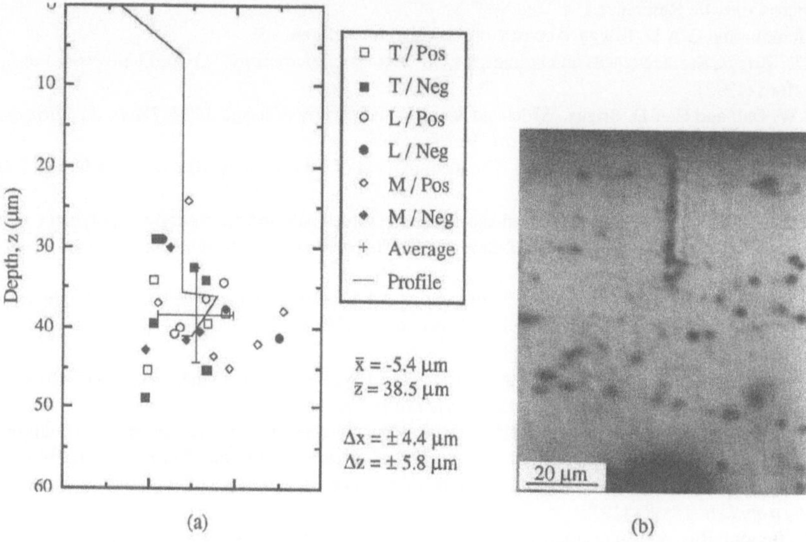

Figure 5. Crack subsurface profile: (a) scattering site indications from L, T and M SART calculations, along with the average location and one standard deviation error bars, and the basic crack profile measured from (b); (b) an acoustic micrograph at 550 MHz and zero defocus of a through section at the scan location.

signals are fully enhanced, pattern recognition, in the form of the SART algorithm of Eq. (2) for example, can be applied and the more extreme local maxima and minima in the computational results taken as possible scattering site indications. The indications must be checked subjectively by eye to ensure that they correspond sensibly to the experimental data.

The capability of TOFD to resolve scattering sites is limited by the pulse width in the time-resolved measurements, and a spread in the indications like that shown in Figure 5a can be expected whenever the crack geometry varies too rapidly with depth. Uncertainties in the wave velocities, and in practice even in the acoustic nature of the material (i.e. anisotropy and inhomogeniety), also result in a spread in the scattering site indications. Yet it is evident from Figure 5 that the TOFD and time-resolved SAM combination can be successfully used for the measurement of crack depth.

ACKNOWLEDGEMENTS

This work was funded through a grant from the SERC and National Power, UK (Grant No. GR/E 76391).

REFERENCES

1. K.J. Miller and E.R. de los Rios, eds. "The Behaviour of Short Fatigue Cracks," Mechanical Engineering Publications Limited, London (1986).
2. D.J. Nicholls and J.W. Martin, The use of ΔK when examining microstructural effects on small fatigue crack growth, *Int. J. Fatigue* 12:469 (1990).
3. K.J. Marsh, R.A. Smith and R.O. Ritchie, eds. "Fatigue Crack Measurement: Techniques and Applications," Engineering Materials Advisory Services Ltd., Warley, UK (1991).

4. G.A.D. Briggs, E.R. de los Rios, and K.J. Miller, How to observe short surface cracks by acoustic microscopy, *in*: Reference 1.

5. P.J. Jenkins and G.A.D. Briggs, Acoustic microscopy, *in*: Reference 3.

6. G.A.D. Briggs, Surface cracks and boundaries, *in*: "Acoustic Microscopy," Oxford University Press, Oxford (1992).

7. C.M.W. Daft and G.A.D. Briggs, Wideband acoustic microscopy of tissue, *IEEE Trans. on Ultrasonics, Ferroelectrics, and Frequency Control* 36:258 (1989).

8. J. Wang, R. Gundle and G.A.D. Briggs, The measurement of acoustic properties of living human cells, *Trans. R. Microsc. Soc.* 1:91 (1990).

9. A.M. Sinton, G.A.D. Briggs and Y. Tsukahara, Time-resolved acoustic microscopy of polymer coatings, *in*: "Acoustical Imaging," Vol. 17, H. Shimizu, N. Chubachi and J. Kushibiki, eds., Plenum Press, New York (1989).

10. J.P. Charlesworth and J.A.G. Temple. "Engineering Applications of Ultrasonic Time-of-Flight Diffraction," Research Studies Press Ltd., Taunton, UK (1989).

11. M.G. Silk, Ultrasonics, *in*: Reference 3.

12. D. Knauss, D.D. Bennink, T. Zhai, G.A.D. Briggs and J.W. Martin, Depth measurements of short cracks with an acoustic microscope, submitted to *J. of Mater. Sci.*

13. J.M.R. Weaver, C.M.W. Daft and G.A.D. Briggs, A quantitative acoustic microscope with multiple detection modes, *IEEE Trans. on Ultrasonics, Ferroelectrics, and Frequency Control* 36:554 (1989).

14. W. Parmon and H.L. Bertoni, Ray interpretation of the material signature in the acoustic microscope, *Electron. Lett.* 15:684 (1979).

15. H.L. Bertoni, Ray-optical evaluation of V(z) in the reflection acoustic microscope, *IEEE Trans. on Sonics and Ultrasonics* 31:105 (1984).

16. T. Zhai, D.D. Bennink, D. Knauss, G.A.D. Briggs and J.W. Martin, Depth measurements of short cracks in perspex with the scanning acoustic microscope, *in*: this proceedings.

DEPTH MEASUREMENTS OF SHORT CRACKS
IN PERSPEX WITH THE SCANNING
ACOUSTIC MICROSCOPE

T. Zhai, D.D .Bennink, D. Knauss, G.A.D. Briggs and J.W. Martin

Department of Materials
University of Oxford
Oxford OX1 3PH, UK

1. INTRODUCTION

The geometry, such as surface length, depth, shape and orientation, is one of the key factors that dominate the propagation behaviour of short fatigue cracks. Thus, for a quantitative understanding of short crack growth behaviour, it is necessary to monitor the crack geometry throughout the fatigue test. However, a technique for the direct measurement of short crack geometry has been lacking, which, together with the difficulty of short crack detection, is one of the major reasons for the slow progress taking place in the study of short fatigue cracks. The growth of short fatigue cracks is commonly monitored only in its surface length with an optical or scanning electron microscope. The depth determination for the crack is then usually made on the basis of an empirical estimation from the crack surface length. For instance, the penny-shape or semi-elliptical estimation by assuming the profile of the crack as a semi-circle infers the depth of a short crack to be half of its surface length[1,2]. Obviously, such an estimation cannot usually provide reliable data on the depth of short cracks in many materials, because of the strong dependence of the short fatigue crack shape on local microstructural characteristics and environment.

Recently, increasing attention has been paid to acoustic techniques in an attempt to overcome the above difficulty. This is because an acoustic wave is able to penetrate into an opaque material and interact with any inhomogeneities inside. They include the photoacoustic microscope[3] and the surface acoustic wave technique[2,4]. They can, however, only be used to estimate the depth of a short crack, combined with an empirical assumption of the crack profile as well.

The scanning acoustic microscope (SAM), combined with the time-of-flight diffraction technique developed in NDT, is used in this paper to directly measure the depths of short cracks in perspex (PMMA). The SAM, reviewed in[5,6], has already been used as a powerful technique for surface crack detection and investigation due to its great sensitivity to surface-breaking cracks[7,8]. It also offers the possibility of measuring the depth of short cracks since it is capable of time-resolved measurement with short pulses (<20 ns wide)[9,10]. The time resolved measurement technique, similar in principle to the time-of-flight diffraction technique in NDT[11], has been made available in the Oxford SAM previously for the measurements of thickness, velocity, impedance and frequency-dependent attenuation in layered samples and thin sections of biological tissue, etc.[12]. Measurements on long cracks in glass were also carried out, showing that there are usually components appearing in the s(t,y) plots (in which one axis is the time-of-flight of returned signals, the other axis is the lens position, and the brightness at each point corresponds to the signal intensity), such as a surface specular reflection, a Rayleigh wave propagation along the surface and Rayleigh waves reflected from the crack mouth edges when the lens is scanned across the crack perpendicularly[9].

For the purpose of building up a theoretical model for interactions of acoustic waves with a short crack and for quantitative interpretation of the s(t,y) plots for short cracks, perspex was chosen as a first, model material. Perspex is a transparent, isotropic and acoustically slow material; it therefore enables one to measure the crack geometry in the optical microscope, and it eliminates the involvement of the Rayleigh surface waves and the influence of wave speed anisotropy on the acoustic measurements. Also, with perspex it is possible that different crack geometries might be made at different temperatures as a result of the great sensitivity of its stress-strain curve to temperature[13].

2. EXPERIMENTAL

2.1 Short Crack Preparation

The specimens used in this work are from commercial perspex sheet. The dimensions of the specimens are 46 X 10 X 3.5 mm. The short cracks were made by bending the specimen slightly and adding a little acetone with a cotton bud to the surface bearing the tensile stress. The bending stress should preferably be small (e.g. by bending the specimen by hand), otherwise the cracks made will be too long and too deep. The crack geometries were first measured in the optical microscope by focusing down into the specimen and following the crack tip. The depth of the crack tip is equal to the defocus multiplied by the refractive index of perspex, 1.59. To determine the correct focus on the crack tip, the smallest possible depth of field of the objective lens should be employed, by using the aperture diaphragm open as wide as possible at as high a magnification as possible. However, at too high a magnification, e.g. higher than 400X, the space between the objective lens and the specimen surface is too small to be able to defocus the lens sufficiently and, in addition, the contrast of the crack tip is usually too weak to be seen.

Two different specimen temperatures, 20°C and 50°C, were used in making the cracks, in order to acquire cracks with different profiles. Fig.1 shows the acoustical microgragh of opened short crack #1, made at 20°C. Its depth to length ratio is 0.72, as can be seen in its profile under bending, in Fig.2, measured with the optical microscope. It should be noticed that the profile of crack #1 measured at 200X is much larger than that measured at 400X. The profile measured at 200X is associated with the craze tip, the profile of the plastic zone ahead of the crack, while the one at 400X is the actual crack tip[13]. At 50°C, short crack #2 with a depth to length ratio of 0.25 was made. Its profile measured with the optical microscope is shown in Fig.3.

2.2 Time Resolved Measurement

The Oxford SAM[9] can be used in either an imaging quasi-CW mode for surface imaging or an impulse spike mode for the time-resolved measurements. The surface wave coupled to by the lens in perspex is only the longitudinal lateral surface wave; the lens half angle is 60° which includes an incident ray at the longitudinal critical angle of 33.75°. The surface cracks were examined in the image mode operated at 350 MHz in order to position the lens so as to scan across the crack perpendicularly at a chosen point along the crack. While scanning the lens across the crack, a narrow acoustic pulse (< 20 ns wide) is sent into the specimen and the intensities of the returned signals received by the lens are recorded in a plot of time-of-flight versus lens position, called an s(t,y) plot.

3. RESULTS AND THEORETICAL INTERPRETATION

3.1 SAM Images of Bent Cracks

As shown in Fig.1, the contrast of cracks in SAM images is significantly enhanced when the specimen is bent to open the cracks. Another feature of the pictures of the cracks when opened by bending is that more fringes appear on both sides of the cracks, parallel to the crack mouth. The fringe spacing is approximately 3.93 μm, reasonably consistent with half a wavelength of the longitudinal lateral wave at 3.86 μm. This demonstrates that, analogous to the fringes caused by interference between the reflection of the Rayleigh wave

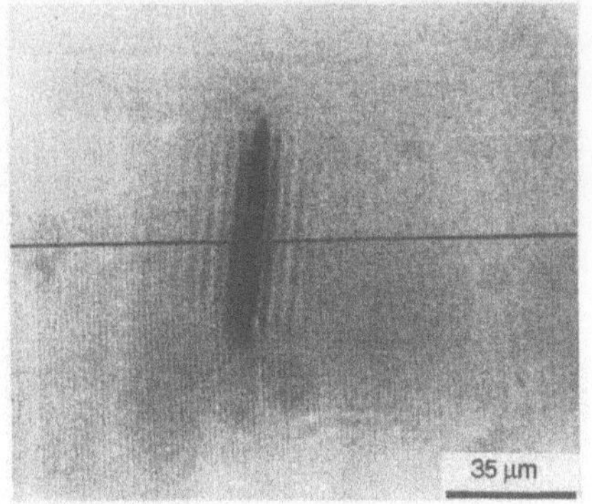

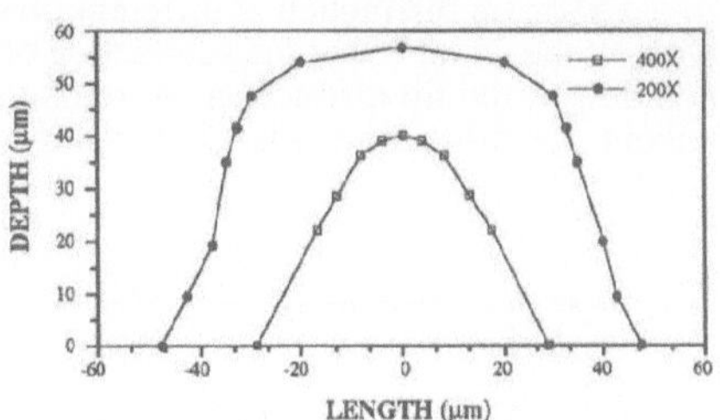

Fig.1 Acoustic image of crack #1 opened, 40 µm deep, in perspex, with the scanning line for the time resolved measurement of Fig.4. Its depth to length ratio is 0.72. More fringes can be seen parallel to the crack, with a spacing of 3.93 µm. Z = - 8 µm, 350 MHz

Fig.2 The profile of crack #1 measured by the optical microscope. The one in line —⊟— is measured at 400X, and the one in line —●— at 200X. The region between the two curves is the craze zone (the plastic zone) ahead of the crack tip

from a crack and the reflection from the surface in a fast material, the fringes here are caused by interference between the reflection of the longitudinal lateral wave from the crack and the reflection from the specimen surface. Thus, such contrast enhancement may be related to the increase of the longitudinal lateral wave reflection from the crack when opened.

3.2 Signal Detection of Crack Tip Diffraction

Similar to the time-of-flight diffraction technique in NDT, the detection of the signal diffracted by a crack tip in the time-resolved measurements using the SAM is to detect a diffraction arc in the s(t,y) plot that gives information about the crack depth. For a tightly closed short crack in perspex, e.g. made in the way mentioned above, the signal corresponding to the crack tip diffraction needs to be enhanced, otherwise it is too weak to be seen in the s(t,y) plot. Several steps were taken to enhance the signal. Firstly, the cracks were opened either by being made at a higher temperature, 50°C, or by bending the specimens while testing in the SAM at the relevant temperatures used in crack preparation. And secondly, signal averaging was applied and the removal of strong signals like the specular surface reflection and the longitudinal lateral surface wave signal, which mask the other weaker signals, was undertaken.

3.3 s(t,y) Plots of the Cracks

The longitudinal lateral surface wave interacts with a crack in perspex in exactly the way the Rayleigh surface wave interacts with a crack in glass, as discussed in 3.1. Therefore, similar signals such as the longitudinal lateral surface wave propagation along the specimen and the reflections of the longitudinal lateral wave from the crack edges, in addition to the strong specular reflection, may exist in the s(t,y) plot for a crack in perspex. In the s(t,y) plot for crack #1, unbent during the test, no diffraction was seen. When crack #1 was bent during the test, a faint arc beyond the two crossing lines can then be seen and, in addition the two crossing lines become stronger than those for the unopened crack. Such a arc can only be observed when scanning the lens across the middle of the crack. The relevant scanning line is shown in Fig.1. This arc corresponds to the crack tip diffraction. In the plot, following components can be seen: the strong specular surface reflection (line E), the longitudinal lateral surface wave propagation (line F), the reflections of the longitudinal lateral wave from the crack mouth edges (the two crossing lines B with the crossing point on the line of the longitudinal lateral wave propagation) and their cable echoes.

After subtraction of the specular surface reflection, the longitudinal lateral surface wave and their cable echoes, the s(t,y) plot in Fig.4 shows clearly all signals scattered by the crack. Three additional curves can now be observed, one (curve A) is a hyperbola tangential to the specular surface reflection line, and the other two (curve C) are the crossing curves between the specular reflection and the surface wave propagation lines. The crack tip diffraction arc (curve D) also becomes clearer.

Crack tip diffraction at different positions along crack #2 can be clearly observed even without the aid of either the subtraction of the strong signals or the opening of the crack. The position of the tip diffraction arc remains the same in the s(t,y) plots for both the bent and unbent specimen when scanning over the same position on the crack.

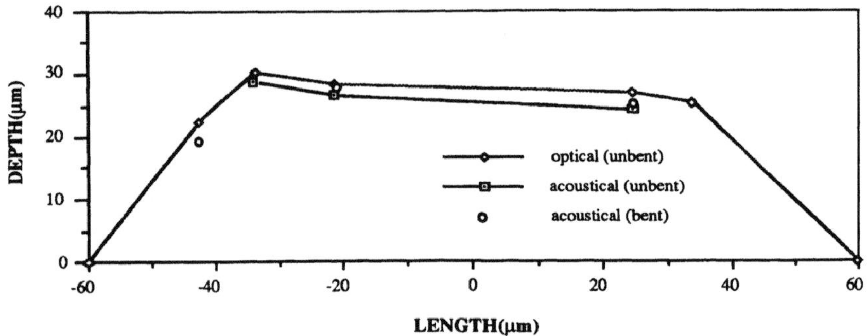

Fig.3 The profiles of crack #2 measured by both the optical microscope (in line ———◇———) at 400X and by the SAM (line ———▫——— is for the unbent specimen and ○ for the bent specimen).

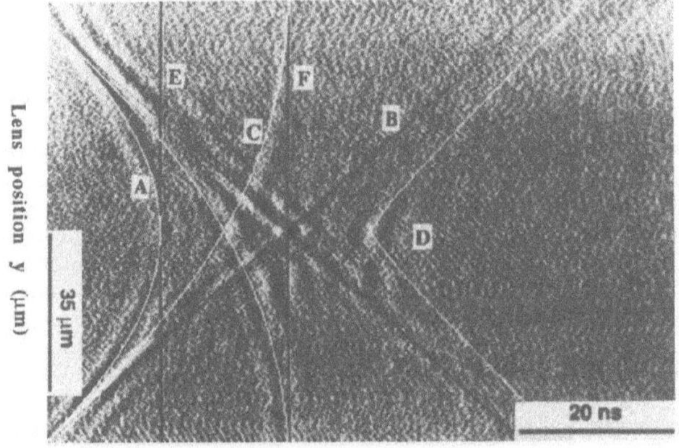

Fig.4 The s(t,y) plot for opened crack #1, with subtraction of the signals of the surface specular reflection, the surface wave propagation and their cable echoes, and imposed on by the calculated results (the black and the white lines). The depth of the crack is measured to be 39.5 μm. The signals in the s(t,y) are interpreted in Fig.7 correspondingly showing the interactions of the acoustic waves with the crack. The relevant scanning line is marked in Fig.1. Z = -80 μm.

3.4 Ray theory interpretation

There are usually seven signal components appearing in an s(t,y) plot for a short crack in perspex. Ray theory can provide a excellent description concerning all these contributions to the s(t,y) plot for the short crack. The schematic diagrams for the interpretation of these contributions by the ray theory are illustrated in Fig.6. Note that the acoustic lens can only

284

accept a returned ray that propagates to the lens back along its own path or along the geometrically symmetric path to its own path, since a ray returned at any other angle will either miss the transducer or will arrive at an angle other than normal incidence, which mostly eliminates the resulting contribution of the ray to the transducer output. By analysing the path difference between the surface specular reflection and the rays responsible for the other signals received by the lens, at a given lens defocus Z, and letting the time-of-flight of the surface specular reflection be equal to zero, the time T taken by the other signal rays can then be derived as a function of the lens position y and the defocus Z (for details see[14]). These components in the s(t,y) plot are: (a) diffraction of the incident beams at the crack mouth edges (Fig.5(a)); (b) reflection of longitudinal lateral surface waves from the crack mouth (Fig.5(b)); (c) conversion of incident waves into longitudinal lateral waves at the crack edges or vice versa (Fig.5(c)); (d) diffraction of longitudinal waves at the crack tip (Fig.5(d)); (e) specular surface reflection of the incident waves; (f) propagation of longitudinal lateral waves along the specimen surface; (g) reflection of longitudinal waves or shear waves from the crack face below the surface, if the crack is oblique (Fig.5(g)). Two of them are relatively important for measuring the geometry of a crack. They are the diffraction of longitudinal waves at the crack tip and the reflection of longitudinal waves or shear waves from the crack face below the surface, if the crack is oblique.

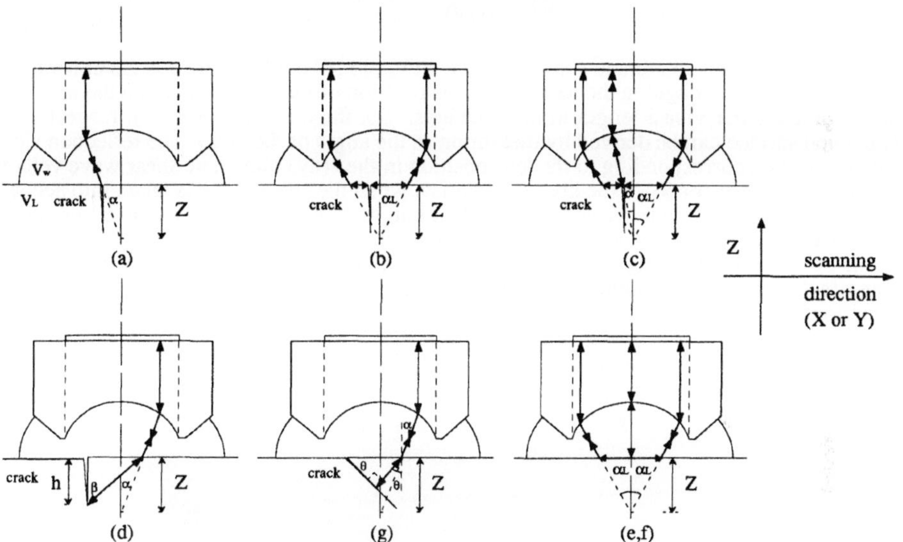

Fig.5 Schematic diagrams for the ray theory interpretation of the components in the s(t,y) plot of a short crack in perspex. (a) crack mouth edge diffraction; (b) longitudinal lateral wave reflection from the crack mouth sides; (c) conversion of incident waves into longitudinal lateral wave at the crack mouth edges, or vice versa; (d) crack tip diffraction; (e) specular surface reflection; (f) longitudinal lateral surface wave propagation along the specimen surface; (g) reflection of the longitudinal or shear waves from the crack face below the surface, if the crack is oblique.

Diffraction occurs when a longitudinal ray refracted at the surface is incident on the crack tip. The diffracted ray then travels back to the lens along the same path as the incident ray (see Fig.5(d)). This ray is described implicitly by the following set of equations:

$$T = \frac{2h}{V_L \cos\beta} - \frac{2}{V_W}\left(\frac{Z}{\cos\alpha} - Z\right)$$

$$h\tan\beta = Z\tan\alpha + y - y_1$$

$$\frac{V_W}{\sin\alpha} = \frac{V_L}{\sin\beta}$$

Where h is the depth of the crack , y_1 is the y shift of the crack tip from the crack mouth ($y_1 = 0$ if the crack is normal to the specimen surface), V_W is the wave speed in water and V_L is the longitudinal wave speed in perspex. The last equation is of cause Snell's law, with α the angle of the ray in the fluid and β the angle of the ray in the solid, both with respect to the normal to the surface. The minimum value of T occurs when $\alpha, \beta = 0$ and is therefore given by $T_{min} = 2h/V_L$ Thus, the depth of the crack can be found from $h = T_{min} V_L/2$.

A longitudinal ray incident on the crack face normally can be reflected back along its own path, and thus be received by the lens (see Fig.5(g)). The expression for the signal in an s(t,y) plot is given by

$$T = \frac{2y\sin\theta}{V_L} - \frac{2Z}{V_W}(\cos\alpha - 1)$$

Where

$$\frac{V_W}{\sin\alpha} = \frac{V_L}{\sin\theta}.$$

This is a straight line in the s(t,y) plot. Its gradient is

$$\frac{dy}{dT} = \frac{V_L}{2\sin\theta} = \cot\varphi \times C$$

Where C is the coordinate constant of the s(t,y) plot, C = a/b, where a is the unit value in microns per visual length in the axis of the lens position of the s(t,y) plot and b the unit value in nanometers per visual length in the time axis. So, the crack angle θ with respect to the specimen surface can be derived by measuring φ, the angle of the crack face reflection signal line to the axis corresponding to the lens position in the s(t,y) plot. The shear wave velocity can be substituted for V_L in the above equations, when the shear wave is taken into account. Besides, because of mode conversion (incident longitudinal or shear waves on a free surface give rise to reflected waves of both longitudinal and shear modes[9]), a reflection signal of mode converted waves may also appear in the s(t,y) plot. The equations for the reflection line are given by substituting its effective velocity $2V_S V_L/(V_L+V_S)$ for V_L in the above equations.

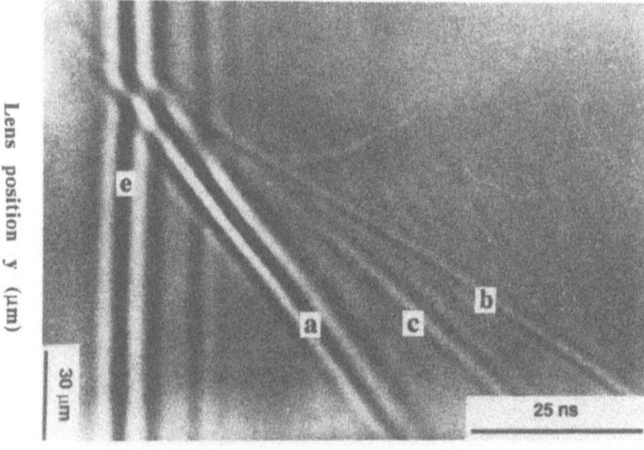

Fig.6 The s(t,y) plot for the oblique crack, showing the reflections of the longitudinal waves (line a), the shear waves (line b) and the converted mode waves (line c) from the crack face below the surface. Line e is the surface specular reflection. Z = -50 µm. The angles φ of these reflection lines to the axis of lens position are 38.4°, 49.0° and 58.5° respectively. $\theta = 25°$.

From the s(t,y) plot for crack #1 (Fig.4), T_{min} is measured to be 29.3 ns, then $h = T_{min}V_L/2 = 39.5$ µm ($V_L = 2.7$ µm/ns). This is in good agreement with the optical result of 40.5 µm. The s(t,y) plot for crack #1 calculated by ray theory, is imposed on Fig.4. This is in very good agreement with the experimental one.

Fig. 6 shows the s(t,y) plot for an oblique, long crack in perspex, in which three straight lines can be seen, but no reflection lines of the longitudinal lateral waves appear. The straight lines correspond to the reflections of the longitudinal waves (line a), shear waves (line b) and mode converted waves (line c) from the crack face respectively. The geometry of the crack was measured by the optical microscope. θ is about 25°. The angles φ in the s(t,y) plot are measured to be 38.4°, 49.0° and 58.5° for the reflections of longitudinal, shear and mode converted waves respectively. The coordinate constant $C = 2.43$ μm/ns. The angles θ are then calculated to be 26.1°, 26.5° and 25° using the above angles φ correspondingly. They are consistent with the optical result of $\theta = 25°$.

3.5 The Geometry Measurement of Crack #2

Depths at different positions along crack #2, with the specimen both bent and unbent, were measured by the technique and the results are illustrated in Fig.3. The accuracy of the acoustical depth measurements as compared with the optical measurements is 93%.

4. DISCUSSION

The above results for the time resolved measurement of short crack depth in perspex have demonstrated the capability of the SAM to measure the depth and geometry of short cracks in perspex with good accuracy. It is the crack tip diffraction received by the acoustic lens that gives information about the crack depth. From the equations given above for the time-of-flight in an s(t,y) plot, we can note that the shape of such a diffraction arc is dependent on the lens position y as well as the lens defocus Z. But its minimum point, given by $T_{min} = 2h/V_L$, and $y_{min} = y_1$, is independent of both. The minimum point occurs for α and $\beta = 0°$, which corresponds to the normal incidence on the specimen surface. The signal of the crack tip diffraction is usually very weak and needs to be enhanced. Opening the crack is one way to achieve this. Cracks in perspex can be opened by both bending the specimen and making the cracks at a higher temperature (50°C, for example). In addition, image processing techniques are also needed in order to enhance the contrast of the crack tip diffraction and to further identify it. A synthetic aperture focusing technique (SAFT)[11], based on ray theory time-of-flight calculations for point diffraction, has recently been used to replace the visual matching procedure and help identify the signals[15].

The technique is obviously also applicable to opaque materials. For metallic materials, however, the s(t,y) plot will be more complicated due to the anisotropy and inhomogeneity of the materials. The anisotropy gives rise to speed differences for acoustic waves propagating along different angles in the anisotropic material and therefore distorts the signal shapes, whereas inhomogeneities such as grain boundaries, inclusions and different phases in the material might cause unwanted reflections and diffractions of the acoustic waves. Therefore, much work needs to be done on the application of the technique to engineering materials. However, the depth measurement of short cracks has been carried out in Al-Li alloy 8090, which has a low elastic anisotropy[16].

5. CONCLUSION

The present work could be summarized as follows:
(1) The scanning acoustic microscope is capable of direct depth measurement for short cracks. It is the crack tip diffraction in an s(t,y) plot that gives information about the depth of a short crack, $h = T_{min}V_L/2$. Furthermore, if the crack is oblique, the angle θ the crack makes with the specimen can be derived by measuring φ, the angle between the crack face reflection signal and the coordinated axis corresponding to the lens position in the s(t,y) plot, with the relation, $\dfrac{dy}{dT} = \dfrac{V_L}{2\sin\theta} = \cotan\varphi \times C$.
(2) It is found that the following components usually appear in the s(t,y) plot for a short crack in perspex:
(a) diffraction of the incident waves at the crack mouth edges;
(b) reflection of longitudinal lateral surface waves from the crack mouth sides;

(c) conversion of incident waves into longitudinal lateral waves at the crack mouth edges, or vice versa;

(d) diffraction of longitudinal waves at the crack tip;

(e) specular surface reflection of the incident waves;

(f) propagation of longitudinal lateral surface waves along the specimen surface;

(g) reflection of the longitudinal waves or shear waves from the crack face below the surface, if the crack is oblique.

(3) Ray theory can provide a excellent description concerning all these contributions to the $s(t,y)$ plot for the short crack

(4) The contrast of the crack tip diffraction signal received by the lens needs to be enhanced in the $s(t,y)$ plot. The following measures could be taken for this purpose:

(i) opening the crack by bending the specimen or making cracks at a higher temperature than room temperature, e.g. 50°C;

(ii) signal averaging;

(iii) subtraction of the strong specular surface reflection and surface wave propagation signals.

6. ACKNOWLEDGEMENT

The authors would like to thank Professor J. Kushibiki for providing the acoustic lens used in these studies, all others who have contributed to the Oxford SAM both in hardware and software which we used in the work, and following bodies for providing fundings: SERC and National Power. T. Zhai is grateful to Sir Run Run Shaw and Oxford University for awarding a Run Run Shaw scholarship for his PhD study at Oxford University, and to Oxford University and Balliol College for partial financial support for attending this conference.

7. REFERENCES

1 S.Suresh and R.O.Ritchie, Propagation of Short Fatigue Cracks, Int. Metals Reviews. 29:445 (1984).

2 B.London, J.C.Shyne and D.V.Nelson, Small Fatigue Crack Behaviour Monitored Using Acoustic Waves in Quenched and Tempered 4140 Steel, in: The Behaviour of Short Fatigue Cracks, K.J.Miller and E.R.de los Rios, ed., Mechanical Engineering Publications, London (1986).

3 W.Arnold, B.Hoffmann, and H.Willems, Crack Depth Estimation by Photoacoustic Microscopy, Zeitschrift fur Physik B-Condensed Matter, 64:31 (1986).

4 M.T.Resch, B.London, G.F.Ramusat, H.H.Yuce, D.V.Nelson, and J.C.Shyne, A Surface Acoustic Wave Technique for Monitoring the Growth Behaviour of Small Surface Cracks, J. Nondestructive Eval., 20:1257 (1989).

5 G.A.D.Briggs, Acoustic Microscopy, Clarendon Press, Oxford (1992).

6 C.F.Quate, A.Atalar, and H.K.Wickramasinghe, Acoustic microscopy with mechanical scanning -- a review, Proc. IEEE, 67:1092 (1979).

7 G.A.D.Briggs, P.J.Jenkins and M.Hoppe, How fine a surface crack can you see in a scanning acoustic microscope? J. Microscopy, 159:15 (1990).

8 P.J.Jenkins and G.A.D.Briggs, Acoustic Microscopy, in: Fatigue Crack Measurement: Techniques and Applications, K.J.Marsh, R.A.Smith, R.O.Ritchie, EMAS, ed., (1991).

9 J.M.R.Weaver, C.M.W.Daft and G.A.D.Briggs, A quantitative acoustic microscope with multiple detection modes, IEEE Trans.UFFC. 36:554 (1989).

10 K.Yamanaka, Surface acoustic wave measurements using an impulsive converging beam, J. Appl. Phys., 54:4323 (1983).

11 J.P.Charlesworth and J.A.G.Temple, Engineering Applications of Ultrasonic Time-of-Flight Diffraction, Research Studies Press Ltd (1989).

12 C.M.W.Daft and G.A.D.Briggs, wideband acoustic microscopy of tissue, IEEE Trans UFFC. 36:258 (1989).

13 Jerold Schultz, Polymer Materials Science, Prentice-Hall, Inc. (1974).

14 T.Zhai, D.D.Bennink, D.Knauss, G.A.D.Briggs and J.W.Martin, Depth Measurements of Short Cracks in Perspex with the Scanning Acoustic Microscope, Materials Characterization (1992), (in press).

15 D.D.Bennink, D.Knauss, T.Zhai, G.A.D.Briggs and J.W.Martin, Image processing for the measurement of crack depth using the scanning acoustic microscope, in: Proc. of the 20th Int. Symposium, Yu Wei, ed., (1992).

16 D.Knauss, D.D.Bennink, T.Zhai, G.A.D.Briggs and J.W.Martin, Depth measurement of short cracks with an acoustic microscope, J. Mater. Sci.(1992), (in press).

SCANNING ACOUSTIC MICROSCOPIC AND BRILLOUIN SCATTERING INVESTIGATION OF SURFACE DAMAGE

S.Sathish[a], M.Mendik[b], A. Kulik[c], G. Gremaud[c], P.Wachter[b]

[a] Analytical Services and Materials Inc., 107, Research Drive
Hampton, Virginia VA 23693,USA

[b] Laboratorium fur Festkorperphysik, ETHZ
CH-8093, Zurich, Switzerland

[c] Institute de Genie Atomique, Departement de Physique, EPFL
CH-1015, Lausanne, Switzerland

INTRODUCTION

Mechanical polishing involves progressive removal of material through repeated rubbing against an another material in presence of abrasives. Near the surface large strains are induced. The existence of surface layer that has been strained to different degree than the bulk has been demonstrated by several techniques like X-ray diffraction[1], electron microscopy[1], positron annihilation[2], hydrogen permeation[3] and mechanical measurements[1]. It is expected that the mechanical properties of material near the surface might be different from the bulk interior. It has been shown that this top layer has large influence on stress-strain curves, creep, fatigue and stress corrosion resistance of metals[1]. Although a variety of techniques have been used to investigate the effect of surface layer on the bulk material properties, the properties of the layer itself have not been investigated in detail, except for some micro hardness and flow stress measurements. These techniques measure the plastic properties of the material. The flow stress of the surface layer of single crystal copper[4] has been found to be lower than the bulk, while in aluminum single crystals[5] the opposite behavior has been reported. The micro hardness measurement of the surface layers has shown two distinctive behaviors. A few investigators have shown that the surface hardness is greater than the bulk, while several others have pointed out that it is low near the surface and increases with the depth and goes through a maximum before reaching the bulk hardness. A discussion about the controversies of the near surface mechanical properties can be found in Ref.[6]. Theoretically both behaviors of flow stress and micro hardness have been explained [7,8,9]. The softness of the layer has been explained on the basis of image

Acoustical Imaging, Volume 20 Edited by Y. Wei
and B. Gu, Plenum Press, New York, 1993

forces on dislocation, while the hard surface layer is attributed to the increase of dislocation density due to deformation.

These techniques that give quantitative results for the mechanical properties of the surface layer have some serious disadvantages. The specimen has to be in contact with the indenter for measurements of hardness and the indentation depth has to be varied to get a profile. During this the mixed hardness is measured which can complicate the interpretation. For flow stress measurements specimens have to be sliced at different levels and then have to be subjected to deformation. In this processes it is not sure whether the surface layer is altered. In this paper we present techniques that are nondestructive and non contacting to study the elastic properties of the surface layer by propagating surface acoustic waves. The results presented here is the summarized form of our recent publications [10,11].

MATERIAL

Single crystal nickel specimens oriented along (100) and (110) were obtained from same ingot. Both sides of the specimen were polished flat and parallel on a rotating wheel. Initial polishing was performed with a continuous flow of water to reduce heat generation due to friction. Final polishing was performed with diamond particles sprayed on a cloth. The pressure on the specimen was kept low. The final polishing with 0.25 μm finish was good even for Brillouin Scattering where surface polishing is extremely important.

EXPERIMENTAL METHODS

Three different techniques were employed in this study to determine the elastic constants of nickel. The bulk elastic constants were measured using continuous wave ultrasound[12]. Longitudinal and shear standing waves of 10 MHz were generated in the specimen and the velocities were measured using a Network analyzer. These were used in determining the velocities and elastic constants, C_{11}, C_{12} and C_{44}.

After the measurements of bulk elastic constants the (100) sample was given a finer polish for further measurements using a Continuous Wave Scanning Acoustic Microscope (SAMCRUW) and Brillouin Scattering(BS). The SAMCRUW that has been used in this study has been described in detail elsewhere[13]. In short it is like a conventional SAM that utilizes CW waves instead of pulses. A Network analyzer treats the transducer-lens-water-sample combination as a two-port device, and measures the input reflection reflection S_{11} parameter of the system. This parameter when plotted as a function of position gives an acoustic image. To obtain CWV(z) that contains the quantitative information S_{11} is plotted as a function of distance between sample and lens. In this investigation CWV(z) was measured at a frequency of 230 MHz, using a cylindrical lens which focuses the acoustic waves on to a line on the specimen. The Fast Fourier Transform (FFT) facility of the Network Analyzer was used to analyze the CWV(z). The peak positions corresponding to the standing waves in water between the lens and sample, and the Rayleigh waves were obtained from the spatial frequency spectrum. These were used to determine the surface wave velocity (V_r) using the equation

$$V_r = V_w \left[1 - (1 - \frac{V_w}{2 F \Delta z})\right]^{-\frac{1}{2}} \tag{1}$$

Where $\Delta z = \dfrac{\dot{z}}{(F_w - F_r)}$ and $V_w = \dfrac{2F\dot{z}}{F_w}$

where $\dot{z}$, is the speed of the specimen stage in the z direction, F_w and F_r are the peak positions for standing waves in water and surface wave in the spatial frequency spectrum, and V_w is the speed of sound in water. The angular dependence of surface wave velocity was measured by rotating the specimen stage in steps of 5° from {100} direction.

A drop of water is used in SAMCRUW measurements. So, the specimen was thoroughly cleaned with acetone or alcohol. Then it was investigated using Brillouin Scattering. Detailed description of the experimental set up for surface acoustic wave BS can be found elsewhere[14]. In our experiments a TM-Polarized 514 nm line of an argon laser was used as the excitation source. Back scattered light is collected by a fast objective and passed through a (3+3) tandem interferometer. The measurements are performed in air. The Rayleigh wave contribution contained in the spectra is analyzed and the velocity is determined. Once again as in the case of SAMCRUW the angular dependance of the Rayleigh wave velocity was measured by rotating the specimen stage by 5° every time from {100} direction.

RESULTS AND DISCUSSIONS

The measurements of angular dispersion of surface wave velocity as determined by the three techniques for the (100) plane of nickel is shown in Fig.1. The continuous line in the figure has been obtained using the model of Farnell[15]. As input parameters into this model bulk elastic constants measured using CW ultrasonic technique were used. Both SAMCRUW and BS see the true Rayleigh wave from 0° to 30° and the Pseudo Rayleigh wave form 30° to 45°. We clearly observe the differences in SAW angular dispersions as obtained by different techniques. Surface waves propagate through a thickness of the material corresponding to at least one wavelength. This wavelength even for BS in which acoustic wave frequency is 10 GHz, corresponds to two tenths of a micron, which spans at least a few thousand atomic layers. So, it is considered a bulk phenomena. That is one should not observe any difference among the three techniques within the experimental resolution. In order to confirm that the differences are arising due to the nature of the surface, we altered the surface of the specimen by annealing it at 930K in vacuum and then sputtering it in argon atmosphere. The surface wave velocities were measured once again. The BS measurements showed a marked increase in SAW velocities. The angular dependence, became closer to the bulk curve. This is shown in fig.1 along with other results. So it is clear that the differences are due to the nature of the surface.

To determine the elastic constants of this layer a nonlinear least square fit was performed to the angular dependance of Rayleigh wave velocity. A procedure to perform this is described in detail in Ref.[14,11]. The results obtained are shown in Table 1. As we can see the elastic constants for the bulk have the highest values and from BS have the lowest values, while SAMCRUW values are just lower than bulk values. The elastic constants from BS after sputtering and annealing are higher than the simply mechanically polished specimen. So we find differences in the elastic constants as well. It is also clear that the sputtering and annealing change the elastic constants bringing them closer to the bulk results. In mechanical polishing the material is removed by force. During this process to obtain a flat surface polishing is performed in all directions. This might affect the position of the atoms and probably randomize a bit. This effect can be seen in the anisotropy factor, η $(= 2 C_{44}/(C_{11} - C_{12})$. In Table1. η is shown for different techniques and surface condition of the specimen. The anisotropy factor is highest for the bulk, while it is least in BS for the mechanically polished surface, and it is slightly lower than bulk in SAMCRUW measurements. This is reasonable because BS sees the top layers of atoms and they are the ones that are randomized to a maximum extent. SAMCRUW generated SAWs propagate through a layer of 25 - 30μm thickness, where in the randomization is not very high but an average over this thickness. The bulk ultrasonic waves do not feel the disturbance because the wavelength is quite large. When the specimen is sputtered and annealed certain amount of atoms close to surface are removed and annealing helps to remove the strain in the specimen and that should increase the anisotropy. This is clearly observed in BS measurements.

An interesting factor to be observed is that the surface layers have lower elastic moduli than the bulk. The reasons for this lowering of elastic constants is not clear but it seems to be related to dislocations within this region.

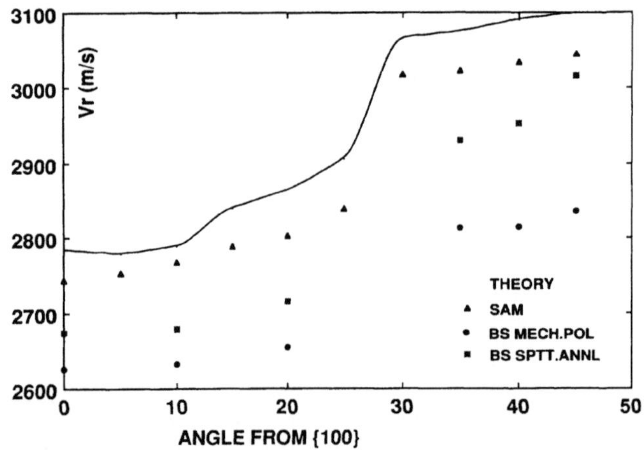

Fig.1. Angular dispersion of Rayleigh wave velocity on Ni (100) plane

TABLE 1. Elastic constants determined by different techniques under different specimen conditions

Technique	Freq MHz	Surface condition	Elastic constants (GPa)			Anisotropy
			C_{11}	C_{12}	C_{44}	
Ultrasonic	10	Mechanical	241	147	126	2.70
SAMCRUW	230	Polish	229±5	134±5	121±5	2.55
BS	2000	Sputtered and	225±5	134±5	100±4	2.20
BS	2000	Annealed	228±5	140±5	118±5	2.56

CONCLUSIONS

In this paper we have shown that Scanning Acoustic Microscopy and Brillouin Scattering can be used to study the surface damage and determine the surface layer elastic constants. The surface layers have much lower elastic moduli compared to the bulk. The anisotropy is least at the top surface and increases to the bulk value as a function of depth. BS is extremely sensitive to the surface damage. SAM is capable of providing surface damage depth because of its deeper penetration capability.

REFERENCES

1. I.R.Kramer. Surface layer effects on mechanical behaviour of metals in "Advances in Mechanics and Physics of Surfaces," . R.M. Latanision and T.E. Fischer. ed, Vol 3, pp109, Hardwood Academic, New.York. (1986)

2. J-L. Lee, J.T. Waber Metall. Trans. 21A : 2037, (1990)

3. J-L. Lee, J.T. Waber and Y.Park, Scripta. Metal. 20 : 823, (1986)

4. J.T. Fourie. Phil. Mag. 17 : 735, (1965)

5. I.R. Kramer. Trans. Metall. Soc. AIME. 233 : 1462, (1965)

6. S.Jahanmir, Laminar wear particle formation, in "Advances in Mechanics and Physics of Surfaces", R.M. Latanision and T.E. Fischer. ed, Vol 3, pp 261, Hardwood Academic, New.York. (1986)

7. F.R.N. Nabarro, "Surface effects in crystal plasticity" NATO Advanced study Series No. 17 (1977)

8. J.P. Hirth and D.A. Rigney. Wear 39 : 133 (1976)

9. N.P. Shu Wear 25 : 111 (1973)

10. S.Sathish, M. Mendik, A.Kulik, G.Gremaud, P. Wachter., Appl. Phys. Lettr. 51 : 167 (1991)

11. M.Mendik, S.Sathish, A.Kulik, G.Gremaud and P. Wachter. J.Appl. Phys. 71, 2830, (1992)

12. A. Kulik, E. Bideaux, G.Gremaud, S.Sathish. Continuous wave ultrasonics : An old method with new applications. In " Ultrasonic Signal Processing", pp 355, A. Alippi, ed, World Scientific, Singapore (1989).

13. A. Kulik, G.Gremaud, S.Sathish Acoustical Imaging Vol 17, pp 71, H.Shimizu, N. Chubachi, J.I. Kushibiki ed, Plenum, New York . (1988)

14. M.W. Elmiger, J.Henz, H.v. Kanel, M.Ospelt and P.Wachter Surfaces. Interface Analysis. 14 : 18, (1989).

15. G.W. Farnell, Properties of elastic surface waves, in " Physical Acoustics" Vol VII, W.P. Mason and R.N. Thurston, ed, pp 109, Academic Press, New York, (1970)

SHEAR WAVE ACOUSTIC IMAGING OF

MICROELECTRONIC CERAMICS

D. Zhang[1] and G. M. Crean[2]

[1]Acoustics Institute, Nanjing University
Nanjing, China
[2]National Microelectronics Research Centre
Lee Maltings, Prospect Row, Cork, Ireland

Annular lenses, used in the practical shear wave imaging of ceramics, are developed in this paper. The theoretical analysis of the distribution of the focused acoustic field in Al_2O_3 shows that the longitudinal wave reflective signal can be suppressed to 20dB below the shear wave signal reflected from the shear wave focus point, if an 8° aperture absorbing disk is placed at the centre of the lens, but the shear wave field are only changed a litter bit. Therefore the shear wave reflective pulse can be esily extracted from reflective pulse train. Several experiment, such as internal voids detection in Al_2O_3 and Al_2O_3-glass sealing interface imaging etc. , are also shown in this paper, which further demonstrates that the annular lens is very useful for realisation and exploitation of shear wave imaging.

INTRODUCTION

Advanced ceramics are of the variety of properties, for example, the low cost and the expectional hardness and wear resistance of aluminium oxide, the high temperature stability of silicon nitride, the widest band gap of silicon carbide and the piezoelectronic properties of PZT etc., are therefore being employed widely as the components of the electronic devices, such as the integrated circuit packages, the optical devices and the electronic ceramic devices and so forth. Therefore the charaterisation of the ceramic materials and the non destructive evaluation of the devices are increasingly improtant for obtaining the high performance devices and the criterion of the device reliability. Acoustic microscopy is an achievement method to observe the internal voids and the other elastic inhomogeneity material[1,2] in the ceramic components. However there are two problem existing in the deep subsurface imaging using a general longitudinal wave mode acoustic microscopy. These concer the effective working distance and the lateral resolution, respectively, in the ceramic material system under characterisation, since the large refraction of the acoustice wave from water. Recently, it has been demonstrated

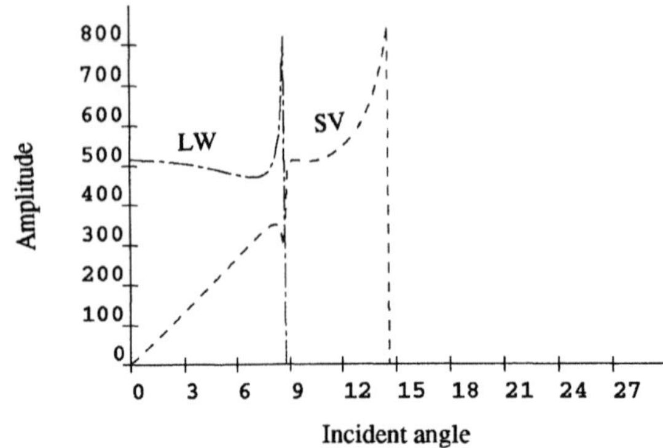

Figure 1 Amplitude of longitudinal (— · — ·) and shear (---------) wave transmitted into Al_2O_3 as a function of incident angle, when an acoustic plan wave impinges on Al_2O_3 surface from water.

that the depth of the penetration and the lateral resolution of the shear wave imaging[3,4], compared to a general longitudinal wave imaging, can be significantly improved, particularly for high acoustic velocity materials, such as ceramics and silicon etc.. However in the practice applications, for example a thin layers system or internal material imaging, the longitudinal and shear wave reflection pulses are too close to be separated by a time gate, since the reflective signal of the longitudinal wave pulse are so strong that it is difficult to pick up a shear wave pulse from the reflective pulse train. This will be a serious problem in the application of the shear wave imaging. Fortunately, the shear wave reflective pulse can be successfully extracted using an annular lens. This paper presents the analysis of a focused shear wave field in ceramic in theory, which shows the longitudinal wave in the sample of Al_2O_3 can be significantly suppressed by a small absorbing disk placed at the centre of the lens, but this disk does not have significant effect on the shear wave field in the sample. Several experiments of the application of the shear wave imaging are also reported in this paper.

THEORETICAL ANALYSIS

The principle of the shear wave acoustic imaging is based on the mode conversion phenomena occurred at the water-solid interface. When an acoustic plane wave impinges on the surface of the solid from water, both longitudinal and shear wave can be transmitted into the solid. the amplitude of the waves in solid, say Al_2O_3, as a function of the incident angle are shown in figure 1, which shows that below the longitudinal wave critical angle, about 9°, the wave principally contributes to longitudinal wave in Al_2O_3, the acoustic wave in this range will be utilised in a general longitudinal wave imaging, However the shear wave imaging uses the acoustic wave in the range between the longitudinal and shear wave critical angles, approximately 9° to 14.5° for Al_2O_3, since the wave transmitted into the material are entirely as the shear wave mode[3].

using the same approach of piston as the analysis of an optical lens to analyse a fuse quartz lens with 28° aperture and 8.5 mm radius spherical surface, the theoretical projections of the focused longitudinal and shear wave field in Al_2O_3 are shown in figure 2 (a) and (b) respectively. In figure 2 (b), a focused quasi annular shear wave field are automatically formed in Al_2O_3. It looks like that the acoustic shear wave field in the Al_2O_3 material is obscured by an opaque disk locating at the centre of the lens.

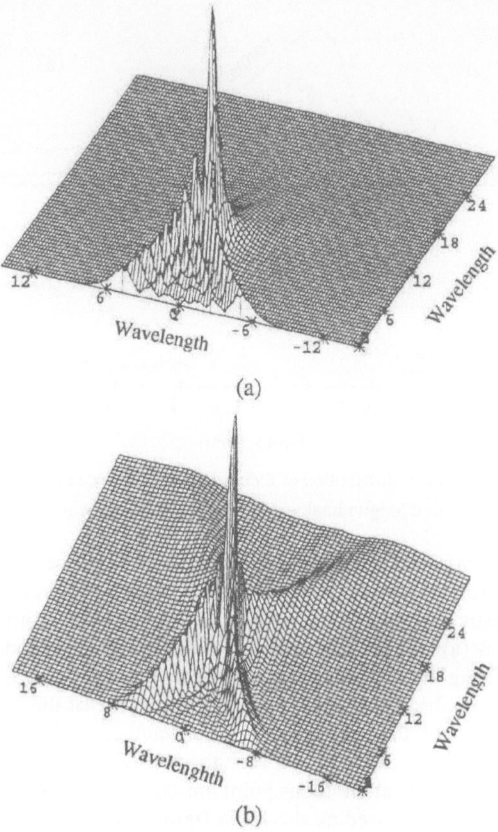

Figure 2 Projections of focused acoustic wave field in Al_2O_3, (a) longitudinal wave mode case, when the A_2O_3 surface is positioned at 2 mm away from the lens, (b) shear wave mode case, when the Al_2O_3 surface is positioned at 5 mm away from the lens.

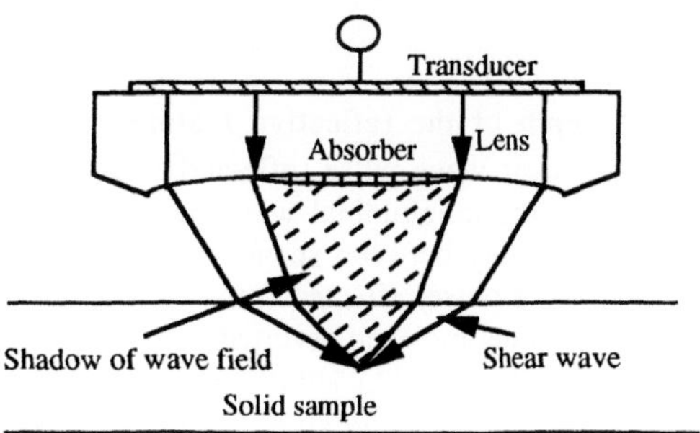

Figure 3 Schematic diagram of an annular lens with an 8° absorbing disk at the centre of the lens.

Therefore it can be imaged that a real absorbing disk, say with 8° aperture, which is put at the centre of the lens and forms a focused acoustic wave shadow in water, as shown in figure 3, does not have much influence on the shear wave field in solid, but the longitudinal wave field in solid can be significantly suppressed by this disk.

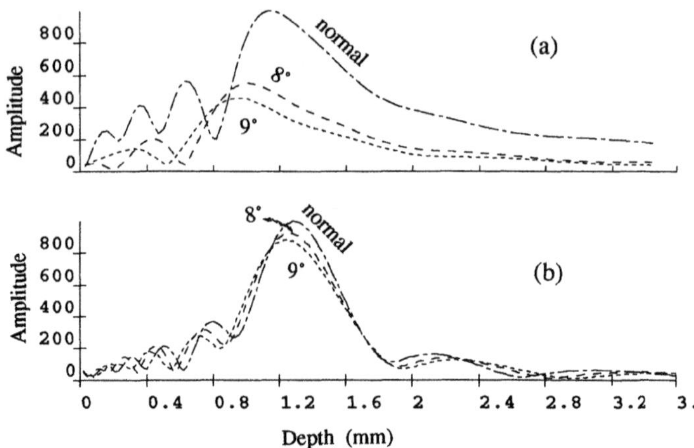

Figure 4 Amplitude distribution of focused acoustic wave along the lens paraxial.
(a) focused longitudinal wave, (b) focused shear wave.

In the normal case, the focus point in the ceramics will be located at the paraxial of the lens which is perpendicular to the surface of the Al_2O_3. Therefore it is very important to analyse the acoustic wave amplitude along the lens paraxial, normalised to focus point, as shown in figure 4 (a) and (b), the curves — · — · ,   — – – and ······· represent the wave amplitudes along the lens paraxial, when using a normal lens, annular lenses with an 8° and a 9° aperture absorbing disks placed at the centre of the lens respectively,. from figure 4 (b), it can be seen that the acoustic shear wave amplitudes at the focus point are only slightly changed as the size of the absorbing disk increased. However the longitudinal wave amplitude at the focus point are decreased quickly and moved ahead as the aperture of the disk increased as shown in figure 4 (a). it makes easy to pick up the shear wave pulse from the reflective pulse train by a time gate.

The further theoretical analysis is for the internal non-destructive evaluation of a Al_2O_3 sample. In the reflective pulse train, the delay time between the acoustic pulse reflected from features and the reference pulse signal reflected from the sample surface can be calculated by the formula,

$$\tau = 2 * L / V \qquad\qquad (1)$$

where L and V are the depth of the reflective features and the acoustic velocity respectively. The τ of the shear wave pulse reflected from the bottom of the sample should be longer than that of the longitudinal wave pulse because of the slower shear wave velocity. However the delay between shear wave pulses reflected from internal voids and the longitudinal wave pulse reflected from the bottom or the interface of the samples will be reduced and sometime even reverted, since the depths of the reflective points are less than the thickness of Al_2O_3 sample. If a time gate is used to pick up the shear wave signal at the focus point, the longitudinal wave signal will be a background noise existing in the time gate. In the case of the scanning acoustic microscope, the reflection signal, received by the lens for imaging, consists of the acoustic wave transmitted into Al_2O_3 and then reflected from the voids or the interfaces in Al_2O_3. After both transmission and reflection are considered, the longitudinal and shear wave signal amplitude along the lens paraxial in Al_2O_3, in decibel scale as a function of the delay time, are shown in figure 5 (a) and (b), when the sample surface is located at 8 mm

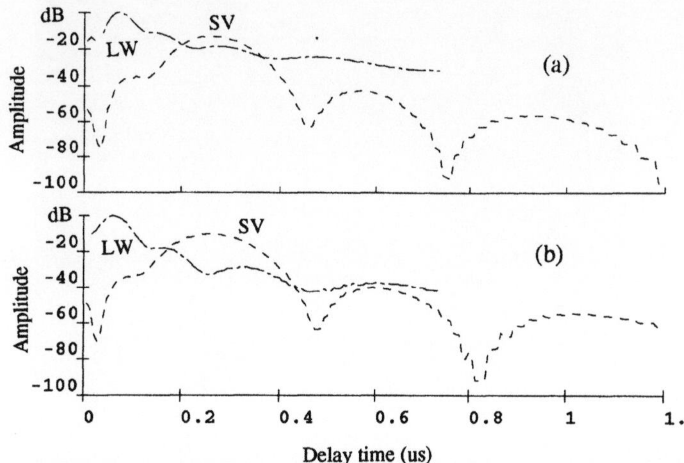

Figure 5 Amplitude of longitudinal and shear wave along the lens paraxial in Al_2O_3, normalised to the amplitude of longitudinal wave, in decibel as a function of delay time. (a) normal lens, (b) 8° annular lens.

away from the lens. In the normal spherical lens case, figure 5(a), the longitudinal wave signal having the same delay time as the shear wave focus point, is only 5 dB below the shear wave signal. However in the case of an 8° aperture absorbing disk at the centre of the lens, figure 5(b), the longitudinal wave signal is suppressed to about 20 dB below the shear wave focus point. Therefore using an annular lens, the shear wave pulse signal, reflected from the focus point, can be easily detected by a normal time gate.

Figure 6 shows the lateral resolutions of focused shear wave beam at focus planes, when the lenses are 5 mm from the surface of Al_2O_3. The curves — · — , — — — — , and · · · · · · represent shear wave fields in Al_2O_3 when focus acoustic beams come from a normal spherical lens, annular lenses with an 8° and a 9° aperture absorbing disks at the centre of the lens respectively. Curves are nearly with the same width in main lobes, but the sidelobes are getting higher when the size of the absorbing disk increasing. However if control the dish aperture and using a proper electrical circuit, the sidelobes does not cause a problem in shear wave imaging, it can be demonstrated in the imaging experiments.

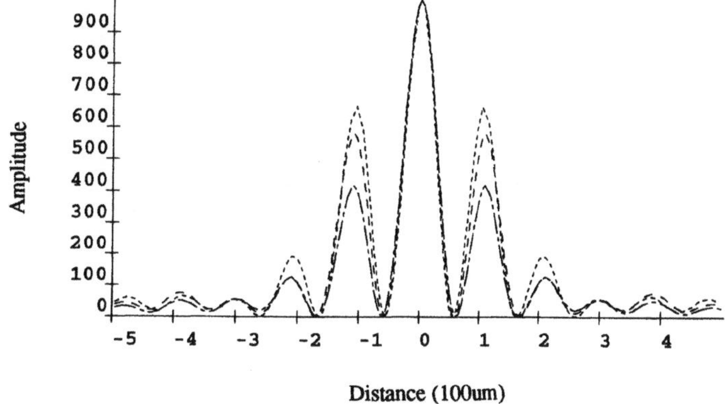

Figure 6 Lateral resolutions of focused shear wave beam at focus planes.

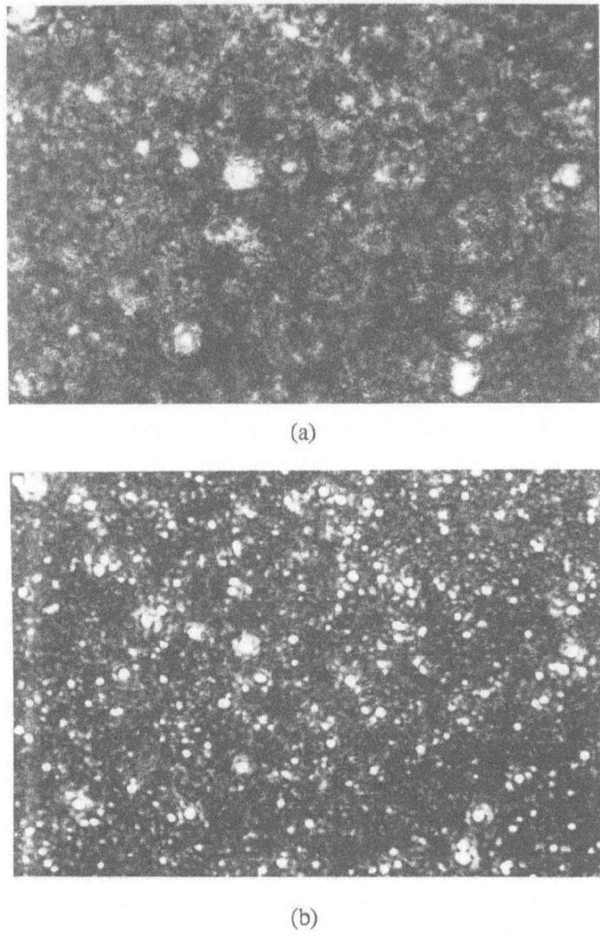

(a)

(b)

Figure 7 Internal voids detection in Al_2O_3. (a) longitudinal wave imaging using a normal lens, (b) shear wave imaging using an 8° annular lens.

EXPERIMENTS AND APPLICATIONS

A 50 MHz fuse quartz acoustic lens with 11.5 mm working distance in water was used in the experiment. The 8° annular lens, used in shear wave imaging, was easy to be made placing an opaque absorbing disk at the centre of the lens. The sample for imaging was two 1.3 mm thick Al_2O_3 layers sealed by glass being part of integrated circuit package. Figure 7 is the internal voids images (with 6 mm width) at 0.65 mm depth of one layer subsurface of Al_2O_3, The interface images (with 15 mm width) of Al_2O_3-glass sealing is shown in figure 8. Compared longitudinal wave imaging, as shown in figure 7 (a) and 8 (a), it can be clear seen that the lateral resolution of shear wave imaging, as shown in figure 7 (b) and figure 8 (b), is significantly improved, and the higher sidelobes of an annular lens does not have much effect on shear wave imaging.

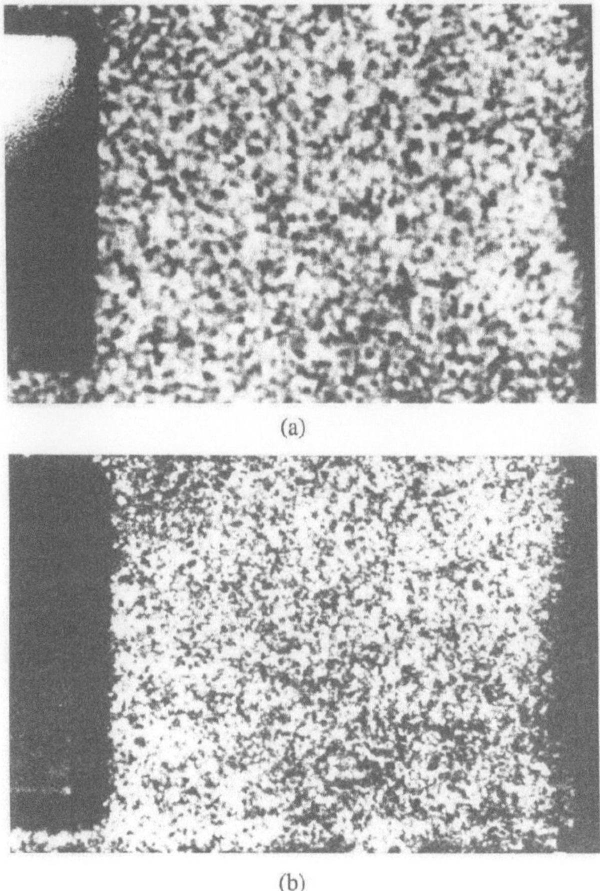

(a)

(b)

Figure 8 Interface images of Al_2O_3-glass sealing of integrated circuit package. (a) longitudinal wave imaging using a normal lens. (b) shear wave imaging using an 8° annular lens

CONCLUSION

An quasi annular focused shear wave beam are automatically formed in ceramics, when a focused acoustic beam impinges on the water-ceramic interface from water, therefore the lateral resolution can be significantly improved. A real annular focused shear wave beam formed placing an opaque absorbing disk at the centre of the lens only slightly influences the shear wave field in ceramics, therefore the annular acoustic lenses are very useful for realisation and exploitation of shear wave imaging.

ACKNOWLEDGEMENT

Authors would like to thanks Mr. E. O'Riain for his graphic software. This work is partially supported by the National Natural Science Foundation of China.

REFERENCE

1. C.W.Lawrence, C.B.Scruby, G.A.D.Briggs and A.Dunhill "Crack detection in silicon nitride by acoustic microscopy". NDT International, Vol. 23, No. 1, pp3-9, 1990
2. G.M.Crean, D.Zhang, J.Flannery and S.C.O'Mathuna "Nondestructive subsurface evaluation of IC ceramic packaging", The proceeding of 14th International Congress on Acoustics, 1992
3. D. Zhang and G. M. Crean "Shear wave imaging for nondestructive evaluation", Elect. Lett., 1991, Vol.27, No.24, pp.2248-2249
4. V. B. Jipson "Acoustic microscopy of interior planes", Appl, Phys. Lett , 1979, 35, pp.385-387

DEFECT SIZING IN CERAMIC MATERIALS BY HIGH-FREQUENCY ULTRASOUND TECHNIQUES

U. Netzelmann[1], H. Stolz[1], W. Arnold[1] and G. Giunta[2]

[1]Fraunhofer-Institute for Nondestructive Testing (IzfP)
University, Bldg. 37
D-6600 Saarbrücken, Germany
[2]Eniricerche S.p.A.
I-00016 Monterotondo (Rome), Italy

INTRODUCTION

Determination of defect sizes of small pores or inclusions in engineering ceramics is a requirement for the prediction of component strength and reliability. Many authors have developed experimental procedures and processing algorithms for sizing defects in the Rayleigh and resonance regime of acoustical scattering. In this work, we have tested some common techniques for defect sizing on small inclusions in SiC, SiSiC and Al_2O_3 ceramics with emphasis on considering practical experimental limitations. Different techniques have been applied to correlate the nondestructive results with the results of other nondestructive and destructive techniques.

EXPERIMENTAL

The experimental arrangement used for the ultrasound measurements has been described in previous publications /1,2/. In short, we are using a broadband spike pulse transmitter for excitation and focussing polymer transducers (Krautkrämer, Cologne) with focal lengths of 7.5 to 50 mm. In the recent years, 80 MHz transducers have become available, which exhibit large bandwidth necessary for most defect sizing strategies. The echo data are detected by fast peak detectors for C-scanning, and by a 400 MHz sample rate digitizer (LeCroy) for B-scanning and volume measurements.

INVERSE BORN APPROXIMATION

A first series of experiments was devoted to the application of Born inversion algorithms to small defects /3/. Here, Al_2O_3 ceramic plates with dimensions of 50 mm x 50 mm and a thickness of 5 mm have been studied. Before sintering, different types of impurities have been added, like particles of Fe_2O_3, small pure iron particles, CrO_2 particles. Pores have been generated by adding a granular plastic material. The impurities were inserted in different depths of the powder and the ceramic was sintered to a final density of 2.62 g/cm^3. After sintering, C-scans of the samples have been measured to locate the scatterers.

An inverse Born scattering algorithm /4,5/ has been implemented, which was tested for different types of inclusions. The defect echo is Fourier transformed and Wiener filtered using the response from a back wall signal of the same ceramic plate. Time-shifting operations were performed in order to bring the imaginary part of the spectrum as close as possible to zero. The result was input into the inversion integral, giving the characteristic function and the defect radius.

The results turned out to be very sensitive to the parameters of the inversion procedure, like selected bandwidth and the zero point of time. To separate experimental and theoretical limitations of the inversion, the real part of calculated backscattering curves /6/ for spherical inclusions for the materials mentioned above have been inserted into the inversion formula. These calculations show, that resonable defect sizes can be obtained for pores, but for most of the solid inclusions the defect size can even not be determined from calculated input data. This is due to the weak scattering assumption inherent to the theoretical model.

MEASUREMENT OF BACKSCATTERING SPECTRA

The theoretical and practical limitations involved with the inverse Born approximation suggest to use an approach omitting the defect reconstruction, where one tries to analyse the backscattering curves of the scatterers directly. The latest polymer probes available exhibit increased bandwidth compared to earlier models, and by making use of real-time average facilities of modern transient digitizers, one is able to average over 1000 ultrasound shots or more in a few seconds. Beside the reduction of the electronic noise, in this procedure the effective resolution of the digitizer is increased by two or

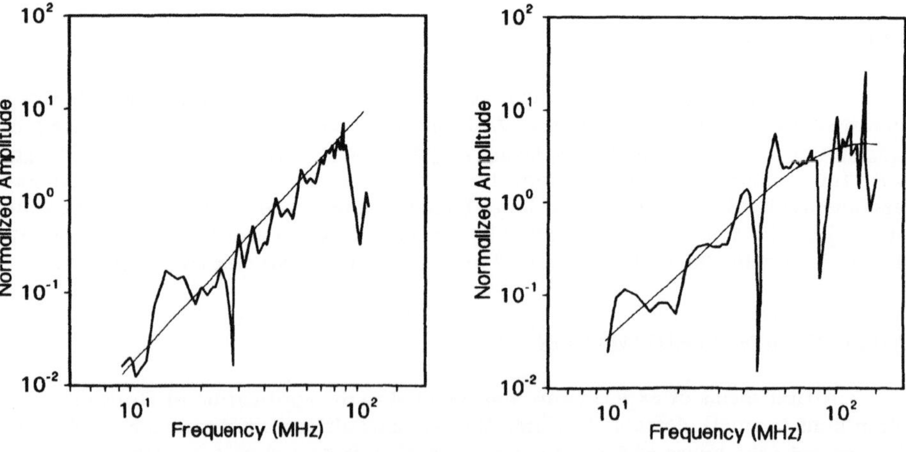

Figure 1. Typical normalized backscattering spectra of iron inclusions inside a SiC ceramic sample. Left part: iron particle size distribution 36-50μm, right part: iron particle size distribution 100-200 μm.

three bits and allows to cover a dynamic range of more than 50 dB. As a result, a usable frequency range of 10 MHz to 120 MHz is available in the backscattering curve.

With this technique, we have determined spectra of small scatterers inside SiC bending specimens with a cross section of 3.5 mm x 4.5 mm. Inclusions of Fe, WC, Si and pores were present in these samples. The nominal inclusion size distributions were peaked around 36-50 µm or 100-200 µm, respectively. The experimental procedure used is as follows: The probe is centered over the defect by maximizing the defect echo amplitude. Then an averaging cycle is performed and the echo data are read to the computer. As many defects are located very close to the front or back wall surface echo of the sample, the defect signal has to be isolated from the wing parts of the neighboring echoes. This is accomplished by performing a second averaging cycle with the probe shifted slightly aside into a region where no defect is present. Both signals are subtracted and a Fourier transform is calculated in the computer. The defect signal is deconvoluted by the system response function by normalization to the frequency spectrum of a back wall echo of the sample.

When plotting the normalized defect response function on a double-logarithmic scale, the Rayleigh scattering regime can be identified as a region of almost constant slope (Fig. 1, left part). When the inclusion size is within the range covered by the available bandwidth, a transition frequency can be observed, where the resonance scattering regime is entered (Fig. 1, right part). The transition frequency is shifted up for the smaller defects and shifted down for the larger defects.

A practical problem of the procedure lies in finding a proper reference signal for deconvolution. We have analysed the frequency spectra of back wall signals from different depths in a test sample. A stepped wedge with thicknesses of 0.5 mm to 4 mm in steps of 0.5 mm has been machined out of an Al_2O_3 bending specimen. Back wall signal spectra obtained with constant focal depth of 2 mm show significant changes of the spectra with depth, which are due to the complicated acoustic field distribution in the near-field of the probe. Amplitude zeroes occuring in the spectra when working out of the focus are presumably also superimposed to the backscattering signals from the defects and are difficult to eliminate in the backscattering spectra.

CORRELATION TO OTHER TECHNIQUES

In a batch of about 10 samples made of SiC-ceramics we first looked for defects using our high-frequency acoustic imaging system. After obtaining the signal/noise ratio (SNR) of each defect indicated, we sectioned the samples destructively in order to get an accurate determination of the shapes, type and size of the defects by Scanning Electron Microscopy. Most of the defects were carbon inclusions surrounded by a cloud of porosity. Their shape was either almost spherical or closely to an ellipsoid. For the smaller defects whose diameter was about 50 µm, the backscattered amplitude turned out to be proportional to a^3, where a is the radius of the defect as expected in the Rayleigh regime /7/. Taking into account the given electronic noise of our system, and the sensitivity of our PVDF-transducer employed, we could deduce the minimal radius of the defects to be detectable as $a_{min} = 25$ µm (SNR = 3dB) in agreement with previous estimates /1/. Additionally, our experiments indicated that a_{min} is almost independent of the type of the defect. This can be understood theoretically because the prefactor in $f_R(\pi)$, the backscattered amplitude in the Rayleigh-regime, contains a complicated combination of the elastic parameters such as E-, and G-modulus and the Poisson-ratio of the defect relative to the host material /6,7/ yielding similar values for defects occuring in practise..

Furthermore, the cloud of porosity and reaction products (see Fig. 2, as a typical example) surrounding a defect makes the excitation of creeping waves and other type of

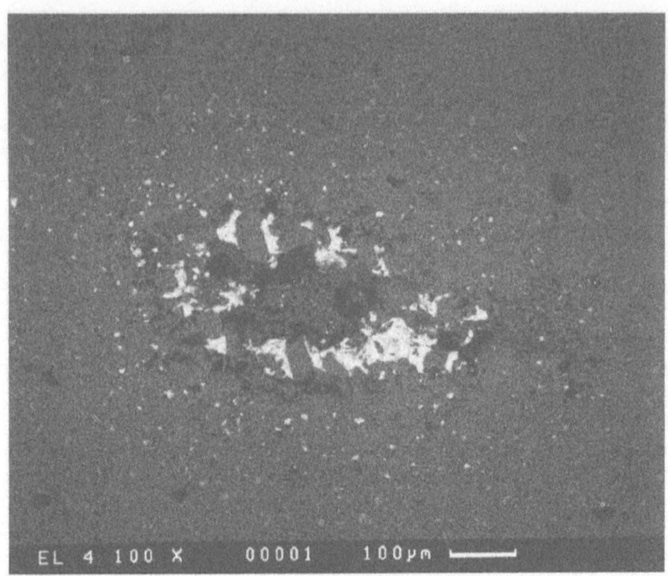

Figure 2. SEM micrograph of a pore in an Al_2O_3 ceramic sample. The iron (white material) originally placed at the position of the pore has diffused into the matrix in the course of the sintering process.

oscillations within the defect difficult, which, however, are at the origin of the characteristic resonances of the backscattered amplitudes. Of course, this applies also in the case where these resonances are excited by narrow-band pulses /8/.

CONCLUSION

In spite of progress in the sensitivity and bandwidth of focussing polymer probes and in data acquisition techniques, our experiments have shown that it is still difficult to obtain reliable size estimations for unknown small defects in engineering ceramics. The results of the inverse Born approximation technique turn out to be very sensitive to a proper selection of the zero point of time in the echo signal, which is basically a consequence of the misuse of the assumption of weak scattering. Obtaining reliable backscattering curves in the near-field of focussing transducers is difficult due to the complicated structure of the sound field. In spite of this, our results show, that the transition between Rayleigh and resonance regime regime can be identified.

Destructive and SEM inspections of samples show, that many inclusions are forming a reaction zone during sintering, which tends to enlarge the observed defect size in the acoustical images and prevents the excitations of resonances usable for defect identification as it was done for defects in metals /9/.

ACKNOWLEDGEMENT

We gratefully acknowledge the financial support of our research by a grant from the German Minister of Research and Technology within the Material Research Programme and the cooperation with Dr. G. Splitt and Dipl.-Ing. M. Rost, Krautkrämer GmbH & Co., Cologne, FRG.

REFERENCES

1. S. Pangraz, H. Simon, R. Herzer, and W. Arnold, Non-destructive evaluation of engineering ceramics by high-frequency acoustic techniques, in: "Acoustical Imaging", **18**, H. Lee and G. Wade, eds., Plenum Press, New-York 1991, p. 189

2. U. Netzelmann, R. Herzer, H. Stolz, and W. Arnold, Volume acquisition and visualization of high-frequency ultrasound data, in: "Acoustical Imaging", **19**, H. Ermert and H.-P. Harjes, eds., Plenum Press, New-York 1992, p. 553

3. H. Stolz, Diploma thesis, Universität and IzfP, Saarbrücken 1992

4. L. J. Bond, Born inversion: application to spherical and spheroidal voids, Proc. Ultrasonics International (1989), p. 768

5. R. B. Thompson and T. A. Gray, Range of applicability of inversion algorithms, Rev. Progr. QNDE (1981), p. 223

6. S. Hirsekorn, IzfP Internal Report No. 780157-TW, unpublished (1978)

7. I. Ermolov , Non-Destructive Testing **5**, 87 (1972)

8. K. Goebbels, H. Reiter, S. Hirsekorn, and W. Arnold, Sci. Ceramics **12**, 483 (1983)

9. G.J. Quentin, M. de Billy, and A. Hayman, J. Acoust. Soc. Am. **70**, 870 (1981)

IMAGING THE 2–D SURFACE ULTRASONIC TRAVELING WAVEFRONT AND THE APPLICATIONS FOR NDE

Xuan-min Yang, Ming Yi, Jin-fu Gan, Ge-xing Yang, Lei-lei Yin

Department of Physics
Nanjing University
Nanjing, China 210008

ABSTRACT

A novel technique for visualization of surface ultrasonic traveling wavefront in real time has been developed. It is based on the correlation theory. The experiments consist of a 12mW He–Ne laser and a 15.5Mc acoustic optical standing wave cell for modulating the laser beam. Imaging of a 31 Mc SAW device with or without flaw in a 4×4 mm area has been investigated by this method.

INTRODUCTION

Various optical techniques have been described in the literatures for visualization of ultrasonic fields. Only few of them were to visualize the surface wavefront for non–destructive evaluation. For instance, the frequency translated holography technique [1] enabled contactless, relatively large area ultrasonic detection. It was based on time average, stroboscopic techniques and phase modulation. This method, however, required exposure time of the order of seconds, the measuring assembly must be extremely carefully protected from external disturbances, such as the building tremors and machine vibrations. H. A. Crostack [2] presented a mobile holographic sound field imaging system by using a ruby double–pulsed laser enabled large area optical detection of insonified sound waves on the tested objects which were not insulated against vibrations. As an other example [3], an instrument for visualizing the SAW on the television screen had been developed. Its key component was a flying spot scanner which contained a laser beam to be used to scan horizontally and vertically at standard television rates. The range of conventional video signals and SAW frequency were limited by the optical resolution of flying spot. The disadvantages of the methods mentioned above are that they are not in real time, and require complex equipments with high cost.

Here, a novel optical correlation method is presented with advantages of simplicity, low cost and real time.

THEORY

A surface wave travels horizontally along X axis with amplitude of h shown in Fig. 1. The surface wave illuminated by a laser beam acts as a moving sinusoidal phase grating represented by

$$E_\alpha = E\exp\{i[\omega t - 2\pi h / \lambda \cos(\Omega t + Kx)]\} \tag{1}$$

which is a spatial–temporal modulated light. Where ω and Ω are the angular frequency of the light and acoustic waves respectively, λ is the wavelength of the light, $K = 2\pi / \wedge$, and $\wedge$ is the wavelength of the sound waves, Assume that the acoustic power is small enough, so that $m = 2\pi h / \lambda < 0.3$, and using identity

$$exp(- im\ cos\Phi) = \sum(- i)^n J_n(m)exp(in\Phi),$$

we can describe the reflected beams mainly with three separated diffraction orders at the far field or focal plane of a lens:

$$E_0 = EJ_0(m)\exp(i\omega t),$$

$$E_{+1} = - EiJ_{+1}(m)\exp\{i[(\omega + \Omega)t + Kx]\},$$

$$E_{-1} = + EiJ_{-1}(m)\exp\{i[(\omega + \Omega)t - Kx]\} \tag{2}$$

According to the Zernike phase contrast method [4], a phase grating can be imaged by spatial filtering. But this is only valid for the steady grating whose diffraction orders have the same temporal frequencies. However, from Equation (2) the temporal frequencies of three diffraction orders are different. Obviously, they are incoherent

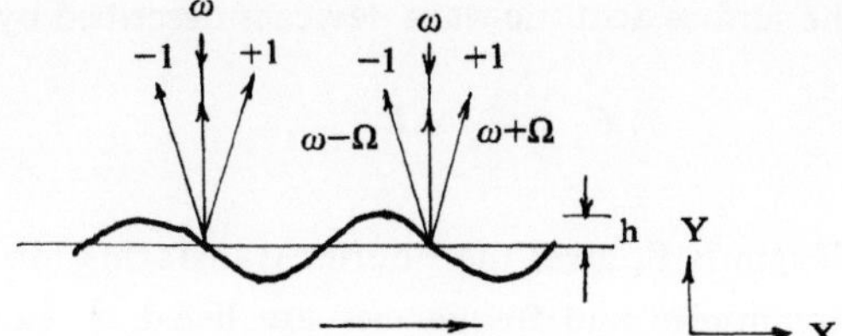

Fig.1 A surface acoustic traveling wave acts as a moving sinusoidal phase grating

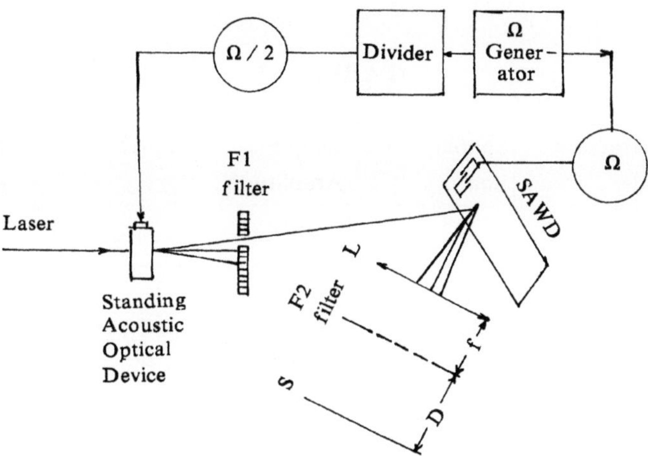

Fig. 2. The imaging setup for surface acoustic traveling wavefront

and therefore couldn't be imaged by Zernike phase contrast method.

A correlation imaging method [5] has been developed and a modulated illuminating laser beam is required. The first order diffracted beam of an acoustic standing wave device which is driven by frequency of $\Omega / 2$ has been used as the modulated laser beam shown in Fig.2. It has been identified that it is simpler than others [6].

The first order diffracted beam of the standing waves used to illuminate the surface waves is given by

$$E_{+1} = E'\{- iJ'_0 J'_1 exp[i(\omega + \Omega / 2)t] + iJ'_{-1} J'_0 exp[- i(\omega - \Omega / 2)t]\} \qquad (3)$$

The output field of the surface acoustic wave device is described by

$$E_s = E_\alpha \times E_{+1} \tag{4}$$

A lens whose focal length is F, gives the Fourier transformation of E_s at the back focal plane. Their amplitudes and frequencies are listed in Table I. The different diffraction orders are coherent with each other now, because of the same temporal frequencies in different diffraction orders. Hence, an image of the surface acoustic wavefront (SAWF) can be formed at far field plane by filtering [6]. The resultant intensity of the image which is also valid in two dimension is given by

$$I \propto (J_0' J_1' J_0)^2 + 4 J_0'^2 J_1'^2 J_0 J_1 sin(\mathbf{K} \cdot \mathbf{x}) \tag{5}$$

TABLE I

Diffraction Order	Spatial Location	Temporal Frequecy Complex Amplitude	
		$\omega + \Omega / 2$	$\omega - \Omega / 2$
+1	Fλ / $\wedge$	$J_0' J_{-1}' J_1$	$- J_{-1}' J_{-2}' J_1$
0	0	$- i J_0' J_1' J_0$	$i J_0' J_{-1}' J_0$
−1	−Fλ / $\wedge$	$- J_1' J_2' J_{-1}$	$J_0' J_1' J_{-1}$

EXPERIMENTS

The optical setup is shown in Fig.2. The power of a He—Ne laser is 12mW, the frequency of the SAW, $f = \Omega / 2\pi = 31$Mc, the focal length of the lens F = 19.5cm, the wavelength $\wedge$ of the SAW = 0.13mm, D = 1.46m is the distance between the screen and the filter.

SUMMARY

A novel technique based on correlation theory to visualize the surface acoustic wavefront has been developed. The equipment is simple and of low cost. It has potential applications for non—destructive evaluation.

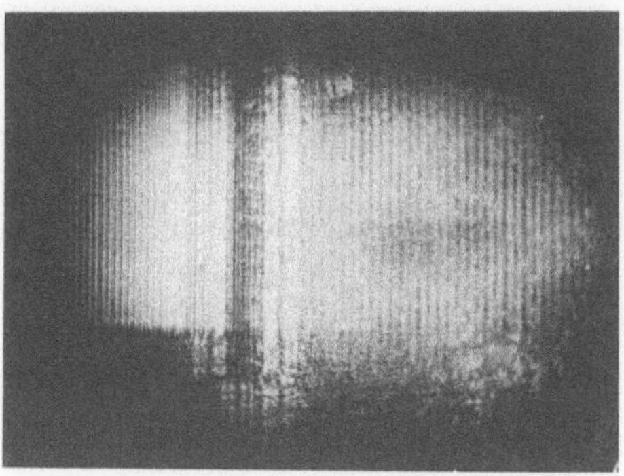

Figure 1. Interdigital transducer and SAWF:
An image of the wavefront of the SAW device. The interdigital transducer is at the left.
The medium is LiNbO$_3$ covered with a thin layer of Aluminum at right.

Figure 2. A partial image of the wavefront of SAW device.

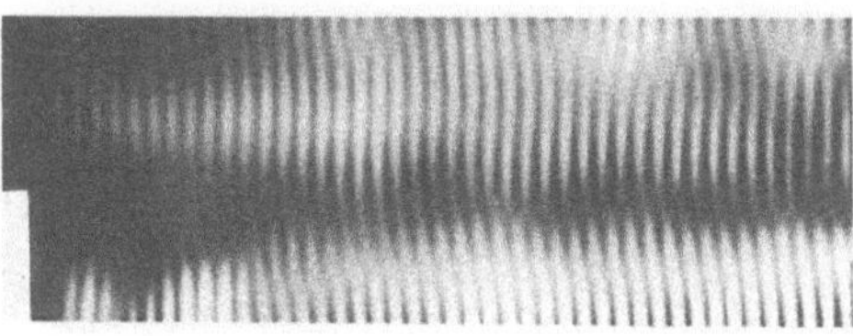

Figure 3. An image of the curved wavefront of SAW device
disturbed by a deep irregular scratch (one wavelength deep).

REFERENCE

[1]   S. Shiokawa, T. Morijzumi and yasuda, Appl. Phy. Lett. Vol. 227, No.8, Oct. 1975, p419.

[2]   H. A. Crostack, A. Kruger and K. J. Pohl., "Non Destructive Testing", p2025-2515.

[3]   R. Adler, A. Korpel, P. Desmares, IEEE Trans. on Sonics and Ultrasonics, Vol. SU-15, No.3, July 1968.

[4]   J. W. Goodman, "Introduction to Fourier Optics", 1968, p145.

[5]   Yi Ming, Yang Xuan-min, etc., Chinese Journal of Acoustics, Vol.7, No.1, 1988, p64-73.

[6]   Yi Ming, Yang Xuan-min, etc., LEOS'90, Proceedings, Nov. 4-9, 1990, p282-283.

NONDESTRUCTIVE   EVALUATION   OF SEMICONDUCTORS
BY PHOTOACOUSTIC IMAGING   WITH HIGH FREQUENCY
AIR TRANSDUCER

Zhao-qiu Wang, Shu-yi Zhang, Jie Zhang and Jing He

Laboratory of Photoacoustic Sciences
Institute of Acoustics
Nanjing University
Nanjing 210008, China

I.INTRODUCTION

The performance of semiconductor devices is related to the chemical
components and crystallographic quality of the semiconductor wafers. The
control of the dose and uniformity of ion implantation, for example, is
critical for good device performance, particularly in MOS Very Large Scale
Integrated Circuit (VLSI) manufacturing. The various thermal and mechanical
processes during device fabrication can generate imperfections and introduce
impurities. The four point probe technique, optical techniques, X ray
techniques and other physical techniques have been used to the characterization of semiconductor materials. Photoacoustic(PA) and Photothermal (PT)
techniques have been recently developed [1] to study the microstructures and
defects in ion-implanted semiconductor wafers, in which the crystal
structure is perturbed. These techniques, being noncontact, nondestructive
and with high spatial resolution, can be used for in-situ monitoring of ion
implantation. Using a PA system with air transducer, we studied previously
the PA signals as the functions of the ion implantation dose for silicon
wafers implanted by boron and phosphorus, before and after annealing [2].
In this paper, we extend the applications of the system to the
imaging of the nonuniformity of ion implantation, and the study of doped
semiconductors and other disturbed crystals. The main purposes of this work
are to improve the previously established photoacoustic system and to find
a simple and efficient method to study the distribution of microstructures
and microdefects of semiconductor and other condensed matters.

II. EXPERIMENTAL SET-UP

The photoacoustic imaging system with high frequency air transducer
is described schematically in Fig.1. An argon ion laser beam, intensity
modulated at 1 MHz by an acousto-optic modulator, is focused onto the
surface of the sample mounted on an X-Y stage. The stage is controlled by
two stepper motors in x and y directions to accomplish the scanning of the
sample under the fixed laser spot. Due to the well known photoacoustic
effect, a sound wave is generated in the air above the sample. It is thus

received by a concave air transducer followed by a lock-in amplifier. The output electric signal is then A/D converted and treated by an IBM 286 computer for imaging. The imaging area of the system can be chosen from $100\mu m \times 100\mu m$ to 20mmx20mm , depending on the desired spatial resolution. The PA image can be displayed in gram scale mode or pseudo-color mode.

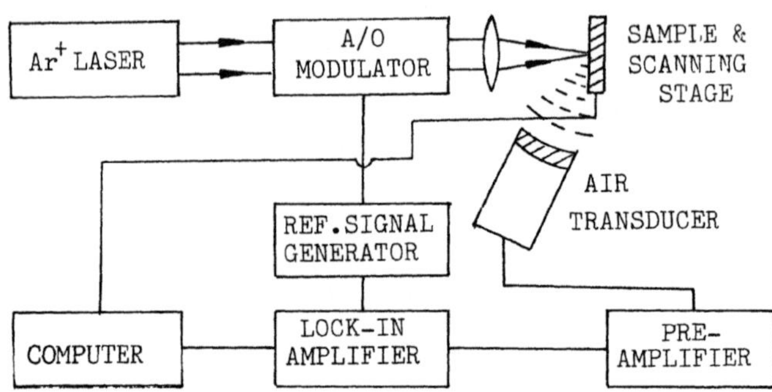

Fig.1 Block diagram of the high frequency photoacoustic imaging system.

The air transducer is the key element of the system, which has not been widely used in photoacoustic detection because of the difficulty in fabricating the transducer. In order to improve the acoustic transmission between the piezoelectric material and the air, a synthetic film is used as matching layer in our design. With an appropriate backing material, the air transducer with a roundtrip insertion loss of 26 dB and a fractional bandwidth of 15% has been fabricated[4].

The air transducer used in our system has a focal length of 20mm and an aperture of 16 mm. The radiation field distribution of the transducer can be estimated by the formula given in reference[5]. The calculated 3 db beam width of the transducer is about 0.9 mm at the focal point. It is much greater than the size of laser spot, so laser spot determines the spatial resolution of the system. The resolution is estimated experimentally by imaging a standard sample with equidistant striae, which is about 1 micron.

The advantages of the photoacoustic technique with air transducer in comparison with other PA and PT techniques are: the simple optical system comparing with the Photothermal Reflectance method[·], the noncontact nature with respect to the PA technique using PZT tranducer bounded on the back of the sample[·], and the higher sensitivity to the surface layer structure than the PA method using electro-mechanical microphones because the higher the frequency, the less the influences of the substrates.

III EXPERIMENTAL RESULTS AND DISCUSSION

A. Ion Implantation Detection

Boron ion implantation is performed in a n-type <100> silicon at 100 keV with doses ranging from 1×10^{10} ions/cm^2 to 5×10^{15} ion/cm^2 in order to obtain the relation between the photoacoustic signal and the implantation dose . The result is given in Fig.2, which shows that the photoacoustic signal increases monotonously with the ion implantation dose. Therefore, the ion implantation dose can be measured by the photoacoustic method, at least for implantations performed under the same conditions.

From these results, we can see that the dose monitoring of boron implantation is possible at about 10^{10} ions/cm^2 to 10^{11} ions/cm^2, which is suitable for MOS device fabrication, for example, the control of the threshold of 64 megabit dynamic random access memory[·].

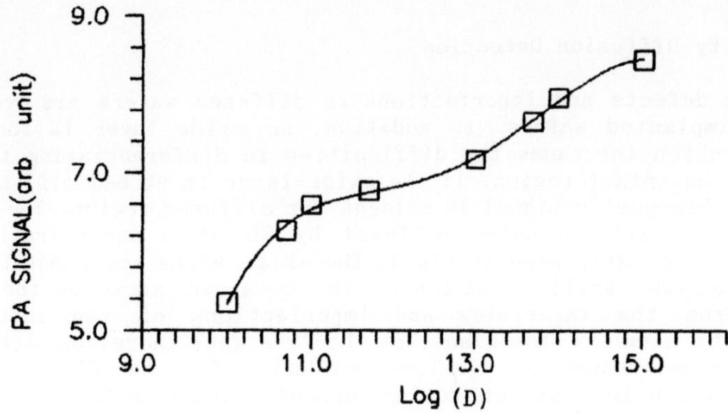

Fig.2 PA signal versus implantation dose (D)
for boron-implanted silicon

The photoacoustic image of a boron implanted wafer at a dose of 10^{11} ions/cm^2 over 1mmx2mm area is presented in Fig.3. Darker regions correspond to greater implantation dose. The inhomogeneity can be seen evidently, the standard deviation of the PA signal is 2.54%, which corresponds approximately to a dose deviation of 6×10^9 ions/cm^2 according to the curve of the PA signal versus implantation doses in fig.2.

In Fig.3, the straight black regions should be related to the dislocations in the crystal wafer, because the PA signal depends on the status of perfection of the initial wafer, on the implantation ions and the implantation conditions. Then the smaller inhomogeneities in Fig.3 may be due to the nonuniformity of the implantation.

Another example is given in fig.3b for GaAs wafer implanted by silicon at a dose of 10^{13} ions/cm^2. Silicon implantation gives much greater photoacoustic signal, because heavier ions disturb the crystal structure more seriously, as we have analyzed previously[3]. In addition, the initial crystal quality seems to be better than that of the wafer in Fig.3a. Two main regions in Fig.3b are supposed to be the result of nonuniform implantation. The standard deviation is about 1.9%.

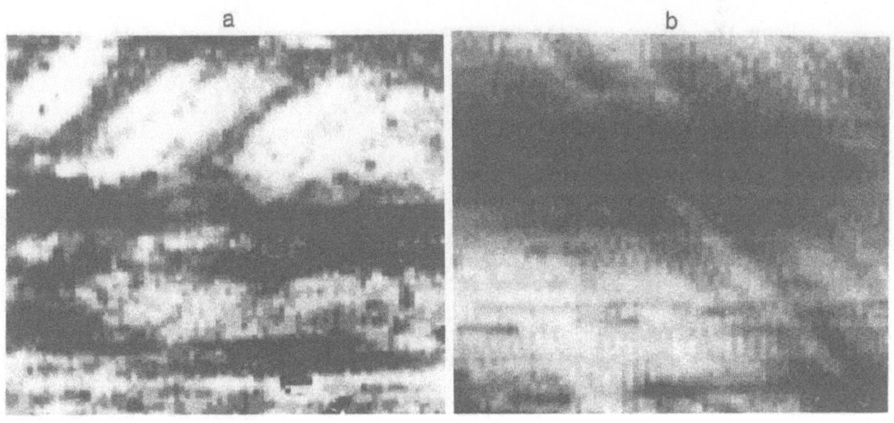

Fig.3 a) Non-unifomity of Boron Implantation, Dose: 10^{11} ions/cm^2, standard derivation: 2.5%   b) Nonuniformity of Si implanted GaAs wafer, dose: 10^{13} ions/cm^2 , standard derivation: 1.9%

B. Impurity Diffusion Detection

The defects and imperfections in diffused wafers are even smaller
than in implanted wafers. In addition, an oxide layer is formed on the
surface, which increases the difficulties in differentiating the diffused
area and the intact region. If the oxide layer is etched off, the increase
of the photoacoustic signal is evident for diffused region. The PA image of
p-type $\langle 111 \rangle$ silicon wafer diffused by Sb at a concentration of 10^{19}
ions/cm^3 is clearly seen in Fig.4. The white areas represent the diffused
regions and the small variation in the remained areas of the photograph
result from the impurities and imperfections of the initial wafer.
Experimental results show also that the PA signal increases with the doping
concentration. Meanwhile, in Te-doped GaSb wafer, the PA signal is 15.4%
greater for 1.6×10^{18} ions/cm^3 doping concentration than for 2.7×10^{17} ions/cm^3
doped sample. The high signal to noise ratio in these preliminary experi-
ments indicates the detection of diffusion-induced imperfections at lower
doping concentration and the monitoring of the diffusion uniformity are
possible.

Fig.4 PA image of Sb diffused GaAs sample.
Dark region is doped at 10^{19} /cm^3.

C. Inhomogeneity Detection of Crystallographic Structure

From the above results on photoacoustic measurement of the implantation and diffusion, it is natural to expect to apply the photoacoustic
technique to other fields of crystal physics. An example is given in fig.5,
where a mediocrely grown silicon crystal is investigated. The nonuniformity
of the photoacoustic signal reflects the inhomogeneity of the crystallo-

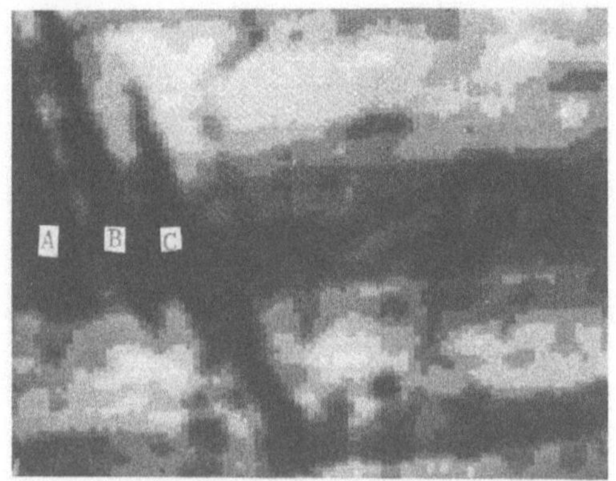

Fig.5 PA image of Silicon with imperfections.

graphic structure and physical properties of the crystal. In the figure, three dark lines (A,B,C) may correspond to three lattice boundaries, while other nonuniform distributions may result from clusters of dislocations and microdefects. The detailed information can be obtained by using other complementary techniques, such as X-ray topography, TEM etc.

IV DISCUSSIONS

The increases of the photoacoustic signal in ion-implanted and diffused semiconductors are related to the imperfections and impurities introduced. In ion implantation samples, a disordered layer is formed in the subsurface region at about 0.1 micron beneath the surface. A multilayer model has been proposed in a previous paper[3], and good explanation of the experimental results has been obtained, particularly for high dose implantation. For implanted wafers with low dose , diffusion wafers, and lightly disturbed crystal samples, it is more difficult to give a rigorous explanation of the experimental phenomena.

It has been pointed out that the photoacoustic signal in a bulk semiconductor can be expressed by[1]:

$$V_{pa} \propto (1-R_i)\,\mu_g\frac{\mu_{th}}{k_{th}}\Big\{\frac{\mu_{th}\beta\,(h\nu-E_g)}{\mu_{th}\beta+(1-i)} + \frac{S_o\mu_{el}E_g}{D_a(1-i)+S_o\mu_{el}} + \frac{(1/2\pi)\,(1-i)\,\mu_{el}^2\mu_{th}E_g}{(\mu_{el}-\mu_{th})\,[D_a(1-i)+S_o\mu_{el}]}\Big\}$$

where R_i is the optical reflectivity of the sample, μ_g, μ_{th} are thermal diffusion length in air and the sample respectively, μ_{el} is the plasma diffusion length given by

$$\mu_{el}^2 = (2D_a/\omega+i\tau)^{1/2}\,,$$

with D_a, the ambipolar diffusion coefficient, τ the lifetime of minority carriers. The parameter β is the optical absorption coefficient, while S_0 is the surface recombination velocity, E_g the energy gap, and k_{th} the thermal conductivity and $h\nu$ the energy of incident photons.

From this expression, it can be seen that any change of above parameters induced by the defects and impurities may influence the photoacoustic signal. The photoacoustic signal must be calculated versus the different parameters. But the parameters are difficult to measure and get the correct values.   In fact, during the fabrication processes, dislocations, microdefects, stacking faults, impurities etc. are introduced. Since the impurities and imperfections can perturb the periodicity of the lattice, the transport properties of the charge carriers will be affected. Particularly, the energy state, the mobility , the lifetime and the surface recombination velocity of the minority carriers and the energy state of the crystal will be changed.

Great attention has been paid to the studies of these parameters. Owing to the different experimental conditions in which experiments have been undertaken , it is not possible to predict how the parameters are modified for a given sample without the complete knowledge of the types and distributions of the defects. In this situation, a numerical fitting of the photoacoustic signal is always possible by assuming such and such a law for certain parameters as a function of ion implantation dose, the diffusion concentration or others. Quantitative analyses of the phenomena need a great deal of systematic experiments by using a variety of analytical techniques to complement each other.

Despite the complexity of the theoretical analysis, the PA signal really reflects the nonuniformity of structure and components of the sample. It provides a complementary, noncontact tool to analyze semiconductor wafers and crystal samples. Particularly, the PA technique is suitable for the

study of thick crystal samples. It is rapid, simple, and with high
resolution, and is thus competitive with the X-ray topography.

REFERENCES

1. W.L.Smith, A.Rosencwaig, and L.Willenborg, Thermal wave measurement of
   ion implanted silicon,Appl. Phys. Lett. 47:584(1985).
2. S.Y.Zhang and L.Chen, Photoacoustic microscopy and detection of
   subsurface features of semiconductor devices, *in* Photoacoustic and
   Photothermal Wave Phenomena in Semiconductors, ed. A.Mandelis, North-
   Holland,(1987).
3. Z.Q.Wang, S.Y.Zhang and Q.S.Gao, Ion implantation measurements with a
   high frequency photoacoustic technique in air, Nucl. Instr. Method. in
   Phys.Res., B62:400(1992).
4. Z.Q.Wang and S.Y.Zhang, High frequency photoacoustic measurements in air,
   Science in China, A34:1273(1991).
5. A.Penttinen and M.Luukkala, The impulse response and pressure nearfield
   of a curved radiator,J. Phys.D: Appl. Phys. 9:1547(1976).
6. G.Washidzu, T.hara, R.Ichikawa et al, Dose and damage measurements in low
   dose ion implantation in silicon by photoacoustic displacement and
   minority carrier lifetime, Jpn. J. Appl. Phys.30:L1025(1991).
7. R.S Sterns, and G.S Kino, Photoacoustic techniques for the measurement
   of electronic properties of semiconductors,*in* Photoacoustic and
   Photothermal Wave Phenomena in Semiconductors, ed. A.Mandelis, North-
   Holland,(1987).

APPLICATIONS OF PHOTOACOUSTIC MICROSCOPY IN
DYNAMIC ANALYSES OF INTEGRATED CIRCUITS

Jian-wen Fang[*], Yao-chun Shen, Yue-sheng Lu,
Jian-chun Cheng, and Shu-yi Zhang

Institute of Acoustics and Lab. of Modern Acoustics
Nanjing University, Nanjing 210008, China

INTRODUCTION

Recently, photoacoustic (PA) and thermal wave techniques have been applied successfully to study the semiconductor materials and devices, especially applied to study the surface and subsurface features of integrated circuits (IC)[1], as well as the PN junctions in which the interface of the PN junction is parallel to the surface[2,3,4]. However, in most previous researches, the IC devices are imaged by the PA microscopy in static state (not in operation), in which some structures having similar properties cannot be distinguished apparently. In this work, we present mainly the PA images of IC devices in dynamic state (in operation), and find out that, as pointed out by Lu et al.[2], PA images of IC devices obtained in dynamic state reflect the electronic structures and logical principles more effectively than those in static state. Besides, a one-dimensional model is proposed to study the effect of the external electric field on the PA signal in the lateral PN junction area, in which the interface of the PN junction is perpendicular to the surface. In theory, the heat absorption and/or heat release of the PN junction are the important factors to affect acoustic wave through thermoelastic coupling. Therefore, the external electric field enhances the influence of the electric properties on the PA signal of the devices. Thus the dynamic PA images (i.e., the PA images of the devices obtained in the operational condition) can show some electric components, which cannot be displayed by the static PA imaging, where the devices are in static condition. The theoretical analysis can properly be used to explain the experimental results.

---

[*]Permanent Address: Dept. of Phys., Zhejing Normal University, Jinhua 321004, China.

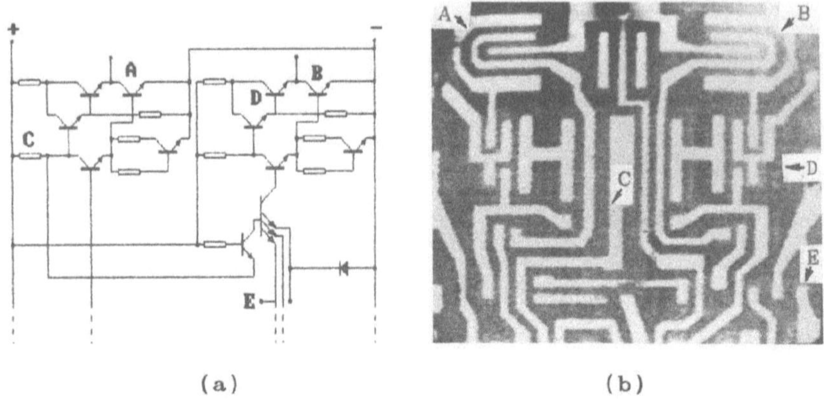

(a)   (b)

Figure 1. (a) A partial logical circuit of the IC device. (b) The static PA image of the IC device.

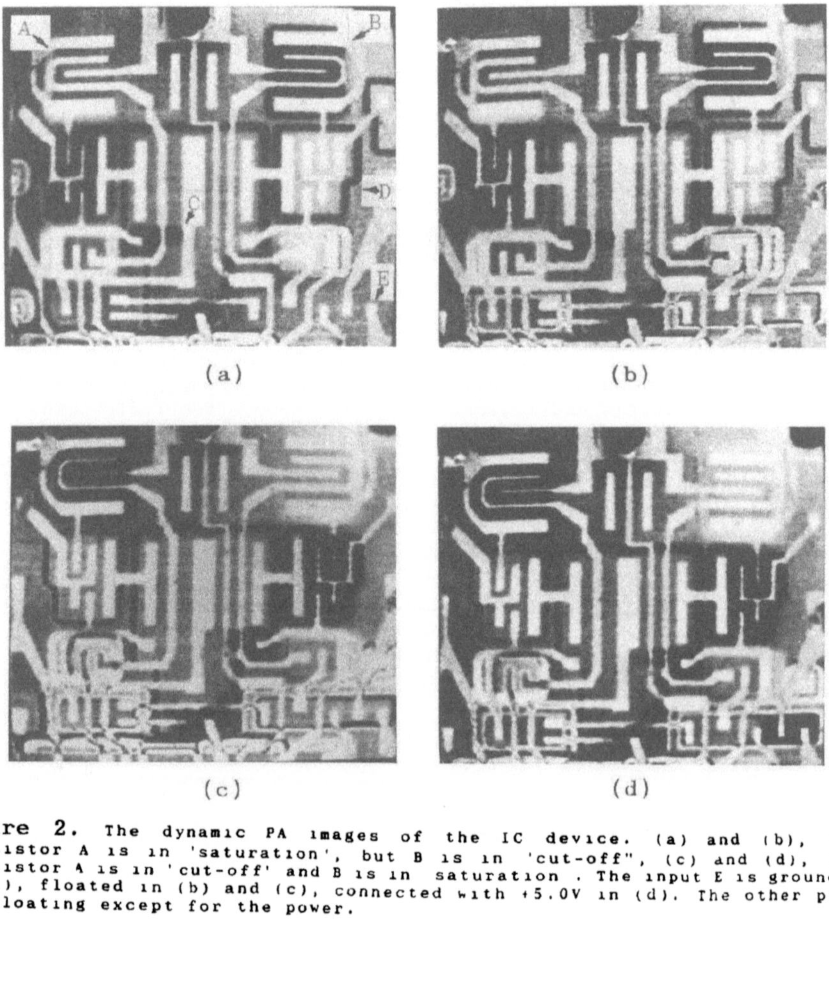

(a)   (b)

(c)   (d)

Figure 2. The dynamic PA images of the IC device. (a) and (b), the transistor A is in 'saturation', but B is in 'cut-off", (c) and (d), the transistor A is in 'cut-off' and B is in saturation . The input E is grounded in (a), floated in (b) and (c), connected with +5.0V in (d). The other pins are floating except for the power.

EXPERIMENT

An IC device (TO76, D trigger) with the logical circuit partly shown in Fig.1(a) is used as a sample. The PA images of the IC device are obtained by the PA microscopy described previously[1]. A static PA image of the investigated areas of the IC is shown in Fig.1(b), which reflects mainly the characteristics of the surface aluminium electrodes. Some dopant areas are also slightly visible because of the difference of optical and thermal properties. Compare Fig.1(a) with Fig.1(b), it can be illustrated that A and B areas of Fig.1(b) are two transistors, C is a resistor which is not visible, D is the compound transistor and E is one of the inputs of IC. As a bias voltage is applied to the devices, a series of dynamic PA images with different biases are obtained and shown in Fig.2. In addition, in order to study the dynamic PA images, the PA images of a lateral PN junction in different biases are also obtained and shown in Fig.3(a). In the dynamic PA images, some components or areas, which are indistinct or even cannot be seen in the static PA imaging, can be shown clearly with different grades of contrast. The essential characterization of dynamic images will be discussed in the following sections.

THEORY

A one-dimensional model of the lateral PN junction is shown in Fig.4, where (Xn-Xp) is the width of the space-charge region (SCR). V is the external D.C. voltage, $V(a,t)$ is the photo-voltage induced by the incident light at $x=a$, V_0 is the diffusion potential across the SCR. In order to simplify the theory, several approximations have been made in the calculation: (i) the light intensity is weak and the nonlinear effects are neglected, (ii) the PN junction is an abrupt junction and the SCR is an exhausted area, (iii) the surface effects are ignored.

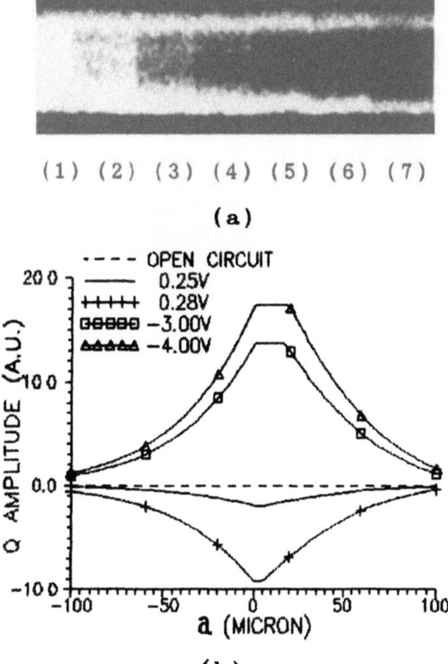

Figure 3. (a) Dynamic PA images of a lateral PN junction under different biases. (1) open, (2)-1.5V, (3)-3.0V, (4)-4.0V, (5)-5.0V, (6)-6.0V, (7)-7.0V, respectively, (b) The amplitude of heat absorption or heat release change with the incident light position in the lateral PN junction under different biases.

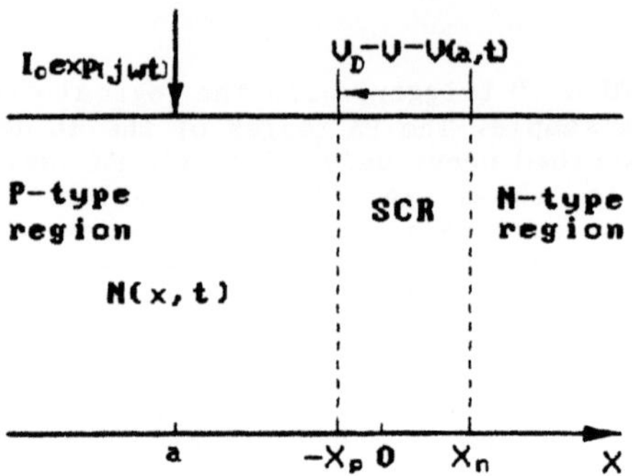

Figure 4. One-dimensional model of a lateral PN junction.

(1) The transportation of Photo-generated-carriers (PGC)

A laser beam with modulation frequency ω is incident upon the point a of the sample. As point a is in the p-type region ($a < -Xp$), the diffusion equation of PGC and the boundary conditions are:

$$\frac{\partial N(x,t)}{\partial t} = D_e \frac{\partial^2 N(x,t)}{\partial x^2} - \frac{N(x,t)}{\tau_e} + N_0 \delta (x-a) e^{j\omega t} \tag{1}$$

$$N(x,t)\big|_{x\to-\infty} = 0$$
$$N(x,t)\big|_{x=-x_p} = 0$$

where D_e and τ_e are the diffusion coefficient and the lifetime of PGC in the p-type region, respectively. $N(x,t)$ is the PGC density and N_0 is the flux of PGC generated by incident laser at x=a plane.

From Eq.(1) we can get the photo-current intensity when the incident light impinges upon the p-type region as

$$J(a,t) = -qD_e \frac{\partial N(x,t)}{\partial x}\Big|_{x=-x_p} \tag{2}$$
$$= qN_0 e^{j\omega t+(a+x_p)/\lambda_e} \qquad (a \leq -x_p)$$

where q is the electron charge and λ_e is the diffusion length of PGC (electron) in p-type region defined as

$$\lambda_e = \sqrt{\frac{D_e \tau_e}{1+j\omega\tau_e}} \tag{3}$$

Similarly, the photo-current intensity in the n-type region is

$$J(a,t) = qN_0 e^{j\omega t - (a-x_n)/\lambda_h} \qquad (a \geq x_n) \tag{4}$$

where λ_h is the diffusion length of PGC (hole) in n-type region defined as:

$$\lambda_h = \sqrt{\frac{D_h \tau_h}{1+j\omega\tau_h}} \tag{5}$$

and D_h and τ_h are the diffusion coefficient and lifetime of PGC in the n-type region, respectively.

When the incident light impinges upon the SCR, the recombination of PGC can be ignored due to the exhausted area approximation. Therefore the photo-current intensity is

$$J(a,t) = qN_0 e^{j\omega t} \qquad (-x_p \leq a \leq x_n) \tag{6}$$

(2) Effect of the external electric field on the thermal wave

When the PN junction under illumination is biased positively or negatively, the current passing through the PN junction includes two parts, i.e., the photo-current and the injection-current, where the latter is induced by the photo-voltage and/or external bias. When carriers pass through the PN junction, its energy changes in the form of absorbing heat or releasing heat because of the energy level difference between the two sides of the PN junction. The absorbed or released heat can be expressed as

$$Q(a,t) = [J(a,t) - J_i][V_D - V - V(a,t)] \tag{7}$$

where J_i is the injection-current induced by the photo-voltage and/or the external electric field. From above equations we can deduced that, although the external electric field is D.C., the alternating current (A.C.) part appears in the thermal effect due to the D.C.-A.C. coupling. Especially, when the PN junction is positively biased, the external electric field reduces the electric field in the SCR, and decreases the heat release induced by the photo-current. On the other hand, the external electric field also enhances the injection-current, and increases the heat absorption resulted from the injection-current. When the positive bias approaches a certain value, the absorbed heat is larger than the released one, thus the total thermal effect appears to be heat absorption. The change of the heat in PN junction can be calculated according to Eq.(7) and the equivalent circuit of PN junction. The numerical results are also shown in Fig.3(b), in which the different curves correspond to different biases. From Fig.3(b) we can see that as the inverse bias increases, the PA signal enlarges and the width of the SCR increases too. This is consistent with the experimental results.

DISCUSSIONS AND CONCLUSIONS

According to the transport properties of semiconductors, there are three thermal sources: (i) instantaneous intraband nonradiative thermalization due to the carrier-phonon collisions within the conduction band, (ii) interband nonradiative bulk recombination of PGC, (iii) interband nonradiative surface recombination of PGC. However, as a semiconductor device is applied external electric field, three additional thermal sources should be considered: (iv) instantaneous Joule thermalization due to PGC drift in the external electric field, (v) Joule thermalization generated when photo-current passes through the load in IC, and (vi) the absorption and/or release of heat caused when carriers pass through the PN junction, due to two sides of the PN junction has different energy levels. The detected PA signal is the combination of the signals caused by above sources. As the IC device is in operation, the total PA signal of IC is the summation of the contributions of the six thermal wave sources discussed above, but the last three sources may play more important role.

Based on the above description, the PA imaging may be discussed as follows:

1. When the laser illuminates a component such as a resistor, PGC drift in the external electric field and release Joule heat. The Joule thermal source enhances the total PA signal of the resistor when the external electric field is sufficiently strong. Thus, the resistance components are distinct in the dynamic PA images[7].

2. If the laser illuminates the transistors, because the modulated laser power in the experiment includes D.C. and A.C. parts, the photo-voltage may change the states of transistors. If the photo-voltage makes the transistor from "cut-off" to "amplification" state, the transistor can amplify the photo-current like phototriode in operating state. While photo-current passes through the loads in IC, it will generate Joule heat. On the other hand, when the photo-current passes through the transistor itself, the potential difference of the electrodes generates heat, so that, the great PA signal is detected (see "B" area in Fig.2(a)). If the photovoltage make the transistor changed from "amplification" or "near saturation" to "saturation" state, the transistor is similar to that in the static image (see "D" area in Fig.2(a)) because the transistor cannot amplify the photo-current and the potential difference of the electrodes is equal to zero. Due to the same reason, the transistor which is originally in "saturation" state is also similar to the static imaging (see "A" area in Fig.2(a)).

3. In Fig.2(a), although the dynamic images of the transistors "A" and "D" are similar mostly to the static image of them, in the edges of the transistors indicated by the arrows A and D, the lateral PN junction lying between the collector (dopant area) and the isolating groove has explicitly different image from those of the static image. The apparent differences between "A" and "D" area are due to the different biases applied to the lateral PN junction. For the transistor "A" in "saturation", the potential of the collector is very low and the lateral PN junction is positively biased. Meanwhile, from Fig.3(b), the heat absorption in the PN junction decreases the total PA signal. As a transistor is not in "saturation", because the incident light impinging upon the edge of the collector does not alter the original state of the transistor, the collector is still in high potential, the lateral PN junction is in inverse biased state, and the total PA signal of the edge is much stronger, where the "dark" areas appear in the image.

  4. For the input part of the IC device, the differences between the four dynamic may be due to the different electrostatic inductions under different electric conditions. However, in the image shown in Fig.2(d) the external voltage applied to the emitter does not alter the working state of the transistor. It makes the PN junction of input part biased inversely, thus the larger PA signal is generated.

  5. On the other hand, it is observed in dynamic image that the aluminum electrodes appear narrower in strong PA signal areas. That is because the diameter of laser spot is not very small. So the width of the aluminum electrodes in dynamic PA image also reflects the intensities of the total PA signal of different areas.

  In general, it is difficult to analyze the dynamic PA images not only due to the complex multilayered structures of IC, but, as discussed above, in addition to the influences of the optical, thermal and elastic properties, the PA signals are also strongly influenced by their electric environment of the components in IC. However, for a transistor, the PA signals are related to its working states, for example, the PA signals of the lateral PN junction lying between the collector and the isolating groove are also related to the collector potential. Therefore, according to the PA imaging of IC devices (especially in dynamic states), we can analyze the working states, such as the "saturation" or "cut-off" conditions of the transistors and the logical circuits, as well as the failure factors of IC devices.

Acknowledgement

This work is supported by the Science Foundation Commission and National Education Commission of China.

REFERENCES

1. S.Y.Zhang and L.Chen, in *Photoacoustic and Thermal Wave Phenomena in Semiconductors*, Edited by A.Mandelis (Elsevier, New York, 1987), P.27.
2. Y.S.Lu, S.Y.Zhang and J.C.Cheng, J.Appl.Phys., Vol.68, P.1088(1990).
3. S.Y.Zhang, J.C.Cheng, Semicond. Sci. technol., Vol.6, P.670(1991).
4. A.Mandelis, J.Appl.Phys., Vol.66, P.5572(1989).

ON THE NEAR-FIELD OPTICAL MICROSCOPY:
A THEORETICAL CONSIDERATION

Ole Keller, Mufei Xiao[1] and Sergey Bozhevolnyi

Institute of Physics, Aalborg University, DK—9220 Aalborg
Denmark, Fax: 98156502, E—mail: 113MUFEL@
VAX87. AUD. AUC. DK

The technique of Near—field Optical Microscopy (NFOM)[2] can be used to obtain microscopic resolutions well below the wavelength of light. In this paper, a theoretical analysis based on a microscopic description of the interaction between the dielectric probe and the tested surface is presented. The probe tip is assumed to be a point—like sphere and the surface is represented by a two—dimensional discrete lattice with subwavelength structure. Using a Green's function technique we established a set of selfconsistent integral equations to describe the local field at the site of the probe tip and the selvedge. All contributions including bulk reflection and many—body interactions have been taken into account. Using point—dipole approach we have solved the selfconsistent equations exactly.

The results have been compared with various approximate solutions. We have found that the approximate solutions are very different from the exact solution in the case of strong interaction. The approximate solutions tend to infinity only when the tip—surface distance decreases to zero, and the results depend strongly on the number of Born iterations performed. However, for the exact solution it is shown that there is a region close to the surface where an enhanced field can be induced by the probe, and when the tip—surface distance decreases to zero the induced field tends to zero. Thus, we have found the resonance interaction between the probe tip and the surface. The qualitative comparison of our results with the experimental data from single — particle plasmon studies[3] revealed the similar features of the system behavior near the resonance. It is shown that in the absence of damping the resolution of the system can be as close to infinite as the tip—surface distance is close to the resonance value, which is essentially nonzero. The induced field at site of the probe as a function of scan coordinate is calculated for different tip—surface distance and other system parameters.

---

[1]   on leave from Nanchang Institute of Aeronautical Tech , China
[2]   R Reddich, R Warmack, T Ferrell Phys Rev B39 767 1989
[3]   U Ch Fischer and D W Pohl Phys Rev Lett 62 458 1989

Acoustical Imaging Volume 20 Edited by Y Wei
and B Gu, Plenum Press New York 1993

FORMATION AND PROPAGATION OF LIMITED DIFFRACTION BEAMS

Jian-yu Lu and James F. Greenleaf

Biodynamics Research Unit, Department of Physiology and Biophysics
Mayo Clinic/Foundation
Rochester, MN 55905

INTRODUCTION

Since the first discovery of localized waves (focused wave mode) in 1983[1] and limited diffraction beams (Bessel beams) in 1987,[2] great advancement in both theory and experiment of these beams have been achieved.[3-6] Generalization of the limited diffraction beams and their applications to medical ultrasonic imaging, biological tissue characterization, and ultrasonic nondestructive evaluation of materials have also been reported.[7-15]

In this paper, formation of the limited diffraction beams by transducer elements and propagation of the resulting two- and three-dimensional waves will be shown. The limited diffraction beams are pencil-like over large depth of field even if they are approximately produced with a finite aperture. In next section we will give a brief review of the theory of the limited diffraction beams. Then, we will show both two- and three-dimensional formation and propagation of the limited diffraction beams. Finally, we have a short discussion and conclusion.

THEORETICAL PRELIMINARIES

One of the special families of solutions of the n-dimensional loss-free, isotropic/homogeneous wave equation is given by[15]

$$\Phi(x_1, x_2, \cdots, x_n; t) = f(s), \tag{1}$$

where

$$s = \sum_{j=1}^{n-1} D_j x_j + D_n(x_n - c_1 t), \quad (n \geq 1), \tag{2}$$

and where

$$c_1 = \pm c \sqrt{1 + \sum_{j=1}^{n-1} D_j^2 / D_n^2}, \quad (n \geq 1). \tag{3}$$

D_j are complex parameters which are independent of spatial and time variables (x_j ($j = 1, 2, \cdots, n$) and t), and $f(s)$ is any complex function (well-behaved) of s.

If c_1 in Eq. (2) is real, f(s) represents a limited diffraction wave propagating along axis, x_n, at the phase velocity of c_1, in an n-dimensional space, i.e., traveling with the wave, one sees a complex wave pattern unchanged in both amplitude and phase. The "$\pm$" term in Eq. (3) represents forward and backward propagating waves, respectively. In the following, we consider forward going waves only. For backward going waves, the results are similar.

Follow the procedures in Reference 15 or 11 (i.e., choose the function forms of the parameters, D_j, and integrate over some of their arguments), we obtain the J_0 Bessel beam[2] and the zeroth-order X wave[11]

$$J_0(\alpha r)e^{i\beta z - i\omega t} \tag{4}$$

and

$$\frac{1}{a_0}b(t) * \frac{a_0}{\sqrt{(r \sin \zeta)^2 + [a_0 - i \cos \zeta(z - c_1 t)]^2}}, \tag{5}$$

respectively, where $J_0(\alpha r)$ is the zeroth-order Bessel function of the first kind, α is a constant, $r = \sqrt{x^2 + y^2}$ represents radial distance, $\beta = \sqrt{(\omega/c)^2 - \alpha^2}$, ω is angular frequency, c is speed of sound or speed of light, t is time, a_0 is a constant, $b(t)$ is an impulse response of a transducer and its associated electronics, ζ is an angle between sidelobe of X wave and surface of transducer, and $c_1 = c/\cos \zeta$ for the X wave. If the J_0 Bessel beam and the zeroth-order X wave are truncated to a finite aperture with a radius, a, the depths of field (propagation distances over which the waves are approximately unchanged) are given by[2,11,15]

$$Z_{Bmax} = a\sqrt{\left(\frac{\omega}{\alpha c}\right)^2 - 1} \tag{6}$$

and

$$Z_{Xmax} = a \cot \zeta, \tag{7}$$

respectively.

In the following, it will be shown how a J_0 Bessel beam and a zeroth-order X wave are formed by an annular array transducer element by element and propagate in water over large distances.

FORMATION AND PROPAGATION OF LIMITED DIFFRACTION BEAMS

Simulations of the formation and propagation of the limited diffraction beams were performed with the Rayleigh-Sommerfeld formulation of diffraction.[16] The aperture geometry of wave sources (transducers) were assumed to be the same as that of an experimental transducer. The driving waveforms[7,12] for producing the J_0 Bessel beam and zeroth-order X wave will be described in the following. The wave fields will be displayed in two- and three-dimensional format.[17]

The experiment setup for measuring acoustic wave fields is shown in Fig. 1. A polynomial waveform generator produced a tone burst or short electric pulses. These electric signals were amplified by a RF power amplifier to excite the transducer element by element. The acoustic wave fields produced in water were measured by a 0.5 mm diameter calibrated hydrophone. Output signals of the hydrophone were amplified and digitized, and then stored in a hard disk. The hydrophone can be scanned along two axes as shown in Fig. 1 to sample either pulse or CW wave fields. It can also be placed at various depths along the wave axis. The transmission of the signals was synchronized to the scan of the hydrophone, and data were transferred to a SUN SPARCstation for further processing and display.

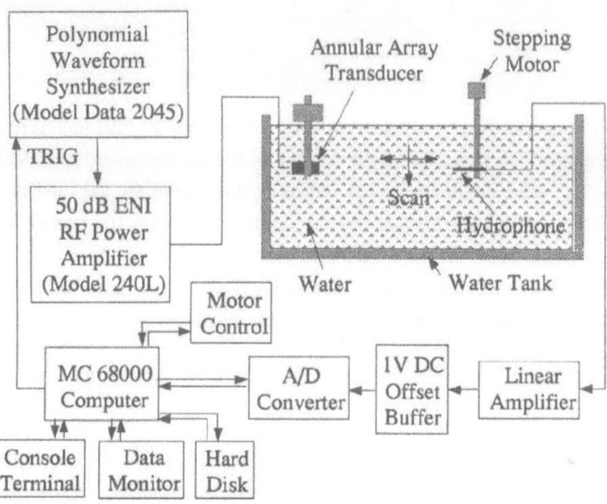

Figure 1. Experiment block diagram for measuring acoustic wave fields in water. The system can measure both CW and pulse wave fields at various depths.

The transducer used in the experiment is a 10–element annular array, which was described in detail in References 9 and 12. The transducer has a diameter of 50 mm, a central frequency of 2.5 MHz, and a −6 dB pulse-echo (two-way) bandwidth of about 50% of the central frequency.

Figure 2 shows analytic envelopes of experimentally measured wave fields in a plane along the axis. The wave fields were obtained by increasing the number of excited transducer elements to form a CW J_0 Bessel beam. Panel (1) represent the wave field when only central element was excited and Panel (10) is the wave field when all the elements were excited simultaneously (Panel (n) shows the wave field when central n elements were excited simultaneously). Each panel was normalized to its maximum and the wave fields were measured from 5 mm to 210 mm away from the surface of the transducer. The elements were driven with alternate voltage polarity and the relative amplitude of the

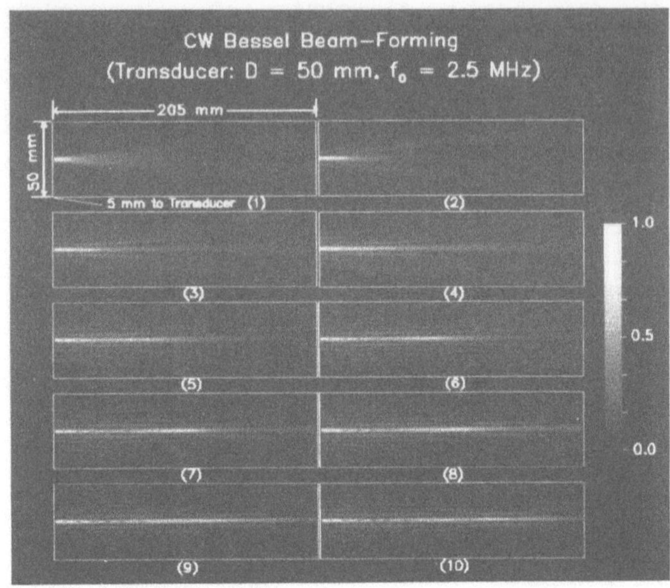

Figure 2. Formation of a CW J_0 Bessel beam in water. Panels (1) to (10) represent the analytic envelope of the wave fields produced when increasing the number of the excited transducer elements, beginning from the central element. The wave fields were measured in a plane along the axis and the image in each panel was normalized to its maximum. The transducer is an annular array with 10 elements. Its diameter is 50 mm and central frequency is 2.5 MHz. The −6 dB pulse-echo (two-way) bandwidth of the transducer is about 50% of its central frequency. All panels of images begin 5 mm away from the surface of the transducer. The panel size of the beams is 205 mm (axial) × 50 mm (lateral).

driving voltages corresponded to the peaks of the lobes of the J_0 Bessel function (with the scaling factor $\alpha = 1202.45\ m^{-1}$).

Figure 3 is the same as Fig. 2 except that a pulse of about one-and-half cycles (2.5 MHz central frequency) was used to drive the transducer. The pulse wave fields were measured at an axial distance of 100 mm (Z/D = 2.0 where Z is axial distance and D = 50 mm is the diameter of the transducer). Figure 4 is a three-dimensional representation of images in Fig. 3. The two-dimensional wave fields in Fig. 3 were rotated around the wave axis to construct three-dimensional wave fields (since the transducer is axially symmetric, the three-dimensional fields should also be symmetric). The three-dimensional wave fields were thresholded at a level of −10 dB down from the maximum of each panel of Fig. 3 to form binary images (anything below −10 dB is "0", and above is "1"). The three-dimensional binary images were contour extracted and surface rendered for the display of the three-dimensional wave objects using commercially available software package ANALYZE^{TM}. Figure 5 shows line plots of images in Fig. 3. The lateral distribution of the maxima of the wave fields in the lines parallel to the wave axis in each panel are shown. This emphasizes the maximum sidelobes of the pulses (the sidelobes are essential to the contrast of medical imaging).

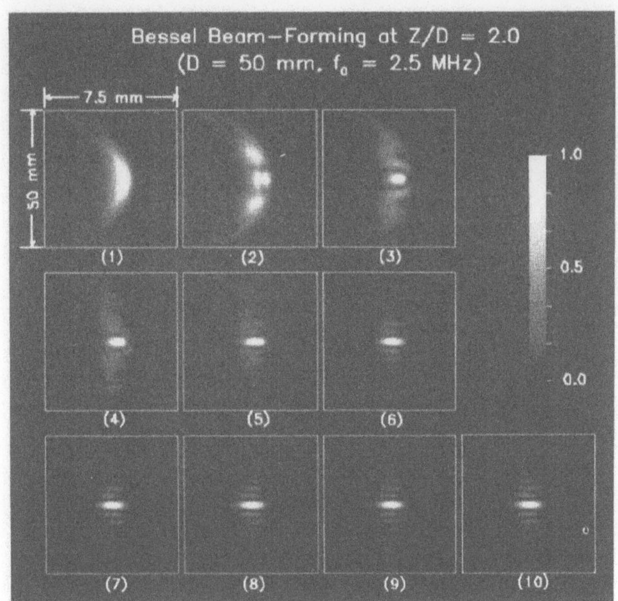

Figure 3. The same format as Fig. 2 for the formation of a pulse J_0 Bessel beam (the transducer was driven by a one-and-half cycle short pulse) obtained at a depth of 100 mm. The panel size of the beams is 7.5 mm (axial) × 50 mm (lateral).

Figures 6 to 8 show the beam-forming of a zeroth-order X wave at a distance of 170 mm (Z/D = 3.4) and correspond in format to Figs. 3 to 5, respectively. The X wave was produced with drive functions calculated from Eq. (5) by setting $b(t)$ as a δ-function, $z = 0$, and r as the mean diameter of each transducer element. The contants a_0 and ζ are 0.05 mm and 4°, respectively.

Figure 9 shows the propagation of the pulse J_0 Bessel beam from 25 mm to 300 mm (Z/D = 0.5 to 6.0), with a depth interval of 25 mm (Z/D = 0.5). The beam was formed with simultaneous excitation of all of the transducer elements. The pulse J_0 Bessel beam stays well in focus (full width at half maximum main beamwidth is about 2.54 mm[7]) over a large depth of field (216 mm calculated from Eq. (6)). Fig. 10 is the three-dimensional representation of images in Fig. 9. It was obtained with the same procedures as those for producing Fig. 4. Figure 11 is the line plots of the images in Fig. 9 and shows sidelobes of the pulse Bessel beam. (Figures 9 to 11 have similar formats as Figs. 3 to 5, respectively.)

Figures 12 to 14 represent the zeroth-order X wave propagating from 25 mm to 400 mm (Z/D = 0.5 to 8.0) with a depth interval of 25 mm (Z/D = 0.5) and correspond in format to Figs. 9 to 11, respectively. The X wave has a depth of field of about 358 mm (calculated from Eq. (7)) with a lateral and axial full width at half maximum of about 4.7 mm and 0.65 mm, respectively.[12]

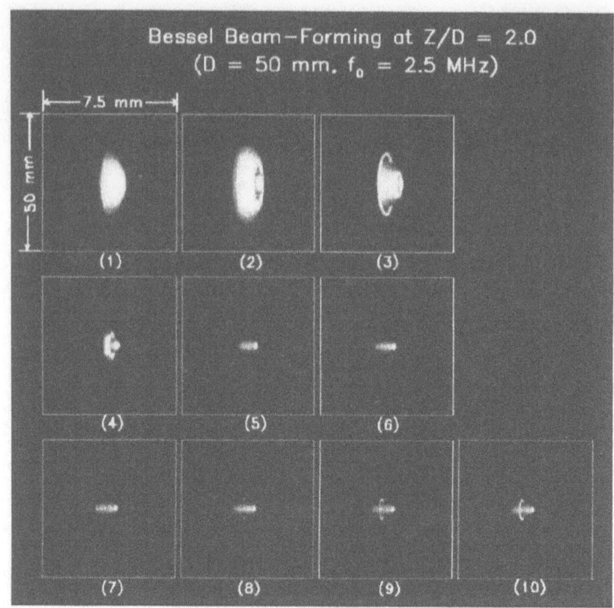

Figure 4. Three-dimensional representation of pulses in Fig. 3. The two-dimensional wave fields in Fig. 3 were rotated around the wave axis to construct three-dimensional wave fields. The three-dimensional wave fields were thresholded at a level of -10 dB down from the maximum of each panel of Fig. 3 to form binary images. The three-dimensional binary images were *contour extracted and surface rendered for the display of three-dimensional wave objects.*

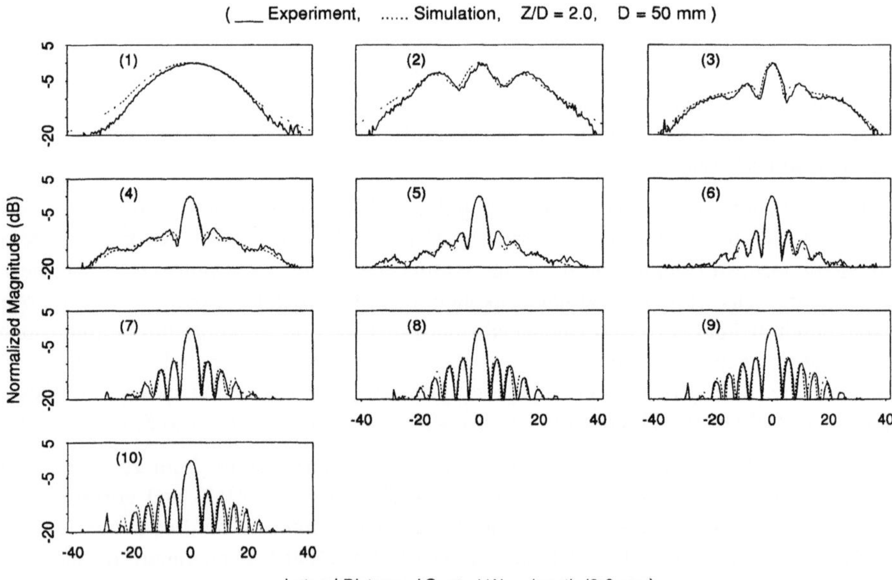

Fig. 5. Line plots of beams in Fig. 3. The maxima of wave fields in lines parallel to the wave axes (axial axes) were plotted over the lateral distance. Full lines and dotted lines represent the experiment and simulation results, respectively.

336

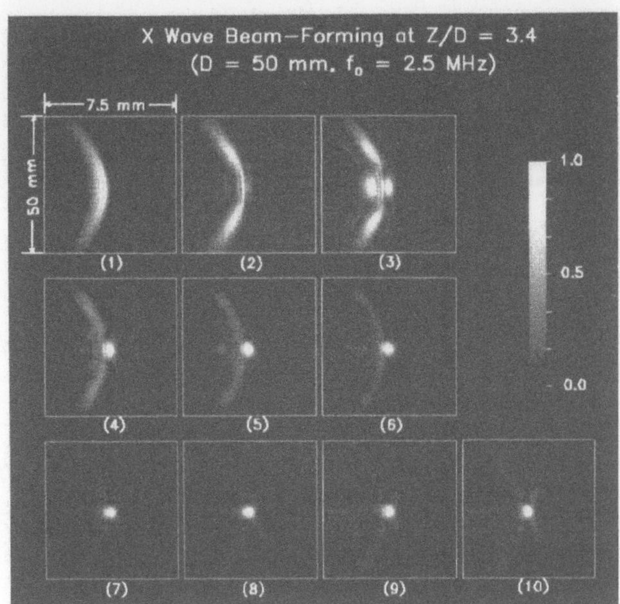

Figure 6. Formation of a zeroth-order X wave in water. The wave fields were measured at a depth of 170 mm. The drive waveforms for the transducer elements are calculated from Eq. (5) by setting $b(t)$ as a δ-function, $z = 0$, and r as the mean diameter of each transducer element. The constants a_0 and ζ are 0.05 mm and 4°, respectively. The display format of each panel is the same as that of Fig. 3.

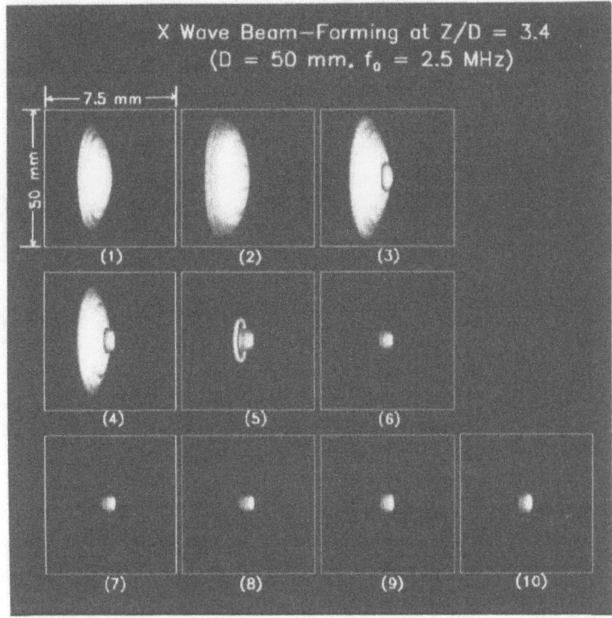

Figure 7. Three-dimensional representation of beams in Fig. 6. The display format of each panel is the same as that of Fig. 4.

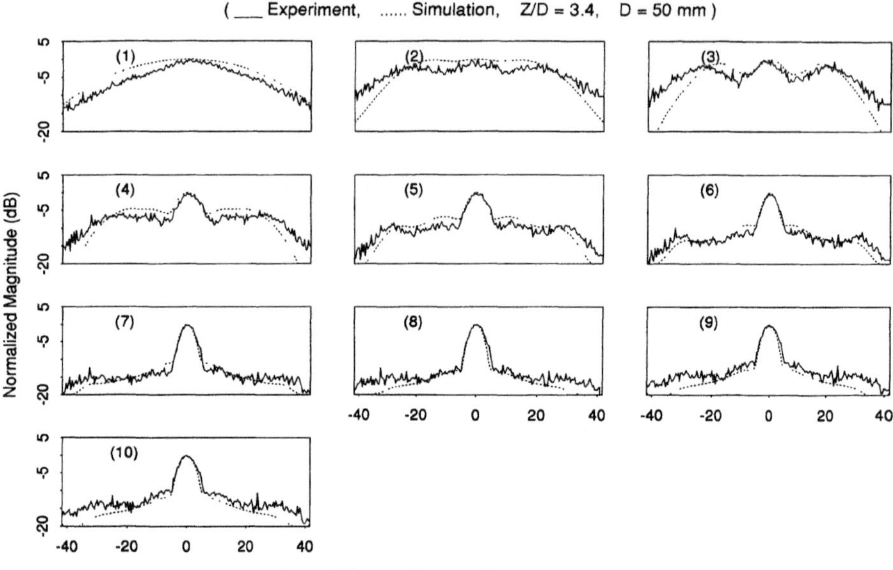

Figure 8. Line plots of images in Fig. 6. The display format of each panel is the same as that of Fig. 5.

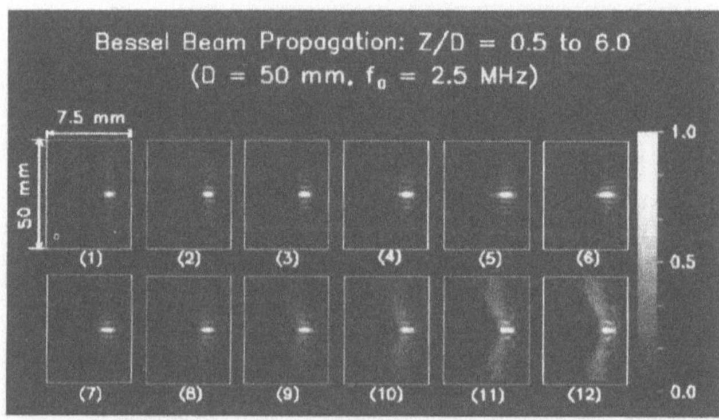

Figure 9. Propagation of the pulse J_0 Bessel beam from $Z = 25$ mm (Panel (1)) to 300 mm (Panel (12)) (Z/D = 0.5 to 6.0, where D = 50 mm is the diameter of the transducer) with a depth interval of 25 mm (Z/D = 0.5). The pulses were produced with the 10 transducer elements excited simultaneously. The display format of each panel is the same as that in Fig. 3.

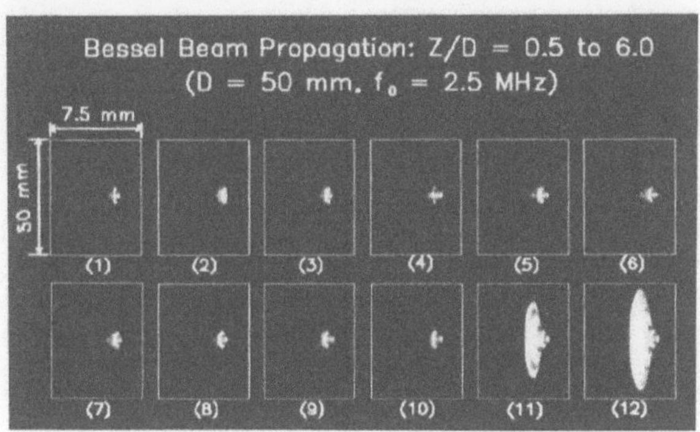

Figure 10. Three-dimensional representation of beams in Fig. 9. The display format of each panel is the same as that of Fig. 4.

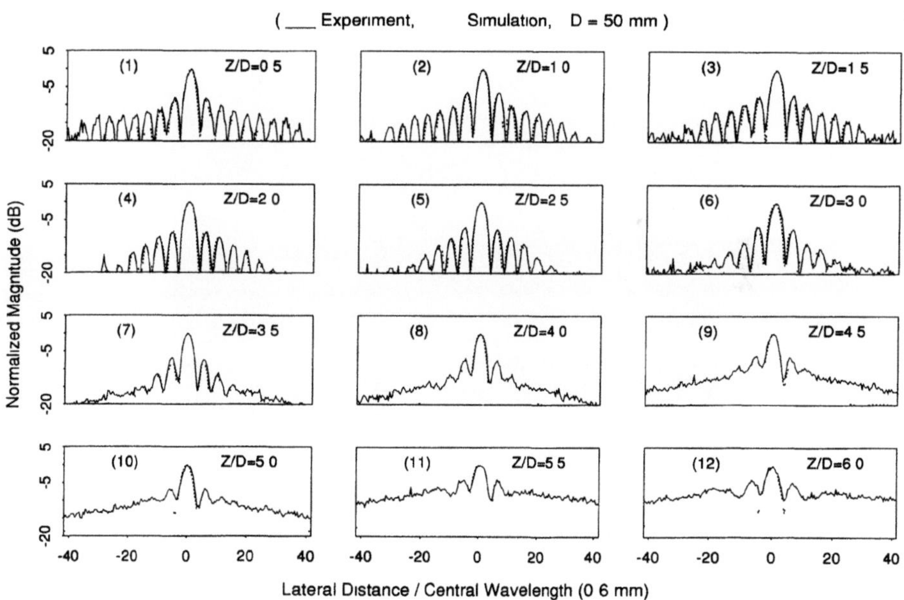

Figure 11. Line plots of images in Fig. 9. The display format of each panel is the same as that of Fig. 5.

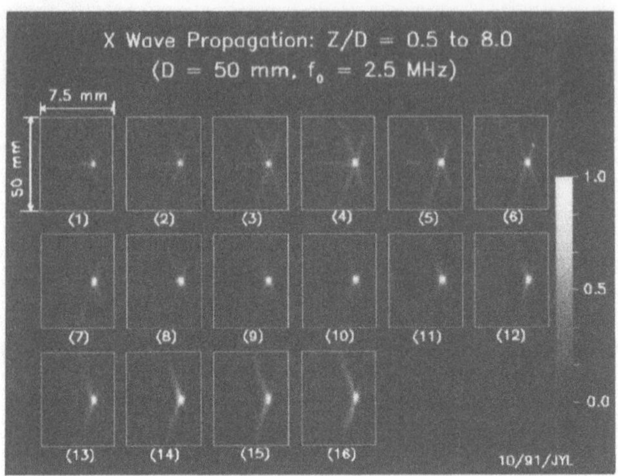

Figure 12. Propagation of the zeroth-order X wave from Z = 25 mm (Panel (1)) to 400 mm (Panel (16)) (Z/D = 0.5 to 8.0, where D = 50 mm is the diameter of the transducer) with a depth interval of 25 mm (Z/D = 0.5). The wave fields were produced with the 10 transducer element excited simultaneously. The display format of each panel is the same as that in Fig. 3.

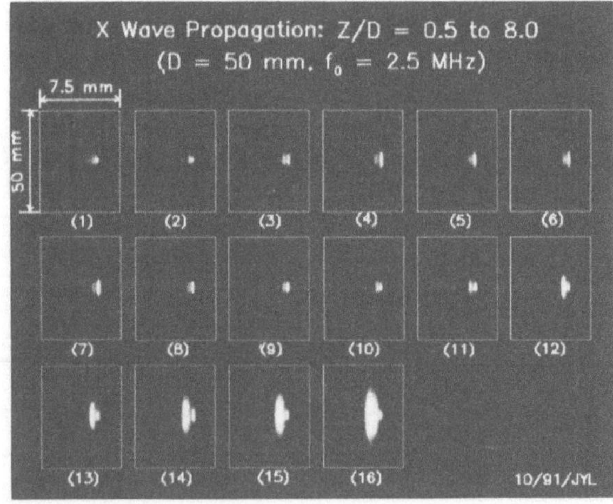

Figure 13. Three-dimensional representation of beams in Fig. 12. The display format of each panel is the same as that of Fig. 4.

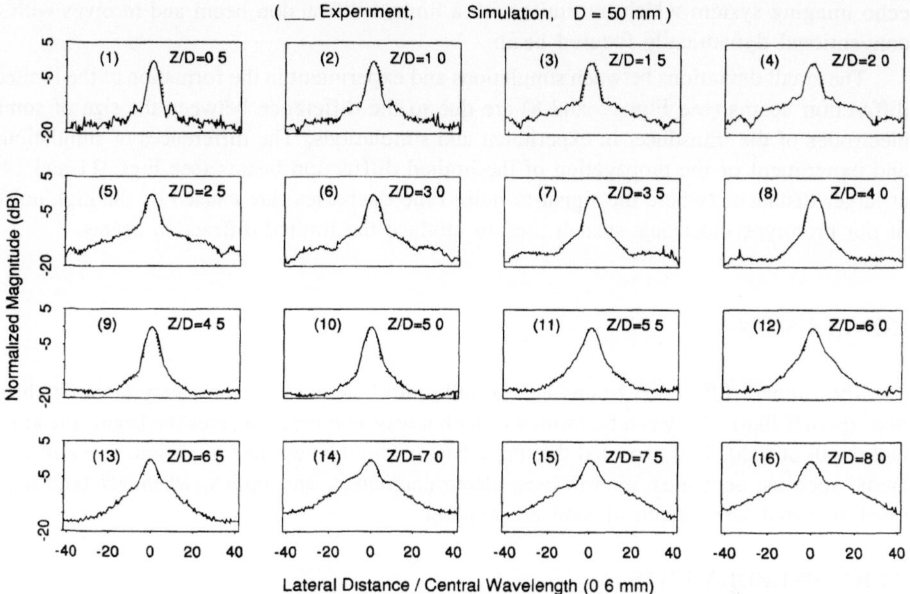

Figure 14. Line plots of images in Fig. 12. The display format of each panel is the same as that of Fig. 5.

DISCUSSION

Figures 2 to 8 show that the limited diffraction beams (J_0 Bessel beam and the zeroth-order X wave) are formed by interference of the wave fields produced from the transducer elements. The center element produces a beam which has a cone-like shape (Panel (1) in Fig. 2). With the central two elements (Panel (2) in Fig. 2), the cone is divided into three sub-cones. Finally, with all the elements (Panel (10) in Fig. 2), a perfect line focusing beam is formed over a large depth of field. The two-dimensional pulse J_0 Bessel beam and the zeroth-order X wave (Figs. 3 and 6) are formed in a similar way as the CW J_0 Bessel beam in Fig. 2. Three-dimensional beams give a clear view of the sizes of the beams (Figs. 4 and 7). Line plots (Figs. 5 and 8) compare the simulations to experiment.

The propagation of the two- and three-dimensional pulse J_0 Bessel beam and the zeroth-order X wave (Figs. 9 to 14) shows that the limited diffraction beams have unique ability to maintain small main beam-widths (2.54 mm and 4.7 mm for the Bessel beam and X wave[7,12], respectively) while propagating over large depths of field (216 mm and 358 mm for the Bessel beam and X wave, respectively). If the medium is loss-free, homogeneous and isotropic, one can expect that the limited diffraction beams produced with an infinite aperture (for example, those given by the mathematical solutions in Eqs. (1), (4) and (5)) will propagate to infinite distance without spreading.

In addition to the main beam component, the limited diffraction beams have higher sidelobes as compared to conventional focused beams at their focuses (sidelobes of a focused Gaussian beam was shown in Reference 9). The sidelobes will degrade the contrast of medical images. One way to overcome the effects of the sidelobes would be a pulse-echo imaging system which transmits with a limited diffraction beam and receives with a conventional dynamically focused beam.

The small deviations between simulations and experiment in the formation of the limited diffraction beams (see Figs. 5 and 8) are due to the difference between the size of some electrodes of the transducer in experiment and simulations. The differences of simulations and experiment of the propagation of the limited diffraction beams (see Figs. 11 and 14) at larger distances (where the signal-to-noise ratio decreases) are caused by the high noise of our prototype electronic system used to produce the limited diffraction beams.

CONCLUSION

The limited diffraction beams have small main beamwidth over a very large depth of field (pencil-like). They can be formed in such a way that best balances the beam spreading and depth of field. It is expected that these beams could be applied to various wave related areas, such as, acoustics, microwaves, electromagnetics, and optics, whenever both high resolution and large depth of field are required.

ACKNOWLEDGMENTS

The authors thank Randall R. Kinnick for constructing the electronic circuits to drive the transducer. The authors appreciate the secretarial assistance of Elaine C. Quarve and the graphic assistance of Christine A. Welch. This work was supported in part by grants CA 43920 and CA 54212 from the National Institutes of Health.

REFERENCES

1. J.B. Brittingham, "Focus wave modes in homogeneous Maxwell's equations: transverse electric mode," *J. Appl. Phys.*, 54(3):1179–1189 (1983).
2. J. Durnin, "Exact solutions for nondiffracting beams. I. The scalar theory," *J. Opt. Soc. Am.*, 4(4):651–654 (1987).
3. R.W. Ziolkowski, D.K. Lewis, and B.D. Cook, "Evidence of localized wave transmission," *Phys. Rev. Lett.*, 62(2):147–150 (Jan. 9, 1989).
4. G. Indebetow, "Nondiffracting optical fields: some remarks on their analysis and synthesis," *J. Opt. Soc. Am. A*, 6(1):150–152 (Jan., 1989).
5. D.K. Hsu, F.J. Margetan, and D.O. Thompson, "Bessel beam ultrasonic transducer: fabrication method and experimental results," *Appl. Phys. Lett.*, 55(20):2066–2068 (Nov. 13, 1989).
6. J.A. Campbell and S. Soloway, "Generation of a nondiffracting beam with frequency independent beam width," *J. Acoust. Soc. Am.*, 88(5):2467–2477 (Nov., 1990).
7. J-y. Lu and J.F. Greenleaf, "Ultrasonic nondiffracting transducer for medical imaging," *IEEE Trans. Ultrason., Ferroelec., Freq. Contr.*, 37(5):438–447 (Sept., 1990).
8. J-y. Lu, and J.F. Greenleaf, "Evaluation of a nondiffracting transducer for tissue characterization," [IEEE 1990 Ultrasonics Symposium, Honolulu, HI, Dec. 4–7, 1990], *IEEE 1990 Ultrason. Symp. Proc.*, 90CH2938-9, 2:795–798 (1990).
9. J-y. Lu and J.F. Greenleaf, "Pulse-echo imaging using a nondiffracting beam transducer," *Ultrasound Med. Biol.*, 17(3):265–281 (May, 1991).

10. J-y. Lu and J.F. Greenleaf, "Theory and acoustic experiments of nondiffracting X waves," [IEEE 1991 Ultrasonics Symposium, Lake Buena Vista, FL, December 8–11, 1991], *IEEE 1991 Ultrason. Symp. Proc.*, 91CH3079–1, 2:1155–1159 (1991).

11. J-y. Lu and J.F. Greenleaf, "Nondiffracting X waves — exact solutions to free-space scalar wave equation and their finite aperture realizations," *IEEE Trans. Ultrason., Ferroelec, Freq. Cont.*, 39(1):19–31 (Jan., 1992).

12. J-y. Lu and J.F. Greenleaf, "Experimental verification of nondiffracting X waves," *IEEE Trans. Ultrason., Ferroelec., Freq. Contr*, 39(3):441–446 (May, 1992).

13. J-y. Lu and J.F. Greenleaf, "Nondestructive evaluation of materials with a J_0 Bessel transducer," [17th International Symposium on Ultrasonic Imaging and Tissue Characterization, Arlington, VA, June 1–3, 1992], *Ultrason. Imaging,* 14(2):203–204 (April, 1992) (abs.).

14. J-y. Lu and J.F. Greenleaf, "Sidelobe reduction of nondiffracting pulse-echo images by deconvolution," [17th International Symposium on Ultrasonic Imaging and Tissue Characterization, Arlington, VA, June 1–3, 1992], *Ultrason. Imaging,* 14(2):203 (April, 1992) (abs.).

15. J-y. Lu and J.F. Greenleaf, "Diffraction-limited beams and their applications for ultrasonic imaging and tissue characterization," [SPIE's 1992 International Symposium on Optical Applied Science and Engineering, San Diego, CA, July 19–24, 1992], *New Developments in Ultrasonic Transducers and Transducer Systems,* F.L. Lizzi, ed. (Invited paper).

16. J.W. Goodman, *Introduction to Fourier Optics.* McGraw-Hill, New York, chs. 2–4 (1968).

17. J-y. Lu, T. Kinter, and J.F. Greenleaf, "Measurement and dynamic display of acoustic wave pulses," [IEEE 1989 Ultrasonics Symposium, Montreal, Quebec, Canada, Oct. 3–6, 1989], *IEEE 1989 Ultrason. Symp. Proc.*, 89CH2791–2, 1:673–676 (1989).

[11] Y. Lu and J.C. Greenleaf, "Theory and accurate experiments of nondiffracting X waves," IEEE 1991 Ultrasonics Symposium, Lake Buena Vista, FL, December 8-11, 1991, IEEE 1991 Ultrason. Symp. Proc., 91CH3079-1, 2:1155-1159 (1991).

[12] Y. Lu and J.P. Greenleaf, "Nondiffracting X waves — exact solutions to free space scalar wave equation and their finite aperture realizations," IEEE Trans. Ultrason. Ferroelec. Freq. Contr., 39(1):19-31 (Jan. 1992).

[13] Y. Lu and J.P. Greenleaf, "Experimental verification of nondiffracting X waves," IEEE Trans. Ultrason. Ferroelec. Freq. Contr., 39(3):441-446 (May 1992).

[14] Y. Lu and J.P. Greenleaf, "Pulse-echo imaging using a nondiffracting beam transducer," Ultrasound in Med. & Biol., 17(3):265-281 (1991).

[15] J.A. Campbell and S. Soloway, "Generation of a nondiffracting beam with frequency independent beam width," J. Acoust. Soc. Am., 88(5):2467-2477 (1990).

[16] D.K. Hsu, F.J. Margetan, and D.O. Thompson, "Bessel beam ultrasonic transducer: fabrication method and experimental results," Appl. Phys. Lett., 55(20):2066-2068 (1989).

AN ACTIVE LOAD BACKING TRANSDUCER
FOR HIGH FREQUENCY ULTRASONIC PROBE :
IMAGING SYSTEMS APPLICATIONS

M. Ravez, M. Ourak, M. Ouaftouh, W. J. Xu and B. Nongaillard

IEMN (UMR CNRS 9929)- Département OAE
Université de Valenciennes- BP311
59304 Valenciennes Cedex, France

INTRODUCTION

Well-known methods for broadband generation of ultrasonic waves in a propagating medium is the so-called backing method. The forward quarter-wavelength ($\lambda/4$) adaptation and the electrical impedance to the internal impedance generator induce (or lead) also to increase the broadband[1,2,3,4].

In a publication [5], we showed there was possible to simulate the mechanical backing medium with an electrically loaded piezoelectric element. This design leaded to decrease the bulky assembly, especially at lower ultrasonic frequencies.

In this paper, we present the realization of a new high ultrasonic frequencies probe (100 MHz), using an active load backing transducer (small thickness) , glued therefore on a substrate. We made the comparison of results obtained with this new active load backing transducer and the classical transducer with mechanical substrate (Fig.1).

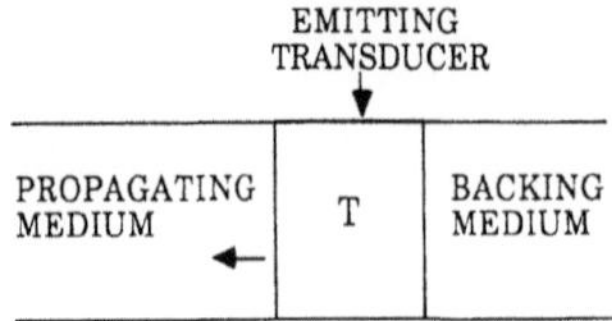

Figure 1 : Classical ultrasonic probe

I- ANALYSIS OF THE NEW PROBE

In the proposed method, the backing medium is composed by a mechanical medium cemented to a piezoelectric element with an electrical circuit across its electrodes cemented to the rear face of the emitting piezoelectric transducer.

The new systeme is shown in Fig.2 where T_1 is the emitting piezoelectric transducer, T_2 is the backing piezoelectric element (behaving as a receiver and electrically loaded by the electrical impedance Z_0), cemented to a mechanical infinite medium, and M is an electrical new-port network and is used to match between generator G with an internal impedance Rg equal to 50Ω , and the transducer T_1. This circuit may comprise reactive part.

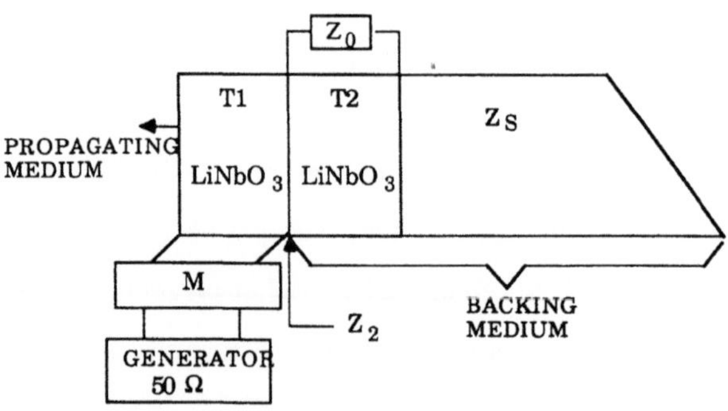

Figure 2: New high ultrasonic frequencies probe

In a general case, we can consider dissimilar thickness transducers; however, for simplicity, we shall assume here that both transducers T_1 and T2 have identical thicknesses, so that their mechanical resonance will occur at the same frequency f_0 (ie, f_0 is the free single isolated transducer half- wave T_1.length resonance frequency). We shall also assume that both transducers T_1 and T_2 are made of the same piezoelectric material with mechanical impe dance Z, and that the active area of each is A.

II- THEORITICAL STUDY

Using the Mason equivalent circuit[6] (shown in fig.3) for the receiving transducer T_2 with port $B_1B'_1$ short circuited, electrical port $B_eB'_e$ charged with the electrical impedance Z_0, the mechanical impedance viewed at port $B_2B'_2$ is $A.Z_2$, and Z_2 is found (shown in fig.4) to be:

$$Z_2=-Z\frac{\left[\frac{2K^2X_c}{f_n}(1-\cos\pi f_n)-\pi f_n(Z_0-j\frac{X_c}{f_n})(\frac{Z_s}{Z}.\cos\pi f_n+j\sin\pi f_n)-j\frac{Z_s}{Z}.\frac{K^2X_c}{f_n}.\sin\pi f_n\right]}{\pi f_n(Z_0-j\frac{X_c}{f_n})(\cos\pi f_n+j\frac{Z_s}{Z}.\sin\pi f_n)+j\frac{K^2X_c}{f_n}.\sin\pi f_n}$$

where $j^2=-1$, K^2 is the electromechanical coupling factor, f_n is the normalized frequency f/f_0 (where f is the electrical frequency), and $-X_c$ is the reactance of the clamped static capacitance C_0 of transducer T_2 at its resonance frequency f_0 (ie., the impedance of the capacitor $\pm C_0$ is jX_c/f_n) .

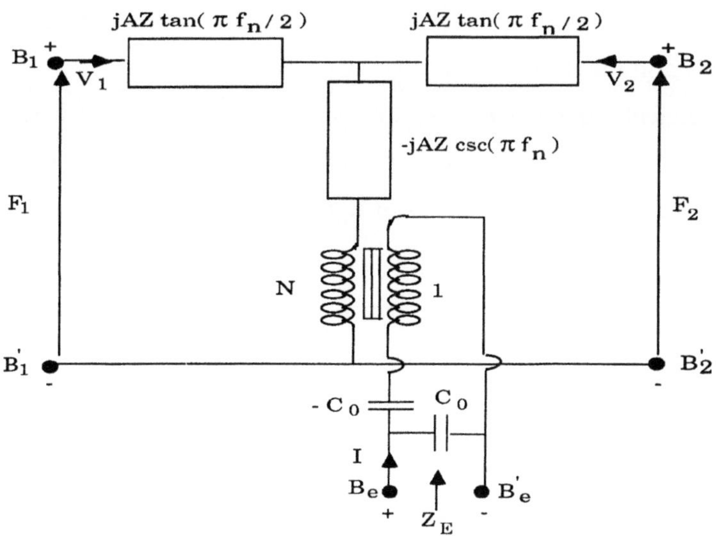

Figure 3 : Mason equivalent circuit for a thin piezoelectric transducer . The ideal transformer's turn ratio is $N = K(AZ/\pi X_c)^{1/2}$. F and V stand for mechanical force and particle velocity at boundaries, respectively.

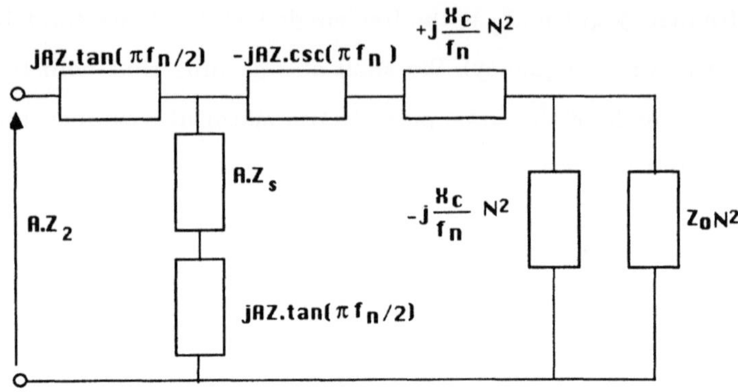

Figure 4 : Equivalent electrical circuit.

Using again the Mason equivalent circuit of Fig.3, now for transducer T_1, with mechanical impedance $A.Z_1$ connected to port $B_1B'_1$, and mechanical impedance $A.Z_2$ at port $B_2B'_2$, the acoustical power radiated at port $B_1B'_1$ into the propagating medium may be written as:

$$P_1 = \frac{K^2 Z |I|^2 X_c}{\pi . f_n^2} . Re(Z_1) \left| \frac{Z(1-\cos\pi f_n)-j.Z_2.\sin\pi f_n}{Z(Z_1+Z_2)\cos\pi f_n + j(Z^2+Z_1Z_2)\sin\pi f_n} \right|^2$$

where Z_1 is the mechanical impedance presented by the propagating medium (eventually seen by the transducer through matching layers) , Re stands for " real part of", and I is the electrical current fed to T_1.

The electrical current I may be knowed the generator parameters, the matching two-port network M characteristics , and the electrical impedance Z_E presented by the transducers T_1 between $B_eB'_e$:

$$Z_E = -j\frac{X_c}{f_n} + \frac{K^2 Z . X_c}{\pi . f_n^2} . \frac{2Z(\cos\pi f_n - 1)+j(Z_1+Z_2)\sin\pi f_n}{Z(Z_1+Z_2)\cos\pi f_n + j(Z^2+Z_1Z_2)\sin\pi f_n}$$

This last expression is also derived from Mason's equivalent circuit.

A computer program has been written to study the new backing method. We have assumed $LiNbO_3$ longitudinal transducers with a coupling coefficient $K^2 = 0.04$ and a mechanical impedance $Z = 35 \times 10^6$ MKSA. The propagating medium is water with mechanical impedance $Z_1 = 1.5 \times 10^6$ MKSA. The resonance frequency f_0 has been chosen equal to 100MHz.

III- RESULTS

1- Theoretical results

Figures 5a and 5b show respectively the variations of the acoustical power versus the frequency and the impulse response of the transducer in the propagating medium in the case of an alone transducer (fig.1) (without active backward damping).

Figures 6a and 6b show the same parameters in the case of the figure 2 (with active backward damping).

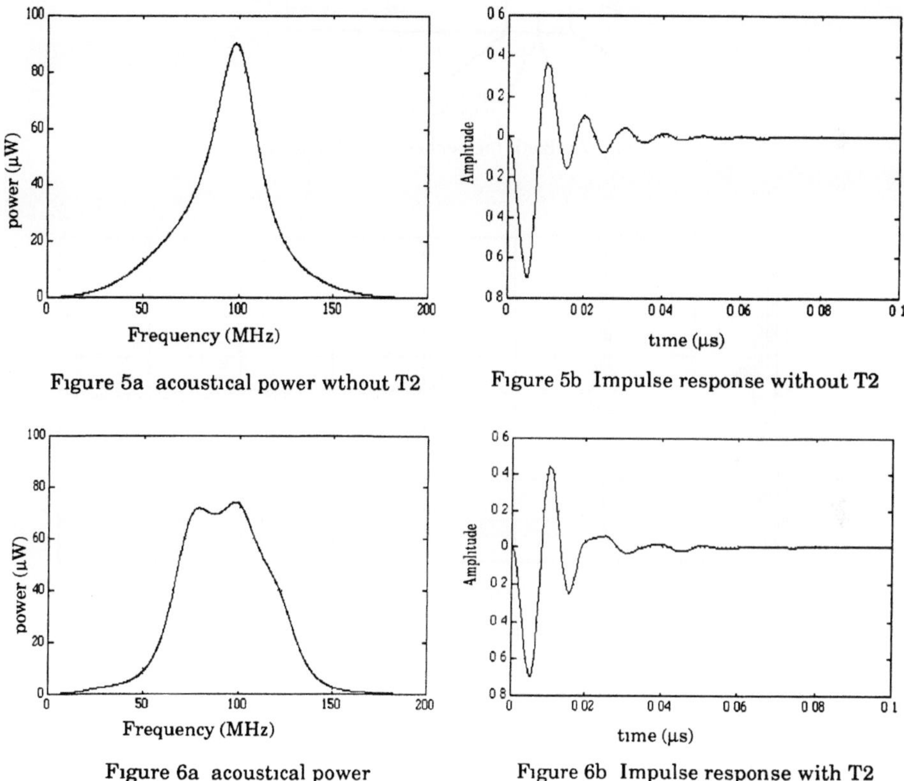

Figure 5a acoustical power wthout T2

Figure 5b Impulse response without T2

Figure 6a acoustical power

Figure 6b Impulse response with T2

We observe an increase of the bandwidth and also a damping of the acoustical signal.

However, the increase of the bandwidth is not very important. This is due to the great mechanical impedance of the substrate (near than the LiNbO3) imepedance); then, the mechanical adaptation is practically realized.

2- Experimental results

Subsequent to our choice of backing material as well as for technical reasons, we have restricted our study to analyse the efficiency of the active backward damping. In this way, we have been interested in the signal reflected by the back face of the substrate (fig. 7) . The figures 8.a and 8.b give the time echograms respectively for T2 unloaded and loaded by the electrical impedance Z_0.

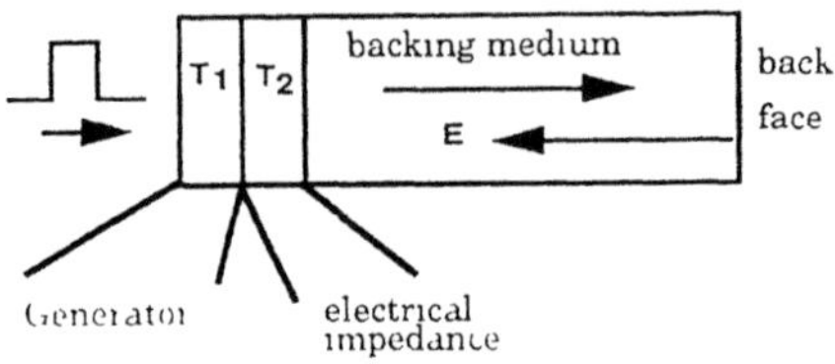

Fig. 7 : Back face echo in the backing

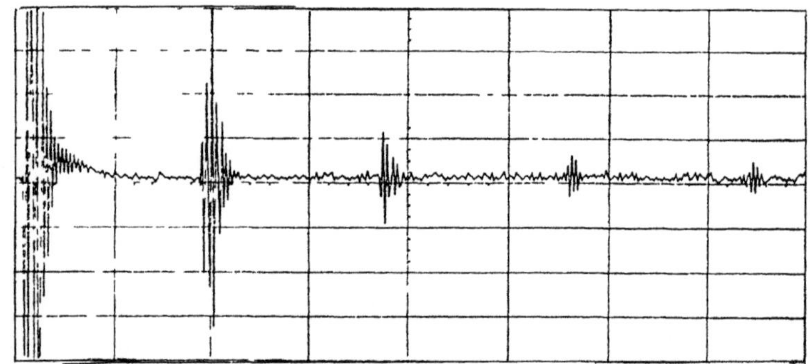

Fig.8.a : Time echogram for T2 unloaded (horiz: 1 µs/div ; vert: 50 mV/div)

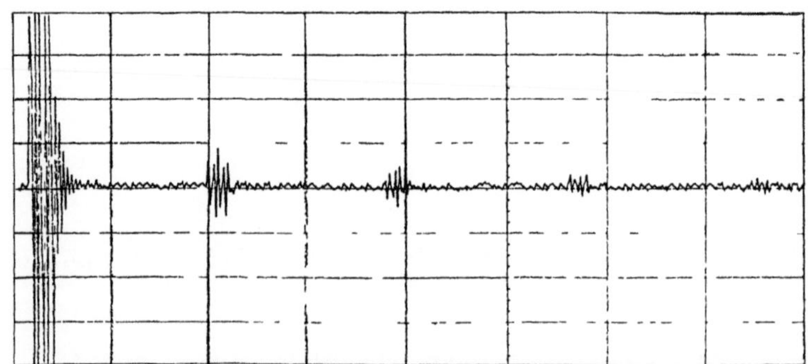

b: Time echogram for T2 loaded by the electrical impedance Z_0 (considerable attenuation of the back face echo) (horiz 1µs/div ; vert: 50 mV/div)

CONCLUSION

This study shows that the transducer T2 loaded by an appropriate electrical impedance can constitute an efficient active damping. Our encouraging results will lead us, in a near future, to the conception of high frequency probes nearly free of acoustic noise caused by multiple reflections in the backing medium.

These probes will be of great use in several applications like biomedical ones which requires a good axial resolution.

REFERENCES

1. J. Souquet, P. Defranould, and J. Desbois, " Design of Low-Loss Wideband Ultrasonic Transducers fot Noninvasive Medical Applications," IEEE Trans. Sonics Ultrason. 26 (2), pp.75-81, 1979.

2. J. H.Goll, " The design of Broad-Band Fluid-Loaded Ultrasonic Transducers, " IEEE Trans.Sonics Ultrason. 26 (6), pp.385-393, 1979.

3. R. H. Coursant, " Les transducteurs ultrasonores", Rev. Electron.Phys. Appl. 22, pp.129-141, 1979.

4. S. H. Van Kervel and J. M. Thijssen, " A Calculation Scheme for the Opti mum Design of Ultrasonic Transducers", Ultrasonics 21, pp.134-140, 1983.

5. M. G. Gazalet, M. Houzé, P. Logette, C. Bruneel, and R. Torguet, " Dummy load backing of ultrasonic transducers for bandwidth enhancement", J. Acoust. Soc. Am. 83 (3), pp.1180-1182, March 1988.

6. E. Dieulesaint, D. Royer, Ondes élastiques dans les solides, Ed. Masson, 1974.

ZnO PLANAR FOCUSING TRANSDUCER ON USING FRESNEL ARRAY

Donghai Qiao, Shunzhou Li
Chenghao Wang, Zheying Zhao

Institute of Acoustics
Academia Sinica
Beijing 100080, P. R. China

ABSTRACT

Fresnel Array can produce focusing of acoustic beam. In this paper we will discuss high frequency Fresnel Arrays which have specially designed electrode structure with ZnO piezoelectric film. The sputtered C-axis oriented piezoelectric film and special electrode structure make longitudinally vibrated transducer arrays. These transducer arrays can generate odd order convergent and divergent beams. But the zero order or even order ones are eliminated. In order to make full use of first order convergent beam we can restrain higher odd order convergent beams by increasing the element width. An important characteristic of Fresnel Array is that its focal-length can change linearly with frequency. And hence its focus can quickly scan linearly along the axis. Two kinds of structures, straight and circular strip arrays, are developed to produce line and point focuses, respectively.

We have developed these two planar transducers with center frequency of 400 MHz and bandwidth of about 200 MHz. For the fused quartz their foci can scan along the depth in the range of 8 mm. Their focusing properties were optically observed. And some prepared experiments were also done for NDT and acoustic imaging.

INTRODUCTION

In fluids, the focusing of acoustic beams has found many applications in different areas. More research work on focusing in fluids has been done than that in solids. There are two possible reasons. One is that there are two kinds of acoustic waves (i. e. the shear wave and the compressional wave), the excitation of one kind can produce another together and one wave mode can change into two modes or three modes at interface. This kind of mode conversion can bring many difficulties to practical application. The other is that the generating of the focusing field is an arduous task. In general, a concave transducer or an acoustic lens transmits the acoustic beam in a liquid and penetrates again into the solid. Such a way leads the great reflect loss on interface and can not quickly scan in depth. But recently, one important application in NDT and acoustic imaging for fine flaw inside solids make it notable. We have investigated the focusing properties of Planar Fresnel Transducer Array (PFTA) and discussed its possible application in NDT and acoustic imaging.[1-3] Now, in this paper, the transducer array structures and some prepared experiments for NDT and acoustic imaging are given.

THE STRUCTURES AND THE FOCUSING PROPERTIES OF PFTA

Figure 1 shows the two geometrical structures of PFTA. Their transducer are sandwiched with C-axis oriented ZnO piezoelectric film between two layers of metal electrodes. The fabrication process can be simply described as follows: ① a thin layer of Au-Cr is evaporated onto the surface of fused quartz in vacuum aparatus as the bottom electrode; ② a layer of ZnO film is deposited onto the bottom electrode by DC-magnetic sputtering; ③ evaporating a layer of Au-Cr onto ZnO piezoelectric film; ④ forming designed top electrode pattern (e. g. straight strip pattern in Figure 1 (a) and circular strip pattern in Figure 1 (b)) by using a conventional photo-lithographic technique. When the eletric signal is used to generate acoustic wave, every strip electrode is inverse in phase with the adjacent one and the bottom electrode is connected to signal ground. In addition, such array structures will be useful for electric increasing impedance. As the focusing properties of the two structures of PFTA are very similar, we only discuss the case of straight strip.

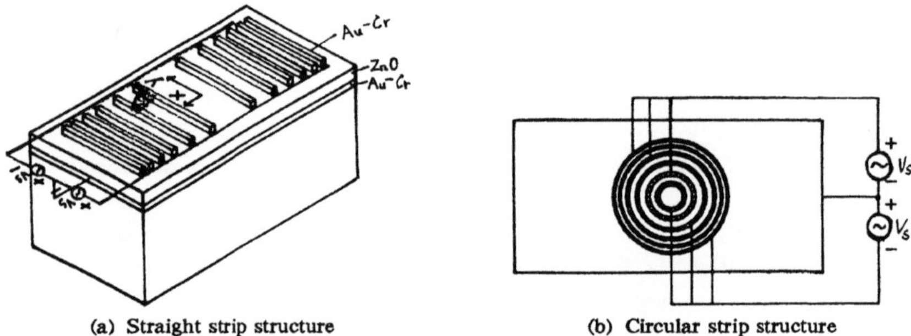

(a) Straight strip structure   (b) Circular strip structure

Figure 1. Two kinds of structures of PFTA

Suppose that the nth strip whose width is $2a$ (Figure 2) on XY-plane is at X_n, and X_n satisfies

$$X_n^2 = (N_S + n) \cdot B \tag{1}$$

where B is a constant; $n = 1, 2, \dots, 2N$ and N_S is a constant.

When the strength of every source element is $A_0 \cdot \exp(j\,n\pi)$ and every groug consists of $2N$ elements, considering the situation at paraxial region and far-field, i. e. $X_n \ll Z$, the element directivity angle $\theta_n \approx \theta_a$, for the amplitude $R_n \approx Z$, but for the phase, R_n can be approximately written as: [3]

$$R_n = \sqrt{Z^2 + X_n^2} \approx Z + (X_n^2/2Z) \tag{2}$$

therefore the acoustic field on axis is

$$U_r(Z, 0) \sim \frac{U_O(Z, \theta_a) \cdot \sin(\eta/2) \cdot \sin(N \cdot \eta)}{N \sin \eta} \tag{3}$$

where $\eta = \pi B/\lambda Z$,

$$U_O(Z, \theta_a) = \frac{1}{\sqrt{2}} U_r(\theta_a) \cdot \cos \theta_a \cdot (4j N A_0)$$

$$\cdot \exp\{j K_r[Z + \frac{B}{2Z}(N_S + N + \frac{1}{2})]\} \tag{4}$$

where K_r and λ are the longitudinal wave number and wave length of solid media, respectively.

From Eq. (3) we know that PFTA will generate convergent and divergent beams with various orders simultaneously as shown in Figure 2. If first order convergent acoustic beams with wavelength λ_0 have focal length Z_0, thus Lth order focus of the beams with wavelength λ will locate at Z which satisfies $\lambda_0 Z_0/\lambda Z = \pm (2m + 1) = \pm L$, where m is a natural number and L is an odd. The symbol $+$ and $-$ correspond to convergence and divergence, respectively. And the even order convergence (or divergence) will not occur because each element is $180°$ of phase difference with the adjacent one. From (3) we also deduce that when the frequency changes, focuses of all orders will linearly scan along axis, i. e. $Z = f Z_0/L Z_0$. Although all the order convergences occur simultaneously, the energy of focusing beams decreases as the other number increases.

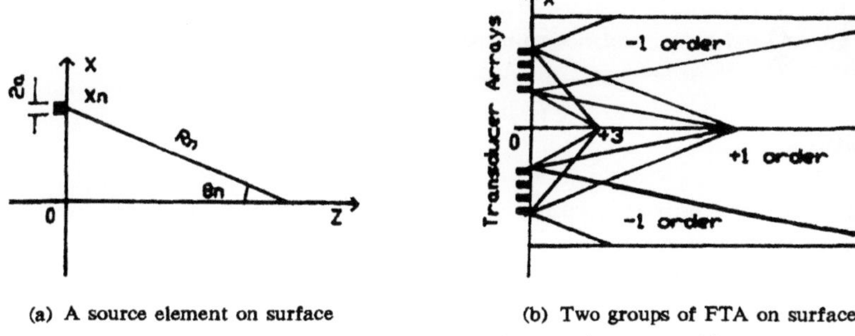

(a) A source element on surface

(b) Two groups of FTA on surface and convergent beams

Figure 2.   Illustration of focusing in solid by PFTA

According to our research, [3] in addition to the above mentioned results, some are breifly mentioned as follows:

1) As the element width increases, 5 order or higher order convergent beams are restrained, and the sound intensity of the focal field between $+1$ order and $+3$ order becomes bigger.

2) $+3$ order focusing beams are finer than the $+1$ order one (also see Figure 3).

3) Longitudinal and transversal resolutions will be enhanced by increasing the whole number of array elements.

EXPERIMENTS AND DISCUSSIONS

We have discussed the properties of the focusing and the scanning in solids by PFTA above. If we put an acoustic scatterer on the focus, when a periodical impulse modulated by high frequency injects to the transducers, a maximum echo signal will be received from the scatterer at focus after a certain time delay. Changing the frequency of the signal, the focus will scan along Z axis linearly, thus the flaw position and size can be given on using the time delay or the frequency corresponding to maximum reflected signal and its peak value.

In order to verify the above mentioned theory of acoustic beam focusing and examine its application further, we have done some experiments of optical observation and electrical measurement. We made two groups of PFTA on one surface of transparent fused quartz substrate. In the experiments, $N_s = 10$, $2N = 50$, $f_o = 400$ MHz, $\lambda_o = 15$ μm, $Z_o = 1000 \lambda_o$. Figure 4 (a) is the optical photograph of acoustical beam and Figure 3 (a) is the focusing in solid taken from acousto-optic diffraction. From Figure 3 (a) we can clearly see the convergence of $+1$ and $+3$ order beams and the divergence of -1 order beam. When $K_r = 5.0$, as the strength of the other order convergent beam is much weaker than that of $+1$ and $+3$ order, the interaction with coherent light is also much weaker and diffcult to be observed by acousto-optic diffraction. Figure 3 (b) shows the situation when the convergent beams travel through the coupling fluid between buffer and sample and foculized at the sample bottom. Figure 3 (c) shows the electrical measurement of Figure 3 (b). From Figure 3 (c) we can observe the reflected signals from the interface and the bottom surface. Figure 4 (a) shows the situation when the convergent beams focalized at a small hole in solid sample. Figure 4 (b) is the delta electric pulse response. From the time delay of Figure 4 (b) we can easily distinguish the reflection from the interface, the small hole deflect and the bottom plane of the sample. Since the beams are focusing at the small hole, the reflected signal from the small hole is greater than that from the bottom. And the time delay are in good correspondance with their position.

CONCLUSIONS

In the experiments of beam focusing in solids by PFTA, the transversal and the longitudinal wave field are excited simultaneously. But the transversal wave field on axis is weaker and its sensitivity of exciting and receiving on transducers is also weaker, so the longitudinal wave field is only considered here when discussing the far-field on the axis. The disturbance of high order modes (especially $+3$ order) is not serious to practical use. One reason is that the reflections from all order focuses can be distinguished by time delay. And the other reason is that the strength of higher order focal field can be restrained by increasing the width of source strip. For instance, from Figure 3 (d) we know that the field strength of $+3$ order convergence is not more than that of the first sidelobe of $+1$ order convergence. And this kind of disturbance can be easily eliminated in practice. Through the above mentioned theoretical and experimental investigation on beam focusing, the focusing produced by PFTA can be expected to be one way of realizing NDT and acoustic imaging for fine flaws inside solids. But there still leaves much arduous work to be done in practice.

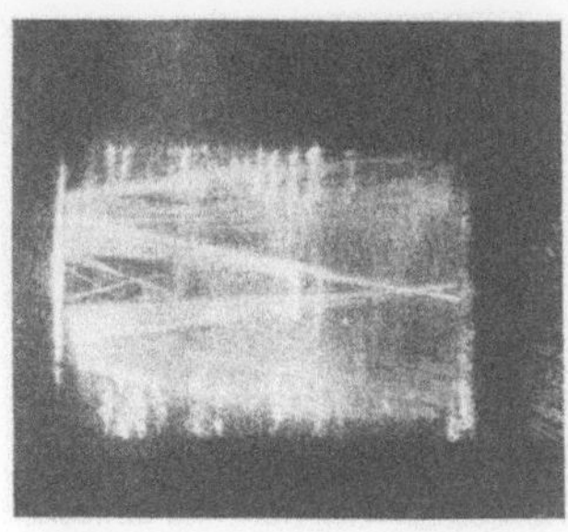

Figure 3 (a) . Photo of focusing

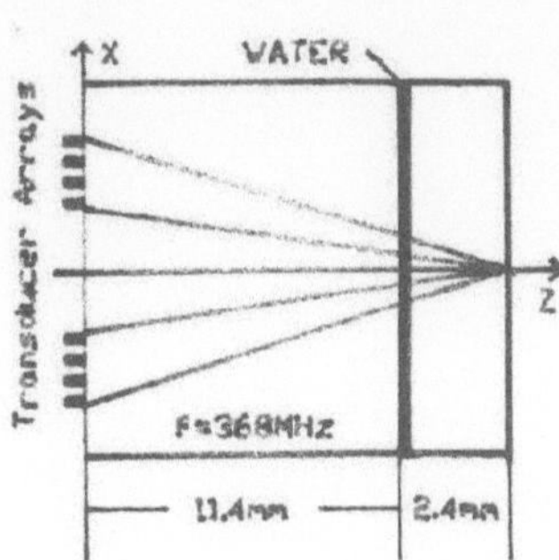

Figure 3 (b) . Focusing in sample

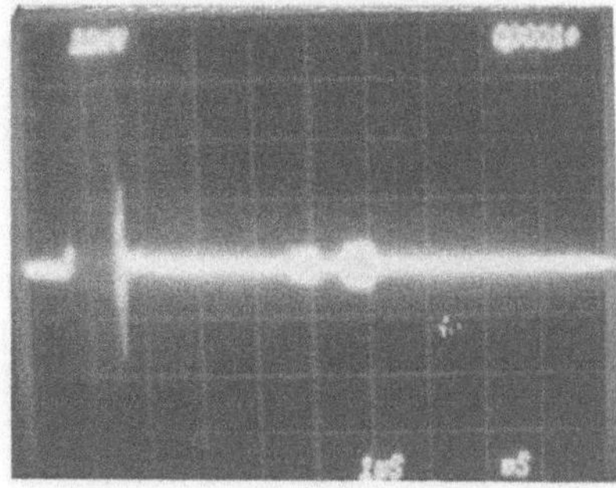

Figure 3 (c) . Received signal of Figure 3 (b)

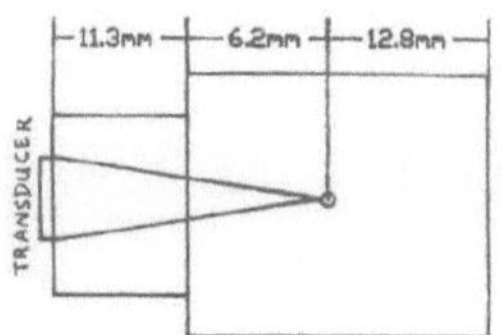

Figure 4 (a) . Focalize at a hole

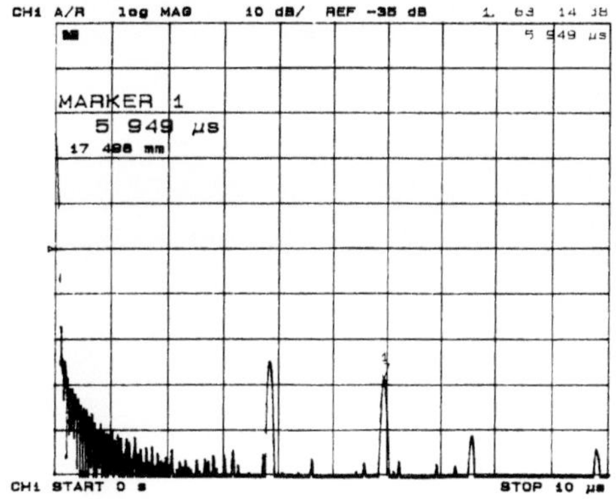

Figure 4 (b) . Electric pulse response of Figure 4 (a)

REFERENCES

1. Wu Jingyuan, Wang Chenghao et al, Chinese Physics Letters, (1987), p257.
2. Qiao Donghai, Wang Chenghao, et al, Proc. Conference on Ultrasonoelectrics, (1990), p57. (in Chinese) .
3. Qiao Donghai, Wang Chenghai, et al, Proc. 14th ICA, C15—2, Beijing, China, (1992) .

AUTOMATIC MODEL IDENTIFICATION OF ULTRASONIC TRANSDUCERS
BY INPUT-OUTPUT BROADBAND MEASUREMENTS

L Capineri, L. Masotti, M. Rinieri, S. Rocchi

Dipartimento di Ingegneria Elettronica, Università degli studi di Firenze
Via Santa Marta 3, 50139 Firenze, Italy

INTRODUCTION

The optimization of ultrasonic transducer performances is generally tackled in the manufacturing stage and defines the upper limit of the obtainable performances. Nevertheless in many practical applications in medical imaging and non destructive testing the transducer is poorly matched with the front-end electronics and the transducers' characteristics are not fully exploited[1,2 3]. The transducer characteristics can be preserved if an effective model of the actual transducer is available to the electronic designer. Aimed to this problem, this paper presents a new method for ultrasonic transducer modelling. This method provides an input-output analytical model of the ultrasonic transducer and the model is built without a priori knowledge of the manufacture characteristics In this way we consider the transducer as a two-port black-box, wholly characterized by the electrical driving-point impedance and the acousto-electric transfer function

Once the transducers' analytical model in the Laplace s-domain is obtained, it can be inserted in analog circuit simulators and the ultrasonic transducers become as other library components. The two-port model is calculated starting from input-output broadband measurements of the transducers' driving point voltage and current and the transducer's output in broadband excitation conditions. With these experimental data sets, the transducer model is carried out by an automatic identification procedure based on the ARMAX technique[4] (Auto Regressive Moving Average eXogenous) that provides a model in the discrete time z domain starting with time series analysis of the digitized broadband signals. Then the discrete time model is converted in a continuous time one (Laplace s domain) but this task make worse the model accuracy due to numerical errors. An efficient optimization algorithm based on the SIMPLEX[8] method is used to fit the continuous time model to the experimental data and to improve the root mean square error. The paper describes the steps of the building process of the transducer model through an example of a 5MHz, 50Ω ultrasonic transducer for medical imaging

TRANSDUCER MODEL AND EXPERIMENTAL SETUP

A preliminary step for the system identification is the definition of characteristic functions that define the transducer model. The transducer model considered in this work is shown in Fig.1.: the function Z represents the driving point impedance and the acousto-electric voltage transfer function defined by a high impedance block tagged as function W In particular we are concerned with a one way model because the transfer function W is the ratio between the transmitted acoustic far field and the driving point voltage, therefore the typical use of this model is in transmitting mode. A work in progress is developing another model for use the transducer as a receiver

The experimental setup for broadband measurements is shown in Fig.2. The characteristics of the transducer under test are summarized in Table 1. The immersion transducer is driven by a one cycle

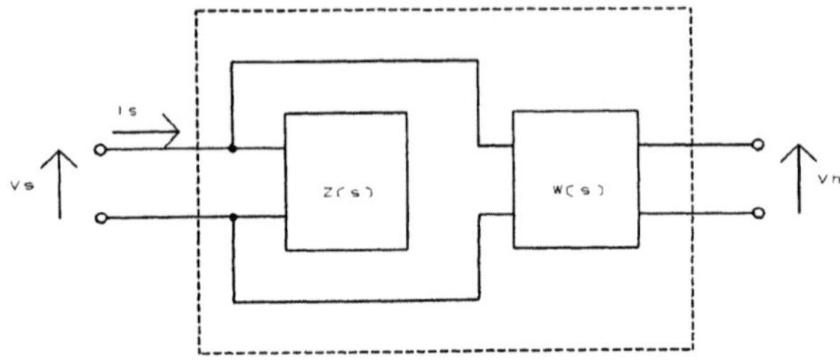

Figure 1. Transducer model: Vs driving point voltage, Is driving point impedance, Vh output voltage corresponding to the acoustic axial field

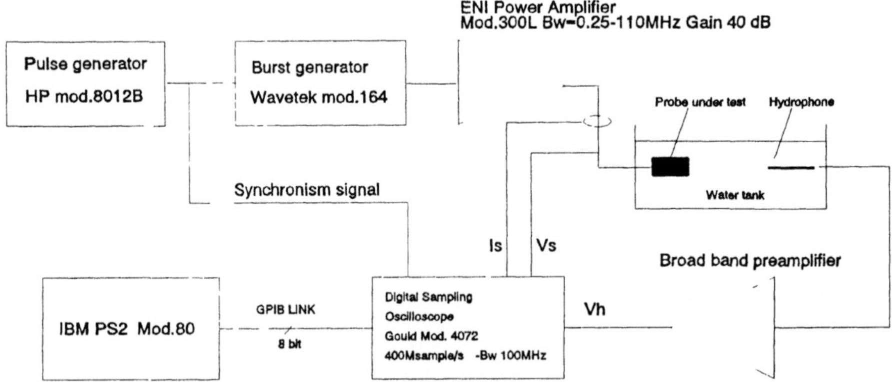

Figure 2. Experimental setup.

broadband power amplifier ENI Mod. 300L with 50Ω output impedance. The driving point voltage Vs and current Is (see Fig.1) are sensed with a voltage probe Tektronix P6108 and a current probe Tektronix P6022.

Table 1. Ultrasonic Transducer Characteristics

Transducer type: GILARDONI BDN10-5
Nominal central frequency: 5 MHz
Bandwidth (-6dB): 2.2 MHz
Characteristic impedance: 50 Ω
Active diameter : 10 mm

Both signals Vs and Is are sampled and digitized at times t_k with 8 bit resolution and with 25 MHz sampling frequency and transferred to the host computer by an IEEE/GPIB link. The ultrasonic axial far field is measured by a 1 mm diameter hydrophone (Medico teknisk institut Bw = 10Mhz) with bandwidth larger than that of the transducer[5,6]. The hydrophone output is amplified with a large bandwidth linear amplifier. We have chosen a high performance linear amplifier to avoid distortions of the output signal. The amplifier Comlinear Corporation Mod. E220 has a linear phase on the -3dB bandwidth 0-190 MHz with a phase deviation of 1.2°, with a voltage gain 20 and output resistance 50Ω. The output voltage Vh is digitized and stored for an off-line processing.

AUTOMATIC TRANSDUCER MODELLING WITH ARMAX TECHNIQUES

The first objective of the model identification is the automatic evaluation of the function Z and W by processing the experimental time series $Vs(t_k)$, $Is(t_k)$ and $Vh(t_k)$ The ARMAX technique was formerly developed for model identification problems and used for design control systems[4,7] but offers also an effective solution for our problem This technique Sources of uncertainty that influence the model approximation are the degree of the polynomial ratio functions and the signal to noise ratio of the input signals The model of the transducer is derived as a "black-box" referring to the fact that some input-output equation is computed on a statistical basis rather than the physical knowledge of the manufacturing characteristics Hence, the ARMAX technique is suitable to solve our problem though it is limited to the discrete time domain The $Vs(t_k)$ - $Is(t_k)$ and $Vh(t_k)$ - $Vs(t_k)$ are considered two discrete-time processes, each of them can be well approximated by a ratio function model in the z domain, called Z'(z) and W'(z) respectively The driving point impedance and the transfer function are evaluated by the Fast Fourier Transform of the sampled signals and the functions $Z(j\omega_k)$ and $W(j\omega_k)$ are calculated Generally in the ARMAX model, the input driving sequence, x(n) and the output sequence y(n) are related by the following linear difference equation

$$y(k) = \sum_{i=1}^{q} b_i x(k-i) + \sum_{j=1}^{p} a_j y(k-j) \tag{1}$$

where the q and p are respectively the order of the moving average and autoregression contributes The identification of the zeroes and poles in the z domain of the ARMAX models that best approximate for $z = e^{j\omega}$ the $Z(j\omega_k)$ and $W(j\omega_k)$ sequences

$$V_s(t_k) = Z'(z) I_s(t_k) \tag{2}$$
$$V_h(t_k) = W'(z) V_s(t_k)$$

Z'(z) and W'(z) can be described as ARMAX models with the following relationship evaluated for $z = e^{j\omega}$ (on the unit circle of the z plane)

$$\frac{\sum_i b_i z^{-i}}{1 + \sum_j a_j z^{-j}} \tag{3}$$

The order of the model must be estimated because is not known a priori Different procedures exist to estimate the correct order Here we decided for an intuitive approach ARMAX models of increasing order are constructed and the computed root mean square errors between the models with different orders and the experimental data are calculated Few trials, applied to different ultrasonic transducers, suggested p and q values that go from 7 to 12 that is a good compromise between model accuracy and complexity These order of magnitudes can be considered as a reference, if an appropriate sampling frequency for the measured signals has been chosen With the chosen order a suitable sampling frequency of Vs, Is, Vh is 25MHz that is slightly higher the Nyquist limit We have observed that an oversampling of the signals avoids the convergence of the ARMAX technique

MODEL CONVERSION FROM DISCRETE TO CONTINUOUS TIME

The final step is to convert the polynomial ratio functions Z'(z) and W'(z) from the z to the Laplace s domain Among different choices we have chosen the *bilinear transformation* because has a straightforward implementation and performed well This transformation replaces the finite differences by differentials The z plane is mapped into the s plane by the relationship
where T is the sampling time The unit circle in the z plane is mapped into the entire $j\omega$ axis of the s plane but there is a nonlinear relationship between the digital and analog frequency The numerical algorithm of the bilinear transformation introduces numerical errors in the poles and zeroes calculations The results obtained by the bilinear transformation need to be feed to an optimization process that fit again the two curves to the experimental data

$$S=\frac{2}{T}\frac{z-1}{z+1} \tag{4}$$

The optimization of the continuous time domain functions Z(s) and W(s) is carried out by a modified version of the SIMPLEX search algorithm[8]. The $Z(j\omega_k)$ and $W(j\omega_k)$ sequences are used as reference curves to calculate a cost function based on the least mean square error. It is an adaptive algorithm in the sense that the parameter search depends not only by the cost function but also by the iteration number. This modified version assures the search of the global minimum avoiding the local minimum trapping, with a better convergence rate than the canonical SIMPLEX algorithm. The optimization process of the real coefficients of the polynomial ratio functions keeps up with two constraints on W(s) and Z(s):

1 -   W(s) must satisfy the stability condition
2 -   Z(s) must be real positive (rp) to allow the equivalent circuit synthesis.

The second condition suggests a cost function MSE made of a mixture of different parameters, as stated in (5):

$$MSE=(MSEm+Cf\times MSEp)\times Crp\times Cpar \tag{5}$$

where MSEm and MSEp are respectively the Mean-Square-Error of the modulus and phase calculated by the difference between the experimental data and the approximating functions. The MSEp term has been introduced to improve the convergence rate and to control the displacement of Z(s) from the minimum phase condition. The Crp and Cpar factors are scalar parameters used to obtain a real positive function and positive coefficients. Eventually, the Cf term adjusts the magnitude order between MSEm and MSEp.

AN EXAMPLE OF TRANSDUCER MODELLING

The results of the modelling process applied to the driving point impedance Z are shown in Fig.3.
On top left of Fig.3 the diagram of modulus approximation with the ARMAX technique Z'(z) is shown with the dashed line and fits well the experimental data (continuous line). With a degree of seven the ARMAX technique is able to reproduce the modulus of the experimental curve with a good accuracy especially in the transducer bandwidth centered on 5.5 MHz. The continuous time domain approximation Z(s) is calculated by the bilinear transformation and is shown in Fig. 3 top left as a dashdot line with a peak shifted on the high frequency side and an unacceptable error. On the top right of Fig.3 the results after the optimization algorithm with the SIMPLEX method applied to the coefficients of the polynomial ratio function Z(s). The optimization algorithm is set to work in the frequency range 3-8MHz and converged after 963 iterations with an accuracy of 0.01 and cost function MSE decreased from 76000 to 550. In Fig.3 bottom are shown the results for the impedance phase diagram. Fig. 3 left shows the good fitting of the final model after the optimization. Fig. 3 bottom left shows another advantage of ARMAX technique that provides a minimum phase approximation: this is essential for the calculation of the impedance equivalent circuit. The presence of linear phase lag lead to not physical all-pass network in the equivalent circuit. The variations of the error terms MSEm and MSEp calculated on 103 points in the bandwidth 3-8MHz, are reported in Table 2. The difference between values MSEm and MSEp after the optimization is less due to the chosen cost function expression. The developed program is able to distribute uniformly the error along all the frequency range instead to have big errors concentrated in few points of the curve as shows the impedance diagrams in Fig 3 left.

Table 2. Error terms in different stages for the transducer impedance.

	MSEm	MSEp
Approximation z domain by ARMAX	533	221
Approximation s domain (after Bilinear Transformation)	52800	27500
Approximation s domain (after optimization with SIMPLEX)	310	377

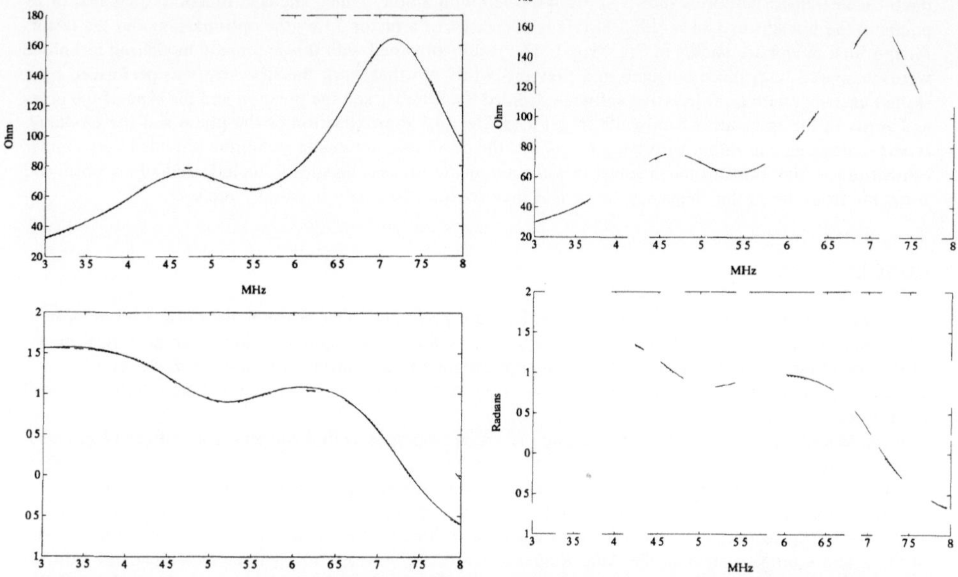

Figure 3 Impedance results relative to a degree 7 Left real z domain with ARMAX s domain after bilinear transformation Right real s domain after optimization with SIMPLEX

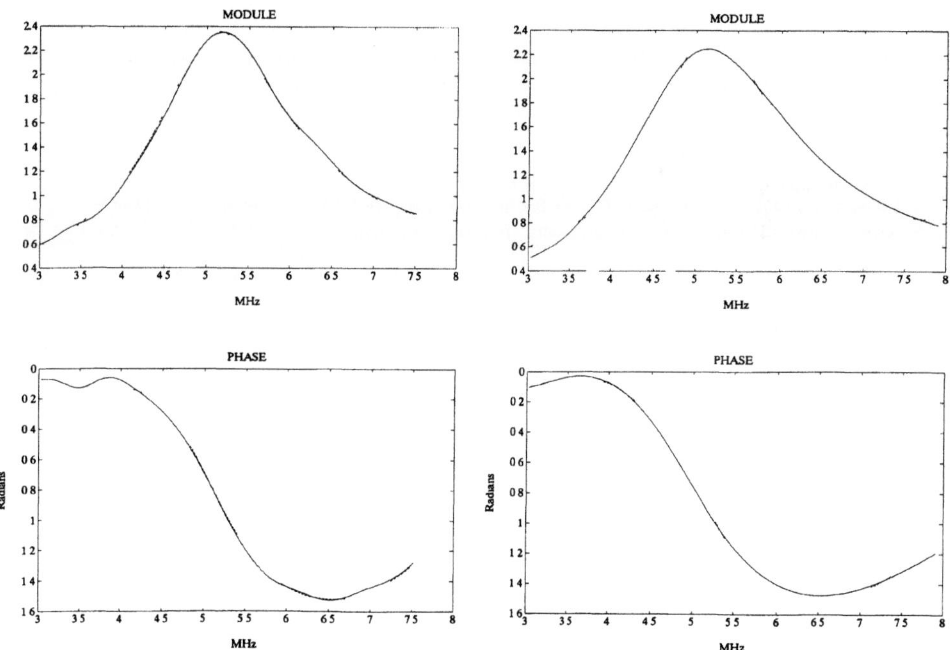

Figure 4 Transfer function results relative to a degree 9 Left real z domain with ARMAX, s domain after bilinear transformation Right real s domain after optimization with SIMPLEX

Similar remarks hold for the transfer function (see Fig 4). A degree equal nine is found and the result after the bilinear transformation is shown in Fig.4 top left with a dotted line. The cost function calculated in 26 points in the transducer bandwidth 3-8MHz is decreased of a factor 33 by the optimization and the results for the final model are shown in Fig 4 right. The results obtained with this automatic modelling technique were compared with those obtained in a previous work[9] In that work the first step was performed by a skilled user that with an interactive software decided the number and the position and the type of the poles and zeros in the transducer bandwidth to get a satisfactory approximation of the phase and the modulus. It was interesting and rather surprising for us that the developed automatic technique provided very similar estimations as the human though sometimes the automatic process decides to include and adjust poles and zeros far from the probe frequency range that they couldn't be easily handle by the user.

CONCLUSIONS

A general method for ultrasonic transducer modelling is presented. This novel method is attractive for both laboratory and industrial applications where a simple technique is required to derive an analytical model of the actual ultrasonic transducer. The main applications of the transducer analytical model are:
1. the analytical model can be inserted in analog circuit CAD system to optimize the design of the front-end electronics
2. to build a library of transducers for testing the actual responses with different error sources (e.g, noise, tolerance etc)
3. Optimization of the transducers' performances by digital signal processing with built in electronics
4. Calculation of equivalent lumped constant networks to design equalization and matching networks
The merit of this technique is to combine different algorithms for the calculation of the input-output model in the z and s domain such as the Auto Regressive Moving Average eXogenous technique, the bilinear transformation and the model optimization with the SIMPLEX method. An example with a 5MHz broadband ultrasonic transducer shows an accurate estimate of the actual driving point impedance and transfer function.

REFERENCES

1. M.K. Schafer, P.A. Lewin, The influence of front-end hardware on digital ultrasonic imaging, IEEE Transaction on Sonics and Ultrasonics, Vol. SU 31, n 4, 295-306 (1984).
2. J.W. Hunt, M. Arditi, and F.S. Foster, Ultrasound Transducers for Pulse-Echo Medical Imaging, IEEE Transactions on Biomedical Engineering, vol. BME-30, n.8, 453-481 (1983).
3. G.S. Kino, Acoustic waves. Devices, Imaging, and analog signal processing. Prentice-Hall Inc. Englewood Cliffs, New Jersey, Cap 1.4 (1987).
4. Ljung L., System Identification. Theory for the user. Prentice Hall, Englewood Cliffs (1987).
5. G.R. Harris, Hydrophone Measurements in Diagnostic Ultrasound Fields, IEEE Transaction on Ultrasonics, Ferroelectrics, and Frequency Control, vol. UFFC-35, 87-101 (1988).
6. S.S. Corbett III, The Influence of Nonlinear Fields on Miniature Hydrophone Calibration Using the Planar Scanning Technique, IEEE Transaction on Ultrasonics, Ferroelectrics, and Frequency Control, vol. UFFC-35, 162-167 (1988).
7. S.L. Marple, Digital Spectral analysis with applications. Prentice-Hall Inc. Englewood Cliffs, New Jersey, (1987).
8. S. Marsili-Libelli, M. Castelli, An Adaptive Algorithm for Numerical Optimization, Applied Mathematics and Computation, vol 23, 341-357 (1987).
9. S.Rocchi, L.Capineri, L.Masotti, M Rinieri, A transducer modelling technique for the identification of the transfer function and driving-point impedance Sensor and Actuators A, 32, 361-365 (1992).

STUDY ON TARGET SURFACE WITH PIEZOELECTRIC ARRAY

Zhao Zheying and Wang Shuhui

Institute of Acoustics
Academia Sinica
P. O. Box 2712
Beijing 100080
P. R. China

INTRODUCTION

Since Sokolov[1] proposed the version of the Ultrasonic Microscope with an ultrasonic converter tube 40 years ago, many studies have been done. Smyth[2] produced the camera with the low-voltage electronic beam scanning. Jacobs[3] improved the effective aperture and increased the field of view of the tube using the discrete matching elements. Addison[4] invented the metal fiber faceplate which acts as a part of envelope and makes the construction of the tube's inner target become outer target so that the chosing of the piezoelectric plate is easier. Nigam[5] adopted the foil-electret to make the condenser-microphone array. His tube has a large aperture up to 60. Brown[6] got the largest area of view of the tube up to $210 \text{ mm} \times 150 \text{ mm}$. Wardley[7] obtained the mechanically strong, long life Sokolov tube. The operating time can be up to 5000 hours. There are still many problems, such as resolution, sensitivity, effective aperture, large field of view, service life and so on, to be improved.

In conventional Sokolov tube the maximum resolution is limited to the thickness of piezoelectric plate which is the acousto-electric convert element of the tube. In order to extend the field of view the thickness of the plate must be increased because the plate also serves as a part of envelope of the tube and must withstand external pressure to prevent the tube from breaking. Thus the resolution of the tube would be decreased. The authors are interested in improving the resolution by means of the array with the independent piezoelectric vibration elements and the sound window that is part of the envelope of the tube.

ANALYSIS

Sokolov tube is a converter. It enables the acoustic signal to be converted into the electric signal.

The tube has a imaging screen which consists of a piezoelectric plate and a part of the tube's envelope, which acts as the sound window and the substrate of the plate (see Figure 1).

The piezoelectric plate is mounted on the inner surface of the envelope and acts as the target surface of the tube's electronic beam. The acoustic image signal passes the sound window to reach the piezoelectric plate. The sound field distribution of the acoustic image on the piezoelectric plate is converted into the electrical signal distribution by means of piezoelectric effect. In order to read the electrical charge distribution of the acoustic image signal on the piezoelectric plate, the electronic beam bombards the target surface. And then the visual picture depended on the acoustic image is displayed on the monitor.

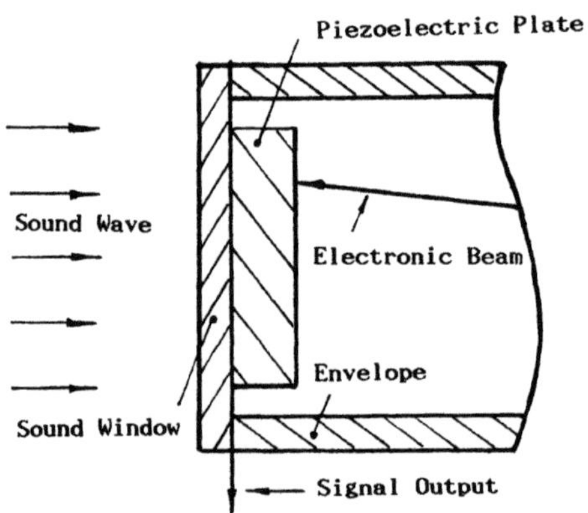

Figure 1.   Schematic cross-sectional view of the construction of the imaging screen

In the conventional convert tube the target surface is a single piezoelectric plate. We consider that the lateral coupling effect exists among the imaging points of the imaging screen of Sokolov tube. The imaging screen is the target surface of electronic beam reading image signal in the tube. When an arbitrary point on the plate is excited by an external ultrasonic beam, the other points around it would be disturbed by the lateral coupling effects such as elastic, piezoelectric and dielectric couplings among the points of the plate as will as the transformation of vibration modes in the plate and so on. For example, the longitudinal wave may be transformed into interface wave or lamb wave, and propagate from the excited point to others. They are regarded as "the lateral coupling effects" here. This kind of lateral coupling effects is mainly mechanical.

Therefore, one would propose the "sound-insulation" imaging screen version to improve its resolution. The version is that the single original piezoelectric plate (imaging screen) is made into the square array, i. e. the array elements do not attach to each other in lateral direction. It makes the sound-insulation between array elements possible. Any array element is an image point. Thus, in principle, the lateral coupling effects among all image points will disappear. The electric signal intensity of every element will be proportional to the acoustic intensity acting directly on that element and disturbed no longer by the lateral coupling effects from other elements.

EXPERIMENT

The piezoelectric array target has been investigated for the construction of the inner target surface before it is mounted into the convert tube.

Experimental Plan

The experimental arrangement is shown in Figure 2.

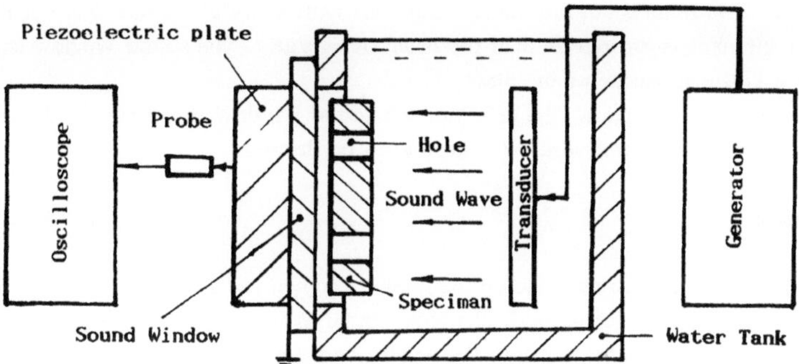

Figure 2.   The experimental equipments and measuring principle

The simulated target consists of the piezoelectrical plate bonded with a sound window (a part of the tube's envelope) which is also the substrate of the piezoelectric plate and mounted over the hole on the wall of the water tank. The piezoeletric plate of the simulated target is set outside the tank. The surface of the sound window is immersed in the water. The specimen observed in the tank is held as close to the sound window as possible. The specimen is the plastic plate with a 2 mm width hole. The sound wave is radiated by a transducer in the water which illuminates it. The experimental data are measured, point by point and line by line, by the probe of an oscilloscope outside the tank, and then the data are sent into the computer. After the data processing, only the shadow pictures of the specimen with some shape hole have been shown on the monitor.

Piezoelectric Array

In order to compare a single plate with a square array, two kinds of arrays have been made. One of them is "the electrode array" which is that the whole electrode on the piezoelectric plate is separated into a square electrode array. The other is "the vibration element array" which is that the whole piezoelectric plate is cut into the square array whose elements are not attached to each other in the lateral direction. They are both square arrays in 11×11 (mm/mm). The area of any element is equal to 1×1 (mm). The number of elements is 121.

Tests

The simulated tests of four imaging screens with the two kinds of arrays have done:

a) The imaging screen is the piezoelectric plate with the electrode array on one of its surfaces. The plate is mounted over the hole on the wall of the tank. The surface with the electrode array is outside the tank. The other electrode of the plate is a whole one connected to ground.

b) The imaging screen is the plate, one surface of which is made into the electrode array, and the other surface with a single electrode is bonded on a piece of the glass (the sound window). The other surface of the sound window is mounted over the hole of the tank.

c) The imaging screen consists of the piezoelectric plate with the electrode array and the sound window. The plate is cut into the vibration element array along the figure of the electrode array. Every element is an independent piezoelectric vibrator. The sound window should not be cut because it is the substrate of the plate.

d) In order to investigate, the effect of the sound window thickness for the resolution of the imaging screen, the sound window is ground to be thinner.

Data Processing

Since the illustrating sound field on the imaging screen radiated by a transducer is inhomogeneous, the data must be normalized. The local signal $S_0(i)$ is measured as the normalized datum before the specimen being put into the water. And then the signal $S_s(i)$ is measured under the same experimental conditions with the specimen in the water.

The relative amplitude distribution $S_r(i)$ of the specimen figure signal on the target surface has been obtained:

$$S_r(i) = S_s(i)/S_0(i),$$

where i indicates the ith element.

In order to get the simple images of the specimen figure the relative amplitudes $S_r(i)$ are separated into five groups:

$$
\begin{aligned}
\text{Group 1:} \quad & S_r(i) > 0.0, \\
\text{Group 2:} \quad & S_r(i) > 0.2, \\
\text{Group 3:} \quad & S_r(i) > 0.4, \\
\text{Group 4:} \quad & S_r(i) > 0.6, \\
\text{Group 5:} \quad & S_r(i) > 0.8.
\end{aligned}
$$

Using the black dots with five different diameters to express the values of five relative amplitudes at the dependent image points. The acoustic images of the specimen have been obtained, and two of the five groups are shown in Figure 3 by this way.

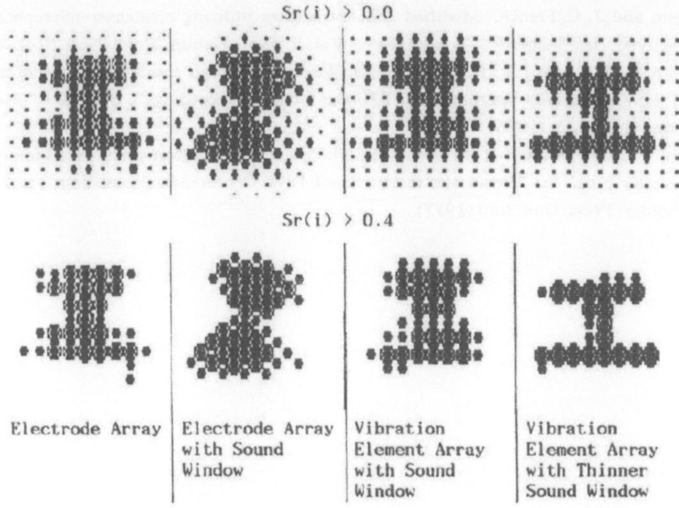

Figure 3.   The acoustic shadow images of the specimen with some hole using four imaging screens.

SUMMARY

The simulating tests have shown that the resolution of the vibration element array is higher than that of the electrode array. The version of the sound-insulation may be possible to improve the resolution of the imaging screen of Sokolov tube.

In addition, the thinner the sound window, the smaller the effect on the resolution is. Since the sound window is a part of the tube's envelope, it must withstand external pressure to prevent the tube from being broken. So the sound window must have a certain thickness.

ACKNOWLEDGEMENTS

The authors wish to thank Senier Engineer Li Zigao and Engineers Chen Shaofeng and Feng Ziying at the Institute of Electronics for their help. This research is supported by Chinese National Natural Scientific Fund No. 8588017. The Institute of Electronics, Academia Sinica is the cooperator.

REFERENCES

1.   S. J. Sokolov, The ultrasonic microscope. Dokl. Adad. Nauk U. S. S. R. , 64: 333 (1949) .

2.   C. N. Smyth, F. Y. Paynton, and U. F. Sayers, The ultrasound image camera, Proc. I. E. E. , 110 (1) : 16 (1963) .

3.   J. E. Jacobs and D. A. Peterson, Advances in the Sokoloff tube, in: "Acoustical holography", Vol. 5: 633, Plenum Press, New York (1974) .

4.   R. C. Addison, A progress report on the Sokolov tube utilizing a metal fiber faceplate, in: "Acoustical holography", Vol. 5: 659, Plenum Press, New York (1974) .

5.   A. K. Nigam and J. C. French, Modified Sokolov camera utilizing condenser-microphone array of the foil-electret type, in: "Acoustical holography", Vol. 5:685, Plenum Press, New York (1974).

6.   P. H. Brown, R. P. Randall, R. F. Sivyer and J. Wardley, A high resolution sensitive ultrasonic image converter, in: "Ultrasonics international 1975 Conference Proceedings", p73, IPC Science and Technology Press, Guildford (1975).

7.   J. Wardley, P. H. Brown and R. C. Croucher, The design and performance of an improved ultrasonic image converter tube, in: "Ultrasonics international 1977 Conference Proceedings", p121, IPC Science and Technology Press, Guildford (1977).

PERFORMANCE OF A COMPOSITE ULTRASONIC TRANSDUCER

COMPOSED OF A PZT PLATE AND A PIEZOELECTRIC POLYMER FILM

G.R. Li,[a] K. Omote, K.S. Park,[b] S. Takahashi,
H. Ohigashi and J.X. Wen[*]

Department of Materials Science and Engineering
Yamagata University, Yonezawa 992, Japan.
[*]Shanghai Institute of Organic Chemistry
Academia Sinica, lingling Lu 345, Shanghai, China.

[a]On leave from Shanghai Institute of Technical
Physics, Academia Sinica, Shanghai, China
present address, Shanghai Institute of Organic
Chemistry, Academia Sinica, China
[b]On leave from Department of Electrical
Engineering, Ninha University, Ninha, Korea

ABSTRACT

The performance of two types of composite transducers designed for simultaneous imaging at two frequencies and for sensitive and low noise imaging is studied. The composite transducer is composed of a PZT plate (or PZT concave) and a piezoelectric polymer film of P(VDF-TrFE), which are stacked together on a backing load. The PZT plate (or PZT concave) and P(VDF-TrFE) film operate independently at the different center frequencies, respectively, as a transmitter and/or receiver. The frequency response and impulse response characteristics measured for various driving/receiving modes are in good agreement with the expected ones evaluated from the Mason's equivalent circuit. The response frequency band is widen in the operation mode where PZT is used as a transmitter and P(VDF-TrFE) as a receiver. The composite transducer has a useful transducer for medical imaging.

INTRODUCTION

Medical acoustic imaging requires ultrasonic transducers operating in a wide bandwidth with high sensitivity and high S/N ratio.[1,2] The transducer operating with different frequencies is also desirable for ultrasonic diagnosis, because ultrasonic waves of low frequency and those of high frequency will complementarily widen the observable range maintaining highest resolution and because the difference in ultrasonic images of a object obtained at different frequency is expected to achieve more precise diagnosis.

Aiming to develop ultrasonic transducer having such features, we fabricated two types of composite transducers, each composed of a PZT plate and a P(VDF-TrFE) film (a vinilidene fluoride and trifluoroethylene copolymer film known as the most effective piezoelectric polymer), and we evaluated the basic properties of the transducer. Since P(VDF-TrFE) has a much lower acoustic impedance compared with that of PZT; these piezoelectric elements in these transducers work almost independently.[2] Furthermore, owing to its low dielectric constant, P(VDF-TrFE) has much higher voltage sensitivity, which is much higher than that of ceramic piezoelectics. (Voltage sensitivity is proportional to piezoelectric g-constant.) The performance of the composite transducers shows that they have a high potential for medical acoustic imaging and other applications.

EXPERIMENTAL

Figure 1 shows the structure of the composite ultrasonic transducers (COMPT-A and COMPT-B) fabricated in this study. In COMPT-A (Fig.1a), a PZT disk plate, a P(VDF-TrFE) film and a matching layer of polyimide (Kapton) film, being 300, 57 and 12 μm in thickness, respectively, were stacked on a brass backing load in this order using epoxy bond. The diameter of active area is 1.0 cm. The polarization directions of PZT plate and P(VDF-TrFE) are opposite to each other. A common ground electrode is placed between the PZT plate and P(VDF-TrFE) film. An electrode on the other side of PZT is connected to a driving source through the brass baking load, and an electrode on the other side of P(VDF-TrFE) (near the Kapton film) is connected to a driving source. COMPT-B (Fig.1b) is com-

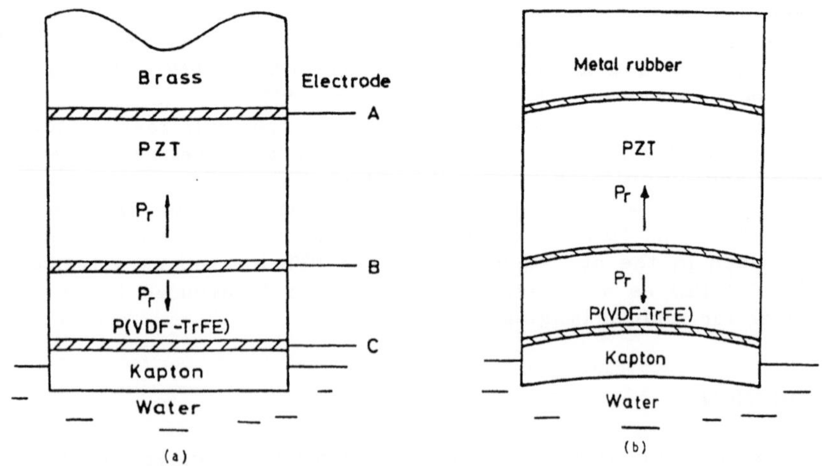

Fig. 1 The structure of the composite ultrasonic transducers: (a) COMPT-A, (b) COMPT-B.

posed of a disk backing load of metal power dispersed in rubber, a concave PZT disk (0.63 mm in thickness, 1.2 cm in diameter, and focal length of 70 mm), a P(VDF-TrFE) (58 μm thick) and a metallized Kapton film (12 μm thick). The connection of electrodes is the same as in COMPT-A.

Frequency and pulse response characteristics of this transducer were studied for the following five driving/receiving operation modes using a 50 ohm driving source, a metal block reflector in water, and a 50 ohm amplifier: (1) the PZT element is used as a transmitter and receiver, (2) the P(VDF-TrFE) element is used as a transmitter and receiver, (3) the PZT element as a transmitter and the P(VDF-TrFE) element as a receiver, (4) the P(VDF-TrFE) element as a transmitter and the PZT element as a receiver, and (5) ultrasonic waves are transmitted and received by the P(VDF-TrFE) and PZT elements in parallel electric connection.

To evaluate the frequency response of a transducer quantitatively, the transducer loss (TL), defined as the ratio of the mechanical output power (P_m) of a transducer to the maximum power (P_0) available from a driving source, and conversion loss (CL), defined as the ratio of P_m to the net electric power (P_t) inputted in the transducer, is calculated by[3,4]

$$TL = -10\log \left(\frac{4\mathrm{Re}(Z_s)\mathrm{Re}(Z_m)|Z_t|^2}{|Z_s + Z_t|^2 |AZ_m + B|^2} \right) , \qquad (1)$$

$$CL = -10\log \left(\frac{\mathrm{Re}(Z_m)|Z_t|^2}{|Z_m + B|^2 \mathrm{Re}(Z_t)} \right) , \qquad (2)$$

where Z_s, Z_t are the electric impedance of the source and transducer, respectively, and Z_m is the acoustic impedance of the propagation medium, A and B are the matrix elements depending on the driving frequency and the structure of the transducer, and A, B and Z_t can be calculated by a modified Mason's equivalent circuit including mechanical and dielectric losses in the materials used.[3] The difference TL – CL is the matching loss ML due to electric impedance mismatching between the source and the transducer, which can be reduced in a given frequency range by using an inductance and a transformer.[3]

TL_n and CL_n (n = 1, 2, 3, 4, 5) are the transducer loss (TL) and conversion loss (CL) for operating mode (n). In mode (1), P(VDF-TrFE) element acts only as a non-piezoelectric film, since the electrodes on the surfaces of P(VDF-TrFE) are opened during the measuring. TL_1 and CL_1, therefore, can be directly calculated by eqs. (1) and (2), respectively. TL_2 and CL_2 can be calculated by eqs. (1) and (2), respectively, as well. TL_3 and TL_4, and CL_3 and CL_4, can be approximately expressed as $(TL_1 + TL_2)/2$ and $(CL_1 + CL_2)/2$, respectively, considering that the PZT and P(VDF-TrFE) elements individually operate in modes (3) and (4). TL_5 and CL_5 can be calculated from eq. (1) and (2), respectively, where calculation of A and B is more complex than that in modes from (1) to (4), because an equivalent circuit of PZT and P(VDF-TrFE) elements are electrically connected in parallel.[4]

Measurement of a transducer losses (TL) were carried by an experimental setup described in elsewhere.[5] The transducer was driven by an RF-burst oscillator having 50Ω-output impedance, and the echo signal (V_2) from a reflector of a brass block placed at a depth of about 3 cm (COMPT-A) or 7 cm (COMPT-B) in water was detected by the same transducer. Then, we insert an attenuator into the circuit between the RF-burst oscillator and the transducer, adjusting a value of the attenuator so that the observed echo signal (V_1) is equal to V_2. The attenuator, therefore, indicates $2(TL+AL)+20\log 2$, where AL is absorption loss in water calculated from $AL=8.86\times(25.5\times f^2)\times10^{-17}\times L$(dB), where f is the frequency and L(cm) is the path length in water. Thus, we obtained the value of TL.

RESULTS AND DISCUSSION

COMPT-A

Figure 2 shows the observed result TL_1, and calculated TL_1 and CL_1. Good agreement between the observed TL_1 and its calculated curve, as shown in Fig. 2, indicates that the COMPT-A is fabricated well. The resonance frequency (7.5 MHz) of COMPT-A is not much lower than but very close to the frequency (7.9 MHz) of the free PZT resonantor ($\lambda/2$ mode), indicating that the PZT plate in the transducer operate approximately in the $\lambda/2$ mode.

Figure 3 shows the observed result TL_2, and calculated TL_2 and CL_2. The observed TL_2 is close to the calculated curve, and has broad frequency response characteristics inherent to a polymer transducer.[3] The resonance frequency (most effective frequency) around 9.5 MHz corresponds to the $\lambda/4$ resonance mode of P(VDF-TrFE) film: Since the acoustic impedance of P(VDF-TrFE) (4.5×10^6kg/m^{2}s) is much lower than that of PZT (35×10^6kg/m^{2}s), the PZT plate acts as a heavy backing for the P(VDF-TrFE) film.

The observed TL_3 and TL_4 are both close to the evaluated curve calculated from $(TL_1 + TL_2)/2$, as shown in Fig.4, where TL_1 and TL_2 are calculated from eq.(1), and CL_3 and CL_4 are plotted as $(CL_1 + CL_2)/2$ as well. This result implies that both elements of PZT and P(VDF-TrFE) work nearly independently, as we mentioned in **EXPERIMENTAL**. The bandwidth of the frequency response, therefore, becomes broader by this composite structure, although the broad frequency response of COMPT-A is not observed so obviously, because the resonance frequencies of the PZT element(TL_1) and P(VDF-TrFE) elements(TL_2) are rather close to each other.

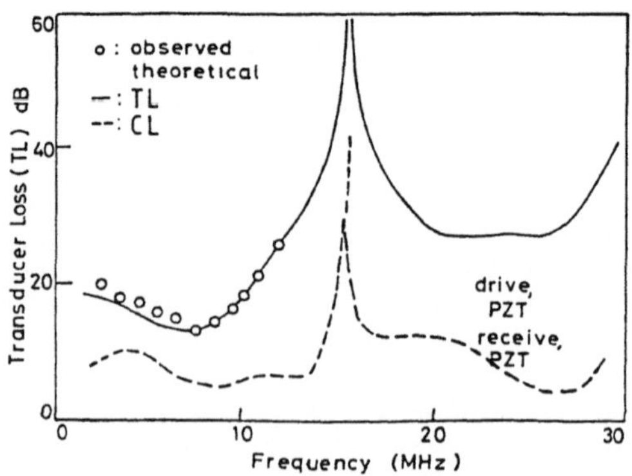

Fig. 2   The observed result for TL_1, and evaluated curves TL_1 and CL_1 calculated from eqs.(1) and (2), respectively.

374

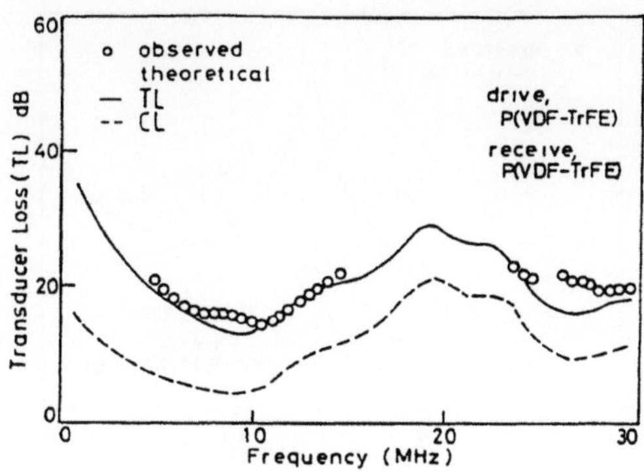

Fig. 3 The observed result for TL_2, and evaluated curves TL_2 and CL_2 calculated from eqs.(1) and (2), respectively.

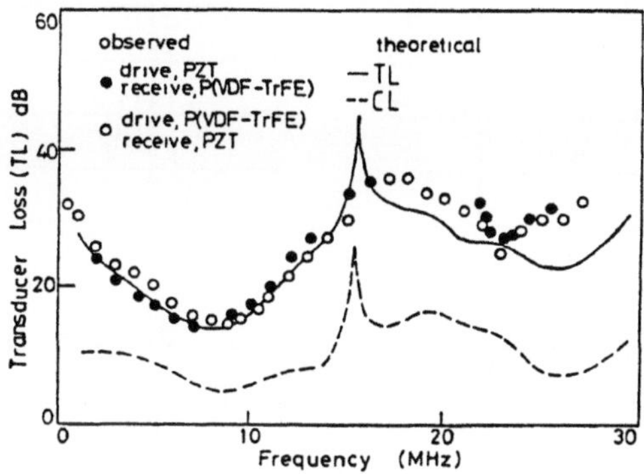

Fig. 4 The observed TL_3 and TL_4, and evaluated curves $(TL_1 + TL_2)/2$ and $(CL_1 + CL_2)/2$ calculated from eqs. (1) and (2), respectively.

In operation mode (5), both PZT plate and P(VDF-TrFE) film are excited simultaneously. In this case the direction of the polarity of these piezoelectric elements reflects on the frequency response characteristics of the transducer. Taking into account the fact that the piezoelectric constant e_{33} for PZT is positive and that of P(VDF-TrFE) is negative,[6] we calculated TL and CL for the composite transducers in which the polarity of PZT and P(VDF-TrFE) are arranged parallel and anti-parallel (opposite). We found that the transducer with the opposite polarity arrangement (the structure shown in Fig.1.) has more sensitive and broader frequency response. The observed result of TL_5 are close to the calculated curve as shown in Fig.5. Thus, different performance is obtained by operating the composite ultrasonic transducer with different operation mode.

Four pulse-echo response waveforms $v_{out}(t)$ corresponding to the driving/receiving modes (1), (2), (3) and (5) are shown in Fig. 6. the transducer was driven by a pulse with rise time less than 10 ns. The operation mode (3) shows a short response waveform with high S/N ratio, indicating the potential of the composite transducer.

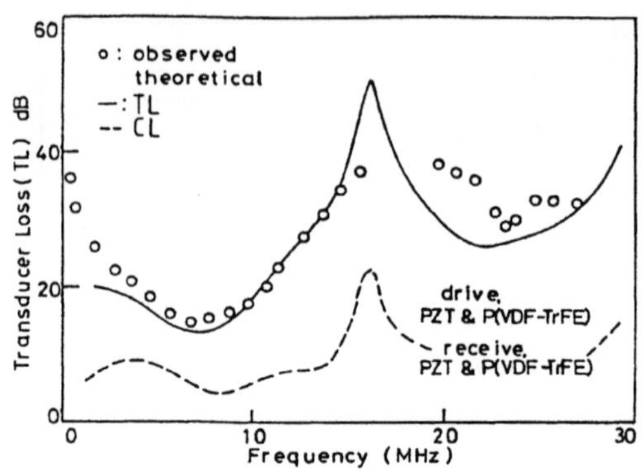

Figure.5 The observed result of TL_5 and the evaluated curves TL_5 and CL_5 calculated from eqs. (1) and (2), respectively.

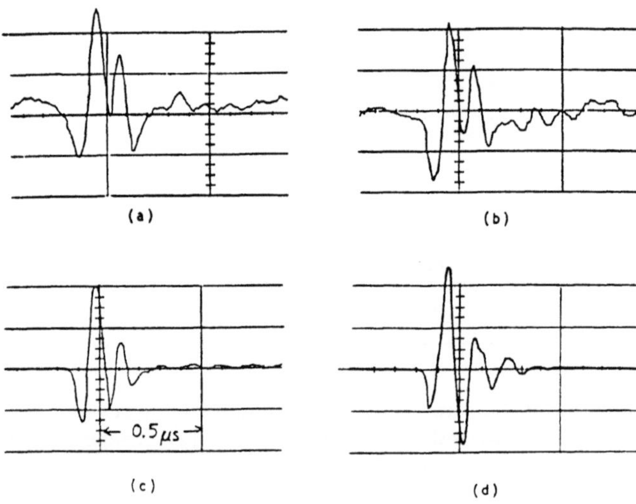

Fig. 6   Four pulse-echo response waveforms $v_{out}(t)$ of (a), (b), (c) and (d), corresponding to modes of (1), (2), (3), and (5), respectively, are observed by the pulse with rise time less than 10ns.

COMPT-B

COMPT-A described above has two problems.   One is that the resonance frequencies of PZT plate and P(VDF-TrFE) film are too close to each other that the merit of the composite transducer is not clearly apparent. The other is the acoustic matching between PZT and backing load is too close, and the acoustic power transmitted into water is not strong enough for practical use, although the wide frequency band characteristics can be achieved.   COMPT-B (Fig.1b) is designed for the purpose of more practical imaging and its performance is evaluated.   The concave PZT disk has a resonance at 3.2 MHz.   The rubber/metal-power backing/absorber disk has an acoustic impedance of about $6 \times 10^5 kg/m^2 s$.

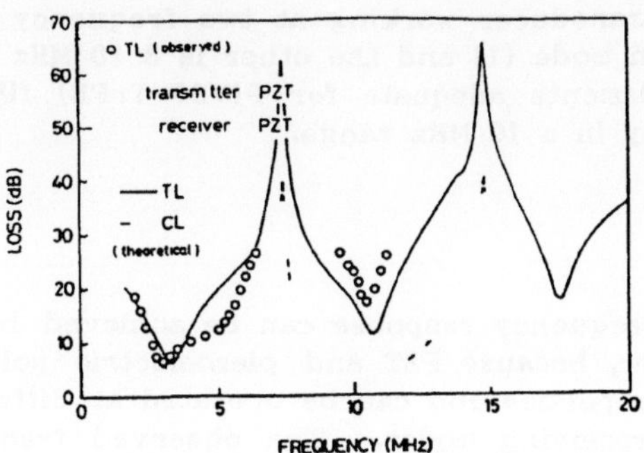

Fig. 7   The observed result for TL_1, and evaluated curves TL_1 and CL_1 calculated from eqs.(1) and (2), respectively.

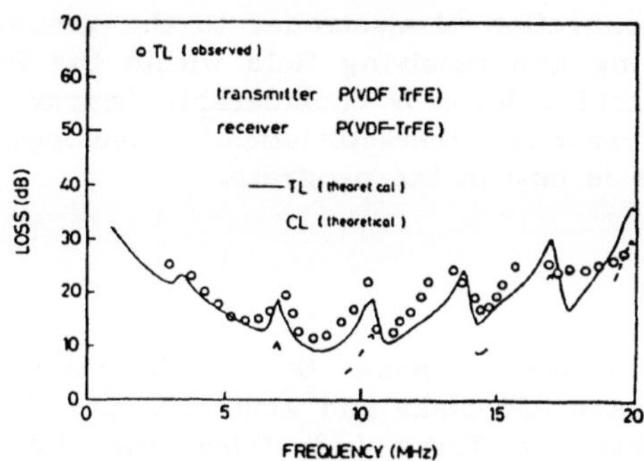

Fig. 8   The observed result for TL_2, and evaluated curves TL_2 and CL_2 calculated from eqs.(1) and (2), respectively.

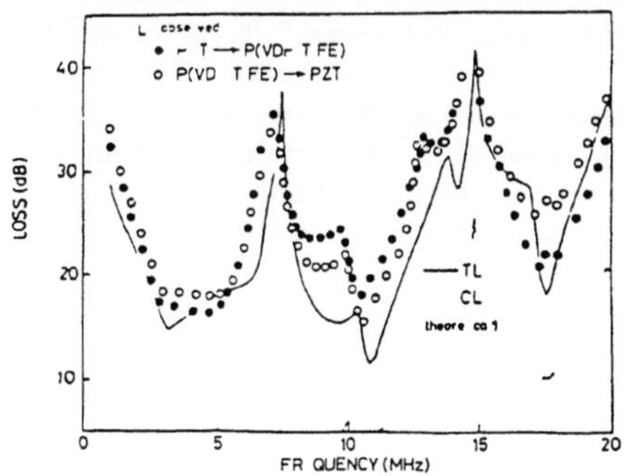

Fig. 9   The observed TL_3 and TL_4, and evaluated curves $(TL_1 + TL_2)/2$ and $(CL_1 + CL_2)/2$ calculated from eqs. (1) and (2), respectively.

Figures 7~9 show the measured frequency dependencies of TL for operation modes (1), (2), (3) and (4).   These results show that COMPT-B can be used as a transducer working at two frequency band: one is at 3.0 MHz by operation mode (1) and the other is 8 10 MHz by mode (2). If we use matching elements adequate for P(VDF-TrFE) film, COMPT-B will have good sensitivity in a 10 MHz range.

CONCLUSION

Broad band of frequency response can be achieved by one composite ultrasonic transducer, because PZT and piezoelectric polymer have quite different acoustic properties and can be operated at different frequencies in various driving/receiving modes. The observed frequency responses for five operation modes are close to curves calculated by Mason's equivalent circuit. The composite transducer shows to have a potential application in medical imaging in the mode of PZT element as a transmitter and P(VDF-TrFE) element as a receiver, since short response time and high amplitude of pulse response waveform with high S/N ratio are observed. However, since the connection of electrodes to the ground is considered to influence the driving and receiving field within the P(VDF-TrFE) film, and $\lambda/4$ acoustic matching layer is considerably improves the frequency response and pulse response characteristics,[7] development of improved composite transducers is now in the progress.

ACKNOWLEDGEMENTS

The authors would like to thank Honda Electronics Co., Ltd. for providing us the concave PZT disks and Aloka Co., Ltd. for providing PZT plate. We are indebted to Toray Industries, Inc. for supporting the present study.

REFERENCES

1. H. Ohigashi, Ultrasonic transducers in the megahertz range, in: "Application of Ferroelectric Polymers," T. T.Wang, J. M. Herbert and A. M. Glass, ed. Blackie & Son, Glasgow (1988).
2. C.T. Lancee, J. Souquet, H. Ohigashi and N. Bom, Ferro-electric ceramics versus polymer piezoelectric materials, Ultrason. :138 (1985).
3. H. Ohigashi, T. Itoh, and K. Kimura, Analysis of frequency response characteristics of polymer ultrasonic transducers, Jpn. J. Appl. Phys. 27: 101 (1988).
4. E. K. Sittig, Transmission parameter of thickness-driven piezoelectric transducers arranged in multilayer configurations, IEEE Trans. Sonics Ultrason. SU-14: 167 (1967).
5. K. Kimura, N. Hashimoto, and H. Ohigashi, Performance of a linear array transducer of vinylidene fluoride and trifluoroethylene copolymer, IEEE Trans. Sonics Ultrason. SU-32: 566 (1985).
6. H. Ohigashi, Electromechanical properties of polarized PVDF films as studied by the piezoelectric resonance method, J. Appl. Phys. 47:949 (1976).
7. K. Kimura and H. Ohigashi, A wide-band polymer ultrasonic transducer using acoustic impedance matching technique, Jpn. J. Appl. Phys. 27: 547 (1988).

2-CH M-SEQUENCE ENCODING ARRAY TRANSDUCER SYSTEM FOR ACOUSTICAL IMAGING

Yorinobu Murata, Kiyohito Koyama and Yasutaka Tamura

Faculty of Engineering, Yamagata University
Jonan 4-3-16, Yonezawa 992, Japan

INTRODUCTION

The "Encoding Array (EA)", proposed by the authors[1], represents a new concept in image sensors. With this EA, an image can be reconstructed without resort to either mechanical or electrical scanning--something that we, the authors, have already proposed the benefits of in previous papers[2-5].

Here in this paper, we describe the findings of an experimental study carried out with an ultrasonic imaging system incorporating a pair of encoding array receivers made of VDF-TrFE copolymer. The aim of the experiment was to determine the extent the use of M-sequence spatial modulation would improve the resolving property of the image sensor. M-sequences are highly sensitive to the delay and time-scaling of detected waveforms caused by the position of a reflecting target.

The frequency used for the experiment was 5MHz and the images were reconstructed by correlation after filtering. A cross sectional image, measuring 512 X 512 pixels, for a single object was then obtained with the imaging system.

PRINCIPLE OF ENCODING ARRAY

Observation of wave front

The schematical representation of the EA is shown in Fig.1. The N receiver elements are positioned to form an equally spaced linear array. The sensitivity of the elements is modulated by means of an appropriate code sequence. A maximal length sequence

(M-sequence) was used with respect to the modulation for the purposes of the present paper.

Objects were illuminated by means of a wide-band ultrasonic wave, so that the ensuing reflected wave could be detected by the array, and recorded. The recorded waveform is the digitized equivalent of the reflected wave front. From the recorded data, an image of the objects can be reconstructed by means of numerical decoding.

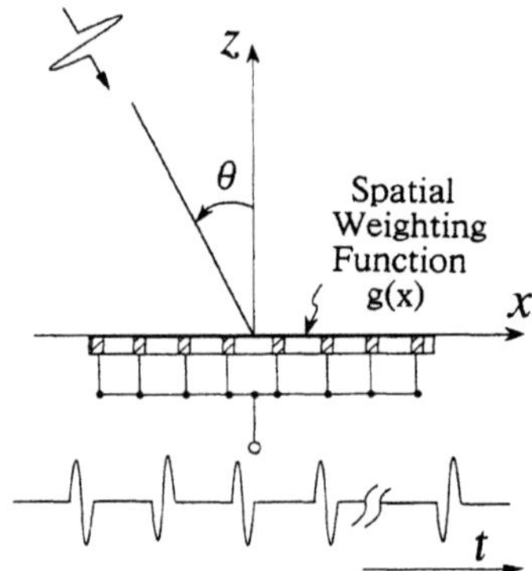

Fig.1. Schematic view of the EA.

Let us Suppose that the transmitted waveform is a pulse of short duration, and that a single point target is being observed diagonally by the array. In such a case, the output waveform of the array would become a train of pulses (pulse train) where each pulse was modulated owing to the coded sensitivity of the receiving element. From the respective intervals and delays of a pulse train, the position of each target point can be determined.

With multiple targets, the pulse trains overlap. However, if the pulses corresponding to distinct target positions are uncorrelated to each other, then a particular pulse train can be discerned and the targets spatially resolved.

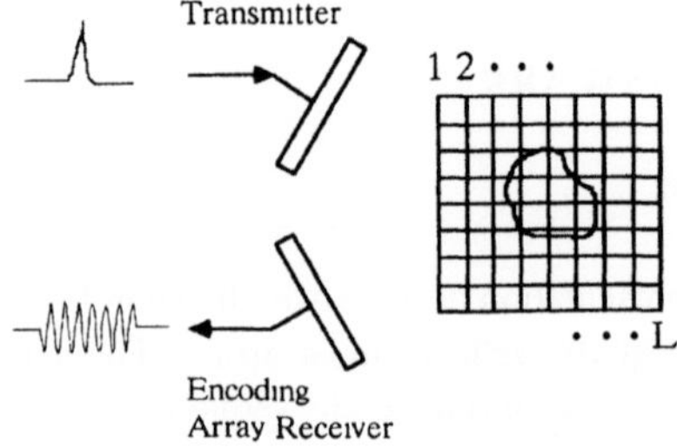

Fig.2. Diagram of the imaging system.

Fig.2 shows, in diagrammatic form, the configuration for the observing a wave front and reconstructing an image. Wide band ultrasonic waves are transmitted from a non-directional transducer so that the ensuing reflected wavefronts can be detected by the array.

With this arrangement, the functions representing the reflectance map generated by an object and the actual waveforms determined by the array can be represented in discrete values by appropriate sampling. A two dimensional reflectance map, represented by a vector, s, can be determined on L grid-points. At each grid-point, the detected waveform, h_i, corresponding to a single point target located on the i-th point, is calculated for all subsequent grid-points, $i = 1 \sim L$. The waveforms, h_i ($i = 1 \sim L$), are the impulse responses determined by the EA system. They conform a transfer matrix, H, of the observation system ; the detected waveform, r, for the objects of reflectance map, s, is given by

$$r = H s \ .\tag{1}$$

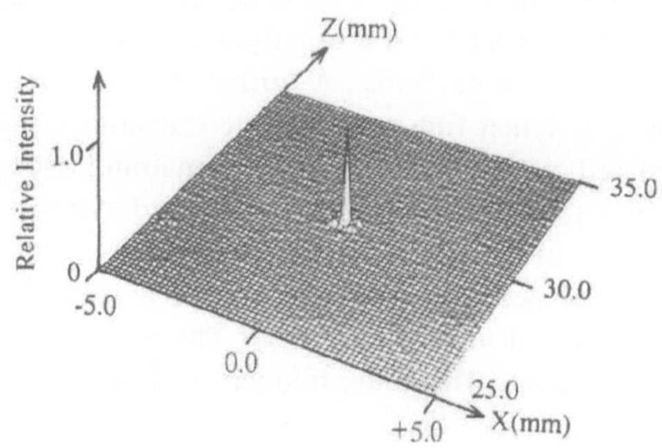

Fig.3. Two dimensional point spread function for the 2-ch M-sequence encoding array transducer system shown by means of computer simulation ; a single wire target exists at $X = 0$ mm and $Z = 30$ mm.

Image reconstruction

We employed a correlation operand in image reconstruction. A complex image of objects, $\hat{s}$, is given by following equation,

$$\hat{s} = H^* r \ .\tag{2}$$

Here, a complex conjugated and transposed matrix H^* is modified by the echo vector r. This correlation operand, when used in image reconstruction, produces a simple and robust image.

Figure 3 shows the result of a computer simulation of two dimensional point spread function of the imaging system. The point spread function for a j-th grid-point is given by

$$P_j = H * h_j \ . \tag{3}$$

A target was assumed to exist at $X = 0$ mm and $Z = 30$ mm, where X is the parallel distance from the center of the transmitter, and Z is the perpendicular distance from the transducer surface. The two dimensional point spread function was expressed on a grid of 64 X 64 points.

EXPERIMENTS

Set up for the experimental system

VDF-TrFE copolymer with a comonomer ratio of 78/22 was used as the piezoelectricitic material. Films of 60 μm thickness were prepared by solution casting. After heat-treatment, the films were polarized by applying an AC electric field. For further details, see reference (1). The electromechanical coupling constant of the films was 0.29. This was estimated from the frequency dependence of electrical impedance.

The arrangement used for the 2-ch M-sequence encoding array transducer system is shown in Fig.4. Three films of VDF-TrFE copolymer were prepared. A gold electrode was placed on one side of each of the films. A further flat gold electrode was then placed on the other side of the copolymer film acting as the transmitter and M-sequence pattern electrodes were then placed on the other side of the remaining two copolymer films. The two transducers, designed to act as receivers, were placed symmetrically about the flat central transmitter. One side of the composite transducers was attached to a thin polyimide film of 25 μm thickness. The polyimide film was used as a matching layer. The other side of the transducers was attached to ferrite rubber. The ferrite rubber acts as a backing material and ultrasonic absorbent. The center frequency of the transducers was 5 MHz.

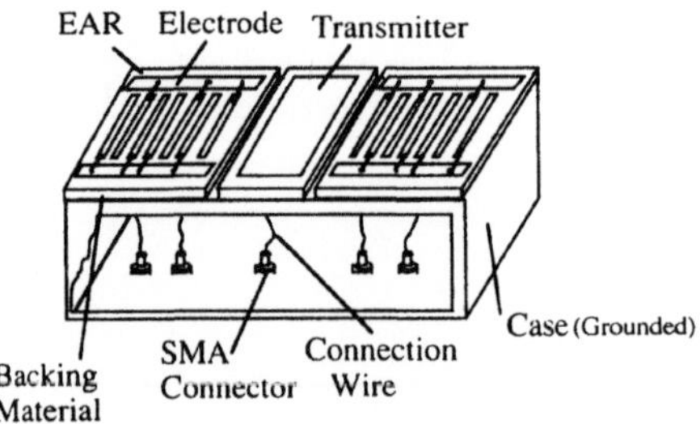

Fig.4. Configuration of the 2-ch M-sequence encoding array transducer system.

The array electrode pattern was designed to have an M-sequence of length 63 bits and binary values of +1 and -1. Details of the M-sequence electrode pattern are shown in Fig. 5. The width of each element in the array pattern was 0.15 mm and total length of the array pattern was 30.00 mm. The counter electrode of the transducer film was a 30.00 mm X 10.00 mm rectangle. Thus, the array transducer could receive a propagated ultrasonic wave with two different polarities encoded by the M-sequence, because the polymer transducer was active only in a sandwiched area between the electrode patterns. The mechanical loss of the piezoelectrical polymer was large. Therefore, the array transducers can be simply made just by forming electrodes from the vapor deposition of the films.

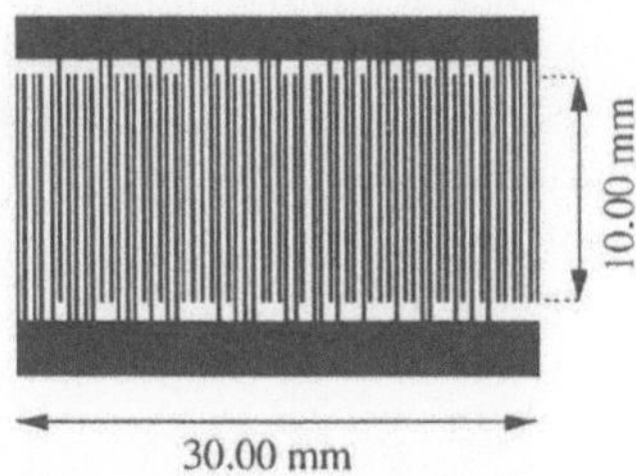

Fig.5. Electrode pattern of the M-sequence encoding array receiver.

A block diagram of the experimental imaging system with the 2-ch M-sequence encoding array transducer is shown in Fig. 6. A wide-band electrical impulse with a peak voltage of 100 V, generated by a pulse generator (AVTECH ELECTOR SYSTEM AVL-AV1CP), was applied to the centrally placed transmitter. A wide-band ultrasonic wave was thus generated radially from the transmitter, and the reflected pulses, from a single target, were received by the pair of M-sequence encoding array receivers. The target was a piece of brass wire of diameter 1.4 mm. The difference between the two signal amplitudes detected by each M-sequence encoding array receiver was calculated using a differential amplifier, and stored temporarily in a digital memory (LeCroy 9400). In order to reconstruct the images, the data was sent to a mini-computer (HP 9000/370).

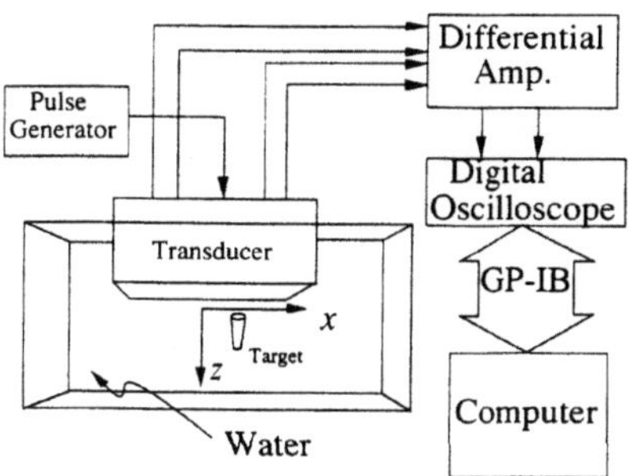

Fig.6. Block diagram of imaging system with the 2-ch M-sequence encoding array transducers .

Filtering

Figure 7 shows a sample of the received waveforms from a single target. Waves (a) and (b) respectively show the wave patterns generated by simulation and in the actual experiment. The simulated waveform was calculated by computer based on the assumption that a cycle of sinusoidal pulse at 5 MHz was being received by an individual electrode on the array receivers. Each waveform was digitized into 2048 points at a sampling rate 100 MHz. By comparing experiment results with simulated results, the experimental waveform was seen to be distorted. The distortion is thought to arise as a result of a distortion of the pulse being detected by each electrode on the array.

To supress this distortion, the received waveforms were therefore filtered.   The transfer function $H(\omega)$ of the filter is defined by

$$H(\omega) = O(\omega) / I(\omega), \tag{4}$$

where $I(\omega)$ is the Fourier transform of the waveform that was received by only one electrode on the array transducers and $O(\omega)$ is the Fourier transform of a cycle of sinusoidal pulses at 5 MHz. To obtain the objective waveform $m(t)$, we calculated by the equation;

$$m(t) = F^{-1}[H(\omega) A(\omega)] \tag{5}$$

where $F^{-1}[\ ]$ is the inverse Fourier transform, and $A(\omega)$ is the Fourier transform of the waveform that was actually detected by the M-sequence encoding array transducer system. The images were reconstructed after the received waveforms were filtered.

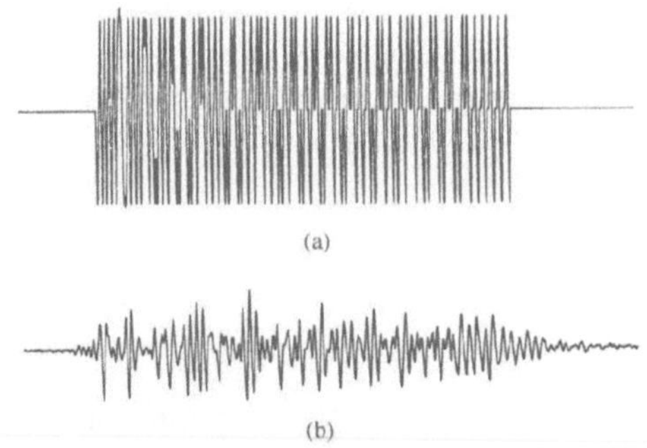

Fig.7. Ultrasonic reflected echo waveform from a single target using the 2-ch M-sequence encoding array transducer system. (a) as simulated by computer, (b) as obtained in the experiment.

RESULTS

Fig.8, shows the image functions of the system before filtering (a) and that after filtering (b). It was confirmed that the sub-peak levels became low following filtering.

Fig.9 shows the gray-scale displayed image of Fig.8(b). A single wire was optionally located as the target in the imaging area.   The imaging area is 10 mm X 10 mm centering at

$X = 0$ mm and $Z = 35$ mm. The number of pixels is 262144 (512×512 mesh points). The time expended on imaging the wire was about 10 minutes.

The bright spot about $(X, Z) = (0$ mm, 36 mm$)$ corresponds to the object. The spatial resolution evaluated from the size of the spot was 0.20 mm (0.13 mm on the

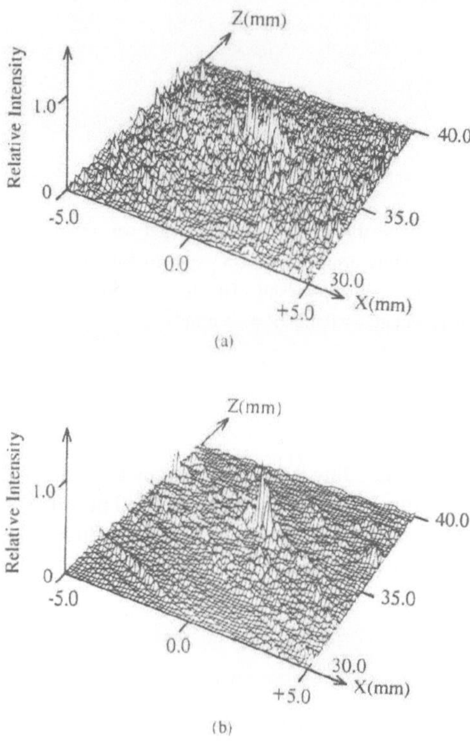

Fig.8   Two dimensional point spread function for a single target with the 2-ch M-sequence encoding array transducer system   (a) pattern before filtering and (b) pattern after filtering.   A correlation method was applied to results.

simulation) along the X axis, and 0.12 mm (0.11 mm on the simulation) along the Z axis, respectively.

Some artifacts could be observed in the image. The dynamic range of the image was 2.5 dB (7 dB on the simulation). It is lower than that of the value obtained by simulation. This probably stems from an error in the positioning of the transducers.

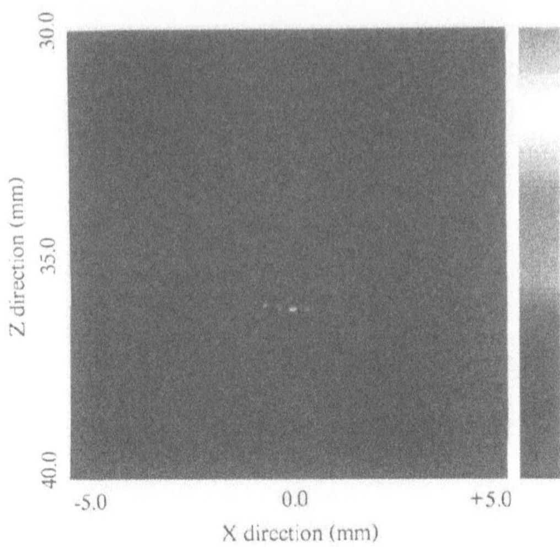

Fig 9   The cross sectional image of a single target was reconstructed with the M-sequence encoding array transducer system following filtering   A correlation method was applied to result The imaging area was 10 mm X 10 mm centered on a point at $X = 0$ mm, $Z = 35$ mm

CONCLUSIONS

The 2-ch M-sequence encoding array transducer system, with its high resolution features, was constructed using VDF-TrFE copolymers. The cross sectional image for a single target was obtained in short time. In the future, to reduce the artifacts generated in imaging, we must produce a feasible operand.

REFERENCES

(1) Y. Tamura, An echo location using spatial modulation and wide-band pulse, Proc. of SICE (Japan), 26 (1987) 717.
(2) K.Koyama, H.Ito, K.Tanaka and Y.Tamura, Jpn.J.Appl.Phys., 28, S28-1(1989)251.
(3) K.Koyama, H.Ito, K.Tanaka and Y.Tamura, Jpn.J.Appl.Phys., 29, S29-1(1990)240.
(4) K.Koyama, Y.Murata, K.Tanaka and Y.Tamura, Jpn.J.Appl.Phys., 30, S30-1(1991) 152.
(5) Y. Tamura and T. Akatsuka, Acoustical Imaging, 18 (1991) 73.

AN ENCODING APERTURE USING AN INCLINED LINEAR ARRAY

Yasutaka Tamura, Takashi Koseki, Takayoshi Sato, and Takao Akatsuka

Faculty of Engineering, Yamagata University
Jyonan 4-3-16 Yonezawa 992, Japan

ABSTRACT

A new echolocation system, based on a concept of an "Encoding Array Receiver (EAR)" is described The system, which requires no mechanical or electrical scanning, determines the 2-D cross-section of images Only one or two signal channels are needed Images can be reconstructed from the data acquired by means of a single transmitting and receiving process

In this paper, we propose a new encoding array receiver system, where the receiver is an inclined linear array coded with Golay codes The equivalence of the EAR to conventional arrays was also shown

The results of simulations and preliminary experiments clearly demonstrate the feasibility of the proposed array to 2-D imaging

INTRODUCTION

The echolocation and imaging system has to handle a wavefield varying spatially and temporarily. An array of transducers and parallel electronic circuits are used in conventional systems.

These parallel features, however, complicate the system hardware and make the cost of the system higher

We have proposed a new configuration of transducers for an echolocating system [1] The proposed system is an alternative to conventional array transducers which use parallel electronic circuits Several improvements have been proposed These include a inear array

Acoustical Imaging, Volume 20 Edited by Y Wei
and B Gu, Plenum Press, New York, 1993

made of piezo-electrical polymer film and coded with random [2] or M-sequence [3] , an inclined 2-dimensional array of wide-band receivers [4] .

The basic idea is to use a coded and inclined array, and to encode generated wavefronts into a serial signal. A single signal channel is connected to an array of ransducers. The array is inclined and the sensitivity of the transducers is modulated by means of an appropriate code sequence.

When a wide-band acoustical signal is used, this array encodes any received wavefront into a serial signal. From the serial signal, an image of multiple targets can be reconstructed using a numerical decoding process. Accordingly, we have dubbed our array receiver an "Encoding Array Receiver".

The impulse response of an EAR varies with the direction of a wavefront. It is known that our ears employ a similar form of angle dependence as a means of vertically locating sound sources. [5]

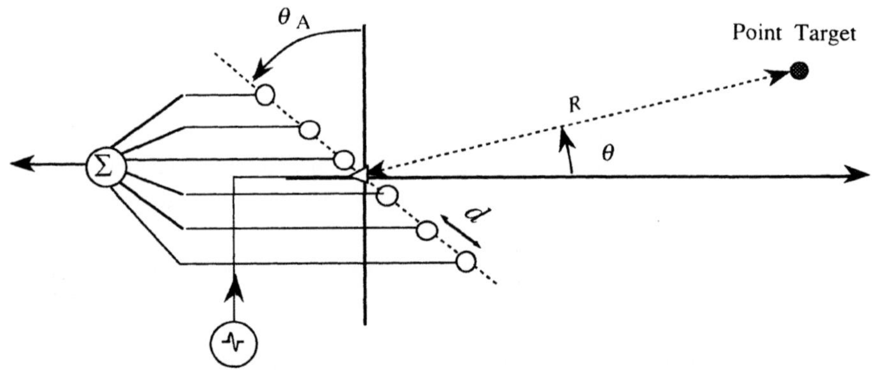

Fig.1 Schematic diagram of an EAR with a linear array

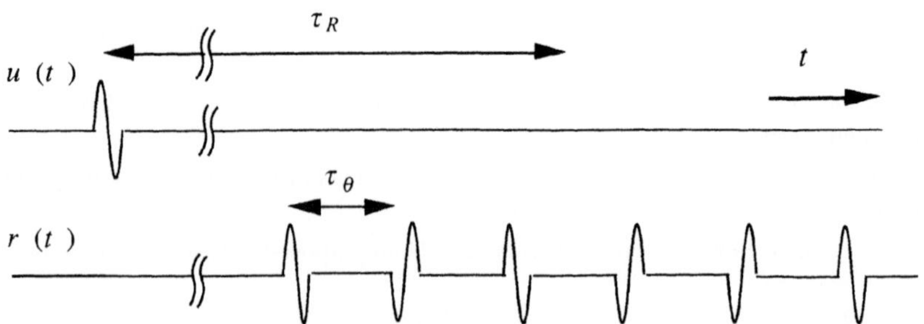

Fig.2 Transmitted pulse $u\ (t\ )$ and waveform r $(t\ )$ detected by an EAR

PRINCIPLE

Detecting Waveforms

Fig.1 shows the schematic diagram of the system. A linear array, of wide-band receivers, is inclined at an angle of θ_A from the direction of observation. N receivers are arranged at equal intervals of d. A wide band pulse is transmitted from the center of the array, and the reflected echo wavefront detected by it.

For the sake of argument, suppose that a target is sufficiently far from the array. Then, the detected waveform becomes a train of pulses as shown in Fig.2. The intervals of the pulses, τ_θ, and the delay time of the pulse train, τ_R are given by the following equations,

$$\tau_\theta = \frac{d \sin (\theta_A - \theta)}{C} \; ,$$

(1)

$$\tau_R = \frac{2R}{C} \; ,$$

(2)

where, θ and R are the angle and the range of the target respectively.

Accordingly, we can determine the direction, θ, and the range, R, from the intervals and delay of the pulse train.

Imaging with Correlation

The system also allows the reconstruction an image, including multiple and distributed objects, following the application of a correlation process.

For instance, where $r(t)$ is a detected waveform, the image function $s(x)$ is obtained from the following equation,

$$s(x) = \int r(t) \cdot h(x,t)^* dt \; ,$$

(3)

where, $h(x,t)$ denotes the detected waveform from a single point target located at a position x, and $*$ denotes the complex conjugation.

In the paraxial region of the array, $h(x,t)$ is given by

$$h(x,t) = \sum_{i=1}^{N} m_i \cdot u\left(t - \frac{|x| + |x - x_i|}{C} \right) ,$$

(4)

where, x_i is a position vector of the i-th receiver, m_i is a code value corresponding to the receiver, and $u(t)$ is the waveform function of the transmitted acoustical pulse. This relationship is a correlation between the received waveform r and the impulse response function h.

Point Spread Function

The point spread function (PSF) of the imaging system is the image function for a single point target. In its general form, the PSF for the position, x_0, is given by

$$\psi(x;x_0) = \int h(x_0,t)h(x,t)^* dt = \sum_{i=1}^{N} \varphi\left(\frac{|x|+|x-x_i|-|x_0|-|x_0-x_i|}{C}\right),\tag{5}$$

where

$$\varphi(t) = \int u(t)u(t-t)^* dt\tag{6}$$

is the auto-correlation function of the transmitted waveform function $u(t)$.

If the imaging region is in the Fraunhofer field, the impulse response function of the system is given by

$$h(\theta,R,t) = \sum_{i=1}^{N} m_i \cdot u\left(t - \frac{2R + \left(i\cdot d - \frac{L}{2}\right)\sin(\theta_A - \theta)}{C}\right),\tag{7}$$

Then, the PSF becomes a function of the differences in range and angle, and can be approximated by

$$\psi(\Delta\theta,\Delta R;\theta_0,R_0) \approx \sum_{i=1}^{N} \varphi\left(\frac{2\Delta R + \left(i\cdot d - \frac{L}{2}\right)\cos(\theta_0 - \theta_A)\sin\Delta\theta}{C}\right).\tag{8}$$

It can be easily shown that the PSF of eq.(8) is equal to the PSF of the linear array with parallel signal channels.

It should be noted that this equivalence holds true in the region of the PSF's peak, $x_0 \approx x$, where

$$\frac{d\sin(\theta_A - \theta_{max})}{C} > \frac{1}{B} , \Delta R\tag{9}$$

is applicable. Equation (9) means that the minimum time interval of the pulse train h is greater than the equivalent pulse width of the transmitted signal u.

For large range differences, however, spurious peaks emerge in the PSF. To eliminate the range side-lobes, caused by the side-lobes of the auto-correlation function of the pulse train $h(x,t)$, we employed the Golay code on the spatial modulating sequence.

Golay Code

Golay code is a binary complementary code consisting of two codes, namely, an A-code and a B-code. By summing the auto-correlation functions of the two codes, a single peak function without side-lobes can be obtained.

In the present system, the codes A and B of $N/2$ bits, were alternatively assigned to the N receivers, see Fig.3.

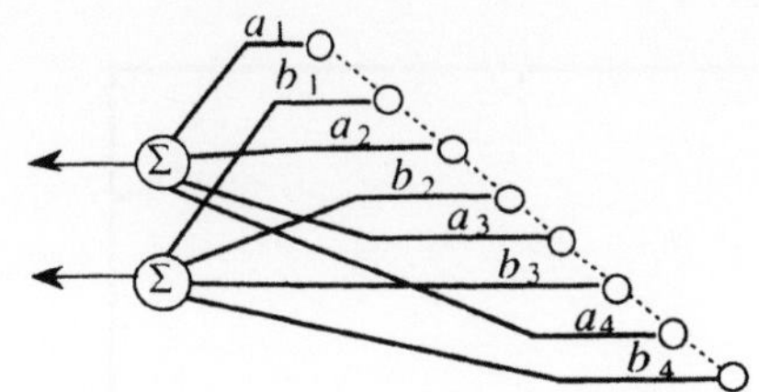

Fig.3 Assignment of codes to the elements

SIMULATION

The PSF of the imaging system was obtained by means of computer simulation.

The simulation assumed a 2-D imaging system operating in mid-air. The width of the array was 44 cm , and 16 elements were used. The angle of the inclination was 45 degrees. An LFM (linear frequency modulated) acoustical pulse was transmitted from the center of the array. The frequency band of the waveform was between 25kHz and 75kHz.

Fig.4 shows a point spread function at the distance of 1 meter.

Fig.5 shows the relation between the level of the spurious peaks and the number of the receivers. The level of the spurious peaks was evaluated as a ratio of the main peak to the maximum spurious peak. A value about 11 dB was obtained when 16 receivers were used, and 16dB for 64 receivers.

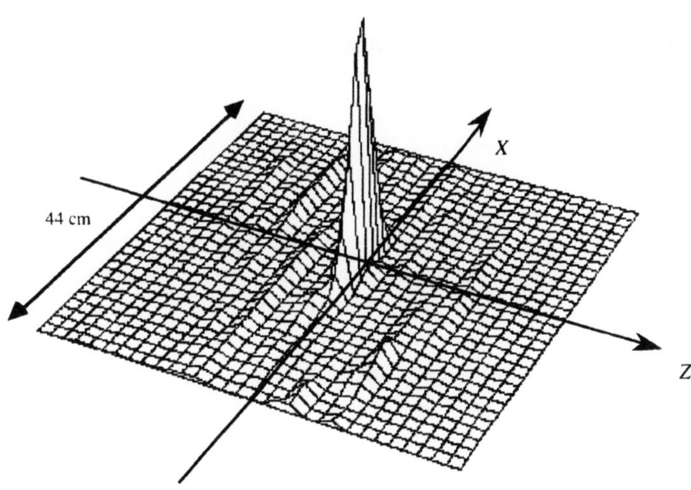

Fig.4 The point spread function of the imaging system obtained by means of computer simulation. The distance was 1 meter, and the imaging area 44 cm · 44cm.

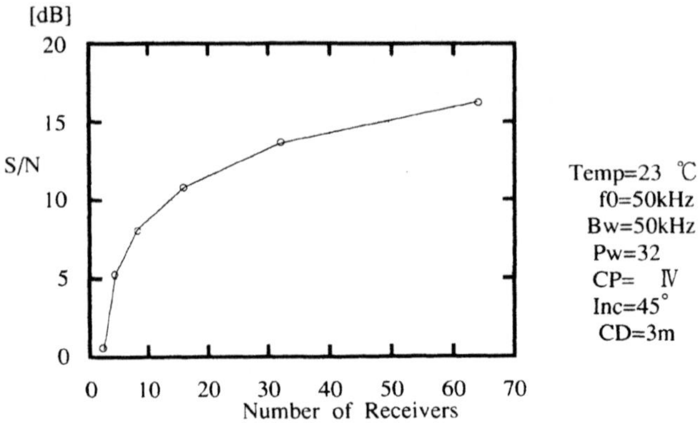

Fig.5 The level of the spurious peaks vs the number of receivers

EXPERIMENTAL RESULTS

Experimental system

Photo. 1 shows a view of the array. The experimental system had a pair of linear arrays configured symmetrically. Each array (linear in form) consisted of 16 microphones. Transmitters, designed to generate a non-directional wavefront , were located in the center between the two receiving arrays. The polarities of the receivers were controlled by means of a GPIB. The transmitted pulses were generated by means of a wave-generator. The center frequency of the wave was 50kHz (Wavelength λ is about 6.9mm). The detected wave forms were digitized by A/D converters , and transferred to a personal computer. Then, the data was sent to an HP9000 workstation so that the image could be reconstructed.

Preliminary experiment

An image of a wide-band sound source was reconstructed as a preliminary test of the experimental system. The linear array on the right-hand side was used. The reconstructed cross-sectional images are shown in Photo. 2-a). The size of the images is $64\lambda \times 64\lambda$. The bright spot , indicating the position of the sound source , could be recognized.

Photo. 2-b) shows the image reconstructed from the echo waveform generated by means of   computer simulation. This assumes the same array configuration as the experimental system.

Many spurious peaks were observed in the image. The level of the peaks were higher

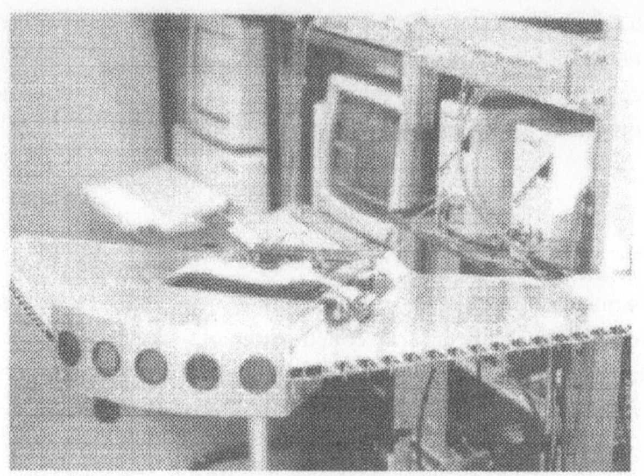

Photo.1 View of the array

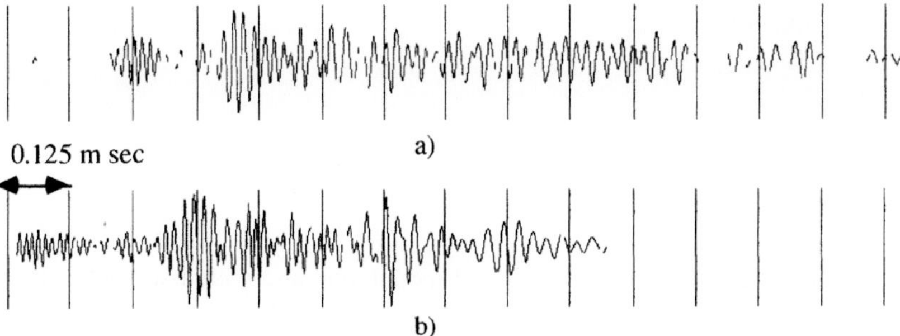

0.125 m sec

a)

b)

Fig.6 Detected waveform : a) Experiment b) Simulation

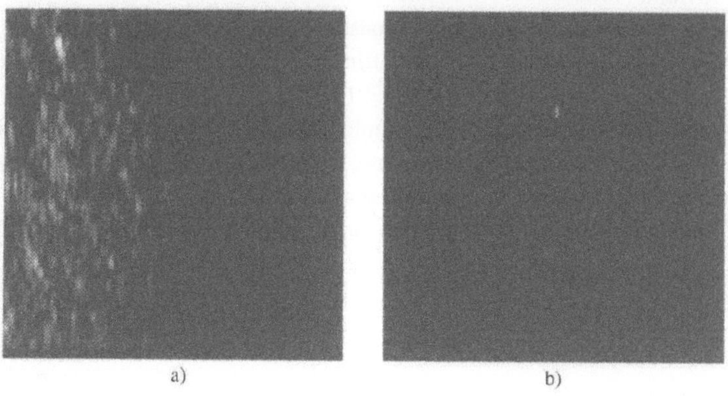

Photo.2 Reconstructed image : a) Experiment b) Simulation, The imaging area is $64\lambda \times 64\lambda$ (44 cm $\times$ 44 cm), the array is located at the left-hand side of the image.

than expected following computer simulation. This probably stems from an error in the positioning of the elements and a distortion of the received waveforms.

We are presently engaged in developing a calibration system designed to produce an imaging operator which would compensate for these effects .

SUMMARY

In conclusion, we propose a new encoding array receiver system, where the receiver array is an inclined linear array coded with Golay codes.

Although the hardware of the imaging system can be simplified considerably, the proposed array is approximately equivalent to a parallel array, in terms of spatial resolution.

The relative level of the spurious peaks is -16 dB when 64 receivers are used in the simulation.

However, in any practical imaging system the spurious images must be under -20 dB. Accordingly, the next goal of our study is to obtain an appropriate operator to reduce artifacts.

REFERENCES

1) Y. Tamura, An echolocation using spatial modulation and wide-band pulse, Proc. of SICE, 26, 717/718 (1987)

2) Y. Tamura, H. Itoh, and K. Koyama, An acoustical imaging sensor with an encoding aperture — Transducer made of piezo-electrical polymer composite film — , Proc. of SICE, 28, 337/338 (1989)

3) Y. Tamura, Y. Murata, K. Koyama, Acoustical image sensor with encoding aperture, —directivity of piezo-electrical polymer film transducer with electrode coded by M-sequence —, Proc. of SICE, 30, 215/216 (1991)

4) Y. Tamura and T. Akatsuka, An echolocation and imaging using transducers of directionally distinguishable impulse response, Acoustical Imaging , 18, 73/82 (1991)

5) P.J. Bloom , Determination of monaural sensitivity changes due to the pinna by use of minimum audible field measurements in the lateral vertical plane, JASA, 61, 820 (1977)

MEMBRANE ACOUSTIC BEAM DEFLECTOR FOR
IMAGING AND MONITORING APPLICATIONS

H. Dabirikhah and C.W. Turner

Dept. of Electronic and Electrical Engineering
King's College, University of London
Strand, London WC2R 2LS, U.K.

ABSTRACT

The performance of an immersed thin membrane acoustic beam deflector has been investigated for a number of materials and immersion fluids. Such beam deflectors are of interest for a range of imaging and monitoring applications, especially for high amplitude focused acoustic beams. Results presented show that it is possible to design membrane deflectors with extremely good fidelity for both C.W. and pulsed acoustic signals and for highly convergent beams. Conditions are established under which no membrane modes are excited by the incident beam, avoiding the problem of non-specular reflection. A Schlieren method has been used to confirm the theoretical results for the transmission and reflection of both focused and collimated incident beams. The practical considerations affecting the design of membrane beam deflectors are presented and possible applications in NDT and medical ultrasound are discussed.

INTRODUCTION

Unloaded plate (free plate) and liquid-loaded plate (immersed plate) phenomena have been investigated for many years. Immersed plate behaviour is of considerable practical importance to non-destructive testing. Lord Rayleigh and Lamb[1,2,3] were the first to study the dispersion relations of elastic waves in a free isotropic plate. Since then many published papers have increased our knowledge and understanding of this subject[4]. Osborne and Hart were the first who derived the appropriate dispersion (characteristic) equations for an immersed plate[5]. Despite the simple appearance of the characteristic equations of a free plate, they are difficult to deal with and become worse by the introduction of liquid loading on the plate. Many authors have tried to solve the characteristic equations of an immersed plate by using approximate methods[5,6]. The zeroth approximation is taken to be the solution of the dispersion equations for a plate with free boundaries. Merkulov obtained solutions by taking the liquid density to plate density ratio to be much smaller than unity. He derived formulas for the damping coefficients for symmetric and antisymmetric waves and calculated them for the A_0, S_0, A_1, and S_1 modes for steel plates in water. In his paper, he showed for a thin plate that when $v_{a_0} < v_0$, where v_{a_0} is the zeroth antisymmetric mode phase velocity and v_0 is the liquid longitudinal wave velocity, the damping coefficient α_{a_0} becomes imaginary, which means that there is no radiation loss. These results were unchallenged for many years and engendered the idea that there is no damping when the plate phase velocity is below the liquid longitudinal wave velocity, known as the subsonic region.

Another way of solving the characteristic equations for plate waves is by using numerical methods. This is straightforward for dispersion relations of the free plate because the roots of these equations are either purely real or purely imaginary. The real roots represent the propagating waves and the imaginary ones indicate evanescent waves. For an immersed plate the roots of the equations are in general complex and, as the equations are also complex, standard numerical root solving methods must be modified to be capable of finding the complex roots of a complex function with acceptable accuracy.

In the following the results obtained by using Muller's method are presented and compared with those of Merkulov. The differences are discussed and the calculations are repeated for several materials such as steel, aluminum, and silicon nitride. As the roots of the characteristic equations are also the poles of the reflection coefficient function for a plane wave, the reflection coefficient can be computed and the effects of the plate modes can be observed. Since, for a very thin plate, the effects of plate modes on the reflection coefficient can be largely suppressed, it is possible to use a thin plate or membrane as an ultrasonic beam deflector for both collimated and focused incident beams. This has been demonstrated experimentally in our laboratories using a Schlieren system.

THEORY

Characteristic equations for an immersed plate have been derived and presented by many authors. Taking into account differences in signs and notation, all forms of representation of these equations describe the same basic physical phenomenon, i.e., normal modes propagating in an unbounded isotropic plate immersed in a liquid. The equations for the symmetric and antisymmetric modes are given by[7]

$$f_s = [k_s^2 - 2k_x^2]^2 \cot\left(\frac{d\kappa_d}{2}\right) + 4k_x^2\kappa_s\kappa_d \cot\left(\frac{d\kappa_s}{2}\right) - \frac{i\rho k_s^4 \kappa_d}{\kappa} \tag{1}$$

for the symmetric case, and

$$f_a = [k_s^2 - 2k_x^2]^2 \tan\left(\frac{d\kappa_d}{2}\right) + 4k_x^2\kappa_s\kappa_d \tan\left(\frac{d\kappa_s}{2}\right) + \frac{i\rho k_s^4 \kappa_d}{\kappa} \tag{2}$$

for the antisymmetric case, where k is the incident wave number of the acoustic wave in the liquid, and k_x, k_s, and k_d are the wave numbers of the plate modes, transverse (shear) waves and dilatational (longitudinal) waves in the plate material respectively. The parameter ρ is the ratio of liquid to solid mass densities. The parameters κ_s, κ_d, and κ are defined as follows: $\kappa^2_s = k^2_s - k^2_x$, $\kappa^2_d = k^2_d - k^2_x$, $\kappa^2 = k^2 - k^2_x$.

These equations are not easy to solve since they are complex transcendental equations with complex roots. Numerical root finding methods can be used to find their roots such as Muller's method[8] which uses a quadratic approximation to the function whose roots are required. This method is able to search very effectively for complex roots of a complex function. There is no need for the calculation of the derivative of the function and no need for any close initial guess to the root. As Muller's method is a global root finding technique it was used to find all roots of interest for the two characteristic equations (1) and (2). In the following the results are presented for plate waves propagating on a very thin membrane immersed in water, a situation of special interest to the design of beam deflectors where only a small percentage of the incident energy is reflected.

NUMERICAL RESULTS

The characteristic equations were solved for the case of an immersed steel plate in water when the frequency times thickness product was varied from 0.01 to 8.0 mm.MHz. The results of these calculations are shown in Figures 1 and 2. Figure 1 shows the zeroth symmetric S_0 and antisymmetric A_0 mode phase velocities for the above range and Figure 2 shows the imaginary parts of the wave numbers.

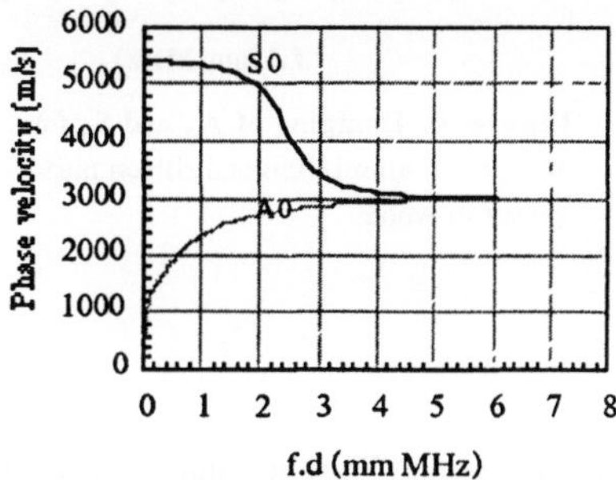

Figure 1. Phase velocities of A_0 and S_0 modes for immersed steel plate in water.

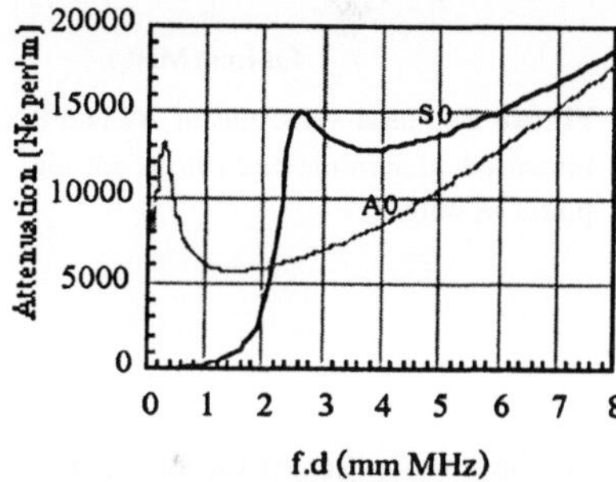

Figure 2. Damping of A_0 and S_0 modes for immersed steel plate in water.

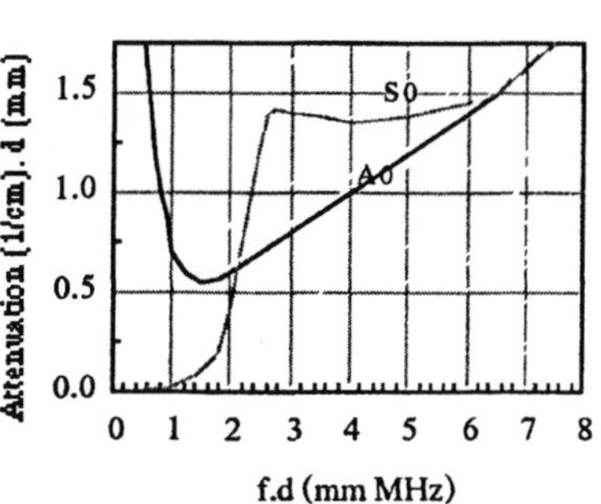

Figure 3. Damping of A_0 and S_0 modes for immersed steel plate in water.(Merkulov)

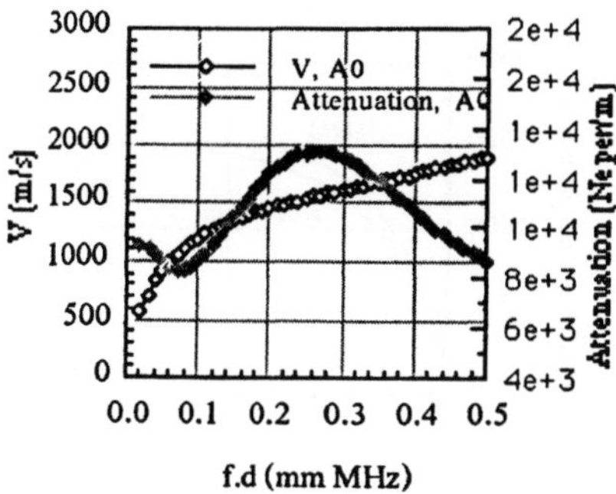

Figure 4. Velocity and damping of A_0 mode for immersed steel plate in water.

Comparison of Figure 2 with Figure 3 which shows Merkulov's approximate solutions[6] reveals a significant difference. We obtain a large imaginary part for A_0 mode wave number in the subsonic range, $v_a < v_0$ while Merkulov's method gives no damping. Figure 4 gives a more detailed presentation of the subsonic range for A_0 mode. As is well known the phase matching condition cannot be satisfied in this region, i.e., the plate wave cannot radiate into the adjacent liquid medium. These calculations are repeated for aluminium and silicon nitride plates and the results are shown in Figures 5 and 6.

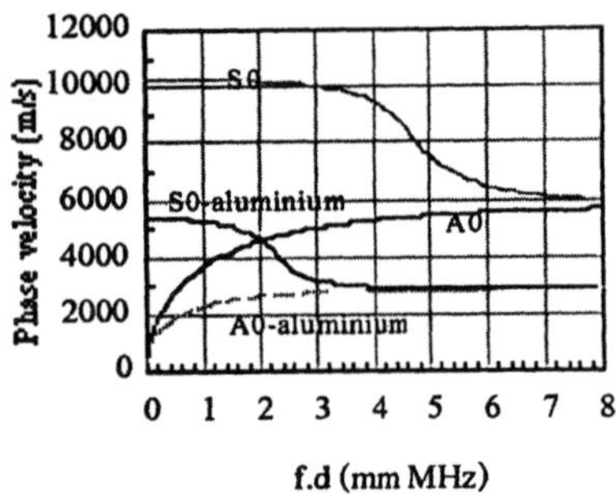

Figure 5. Phase velocities of A_0 and S_0 for immersed aluminium and silicon nitride plates in water.

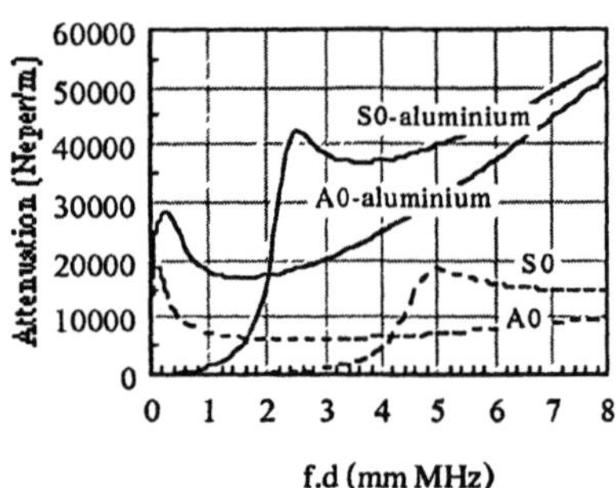

Figure 6. Damping of A_0 and S_0 for immersed aluminium and silicon nitride plates in water.

The phase velocities of the antisymmetric and symmetric undamped Scholte (Stoneley) waves which are also solutions of the characteristic equations are presented in Figures 7 and 8.

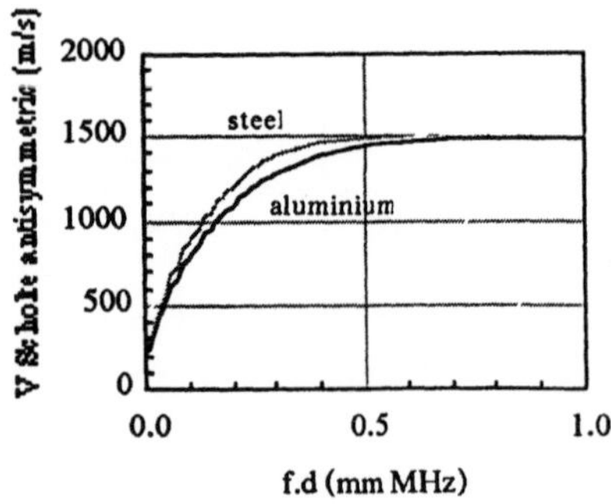

Figure 7. Phase velocity of antisymmetric Scholte wave for immersed steel and aluminium plates in water.

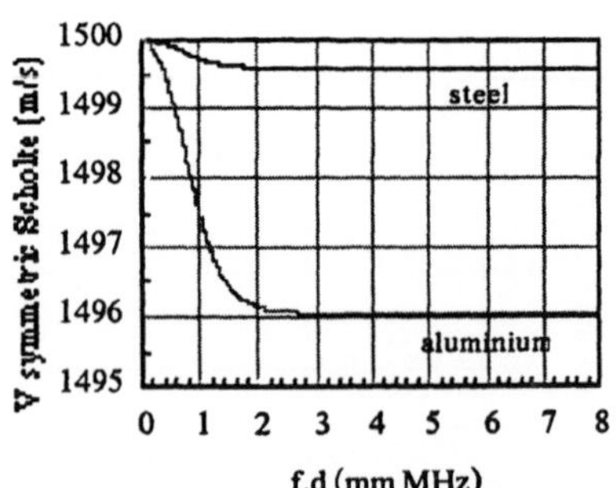

Figure 8. Phase velocity of symmetric Scholte wave for immersed steel and aluminium plates in water.

REFLECTION COEFFICIENT

The plane wave reflection coefficient of an immersed isotropic plate is given[7,9] as

$$R(k_x) = \frac{N}{f_s f_a} \tag{3}$$

where f_a and f_s are the antisymmetrical and symmetrical characteristic equations of an immersed plate as in (1) and (2) and

$$N = [k_s^2 - 2k_x^2]^4 + 16k_x^4 \kappa_s^2 \kappa_d^2 - \frac{\rho^2 k_s^8 \kappa_d^2}{\kappa^2} + \frac{8k_x^2 \kappa_s \kappa_d [k_s^2 - 2k_x^2]^2 [1 - \cos(d\kappa_d) \cos(d\kappa_s)]}{\sin(d\kappa_d) \sin(d\kappa_s)}$$

$$\tag{4}$$

It can be seen that the poles of the reflection coefficient are the roots of the characteristic equations of the immersed plate. The reflection coefficient is a function of the frequency thickness product, f·d, and the angle of incidence θ and it is known that its absolute value is bounded between zero and one. By keeping the f·d product constant equal to 0.01 mm MHz (d=10 μm, f=1.0 MHz) and varying θ, R was computed and plotted as shown in Figures 9 and 10 for an aluminum plate in the lossless and lossy cases. The effect of the S_0 mode appears as a spike at the angle equal to $\theta = \sin^{-1}(v_0 / v_{s_0})$ and the rest of the graph is relatively flat. By introducing a realistic value of acoustic loss for the plate material the spike is suppressed significantly. Figure 11 shows the same calculation for an immersed Teflon membrane.

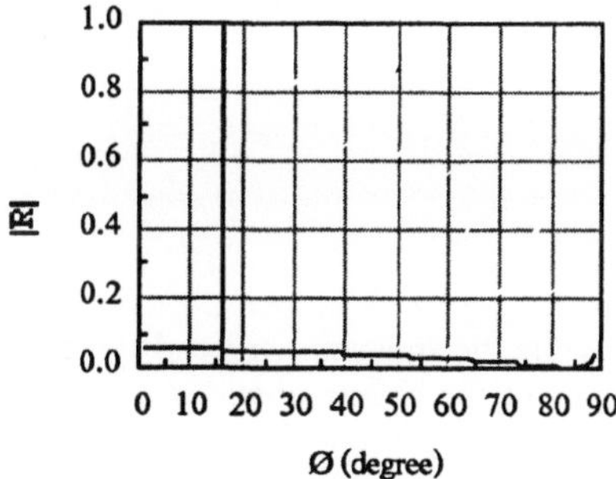

Figure 9. Refelction coefficient of immersed lossless aluminium plate in water, f.d = 0.01 mm.MHz.

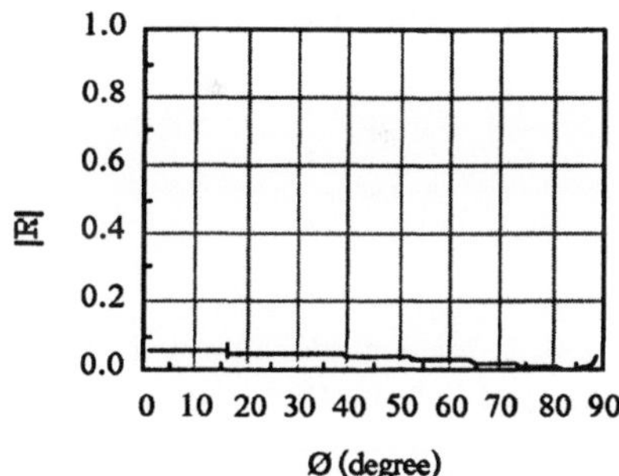

Figure 10. Refelction coefficient of immersed lossy aluminium plate in water, f.d = 0.01 mm.MHz.

This shows that the plate (membrane) can be used as a semi-reflector that allows only a small percentage of the incident energy to be reflected whilst the rest is transmitted with minimal distortion. This could be useful in cases where measurement of very high intensity incident beams is attempted using miniature hydrophones[10]. This type of transducer cannot tolerate very high pressures and would be severely damaged in the process of measurement of a large number of pulses.

DISCUSSION

Using a Schlieren[11] visualization system, we have been able to confirm the reflection from and transmission through some thin plates for both collimated and focused incident beams. We have also demonstrated that plane wave reflection and transmission phenomena in thin membranes are in agreement with the numerical results given in this paper[10]. The results of numerical calculation, using Muller's method, for solving the immersed plate characteristic equations have been presented and compared with the results of Merkulov. It has been shown that for the zeroth antisymmetric mode A_0 there is a large damping in the subsonic case where the coincidence condition cannot be satisfied. The A_0 mode is essentially non-propagating under these conditions and is not only non-radiative but also cannot be excited by an

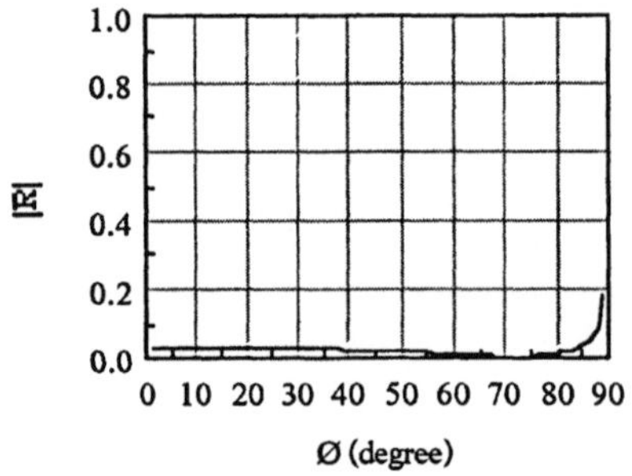

Figure 11. Refelction coefficient of immersed lossy aluminium plate in water, f.d = 0.01 mm.MHz.

externally incident plane wave in the fluid. It is shown that the antisymmetric and symmetric undamped Scholte (Stoneley) waves are also solutions of the characteristic equations and their effects on reflection and transmission coefficients have to be considered. The effects of the A_0 and S_0 modes on the reflection coefficient of an immersed plate have been discussed by Freedman[12]. Our results confirms his general conclusion. It is shown that the introduction of finite acoustic loss helps to minimize the unwanted effects of plate mode excitation in thin membrane acoustic beam deflectors which could be useful in medical ultrasound and NDT applications. Ideally a beam deflector should reflect a specified proportion of the incident beam for a given range of angle of incidence (to deal with convergent beams), as in Figures 12 and 13, and over a given bandwidth (for pulsed acoustic signals). In this paper we have shown that thin membranes, operated with f·d << 1, can achieve satisfactory performance using either metal foils or polymer membranes.

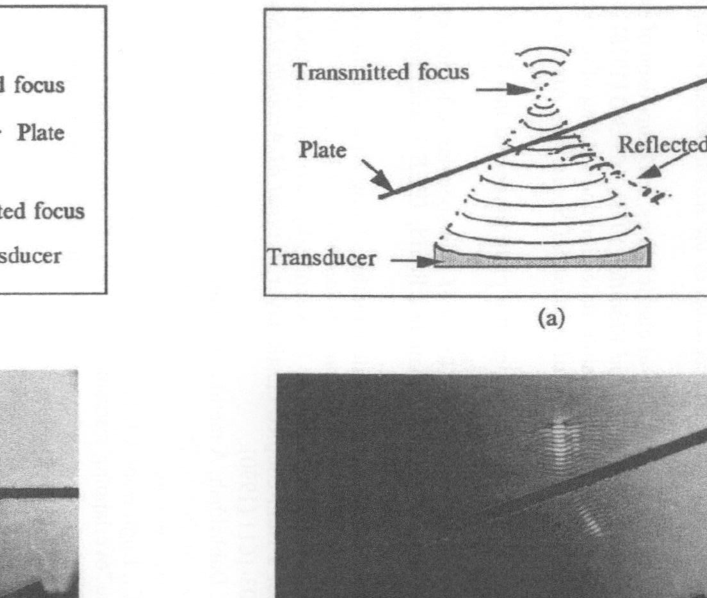

Figure 12. Schematic diagram (a) and Schlieren photograph (b) for incident focused beam and its reflection from the reflector when the plate is placed normal to the beam axis.

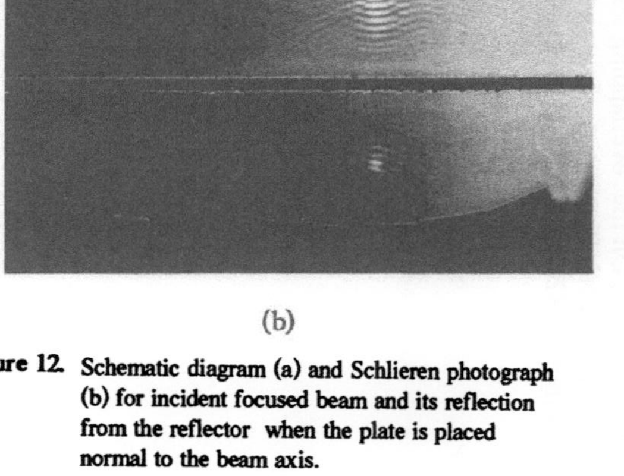

Figure 13. Schematic diagram (a) and Schlieren photograph (b) for incident focused beam and its reflection from the reflector when the plate is placed obliquely to the beam axis.

REFERENCES

1. Lord Rayleigh, On the free vibration of an infinite plate of homogeneous isotropic elastic matter, *Proceedings of the London Mathematical Society* 20,225-234 (1889).
2. H. Lamb, On the flexure of an elastic plate, *Proceedings of the London Mathematical Society* 21,70-90 (1889).
3. H. Lamb, On waves in an elastic plate, *Proceedings of the Royal Society* A 93, 114-128 (1917).
4. A. Freedman, Reflectivity and transmitivity of elastic plates, I: Comparison of exact and approximate theories, *Journal of Sound and Vibration* 59(3), 369-393 (1978)
5. M. F. M. Osborne and S. D. Hart, Transmission, reflection and guiding of an exponential pulse by a steel plate in water, *Journal of Acoustical Society of America* 17, 1-18 (1945).
6. L.G. Merkulov, Damping of normal modes in a plate immersed in a liquid, *Soviet Physics-Acoustics* 10(2), 169-173 (1964).
7. L. E. Pitts, T. J. Plona, and W. G. Mayer, Theory of Non specular reflection effects for an ultrasonics beam incident on a solid plate in a liquid, *IEEE Transaction on Sonics and Ultrasonics* SU24(2), 101-109 (1977).
8. D. E. Muller, A method of solving algebraic equations using an automatic computer, *Mathematical Tables and Other Aids to Computation* (MTAC) 10, 208-215 (1956).
9. D. E. Chimenti, S. I. Rokhlin, Relationship between leaky Lamb modes and reflection coefficient zeros for a fluid-coupled elastic layer, *Journal of Acoustical Society of America* 88(3), 1603-1611 (1990).
10. C. W. Turner and H. Dabirikhah, A membrane sampler for lithotripter monitoring, *IEEE Ultrasonics Symposium: Proceedings* 1-3, 1627-1630 (1990).
11. G. S. Kino, "Acoustic Waves Devices, Imaging, & Analog Signal Processing", Prentice-Hall , INC., Englewood Cliffs, New Jersy, (1987).
12. A. Freedman, On the "overlapping resonance" concept of acoustic transmission through an elastic plate, II: Numerical examples and physical implications, *Journal of Sound and Vibration* 82(2), 197-213 (1982).

MULTIDIMENSIONAL CARDIAC IMAGING

James F. Greenleaf, Marek Belohlavek, and Thomas C. Gerber

Biodynamics Research Unit
Department of Physiology and Biophysics
Mayo Clinic and Foundation
Rochester, MN 55905 U.S.A.

INTRODUCTION

There are many ways to obtain three-dimensional images with ultrasound scanners. Some investigators are developing two-dimensional arrays and associated beam formers to obtain ultrasonic data from a three-dimensional volume in real time. Other investigators are working on methods of acquiring data using gating so that only a small part of the heart is scanned at a time, but by moving the scanned plane, the entire heart can be imaged over several heart and respiratory cycles. It is the latter method that is described in this paper.

True Three-Dimensional Acquisition

True three-dimensional acquisition implies that the entire volume of tissue is interrogated about 30 times per second. The goal is to display a perspective or stereo image of the beating heart in three dimensions in real time. This results in strict speed requirements and resulting compromises.[1]

Speed Requirements. Three-dimensional acquisition in real time requires a very complicated beam forming system. A speed up of some 30 times as many image lines as are normally acquired must be accomplished. The scanner must transmit relatively broad beams and receive on multiple beams simultaneously to get a great enough data acquisition rate for real-time three-dimensional images.[2]

Compromises. The compromises required by very high-speed acquisition for real-time three-dimensional display result in lower contrast due to the broad transmit beams and in beam forming compromises due to the large number of required receive beam formers. The result is very fast but low resolution and low contrast images. Whether these trade-offs are worth the speedup in image acquisition and whether real-time three-dimensional acquisition and display increases comprehension of cardiac geometry and function, is yet to be proven.

Need for Surfaces. We have shown previously that comprehension of three-dimensional objects is greatly enhanced if the objects have surfaces as do those of every day experience.[3] The human has difficulty comprehending objects that are made of puffs of brightness distributed in space. Therefore, we have embarked on a program to study the methods and efficacity of shaded surface imaging from three-dimensional ultrasound data.[4] This requires the acquisition of three-dimensional data, its processing, and computer visualization methods to produce suitable three-dimensional depictions of the anatomic components of the heart and associated vessels.[5]

Post-Acquisition Three-Dimensional Analysis

Post-acquisition three–dimensional analysis is a description our method of making three-dimensional images off line after the acquisition of three-dimensional data. Many cardiac ultrasound examinations in the U.S.A. are evaluated after the data are obtained so the delay in analysis of the data for three-dimensional images may not be a serious time problem.

Applications

Multidimensional images of the anatomy and function of the cardiac structures have applications in education, surgical planning, structural measurements, and other areas where the current analysis of two-dimensional images may produce limited understanding of the volumetric data.

METHODS

Acquisition

The first step in multidimensional cardiac imaging is the acquisition of image data. This requires a transducer and associated scanner hardware for obtaining high-speed two-dimensional images. Typical scanners obtain 30 ultrasound images per second in the television format. These images represent ultrasonic backscatter in a plane. To get three-dimensional data the imaging plane must be swept through the tissue volume under investigation. The two-dimensional planes are digitized and stored to be interpolated later into a three-dimensional image in computer memory. The data must be gated in order to have a complete set of data for one physiologic state of the tissue.

Physiologic Gating. The heart is a dynamic organ which is beating at a rate of about one beat per second. In addition, the heart lies between the lungs which are varying in size and position at the rate of about twelve times per minute. Therefore, data acquired by the scanner must include the cardiac and respiratory state of the patient. Patients cannot hold their breath more than several seconds and seldom repeatedly at the same lung volume. The respiratory motion and the EKG must be recorded along with the ultrasonic images so that the data can be retrospectively gated. Retrospective gating means that the final set of images making up a image volume are selected from a large set of data so that only one respiratory and one cardiac state are represented. For instance, the images may be selected to represent diastole at end expiration. Thus, the entire set of digitized images is searched for those images obtained at end expiration and at diastole and this subset of images is used for constructing the final three-dimensional image.

Need for Position Control. Because the scanner must be swept over the volume of interest, the precise position of the planes must be known to reconstruct the volume in computer memory. This is done in several ways.

Free Motion—Some investigators have placed position sensing devices such as spark gaps or potentiometers on the handheld transducer to record, along with the digitized image, the position of the imaged plane in space.[6,7] The computer then uses this information to reconstruct the volume image from a set of randomly oriented planes. The advantage of this method is that the user can orient the transducer in any anatomic window such as the apical or parasternal views. The disadvantage is that some regions of the volume may be undersampled and not represented in the complete image.

One Degree of Freedom—To overcome the sampling problem of the free motion approaches several investigators have used methods in which only one degree of freedom is allowed by the mechanics of the scan system. This way sequential samples along the single degree of freedom can be better controlled (Fig. 1).

Figure 1. Schematic of "propeller scan", pullback scan and fan scan. Each method has advantages and disadvantages depending on the scanner and window position.

Rotation—An example of one degree of freedom imaging is the 'propeller scan'. This method uses a stepper motor to rotate the scan head around its central beam resulting in a sampled volume of conical shape. Only 180 degrees of rotation are required to completely sample the volume. We have reported such methods in which very good sampling has been obtained using 1.8 to 3.6 degree steps (100 to 50 planes).[4]

Pullback—A method which requires very simple computer interpolation routines is the pullback method. The transducer is translated in such a way as to obtain a sequence of parallel planes resulting in a wedge-shaped scanned volume. The disadvantage of this method is that the window along which the transducer is translated must be as long as the volume being imaged. Another disadvantage is that this method usually takes longer to perform.

Fan—A scan method that is especially conducive to transesophageal imaging is the fan method. The transducer is rotated about an axis orthogonal to the central beam of the transducer sector scan and at the apex of the sector. This results in a wedge-shaped segment representing the scanned volume. Transesophageal probes which image a plane parallel to the endoprobe can be rotated around their long axis to obtain this type of scan. Transthoracic probes can be rotated this way also.

Preprocessing and Analysis

The raw data represented by digitized ultrasound images have artifacts including misregistration due to tissue motion, and noise due to drop outs and spurious reflections etc. To produce a clean representation of a desired anatomic component of the heart or vessels the data must be reregistered, filtered, and the desired region must be isolated (segmented) from the remainder of the heart.

Interpolation. The data acquisition methods described previously result in two-dimensional images that are at angles to one another and are not in a cartesian coordinate system as required by computer imaging methods. Therefore a method of interpolating these planes through space onto a preferably cubic coordinate system must be implemented. From knowledge of the exact position in space of the plane of each digitized image, the interpolation program produces a three-dimensional volume image of the ultrasonic data acquired for a specific point in the cardiac and respiratory cycles.

Registration. Certain regions of the heart such as the valves move very quickly and therefore the image rate of 1/30 second is not quite fast enough to freeze the motion. Therefore, regions such as the valves are slightly misregistered in the sequence of images. These images can be reregistered using either automatic or manual methods. This is found to be necessary when imaging the atrial wall, aneurysms, and valves.

Segmentation

Usually the anatomic object of interest is combined in the images with other objects. This necessitates the separation of the object of interest from the remainder of the image. This process is called segmentation. Usually the process is performed manually by tracing the object in many cross sections, but automatic methods exist for some anatomic regions such as the ventricles.[8]

Rendering

The final goal of data acquisition, filtering, registration, and segmentation is to obtain a volumetric image that can be displayed in a comprehensible, useful manner. The process used to convert the three-dimensional data set into a comprehensible two-dimensional image that can be viewed on a computer monitor or in a published photograph is called rendering. The images to be shown in this paper are volume rendered.[9] This is done by calculating the gray scale value of each pixel on the two-dimensional display image as if one projected a light ray from a light source onto the object and back to the image plane. Specular and diffuse reflections are taken into account and, if required, opacity and color can also be accounted for. The resulting images appear as images of an anatomic object that has been extracted from the live beating heart of the examined patient.

EXAMPLES

Scatter Image

A projection of a three-dimensional distribution of scatter on a two-dimensional plane or shadowgram produces an image that might be called a three-dimensional scatter image (Fig. 2). These images are surprisingly difficult to comprehend unless they are viewed in stereo or are rotated to provide depth cues. It is for this reason that we have developed other methods of evaluating these volumes of ultrasound scatter data.

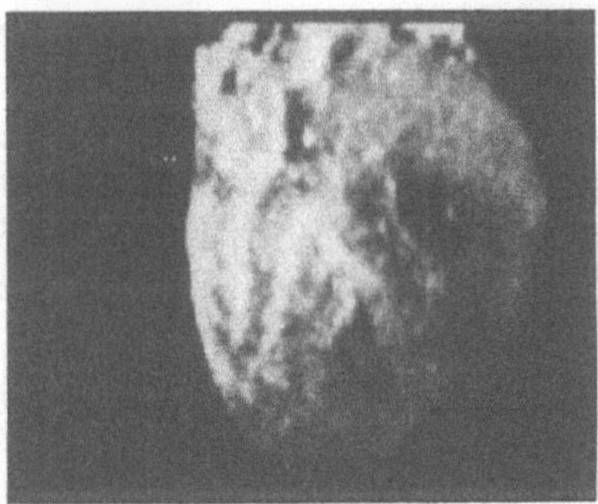

Figure 2. A two-dimensional image produced by projecting three-dimensional data onto a two-dimensional screen. Comprehension of these data is difficult because of the absence of surfaces. Dynamic rotation and stereo imaging do help somewhat.

Oblique Sections

An effective method of evaluating scatter data is to "slice" the data in two-dimensional planes through the volume and to view the scatter within these planes. Interactive movement of the plane is effective in evaluating the object if the position of the plane is clearly depicted in some effective graphical manner (Fig. 3).

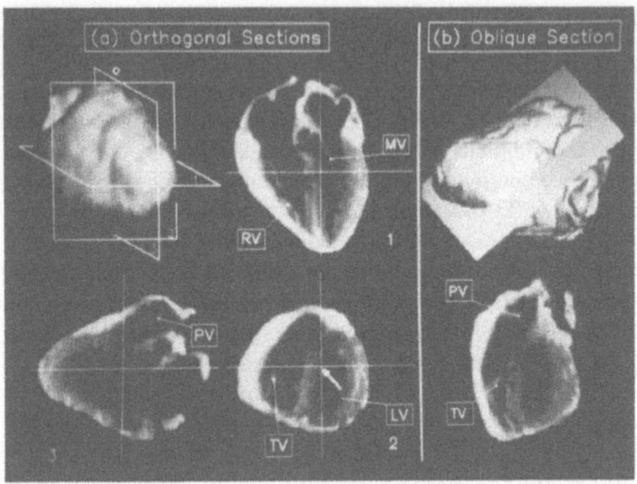

Figure 3. Both orthogonal (a) and oblique (b) planes can be passed through the data shown in Fig. 2 and the scatter within the plane shown in a two-dimensional image. Real-time interaction with the image greatly aids comprehension of the data. Images at angles that cannot be obtained using a scanner can be obtained using this method. Dynamic images can be obtained if the data are four-dimensional. RV = right ventricle, LV = left ventricle, TV = tricuspid valve, PV = pulmonary valve. [Reproduced with permission from M. Belohlavek and J. F. Greenleaf, *1993 McGraw-Hill Yearbook of Science and Technology*]

Surface Images

As mentioned before, images of three-dimensional distributions of scatter are not easily comprehended. Therefore, we have developed methods of displaying surface images representing anatomical and function components of the heart. These components can be displayed as a static image or as a dynamic sequence of images representing the motion of the viewed object through the cardiac or respiratory cycle. In this paper we can only display static images, however sequences of images can be obtained in a similar manner for cinematic display.

Atrial Septum. The base of the heart is the site for many abnormalities and the atrial septum accounts for a large fraction of the nonlethal abnormalities. Septal defects (Fig. 4) and aneurysms (Fig. 5) are examples of lesions that can be well depicted using the multidimensional visualization methods described in this paper.

Ventricle. The ventricles are more difficult to image using transesophageal probes because they are more distant from the esophagus than the base of the heart. However, we have imaged ventricles in vitro using a scanner and a water bath (Fig. 6), and are working on methods to depict the regional function of the ventricle using visualization methods.

Flow Image. An example of a functional image is that of blood flow jets escaping from the mitral valve. Data acquisition of Doppler shifts is complicated by the extreme dynamic nature of the flow and by the fact that the true velocity is not measured by Doppler. However the distribution of measured velocities can result in effective images (Fig. 7).

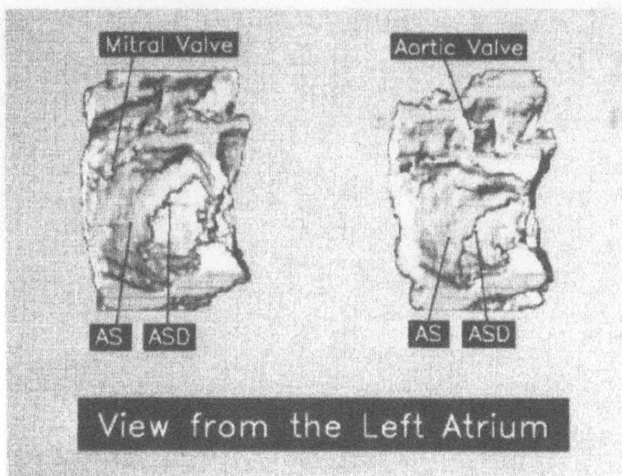

Figure 4   Surface shaded display of ultrasound data representing a septal defect as viewed from the left atrium   [Modified from M   Belohlavek, et al , *1991 Ultrasonics Symposium Proceedings*]

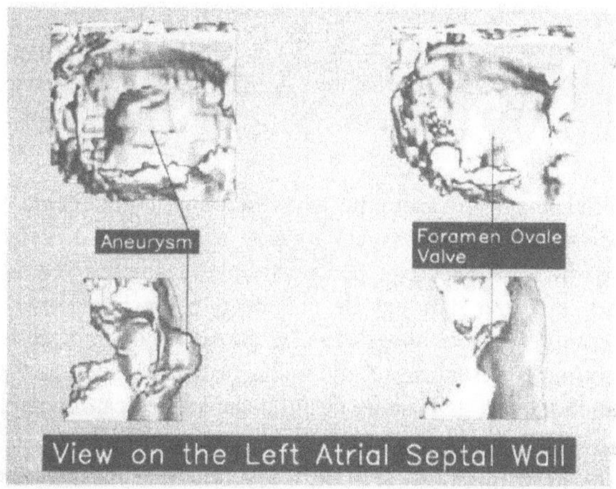

Figure 5   Surface shaded display of an atrial aneurysm   A sequence of these images through the heart cycle result in a cinematic depiction of the dynamics of the lesion

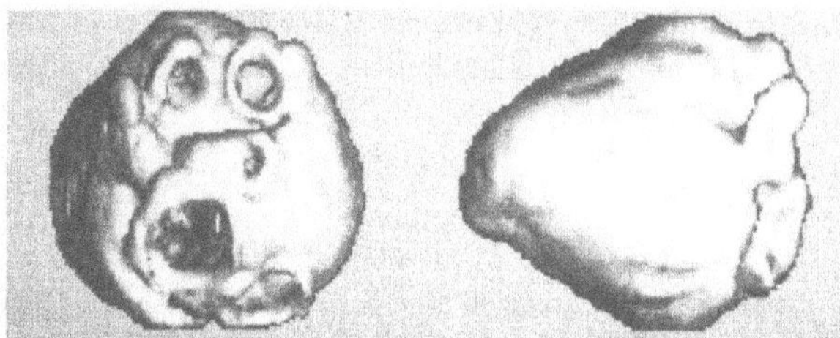

Figure 6   Shaded surface display of ultrasound data representing the entire heart and ventricles   Although obtained *in vitro* these images demonstrate that ultrasound data has the potential to present detailed images of the cardiac structures

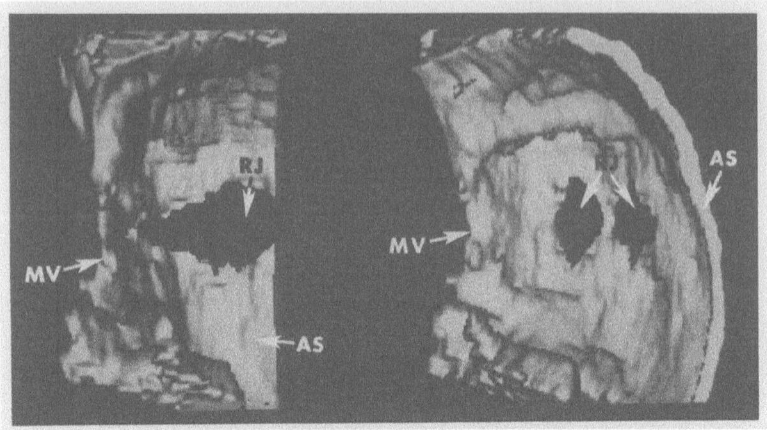

Figure 7. Shaded surface display of the mitral valve as viewed from two angles in the left atrium. Doppler jets representing blood leaking from the closed valve are shown as dark plumes. MV = mitral valve, AS = atrial septum, RJ = regurgitant jet.

DISCUSSION

Multidimensional images of anatomic and functional components of the heart can be effectively obtained from transesophageal images of the cardiac structures. Both static snapshots and dynamic cineloops can be constructed from the acquired data. Comprehension of volumetric scatter data can be enhanced using interactive oblique sectioning methods. Highly comprehensible images can be constructed from scatter data which have been shaded by computer visualization methods. Function such as Doppler flow in the chambers or disfunction such as leakage through the valves can be depicted effectively.

Accurate images of anatomic components of the heart require physiologic stationarity. This means that the heart must be cyclic in a reproducible manner, otherwise the gating required to obtain the data will result in an incoherent set of measurements. Minimization of motion of the transducer and accuracy of its known position for each slice are also requirements for high-quality data.

The extremely high-quality data that are consistently obtained from patients using transesophageal images imply that multidimensional visualization of anatomic and functional components of the cardiovascular system can be commonly obtained in the clinical laboratory.

ACKNOWLEDGMENTS

The authors appreciate the secretarial assistance of Elaine C. Quarve and graphic assistance of Christine A. Welch. This work was supported in part by grant HL 41046 from the National Institutes of Health.

REFERENCES

1. S.W. Smith, H. Pavy, Jr., and O. T. von Ramm, High-speed ultrasound volumetric imaging system—Part I: Transducer design and beam steering, *IEEE Trans. Ultrason., Ferroelec., Freq. Cont.* 38(2):100–108 (March, 1991).
2. O.T. von Ramm, S.W. Smith, and H.E. Pavy, High-speed ultrasound volumetric imaging system—Part II: Parallel processing and image display, *IEEE Trans. Ultrason., Ferroelec., Freq. Cont.* 38(2):109–115 (March, 1991).
3. J-y. Lu, R. Kinnick, J.F. Greenleaf, and C.M. Sehgal, Multi-electrical excitation of a transducer for ultrasonic imaging, *in*: "Acoustical Imaging," H. Lee and G. Wade, eds., Plenum Press, New York, vol. 18, pp. 511–519 (1991).
4. T. Kuroda, T.M. Kinter, J.B. Seward, H. Yanagi, and J.F. Greenleaf, Accuracy of three-dimensional volume measurement, *J. Amer. Soc. Echocardiogr.* 4(5):475–484 (Sept., 1991).
5. J.F. Greenleaf, M. Belohlavek, T.C. Gerber, D.A. Foley, and J.B. Seward, Part I: Introduction—Multidimensional visualization with ultrasound, *Mayo Clin. Proc.* (Submitted).
6. J.F. Brinkley, W.E. Moritz, and D.W. Baker, Ultrasonic three-dimensional imaging and volume from a series of arbitrary sector scans, *Ultrasound Med. Biol.* 4:317–327 (1978).
7. W.E. Moritz, A.S. Pearlman, D.H. McCabe, D.K. Medema, M.E. Ainsworth, and M.S. Boles, An ultrasonic technique for imaging the ventricle in three dimensions and calculating its volume, *IEEE Trans. Biomed. Eng.* 30:482–492 (1983).
8. W.E. Higgins, N. Chung, and E.L. Ritman, Extraction of left-ventricular chamber from 3D CT images of the heart, *IEEE Trans. Med. Imaging* 9:384–395 (1990).
9. R.A. Robb and C. Barillot, Interactive display and analysis of 3–D medical images. *IEEE Trans. Med. Imaging* 8:217–226 (1989).

ACOUSTICAL FRACTAL IMAGES APPLIED
TO MEDICAL IMAGING

Woon S. Gan and Cheong K. Gan

Acoustical Services Pte Ltd
29 Telok Ayer Street
Singapore 0104
Republic of Singapore

INTRODUCTION

Over the last few years, the development of the science of chaos have led to new insights and understanding of nonlinear dynamics. In this paper we study ultrasonic wave propagation in human tissue which is an inhomogeneous medium and causes nonlinearity and strong multiple scattering. Environments that produce strong multiple scattering by irregularities with a large length scale are most likely to cause chaotic propagation. This gives rise to fractal images. Two approaches will be used to obtain fractal images: the rigorous approach and the phenomenological or empirical approach.

RIGOROUS APPROACH

Two methods will be used: mapping technique and stochastic approach.

Mapping Technique

For imaging in human tissue, an inhomogeneous media, the inhomogeneous Helmholtz wave equation is used. The inhomogeneous Helmholtz wave equation is a partial differential equation and can be represented by the nonlinear oscillator equation perturbed by an array of impulses:

$$(\nabla^2 + k_0^2)u_s(\bar{r}) = -u(\bar{r})O(\bar{r})$$

$$= Af(\bar{r})\sum_{n=1}^{\infty}\delta(t-nT) \tag{1}$$

where $k_0 = \dfrac{2\pi}{\lambda}$ and λ=sound wavelength, $u_s(\bar{r})$ =scattered field, $u(\bar{r})$ =total field

$= u_0(\bar{r}) + u_s(\bar{r})$, $u_0(\bar{r})$ = incident field, $O(\bar{r})$ = object function $= k_0^2[n_0^2(\bar{r})-1]$, $n_0(\bar{r})=$

complex refractive index at position $\bar{r}$ and $f(\bar{r}) =$ arbitrary function of $\bar{r}$. This can be shown to be equivalent to the nonlinear oscillator equation:

$$\ddot{x} + \omega_0^2 x = Af(x) \sum_{n=1}^{\infty} \delta(t - nT) \tag{2}$$

where $\omega_0 =$ angular frequency.

The solution for Eq.(2) can be given in two-dimensional mapping form as:

$$\begin{bmatrix} x_{k+1} \\ \dot{x}_{k+1} \end{bmatrix} = \begin{bmatrix} \cos\omega_0 T & \omega_0^{-1} \sin\omega_0 T \\ -\omega_0 \sin\omega_0 T & \cos\omega_0 T \end{bmatrix} \begin{bmatrix} x_k \\ \dot{x}_k + Af(x_k) \end{bmatrix} \tag{3}$$

which gives the position and velocity just before the time (k+1)T in terms of the position and velocity just before the time kT.

Expressing the solution of Eq.(1) in mapping form, we have

$$\begin{bmatrix} u_{s_{k+1}}(\bar{r}) \\ \dot{u}_{s_{k+1}}(\bar{r}) \end{bmatrix} = \begin{bmatrix} \cos\omega_0 T & \omega_0^{-1} \sin\omega_0 T \\ -\omega_0 \sin\omega_0 T & \cos\omega_0 T \end{bmatrix} \begin{bmatrix} u_{s_k}(\bar{r}) \\ \dot{u}_{s_k}(\bar{r}) + Af(r_k) \end{bmatrix} \tag{4}$$

The $Af(\bar{r}_k)$ term can be expressed as human tissue parameters as follows:

$$Af(r_k) = O(\bar{r})u(\bar{r})$$

$$= O(\bar{r})[u_0(\bar{r}) + u_s(\bar{r})]$$

$$= k_0^2[n^2(\bar{r}) - 1][u_0(\bar{r}) + u_s(\bar{r})] \tag{5}$$

$$= k_0^2[\frac{c_0}{c(\bar{r})} - 1][u_0(\bar{r}) + u_s(\bar{r})]$$

where $c_0 =$ propagation velocity in the medium in which the object is immersed which is air in our case and $c(\bar{r}) = \dfrac{1}{\sqrt{\rho k}}$ where ρ and k are the local density and the complex compressibility at location $\bar{r}$ of the human tissue. Eq.(5) is a first order dependence on $u_s(\bar{r})$. A plot of Eq.(5), $u_s(\bar{r})$ versus $\dot{u}_s(\bar{r})$ by iteration will produce the fractal images.

Stochastic Approach

In a previous paper[1] , the stochastic approach is applied to sound propagation in random media. The statistical properties of the fluctuations of the wavefield can be characterized more completely by using correlation functions. The concept of chaos is useful to the statistical approach of sound propagation in random media. The chaotic behaviour of a system can be identified from its probability density function plot. Autocorrelation function and probability density function are related by:

$$C_\gamma = \frac{1}{N} \sum_{\beta=1}^{N} P_\beta e^{-i2\pi\beta(\gamma/N)} \tag{6}$$

$$P_\beta = \sum_{\gamma=1}^{N} C_\gamma e^{i2\pi\gamma(\beta/N)} \tag{7}$$

where C_γ = autocorrelation function and P_β = probability density function.

To identify chaotic behaviour of the system from the probability density function, it is necessary to make a map of P(x,t) versus P(x,t+τ) for constant x and τ where x is ordinate of phase space. In the case of chaotic behaviour, the numerical results indicate that the maps have a Canton set structure. For regular behaviour, the probability density function does not depend on time and in this case the map P(x,t) versus P(x,t+τ) consists of a single point. The map of P(x,t) versus P(x,t+τ) is called Poincaré section. Poincaré section is related to fractal image in the following way:

The Poincaré section is a section plane chosen to intersect the trajectories of a dynamical system in phase space and it will become the fractal image if the path followed by the point is expressed in fractal dimension.

PHENOMENOLOGICAL OR EMPIRICAL APPROACH

In the phenomenological or empirical approach, one assumes that chaos exist in the image, and the image is transformed into a fractal image by calculating the fractal dimension of each pixel over the whole image[2] .

One of the hallmarks of chaotic behaviour has been the manifestation of fractal geometry, particularly for strange attractors in dissipative systems. For a practical definition, one takes a strange attractor to be an attracting set with fractal dimension. One important characterization of a self-similar curve or point set is its fractal dimension. There are four different measures of fractal dimension: (a) Hausdorff dimension, (b) information dimension, (c) correlation dimension and (d) Lyapunov dimension. In this paper, the Lyapunov dimension will be used. The Lyapunov characteristic exponents have been made the basis of a measure of fractal dimension. For two exponents, the Lyapunov dimension is defined as

$$d_L = 1 - \frac{L_1}{L_2} \tag{8}$$

where L=Lyapunov characteristic exponent.
For the case of m dimensions,

$$d_L = k - \frac{\sum_{i=1}^{k} L_i}{L_{k+1}} \tag{9}$$

where k labels the last L_k for which

$$L_1 + L_2 + ... + L_k \geq 0$$

This approach is useful for fractal edge enhancement of image. The Lyapunov characteristic exponent is used to get the average fractal dimension along all directions. The result is a transformed image in which each pixel value is a fractal dimension. In this paper, ultrasonic liver images are used. The fractal dimension value of each pixel is calculated which is done by calculating the fractal dimension of a 7x7 pixel block centred on this pixel which is then transformed over the entire image surface. The fractal dimension of each pixel is then converted into an intensith gray level. Finally a fractal dimension distribution transformed image is obtained. The fractal dimension distribution transformation image can enhance and detect edges of the original image. The reason why the fractal dimension can detect and enhance the edges is that the fractal dimension has lower values on the boundary of different textures and tends to be constant over uniform texture regions.

CONCLUSION

The fractal images give insight into the nonlinear property of the human tissues. They are useful for studying image texture. Fractal feature extraction can be used to classify different image patterns and thus enhance edges in medical images.

REFERENCES

1   W S Gan. "Application of Chaos to Sound Propagation in Random Media", paper presented at *the 19th International Symposium on Acoustical Imaging,* Bochum, Germany, April 1991.

2   A Pentland, "Fractal-based Description of Natural Scenes", *IEEE Trans. Pattern Anal. Machine Intell*, Vol. PAMI-6, 666(1984).

NONLINEAR PARAMETER IMAGING

Dong Zhang, Xiufen Gong, and Shigong Ye

Institute of Acoustics
Nanjing University
Nanjing, 210008
P.R.China

I . INTRODUCTION

The use of ultrasound to detect and display interfaces between tissues is increasing in medical imaging. The propagation of ultrasound through biological media has been modeled as linear process until recently. But at biological frequencies and intensities, the nonlinear acoustic effect is very important[1].

The nonlinear parameter B/A is a measure of the nonlinearity of the pressure-density relation for a medium. It describes the distortion of a finite amplitude wave as it propagates through the medium. The nonlinear parameter also describes the dependence of ultrasonic velocity on pressure and may be one of the significant parameters for ultrasonic tissue characterization, since this parameter has dynamical information of the media.

Until recently, very few efforts have been made to use nonlinear parameter in medical imaging. Ichida et al proposed a method using phase shifts in high-frequency acoustic probe by an impulsive acoustic pump[2]. Cain reported a reflection-mode imaging technique which uses a pump wave propagating in the opposite direction to a high-frequency probe wave[3]. Nakagawa et al proposed a nonlinear parameter imaging CT using difference-frequency wave generated by parametric acoustic array[4]. When two high-frequency waves in the same direction are projected as primary waves, the secondary wave with low frequency is generated by nonlinear interaction of primary sound waves. This wave depends on nonlinear parameter $\beta=1+B/2A$ and on attenuation of the media, so nonlinear tomography can be reconstructed by attenuation compensation. Instead of the difference-frequency wave, Nakagawa extend the analysis to nonlinear parameter imaging involving the second-harmonic wave[5].

In this paper, we propose a method based on second-harmonic wave to image the nonlinear parameter by using comparative, insert-substitution method[1]. In order to study nonlinear parameter B/A imaging, an experimental system was established in our Lab. based on IBM-PC 386 computer. Some simulation and experimental images on this system are presented in this paper.

II. THEORY

1.The Nonlinear Parameter B/A

The nonlinearity of the pressure-density relation of a medium is expressed quantitatively in terms of a parameter called B/A, which is the ratio of the coefficients of quadratic and linear terms of a Taylor series used to express the pressure-density equation of the state of a medium. The Taylor series is:

$$P-P_0=(\rho-\rho_0)\left(\frac{\partial p}{\partial \rho}\right)_{0,s}+\left[\frac{(\rho-\rho_0)^2}{2}\right]\left(\frac{\partial^2 p}{\partial \rho^2}\right)_{0,s}+\cdots \tag{1}$$

where p and p_0 are instantaneous and hydrostatic pressure, respectively, ρ and ρ_0 are the respective instantaneous and equilibrium densities, and the subscripts 0 and s on partial derivatives represent equilibrium density and constant entropy, respectively.

Defining

$$A=\rho_0\left(\frac{\partial p}{\partial \rho}\right)_{0,s}=\rho_0 c_0^2$$
$$B=\rho_0^2\left(\frac{\partial^2 p}{\partial \rho^2}\right)_{0,s} \tag{2}$$

the nonlinear parameter B/A can be expressed by the following relation[6].

$$B/A=2\rho_0 c_0\left(\frac{\partial C}{\partial p}\right)_{0,s} \tag{3}$$

2.2 Propagation of A Finite-Amplitude Sound Wave

When a finite amplitude wave is transmitted in a medium, it will be distorted along the propagation path. The following Burgers' equation can describe this nonlinear propagation for lossless medium:

$$\frac{\partial W}{\partial \sigma}-W\frac{\partial W}{\partial Y}=0 \tag{4}$$

Where $W=u/u_0, \sigma=x/1, y=\omega\tau, \tau=t-x/c_0, u$ and u_0 are the instantaneous and peak values of particle velocity, respectively. 1 is discontinuity distance. $1=c_0^2/(\beta u_0\omega)$, $\beta=1+B/2A$.

If primary wave at the source is described by:

$$p_1 = p_0 \sin(\omega \tau) \tag{5}$$

then from Eq(4), the pressure amplitude of second harmonics at distance x=L can be expressed by

$$p_2(L) = \frac{\omega_2}{4} p_1^2(0) \int_0^L \beta_1(x)\, dx \tag{6}$$

Considering the sound attenuation, we obtain the following differential equation describing the second harmonic sound wave[8].

$$\frac{dp_2}{dx} = \frac{\beta(x)\, \omega_2 p_1^2(x)}{2\rho_0(x)\, c_0^3(x)} - \alpha_2(x)\, p_2(x) \tag{7}$$

Where $p_1(x)$ is the sound source pressure amplitude; ρ_0, c_0 are the density and the sound velocity ,respectively; $\omega_2 = 2\omega_1$ and $\alpha_2(x)$ is the attenuation coefficient of second harmonic frequency, and $p_1(x)$ is described by:

$$p_1(x) = p_1(0) \exp\left[-\int_0^x \alpha_1(x)\, dx\right] \tag{8}$$

The second-harmonic wave sound pressure amplitude after propagating a distance L is given by Eq(9)

$$p_2(L) = \frac{\omega_2}{4} p_1^2(L) \int_0^L \beta_1(x) \exp\left[\int_x^L (2\alpha_1(x) - \alpha_2(x))\, dx\right] dx \tag{9}$$

Where

$$\beta_1(x) = \frac{\beta(x)}{\rho_0(x)\, c_0^3(x)} \tag{10}$$

α_1 is the attenuation coefficient of primary wave frequency and α_2 is the attenuation coefficient at the second-harmonic frequency. Since the attenuation of biological soft tissue is proportional to frequency, we can assume $2\alpha_1 = \alpha_2$, the Eq(9) is therefore simplified.

$$p_2(L) = \frac{\omega_2}{4} p_1^2(L) \int_0^L \beta_1(x)\, dx \tag{11}$$

Eq(11) shows that $p_2(L)$ is a line integral of the nonlinear parameter. The primary wave amplitude $p_1(L)$ can be measured simultaneously with $p_2(L)$ so that the nonlinear parameter image can be reconstructed by using CT method. Nakagawa et al successfully did it in 1986[5].

2.3 Comparative, Insert-Substitution Imaging

For avoiding to measure the absolute values of sound pressure of primary wave and second-harmonic wave at distance L, the finite amplitude insert-substitution method is used here. Eq(9) can be expressed by Eq(12), too.

$$p_2(L) = \frac{\omega_2}{4} p_1^2(0) \int_0^L \beta_1(x) \exp\left[\int_0^x (-2\alpha_1(x)\,dx - \int_x^L \alpha_2(x)\,dx\right]dx \quad (12)$$

Defining the Modified Correction Matrix $C(x,y)$

$$C(x,y) = \left[\frac{1}{2\pi}\int_0^{2\pi} \exp\left[\int_0^{l_0} -2\alpha_1(x,y)\,dl - \int_{l_0}^L \alpha_2(x,y)\,dl\right]d\theta\right] \quad (13)$$

Now if the sample is absent in the distilled water, we can obtain

$$p_{20}(L) = \frac{\omega_2 L}{4} p_1^2(0)\beta_{10} \quad (14)$$

where $p_{20}(L)$ is the sound pressure amplitude of the second-harmonic wave at distance L and β_{i0} is the β_i of the distilled water. By Eq(12) and Eq(14) we get

$$\frac{p_2(L)}{p_{20}(L)} = \frac{1}{\beta_{10}L}\int_0^L \beta'_i\,dl \quad (15)$$

where $\beta_i' = \beta_i(x,y)C(x,y)$.

We can obtain β_i' image by using CT method. To obtain a good image from the measurement of the second-harmonic wave, we must compensate for attenuation.

2.4 CT Method

There are many tomographic computation ways. Here, we use convolution algorithm[9] to get nonlinear parameter image. The way consists of four parts: (1) getting projection data; (2) constructing filter function; (3) convolving and interpolating; (4) imaging on VGA monitor.

In order to get satisfied image, three filter functions are used in this paper, they are B-R function, R-L function, and S-L function[9]. Of all images, the linear interpolating way is used[9].

III. EXPERIMENTAL SYSTEM

Fig.1 is the block diagram of the experimental system, which consists of four parts: (1) acoustic system; (2) mechanical scanning system; (3) signal processing of received second-harmonic wave; (4) tomographic computation using convolution algorithm for reconstructing nonlinear parameter images. All parts are controlled by PC-386 computer.

Signal generator 1 is the primary wave source (2MHz), the signal from it mixed with a pulse from generator 2 by mixer 3. The transmitting transducer S is a PZT-ceramic disk with resonance frequency 2MHz. This transducer is driven by broadband power amplifier 4. The resonance frequency of receiver R is 4MHz sensitized to the second-harmonic frequency signal. The selective amplifier 5 and PHILIP PM3315 digital oscilloscope are used to detect and measure the amplitude of second-harmonic frequency signal. The computer 10 controls the digital oscilloscope and motor 7 by use of IEEE-488 interface 9 and parallel port 8, respectively. 11 represent the mechanical scanning system.

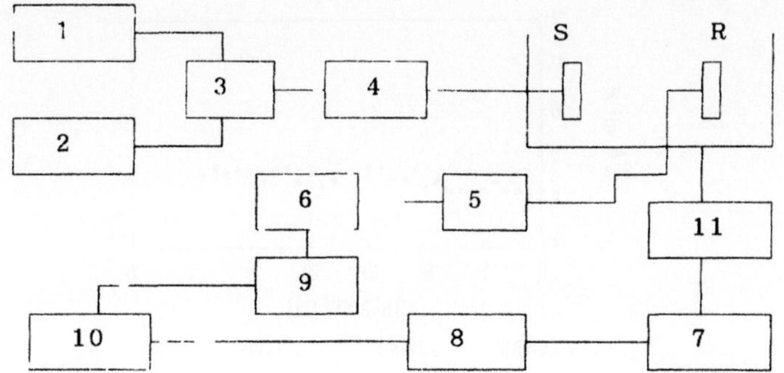

Figure 1. The block diagram of experimental system

IV. EXPERIMENTAL RESULTS

In this paper, we study one dimensional and two dimensional imaging. 1.Some fluids and animal soft tissues such as water, ethylene glycol and the fresh porcine liver, tongue, heart are conducted for one dimensional imaging, several reconstructed images are obtained and demonstrated. Fig.2, Fig.3 and Fig.4 show the results of one dimensional imaging.

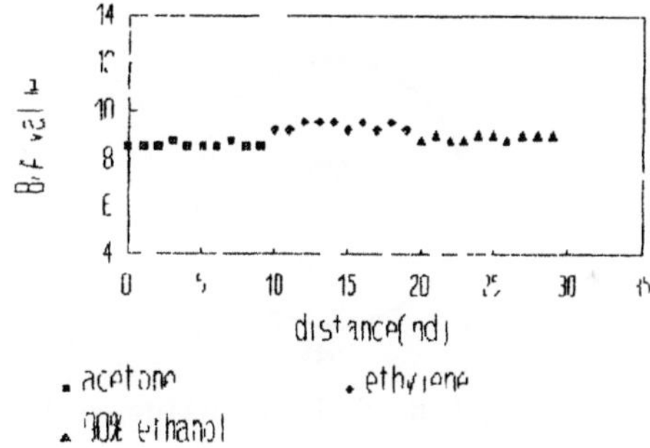

Figure 2. One dimensional B/A images for some fluids.

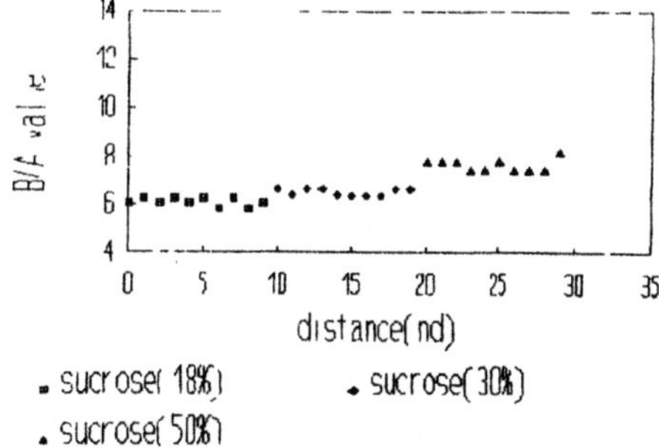

Figure 3. One dimensional B/A images for aqueous solution of sucrose with different concentrations

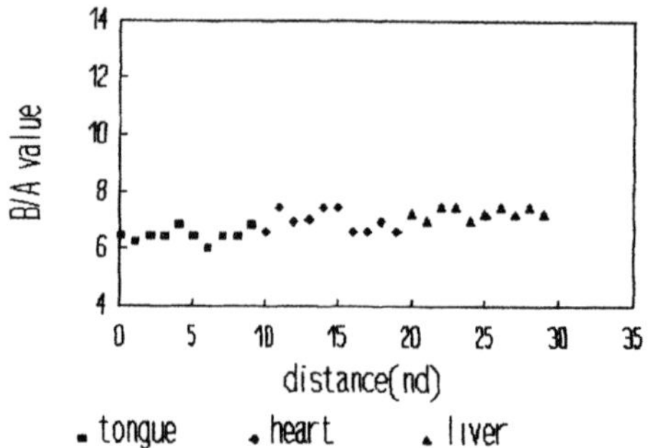

Figure 4. One dimensional B/A images for animal soft tissues

2.In our experiment, because only fluids are conducted to two dimensional imaging, the Modified Correction Matrix C(x,y) can be ignored. Our work consists of two parts (1) simulation study; (2) experimental study.

In the simulation study, we build a model showed in Fig.5, Fig.6 shows the nonlinear parameter images using different filter functions.

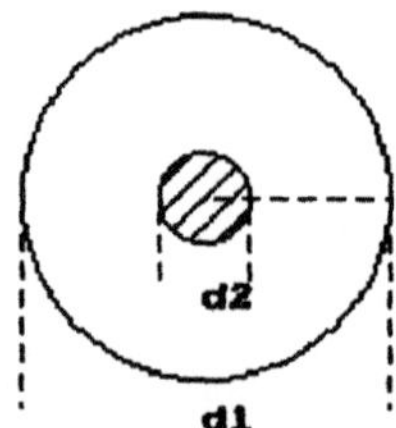

Figure 5. A sample model. d_1=8cm, d_2=2cm, inside--ethanol, outside--water

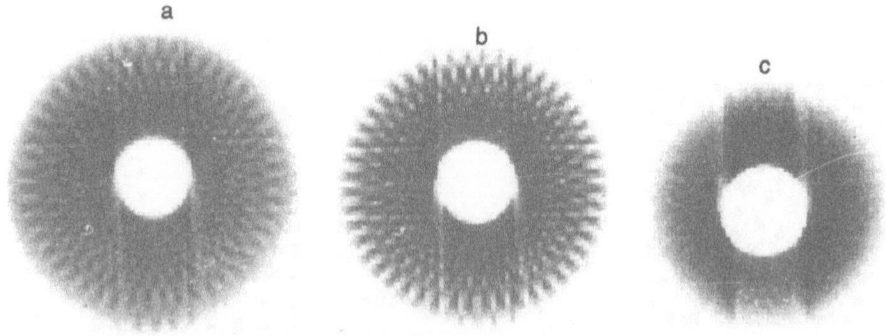

Figure 6. Pictures of the model B/A imaging. (a) B-R function (b) R-L function (c) S-L function

Fig.7 shows the experimental sample. The nonlinear parameter images of the experimental sample are shown in Fig.8. These images show the effectiveness of the new CT system.

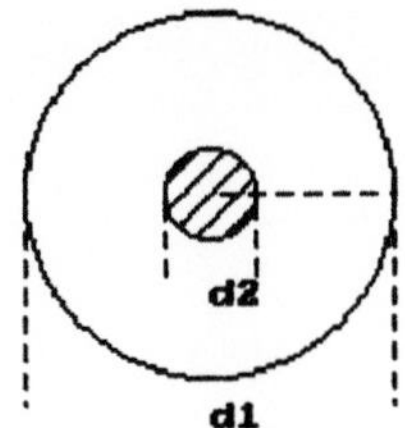

Figure 7. Experimental sample. d_1=8cm, d_2=2.5cm, inside--ethanol, outside--water

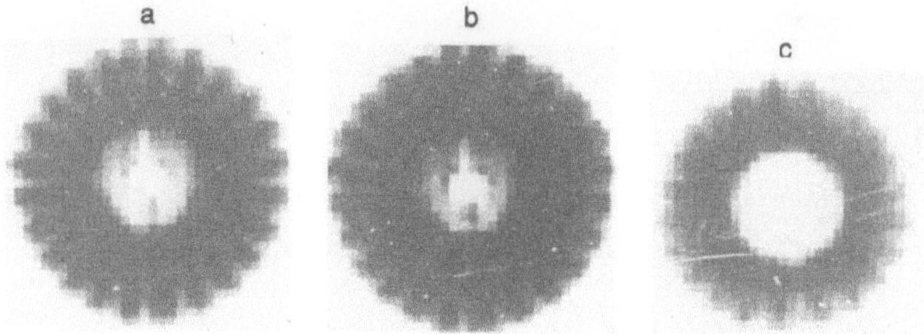

Figure 8. Pictures of experimental sample. (a) B-R function (b) R-L function (c) S-L function.

V.REFERENCES

1. E.L. Carstensen, W.K.Law, N.D.Mckay, J.G.Muir,"Demonstration of nonlinear acoustic effects at biological frequencies and intensities", Ultrasound Med. Biol. 6;359-368; 1980
2. N.Ichida, T.Sato, M.Linzer, "Imaging the nonlinear ultrasonic parameter of a medium", Ultrasonic Imaging 5; 295-299; 1983
3. C.A.Cain, J.Acoust.Soc.Am. Bo(1), pp.28-32, 1986
4. Y.Nakagawa et al, Proc. IEEE 1984 Ultrasonic Symposium, pp 637-676, 1984
5. Y.Nakagawa, W.Hou, A.Cai, N.Arnold and G.Wade, "Nonlinear parameter imaging with finite amplitude sound waves", Ultrasonics Symposium, 1986
6. R.T.Beyer, "Parameter of nonlinear in fluids" J.Acoust.Soc.Am. 32, 719-721, 1960

7. X.F.Gong, Z.M.Zhu, T.Shi and J.H.Huang, "Determination of acoustic nonlinearity parameter B/A in biological medical using FAIS and ITD methods", J.Acoust.Soc.Am., 86;1(1989)
8. F.Dunn et al, Proc, IEEE Ultrasonics Symposium, pp.527-532, 1981
9. G.T.Herman, "Image reconstruction from projection" Spingerverlay Berlin Heidelberg, New York, 1979

THREE-DIMENSIONAL (3D) ACQUISITION AND DISPLAY OF BEATING HEART ECHO IMAGES

Riccardo Pini,[1] Magda Costi,[2] Leonardo Masotti,[2] Kevin L. Novins,[3] Donald P. Greenberg,[3] Barbara Greppi,[4] Marino Carofolini,[4] and Richard B. Devereux[5]

[1]Institute of Gerontology and Geriatrics-University of Firenze, Firenze, Italy
[2]Department of Electronic Engineering-University of Firenze, Firenze, Italy
[3]Program of Computer Graphics-Cornell University, Ithaca, NY, U.S.A.
[4]ESAOTE Biomedica, Firenze, Italy
[5]Division of Cardiology, The New York Hospital-Cornell Medical Center, New York, NY, U.S.A.

INTRODUCTION

Conventional two-dimensional (2D) echocardiography provides high-resolution images of the beating heart that allow accurate qualitative and quantitative estimation of structure and function[1]. However, 2D echocardiography visualizes only one tomographic plane at a time and then it requires the operator to reconstruct in his mind a composite view of the heart using multiple 2D images acquired from different acoustic windows. In fact, especially in the presence of regional abnormalities as induced by ischemic heart disease, a single tomographic plane often does not allow a correct diagnosis. Furthermore, we have shown that standard 2D echocardiographic imaging planes that are assumed to be orthogonal to each other commonly are not[2], undermining the accuracy of both qualitative diagnosis and of quantitative measurements that rely on assumed geometric models.

The three-dimensional (3D) reconstruction of the heart from multiple 2D images represents a solution to this technical limitation of conventional echocardiography, preserving the same spatial and temporal resolution as the original 2D images. Thus, to obtain 3D images of the beating heart, we developed a new 3D echocardiographic system that, with computer assistance, allows the acquisition and display of 3D echocardiograms.

METHODS

We developed a new 3D echocardiographic system based on a 3.5 MHz dynamically focused anular array transducer[3] rotating around its central axis. The ultrasonic beam profile exhibited a regular shape along the entire depth used to visualize the heart with a lateral resolution of 1.3 mm at -6dB   This transducer allows the acquisition of 51 standard fan shaped 2D echocardiograms at 3.6 degree increments of rotation. Since the rotation axis is in the center of the 2D fan, a transducer rotation of 180 degrees permits the 3D visualization of a solid cone encompassing the heart through the chest wall. Comparing the 0 degree (first 2D scanning plane) and the 180 degree (51st scanning plane) images provides an immediate check of the rotation axis's stability during the recording since, if the transducer maintains the same relative position to the heart during the examination, the 0 and 180 degree images are mirror images.   The transducer is connected to an echocardiographic system (ESAOTE Biomedica, Firenze, Italy) modified by adding an electronic circuit that, with computer assistance (8086 personal computer), controls the acquisition process and provides the synchronization with the ECG signal to reconstruct the beating heart (Figure 1). An entire cardiac cycle was recorded from each transducer

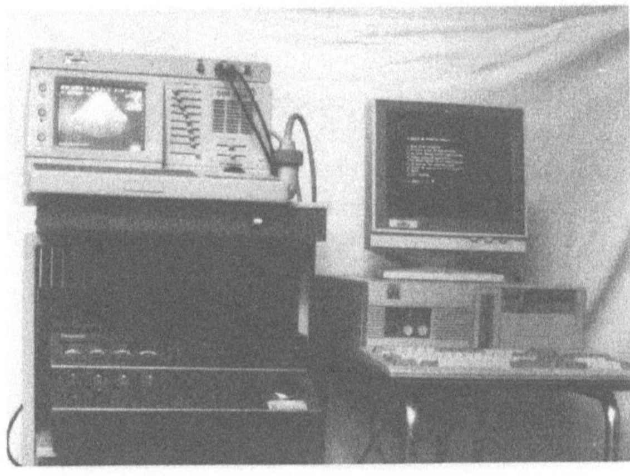

Figure 1. Three-dimensional echocardiograph (left) connected to the personal computer (right) which controls the acquisition process.

position while the subsequent cycle was utilized to rotate the transducer; thus, the system acquired images over 100 cardiac cycles, or 60 to 100 seconds in normal sinus rhythm. To reduce artifacts derived from heart rate variability and arrhythmias, the software repeats the cardiac cycle acquisition if the actual RR interval differs from the mean value by more then a percentage threshold selected by the operator. The program can also be used, at the option of the operator, to rotate the scanning plane without the check on heart rate stability for patients in an irregular rhythm such as atrial fibrillation or sinus arrhythmia. The echocardiograph displays the 2D images on the screen in real time and the same images are stored on video-tape at TV frame rate (25 frames/sec).

The videotaped 2D echocardiographic images were digitized using a 80386 personal computer (Compaq Computer Corporation, Houston, Texas, U.S.A.) with a 768x576 pixel resolution and 256 gray level frame grabber (GTI/Freeland Medical Division, Louisville, Colorado, U.S.A.). Only the first 50 2D images (scanning plane 0 to 49) were digitized because the image at the 50th increment of rotation (180 degrees) was used exclusively to check the rotation axis stability; up to 1250 images (25 frames throughout a cardiac cycle in each of the 50 scanning planes) were used to reconstruct the 3D images of the beating heart.

For each frame in the entire cardiac cycle, the 50 images acquired in cylindrical coordinates were processed by an algorithm for linear interpolation to reconstruct a 3D cone of information in cartesian coordinates. Each of the up to 25 successive 3D images were stored in matrices of 256x256x576 pixels and 256 gray levels. From the 3D matrices stored in the computer, 2D echocardiographic images in any plane at specified times in the cardiac cycle, or throughout the cardiac cycle, can be derived and visualized.

To create perspective projection of the 3D matrices, a computer work station based system (Hewlett-Packard 9000/825 with 50 MB of main memory and 1 GB of disk storage, Fort Collins, Colorado, U.S.A.) was developed. The algorithm used to create the perspective images[4] is based on ray-tracing technique[5] adapted for use in visualizing 3D scalar fields; 3D spatial coordinates are calculated for each pixel, and a light source and camera position (or viewer's location) are specified. A simulation of the light propagation in the environment is then executed, revealing how the scene appears from the camera's point of view[6]. To assign the material properties (i.e. reflectiveness, transparency) to each element in the 3D matrix, a threshold value to screen out low intensity echoes was selected in order to visualize a single isovalue surface in the dataset. Using isovalues, a crisp surface may be extracted, at the expense of filtering out some of the information present in the raw data. In order to give the physician some interactive control over the viewing of the data, a full rotation sequence is computed; these pre-computed images are then displayed at video rates with user control of speed and direction of rotation.

The accuracy of the 3D reconstruction was tested with a standard ultrasound phantom (RMI, model 412A, Middleton, Wisconsin, U.S.A.) and balloons and surgical gloves filled with water and immersed in a fish tank.

To test the ability to reconstruct precisely details of complex anatomical structures, we visualized 4 excised animal hearts immersed in a fish tank filled with water.

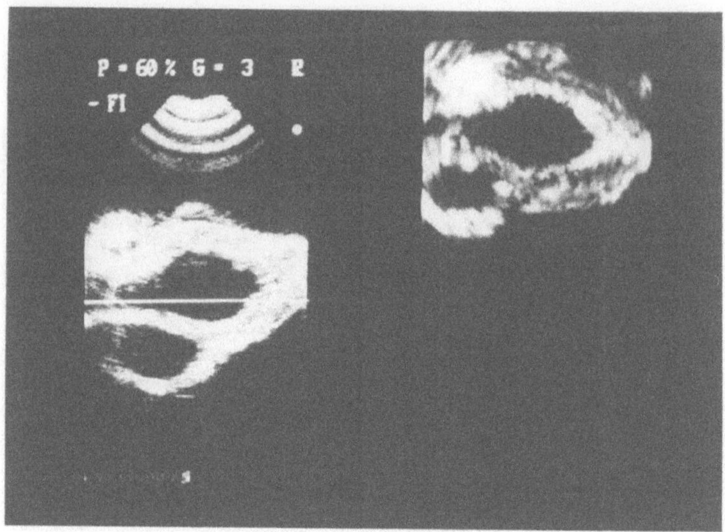

Figure 2. Original 2D echocardiogram of an excised lamb heart immersed in a fish tank (left) with an horizontal line indicating the plane selected at the level of the left ventricle to derive an image perpendicular to the ultrasound beam (right) from the 3D matrix. The reconstructed tomographic image exhibits good definition of the left ventricular endocardium and similar thickness of the left ventricular wall to that in the original image.

To verify the clinical applicability of our system, we studied 40 subjects referred to the Echocardiographic Laboratory at The New York Hospital-Cornell Medical Center.

RESULTS

Two-dimensional images derived from the 3D matrix demonstrated that wires and tubes contained in the ultrasound phantom were accurately reconstructed at any depth in the visualized solid cone. These structures were well visualized in reconstructed 2D tomographic slices of planes perpendicular to the ultrasound beam, which could not be imaged by any conventional 2D ultrasound method. As shown in Figures 2 through 5 , 2D images derived from the 3D matrices exhibit the same spatial resolution as standard echocardiograms both in studies performed with excised animal hearts as well as in patients.

The quantitative validation of the volumes derived from the 3D matrices was obtained by visualizing balloons and surgical gloves filled with 28 to 307 ml of water and immersed in a fish tank. There was an excellent correlation between the measured and the true volumes (r=0.995, p<0.001) with a standard error of estimate of 8.7 ml[7].

In our clinical study, the system allowed the acquisition of a technically adequate parasternal rotation in 35 subjects (87.5%) but in only 27 patients (67.5%) were adequate apical rotations obtained[2].

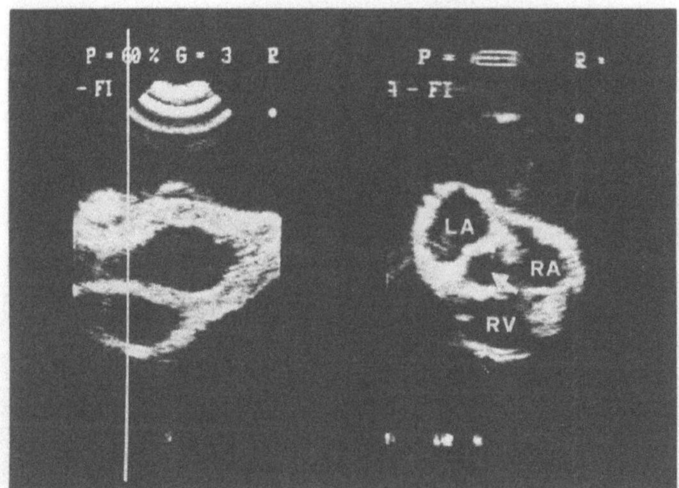

Figure 3. Original 2D echocardiogram of an excised lamb heart immersed in a fish tank (left) with a vertical line indicating the plane selected at the level of the base of the heart to derive a short-axis image (right) from the 3D matrix. The reconstructed tomographic image visualizes the left atrium (LA), the right atrium (RA), the right ventricle (RV), and the aorta (arrow). Of note, the left and right atria are not present in the original image.

DISCUSSION

To perform 3D reconstruction of the beating heart by ultrasound, multiple 2D images with known spatial orientation must be acquired. To define the transducer position, different transthoracic echocardiographic systems have been proposed based on either a

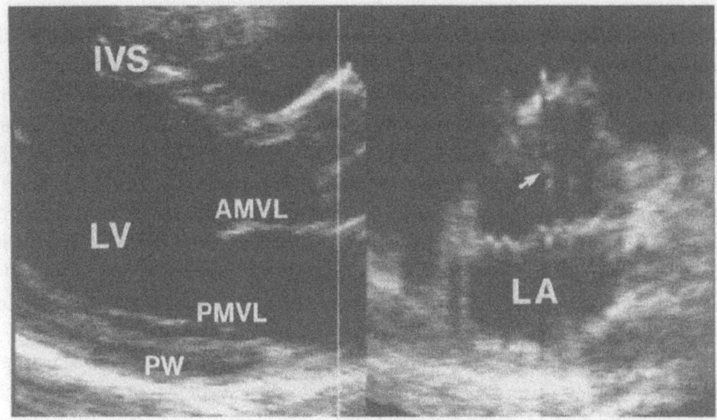

Figure 4. Original 2D echocardiogram of a normal subject acquired from the parasternal long-axis view at end-diastole (left) with a vertical line indicating the plane selected at the level of the aortic valve to derive a short-axis image (right) from the 3D matrix. The tomographic image exhibits good resolution of the three aortic valve cusps at their point of opposition (arrow) as well as the left atrium (LA). LV=left ventricle; IVS=interventricular septum; PW=left ventricular posterior wall; AMVL=anterior mitral valve leaflet; PMVL=posterior mitral valve leaflet.

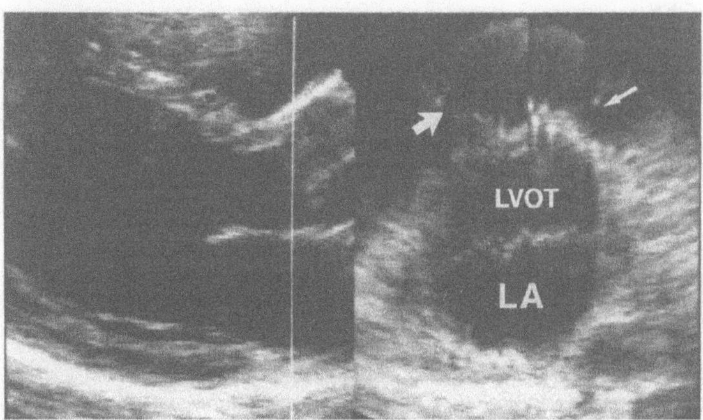

Figure 5. Original 2D echocardiogram of a normal subject acquired from the parasternal long-axis view at end-diastole (left) with a vertical line indicating the plane selected to derive a short-axis image (right) from the 3D matrix. The tomographic image visualizes the left ventricular outflow tract (LVOT) and left atrium (LA), the tricuspid valve (thick arrow) and the pulmonic valve (thin arrow).

mechanical arm with position sensors[8,9] or an acoustic ranging device with multiple fixed microphones and spark gasps affixed to a freely movable ultrasound transducer[10-12]. These systems have demonstrated the possibility of obtaining accurate measurements of the volume of heart chambers but they have limited clinical applicability because they do not permit the acquisition of a sufficient number of 2D images for realistic echocardiographic 3D reconstruction. Recently, the in vitro accuracy of the 3D reconstruction of the heart using a rotating transducer similar to ours has been demonstrated both with transthoracic[13] and with transesophageal[14] probes.

Our system, based on a 2D ultrasound transducer rotating around its central axis, overcomes the limitations of the previous methods and allows realistic echocardiographic tomography of the heart. In fact, because the dynamically focused annular array transducer has an uniform lateral resolution along the entire depth used to visualize the heart[3] and the 50 2D images are separated by angles narrower than the lateral resolution of the transducer, from the 3D matrices we can derive 2D images with the same spatial resolution as the original 2D echocardiograms. Volume measurements derived from the 3D matrices reconstructed with our system exhibited excellent correlation with the true volumes even with test-objects with an irregular shape that could not be described with a simple geometric model. Thus, this system may allow accurate estimation of cardiac chamber sizes and myocardial mass in patients with distorted ventricular shape. Finally, data from our preliminary clinical trial demonstrated the our system allows the acquisition of technically adequate studies in a majority of patients thus suggesting that 3D echocardiography could have wide clinical applicability.

In conclusion, we developed a new system for 3D echocardiography which allows 3D reconstruction of the beating heart with the same spatial and temporal resolution of the standard 2D images without cumbersome external reference systems, expensive or immobile equipment, or ionizing radiation exposure.

REFERENCES

1. D.C. Wallerson, and R.B. Devereux, Reproducibility of quantitative echocardiography: factors affecting variability of imaging and Doppler measurements, *Echocardiography* 3:219 (1986).
2. A.S. Katz, D.C. Wallerson, R. Pini, and R.B. Devereux, Visually determined long-axis and short-axis parasternal and four-chamber and two-chamber apical views do not represent paired orthogonal projections, *J. Am. Coll. Cardiol.* 13:94A (1990).
3. R. Pini, L. Ferrucci, M. DiBari, B. Greppi, M. Cerofolini, L. Masotti, and R.B. Devereux, Two-dimensional echocardiographic imaging: in-vitro comparison of conventional and dynamically focused annular array transducers, *Ultrasound in Med. and Biol.* 13:643 (1987).
4. M. Levoy, Display of surfaces from volume data, *IEEE Computer Graphics and Applications* 8:29 (1988).
5. T. Whitted, An improved illumination model for shaped display, *Communications of the ACM* 23:343 (1980).
6. D.P. Greenberg, Light reflection models for computer graphics, *Science* 244:166 (1989).
7. G.A. Mensah , R. Pini, E. Monnini, L. Masotti, K.L. Novins, D.P. Greenberg, B. Greppi, M. Cerofolini, and R.B. Devereux, Three dimensional echocardiographic reconstruction: experimental validation of volume measurements, *J. Am. Coll. Cardiol.* 17:291A (1991).
8. E.A. Geiser, L.G. Christie, D.A. Conetta, R. Conti, and G.S. Grossman, A mechanical arm for spatial registration of two-dimensional echocardiographic sections, *Cathet. Cardiovasc. Diagn.* 8:89 (1982).
9. J.S. Raichlen, S.S. Trivedi, G.T. Herman, M.G.J. Sutton, and N. Reichek, Dynamic three-dimensional reconstruction of the left ventricle from two-dimensional echocardiograms, *J. Am. Coll. Cardiol.* 8:364 (1986).
10. D.L. King, D.L. Jr King, and M. Y. Shao, Three-dimensional spatial registration and interactive display of position and orientation of real-time ultrasound images, *J. Ultrasound Med.* 9:525 (1990).
11. R.A. Levine, M.D. Handshumacher, A.J. Sanfilippo, A.A. Hagege, P. Harrigan, J.E. Warshall, and A.E. Weyman, Three-dimensional echocardiographic reconstruction of the mitral valve, with implications for the diagnosis of mitral valve prolapse, *Circulation* 80:589 (1989).
12. W.E. Moritz, A.S. Pearlman, D.H. McCabe, D.K. Medema, M.E. Ainsworth, and M.S. Boles, An ultrasonic technique for imaging the ventricle in three-dimensions and calculating its volume, *IEEE Trans. Biomed. Eng.* 30:482 (1983).
13. H.A. McCann, J.C. Sharp, T.M. Kinter, C.N. McEwan, C. Barillot, and J.F. Greenleaf, Multidimensional ultrasonic imaging for cardiology, *Proc. IEEE* 76:1063 (1988).

14. T. Kuroda, T.M. Kinter, J.B. Seward, H. Yanagi, and J.F. Greenleaf, Accuracy of three-dimensional volume measurement using biplane transesophageal echocardiographic probe: in vitro experiment, *J. Am. Soc. Echocardiogr.* 4:475 (1991).

AN INTRAVASCULAR ULTRASONIC IMAGING TECHNIQUE FOR MEASUREMENT OF ELASTIC PROPERTIES OF THE ARTERY

Katsuyuki Yamamoto, Masahiro Shikutani,
Toshiyuki Amano, and Eiji Okamoto

Division of Biomedical Engineering
Faculty of Engineering
Hokkaido University
Sapporo, 060 Japan

INTRODUCTION

In recent years, intravascular ultrasonic imaging has become important as this technique provides cross-sectional images of the arterial wall. Several in vitro studies[1-3] demonstrate good correlation for the lumen or the wall thickness between ultrasound images and histological sections. This unique technique has also been applied to clinical observations of the atherosclerotic plaque.[4-6] Thus, the intravascular ultrasonic imaging has been well placed as a new modality allowing to characterize the morphological features of the blood vessels. In addition to the morphological data, the intravascular ultrasonic imaging has the potential of obtaining elastic properties of the artery. The present study aims to find the feasibility of this new application of the technique.

Since the elastic properties are reflected in a pulsatile motion of the artery, the arterial elasticity can be assessed with the pulsatile change of the diameter in response to the pulsation of blood pressure. Based on this principle, we first developed a noninvasive technique for measuring the arterial elasticity using ultrasound and demonstrated the usefulness of this technique in clinical applications.[7,8] In a later study,[9] we also developed an intravascular transducer, which was superior to the noninvasive one with respect to accuracy, because it incorporated a manometer for simultaneous blood pressure measurement.

The objectives of the present study were to develope a prototype of intravascular ultrasonic imaging for the measurement of the arterial elasticity and to evaluate the feasibility of this technique.

ELASTICITY OF THE ARTERY

Many kinds of elasticities have been proposed in the field of blood vessel mechanics. Among them, a pressure strain elastic modulus E_p has widely been used for the evaluation of the elastic properties of the artery. E_p is defined by the following equation.

$$E_p = \Delta P/(\Delta D/D) \tag{1}$$

where ΔP, ΔD and D are pulse pressure, pulsatile change of the arterial diameter and the mean diameter, respectively. Volume elasticity E_v also has been used, which is similar to E_p. It is defined by the volumetric change of the lumen, instead of the diameter change, as the following equation shows.

$$E_v = \Delta P/(\Delta V/V) \tag{2}$$

We may also define E_v by employing the area S of the lumen.

$$E_v = \Delta P/(\Delta S/S) \tag{3}$$

Assuming the roundness of the arterial lumen, we obtain the following approximate relationship.

$$E_p = 2E_v \tag{4}$$

In the present study, E_p is calculated from Eqs. (3) and (4), where the luminal area is measured by means of the intravascular ultrasound imaging.

MEASUREMENT SYSTEM

Data Acquisition for Intravascular Imaging

The measurement system developed in the present study is composed of an intravascular probe, a pulser and receiver unit, an A/D converter, a personal computer, and a catheter–tip manometer (Fig. 1). As shown in Fig. 2 (A), the intravascular probe is about 80 cm in length and 2.1 mm in diameter, except for the tip portion which is 2.8 mm. Fig. 2 (B) shows the tip portion of the probe, in which a single PZT transducer (1 by 1.5 mm) is mounted and connected to a flexible shaft for mechanical scanning. The frame rate, which is variable, was usually fixed at 30 frames per second in the present study. The system works in the following manner.

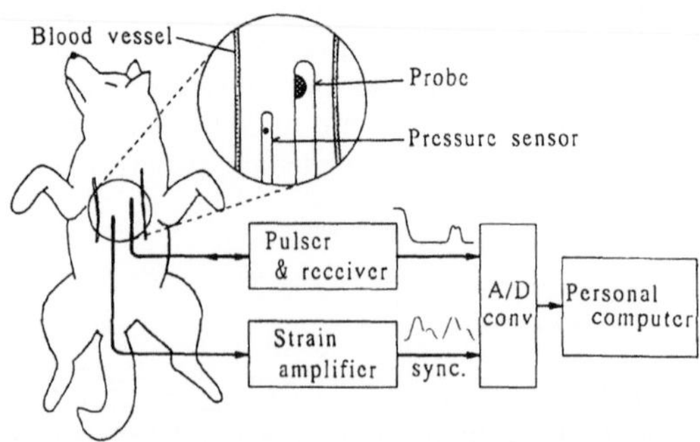

Figure 1. A block diagram of the system and the experimental setup.

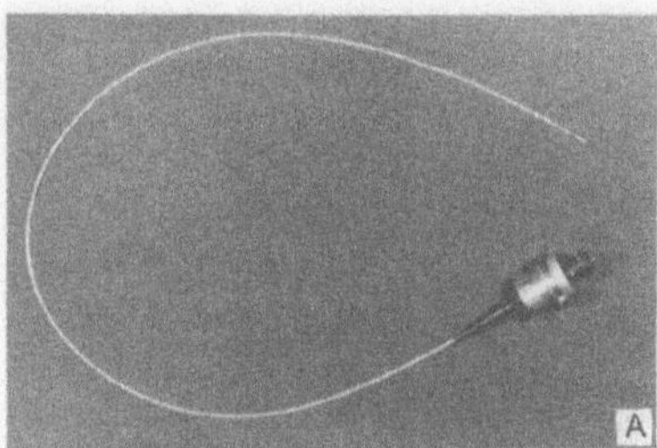 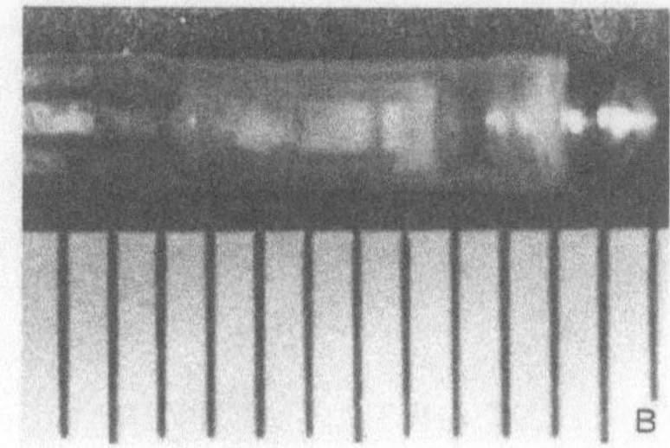

Figure 2. An intravascular ultrasound probe (A) and the tip portion of the probe (B). The probe is 80 cm in length and 2.1 mm in diameter, except the tip of 2.8–mm diameter. A 1 by 1.5–mm element is mounted within the tip and mechanically rotated by a flexible shaft.

The probe emits 256 pulses during a rotation and receives the echo signals followed by logarithmic amplification. Amplified and rectified echo signals are transferred to the computer via the A/D converter at the sampling frequency of 25 MHz. The computer processes the stored data to construct the image and to calculate the elasticity. The present system is not capable of real time imaging, but is designed to gain the echo data, from which only two frames are constructed, synchronously with the blood pressure waveform. When triggered by the pressure waveform at the peak–systolic and the end–diastolic phase, the computer starts the A/D conversion. After that, the freeze images at these phases are obtained.

Calculation of the Area of the Arterial Lumen

Figure 3 shows how to calculate the area of the lumen from the intravascular image of the artery. Prior to the calculation, the contour of the inner wall of the artery was detected by the following procedure. The echo signal corresponding to the inner wall is searched for, in an axial direction, from the beginning of the echo data on each beam. The inner wall is detected by the derivative value of the echo signal which exceeds a certain threshold level. In order to prevent the derivation from being effected by noise, the original image is smoothed by 2–D filtering, prior to the derivation. When the contour line is not complete, because the echo signal from the wall is of an insufficient intensity, the data is interpolated so to perfect the circle.

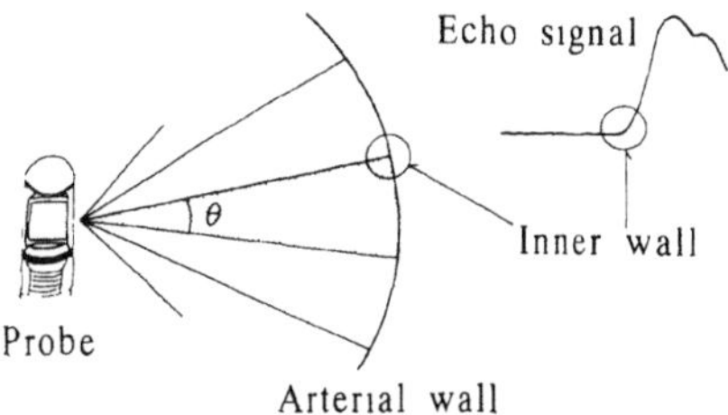

Figure 3. Detection of the arterial wall and the calculation of the luminal area. A inner wall was detected as the rising part of echo signals (right panel).

The area of the lumen S is calculated with Eq. (5).

$$S = (\theta/2) \, \Sigma \, r_i \cdot r_{i+1} \tag{5}$$

where θ is the angle made by two beams adjacent and r is the distance from the center of the probe to the inner wall.

ASSESSMENT OF THE MEASUREMENT ACCURACY

Measurement of the Luminal Area

The accuracy of the system was evaluated by measuring the luminal area of acrylic pipes whose diameters are 6–24 mm. Care was taken so as to position the probe at the center of the lumen and to make the ultrasound beam perpendicular against the wall. The luminal area was calculated by the procedure described in the previous section.

The measured area was compared with the control values which were calculated from the internal diameter measured with a caliper. As shown in Fig. 4, a good correlation, with less than 3% error, was obtained.

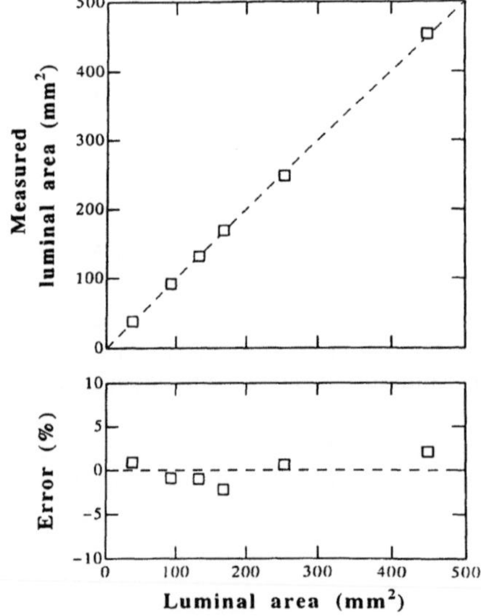

Figure 4. Measurements of luminal area in acrylic pipes. Internal diameters of the pipes are 6–24 mm. Measurements from the present method are compared with the area calculated from the diameters.

Measurement of the Luminal Area Change

The measurement accuracy of the total system is subject mostly to that of the pulsatile change in the luminal area, because the pulsatile change is as small as 20% at most. There-fore, the experiment shown in Fig. 5 was carried out in order to evaluate the accuracy. In the experiment we measured small changes of in the luminal area of a latex tube. The latex

present area[7] and from the diameter[8]. The 1.9% mm² mis-change in the area was measured with error of less than 1.2%.

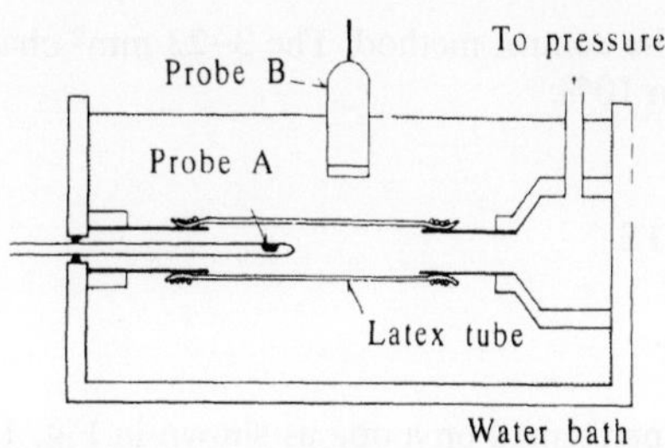

Figure 5. Experimental setup for evaluating the measurement accuracy of luminal area changes. A latex tube is mounted to holders, from the end of which an intravascular probe A is inserted, to provide the cross–sectional image. As a control, probe B for measuring the diameter is placed above the tube. Internal pressure of the tube changes its diameter.

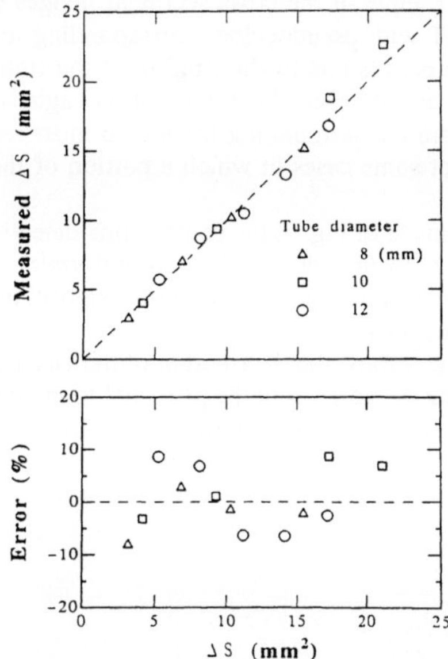

Figure 6 Measurements of the luminal area change in latex tubes. Measurements from the present method are compared with the area change which were calculated from the diameter measured with the probe B in Fig. 5.

tube was mounted to holders in a water bath. The inside of the tube was pressurized by water to cause certain changes of the luminal area. The diameter of the tube used was 8–12 mm. The luminal area was changed by 5–30%.

For comparison, the diameter of the tubes was measured by means of a 7.5–MHz single element transducer placed above the tube. This measurement is based on the conventional method of observing the echoes from the anterior and posterior walls on an oscilloscope and reading the distance between the walls by using the caliper function of the scope. Assuming the roundness of the tube, we calculated the area of the lumen from the diameter.

Figure 6 shows a good correlation between the measurements obtained from the present system and from the control method. The 3–23 mm² change in the area was measured with error of less than 10%.

IN VIVO EXPERIMENTS

Methods

The experiment was performed on a dog as shown in Fig. 1. The mongrel dog weighing 10 kg was anesthetized with an intravenous infusion of pentobarbital. The intravascular transducer and the catheter–tip manometer were inserted into the aorta through the right and the left femoral artery, respectively. Changing the measured sites along the aorta, we obtained the cross sectional images and the pressure waveforms to calculate the elasticity.

Results

Figure 7 shows an example of the cross sectional images of the aorta. The original image is displayed on CRT with pseudocolors corresponding to the echo intensities. The strong signal within the lumen is due to the ringing of the transducer. The image of the inner wall is somewhat clear; however, the outer wall is vaguely identified because of the existence of the echoes from the surrounding tissues. In most regions of the aorta, similar images were gained, except some cases in which a portion of the wall image was incomplete.

When applied to the image of Fig. 7, the contour line detection algorithm results in the white curve shown in Fig. 8. Even if there is lack of uniformity of the detected wall owing to variations of the echo intensities, the wall detection algorithm provides the entire circle necessary for calculating the area.

The plotted data in Fig. 9 show the distribution of the elasticity E_p along the aorta. The E_p increases gradually along the aorta from the proximal to the distal site. In Fig. 9, another distribution of E_p obtained by our previous study[9] is also illustrated by the shaded area. The plots of the present study distribute in the same region as the previous ones. However, the plots of the present study disperse to some extent even at the same measurement site of a dog.

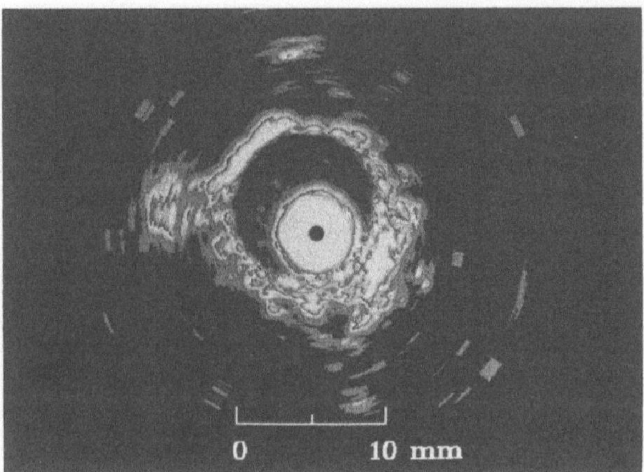

Figure 7 A cross–sectional image of the aorta. The original image is displayed on CRT by pseudocolors corresponding to echo intensities.

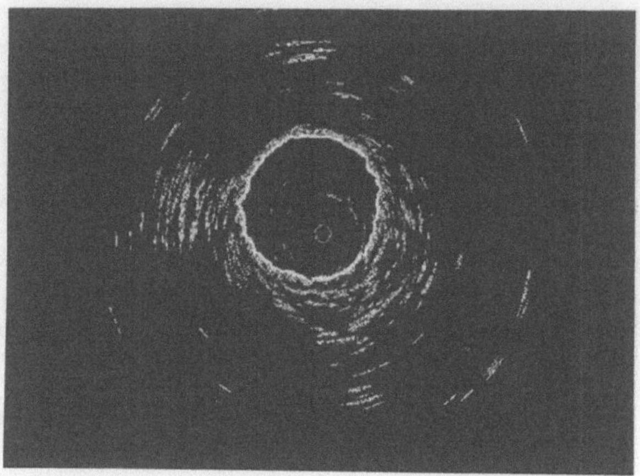

Figure 8 A detected inner wall contour displayed in the highest brightness. The cross–sectional image was obtained from contour detection filtering on the original image in Fig. 7.

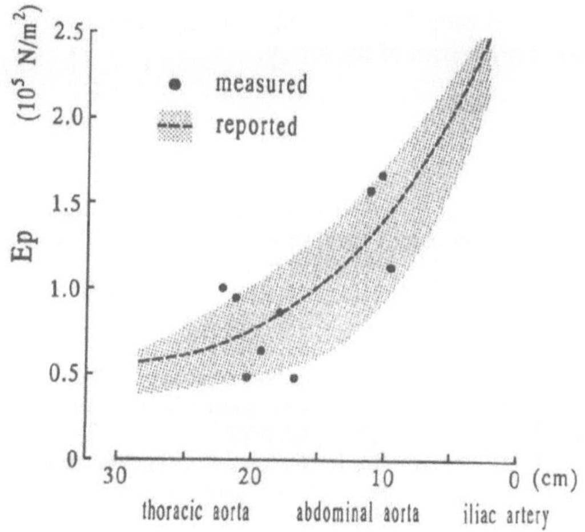

Figure 9 Distribution of the elasticity E_p along the aorta. Plots indicate the measurements from the present study, whereas the measurement distribution from our previous study are shown by the shaded area.

DISCUSSIONS

In this study, we presented the elasticity measurement using the intravascular ultrasound imaging and demonstrated the feasibility of this technique. The measurement is based on the detection of the small pulsatile motion of the arterial wall. The pulsatile change of the luminal area of the artery is about 10% at the thoracic aorta and is much less at other peripheral and the sclerotic arteries. Therefore, the measurement accuracy depends on whether the intravascular ultrasound imaging can detect a change as small as the pulsatile change.

To determine the accuracy of the intravascular ultrasound measurement of the luminal area, the model experiments using the acrylic pipes and the latex tubes were performed. The results showed that the present system has sufficient accuracy of measuring the luminal area and its small changes, as shown in Figs. 4 and 6. However, the elasticity measurements from the in vivo experiments varied widely even at the same site of the aorta. It is partly because the intravascular image of the present system using a 15–MHz frequency does not have high resolution. It is estimated that the prototype probe for this study has the 0.5-mm axial and the 1-mm lateral resolutions, which are not enough to visualize clearly the arterial cross sections. A higher frequency intravascular transducer[2,5] of 30–40 MHz is now available; therefore, the error due to poor resolution may be greatly reduced. In addition, a more advanced contour detection algorithm such as smooth curve fitting would also much improve the accuracy of the measurement.

There may be other factors causing the image distortion; e.g., a low frame rate against a rapid motion of the artery, the mechanical vibration of the probe tip, or the movement of the tip by blood flow. These may have a great influence on the measurement accuracy. It is now under investigation.

The purpose of this study was to demonstrate that the intravascular ultrasound imaging with the simultaneous blood pressure measurement also had the potential of measuring the elastic properties of the artery. Although there may be technical difficulties such as incorporating a micromanometer into the probe for practical use, the results from this study suggest that this technique is promising not only for the morphological studies, but also for characterizing the elastic properties of the artery.

REFERENCES

1. J. M. Tobis, J. A. Mallery, J. Gessert, J. Griffith, D. Mahon, M. Bessen, M. Moriuchi, L. McLeay, M. McRae, and W. L. Henry, Intravascular ultrasound cross–sectional arterial imaging before and after balloon angioplasty in vitro, *Circulation*, 80:873–882(1989).

2. E. J. Gussenhoven, C. E. Essed, C. T. Lancee, F. Mastik, P. Frietman, F. C. van Egmond, J. Reiber, H. Bosch, H. van Urk, J. Roelandt, N. Bom, Arterial wall characteristics determined by intravascular ultrasound imaging: An in vitro study, *J. Am. Coll. Cardiol.* 14:947–952(1989).

3. J. A. Mallery, J. M. Tobis, J. Griffith, J. Gessert, M. McRae, O. Moussabeck, M. Besen, M. Moriuchi, and W. L. Henry, Assessment of normal and atherosclerotic arterial wall thickness with an intravascular ultrasound imaging catheter, *Am. Heart J.*, 119:1392–1400(1990).

4. J. M. Isner, K. Rosenfield, D. W. Losordo, L. Rose, R. E. Langevin, S. Razvi, and B. D. Kosowsky, Combination balloon–ultrasound imaging catheter percutaneous transluminal angioplasty, *Circulation*, 84:739–754(1991).

5. F. G. St. Goar, F. J. Pinto, E. L. Alderman, H. A. Valantine, J. S. Schroeder, S. Gao, E. B. Stinson, and R. L. Popp, Intracoronary ultrasound in cardiac transplant recipients, *Circulation*, 85:979–987(1992).

6. J. Honey, D. J. Mahon, A. Jain, C. J. White, S. R. Ramee, J. B. Wallis, A. Al–Zarka, and J. M. Tobis, Morphological effects of coronary balloon angioplasty in vivo assessed by intravascular ultrasound imaging, *Circulation*, 85:1012–1025(1992).

7. T. Imura, K. Yamamoto, K. Kanamori, T. Mikami, and H. Yasuda, Non–invasive ultrasonic measurement of the elastic properties of the human abdominal aorta, *Cardiovasc. Res.* 20:208–214(1986).

8. T. Imura, K. Yamamoto, T. Satoh, T. Mikami, and H. Yasuda, Arteriosclerotic change in the human abdominal aorta in vivo in relation to coronary heart disease and risk factors, *Atherosclerosis*, 73:149–155(1988).

9. K. Yamamoto, M. Shikutani, T. Imura, K. Itoh, T. Sakamoto, and T. Mikami, Measurement of arterial elasticity using an intravascular ultrasonic transducer, *Med. Biol. Eng. Comput.*, 29 Suppl.:484(1991)

A NEW SPECTRUM-AMPLITUDE COMBINED IMAGING
TECHNIQUE FOR INTRAOCULAR TUMOR CHARACTERIZATION

S.Rocchi, A.Fort, C.Manfredi, L.Masotti

Department of Electronic Engineering- University of Florence-
Via Santa Marta 3, 50139 Firenze, Italy

INTRODUCTION

The interest in evaluating the histological features of uveal lesions by non-invasive techniques raises mainly from two issues.

There is an obvious interest in obtaining a differential diagnosis for the intraocular pathologies that may resemble intraocular tumors. Moreover, due to the continuous therapeutic advances achieved by conservative treatments, the ophthalmologist uses them with increasing frequency. Nevertheless this therapy cannot be supported by the information from the histological analysis on the pathology specimens. Therefore the non-invasive evaluation of the intraocular tumor histological features during and after the conservative treatments becomes of the utmost importance. In this context the ultrasonic investigation techniques play a major role.

It is well known that the ultrasonic echoes backscattered by the biological tissues convey information on the average features of the investigated microstructure. A pathologic region behaves with respect to the ultrasonic investigating waves as a random scattering medium, consisting of a distribution of particles with different size and shape immersed in an homogeneous fluid. In general the dimension of the particles are well below the wavelength used in ophthalmologic analysis. Therefore the tissue microstructure information is embedded in random components due to the stochastic nature of biological media. To obtain a characterization of the tissue, a statistical analysis of the backscattered signal is required. A common approach is based on the estimation of the Power Spectral Density (PSD) of the ultrasonic echoes, since it is related to the average tissue properties such as scatterer density, size shape and strength [1] [2].

However the signals backscattered by a large region of the tumoral tissue can't in general be considered stationary, since the tumoral tissue is usually formed by zones with different nature: necrosis, small vessels, calcification regions etc. The backscattered echoes must be considered stationary only in little regions (with diameter of less than 1 mm). Each zone must be characterized by different PSDs. This signal processing problem can be tackled as a time or space (a propagating phenomenon is handled) frequency characterization where an intrinsic trade-off between the time and frequency resolution must be met.

In the specific application the tissue spatial characterization is based on the PSD centroid value related to the scatterer mean size: this spectral parameter represents the spatial development of the backscattered echo frequency content.

The spectral PSD estimators which are commonly used, are based on an averaging process of the FFT (Fast Fourier Transform) of windowed signals, but they are not effective for space-frequency characterization, due to the poor frequency resolution that can be achieved and to the residual randomness that affects the spectral parameter estimate when the space resolution must increase.
In the present paper the frequency content of the ultrasonic signals backscattered by small regions of the investigated tissue is extracted by means of spectral estimators based on the autoregressive (AR) modelling technique [3],[4]. In particular, low-order AR models are used as they smooth the random peaks and pits of the backscattered signal spectrum while being highly sensitive to the spectral shift trend. Moreover this technique is intrinsically insensitive to the windowing side-lobe effects; as a consequence, the centroid frequency estimate is meaningful also when the number of samples of the radiofrequency signal is drastically reduced, allowing a local characterization of inhomogeneous media.

The PSD centroid is displayed in a color coded version to form spectral images that present the spatial development of the backscattered signal frequency content. The technique aims to emphasize local inhomogeneities, in terms of particle mean size. The combination of these spectral images with the conventional B-Mode display, offers a complete and integrated diagnostic tool. It allows the local characterization of the investigated region in terms of both amplitude and frequency content of the corresponding echoes. This approach has been tested on a gel suspension of calibrated latex spheres and then applied to the backscattered signals from in-vitro eye specimens.

ULTRASONIC SCATTERING FROM PATHOLOGICAL TISSUES

A tumoral region behaves as a discrete random scattering medium. A simplified physical description of the ultrasonic signals coming from the pathologic tissues can be obtained by considering the case of scatterer concentration weak enough to ignore multiple scattering effects, so that the ultrasonic echoes are formed by the superimposition of single scatterer contributions.

A further simplification can be achieved when examining isotropic media in the focal zone of a narrow ultrasonic beam so that plane-wave propagation can be assumed. This is the case in medical analysis when a weakly focussed ultrasonic transducer is used and the region of interest lies in the focal zone.
Under these assumptions, considering an impinging plane wave, the backscattered pressure p_s, is a stochastic process that can be written as follows [5]:

$$p_s(r) = e^{jK \cdot R} \sum_i \Phi_i(2\,K) e^{jK \cdot r_i}$$

$$(1)$$

In Eq. 1, f is the frequency; K is the wave vector of the impinging field, $\Phi_i(2K)$ is the complex amplitude of the backscattered field for the i-th scatterer which depends on its shape and size, z_i is its axial position.

Assuming the validity of the Rayleigh-Born approximation, $\Phi_i(2k)$ can be derived for a fluid spherical scatterer that is explicated by the following function of frequency [6]:

$$\Phi(2K) = \mu \left[\frac{\sin(2Ka)}{8K} - a \frac{\cos(2Ka)}{4} \right] \qquad (2)$$

where μ is a parameter related to the particle acoustic impedance and a is the particle radius.

It is evident, also from this heavily simplified physical model that the desired information concerning the scatterer size contained in the backscattered ultrasonic signals, is embedded in random components due to the tissue microstructure, represented in this model by the scatterer random positions.

The function $\Phi(2\,K)$, in Eq. 1, is the only term dependent on the scatterer size and, the average size of the scatterers can be obtained by evaluating the stochastic process PSD, thus eliminating random phase shift effects.

For a tissue characterized by uniform spatial distribution the PSD results:

$$PSD \propto |\Phi(2K)|^2$$

$$(3)$$

and the PSD centroid results a good spectral parameter able to represent the scatterer mean size.

THE POWER SPECTRAL DENSITY ESTIMATION OF BACKSCATTERED ULTRASONIC SIGNALS AND SPECTRAL MAPS PRODUCTION

Usually for ultrasound tissue characterization, the PSD estimation of the backscattered ultrasonic signal, is derived by the periodogram technique via FFT [2]. Despite the wide range of its possible applications, this approach bears two main limitations: an increase in the PSD estimation quality decreases the frequency resolution; and the implicit windowing of the data causes the 'leakage effect' in the spectral domain that may obscure and distort weak signal spectral responses.

The above mentioned limitations are particularly troublesome when analyzing short data records, as it is the case for a local tissue characterization. Therefore an alternative spectral estimation procedure was used, based on the fitting of the measured data to an assumed model. The PSD estimation, in the context of modelling, can be summarized in three steps: the first is the choice of a model describing the process under study; the second is the estimation of the assumed model parameters, by minimizing a cost functional representing the quality of the known data fitting on the model. The third step is the substitution of the estimated parameters into the theoretical PSD implied by the model.

With this approach, more realistic assumptions are made on the measured process outside the considered interval, with respect to the implicit assumptions made by the periodogram approach that sets all the unknown samples (outside the considered interval) equal to zero. The modelling procedure extrapolates the data behavior outside the considered interval, so that the data windowing effect is intrinsically eliminated. Moreover in the last step a function describing the PSD is found, thereby removing the theoretical limits for the frequency resolution. For these reasons the improvement over the conventional periodogram approach can be quite noticeable

especially for short data records.
In this paper the backscattered echoes are fitted on an AR model of order p:

$$x(n) = -\sum_{k=1}^{p} a_k x(n-k) + u(n) \tag{4}$$

where x(n) represents the measured data, u(n) is the unknown input sequence and a_k are the coefficients of the model to be determined starting from the known data. The choice of an AR model is particularly convenient since the estimation of the a_k coefficients results in solving a set of linear equations.
The power spectral density (AR PSD) of an AR process is:

$$P_{AR}(f) = \frac{T\sigma_u^2}{\left| 1 + \sum_{k=1}^{p} a_k e^{-j2\pi f kT} \right|^2} \tag{5}$$

where T is the signal sampling period and σ_u^2 is the variance of u(n), assumed to be a zero mean, white noise process.
The filter order p is not known, and can be selected either through an analysis of the data or through a priori considerations related to the specific application. In the present work, the order p was fixed to a small value, namely p=2 or 3, the main interest being in emphasizing the signal high energy frequency, regardless of the spectrum details [8].
To evaluate the AR parameters the Modified Covariance Method (MCM) was used [7] which, independently of the selected AR order, exploits the information of the whole data sequence. In fact, this method leads to a linear system of order p+1, whose coefficient matrix has entries which are combinations of all the available data. By expressing it as sums and products of Toeplitz and Hankel matrices, a fast computational solution of the MCM exists, only requiring $Np+6p^2$ operations and N+4p data storage, N being the number of data samples. This aspect makes the algorithm particularly suitable for real-time applications in the biomedical field.
The spectral maps based on the processing technique presented above are produced by considering a set of RF backscattered signals obtained by linear or sectorial scan. Each echo line is divided into segments corresponding to small tissue regions. Second and third order AR models are used to estimate the PSD centroid. Finally a bidimensional spectral image is produced by associating the centroid value to a gray level or to a color coded legend.

EXPERIMENTAL RESULTS

For ophthalmological diagnosis, the differentiation is required among tissues made up of particles whose mean dimensions range from about 20μm to 100μm. Commonly these tissues are investigated with ultrasonic transducers characterized by a central frequency of 10 MHz, which implies a wavelength of 150μm, well above the mean particle diameters. This choice is determined by the trade-off between good resolution and low attenuation of the ultrasonic waves in tissues.
The presented spectral imaging technique was previously tested on signals provided

by a simulation model of the scattering tissue derived from eqs. 1-2 [8]. For an experimental validation of the proposed technique a laboratory set-up was carried out as follows. Test objects consisting of gel suspensions of calibrated latex spheres (with diameters of $20\mu m$, $50\mu m$, $80\mu m$, and 100μ) were produced. A linear scanning was performed on the test objects to gather the Radio-Frequency (RF) echo signals that were digitized and stored in a computer memory for subsequent processing. A wideband focused transducer (PANAMETRICS V311) was used, whose main characteristics are: -6 dB bandwidth of 7MHz, central frequency equal to 9.9MHz and -12 dB focal spot of 0.8 mm at the distance of 35mm.

Fig. 1 shows the spectral maps obtained from simulated signals (a) and echoes coming from the test objects (b) by considering $250\mu m$ (right hand side) and $500\mu m$ (left hand side) tissue segments per pixel. The maps represent three adjacent regions characterized by scatterer sizes respectively of $20\mu m$, $50\mu m$ and $80\mu m$. The gray scale associates the brighter gray level to the PSD centroid value relative to scatterers with $20\mu m$ diameters, while the darker one is related to $80\mu m$ diameter scatterers. In the simulations the actual frequency response of the ultrasonic equipment was considered.

The experimental results were in good agreement with the theoretical predictions and were used to adjust the gray scale.

Finally, measurements were performed with the same experimental set-up on in-vitro eyes, using a linear scanning with a 0.4 mm step. A B-mode image, was generated from the set of RF signals to detect the signal portions concerning the pathological region on which the spectral analysis was to be performed. The spectral maps are simultaneously displayed with the B-mode images, preserving the topological references. A complete diagnostic tool is thus obtained since the B-mode image exploits the information related to the echo amplitude and, by outlining the presence of interfaces, allows the spatial locating of the tumoral mass, while the spectral maps outline the backscattered echoes frequency dependence strictly related to the tissue microstructure. This enables the differentiation of various pathologic zones that are not solved with the conventional imaging technique.

Figure 2 represents the spectral maps and the relative B- mode images of two intraocular pathologies' portions (5.m x 3.m) both coming from human eyes retinoblastoma. The tumor represented in the images at the left side of Fig.2 was surrounded by inflammation pattern, while the images at the right side of Fig.2 concern a homogeneous tumoral mass. In this case only the use of the spectral maps enables to distinguish between the inflammation and the tumoral lesion. Each pixel in the spectral maps corresponds to a tissue segment with 250 μm length.

CONCLUSIONS

An automatic detection of a tissue signature based on spectral analysis of ultrasonic signals is very troublesome due to the complex nature of the biological media. Hence a spectral imaging technique is proposed that yields the improvement of the information offered to the physicians. The diagnosis in this case is supported by the filtering and the interpretation of the physicians' experience. The spectral images are based on the local estimate of PSD centroid frequency, which is extracted from data related to region with small size (diameter smaller than 1mm, typically 250 μm). The spectral maps emphasize both the average value of the scatterers size and the degree of local inhomogeneity of the investigated pathology.

ACKNOWLEDGEMENTS

The authors acknowledge the contribution of Prof. R.Frezzotti and his co-workers for their support in the clinical field.

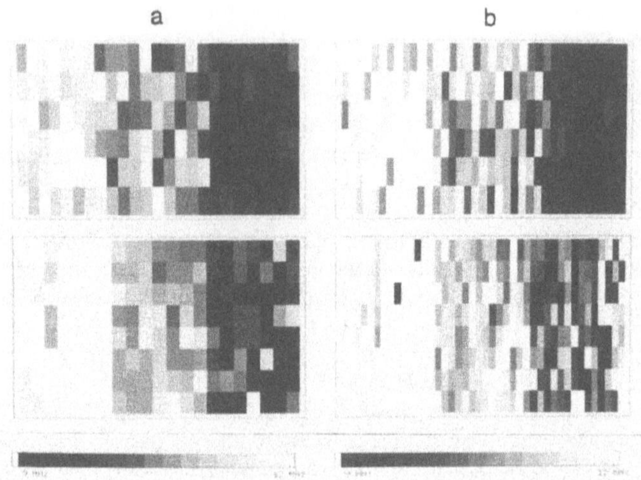

Figure 1. Spectral maps based on frequency centroid evaluation by AR estimators. The maps are relative to a fictitious tissue, made up of three zones containing scatterers with diameters of $20\mu m$, $50\mu m$ and $80\mu m$, respectively. Maps a) are obtained analyzing simulated signals, maps b) are obtained by assembling spectral maps coming from 3 different test objects, consisting of a gel suspension of latex spheres of respectively $20\mu m$, $50\mu m$, $80\mu m$. In the maps at the left one pixel corresponds to a $500\mu m$ tissue segment (sequence of 64 data samples), at the right it corresponds to $250\mu m$ (32 data samples).

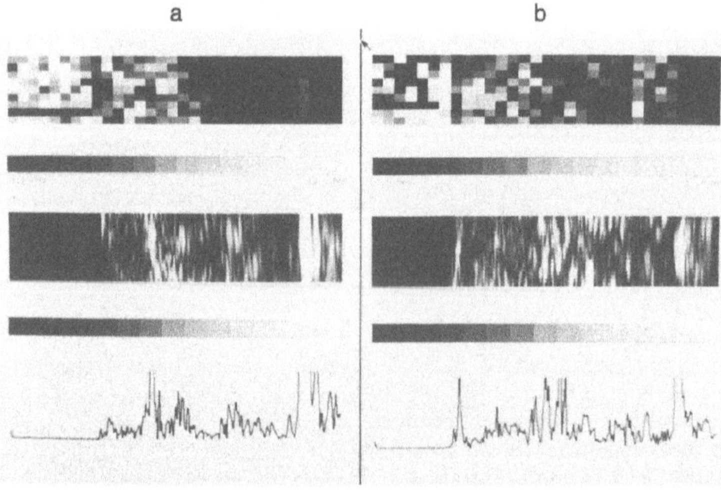

Figure 2. Comparisons between the spectral maps and B_mode images concerning a) a retinoblastoma b) a retinoblastoma surrounded by inflammation. Each pixel in the spectral map corresponds to tissue regions with a length of 250 μm.

REFERENCES

1. F.L.Lizzi, M.Greenbaum, E.J.Feleppa, M.Elbaum, *"Theoretical framework for spectrum analysis in ultrasonic tissue characterization,"* J. Acoust. Soc. Am. 73-4, pp. 1366-1373, 1983.
2. F.L.Lizzi, M.Ostromogilisky, E.J.Feleppa,M.Rorke,M.M.Yaremko, *"Relationship of ultrasonic spectral parmaters to features of tissue Microstructure,"* IEEE Trans. on UFFC 33-3, pp. 319-328, May 1986.
3. S.M.Kay, S.L.Marple, *"Spectrum analysis- A modern perspective,"* IEEE Proc. 69-11, pp. 1380, Nov. 1981.
4. T.Wang, J.Saniie, X.Jin, *"Analysis of low-order autoregressive models for ultrasonic grain signal characterization,"* IEEE Trans. on UFFC 38-2 pp. 116-124, March 1991.
5. M.F.Insana, R.F,Wagner, D.G.Brown, T.J.Hall, *"Describing small-scale structure in random media using pulse-echo ultrasound,"* J. Acoust. Soc. Am. 87-1, pp. 179-191, Jan. 1990.
6. A.T.Cheung, P.W.Buchen, C.Macaskill, D.E.Robinson, *"Backscatter spectrum of a random perturbed regular array of discrete scatterers,"* J. Acoust. Soc. Am. 86-1, pp. 407-413, July 1989.
7. S.L.Marple, "Digital spectral analysis with applications" Prentice Hall, Englewood Cliffs, N.J., 1987.
8. A.Fort, C.Manfredi, L.Masotti, S.Rocchi, *"Spectral maps by AR models as a support to conventional ultrasonic imaging,"* EUSIPCO 1992 - Bruxelles 24-27 August, 1992.

QUANTITATIVE DIAGNOSIS OF HEPATOCELLULAR CARCINOMA AND HEPATIC CAVERNOUS HAEMANGIOMA ULTRASONIC IMAGES

Yao Sun,[1] Zhen Yao,[2] Feng Sun[1]

[1]Department of Automatic Control
Harbin Shipbuiding Engineering Institute
[2]The Second Hospital
Harbin Medical University
Harbin 150001, P. R. China

ABSTRACT

A new method of distinguishing hepatocellular carcinoma and hepatic cavernous haemangioma ultrasonic images has been proposed in this paper. The recognition method of Gradient-Square Deviation Co-occurrence Matrix is effective on discriminating HCC ultrasonograph and HCH ultrasonograph, and 92 cases of HCC and HCH ultrasonic image have been analyzed quantitatively by computer image processing system. The experimental results obtained are satisfactory.

INTRODUCTION

It is well known that primary hepatocellular carcinoma (HCC) is a kind of harmful disease to human body, and it is difficult to distinguish early primary hepatocellular carcinoma from hepatic cavernous haemangioma (HCH) ultrasonic images, and doctors can not judge the similar images accurately. With recent advances of imaging diagnostic modalities, many lesions of small liver cancer become detectable during clinical observation of liver disease. Of noninvasive modalities, such as ultrasound (US), X-ray computed tomography (X-CT) and liver scintigraphy (RI). In general, US is one of the most effective methods to detect small liver cancer. But it is very difficult to distinguish small early primary hepatocellular carcinoma from small hepatic cavernous haemangioma ultrasonic images by doctros' eyes. When the tumor is less than 2 cm in diameter, doctors can not differentiate them accurately on the screen of US. X-ray computed tomography and magnetic resonance imaging have enhanced the possibility, but there remain some difficulties for differentiation. All of the advanced imaging techniques have some weaknesses and limits in identifying cancer from other lesions, and the resolution of human eyes is limited, so it is necessary to seek new measure to distinguish HCC ultrasonograph from HCH ultrasonograph.

In our experiment, computer image processing technique is used to analyze ultrasonograph in order to recognize HCC and HCH quantitatively and to increase the diagnosis accuracy of small liver tumor.

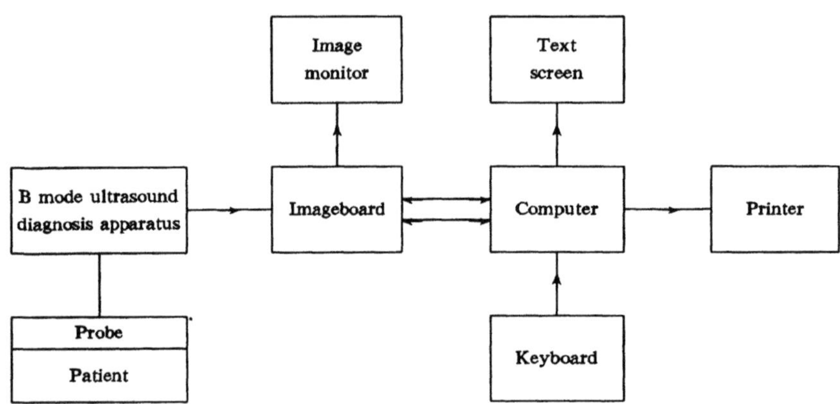

Figure 1.   Block diagram of HCC and HCH ultrasonograph recognition system

MATERIALS AND METHODS

TOSHIBA SSA-90A B mode ultrasound diagnosis apparatus was used to obtain HCC and HCH images. The ultrasonic images were transmitted into the GWPVP computer image processing system through the video interface. Video signal was converted into digital signal, and digitalized ultrasonograph matrix was stored in harddisk. The GWPVP computer image processing system includes a GW286BH microcomputer, a CEGA display screen, a PCVISION PLUS frame grabbing memory of imageboard, a SAMSUNG high resolution color monitor, an NEC P5 color printer, a National L15 video recorder and a RCA high resolution camera. The frame memory of imageboard can hold two images of $512 \times 512 \times 8$ bits, and processing image matrices at high speed. Figure 1 shows the block diagram of HCC and HCH ultrasonograph recognition system.

For liver ultrasonograph, one of the most important features is texture feature. Because texture has the space distribution pattern of gray level of images, the texture analysis method can extract the essential features of liver ultrasonic images. In this paper, some important image analysis methods are reviewed. For overcoming the weaknesses of various traditional analysis methods and fitting to the characteristics of liver ultrasonograph, a new texture analysis method of Gradient-Square Deviation Co-occurrence Matrix is proposed. The new method executes in four steps: the first step is to calculate the gradient matrix from gray level image, the second is to calculate the square deviation matrix from gray level image, the third is to achieve the gradient-square deviation co-occurrence matrix from gradient matrix and square deviation matrix; and the last is to extract five texture feature parameters. The five feature parameters represent the characters of HCC and HCH ultrasonic images. Figure 2 shows the flow chart of analysis-program. The gradient matrix reflects the relative stability of pixels value. The square deviation matrix reflects the roughness and fineness of image texture. The 3×3 masks are used to calculate the gradient matrix and square deviation matrix in this new method. The co-occurrence matrix obtained from gradient matrix and square deviation matrix is very helpful to extract the effective texture feature parameters.

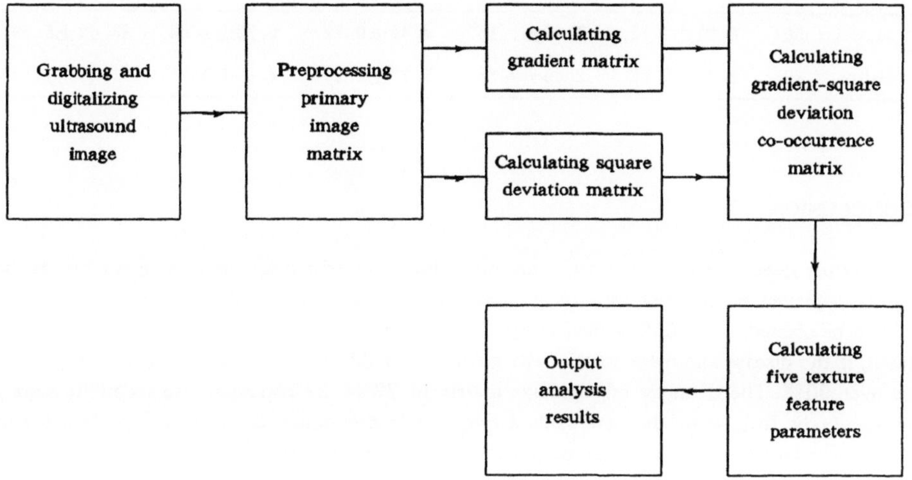

Figure 2. Flow chart of analysis program

RESULTS AND DISCUSSION

In this paper, 50 cases of HCC ultrasonic image and 42 cases of HCH ultrasonic image are analyzed quantitatively by using the new Gradient-Square Deviation Co-Occurrence Matrix method on the computer image processing system. Table 1 shows the experimental results. The formulas of five feature parameters are:

(1) power: $\quad W_1 = \sum\limits_{m=1}^{32} \sum\limits_{n=1}^{32} [B(m, n)/M]^2$

(2) contrast: $\quad W_2 = \sum\limits_{m=1}^{32} \sum\limits_{n=1}^{32} (m - n)^2 B(m, n)/M$

(3) absolute: $\quad W_3 = \sum\limits_{m=1}^{32} \sum\limits_{n=1}^{32} |m - n| B(m, n)/M$

(4) counter moment: $\quad W_4 = \sum\limits_{m=1}^{32} \sum\limits_{n=1}^{32} \frac{1}{1 + (m - n)^2} B(m, n)/M$

(5) entropy: $\quad W_5 = - \sum\limits_{m=1}^{32} \sum\limits_{n=1}^{32} B(m, n) \log_2 B(m, n)/M$

where $M = 32 \times 32$.

The experimental results show that the new method is effective for distin guishing HCC from HCH ultrasonic images. The diagnosis reach an accuracy of 91. 3%, the smallest diameter of nodules was 0. 6 cm.

Table 1.   HCC and HCH ultrasonograph texture feature parameters

Disease	Case	W_1	W_2	W_3	W_4	W_5
HCC	50	8. 21±1. 42	7. 32±1. 15	4. 46±0. 77	5. 39±1. 03	88. 62±5. 69
HCH	42	2. 45±0. 76	2. 45±0. 76	9. 73±2. 21	4. 05±0. 71	191. 36±14. 38

CONCLUSION

When liver ultrasonograph was transmitted into computer image processing system, the image is displayed on the monitor. Usually, we are interested in the disease areas, so a window should be opened in the 512×512 image. The window is also called subimage. In our experiment, if the disease subimage matrix was greater than 22×22, the diagnosis accuracy would be over 90%. The diameter of subimage matrix of 22×22 correspond to six millimeters in human body. In a word, the new method proposed in this paper can extract important texture features of HCC and HCH ultrasonic images, and is helpful to recognize early liver tumor.

REFERENCES

Horiguchi Yuji, et al. 1989. Imaging diagnosis of hepatocellular carcinoma, *Medical Imaging*, Vol. 2, No. 2, pp. 35—42.

Laws, K. I. , 1980, "Textured Image Segmentation", Dept. of EE. Univ. of Southern California.

Haralick, R. M. et al, 1973, Textural features for image classification, *IEEE Trans. on System, Man and Cybernetics*, SMC—3: 610—621.

NONLINEAR ULTRASONIC PARAMETER IN TISSUE CHARACTERIZATION AND IMAGING*

Xiufen Gong

Institute of Acoustics
Najing University
Nanjing 210008, P. R. China

INTRODUCTION

In the wide applications of ultrasound in medicine, the ultrasonic imaging is used extensively in the medical diagnosis. The linear parameters such as acoustic impedance, sound velocity and attenuation are used for human tissue characterization in the conventional clinical ultrasonic imaging systems. However, in recent ten years many works have shown that the important nonlinear acoustic effects can be produced at the biomedical frequencies and intensities. [1,2] These include the finite amplitude wave distortion, harmonic generation, extra-attenuation and sound saturation. The acoustic nonlinearity parameter B/A is the basic parameter to describe these nonlinear phenomena. The existence of nonlinear effects of ultrasound waves on biological media can accelerate the sound losses and limit the amount of sound power diliverable to the media. Especially, our previous work indicated that the value of the nonlinear parameter of tissues was found to be related to their structural characteristics. Therefore, nonlinear ultrasonic parameter B/A of biological media may become a new parameter in tissue characterization and imaging, thus a valuable tool for medical dignosis.

This paper is devoted to an experimental demonstration of this nonlinearity parameter in biological tissues. Starting from the definition of acoustic nonlinearities in a medium, some efforts have been made recently to develop techniques for measuring the nonlinear parameter B/A.

Three methods of B/A determination for tissues established in our laboratory have been presented: finite amplitude insert-substitution method, improved thermodynamic method (phase shift method) and method of parametric array. The obtained experimental results of B/A values for various biological samples (including normal and pathological tissues) indicate the importance and prospects of nonlinearity parameter B/A in the tissue characterization and imaging.

---

* This work was supported by National Natural Science Foundation of China.

MEASUREMENT OF NONLINEARITY PARAMETER

1. Definition of Nonlinearity Parameter

The state equation can be expressed by the nonlinear relationship between pressure and density given to second order as follows:

$$p - p_0 = A\left(\frac{\rho - \rho_0}{\rho_0}\right) + \frac{B}{2}\left(\frac{\rho - \rho_0}{\rho_0}\right)^2$$

$$A = \rho_0 c_0^2 \ , \quad B = \rho_0^2\left(\frac{\partial^2 p}{\partial \rho^2}\right)_{0,\,s}$$

B/A is a measure of the quadratic term in pressure-density relation. As a result of this nonlinearity, a finite amplitude sine plane wave with frequency ω travelling through a medium will result in wave distortion and harmonic generation. based on the Burgers' equation and its Fubini solution, the pressure amplitude of the second harmonics at distance x from the source for lossless medium is given by:

$$p_2 = \frac{\omega p_1^2 x}{4\rho_0 c_0^3}\left(\frac{B}{A} + 2\right) \tag{1}$$

where p_1 is the pressure amplitude at the source. Therefore, B/A is a very basic parameter for determining the degree of wave form distortion.

2. Finite Amplitude Insert-Substitution Method (FAIS)

The FAIS method is a comparative method. Its main point is to measure the amplitude of the second harmonics p_{20} of a comparative liquid with the known B/A (e. g. $(B/A)_0 = 5.2$ for degassed distilled water) and p_{2x} of the sample whose B/A is unknown. If a sound permeable box of thickness d filled with the samples or the comparative liquid is inserted in the water between the source and receiver, locating the box close to the receiver, then a simplified formula can be obtained:[3]

$$\left(\frac{B}{A}\right)_x = \left[\left(\frac{p_{2x}}{p_{20}}\frac{L}{dD_{x0}D_{0x}} - \frac{L}{d} + 1\right)\frac{(\rho c^3)_x}{(\rho c^3)_0}\frac{1}{D_{x0}}\left(\frac{B}{A} + 2\right)_0\right] - 2 \tag{2}$$

Considering the high sound attenuation of tissues, a modified relation can be rewritten as follows:[4]

$$\left(\frac{B}{A}\right)_x = \left[\frac{p_{2x}}{p_{20}}\frac{L}{dI_1 I_2} - \left(\frac{L}{d} - 1\right)\frac{I_2}{I_1}D_{x0}D_{0x}\right]\frac{(\rho c^3)_x}{(\rho c^3)_0}\frac{(B/A)_0 + 2}{D_{0x}D_{x0}^2} - 2 \tag{3}$$

where L is the length of the path between the source and receiver, $D_{x0} = 2(\rho c)_x/[(\rho c)_0 + (\rho c)_x]$, $D_{0x} = 2(\rho c)_0/[(\rho c)_0 + (\rho c)_x]$, $I_1(\alpha_1) = \exp[-\alpha_1 d]$, $I_2(\alpha_2) = \exp[-\alpha_2 d/2]$, α_1 and

α_2 are the attenuation coefficients of the fundamental and 2nd harmonic components. So, if the ratio p_{2x}/p_{20} is measured, then $(B/A)_x$ can be easily calculated by this formula. Advantages of this method are that only a small volume of sample material is needed and absolute measuerment of sound pressure is unnecessary.

3. Improved Thermodynamic Method (ITD)

The idea of this method is to make direct isentropic measurement of sound speed variation rather than isothermal process. In this case, B/A can be expressed as:

$$\frac{B}{A} = 2\rho_0 c_0 \left(\frac{\partial c}{\partial p}\right)_s \tag{4}$$

where ρ_0 is the undisturbed density, c_0 is the infinitesimal wave velocity, and p is the static pressure. Eq. (4) shows that B/A depends directly on the change of sound speed due to the change of static pressure under adiabatic condition. A test vessel contains two 4.6MHz acoustic transmitter and receiver saparated by a fixed distance d and immersed in the medium of sound speed c. A sound wave travels from one transducer to the other in time t, then Eq. (4) can be rewritten as:

$$\frac{B}{A} = -\frac{2\rho_0 c_0^3}{d}\left(\frac{\Delta t}{\Delta p}\right)_s \tag{5}$$

where Δt is the change of transit time due to the change of ambient static pressure $\Delta p = 1$ atm (from 1 to 2 atm). As $\Delta\varphi = 2\pi f\Delta t$, using high sensitive phase detect technique we can measure this phase shift $\Delta\varphi$ and then determine the B/A value. The main advantages of this method are that it does not require calibration of sound field and the phase detection technique allows the experiment to be carried out under small variation of static pressure.

4. Method of Parametric Array

When two waves with different frequencies ω_1 and ω_2 are projected as primary wave, the secondary wave with sum frequency (SF) and difference frequency (DF) generated as parametric array are received at distance x from the source. On the basis of Fenlon's theory, the pressure amplitude of the secondary wave due to the nonlinear interaction of a low frequency pump wave and a high frequency small wave can be expressed as:

$$p_{\omega_1 \pm \omega_2} = p_s(x) = \frac{xp_1(0)\,p_2(0)}{4\rho_0 c_0^3}(\omega_1 \pm \omega_2)\left(\frac{B}{A} + 2\right) \tag{6}$$

Considering the correction of sound attenuation of the medium, $p_s(x)$ can be rewritten as:

$$p_s = p_s(x) = \frac{p_1(0)\,p_2(0)}{4\rho_0 c_0^3}(\omega_1 \pm \omega_2)\left(\frac{B}{A} + 2\right)I(x)$$

$$I(x) = \frac{e^{-\alpha_s x} - e^{-(\alpha_1 + \alpha_2)x}}{\alpha_1 + \alpha_2 - \alpha_s}$$

where α_1, α_2 and α_s are the attenuation coefficients of two primary waves with ω_1, ω_2 and secondary wave with $\omega_s = \omega_1 \pm \omega_2$, respectively. Using the principle of comparative method, similar to FAIS, the B/A value for sample can be calculated by measuring p_{sx}/p_{s0} as follows:

$$\left(\frac{B}{A}\right)_x + 2 = \left[\frac{p_{sx}}{p_{s0}}\frac{L}{dA_1A_2} - \left(\frac{L}{d} - 1\right)\frac{A_1}{A_2}D_{x0}D_{0x}\right]\frac{(\rho c^3)_x}{(\rho c^3)_0}\frac{1}{D_{0x}D_{x0}^2}\left(\frac{B}{A} + 2\right)_0$$

where $A_1 = \exp[-(\alpha_1 + \alpha_2)d/2]$, $A_2 = \exp[-\alpha_s d/2]$.

EXPERIMENTAL RESULTS

1. B/A Values for Biological Soft Tissues

B/A values for biological soft tissues using the above-mentioned methods are listed in Table 1. The fresh porcine soft tissues taken from the slaughter house were kept in physiological saline before measurement. All the experiments were carried out at 26 °C.

Table 1. B/A values for fresh porcine soft tissues

Specimen	Density (g/cm³)	Velocity (m/s)	B/A (FAIS)	B/A (ITD)	B/A (SF)	B/A (DF)	B/A [2]
Water	0. 997	1497	5. 2*	5. 1			Beyer*
Liver	1. 05	1588	7. 1	6. 9	7. 1	6. 9	7. 2—7. 9[2], beef
Kidney	1. 05	1569	6. 9	6. 3	6. 6	6. 8	7. 4[2], Canine
Heart	1. 06	1572	7. 1	6. 8	7. 5	7. 3	6. 8—7. 4[2]
Spleen	1. 05	1575	6. 9	6. 3	6. 7	6. 7	6. 8[2]
Brain	1. 05	1550	—	—	6. 2	6. 3	—
Tongue	1. 05	1551	6. 6	6. 8	7. 4	7. 2	—
Fatty tissue	0. 95	1445	10. 8	10. 9	10. 3	10. 6	11. 0[2]

2. Structural Characteristics of Nonlinearity Parameter B/A

The dependence of B/A values on specimen structural hierarchy for normal tissue is listed in Table 2. Fresh human whole blood taken from vein was treated with anticoagulant. Coagulative human blood was coagulated naturally under the room temperature 26 °C. The homogenized porcine liver was prepared by liquifying a small piece of fresh in a blender, removed the vessels and gas babble.

It can be found from Table 2 that B/A value for whole liver is 7. 1, and for homogenized liver is 6. 6. Such structure is destroyed by homogenization and the value B/A decreases. Also, the B/A values of water, fresh human blood and coagulative human blood increase progressively with the increase of complication in structure.

Table 2. B/A value of blood and liver

Specimen	Density (g/cm³)	Velocity (m/s)	B/A (FAISM)
Fresh porcine liver	1. 05	1588	7. 1
Homogenized liver	1. 06	1551	6. 6
Distilled water	0. 997	1497	5. 2 (Beyer)
Fresh human blood	1. 075	1561	6. 1
Coagulative human blood	1. 075	1586	7. 3

3. B/A Values for Normal and Pathological Tissues

Since structural changes in tissues are often related to pathological states, it can be predicted that B/A for pathological tissue may be very different from B/A value of normal tissue. More recently we studied the nonlinearity parameter of pathological tissue and the relation between the B/A value and the structures and pathological state of tissues. The B/A values of four kinds of pathological porcine liver tissue were studied by using FAIS method. In order to make comparison the B/A value of normal tissue was made simultaneously. The liver tissues obtained from slaughthouse freshly were kept in 0. 9% physiological saline and studied in 3 hours. At the same time, the pathological analysis of tissues was carried out by the microphotography with slice piece of corresponding samples. The results of sound velocity, attenuation and nonlinear parameter B/A for normal and four kinds of pathological tissues are listed in Table 3.

Table 3. B/A value for normal and pathological livers

Specimen	Velocity (m/s)	Attenuation (2 MHz) dB/cm (4 MHz)		B/A
1. Normal liver	1588	1. 10	2. 16	6. 9
2. Hepatocirrhosis liver (mild)	1573	1. 13	2. 15	7. 8
3. Hepatocirrhosis liver (moderate)	1574	1. 10	2. 12	8. 3
4. Hepatocirrhosis liver (severe)	1580	1. 05	2. 24	8. 5
5. Hepatonecrosis liver	1494	1. 11	2. 06	8. 8

The microphotographs for liver 2 to liver 5 show the structural changes of tissues. The experimental results listed in Table 3 indicate that: (1) nonlinearity parameter is senstive to the change of structure and to the pathological state of tissue. Liver 2 to liver 4 are all hepatocirrhosis but with degree from mild to severe and their corresponding nonlinearity parameters increase gradually; (2) the pathological degree of hepatonecrosis is more serious than that of general pathological livers, so its B/A value increases clearly.

The above experimental results show that compared with those of the normal liver tissue

1, the values of pathological tissues (livers 2 to 5) increase obviously, but at the same time, the values of the other linear parameters (velocity, attenuation) change slightly.

CONCLUSION

The analysis of experimental results indicates that the acoustic nonlinearity parameter B/A is evidently dependent on the change of structure and sensitive to the pathological state of tissues. These results offer the important information and physical basis for using this nonlinearity parameter as a new parameter for tissue characterization and imaging. [6,7] The experimental research work concerning the nonlinear parameter imaging in our laboratory was reported in another paper. [8]

REFERENCES

1. T. G. Muir and E. L. Carstensen, Prediction of nonlinear effects at biological frequencies and intensities, *Ultrasound in Med & Biol*, 6: 345 (1980).
2. W. K. Law, L. A. Frizzell and F. Dunn, Determination of nonlinearity parameter B/A of biological media, *Ultrasound in Med & Biol*, 11: 307 (1985).
3. X. F. Gong, R. Feng, C. Y. Zhu and T. Shi, Ultrasonic investigation of nonlinearity parameter B/A in biological media, *J Acoust Soc Am*, 76: 949 (1984).
4. X. F. Gong, Z. M. Zhu, T. Shi and J. H. Huang, Determination of acoustic nonlinearity parameter in biological media using FAIS and ITD methods, *J Acoust Soc Am*, 86: 1 (1989).
5. X. F. Gong, Z. M. Zhu, B. Fan and D. Zhang, Parameter effect and nonlinear parameter in biological media, Fronties of Nonlinear Acoustics, in: "Proc of 12th ISNA", 397 (1990).
6. T. Sato, Y. Yamakoshi and T. Nakamura, Nonlinear tissue imaging, in: "Proc of Ultrasonic Symposium", 889 (1986).
7. Y. Nakagawa, W. Hou, A. Cai, N. Anold and G. Wage, Nonlinear parameter imaging with finite amplitude sound waves, in: "Proc of Ultrasonic Symposium", 901 (1986).
8. D. Zhang and X. F. Gong, Nonlinear parameter B/A imaging, in: "Proc of Acoustical Imaging", 1992.

SOFT TISSUE ATTENUATION ESTIMATION BASED ON THE ENTROPY DIFFERENCE METHOD AND BY APPLYING A WIDE-BAND PULSE AND DYNAMIC FOCUSING TO AN ANNULAR ARRAY

Jong Cheul Park and Song Bai Park

Department of Electrical Engineering
Korea Advanced Institute of Science and Technology
P. O. Box 201, Chongyangni, Seoul, Korea

ABSTRACT

In the entropy difference method for estimating the ultrasonic attenuation coefficient in soft tissue, the entropies of speckle-contaminated echo envelopes returning from two adjacent rectangular regions in the axial direction are compared as the attenuation is continuously compensated for until the two entropy values become identical; the amount of compensated attenuation is regarded as the average attenuation coefficient in the two regions. In the past CW burst pulses were generated for this purpose by a loosely focused disk-type transducer which was scanned in the lateral direction to obtain a large number of scan data. It turned out that this method is by far superior to the well-known log spectrum difference method in the respect of accuracy with the same amount of data.

In order to improve the resolution in the two-dimensional attenuation mapping we propose in this paper to use an annular array sector scanning system, that is, (1) to scan an annular array in the azimuthal as well as lateral direction such that each scan line is equidistant from its neighboring six scan lines, and (2) to apply a wide-band pulse to one annular element at a time and receive the echo by all annuli and, using these data for a given scan direction, apply dynamic focusing both in the transmit and in the receive modes.

Computer simulation and experimental results with a tissue-mimicked phantom were performed; in both cases the proposed method was capable of estimating the attenuation coefficient within about 10% accuracy for two adjacent regions of $(6mm)^3$ each, using a 3.5MHz, 6-element annular array. To the authors' knowledge, this resolution has never been achieved so far.

I. INTRODUCTION

The area of tissue characterization is concerned with extracting diagnostically useful informations from the soft tissue of human body by applying ultrasonic pulses. A number of ultrasonic tissue parameters are conceivable - reflectivity, attenuation, velocity, nonlinearity, ultrasonic impedance, density, elasticity, movability, etc. Ultrasonic pulse echoes reflected from the interfaces of objects with different acoustic impedances provide the most valuable information about the internal structure of the human body and are used extensively in the ultrasonic diagnostic equipments in different image modalities - most commonly B mode, A mode, M mode and rarely C mode. However, the pulse echoes from the soft tissue, when displayed as a B-mode image, fail to distinguish the malignant region from the wealthy region unless the disease has progressed severely so as to show distinctively different reflectivities in the two regions. This is primarily because the echoes are usually heavily contaminated by speckle noises which obscure a small difference of reflectivities. Among the parameters listed above, attenuation has been the subject of most intensive research in the tissue characterization area in the past 20 years and is regarded as the most promising although commercial equipments for attenuation mapping have not appeared on the markets as yet.

The ultrasonic attenuation coefficient β can be measured either in the transmission mode or in the reflection mode. The first mode is very accurate and convenient to use [1] but can be applied only to a limited region of the human body, e.g., the breast. In contrast, the reflection mode can be applied to any region but is difficult to obtain an accurate result. Measurement of the β value by the reflection mode can be classified into two categories - the frequency domain approach [1-8] and the time domain approach [9-11]. We will first review the frequency domain approaches. Kuc [2] proposed the spectral difference approach, which estimates β, using wide-band pulses, from the slope of the difference between log spectra of echoes from two regions in the soft tissue. Fink et al. [3] proposed short time Fourier analysis of the backscatterered signal to obtain local unbiased estimation of the attenuation slope, which may be considered as an improvement of the Kuc's method. Both methods assume a one-dimensional wave propagation through the medium, neglecting the diffraction effect which is known to cause a change in the spectra [12-14]. The central frequency shift method [6, 11] using wide-band pulse is the simplest but most error-prone. To improve this method, Kuc [6] proposed the two-dimensional maximum entropy method. All of these frequency domain approaches have one drawback in common, that is, they require an enormous echo data to achieve some an acceptable degree of accuracy in the estimation of the attenuation and hence are not suitable for two-dimensional imaging of the parameter where the spatial resolution is of major concern.

Two methods have been proposed in the time domain for attenuation estimation; one is the noise-to-signal ratio (NSR) method proposed by Greenleaf [9] and the other is the entropy difference method proposed by our research group [10]. In the first method, the variance-to-mean ratio of the echo "envelope peaks" in the region of interest (ROI) is measured as the attenuation is continuously compensated for until the ratio reaches the minimum value — the amount of attenuation compensation is then regarded as the average attenuation coefficient in the ROI. In the entropy difference method the entropy difference of the echo envelope data from two adjacent regions is measured as the attenuation is continuously compensated for until the difference becomes zero — the amount of attenuation compensation is then regarded as the average attenuation coefficient in the two regions. Both methods have an advantage in common over the frequency domain approach, that is, they require far less echo data for the same degree of accuracy in the attenuation estimation and hence are suitable for attenuation imaging.

In this paper, in order to improve the spatial resolution of attenuation mapping we use wide-band pulses, instead of burst pulses, and sector scanning of an annular array for dynamic focusing in both and receive modes, instead of a disk-type transducer or a linear array. The use of wide-band pulses requires the introduction of inverse attenuation filtering and the diffraction effect becomes a more serious problem. Both of these problems will be discussed in this paper.

Justification of the approach taken in this paper will be confirmed by comparing the simulation results with the experimental results.

II. REVIEW OF THE ENTROPY METHOD

The time-domain estimation of the ultrasonic attenuation coefficient starts from the statistical analysis of the reflected waveforms. It is known that the probability density function (pdf) of the envelope of narrow-band pulses reflected from a homogeneous and non-attenuating medium, consisting of randomly distributed scatterers follows the Rayleigh distribution:

$$p(R) = \frac{R}{\psi^2}e^{-R^2/2\psi^2}, \quad R > 0 \tag{1}$$

where ψ is the average value of the echo envelope divided by $\pi/2$ and takes a different value depending on the property of the soft tissue. When the wave undergoes attenuation while propagating through the medium, the envelope of the echo waveform is expressed as

$$R(d) = \frac{Ro}{d}\sigma(d)e^{-\int 2\alpha(f)d} \tag{2}$$

where σ is the backscattering coefficient and $\alpha(f)$ represents the frequency-dependent attenuation, d is the one-way travelling distance, and R_o is a constant of proportion. Eq. (2) suggests that the pdf of the echo envelope in the case of attenuation will change from the Rayleigh distribution given by Eq. (1), and this change can be utilized in estimating the attenuation in the medium. It is known from the measurements that $\alpha(f)$ changes almost linearly with the frequency in the biological soft tissue in the frequency range from 1 MHz to 10 MHz. In this case

$$\alpha(f) = \beta f \tag{3}$$

where β is called the attenuation coefficient and measured in the unit of [dB/MHz/cm].

Due to the attenuation, the pdf of the echo envelope skews toward the lower gray level [10] from the Rayleigh pdf. Figure 1 skews the change of the pdf with βd. The curve for $\beta d = 0$ corresponds to the Rayleigh distribution. As βd increases with a positive value, the curve skews towards the left, while the curve skews towards the right as βd decreases with a negative value. Now suppose a rectangular region A perpendicular to the beam propagation and with a width D as shown in figure 2. The pdf of the echo from this region skews

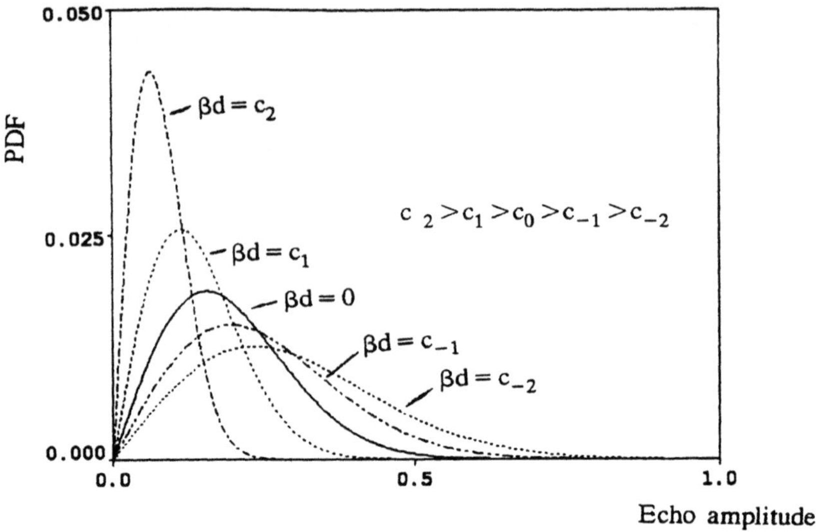

Fig. 1   Pdf's of the echo envelope for various values of βd.

toward the left, as mentioned above. Suppose we multiply the echo waveform by $e^{+\beta_c f_0 d}$ to compensate for the attenuation, where f_0 is the center frequency of the narrow-band pulse and d is the penetration from the front surface of region A. As β_c increased, the resulting pdf will approach, and eventually become identical to, the Rayleigh distribution. The value of β_c for this exact compensation can be regarded as the β value of region A. The time-domain approaches of the attenuation are based on this reasoning.

One immediately conceivable algorithm is then to compare some statistical parameters of the successively compensated echo envelope with the known statistical parameters of the Rayleigh distribution. Since the Rayleigh distribution is characterized by one parameter, namely, ψ in Eq. (1), or the ratio of the mean to the variance takes a constant value, the simplest algorithm is to compute this ratio for the successively compensated echoes until the two ratios become identical. However, it turns out that this approach is not very accurate when the data is limited in number as in the case where a high spatial resolution in attenuation imaging is intended. Greenleaf [9] utilized the fact that the NSR takes a minimum value at the exact compensation.

Another statistical parameter found useful in the β estimation is the entropy, which is defined, as a function of any pdf p(x), by

$$H(p) = -\int_{-\infty}^{\infty} p(x) \ln p(x) \, dx \qquad (4)$$

Since p(x)=1 over the domain of the random variable x, the entropy takes a positive value. It is known that the entropy serves a good representation of the statistical properties, especially when the available samples are limited in number. Broadly speaking, the entropy increases as the pdf curve broadens, and hence we expect that the entropy will increase continuously with the attenuation

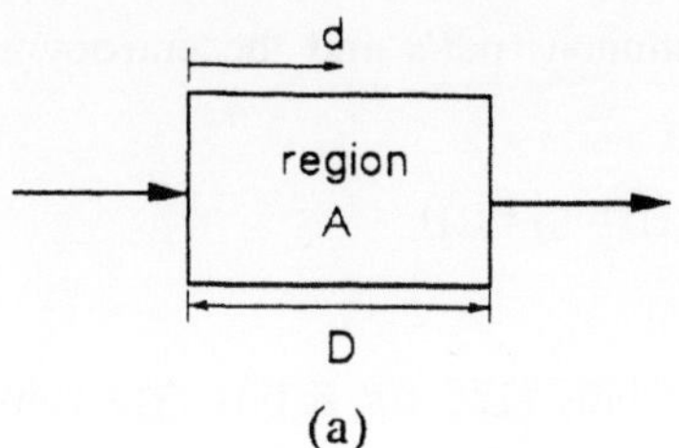

(a)

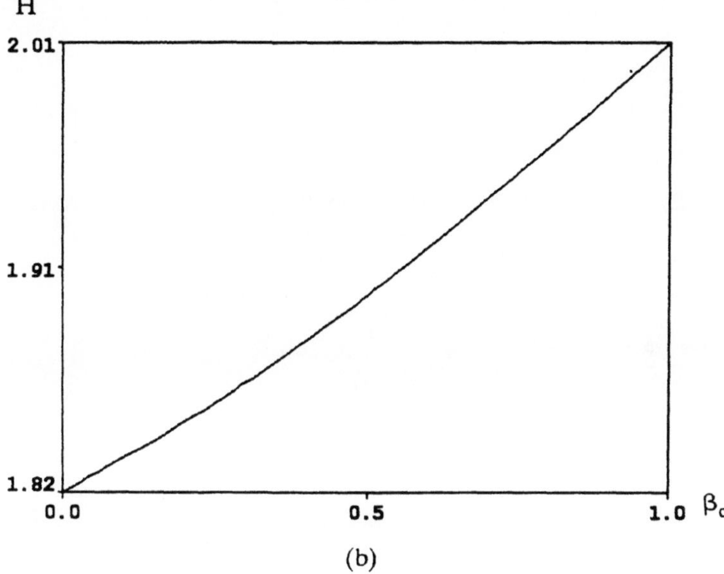

(b)

Fig. 2   (a) ROI (b) entropy of the echo envelope as a function of attenuation compensation (β_c).

compensation as expected from figure 1 or as shown in figure 2(b). To alleviate this difficulty we consider two adjacent regions A and B with the same width D, as shown in figure 3(a). The pdf of the echoes from the region B skews more severly to the left with a correspondingly smaller entropy than that from the region A, as shown in figure 3(b), since the former echo travels a lager distance. For the same reason, as the attenuation compensation proceeds, the entropy of region B, H_B, increases more rapidly than that of region A, H_A, as shown in figure 3(c); the two entropy curves cross each other at the extract compensation, where the pdf becomes Rayleigh. In figure 3(d) $H_B - H_A$ is plotted as a function of β_c and the average β value in regions A and B is found at the zero crossing point, which is a more practical way to do. We note that the entropy difference changes very linearly in the vicinity of the zero crossing. This suggests that linear fitting to an actual fluctuating curve would yield a satisfactory result when dealing with a finite number of data. In the actual estimation procedure, we generate the normalized histograms of the echo amplitude, instead of continuous pdf's and the entropy expression is modified as

$$H(P_h) = -\sum_{j=1}^{B} P_n(j) \log_2 P_n(j) \tag{5}$$

where B is the number of bins (128 for 8 bits AD conversion).

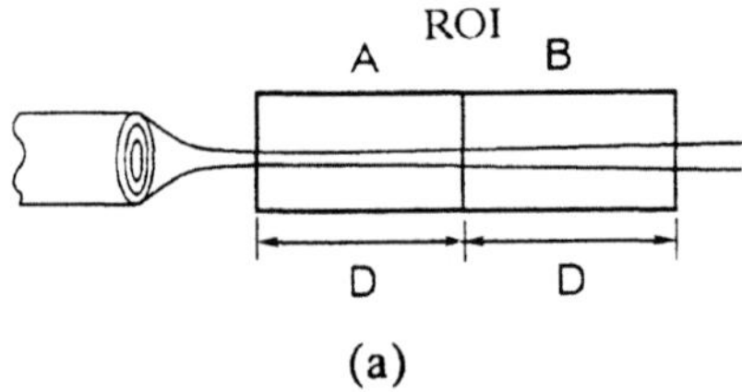

(a)

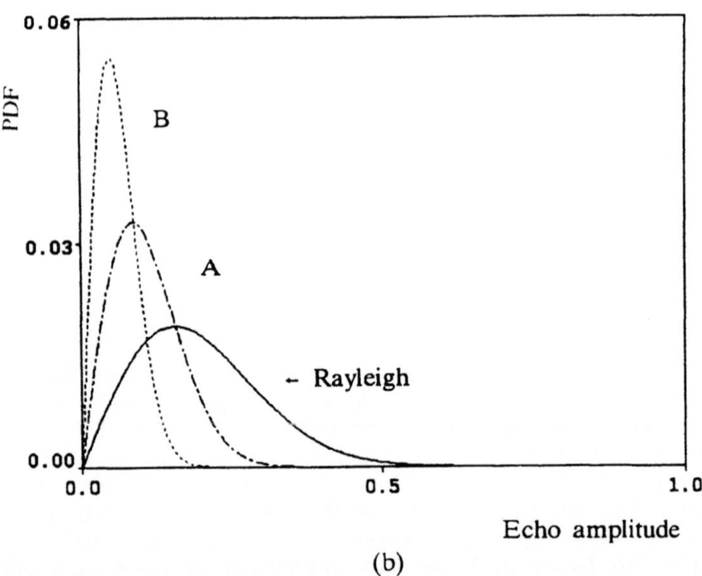

(b)

Fig. 3   Illustration of the entropy difference method.   (a) ROI consisting of two adjacent regions A and B. (b) Pdf's of the echo envelope of the two regions and Rayleigh distribution.   (c) Entropy in each region. (d) Entropy differences of the two regions.

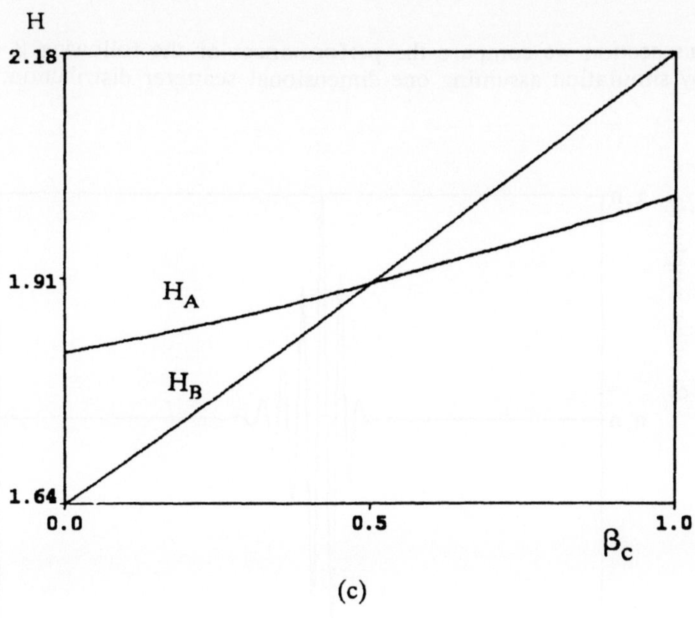

(c)

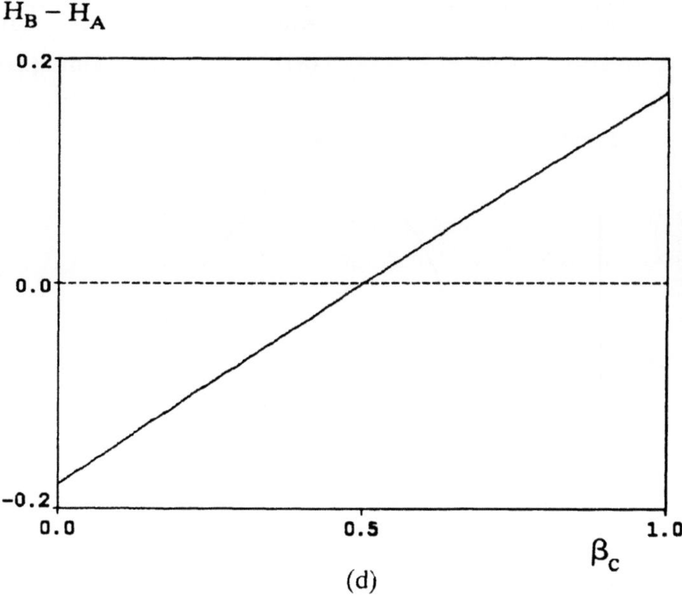

(d)

Fig. 3 (continued)

III. COMPARISONS OF VARIOUS β ESTIMATION
BY SIMULATION WITH 1-D SCATTERER DISTRIBUTION

In this section we compare the performances of the following β estimation methods by simulation assuming one-dimensional scatterer distribution:

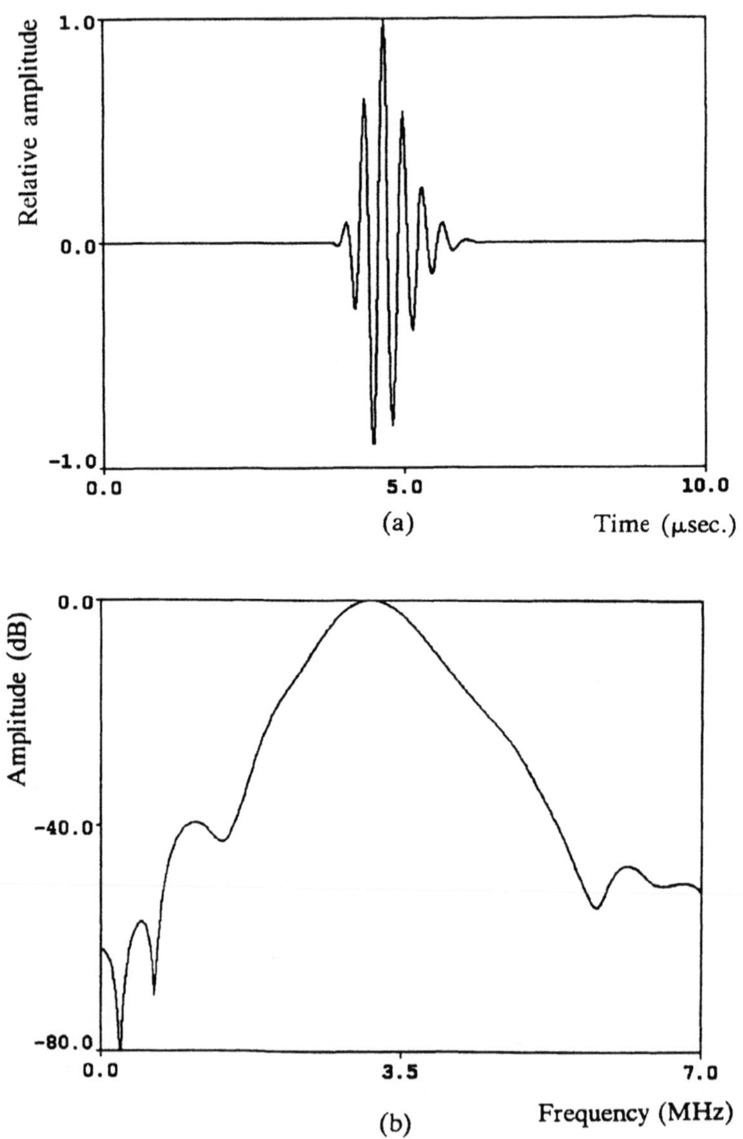

Fig. 4 Measured pulse echo waveforms and its power spectrum. (a) and (b): wide-band pulse, (c) and (d): narrow-band pulse.

(1) log spectral difference method (LSD)

(2) center frequency shift method (CFS)

(3) entropy difference method(ED) using a narrow-band pulse (EDNB)

(4) entropy difference method(ED) using a wide-band pulse and no inverse filter (EDWB)

(5) entropy difference method(ED) using a wide-band pulse and an inverse filter (EDWBIF).

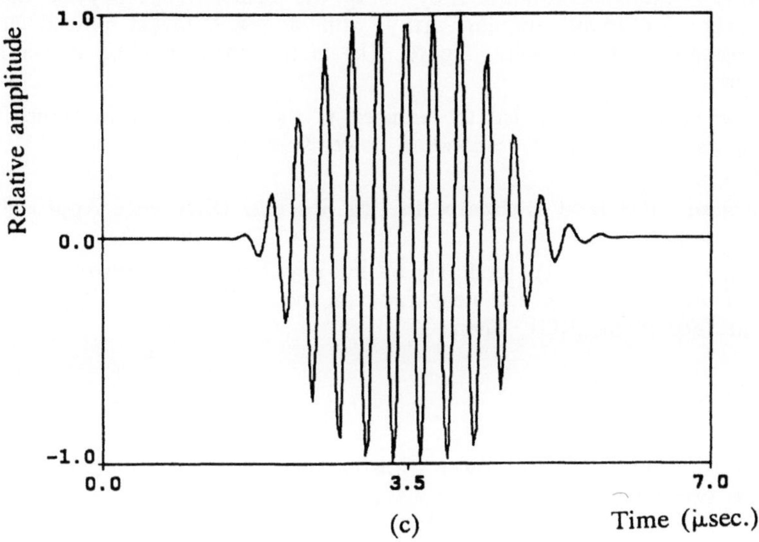

(c)

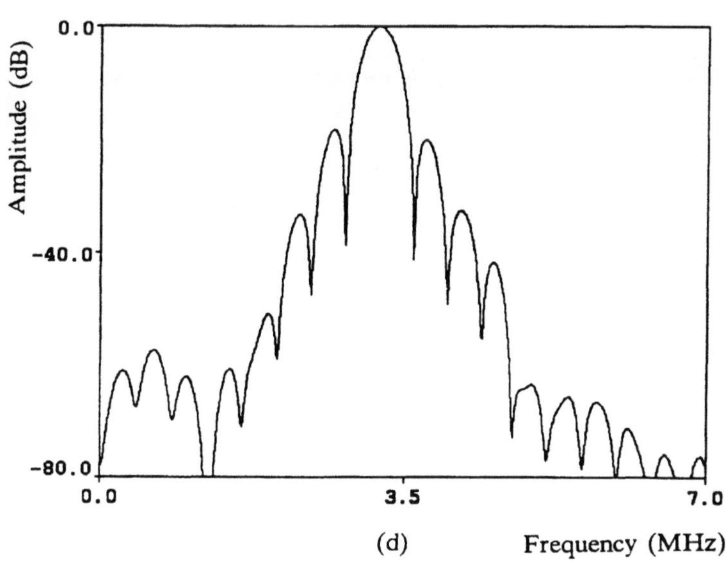

(d)

Fig. 4 (continued)

In all simulations we assume that

(1) 366 scatterers in total are randomly distributed on two adjacent line segments A and B, 1 cm long each,

(2) the reflectivities of the scatterers are also randomly and uniformly distributed,

(3) the attenuation coefficient β of the medium is assumed to be 0.5 dB/MHz/cm,

(4) the wide-band echo waveform to be used is the one shown in figure 4(a), which was obtained by using the center frequency of the 3.5 MHz, 6-element annular array with a point target placed at the geometrical focal point; figure 4(b) is the corresponding power spectrum.

(5) The narrow-band pulse to be used is the one shown in figure 4(c); figure 4(d) is the corresponding power spectrum.

III.1 Maximum Likelihood Estimator in Log Spectral Difference Approach

Referring to figure 3(a), let $P_N(f)$ be the power spectral density (or simply spectrum) of the echo from the near face of the ROI, and let $P_F(f)$ be that from the far face of the ROI. Then

$$P_F(f) = |H(f)|^2 P_N(f) \tag{6}$$

where $|H(f)|^2$ is the two-way power transfer function of the medium between the surfaces. From Eq. 6,

$$20\log_{10}|H(f)|^2 = 20\log P_F(f) - 20\log_{10}P_N(f) = -2\beta f D \tag{7}$$

where D is the width of the ROI. When this spectral difference is plotted as a function of f, the slope denoted by S_f is related to the β value by

$$\beta = \frac{S_f}{2D} \ \ dB/cm/MHz. \tag{8}$$

$$\overline{\beta} = \frac{\overline{S_f}}{2D} \tag{9}$$

where

$$\overline{S_f} = \frac{\displaystyle\sum_{k=0}^{K} f_k L(f_k) - \left(\sum_{k=0}^{K} f_k \sum_{k=0}^{K} L(f_k)\right)/(K+1)}{\displaystyle\sum_{k=0}^{K} (f_k - \overline{f})^2} \tag{10}$$

with $L(f_K) = LR_N(f_K) - LR_F(f_R)$, $f_K = K\Delta f + f_0 = K/T + f_0$ and $\bar{f}$ is the mean frequency is the useful bandwidth of the NAC (normalized attenuation coefficient) curve defined in [15].

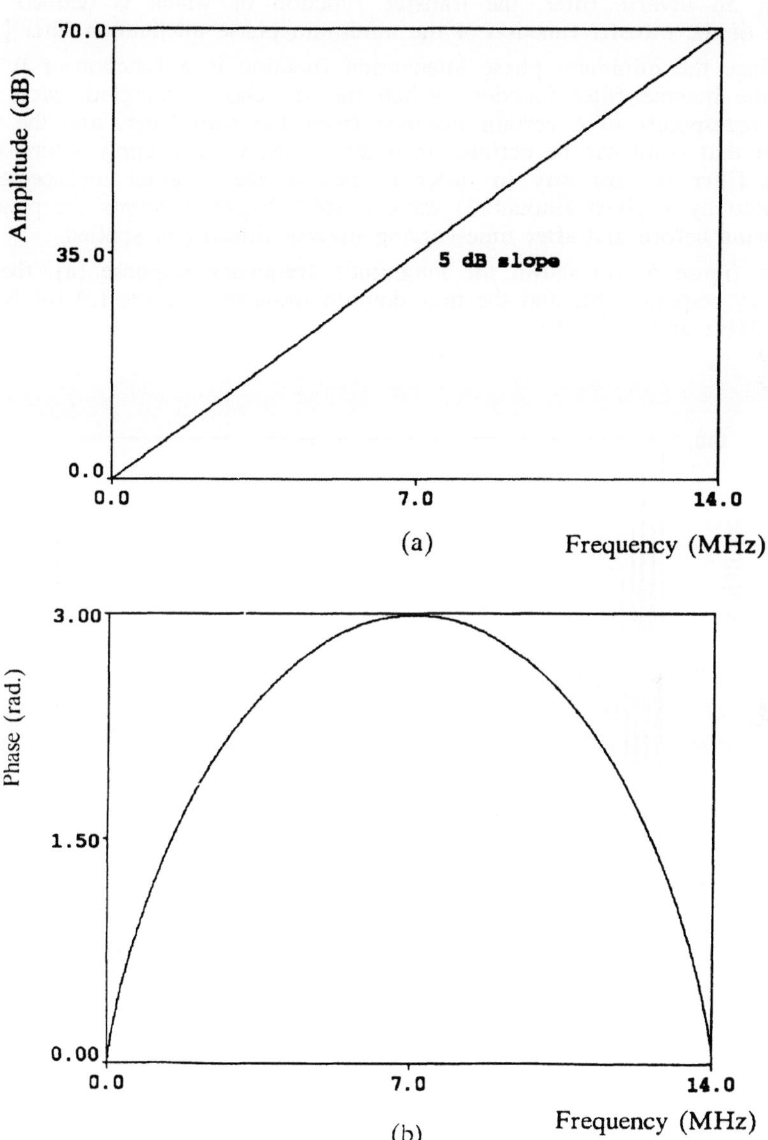

Fig. 5   The inverse filter characteristics (β = 0.5 dB/MHz/cm and for d = 10 cm).   (a) The magnitude frequency characteristic, (b) the phase frequency characteristic, and (c) impulse response.

III.2 Inverse Filter

When a wide-band pulse [figure 4(a)] propagates through an attenuating medium, not only the amplitude but also the wave shape itself changes (widens) due to the frequency-dependent attenuation − this is not the case for the narrow-band signal. Therefore, we have to restore the original waveform before applying any β estimation method. This can be done by passing the echo through an inverse filter, the transfer function of which is defined by the inverse of the transfer function of the minimum phase attenuation filter [16].

Since the minimum phase attenuation function is a function of β and d, so is the inverse filter function. When the rf echo is sampled, each sample point corresponds to a certain distance from the transducer, and the inverse filter at that point can be defined; in other words, we can apply a time-varying inverse filter in this way in order to restore the original undispersed (but attenuated by a given amount β) wave shape. Figure 6 shows the pulse echo waveforms before and after time-varying inverse filtering is applied.

In figure 5 are shown the magnitude frequency response (a), the phase frequency response (b), and the time-domain impulse response (c) for β = 0.5 dB/cm/MHz and d = 10 cm.

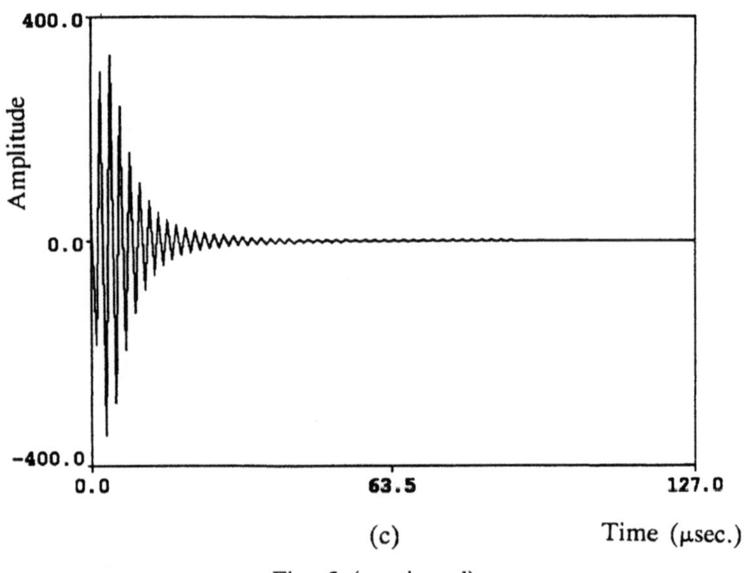

Fig. 5 (continued)

III.3 1-D Simulation Results

Figure 7 shows the histograms obtained by (i) EDWBIF (entropy difference method using wide-band pulses and inverse filtering), (ii) EDNB (entropy difference method using narrow-band pulses), (iii) LSD (log spectral difference

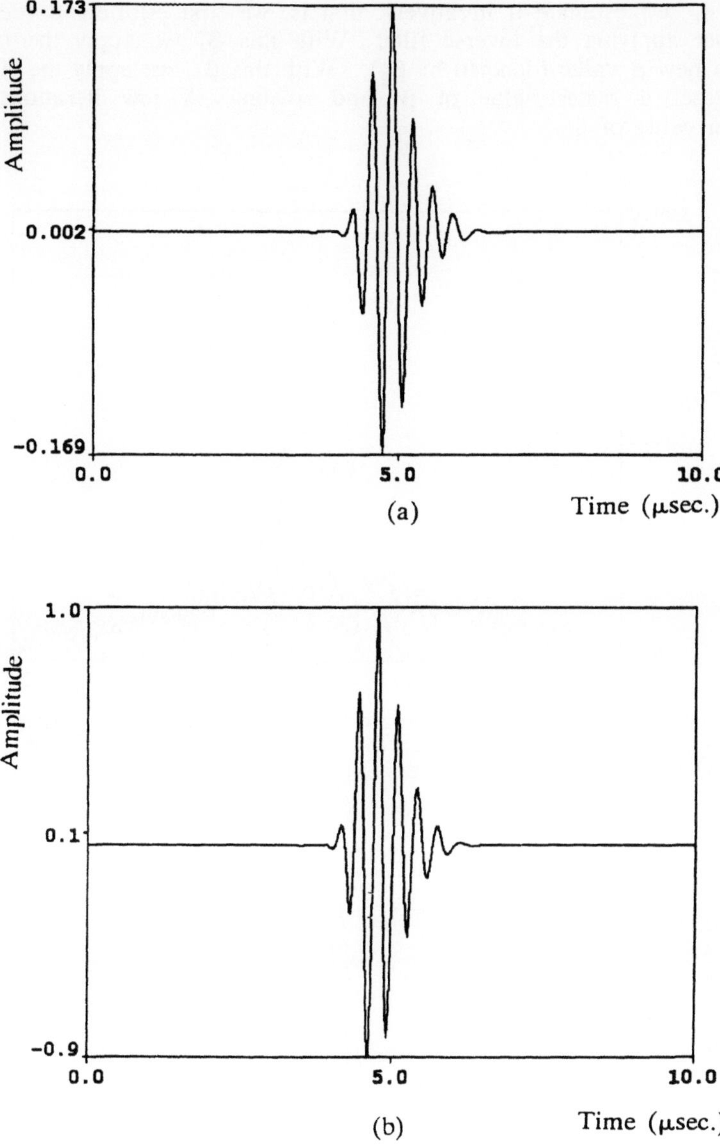

Fig. 6   Pulse echo waveforms (a) before and (b) after time-varying inverse filtering.   (β =0.5 dB/MHz/cm and d = 10 cm.)

method), and (iv) CFS (center frequency shift method), all under the assumptions listed at the beginning of this section. In the inverse filter β was assumed to be a known value of 0.5 dB/MHz/cm.   Figure 9 shows the histograms obtained by (i) EDWBIF and (ii) EDWB; we see that a down-shift of the histogram of (iii), which shows the necessity of applying the inverse filter when using wide-band pulses. Since in actual situations we do not know the β value in advance, we estimate it iteratively, that is, we first estimate β (denoted by β_1) without applying the inverse filter. With this β_1 we apply the filter and estimate a new β value (denoted by β_2). With this β_2, we apply the filter once again to get a better value of β, and so on. A few iterations give a convergent value of β .

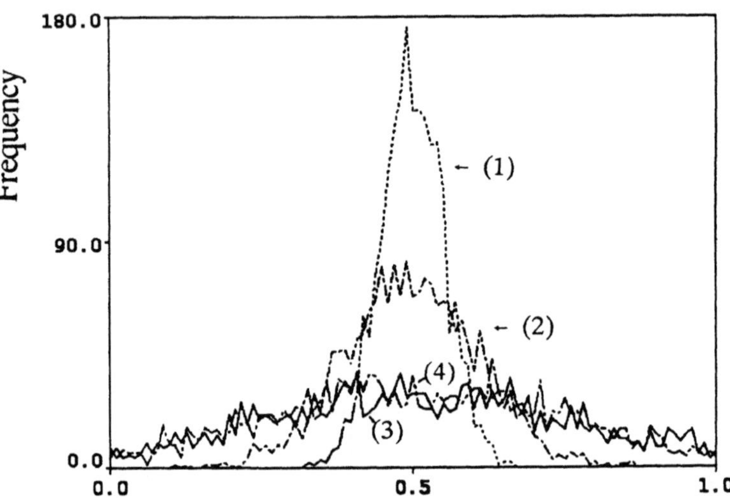

Fig. 7   Histograms of various estimators obtained with 2000 ensembles each using 10 scan data (d = 1 cm).   (1) EDWBIF, (2) EDNB, (3) LSD, (4) CFS.

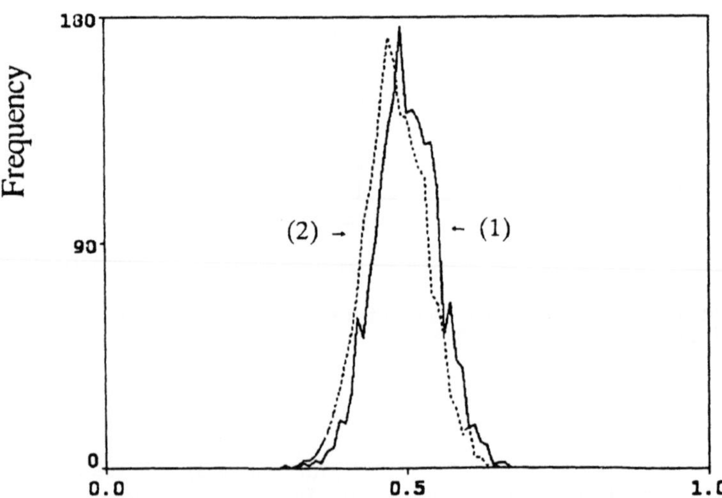

Fig. 8   Histograms obtained by (1) EDWBIF and (2) EDWB estimators.

Table I.   Statistical results for various estimators (1-D simulation)

(Units dB/MHz/cm)

	EDWBIF	EDNB	LSD	CFS	EDWB
$\hat{\beta}$	0 4997	0 497	0 5033	0 5032	0 4812
σ_β	0 0508	0 107	0 276	0 261	0 0502

Table I summarizes the statistical results for various estimators. We see that, although the mean values do not differ very much, the standard deviations of the entropy difference methods are much smaller than the frequency domain methods. In all we notice that EDWBIF is by far the best.

IV. 3-DIMENSIONAL SIMULATION AND EXPERIMENTAL RESULTS FOR β ESTIMATION AND DYNAMIC FOCUSING

In this section we present 3-D simulation and experimental results for the β estimation using the annular system and dynamic focusing. The focused waveform was synthesized such that two-way focusing was attained by using all of the rf echo data by all annuli for each transmitted pulse from one annulus after another. Figure 9 shows the geometry of ROI and the hexagonal pattern of scan lines. The raw data (annulus-pair echo data) are obtained by the hexagonal scanning with a scan line separation of about 0.45 mm on the near spherical surface of the ROI. As preliminary studies, we will first examine the correlations among the synthesized waveforms of neighboring scan lines which are separated by 1 mm in the attenuation estimation. Then, the β value of the tissue-mimicked phantom to be used in the experiments was measured by the transmission mode and also the characteristic of the TGC (time varying gain control) circuit was measured for the purpose of calibration.

The diffraction effect will be then studied by comparing the calibrated amplitude curve with the actually measured amplitude curve as a function of penetration. We will show a good agreement between the diffraction effect computed by simulation and that obtained by measurements. The synthesized echo waveforms are calibrated in amplitude by taking into account of the diffraction effect of our data acquisition system.

After these preprocessings for the synthesized amplitude of each scan line, that is, calibration of the TGC curve and the diffraction effect correction, the β value of the medium is estimated by simulation and experiments using two methods, namely, entropy difference method using narrow-band pulses (EDNB) the entropy difference method using wide-band pulses and inverse filtering (EDWBIF). Thirty-seven scan line data will be used to estimate the β value,

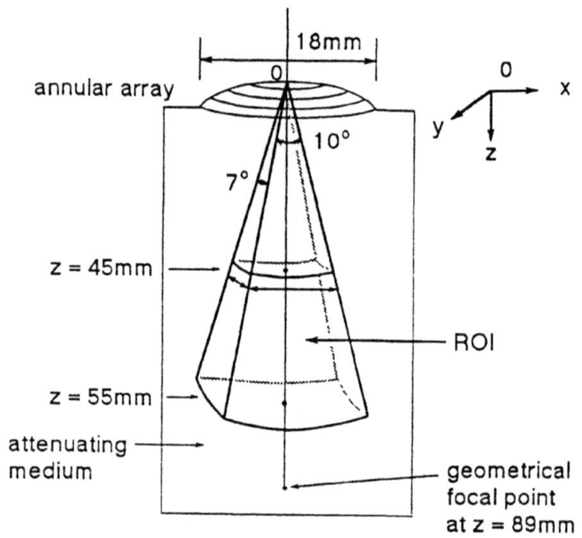

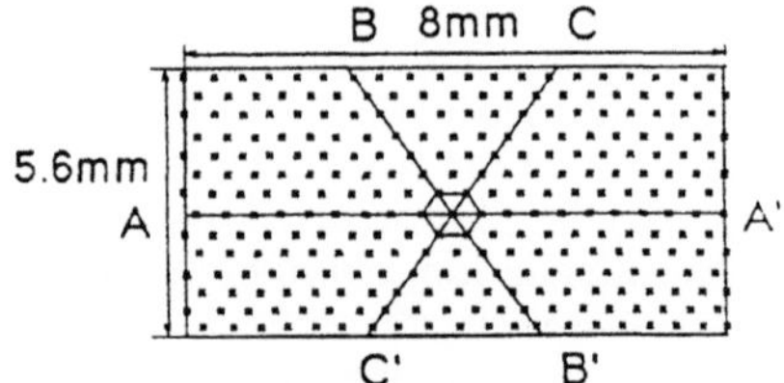

Intersections of scan lines
with the upper spherical
surface (flattened for easy
visualization)

Fig. 9   Geometry for the data acquisition.

and the estimation will be repeated 100 times for different sets of scan lines to plot the histogram to show again the outstanding performance of the EDWBIF method as compared with the EDNB method which is the most reputed method published so far.

IV.1 Scan Line Correlation

For the sake of spatial resolution and accuracy we need dense scan lines with little correlations. However, these two requirements are contradictory and we have to make some compromise between the scan line density and the scan data correlation. Figure 10 shows 7 scan lines positions chosen for the correlation study. The separation of adjacent scan lines is 1 mm on the near spherical surface of the ROI. The correlation coefficient r_{ij} between the synthesized waveform of the i th scan line and that of the j th scan line (hereafter to be called the i th scan data and the j th scan data) is defined as

$$r_{ij} = \frac{\sum_{k}^{n} (x_{ik} - m_i)(x_{jk} - m_j)}{\sigma x_i \sigma x_j} , \qquad r_{ij} = r_{ji} \qquad (11)$$

where m is the mean of the envelope and σ is the standard deviation. And the correlation matrix is defined as

$$R = \begin{pmatrix} 1 & r_{12} & \cdots & \cdots & r_{1n} \\ r_{21} & 1 & \cdots & \cdots & r_{1n} \\ & & \cdot & & \\ & & \cdot & & \\ & & \cdot & & \\ & & \cdot & & \\ r_{n1} & r_{n2} & \cdots & \cdots & 1 \end{pmatrix} . \qquad (12)$$

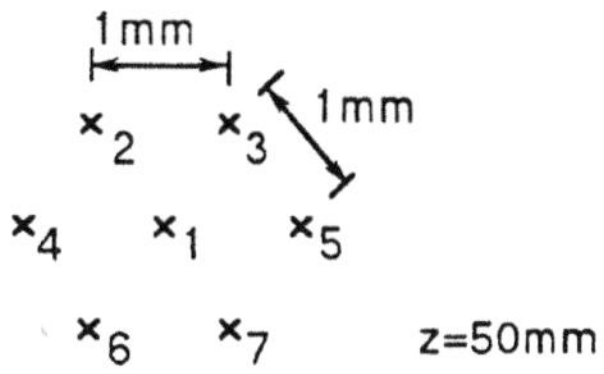

Fig. 10 Seven scan lines for the correlation study.

$$\begin{pmatrix}
1 & 0.25 & 0.14 & 0.16 & 0.19 & 0.09 & 0.21 \\
 & 1 & 0.19 & 0.25 & 0.08 & 0.14 & 0.09 \\
 & & 1 & 0.19 & 0.22 & 0.36 & 0.07 \\
 & & & 1 & 0.09 & 0.06 & 0.09 \\
 & & & & 1 & 0.03 & 0.24 \\
 & & & & & 1 & 0.10 \\
 & & & & & & 1
\end{pmatrix}$$

(a)

$$\begin{pmatrix}
1 & 0.19 & 0.12 & 0.18 & 0.18 & 0.13 & 0.27 \\
 & 1 & 0.09 & 0.19 & 0.16 & 0.09 & 0.13 \\
 & & 1 & 0.13 & 0.24 & 0.04 & 0.13 \\
 & & & 1 & 0.11 & 0.14 & 0.18 \\
 & & & & 1 & 0.05 & 0.17 \\
 & & & & & 1 & 0.04 \\
 & & & & & & 1
\end{pmatrix}$$

(b)

Fig. 11 The correlation matrices with (a) simulated data and (b) for experimental data.

Figure 11 shows the correlation matrices obtained by simulation (scatterer density $= 10^4/cm^3$) and experiments. We see that the simulation and experimental mean values of the correlation agree well with each other, and the correlation between closest scan line data is not very large. Thus, we will use scan line separation of 1 mm in the β estimation in section IV. 4.

IV.2 Measured β Value by the Transmission Mode

The attenuation coefficient β of the tissue-mimicked homogeneous phantom was measured by the transmission mode using a disk-type transducer of 3.5 MHz and a wide-band hydrophone placed on the opposite surface of the phantom. A wide-band pulse was transmitted and received first with the phantom inserted and next with water filled in between. The log spectral difference of the two received signals is plotted as a function of frequency, as shown in figure 12, the slope of which gives the β value as $0.2 \ f^{0.95}$ in most of the frequency range of 2.8 MHz - 3.0 MHz at z = 45 mm - 75 mm. The linealized attenuation coefficient is approximately calculated as 0.54 dB/MHz/cm. Repeated measurements with different transducer locations gave the same result.

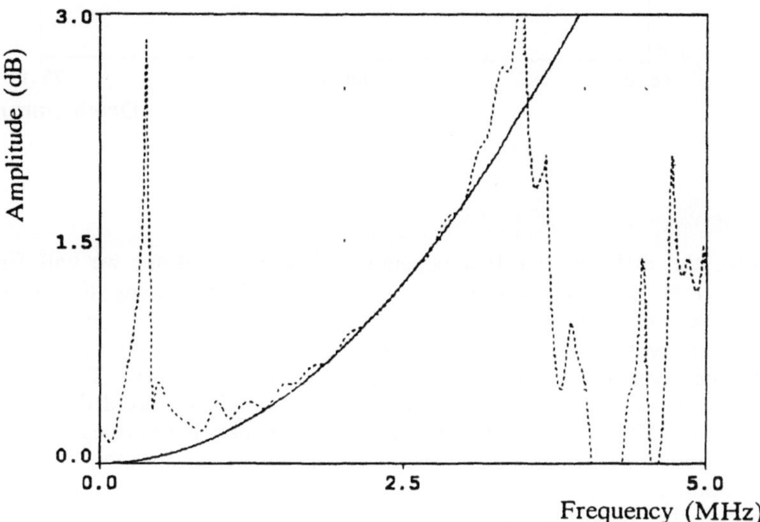

Fig. 12 β estimation by the transmission mode.

IV.3 TGC Curve Calibration

The gain of the TGC circuit in the receiver system was measured using a sinusoid of 3.5 MHz as a function of time (or the depth). The result is plotted in figure 13 — ideally this curve should be exponential. We divide the synthesized rf data by the TGC value to recover the original attenuated data. The result will be call attenuation calibrated data.

IV.4 Diffraction Effect

Since the beam spreads as it propagates through the ROI, the echo amplitude and hence the synthesized waveform decreases in excess of mere attenuation. This diffraction effect must be taken into account and the amplitude of the synthesized scan data must be calibrated accordingly before any β estimation is started.

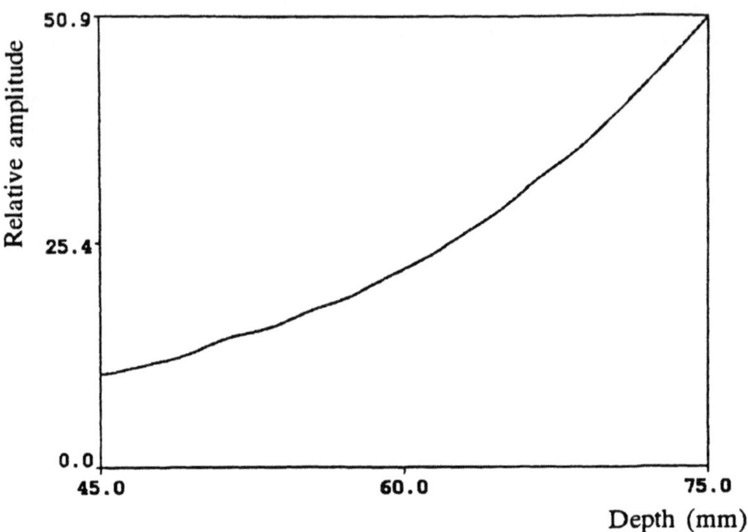

Fig. 13 Actual TGC curve.

Since the diffraction effect depends on the attenuation, we will first compare the simulated and expermental diffraction effects using the measured β value of the phantom, that is, $\beta = 0.54$ dB/MHz/cm. In the simulation we averaged 280 synthesized scan data; the result is plotted in figure 14(a). Next, we take the average of the original echo envelopes before TGC is applied (which can be obtained from the envelopes after TGC divided by the TGC factor), using 280 synthesized scan data expermentally obtained; the result is plotted in figure 14(b). In both the fluctuating curve is smoothed and the ratio of the smoothed amplitude to that of the expected attenuation is plotted as a function of the depth in figure 14(c), which shows that the simulation and experimental results agree very well and that the diffraction effect increases with depth.

Figure 15 shows the simulated diffraction effect as a function of depth in an nonattenuating medium for four different center frequencies of the annular array with otherwise the same geometry. We note that the curves are almost uniformly shifted as the center frequency is shifted. Numerically it was found that the following approximation holds:

$$\frac{A_{f_0}}{A_{f_0'}} = (\frac{f_0'}{f_0})^{3/2} \tag{13}$$

where A_{f_0} and $A_{f_0'}$ are the diffraction factors at center frequencis of $f_0 = 3.2$MHz and f_0'. Therefore, we can tell the diffraction effect at an arbitrary depth, since the shifted center frequency at that depth can be measured by, say, the autocorrelation method [17]. This will be utilized in steps 1 and 4 of the β estimation procedure described in the following.

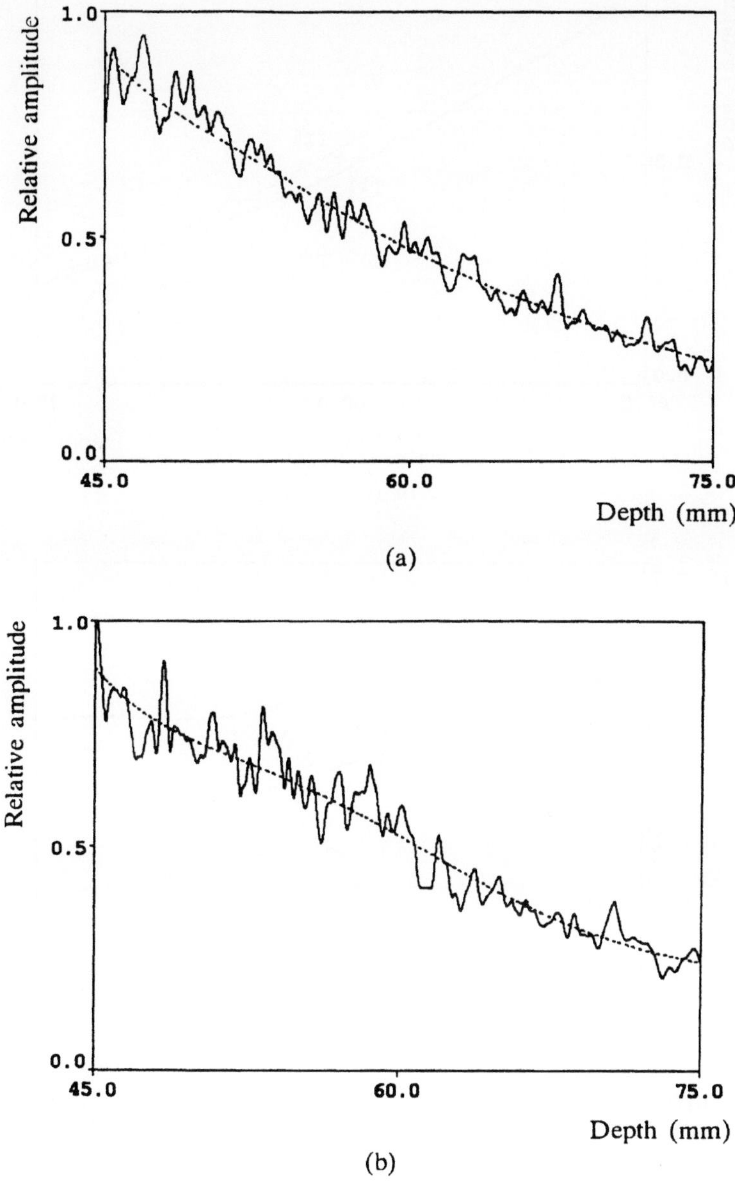

Fig. 14 Diffraction curves. (a) Averaged over 280 scan data from the simulated envelopes, (b) averaged over 280 scan data from the experimently obtained envelopes, and (c) smoothed curves (1) simulated, (2) experimental.

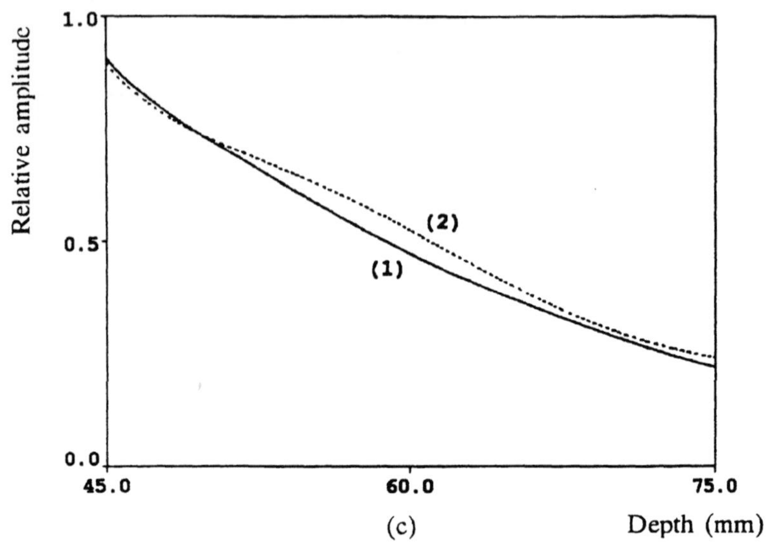

(c)

Fig. 14 (continued)

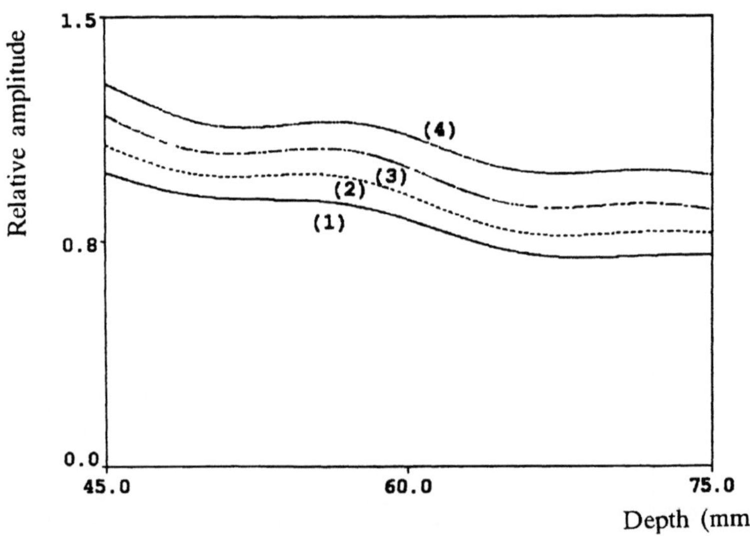

Fig. 15 Diffraction curves for various frequencies. (1) f_0 = 3.2 MHz, (2) f_0' = 3.0 MHz, (3) f_0' = 2.8 MHz, and (4) f_0' = 2.6 MHz.

IV.5 Experimental Results

The experimental procedure of the β estimation follows:
step 1. Obtain the TGC and diffraction effect calibrated scan data.
step 2. Estimate the β value by the ED method.
step 3. Apply the inverse filter with the estimated β value.
step 4. Calibrate the diffraction effect.
Repeat the step 2, 3, 4 until a convergent value of β is obtained.

Both simulation and experiments were performed using 37 scan lines (in a volume of 6 mm x 6 mm x 6 mm (depth)) to get one estimated β value. The estimation was repeated 100 times using different sets of scan line data. In the simulations the scan line data were generated by using the β value of the phantom, 0.54 dB/MHz/cm. Table 2 summarizes the mean and standard deviation of estimated β values obtained by simulation and experiment using two different ED methods, EDWBIF and EDNB. We see that the use of wide-band pulses and inverse filtering results in much better estimation than that of narrow-band pulses.

In actual calculations, the algorithm stated above always yield a convergent value of β (within a few %) after 2 iterations, for the initially estimated value of β is quite close to the actual value. It would be very difficult, however, to prove the convergence theoretically since the objective function in this case is a statistical measure (entropy difference) and its behavior is quite unpredictable because of the inverse filtering (a frequency domain operation) and the β compensation (a time operation) involved in each iteration.

Table II. Statistical results for various estimators (3-D simulation and experiment)

Units: dB/MHz/cm

	Simulation		Experiment	
	EDWBIF	EDNB	EDWBIF	EDNB
$\hat{\beta}$	0.543	0.541	0.532	0.533
σ_β	0.032	0.078	0.041	0.093

V. CONCLUSIONS

We proposed a new entropy difference estimator for ultrasonic attenuation coefficient in soft tissue, in which we used wide-band pulses instead of narrow-band pulses. We have shown the necessity of inverse filtering in this case. And we applied the diffraction effect correction which has not been taken into account in other estimators so far. The superiority of the proposed estimator over other estimators was confirmed by simulation and experiments.

It is expected that the spatial resolution of attenuation mapping can further be improved by using pulses of higher frequencies and an annular array having more elements. The in-vivo attenuation images will be obtained in the future.

REFERENCE

1. A. C. Kak and K. A. Dines, Signal processing of broadband pulsed ultrasound: measurement of attenuation coefficient of soft biological tissues, *IEEE Trans. Sonics* and *Ultrasonics* SU-31, 313-330, 1978.

2. R. Kuc and M. Schwartz, Estimating the attenuation coefficient slope for the liver from the reflected ultrasound signals, *IEEE Trans. Sonics* and *Ultrasonics* SU-26, 353-362, 1979.

3. M. Fink, F. Hottier, and J. F. Cardoso, Ultrasonic signal processing for in vivo attenuation measurement: short time Fourier analysis, *Ultrasonic Imaging* 5, 117-135, 1983.

4. S. Shaffer, D. W. Pettibone, J. F. Havlice, and M. Nassi, Estimation of the slope of the acoustic attenuation coefficient, *Ultrasonic Imaging* 6, 126-138, 1984.

5. R. Kuc, Bounds on estimating the acoustic attenuation of small tissue regions from reflected ultrasound, *Proceedings of IEEE* Vol. 73, NO. 7, 1159-1168, 1985.

6. R. Kuc and H. Li, Reduced-order autoregressive modelling for center-frequency estimation, *Ultrasonic Imaging* 7, 224-251, 1985.

7. J. Ophir, T. H. Shawker, N. F. Maklad, J. G. Miller, S. W. Flax, P. A. Narayana, and J. P. Jones, Attenuation estimation in reflection: progress and prospects, *Ultrasonic Imaging* 6, 349-395, 1984.

8. M. Insana, J. Zagzebski, and E. Madsen, Improvements in the spectral difference method for measuring ultrasonic attenuation, *Ultrasonic Imaging* 5, 331-345, 1983.

9. P. He and J. F. Greenleaf, Application of stochastic analysis to ultrasonic echoes-estimation of attenuation and tissue heterogeneity from peaks echo of envelope, *J. Acoust. Soc. Am.* 79, 526-534, 1986.

10. H. S. Jang, T. K. Song, and S. B. Park, Ultrasound attenuation estimation in soft tissue using the entropy difference of pulsed echoes between two adjacent envelope segments, *Ultrasonic Imaging* 10, 248-264, 1988.

11. J. Ophir, M. A. Ghouse, and L. A. Ferrari, Attenuation estimation with the zero-crossing technique: phantom studies, *Ulatrasonic Imaging* 7, 122-132, 1985.

12. S. Leeman, L. Ferrari, J. P. Jones, and M. Fink, Perspectives on attenuation estimation from pulse-echo signals, *IEEE Trans. Sonics* and *Ultrasonics* SU-31, 352-361, 1984.

13. M. Fink and J. F. Cardoso, Diffraction effect in pulse echo measurement, *IEEE Trans. Sonics* and *Ultrasonics* SU-31, 313-330, 1984.

14. M. O'Donnell, Effects of diffraction on measurements at the frequency-dependent ultrasonic attenuation, *IEEE Trans. Biomedical Engineering* BME-30, 320-326, 1983.

15. R. Kuc, Estimating reflected ultrasound spectra from quantized signals *IEEE Trans. Biomedical Engineering* BME-32, 105-112, 1985.

16. R. Kuc, Modeling acoustic attenuation of soft tissue with a minimum-phase filter, *Ultrasonic Imaging* 6, 24-36, 1984.

17. C. Kasai, K. Namekawa, A. Koyano, and R. Omoto, Real time two-dimensional flow imaging using an autocorrelation technique, *IEEE Trans. on Sonics* and *Ultrasonics*, SU-32, NO. 3, 458-464, 1985.

HIGH FREQUENCY ULTRASONIC CHARACTERIZATION
OF BIOLOGICAL MEDIA

M. Toubal, E. Radziszewski, M. Asmani, M. Ourak, F. Lefebvre

IEMN (UMR CNRS 9929)- Département OAE
Université de Valenciennes- BP311
59304 Valenciennes Cedex
France

INTRODUCTION

Although acoustic attenuation phenomena play an important role in many medical applications, we still have only a rudimentary understanding of the interaction mechanisms fifty years after the first work in the field[1].

A particular attention has been devoted to describe these phenomena at medical frequencies (1 to 30 MHz), proving detrimental to the microscopic information carried by higher ones.

The results we present here constitute a continuation to previous ones[2] dealing with the study of ultrasonic attenuation in blood in the frequency range 50 to 150 MHz. Our interest is in the quantification of the scatterers' influence and particularly their membranes in the attenuation phenomena by comparing media with and without scatterers.

We made use of liposomes to approach our results; they represent a simpler biological model than blood, as they appear as membranes which can separate two identical media.

I- MATERIAL AND METHOD.

1- Experimental device

Many methods can be used for the experimental determination of the ultrasonic attenuation. Radiation force insertion loss measurements are employed to determine amplitude attenuation coefficients under continuous wave (C.W.) irradiation[3]. Care must be taken however to use power levels low enough to avoid nonlinear effects. The acousto-optic method concernes only transparent media.

Moreover, considering the high attenuation in blood, we opted for the transmission and pulse decay technique, using two pairs of lithium niobate transducers to cover the whole frequency range.

We got rid of nonlinear effects, owing to the low power level (less than 10 mW/cm^2).

The system's calibration was checked with regard to water. The measurements' precision is about 5%.

2- Biological media

a- Blood as a random medium

Human blood is composed of a liquid (the plasma) in which are suspended Red Blood Cells (RBC's) or erythrocytes, White Blood Cells, and platelets.

The number of RBC's is much larger than that of platelets, and the volume of a red cell is much larger than that of a platelet[4]. Thus the scattering of ultrasound by blood is presumably due to RBC's[5].

While propagating in blood, the ultrasonic wave interacts with three different media: plasma, hemoglobin and the membrane containing it; showing each peculiar ultrasonic characteristics. Cartensen's previous work[6] showed that the acoustic properties of blood are largely determined by the proteins which it contains.

The values of wave velocity c and acoustic impedance ρc for blood[7], plasma[8], erythrocyte[8], and membrane[9], are shown in table 1.

On another hand, to study the influence of blood parameters related to the attenuation (intra and extra-cellular media, membrane, RBC's shape...), it is important to be able to vary them.

Table 1. Velocity and acoustic impedance data

	Blood[7]	Plasma[8]	Erythrocyte[8]	Membrane[9]
Velocity m.s^{-1}	1570	1547	1639	1520
Acoustic impedance x 10^6 Kg.m-^{2}.s^{-1}	1,61	1,58	1,79	1,67

Liposomes provide a suitable biological model in which these operations are easy to make.

b- Liposomes

Liposomes are small vesicles with diameters between 0.02 to 10μm and 4nm thick whose membranes consist of a bilayer of phospholipid molecules. This bilayered structure make them approximate to cell membranes[10].

The liposomes used are multilamellar single bilayers, 2μm in diameter with a distribution of 72%.

They were prepared from egg phosphatidyl choline, and suspended in Physiological Buffered Saline (PBS 5mM, pH 7.4) which absorption is like that of water.

II- EXPERIMENTAL RESULTS

When interacting with a medium containing suspending particles, the acoustic wave, over and above absorption, undergoes a scattering of its energy.

The scattering by erythrocytes of blood is widely studied[8,11,12] in the medical frequency range.

The developed models are however hardly transposable in higher frequency range because of the observation's scale that modifies the proportion between wavelengths and interacting media.

In a previous work[2], we showed that about half of the attenuation we measured in blood was due to the presence of RBC's in the frequency range 50 to 400 MHz. We aimed here at seeing the influence of the membrane in the attenuation. We used for that RBC's alone (obtained by blood centrifugation) and liposomes.

1- Comparison between RBC's and haemolysed RBC's

In RBC's, the wavelength interacts with only the membrane and the hemoglobin. When haemolysing them, we suppress the membrane and we just measure then the absorption in hemoglobin.

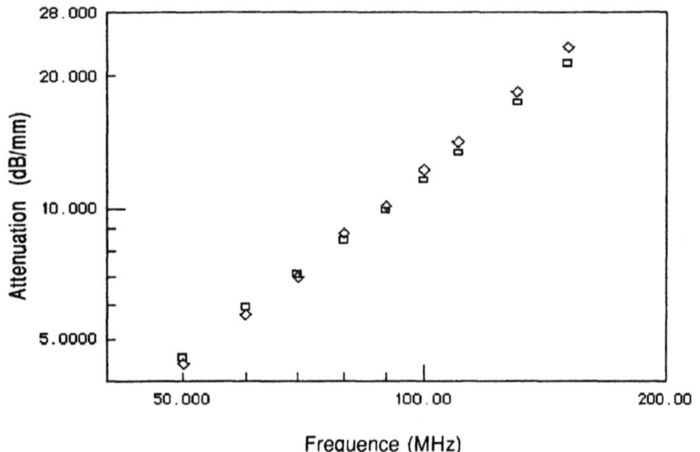

Figure 1. Membrane's influence on RBC's
◊ Centrifuged RBC's □ Haemolysed RBC's

Figure 1 shows that by comparing both of the curves we can deduce that the membrane's participation is about 10% in the frequency range from 50 to 150 MHz.

2- Attenuation in a liposomal solution

The suspending medium, being from the acoustical position, identical to water, we compared in figure 2, the attenuation measured in a solution of liposomes with the absorption in water.

Here again, we can see that in the frequency range from 50 to 150MHz, about 10% of the attenuation is due to the membrane.

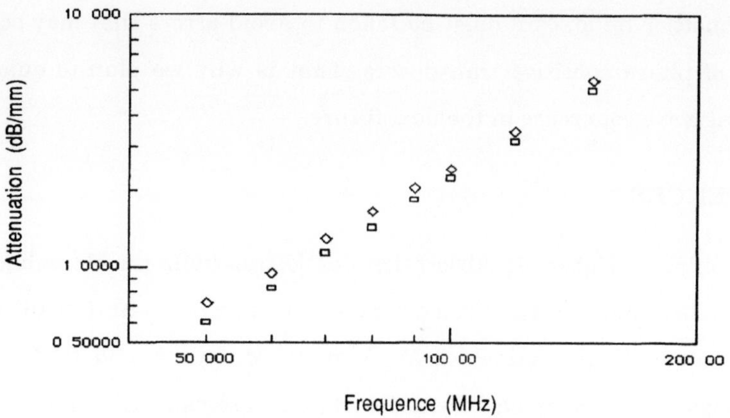

Figure 2. Membrane's influence on liposomes

◇ Liposomes   □ Water

III- DISCUSSION

In spite of the proportion difference between wavelengths and scatterers'size, (7µm for RBC's and 2µm for liposomes), and although the membranes are not the same thickness in RBC's (100nm) and liposomes (4nm), we could measure a membrane's contribution of about 10% in both cases.

This effect has previously been put forward by Anson and Chivers[12] who showed in their model of ultrasonic propagation in mammalian cell suspensions, that below 0,5 MHz and 30 MHz, the effect is virtually negligible. For all intermediate frequencies, it was found to be less than 14%.

As regards, Cartensen[6] experimentally checked this effect. Between 1 and 20 MHz, he concluded that one tenth of the attenuation he measured in blood is due to the membrane.

IV- CONCLUSION

Scattering of the acoustical energy by RBC's is a wide prospection subject that needs to take into consideration many physical and chemical parameters such as shape, size, concentration, compressibility, density...

The use of liposomes will help us to understand some of the mechanisms ruling this phenomenon.

On another hand, care must be taken to avoid errors that may occur from the use of phase sensitive transducers. That is why we plan to quantify the acoustical wave coherence in the next future.

REFERENCES

1. R. Pohlman, Ueber die Absorption des Ultraschalls im Menschlichen Gewebe und ihre Abhaengigkeit von der Frequenz, Physik. Z.40, 159-161 (1939).

2. M. Toubal, E. Radziszewski, M. Asmani, M. Ourak and B. Nongaillard, Caractérisation du Sang par Ultrasons Hautes Fréquences, Colloque C1, supplément au journal de Physique III, Volume 2, 771-774, Avril 1992.

3. M. J. Lyons and K. J. Parker, Absorption in Soft Tissues II- Experimental Results, IEEE Transactions on Ultrasonics, Ferroelectrics, and Frequency Control, Vol. 35, N° 4, July 1988.

4. W. R. Platt, Color Atlas and Textbook of Hematology. Philadelphia, PA: Lippincott, 1969.

5. J. M. Reid and R.A. Siegelman, private communication.

6. E. L. Cartensen, H. P. Schwan, Determination of the Acoustic Properties of Blood and its Components, J.A.S.A, Vol.25, 1953.

7. J. P. Jones, S. Leeman, La Caracterisation des Tissus Biologiques par Ultrasons, Acta Electronica, 26, 1-2, 1984, 3-31.

8. K. K. Shung, R. A. Sigelmann, and J. M. Reid, Angular Dependence of Scattering of Ultrasound from Blood, Ibid., BME-24, 325-331.

9. L. W. Anson, and R. C. Chivers, Ultrasonic Propagation in Mammalian Cell Suspensions Based on a Shell Model, Phys. Med. Bio., 1989, Vol. 34, N° 9, 1153-1167.

10. D. D. Lasic, Les Liposomes, La Recherche, Vol. 20, N° 212, 905-913,Juillet-Août 1989.

11. K. K. Shung, A. Rubens, and J. Reid, Scattering of Ultrasound by Blood, IEEE Transactions on Biomedical Engineering, Vol. BME-23 N° 6, Nov. 76.

12. B. A. J. Angelsen, A Theoretical Study of The Scattering of Ultrasound from Blood, IEEE Trans. on Biomed. Engin., Vol. BME-27, N° 2, Feb. 80.

IMAGE SEGMENTATION AND 3D RECONSTRUCTION
OF INTRAVASCULAR ULTRASOUND IMAGES

W. Li[1,4], J.G. Bosch[3], Y. Zhong[1], H. v. Urk[2], E.J. Gussenhoven[1,4],
F. Mastik[1], F. v. Egmond[1], H. Rijsterborgh[1,4], J.H.C. Reiber[3], N. Bom[1,4]

[1]Thoraxcenter, Erasmus University Rotterdam
[2]Department of Vascular Surgery, University Hospital Rotterdam-Dijkzigt
[3]Department of Diagnostic Radiology, University of Leiden
[4]Interuniversity Cardiology Institute
The Netherlands

ABSTRACT

Intravascular ultrasound is a new imaging modality which provides real-time, high-resolution, cross-sectional images of the arterial lumen and structures of the vessel wall. An insight into the spatial distribution of atherosclerosis can be obtained by three-dimensional (3D) reconstruction of intravascular ultrasound. A method is described for image segmentation and subsequent 3D modelling of sequential ultrasound data. The segmentation procedure detects the contours of the arterial lumen and media interface; the voxel modelling method is then applied to create the arterial objects in a 3D space.

INTRODUCTION

Development of the intra-luminal ultrasonic imaging technique provides an opportunity to visualize the arterial lumen and the lesion in humans. By providing real-time, cross-sectional images of the vessel, this new technique enables additional assessment of vascular atherosclerotic diseases prior to and following intervention [1-4]. One limitation of intravascular ultrasound is that only one cross-sectional view is available at each moment. Interpretation of the longitudinal distribution of the vessel pathology is only obtainable after reviewing all the ultrasonic cross-sections from videotapes. With computerized image processing techniques, three-dimensional (3D) images of the vessel length can be reconstructed from a set of longitudinally stacked intravascular ultrasonic images. The 3D presentation of echographic data may facilitate interpretation of the complex nature of diseased arteries.

This paper describes an image segmentation method developed at our laboratory for 3D reconstruction of intravascular ultrasonic data.

Acoustical Imaging, Volume 20 Edited by Y. Wei
and B. Gu, Plenum Press, New York, 1993

INTRAVASCULAR ULTRASOUND

Intravascular ultrasound images are generated using a 32 MHz single element system (Du-Med, Rotterdam, The Netherlands). The ultrasonic transducer is mounted on the tip of a 5F catheter and rotated by a drive-shaft at a high speed (1000 rpm). The system is capable of producing 16 ultrasonic scans (360°) per second. Axial resolution of the system is 0.75 mm and lateral resolution is better than 2.25 mm at a depth of 2-4 mm. The ultrasound data are displayed in a standard video format via a real-time digital scan convertor.

COMPUTER SYSTEM

Ultrasound data are processed with an IBM PC/AT compatible system. The computer is equipped with a DT 2851 framegrabber to digitize the video signals. The digital image data are stored on the hard disk in a 512 x 512 x 8 bits resolution. The results of image segmentation and 3D reconstruction are displayed on a colour videomonitor connected to the output of the framegrabber. Manual interaction is performed with a PC mouse device.

DATA ACQUISITION

Manual pull-back of the echocatheter at a steady speed is a commonly-used method to obtain a sequence of echographic slices from different vessel positions[5,6]. One disadvantage of this method is that speed variations during the manual pull-back processing may produce errors in the estimation of the spatial distance between the slices, resulting in inaccurate sampling of the vessel along the Z axis. To solve this problem, a dedicated displacement sensor has been developed at our laboratory to determine the longitudinal position of each echographic cross-section. The system consists of an optical sensing device

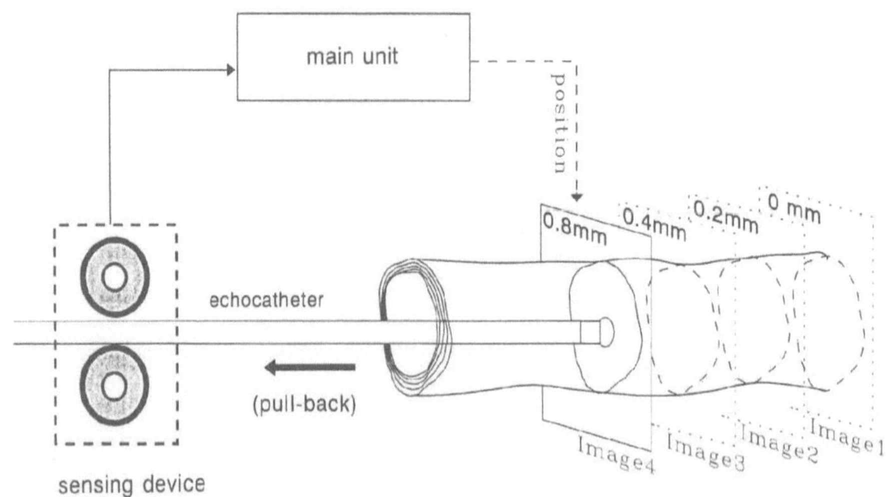

Figure 1. Diagram illustrating the ultrasound data acquisition from the artery specimen using a dedicated sensing device.

and a main-unit. The catheter is pushed through the sensing device prior to being advanced into the arterial specimen. The mechanical movement by advancing/withdrawing the catheter is detected and encoded as digital signals. These digital data are received and processed by the main-unit to provide the position measurement in each 0.1 mm displacement. The position data are simultaneously superimposed on ultrasonic images and recorded on a videotape. With the position information ultrasound data can be accurately acquired at a given spatial resolution. The experimental setup for data acquisition is illustrated in Figure 1.

In this experiment ultrasonic slices were resampled at 0.2 mm intervals from videotape playback and a total of 80 cross-sections were obtained for a length of 1.6 cm coronary specimen.

IMAGE SEGMENTATION

Image segmentation is performed to define the regions of the vessel lumen and lesion on each ultrasonic cross-section. The luminal area is found by detecting the leading edge of the arterial tissues. To estimate the size of the *original native lumen,* the interface of the echolucent media is derived (Figure 2). The lesion area is extracted by subtracting the free lumen from the media-bounded region[7].

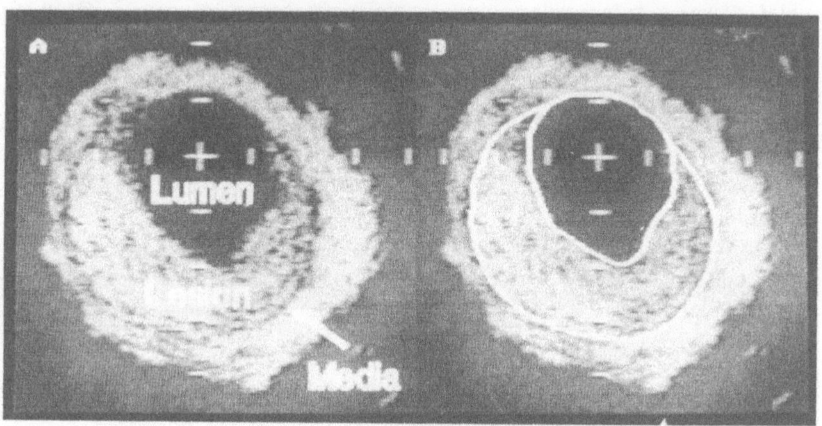

Figure 2. Intravascular ultrasound cross-sections of a pressurized coronary artery specimen. (A) The media interface (arrow) is interposed between the lesion and the adventitia. (B) The lesion area is bounded by the contours of the lumen and media.

The borders of the free lumen and media-bounded area are extracted using a contour detection method developed from the minimum cost algorithm[8,9]. This technique consists of 4 steps: 1) image resampling; 2) cost computation; 3) determining the minimum cost path; 4) converting the path to a contour.

The segmentation of the image sequence is initialized by the manual indications of the lumen and media contours on the first frame. The manual contours are then used as model to resample the following image into a polar format. From the resampled data a cost matrix is yielded, in which each element represents the edge strength or the possibility to be an edge point. For detection of the luminal boundary, the cost value is defined by the spatial first-derivative method; large changes in the echo intensity will produce low cost values in the matrix. This method permits detection of the luminal edge which contains a significant transaction in grey levels. The media interface, however, is usually recognized by its black-ring appearance rather than by the changes in image intensity. Edge enhancement based on the gradient of grey levels is unable to produce sufficient contrast for discrimination of the media boundary. To detect this interface, a pattern matching

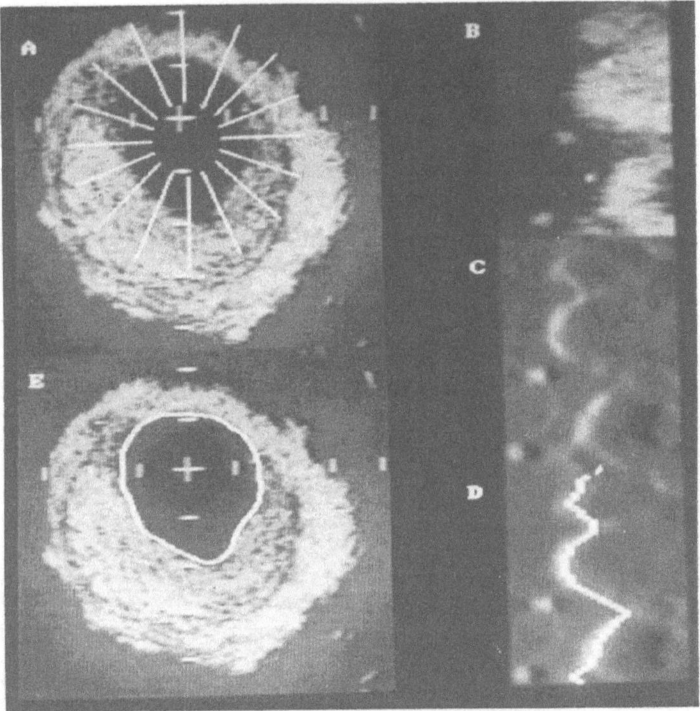

Figure 3. (A) Resampling of the image along the scan lines. (B) The ultrasound data in a polar format. (C) The cost matrix displaying low cost values with high grey levels. (D) The path as determined by the minimum cost algorithm. (E) The luminal contour derived by transferring the path back to the image coordinates.

processing by cross-correlation is adopted for the cost calculations. The value of the cost element is defined by the normalized cross-correlation coefficient between a line pattern and the resampled data[10]. Low cost values will be assigned to the points with high similarity levels. Next, through the cost matrix a path with the smallest accumulated cost is determined by the minimum cost algorithm. This path is globally optimized in terms of connectivity and smoothness. Finally, the points in the path are transformed back to the image coordinates, and interpolated to generate a circumferential outline. Figure 3 illustrates each step of the contour detection procedure.

THREE-DIMENSIONAL RECONSTRUCTION

The 3D arterial objects are reconstructed using the voxel modelling method[11]. In this modelling approach, a pixel on the 2D echo image is extended to a 3D volume element or so-called "voxel". Each voxel is classified as a member of either the lumen or the wall structures by the detected contours. The reconstructed data sets are rendered by the depth-gradient shading method. The brightness of the voxel is computed from both the depth of the voxel to provide distance perception, and the gradient vector of the voxel to enhance the orientation of the voxel surface. The resulting 3D reconstruction is displayed on the color monitor via the video memory in the framegrabber.

To visualize the interior structure of the vessel segment, the reconstructed arterial wall can be cut and opened longitudinally (Figure 4). The lesion volume is encoded with different colours to allow optimal differentiation of the lesion from the normal arterial

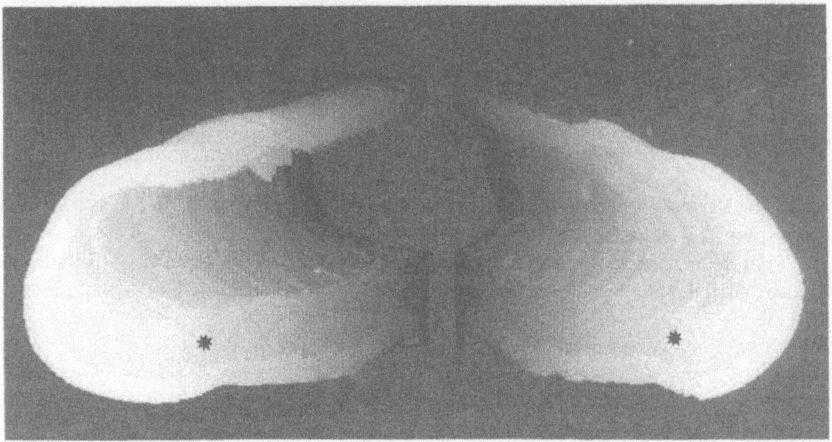

Figure 4. The cut-open view of a coronary artery showing an eccentric lesion (*) distribution along the segment.

wall. Figure 5 shows the 3D images of the arterial lumen reconstructed as a solid model and visualized from X and Y planes. The separate 3D display of the vessel lumen has the advantage that it provides a similar image to that obtained by angiography and thus facilitates analysis of the luminal geometry.

With the detected contours the cross-sectional areas of the lumen and lesion can be automatically computed. The area measurement is obtained by converting the pixels enclosed by the contour into square millimeters with calibration. The resulting quantitative data from all slices can be plotted as a function of vessel position (Figure 6). This display format provides immediate assessment of the lesion distribution along the length of the vessel.

DISCUSSION

This study has demonstrated the feasibility of a computerized method for image segmentation and 3D reconstruction of intravascular ultrasound images. Preliminary results

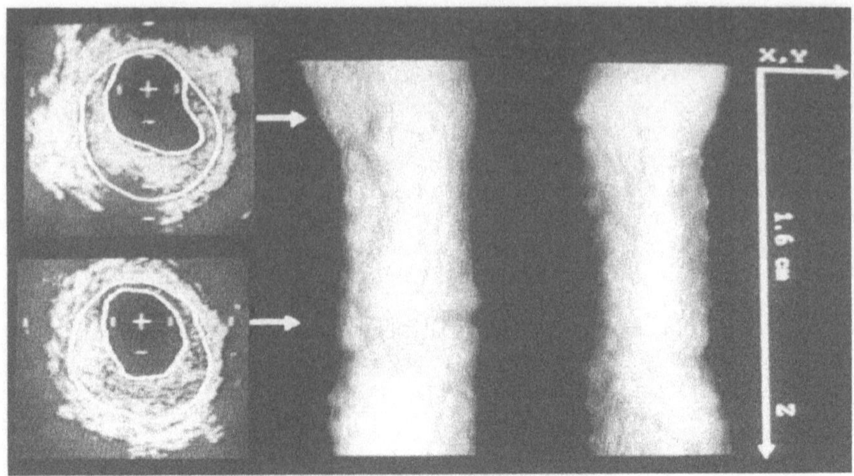

Figure 5. 3D reconstruction of the vessel lumen as a solid model and viewed from the X and Y directions. Two ultrasonic cross-sections are shown with the arrows to indicate the corresponding longitudinal positions.

have indicated that reconstruction of sequential echographic slices can reproduce in detail the complex 3D geometry of the vessel segment.

The development of the catheter displacement sensor enables to obtain accurate measures of the vessel position for each ultrasonic cross-section. Further implementation

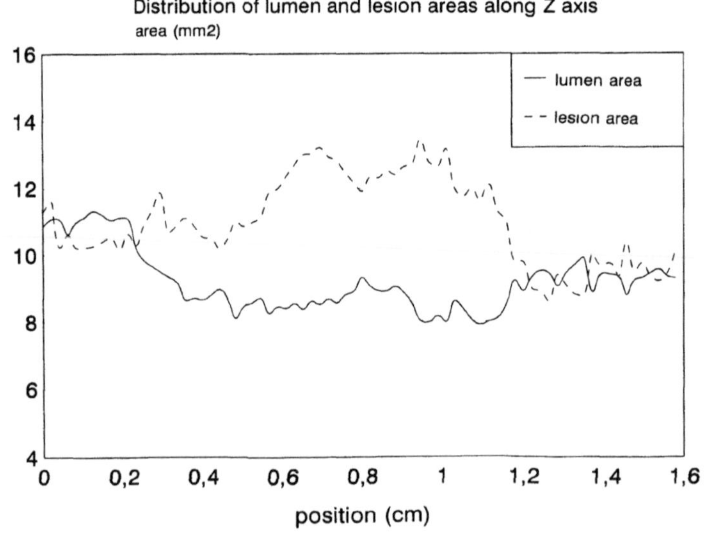

Figure 6. Lumen and lesion areas plotted as a function of the vessel position.

of a computer interface in this device may automatize the data acquisition procedure and allow real-time storage of a complete image sequence.

Image segmentation is an important step for the accuracy of 3D reconstruction. The threshold processing as described by other research groups is a simple and fast technique[5,6,12]. This method, however, is unable to derive the boundary of the media interface for the color-encoded display or for quantification of the lesion volume. The use of the pattern matching technique in combination with the minimum cost algorithm provides possibilities to detect the media boundary semi-automatically and thus allows classification of the lesion voxel in 3D reconstruction. Application of this contour detection technique will require further validation studies to determine its accuracy and reproducibility.

In conclusion, 3D reconstruction of 2D ultrasonic data offers potential as an adjunct to enhance the diagnostic capability of intravascular ultrasonic imaging techniques. The quantitative assessment of the lumen and lesion dimensions along the longitudinal direction may play an important role in assessing the clinical outcome of therapeutic intervention.

REFERENCES

1. Gussenhoven EJ, Essed CE, Frietman P, Mastik F, Lancée C, Slager C, Serruys P, Gerritsen P, Pieterman H, Bom N. Intravascular echographic assessment of vessel wall characteristics. *Int J Card imag* 1989; 4: 105-116.
2. Potkin BN, Bartorelli AL, Gessert KM, Neville RF, Almagor Y, Roberts WC, Leon MB. Coronary artery imaging with intravascular high-frequency ultrasound. *Circulation* 1990; 81: 1575-1585.
3. Pandian NG, Kreis A, Weintraub A, Motarjeme A, Desnoyers M, Isner JM, Konstam M, Salem DN. Real-time intravascular ultrasound imaging in humans. *Am J Cardiol* 1990; 65: 1392-1396.
4. The SHK, Gussenhoven EJ, Zhong Y, Li W, van Egmond F, Pieterman H, van Urk H, Gerritsen GP, Borst C, Wilson RA, Bom N. The effect of balloon angioplasty on the femoral artery evaluated with intravascular ultrasound imaging. *Circulation* 1992; 86: 483-493.
5. Rosenfield K, Losordo DW, Ramaswamy K, Pastore JO, Langevin RE, Razvi S, Kosowsky BD, Isner JM. Three-dimensional reconstruction of human coronary and peripheral arteries from images recorded during two-dimensional intravascular ultrasound examination. *Circulation* 1991; 84: 1938-1956.
6. Cavaye DM, Tabbara MR, Kopchok G, Laas T, White RA. Three-dimensional vascular ultrasound imaging. *Am Surg* 1991; 57:751-755.
7. Li W, Gussenhoven WJ, Zhong Y, The SHK, Di Mario C, Madretsma S, Egmond F van, Feyter P dd, Urk H van, Rijsterborgh H, Bom N. Validation of quantitative analysis of intravascular ultrasound images. *Int J Cardiac Imag* 1991; 6: 247-253.
8. Gerbrands JJ, Hoek C, Reiber JHC, Lie SP, Simoons ML. Minimum cost contour detection in technetium-99m gated cardiac blood pool scintigrams. In *Proc. Comput Cardiol Conf* 1982: 253-256.
9. Bosch JG, Reiber JHC, Burken G van, Gerbrands JJ, Gussenhoven WJ, Bom N, Roelandt JRTC. Automated endocardial contour detection in short-axis 2-D echocardiograms: methodology and assessment of variability. In *Proc Comput Cardiol Conf* 1988: 137-140.
10. Li W, Bosch JG, Zhong Y, Gussenhoven WJ, Rijsterborgh H, Reiber JHC, Bom N. Semiautomatic frame-to-frame tracking of the luminal border from intravascular ultrasound. In *Proc Comput Cardiol Conf* 1991: 353-356.

11. Kitney R, Moura L, Straughan K. 3-D visualization of arterial structures using ultrasound and voxel modelling. *Int J Cardiac Imag* 1989; 4: 135-143.
12. Franceschi D, Bondi JA, Rubin JR. A new approach for three-dimensional reconstruction of arterial ultrasonography. *J Vascular Surgery* 1992; 15: 800-805.

TEXTURE SEGMENTATION OF ECHOGRAMS FOR TUMOUR DETECTION

Johan M. Thijssen and Hans J.T.M. Verhoeven

Biophysics Laboratory
Department of Ophthalmology
University Hospital
6500 HB Nijmegen, The Netherlands

INTRODUCTION

Echography is a non-invasive diagnostic method, using ultrasound frequencies in the range of 2 to 25 MHz and producing real-time images. The latter facility enables interactive use of this medical imaging modality. This is one of the reasons of its wide spread in the clinical environment: 20 percent of medical imaging diagnoses are made with echographic devices of various kinds. The market share of echographic devices is also of the order of 20 percent of the mondial market for medical imaging devices, whereas, its growth is 15 percent per year.

One of the essential limitations of echographic imaging is the "speckle" formation, which is due to the coherent summation of backscattered echo signals at the receiving transducer. This backscattering is caused by small inhomogeneities and structures (e.g. microvasculature, stroma) of parenchymal tissues. The speckle in echographic images represents a kind of spatial noise, or "clutter", which significantly limits the detectability of small, or low contrast, lesions like tumors. Furthermore, the structural characteristics of the (micro)anatomy of tissues are masked by this noise thereby hindering the detection of diffuse pathological changes which may occur in tissues. Since, the reflectivity contract (i.e. grey level contrast of the image) produced

Acoustical Imaging, Volume 20 Edited by Y. Wei
and B. Gu, Plenum Press, New York, 1993

by tumors depends on the kind of tumour and on the growth rate, tumour detection algorithms should not only be designed to enhance the grey level contrast, but also to improve the display of other texture features which are more related to the tissue structure.

By realistic 3-D simulations it could be shown that the 1st and 2nd order statistical parameters of the speckle pattern (texture) in echographic images are systematically dependent on the number density of scattering sites within the tissue (Thijssen and Oosterveld, 1985; Oosterveld et al. 1985). Moreover, the effects of a structural component of the scattering on the texture were investigated quantitatively and the limits of the analysis methods to extract the structure characteristics from the image texture were assessed (Jacobs and Thijssen, 1991). The significance of the statistical parameters was investigated by in vitro (Cloostermans et al., 1986) and in vivo studies (Romijn et al., 1991; Thijssen et al., 1991; Oosterveld et al., 1991; Hartman et al., 1992; Oosterveld et al., 1992; Thijssen et al., 1992).

Based on these promising clinical results, a new approach was initiated which is aiming at image processing rather than image analysis. In other words the goal is to either enhance the images, or to produce new kinds of images by local analysis of the data. In a recent study (Verhoeven et al., 1991) the authors investigated the potentials of SNR images for improving the detectability of lesions. The SNR stands for "signal-to-noise ratio" of the gray level histogram, or the inverse of the speckle contrast. It could be shown that lesions due to a reduction of the scatterer number density became detectable, even in case of zero grey level contrast!

A second approach of image processing for the texture segmentation in lesion detection is based on a scale-space approach (Witkin, 1983; Babaud et al., 1986) to estimate the fractal dimension of textures (Muessigmann, 1989,1990; Rueff, 1989). This fractal dimension is derived by systematically varying a scale variable. It could be shown that the fractal dimension images produced by a sliding window technique yield a possibility of segmentation of lesions.

STATISTICS OF A SPECKLE IMAGE

The generation of a radiofrequency (rf-)echogram from an isotopically scattering medium can be described as a random walk in two dimensions (Goodman, 1975; Burckhardt, 1978). The joint probability density function (p.d.f.) of the statistics of the rf-signal is therefore a circular Gaussian p.d.f. It can easily be shown (e.g. Papoulis, 1965) that after demodulation the statistics of the envelope, p, are governed by a Rayleigh p.d.f. and of the intensity, I (=squared envelope), by a χ-square p.d.f. with two degrees of freedom (i.e. negative exponential p.d.f.):

$$f(p) = \frac{p}{\phi}\, e^{\frac{-p^2}{2\phi}} \qquad p \geq 0 \tag{1}$$

p = magnitude of backscattered pressure

ϕ = dependent on the signal power

$$f(I) = \frac{1}{2\phi}\, e^{\frac{-I}{2\phi}} \qquad I \geq 0 \tag{2}$$

It can be shown that the "point" signal-to-noise ratios corresponding to these two p.d.f.s. are:

$$\mathrm{SNR}_p = \frac{\mu_p}{\sigma_p} = 1.91 \tag{3}$$

$$\mathrm{SNR}_I = \frac{\mu_I}{\sigma_I} = 1.0 \tag{4}$$

Equations (1) and (2) are applicable to the first order speckle statistics in case of a high number density of the scatterers. This condition is characterized as "fully developed" speckle. If, however, the number density decreases to such a value that the number of scatterers within the resolution volume of the echographic transducer is lower than about ten, a much more complicated p.d.f. is valid. Jakeman (1984) reproduced a formula for the SNR_I for this condition:

$$\mathrm{SNR}_I = \left[1 + \frac{<a^4>}{n\, V_s\, <a^2>^2} \right]^{-\frac{1}{2}} \tag{5}$$

where n = number density

     V_s = resolution volume

     a = scatterer strength

     $<.>$ = ensemble averaging operator.

If the scatterers actually have (almost) identical strengths this latter formula reduces to:

$$\mathrm{SNR}_I = \left[1 + \frac{k}{n} \right]^{-\frac{1}{2}} \tag{6}$$

where k is a constant.

The increase of SNR_I with increasing number density is shown in Fig. 1. The data obtained by Oosterveld et al. (1985) for the SNR_p are also displayed in this figure.

The mean grey level of an envelope image, μ_p, is strongly dependent on the number density. This relation is described by a square root. However, because the

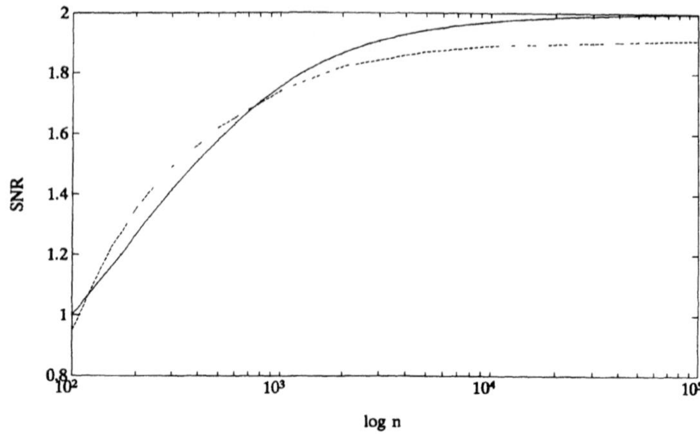

Figure 1. SNR vs. number density of scatterers. Drawn curve: $SNR_I \times 2$, according to Eq. 5; dashed curve: SNR_p, data from Oosterveld et al. (1985).

mean grey level is greatly dependent on the instrument settings it is not feasible to employ the mean as an absolute measure of reflectivity in a clinical routine setting. Moreover, it has been repeatedly shown that tumors can be hyper-, hypo- or iso-echoic, i.e. the reflectivity of a lesion is not consistently and very often not significantly different from healthy tissues. For these reasons the SNR_p, or SNR_I, might be used to produce a new kind of "parametric" images.

SEGMENTATION BY SNR IMAGING

The SNR image is obtained from the conventional envelope image by calculating the local SNR_p in sliding windows and assigning this value to the central pixel of each window. Defining the new variable:

$$r_{i,j} \equiv \frac{\overline{p}_{i,j}}{s_{p_{i,j}}} \tag{7}$$

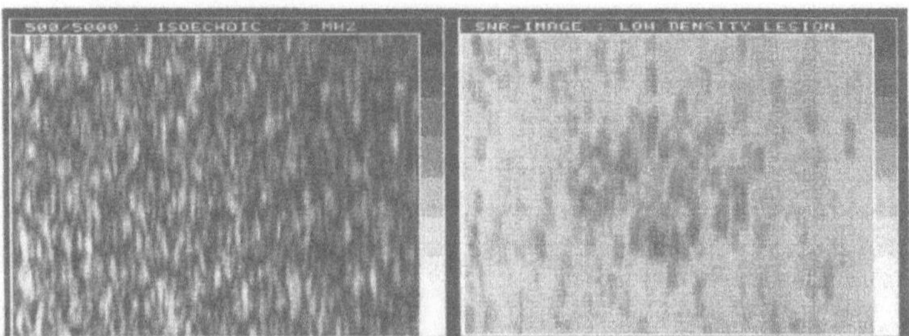

Figure 2. Left: simulated echographic image, containing iso-echoic central "lesion", with density 500/cm^3, in background 5000/cm^3. Right: corresponding SNR-image, showing low contrsat lesion (Verhoeven et al., 1991).

where p and s are the local mean and standard deviation of a window and i,j are the coordinates of the central pixel in a Cartesian grid. The approximate statistics of the SNR-image have been derived by Verhoeven et al. (1991) in case of fully developed speckle (i.e a Rayleigh p.d.f. applies to the envelope image):

$$E\{r\} \approx 1.91 \left(1+\frac{0.68}{N}\right) \tag{8}$$

$$\mathrm{var}\{r\} = \frac{1.85}{N} \tag{9}$$

where N = number of pixels in a window.
Therefore, the signal-to-noise ratio, SNR_r, of the SNR image, i.e.

$$\mathrm{SNR}_r \approx \sqrt{N}\left[1.40+\frac{0.95}{N}\right] \tag{10}$$

This formula was tested by using simulated echographic images with fully developed speckle. It appeared, that the SNR_r obtained from images corresponds only to the SNR_r from Eq. (10) when the number of speckles N_s is inserted. In other words the number of independent information "grains" determines the effective signal-to-noise ratio of the SNR image.

An illustration of the potential of SNR-imaging is given in Fig. 2. In the left image the result of a simulation is shown. A 3.5 MHz single element transducer (for details see Verhoeven et al., 1991) was linearly scanned along a cloud of point scatterers, number density 500 cm^{-3}, and reflectivity $\sqrt{10}$ times the reflectivity of the scatterers in the surrounding medium. The resulting mean grey level of the disc in the image was thereby identical to the mean of the surrounding part of the image. In other words, the detectability became equal to zero when based on the (mean) grey level. The right image shows the SNR image obtained by a sliding window of 5 x 15 pixels. The lesion has become visible here as a hypoechoic region. Further improvement of lesion detectability can be achieved by combining grey level information with SNR information in a synergistic manner (Verhoeven et al., 1991).

SEGMENTATION BY FRACTAL ANALYSIS

The concept of fractal analysis of image texture was combined with that of scale space filtering (Witkin, 1983; Baband et al., 1986) in recent literature. Scale space filtering implies the convolution of the image with a kernel over a certain range of resolution. It was shown by Witkin (1983) that a Gaussian kernel has the unique property that the extrema of the surface never disappear when the scale parameter decreases. Rather than counting the number of extrema and plotting this number vs. scale, Rueff (1989) and Muessigmann (1989) proposed to employ the surface spanned by the grey-levels of an image (or the length in the one dimensional case). The assumption made now is that over a certain range of the scale parameter a scaling law applies, which is related to the fractal self-similarity property of the image.

Defining the "surface" of the image by:

$$S(\sigma) = \int \int \sqrt{1 + I_x^2(x,y;\sigma) + I_y^2(x,y;\sigma)} \; dxdy \qquad (11)$$

where:
$$I_x(x,y;\sigma) = \frac{\partial I(x,y;\sigma)}{\partial x}$$

$$I_y(x,y;\sigma) = \frac{\partial I(x,y;\sigma)}{\partial y}$$

and:

$$\begin{aligned} I(x,y;\sigma) &= E(x,y) \otimes G(x,y;\sigma) \\ &= \int \int E(\xi,\eta) G((\xi-x),(\eta-y);\sigma) d\xi d\eta \end{aligned} \qquad (12)$$

where $E(x,y)$ is the original echographic image and

$$G(x,y;\sigma) = \frac{1}{2\sigma^2 \pi} \, e^{\frac{-(x^2+y^2)}{2\sigma^2}} \qquad (13)$$

is the Gaussian Kernel.

The analysis is based on the assumption that the scaling property leads to:

$$S(\sigma) \propto \sigma^P \qquad \sigma \in \{\min,\max\} \qquad (14)$$

where P = scaling dimension,

and the Minkowski dimension M (Muessigmann, 1990):

$$M = 2 - P \qquad (15)$$

This method was applied to simulated echographic images (Wijshoff, 1991). The results of a one dimensional simulation, i.e. echolines rather than echo-images, are shown in Fig. 3. The simulation consisted of 64 lines of 1024 data points. The left 512 samples were made by using a tissue model equivalent to 50 scatterers per cm in the left half and in the right half equivalent to 500 scatterers per cm. The amplitude level on both sides of the transition was made equal. Due to the finite window size the transition step in the value of P is rather gradual. Plots of log $S(\sigma)$ vs. log σ are shown in Fig. 4. Each curve was based on 16 segments of 64 data points on both sides of the

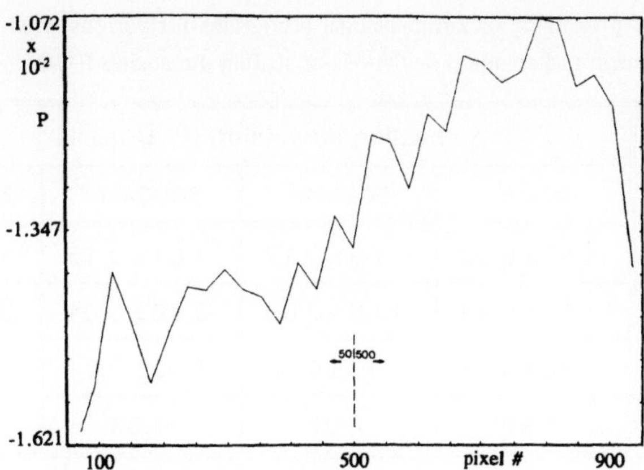

Figure 3. Scaling dimension image of 1-dimensional simulated echosignal. Mean of 64 independent lines. Left part equivalent to 500 scatterers per cm^3, right part to 5000/cm^3. Isoechoic transition.

in Fig. 4. Each curve was based on 16 segments of 64 data points on both sides of the transition. The value of P was estimated from the slope of these lines at the bending point, i.e. it was assumed that a multiscale fractal property applied to the image in the central part of the σ-range. The differences between the curves are indicative of the variability due to stochastic nature of the speckle. The standard deviation of the estimates of P is of the order of 6 percent.

Results of 2-D simulations are shown in Table 1. Here the window size was varied from (32x32) to (241x241). The effect of increasing the scatterer density from 100 to 25,000 per cm^3 is clearly illustrated. The slope parameter P changes from -1.5 to -1.0 over this range, which is considered to be an appropriate starting point for image segmentation.

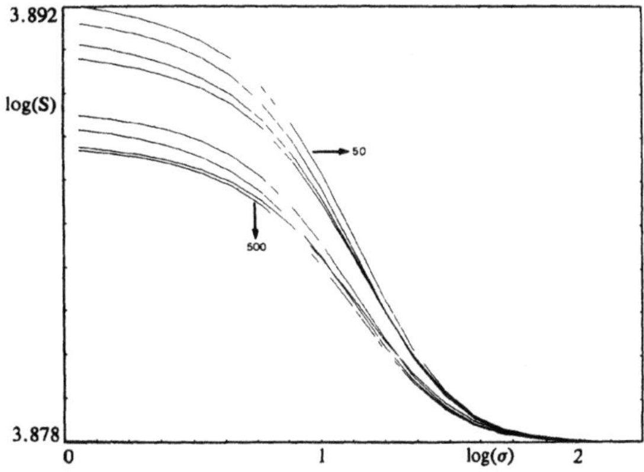

Figure 4. Log-log plot of image "surface" $S(\sigma)$ vs. scaling parameter σ. Simulations of 1-D echo-lines; each curve is average of 16 lines of 1024 data points. Slope at bending point equals the scaling dimension P.

Table 1. Results of processing of 2-dimensional echograms for various windowsizes (rows) and number densities: mean and standard deviations of scaling dimension P*10² in columns.

Window size	scaling dimension (P ± σ_P) 10²			
	100/cm³	500/cm³	5000/cm³	25000/cm³
32x32	-1.52±0.22	-1.09±0.17	-1.01±0.15	-1.00±0.20
64x64	-1.56±0.14	-1.08±0.05	-0.96±0.04	-0.97±0.03
120x120	-1.42±0.18	-1.13±0.09	-1.00±0.04	-1.00±0.10
241x241	-1.50	-1.07	-0.93	-0.93

CONCLUSION

In this paper two methods of parametric imaging of echographic data are discussed. It is concluded that both have the potential to reveal a local change in the image texture which is related to the number density of the scatterers. So, the detectability of isoechoic lesions will become possible, whereas combination of these parameters with the conventional echolevel information will be even more useful.

Acknowledgement

This work was supported by a grant from the Technical Science Branch (STW) of the Netherlands' Organization for Scientific Research (NWO), Grant NGN70.1364.

REFERENCES

Babaud, J., Witkin, A.P., Baudin, M., and Duda, R.O., 1986, Uniquiness Gaussian kernel for scale-space filtering, *IEEE Trans. Patt. Anal. Mach. Intell. PAMI-8*: 26-33.

Burckhardt, C.B., 1978, Speckle in Ultrasound B-mode scans, *IEEE Trans. Sonics Ultrasonics. SU-25*: 1-6.

Cloostermans, M.J.T.M., Mol, H., Verhoef, W.A., and Thijssen, J.M., 1986, In vitro estimation of acoustic parameters of the liver and correlations with histology, *Ultrasound Med. Biol. 12*: 39-51.

Goodman, J.W., 1975, Statistical properties of laser speckle patterns, *in*: "Laser Speckle and Related Phenomena", J.C. Dainty, ed., Springer, Berlin.

Hartman, P.C., Oosterveld, B.J., Thijssen, J.M., Rosenbusch, G.J.E., and van den Berg, J., 1992, Detection and differentiation of diffuse liver disease by quantitative echography: a retrospective assessment, *Invest. Radiol.* in press.

Jacobs, E.M.G., and Thijssen, J.M., 1991, A simulation study of echographic imaging of diffuse and structurally scattering media. *Ultrasonic Imag. 13*: 316-333.

Muessigmann, U., 1989, Texture analysis, fractals and scale space filtering, in: "Proceedings 6th Scandinavian Conf. on Image Analysis, Oulo, 1989", 987-994.

Muessigmann, U., 1990, Homogeneous fractals and their application in texture analysis, *in*: "Proceedings of the 1st IFIP Conference on Fractals, Lisbon, 1990, "H.O. Peitgen, J.M. Henriques, and L.F. Penedo, eds., Elsevier, Amsterdam.

Oosterveld, B.J., Thijssen, J.M., and Verhoef, W.A., 1985, Texture of B-mode echograms: 3-D simulations and experiments of the effects of diffraction and scatterer density, *Ultrasonic Imag. 7*: 142-160.

Oosterveld, B.J., Thijssen, J.M., Hartman, P.C., Romijn, R.L. and Rosenbusch, G.J.E., 1991, Ultrasound attenuation and texture analysis of diffuse liver disease: methods and preliminary results, *Phys. Med. Biol. 36*: 1039-1064.

Oosterveld, B.J., Thijssen, J.M., Hartman, P.C., and Rosenbusch, G.J.E., 1992, Detection of diffuse liver disease by quantitative echography: dependence on apriori choice of parameters, *Ultrasound Med. Biol.* in press.

Papoulis, A., 1965, "Probability, Random Variables and Stochastic Processes", MacGraw Hill, New York.

Romijn, R.L., Thijssen, J.M., Oosterveld, B.J. and Verbeek, A.M., 1991, Ultrasonic differentiation of intraocular melanomas: parameters and estimation methods, *Ultrasonic Imag. 13*: 27-55.

Rueff, M., 1989, Scale space filtering and the scaling regions of fractals, *in*: "From Pixels to Features", J.C. Simons, ed., Elsevier, Amsterdam, 49-60.

Thijssen, J.M., and Oosterveld, B.J., 1985, Texture of B-mode echograms. A simulation study of the effects of diffraction and of scatterer density on gray scale statistics, *in*: "Acoustical Imaging, vol. 14", A.J. Berkhout, J. Ridder, and L. van der Wal, eds., Plenum, New York,481-485.

Thijssen, J.M., Verbeek, A.M., Romijn, R.L., De Wolff-Rouendaal, D., and Oosterhuis, J.A., 1991, Echographic differentiation of histological types of intraocular melanoma, *Ultrasound Med. Biol. 17*: 127-138.

Thijssen, J.M., Oosterveld, B.J., Hartman, P.C. and Rosenbusch G.J.E., 1992, Correlations between acoustic and texture parameters from RF and B-mode liver echograms, *Ultrasound Med. Biol.*, in press.

Verhoeven, J.T.M., Thijssen, J.M., and Theeuwes, A.G.M., 1991, Improvement of lesion detection by echographic image processing: signal-to-noise-ratio imaging, *Ultrasonic Imag. 13*: 235-251.

Witkin, A.P., 1983, Scale-space filtering, *in* "Proceedings Joint Conference on Artificial Intelligence", Karlsruhe, 329-332.

Wijshoff, M.G.J., 1991, "Fractal Analysis of Echographic Images", Internal Report Biophysics Laboratory, Nijmegen (in Dutch).

POTENTIAL FUNCTION MATRIX ANALYSIS
OF TWO-DIMENSIONAL ULTRASOUND CARDIOGRAPHY

Yao Sun and Feng Sun

Department of Automatic Control
Harbin Shipbuilding Engineering Instittute
Harbin 150001, P. R. China

ABSTRACT

Catastrophe theory and some mathematical models have been studied in this paper, an important fact has been discovered, i. e. human heart beating has obvious catastrophic property. A new potential function matrix analysis method is proposed, so as to analyze ultrasound cardiography (UCG) and to judge the state of coronary artery disease quantitatively.

INTRODUCTION

Catastrophe theory is a new branch of mathematics. In 1968, Thom, a French mathematician published the first paper about catastrophe theory. Many problems that had not been solved by the traditional mathematical methods were solved by catastrophe theory. In the area of biomedical engineering, mathematical modelling of human heart movement is an important problem. But the traditional methods, such as infinitesimal calculus and general nonlinear function methods can not provide a mathematical model of heart beating exactly. The human heart is not a simple biological organ, its movement is very complex, especially ill heart has irregular motion. slow diastole and fast systole are the main characters of human heart beating. So we may think human heart beating has catastrophic property, the critical point is on the end diastole of heart.

Two-dimensional echocardiography (2DE) is very useful in visualizing cardiac wall dynamics and in detecting regional left ventricular wall motion abnormalities (RWMA). Some studies have demonstrated the value of UCG in comparison with contrast ventriculography in the evaluation of RWMA. By using computer image processing system, ultrasound cardiography can be analyzed quantitatively, and the noninvasive diagnostic method is more suitable for evaluation of regional cardiac function. In this paper, catastrophe theory is introduced to analyze UCG, and a new catastrophic analysis method is proposed.

POTENTIAL FUNCTION MATRIX ANALYSIS METHOD

On the 2DE images, all the edges between different tissues and cardiac wall are the areas whose gray level appears catastrophe. When cardiac muscles lack blood because of coronary artery being clogged, the expression of UCG is that catastrophe occurs in gray level distribution. Potential function is the basis of establishing catastrophic model of UCG. In this paper, a new concept of potential function matrix (PFM) is proposed, i. e. every element value of PFM is equal to its potential function value which is calculated from gray level matrix of UCG with a kind of potential function model. The optimal potential function matrix is calculated by using hyperbolic umbilic model, the feature parameters are extracted for analyzing and diagnosing UCG quantitatively. Figure 1 shows the flow chart of potential function matrix analysis. The abscissa and ordinate are taken as control variables, the gray level value of UCG and histogram are taken as state variables. Gray level value and its distribution patern are deciding factors of some feature parameters. the hyperbolic umbilic model is:

$$V(x, y) = x^3 + y^3 + ux + vy + wxy$$

Where $V(x, y)$ is the potential function, x and y are the state variables, u, v and w are the control variables. x is corresponding to gray level value of UCG image; y is corresponding to gray level histogram; u is corresponding to abscissa; v is corresponding to ordinate; w is corresponding to absolute value of difference between abscissa and ordinate. PFM represents the catastrophic property of UCG image, the feature parameters extracted from PFM would be helpful to the diagnosis of early coronary artery disease quantitatively. Some catastrophic feature parameters extracted from the potential function matrix are:

(1)   catastrophic power:

$$S_1 = \sum_{i=1}^{M} \sum_{j=1}^{N} D^2(i, j)$$

(2)   catastrophic contrast:

$$S_2 = \sum_{i=1}^{M} \sum_{j=1}^{N} (i - j)^2 D(i, j)$$

(3)   catastrophic entropy:

$$S_3 = \sum_{i=1}^{M} \sum_{j=1}^{N} D(i, j) \log_2 D(i, j)$$

(4)   catastrophic correlation coefficient:

$$S_4 = \sum_{i=1}^{M} \sum_{j=1}^{N} i \cdot j \cdot D(i, j)$$

EXPERIMENTS AND DISCUSSION

A series of cross section image from short-axis, apical four-chamber and two-chamber view are obtained to measure sector area difference of endocardial motion between the end diastole and end systole. In our experiment, TOSHIBA SSA-40A B mode ultrasound diagnosis apparatus was used to obtain2DE images, its probe is phased array sector one. The ultrasound wave ferquency is 3. 75MHz, the sector angle of phased array probe is 90°.

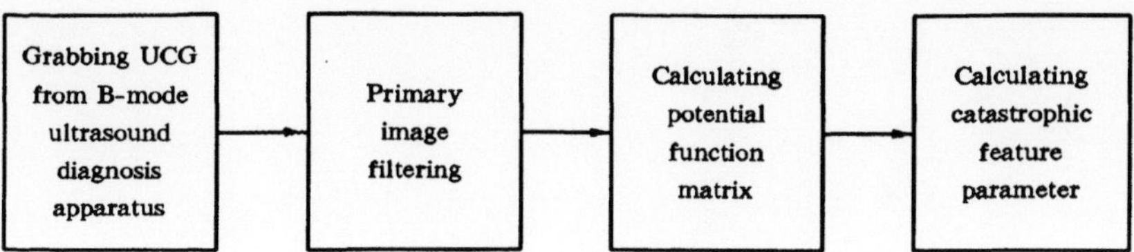

Figure 1.   The flow chart of potential function matrix analysis

Twenty cases of coronary artery disease UCG were examined by the CAG computer processing system, The CAG system includes a compatible 286 microcomputer, a CA5300 frame grabbing imageboard, a high resolution color monitor and a LQ1600 printer. The UCG images were transmitted into the imageboard through the video interface, and the digitalized UCG matrices were stored in harddisk. The imageboard can hold four frames of $512 \times 512 \times 8$ bits, and its processing speed is very high.

The coronary artery disease UCG images were analyzed by using potential function matrix analysis method on CAG system, and the computer analysis results are in agreement with the clinical diagnosis results. The experimental results show that the method is helpful to increasing the diagnosis accuracy of coronary artery disease and analyzing UCG images automatically.

CONCLUSION

It is important to realize the quantitative analysis and measurement of regional left ventricular function. The new method of potential function matrix analysis put forward in this paper is effective for analyzing coronary artery disease UCG images. The sharp potential function and swallow-tailed potential function are helpful to static UCG images, and the ellipse potential function is helpful to analyzing dynamic UCG images. The hyperbolic umbilic potential function is optimal for analyzing the left ventricular wall of two-dimensional echocardiography and other gray level catastrophe of UCG images.

REFERENCES

1.   J. F. Ren, et al, Quantitation of regional left ventricular function by two-dimensional echocardiography in normals and patients with coronary artery disease, *American Heart Journal*, Sept. 1985.
2.   M. Linzer, et al, Ultrasonic tissue characterization, *Ann. Rev. Biophys. Bioeng.*, 11: 303—329 (1982).
3.   S. A. Goss, et al, Compilation of empirical ultrasound properties of mammalian tissues, *J. Acoust. Soc. Am.* 68 (1): 93—108 (1980).

A RESTORATION METHOD OF ULTRASONIC MEDICAL TOMOGRAM

Kangyuan Zhou,[1] Donglin Li,[2] Min Wang,[1] L. G. Wang[1]

[1]Radio-Electronics Department
[2]Electrical Engineering and Electronics Department
University of Science and Technology of China
Hefei 230027, P. R. China

INTRODUCTION

Recently, the technology of ultrasonic medical imaging has played a quite important role in biomedical region. However , in a practical imaging system the echo signals from tissue are affected by many different causes which lead to distortion of images. Therefore it is necessary to restore the medical image for ultrasonic tissue characterization.

In this paper a restoration method of B mode tomogram is proposed. It is based on the principle of ultrasonic attenuation estimation in vivo of DRC method (Difference Ratio Correction Method) proposed by the authors[1].

As stated above, the tomogram is distorted by many causes, which are the dispersion of ultrasonic attenuation in propogating, the convergence and divergence of beams, the sensetivity and bandwidth of transduser, the gain and TGC of amplifiers, and the logrithmatic effect and detection. If TGC compensation and gain of a B scanner set in a certain position, the effects mentioned above will remain unchanged except the tissue characters. Therefore we can estimate the results of those effects and treat them as a function of diagnostic depth called correction curve $T(z)$ using the tomograms of three reference tissue mimicking phantoms, whose attenuation slopes are known. Then the attenuation slopes in small segments on an A line of the image to be restored can be estimated by DRC sequentially.

If all the attenuation slopes corresponding to the depth segments along an A line of a tomogram have been estimated, the restored brightness function along the A line can be obtained.

THEORY

We first consider the ultrasound propogation in homogeneous medium with the same linear attenuation slope everywhere. In the acoustic focusing field, the backscattered signals are wideband and they are dependent on the emitted signal, the acoustic attenuation and the properties and position of scatters in the tissue. The power spectrum of the backscattered ultrasound is[2]:

$$S(f) = C_1 \frac{G(f)}{z^2} \exp\left[-4a(f)z\right] Z(z, x) \qquad z > 0, \qquad x > 0, \tag{1}$$

where $G(f)$ is the power spectrum of the emitted pulse, C_1 is a constant, $a(f)$ is the attenuation coefficient which depends on the frequency f, and $Z(z, x)$ is the fluctuation caused by the backscattering of the resolution cell at lateral position x and axial position z, which is a random variable.

Because the technique of multi-section dynamic focusing is widely used in modern B scanners, the beam patterns of the B scanners are rather complicated. More generally, the power spectrum of the backscattered ultrasound can be expressed as:

$$S(f) = C_1 G(f) D(f, z) \exp\left[-4a(f)z\right] Z(z, x), \tag{2}$$

where $D(f, z)$ is a correction factor for the diffraction of the beam. Eq. (2) is suitable for any kind of focusing at any axial position z.

Assuming that the envelope of emitted pulse is Gaussion shaped, then its power spectrum is also Gaussion shaped, i. e.

$$G(f) = C_2 \exp\left[-(f - f_c)^2/2\sigma^2\right] \qquad f_c > \sigma, \tag{3}$$

where f_c is the central frequency and σ is the bandwidth of the emitted pulse.

For tissue mimicking phantom or some soft tissue, e. g. normal liver, the attenuation coefficient is a linear function of frequency[3]

$$a(f) = \beta f, \tag{4}$$

where β is called the attenuation slope.

From Eqs. (2), (3) and (4), the backscattered energy from the resolution cell at (z, x) is:

$$\Psi(z, x) = \int_0^\infty S(f)\,df$$

$$= C_1 C_2 Z(z, x) \int_0^\infty D(f, z) \exp\left[-(f - f_c)^2/2\sigma^2\right] \exp\left(-4\beta f z\right) df. \tag{5}$$

In general, $D(f, z)$ changes with frequency more slowly than the exponential attenuation function does and $D(f, z)$ can be approximated by $D(f_c, z)$, where f_c is the central frequency. Hence,

$$\Psi(z, x) = C_3 D(f_c, z) \exp\left[-(4f_c\beta z - 8\sigma^2\beta^2 z^2)\right] \int_y^\infty \exp\left(-u\right)^2 du, \tag{6}$$

$$C_3 = \sqrt{2}\,\sigma C_1 C_2, \tag{7}$$

$$y = (4\sigma^2\beta z - f_c)/\sqrt{2}\,\sigma, \tag{8}$$

where C_2 and C_3 are constants.

The integral in Eq. (6) is the well known error function and it has the following proper-
ties: The function $\mathrm{erf}(y)$ is an odd function of y. $|\mathrm{erf}(y)|\cong 1$ when $|y|$ is large enough.
In fact, when $y=-1.2$, $\mathrm{erf}(y)=-0.9745$ which is very close to -1. Hence, if $y\leqslant-1.2$, then $1-\mathrm{erf}(y)\cong 2$.

In Eq. (6), let $y\leqslant-1.2$, then we can get

$$\Psi(z,\ x) = C_3 D(f_c,\ z)\ \exp\left[-\ (4f_c\beta z-8\sigma^2\beta^2 z^2)\right]\left[1-\mathrm{erf}(y)\right]$$

$$\cong 2C_3 D(f_c,\ z)\ Z(z,\ x)\ \exp\left[-\ (4f_c\beta z-8\sigma^2\beta^2 z^2)\right]. \tag{9}$$

The condition under which Eq. (9) is derived is $y\leqslant-1.2$, i. e.

$$\beta z\leqslant (f_c-1.2\sqrt{2}\ \sigma)/4\sigma^2 \tag{10}$$

Eq. (10) determines the limits of application of the DRC method. For instance, if $f_c=3.5$
MHz, $\sigma=7.5$ MHz, then from Eq. (10), $\beta z\leqslant 1$. Because the attenuation slopes of normal
livers are near 0.058 Np/cm/MHz or 0.5 dB/cm/MHz, the useful measurement depth for a
normal liver is about 17 cm. The useful measurement depth z will be smaller when measuring
such soft tissue as fatty liver which has a higher attenuation slope. It should be noted that the
useful measurement depth decreases rapidly with the increase of bandwidth.
From Eq. (9), the amplitude of the backscattered wave from the resolution cell at (z, x) is

$$R(z,x) = \sqrt{\Psi(z,x)} = CD_1(f_c,z)\,E(z,x)\,\exp\left[-\ (2f_c\beta z-4\sigma^2\beta^2 z^2)\right], \tag{11}$$

where $C=\sqrt{2C_3}$ is a constant, $D_1(f_c,z)=\sqrt{D(f_c,z)}$ is the correction factor for the convergence and divergence of the beam, $E(z,x)=\sqrt{Z(z,x)}$ is a random variable and its envelope is
also the brightness function of the image to be restored.

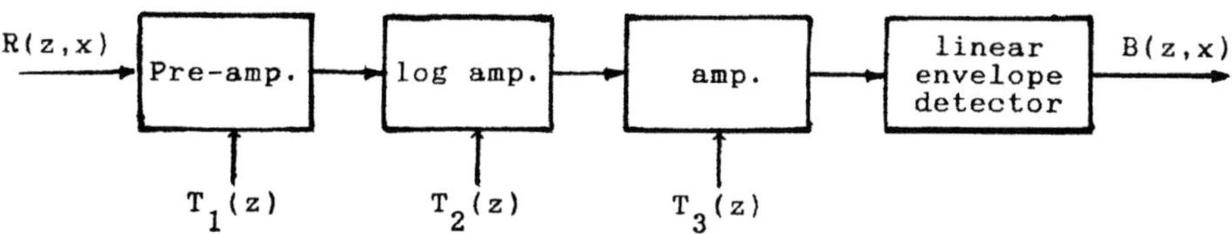

Figure 1.　　The processing of echo signal in a B scanner

As shown in Figure 1, the echo signal is influenced by various factors in a B scanner
where $T_1(z)$, $T_2(z)$, and $T_3(z)$ represent Time-Varying Gain Compensations (TGCs) added to
the preamplifier, the logrithmic amplifier and the following amplifier, respectively. Practically,
in a commercial B scanner, one or two of them may be constants which are independent of z.

If the linear envelope detector can detect the envelope of $E(z, x)$ without distortion, then the output after the linear envelope detector is

$$B(z, x) = T(z) \{ \ln[CT_1(z) D(f_c, z) E(z, x)] - (2f_c\beta z - 4\sigma^2\beta^2 z^2) \}, \qquad (12)$$

where $E(z, x)$ is the envelope of $E(z, x)$, $B(z, x)$ is the video output of the B scanner, and $T(z) = T_2(z) T_3(z)$ is named the correction curve of the B scanner under certain conditions.

Now using the tree reference phantoms, the correction curve $T(z)$ can be obtained as follows.

Because all the parts of a reference phantom are made up of one kind of material and are homogeneous, $E(z, x)$ will be a stationary random process. Hence, if the number of independent scan lines of the region of interest (ROI) is large enough, the average of $E(z, x)$ over lateral position x will be independent of z. By averaging the scan lines over the lateral position x of the tree phantoms, respectively, we can get the averaged scan lines of the tree phantoms $\bar{B}_1(z)$, $\bar{B}_2(z)$, and $\bar{B}_3(z)$. Define

$$F_{ij}(z) = \bar{B}_i(z) - \bar{B}_j(z) \qquad i, j = 1, 2, 3, \qquad (13)$$

then from Eqs. (12) and (13), finally we can get

$$T(z) = \frac{F_{ij}(z)}{2(\beta_j - \beta_i) f_c z - 4(\beta_j^2 - \beta_i^2)\sigma^2 z^2 + C_{ij}}, \qquad (14)$$

where $T(z)$ is the correction curve for a given combination of TGC and the gain of B scanner. Because of the near field effects, $T(z)$ can be used only when z is large enough, generally, $z > 3$ cm. However, the maximum measurable distance z_{max} is determined by Eq. (10), which is relevent to the center frequency f_c and the bandwidth σ of B scanner as well as the value of β. C_{ij} can be obtained by solving simultaneous equations[1]. An example of $T(z)$ for a certain combination of TGC and gain of a type of B scanner is shown in Figure 2.

Assume that there is an ROI which is relatively homogeneous, the averaging attenuation slope of this area is β_0, and the sound speed of the medium within the area is the same as that of the reference phantoms. Also assume that this area includes n independent scan lines while each scan line has m samples, define

$$Y_{01}(z) = \frac{F_{01}(z)}{T(z)}, \qquad (15)$$

then, according to Eq. (14) we have

$$Y_{01}(z) = -4(\beta_1^2 - \beta_0^2)\sigma^2 z^2 + 2f_c(\beta_1 - \beta_0) z + C_{01} = az^2 + bz + c, \qquad (16)$$

where

$$c = C_{01}, \qquad (17)$$

$$b = 2f_c(\beta_1 - \beta_0), \qquad (18)$$

$$a = -4(\beta_1^2 - \beta_0^2)\sigma^2. \qquad (19)$$

Using the least-square law[4] to fit Eq. (16) from the m samples, we can get the value of b.

Finally, the attenuation slope β_0 of the detected area is

$$\beta_0 = -\frac{b}{2f_c} + \beta_1, \tag{20}$$

where β_1 is the attenuation slope of one of the reference phantoms. Or in dB units,

$$\beta_0 = -\frac{4.343b}{f_c} + \beta_1. \tag{21}$$

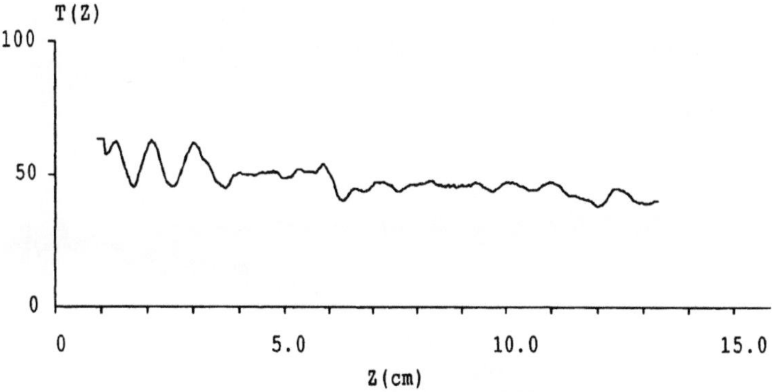

Figure 2. The correction curve with a certain combination TGC
and gain for Aloka SSD 210-2z B scanner

For restoration of a B mode tomogram which is obtained in certain combination of TGC and gain of a B scanner, it is necessary to calculate the correction curve $T(z)$ corresponding to the combination using the DRC method firstly. Then from $T(z)$ and the output of one of the tree reference phantoms the attenuation slope $\beta_0(z_i, x_j)$ of a small homogeneous region in a certain position (z_i, x_j) on the tomogram can be derived. When all of the $\beta_0(z_i, x_j)$ corresponding to every i and j on the tomogram are got sequantially, from Eq. (12) and the outputs of the tree reference phantoms, the brightness function $E(z_i, x_j)$ on the whole tomogram can be calculated. Thus the tomogram will be restored.

EXPERIMENTAL RESULTS

For verifying this restoration method, four ultrasonic tissue mimicking phantoms, which were specially made by the Institute of Acoustics, Academia Sinica, were used. The base of four phantoms was agar, the scatters were uniformly distributed in the base. The scattering properties of the phantoms were analogous to that of soft tissues, e. g. human liver, the ultrasonic velocity was about 1540 m/s. Their attenuations were measured by the transmission method. Each phantom was measured at 10 different positions at a certain frequency and the S. D. of the measured attenuation slopes was less than 0. 005 dB/cm/MHz. As a result, they have a very good linear attenuation-frequency dependency. In our experiment, the four phantoms employed are numbered 0, 1, 2, 3, their attenuation slopes are $\beta_0 = 0.620$, $\beta_1 = 0.292$, $\beta_2 = 0.399$ and $\beta_3 = 0.500$ dB/cm/MHz, respectively. Phantoms Nos. 1, 2 and 3 are used to calculate the correction curve, so these phantoms are named reference phantoms, while phantom No. 1 used as test sample. The size of each phantom is 20. 0 cm long, 18 cm wide and 5. 0 cm thick. The B scanner used in the experiment is Aloka SSD-210 2z. The center frequency of the transducer is 3. 5 MHz. The shape of the power spectrum of the transmitting pulse can be approximated by Gaussian function with center frequency at 3. 50 MHz and $\sigma = 0.78$ MHz.

Figure 3 shows an A line on the tomogram of No. 0 phantom with a combination of TGC and gain: gain = 4. 0, far = 5. 0 and near = 1. 0. Figure 4 shoes a restored A line of the same phantom with the same combination of TGC and gain. From Figure 4 it is obvious that the DRC method can restore the B mode tomogram of phantom effectively.

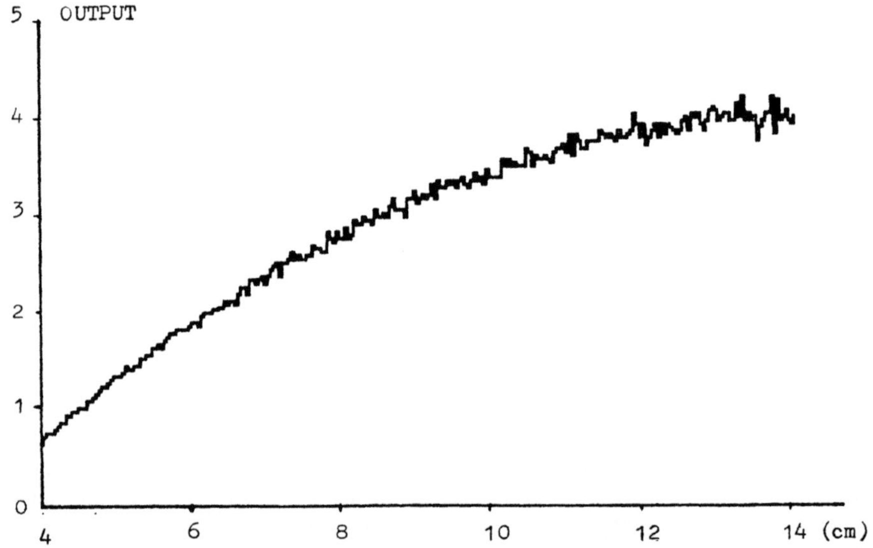

Figure 3. An A line on the tomogram of No. 0 phantom with a combination
of TGC and gain: gain = 4. 0, far = 5. 0, and near = 1. 0

CONCLUSION

If the conditions in which the tomogram was obtained are known, using tree reference phantoms the DRC method can decrease the effects causing the distortion of the tomogram and restore it. When the conditions are unchanged, namely the B scanner and the combination of TGC and gain are unchanged, it is not necessary to repeat the procedure of calculating $T(z)$ and another data before the restoration. We can store all the data which are unchanged in restoring in the memories of a computer so that the restoration processing can become more simply and rapidly.

516

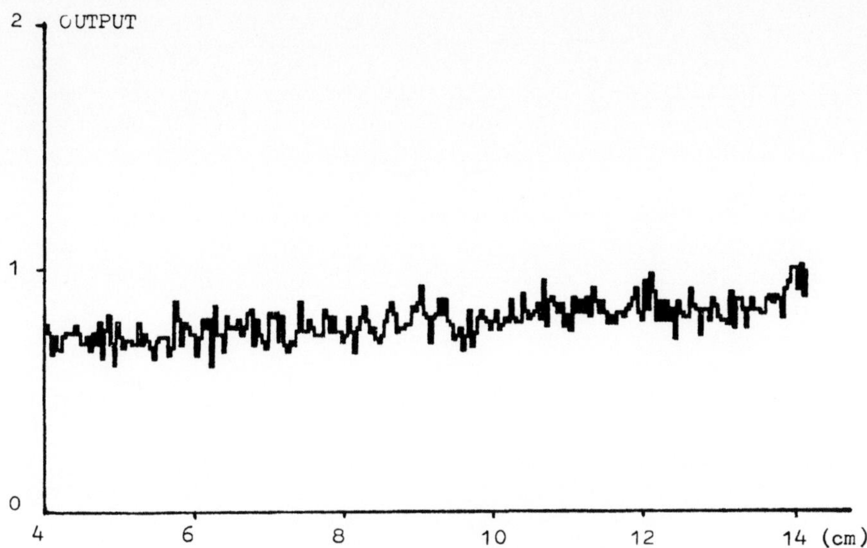

Figure 4.   A restored A line on the tomogram of No. 0 phantom with a combination of TGC and gain: gain＝4. 0, far＝5. 0, and near＝1. 0

It is much more complicated to restore the tomograms clinically than to restore those of phantoms. The main difficulties are that the boundaries in tissues are rather complicated and sometimes the backscattered signals of tissue in front of a boundary are influenced by the reflection of the boundary. [5]

The irregular distribution of veins and arteries contributes to the difficulties of finding relatively large homogeneous area. But a suitable area for restoration of DRC method could be found in large organs such as livers and spleens. The progress of this method is to enhance the resolution of estimating the attenuation slope so that the method can be widely used in clinical application.

REFERENCES

1.   K. Y. Zhou, D. F. Zhang, C. Lin and S. H. Zhu, Ultrasonic attenuation estimation in-vivo using the difference ratio correction method, *J. Acoust. Soc. Am.*, (1992).

2.   K. J. Parker , R. C. Waag, Measurement of ultrasonic attenuation within regions selected from B-scan images, *IEEE Trans. Biomed. Engr.*, BME—30, 431—437 (1983).

3.   P. He, Acoustic attenuation estimation for soft tissue from ultrasound echo envelope peak, *IEEE Trans. on UFFC*, Vol. 36, No. 2, 197—203 (1989).

4.   P. R. Bevington, "Data Reduction and Error Analysis for the Physical Sciences", Chapter 8, McGraw-Hill Book Company (1969).

5.   K. Y. Zhou, D. F. Zhang, The effect to ultrasonic image by tissue boundary, Proc. of China-Japan Joint Conference on Nltrasonics, 411—414 (1987).

IN VITRO ULTRASONIC MONITORING OF BIOLOGICAL TISSUE CRYODESTRUCTION

P. Laugier, G. Berger

URA CNRS 1458 Imagerie Paramétrique
15 rue de l'Ecole de Médecine 75006 Paris, France

INTRODUCTION

The aim of cryosurgery is the cryodestruction of tumour cells within a delimited region of interest. The destruction of tissue by cooling to cryogenic temperature is usually effected by means of a surgical cryoprobe which is continuously cooled and placed on or into the tissue to be destroyed. Cryolesions are the result of phase changes occuring in biological tissues within the "ice ball" centered on the cryoprobe tip. The cryoprobe tip produces a fast cooling effect that makes an ice ball grow, first rapidly, then more slowly. Finally the growing stops when a thermal equilibrium is reached. Thermal exchanges depend on tissue thermal conduction and are under control of blood circulation. The size of the frozen region depends on several parameters: the cryoprobe temperature, the cryoprobe tip shape, the area of contact between the cryoprobe and the tissue, the nature of the tissue, its blood circulation and its metabolism heat generation.
Resulting cryonecrosis is a sharp edged, little hemorrhagic, little painful and little inflammatory tissue lesion. Cryosurgery has established itself as an important clinical technique. Main clinical field of cryosurgery is dermatology (benign tumors and also basocellular carcinoma), but it is used also in pneumology for bronchoscopic airways stricture caused by tumors.
One of the major problem limiting the more widespread use of the cryosurgery is the difficulty of controlling the amount of tissue that is destroyed during the treatment. No satisfactory technique has so far been developped. However, it seems that the control of the ice ball growth using some form of ice ball monitoring is necessary to ensure precise control of cryosurgery (see for example Gill and Long, 1977 or Orpwood, 1981).
Cryonecrosis is located inside the frozen region of the tissue. The extent of the cryonecrosis is directly determined by the accurate knowledge of where the deepest boundary of the tumor lies. It is therefore necessary for the clinician to know the tumor extent and to be able to dynamically monitor the growing cryolesion boundary during the freezing process. When freezing is generated from the surface of the tissue, the extent of cryodestruction is usually assessed by the extent of the white frozen area. But there is no

control of depth extent if deep seated volumes of tissue are exposed to the freeze-thaw cycle without direct vision.

The aim of this study is to evaluate echography possibilities in monitoring the freezing process by accurately measuring the frozen region size and controlling the advancing cold front. The reason for starting this investigation on echographic monitoring of cryosurgery is related to the new diagnostic possibilities of high frequency ultrasound imaging systems in the in vivo determination of invasion depth of skin and parietal tumors (Berson, 1992; Kimmig et al., 1990; Martin et al., 1989; Murashami and Yoshiharu, 1989). It should be useful to the clinician to know by means of ultrasound both the tumor depth extent and the frozen region boundary. In this paper we describe the first experimental results of freezing monitoring by ultrasound obtained in 5 different in vitro skin samples of domestic pig.

METHOD

Principle of the Method

Phase changes occuring during the freezing process can be visualized by ultrasound imaging technique . An interesting attempt was already presented by Tanahashi et al. (1977) to monitor cryosurgery of prostate by use of transrectal ultrasonography. They used a 3.5 MHz transrectal ultrasonic probe to control the formation of the ice ball during the freezing period.

Ultrasound pulses are reflected at the boundary of the frozen region (or "ice ball") due to the difference of acoustic impedance between the unfrozen soft tissue and the frozen tissue. The change in acoustic impedance is mainly explained by the different values of ultrasound velocity in water (1530 m/s) and ice (3980 m/s, Krautkrämer, 1990).As the meaningful information during freezing is the value of the iceball size, the ultrasonic measurement consists in estimating the transit time of an ultrasonic pulse travelling through the skin and reflected at the frozen region boundary. The accuracy of the measurement depends on the axial resolution of the ultrasound imaging system.

However, the ice ball size can be reached only if the value of the ultrasound velocity in the frozen tissue is known. It is well known that the ultrasound velocity is a function of the temperature. For example in water, it increases by about 5 m/s per degree in the range around 0°C and by 3 m/s per degree in the range around 30 °C (Del Grosso and Mader, 1972). Due to the non-uniform temperature distribution in the tissue, the ultrasound velocity is likely to be a complex function of the spatial coordinates. Nevertheless, the ultrasound velocity in the frozen tissue can be easily estimated after some simplifying assumptions. The method uses the distortion of the echogram due to the velocity variations and compares the transit time t_0 of the cryoprobe echo before freezing with the transit time t_2 during freezing. Only two different regions are considered in the sample: a frozen region with constant velocity c_2 and an unfrozen one with constant velocity c_1. As the temperature gradient preceding the cold front is neglected the value of velocity c_1 is derived from known measurements from litterature. Cantrell et al. (1978) give 1720 m/s in porcine skin. The velocity in the frozen tissue may easily be derived from the knowledge of the echo round trip times following relation:

$$e=c_2(t_2-t_1)=c_1(t_0-t_1) \qquad (1)$$

where e is the ice ball size, t_0 is the round-trip time of the cryoprobe echo before freezing, t_1 and t_2 are the round-trip time respectively of the freezing front and the cryoprobe echoes during freezing.

Signal-to-Noise Ratio Enhancement

Transit time measurement is possible only if the ice ball boundary echo can be identified unambiguously. The usual way of identifying an echo in ultrasound imaging is the amplitude detection. It may happen that the meaningful echo arrives on the transducer at the same time than other tissue structure echoes thus producing interference and random fluctuations of the amplitude (i.e. speckle pattern). Under these conditions the echo detection may prove quite difficult. Amplitude detection improvement can be attempted in two different ways : either by a speckle reduction technique consisting of decreasing the amplitude fluctuations in the echogram with incoherent summation or by enhancing the target echo amplitude. When one have some a priori knowledge on the target (size or form for exemple) the second method seems more adequate. One interesting way of increasing the signal-to-noise ratio (SNR) of the target echo is to use the frequency response of the echogram. Previous research indicated that specular reflector echoes in biological tissues (Laugier et al., 1985) or flaw echoes in materials (Shankar et al., 1988) usually occupy the lower frequency of the transducer passband compared to the speckle noise. As expected from the scattering theory, the frequency response of a small scale scatterer (size compared to the wavelength) increases with frequency, while the response of a a specular reflector is frequency independant. The freezing front is expected to be such a specular reflector. Thus SNR enhancement should be possible by using conventional linear low-pass filtering to isolate the target echo. In this particular case, the SNR is defined as:

$$ SNR = \frac{S}{\sigma} \qquad \text{with} \qquad \sigma = \sqrt{\frac{1}{N}\Sigma S_i^2} \qquad (2) $$

where S is the amplitude of the meaningful echo and S_i being the value of the ith sample of the windowed signal of duration N. The windowed signal is an echogram segment just preceding the detectedfreezing front echo.

Experimental Setup

The experimental setup is represented in figure 1. The transducer is immersed in a water bath at 37 °C with 9 g/l NaCl. A skin sample of 5 cm in diameter and 4 mm thickness is placed horizontally at the water surface. It is partially immersed in the bath so that the outer layer of the skin is in contact with the air except for its central part which is in contact with the cryoprobe tip. There is a 1 cm circular contact between the outer layer of the skin and the cryoprobe tip.

The transducer and the cryoprobe are placed on opposite sides of the tissue sample, so that the growing cold front is going away from the cryoprobe and moving towards the transducer. Although this disposal is not adequate for in vivo studies, it has been chosen for in vitro measurements because of experimental facility.

We used a high pressure cryoprobe manufactured by the company Date. Low temperature of the tip is reached by Joule-Thomson expansion of nitrous oxyde (N_2O). The lowest possible tip temperature that can be obtained with N_2O after an adiabatic expansion is -88.5°C. Temperature ranging from +37°C to the tip temperature can be recorded in in vivo tissue.

A 13 mm diameter unfocused Panametrics transducer of center frequency 15 MHz is used to collect echographic signals. Waveform duration is 0.2 μs at -20 dB, therefore axial resolution is 300μm.

The data acquisition system uses a Data 6000 A/D convertor. An PC computer is used to control the data acquisition process and to perform signal processing. A sampling frequency of 100 MHz is used to collect data. In order to follow the evolution of the cold front during one freeze-thaw cycle, echographic data are recorded every 1 second during the whole duration of the cycle, both the transducer and the cryoprobe being at a fixed position, and facing each other. 256 echographic lines are recorded during a freeze-thaw cycle. The freezing period starts at t=10 seconds and stops at t=110 seconds. Five different skin samples were studied.

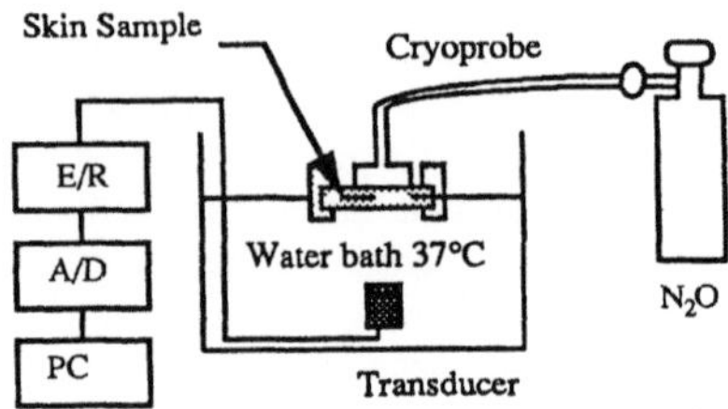

Fig 1 Experimental setup. G: Pulse generator; A/D: Analogic to digital convertor, PC: personal computer.

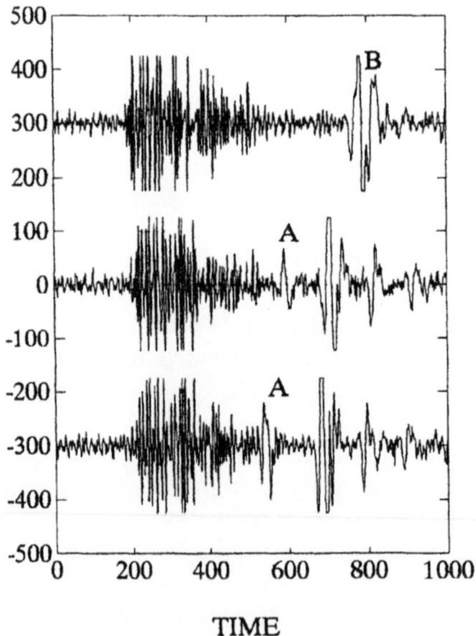

Fig 2 Echographic signals at three different freezing times (respectively from top to bottom 0 s; 10 s and 15 s.) showing the time moving cold front echo (A) and cryoprobe echo (B).

EXPERIMENTAL RESULTS

Figure 2 represents the echographic signals recorded in the biological tissue at three different times of the freezing process. Top signal is recorded at the very beginning of the freezing process. A high amplitude cryoprobe echo (echo B) is visible at the round trip time t0. As further signals are collected during the freezing process the cryoprobe echo is still visible at a different round trip time. Moreover, during the freezing process, another high amplitude echo (echo A) is visible at a variable round trip time t1. This echo is likely to be the advancing freezing front echo. Note that the round trip time t1 of the freezing front echo decreases during the freezing period, as the distance between the transducer and the moving cold front decreases. Although the cryoprobe is at a fixed distance from the transducer, the round trip time of the cryoprobe echo also decreases during the freezing period, meaning that ultrasound velocity in the frozen region surrounding the cryoprobe changes.

The whole data collected during a freeze-thaw cycle in steps of 1 seconds is represented in figure 3. Signal amplitude is plotted versus time (vertical axis) in a TM (Time-Motion) echographic mode. The growing ice ball echo is visible. The figure also shows the cryoprobe in a wrong (time varying) position because of echogram distortion due to velocity variation inside the ice ball. Similar observations were made on all of the five skin samples. Note also that the tissue structure echoes are still visible, even if they are attenuated, within the ice ball.

Fig 3 Echographic TM image of the signal amplitude (freezing time on vertical axis) recorded during a freeze-thaw cycle.

SNR enhancement is performed by low-pass filtering. The effect of different low-pass filters is shown in figure 4 (Cutoff frequency top left: 10 MHz; top right: 8 MHz; bottom left: 6 MHz; bottom right: 4 MHz). It clearly demonstrates SNR enhancement as far as the freezing front echo is concerned. The amplitude of the tissue structure echoes decreases more rapidly than the freezing front echo amplitude. Respective SNR are calculated : top left: 1.75; top right: 1.9; bottom left: 3.2 ; bottom right: 3.8. Because of the specular reflection of the ultrasonic pulse on the ice ball boundary and because of large bandwidth of the transducer, even a low cutoff frequency filter is efficient for SNR enhancement. The lower the cutoff frequency, the better the SNR enhancement. The filtering technique can be used to accurately determine the location of the ice ball boundary.

The ultrasound speed in the frozen tissue is estimated following relation (1). Round trip time are measured from the envelope of the signal by detection of the maximum amplitude. We found $c_2 = 2900 \pm 150$ m/s. The standard deviation value is derived from the velocity values obtained on the whole echographic data set recorded during the freezing process. The velocity value is lower than the known velocity in ice (3980 m/s) which can be

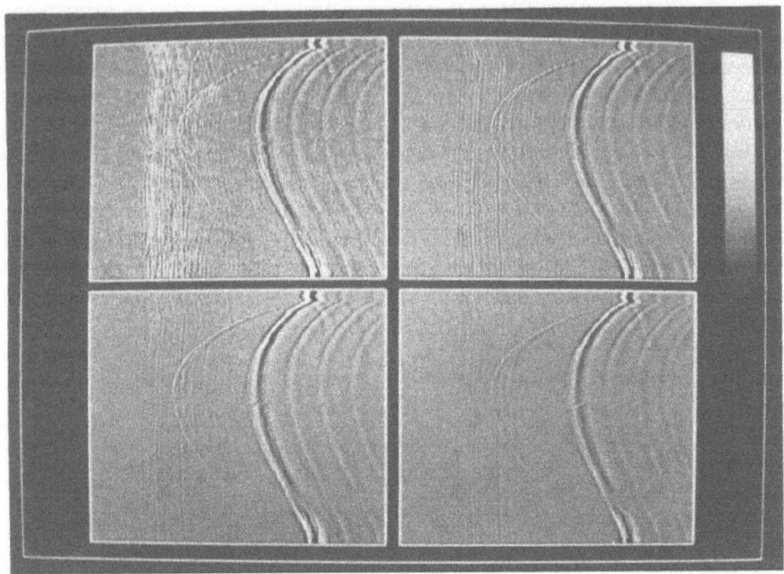

Fig 4 Effect of different low-pass filters (Cutoff frequency top left: 10 MHz; top right: 8 MHz; bottom left: 6 MHz; bottom right: 4 MHz) on the echographic TM image of a freeze-thaw cycle .

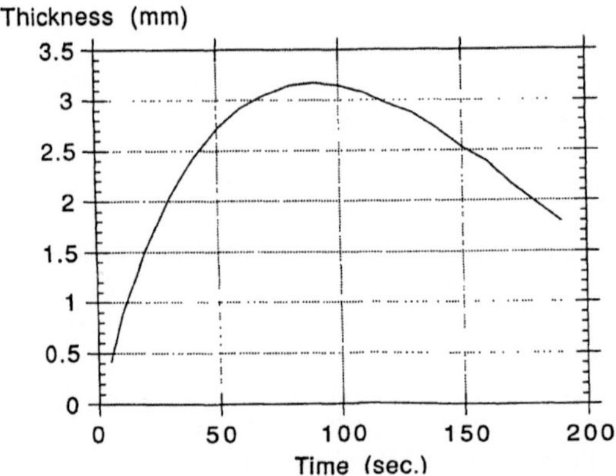

Fig 5 Thickness of the frozen biological tissue during the freeze-thaw cycle.

accounted for by the fact that the ultrasound pulse is not travelling in pure ice but in a complex medium made of ice and of tissue structures. Besides, let us recall that, due to non uniform temperature distribution within the tissue, the velocity estimation can only be an approximate measurement.

Figure 5 shows the frozen region thickness versus freezing time during a freeze thaw cycle. It shows rapid advance of the freezing front at the beginning of the freezing process. Then the frozen region growth gradually slows down. A one millimeter thickness frozen region is obtained in a few seconds. The freezing process duration is 100 seconds.

DISCUSSION AND CONCLUSION

Monitoring the volume of tissue destroyed during freezing is very important if cryosurgery is to be a precise and controlled process. We propose an ultrasound monitoring method to control the freezing process in cryotherapy. Ultrasound monitoring presents several advantages: it is a reliable, simple, non invasive technique of accurate control in real time of the ice ball boundary. It is possible to easily detect the frozen region depth boundary. Conventional linear filtering of the echographic signals can be very useful for SNR enhancement in the detection of a target signal drowned in speckle noise. Once the ultrasound velocity in frozen tissue is known, the kinetics of the freeze-thaw cycle is also measurable (size of the ice ball versus time), which is of major importance as it allows to exactly adjust the freezing time as a function of the tumor depth extent. At present we proposed in this paper a rather simplistic model for the velocity estimation based on the correction of the echogram distorsion only for the cryoprobe echo. From a finer analysis of the whole echogram distorsion during freezing could be derived the ultrasonic velocity changes along the depth axis (Ophir and Yazdi, 1990). As the tissue structure echoes are still visible within the ice ball this analysis could be performed.A more precise analysis of the velocity variations along the ultrasonic axis which are themselves related to temperature variations would give a more relevant information to the surgeon than a simple measurement of ice ball size. Another advantage is the new diagnostic possibilities of high frequency ultrasound imaging systems in dermatology or in endo-echography for determining the depth extent of cutaneous or parietal tumour. A combined system (cryoprobe together with an ultrasound imaging system) could be very usefull to determine both the tumor depth extent and the ice ball size.

ACKNOWLEDGEMENTS

We would like to thank the company Date for lending their cryosurgery material.

REFERENCES

Berson, M., Vaillant, L., Patat, F., and Pourcelot, L., 1992, High-Resolution real-time ultrasonic scanner, *Ultrasound in Med. and Biol.* 18 (5): 471-478.

Cantrell, J., H., Goans, R., E., and Roswell, R., 1978, Acoustic impedance vriations at burn-nonburn interfaces in porcine skin, *J. Acoust. Soc. Am.* 64 (3): 731-735.

Del Grosso, V., A., and Mader, W., 1972, Speed of sound in pure water, *J. Acoust. Soc. Am.* 52 (5): 1442-1446.

Gill, W., and Long,W.B., 1971, A critical look at cryosurgery. *International Surgery.* 56 (5): 344-351.

Kimming, W., Hicks, R., and Breitbart,E.W., 1971, Ultrasound in cryosurgery. *Clinics in Dermatology.* 8 (1): 65-68.

Krautkrämer, J., and Krautkrämer, H., 1990, in "Ultrasonic Testing of Materials", Spinger-Verlag, Berlin.

Laugier P., Berger G., Fink M., and Perrin. J., 1985, Specular reflector noise : effect and correction for In Vivo attenuation estimation, *Ultrason. Imaging* 7: 277-292.

Le Pivert, P., Binder,P. , and Ougier, T., 1977, Measurement of intratissue bioelectrical low frequency impedance: a new method to predict per-operatively the destructive effect of cryosurgery, *Cryobiology* 14: 245-250.

Martin, R.W., Silverstein, E.F., and Kimmey, M.B., 1989, A 20MHz ultrasound system for imaging the intestinal wall, *Ultrasound in Med. and Biol.*15: 273-280.

Mazur, P., 1966, Physical and chemical basis of injury in single-celled micro-organisms subjected to freezing and thawing. In "Cryobiology", Meryman, H.T., Ed., Academic Press Publ., London.

Murakami,S., and Yoshiharu, M., 1989, Human skin histology using high resolution echography. *J. Clin. Ultrasound.* 17: 77-82.

Orpwood, R., D., 1981, Byophysical and engineering aspects of cryosurgery, *Phys. Med. Biol.* 26(4): 555-575.

Ophir, J., and Yazdi, Y., 1990, A transaxial compression technique (TACT) for localized pulse--echo estimation of sound speed in biological tissues,*Ultrason. Imaging* 12: 35-46.

Shankar, P., M., Bencherit, U., Bilgutay, N., M., and Saniie, J., 1988, Grain noise suppression through bandpass filter, *Material Evaluation.* 46: 1100-1104.

Tanahashi, Y., Igari, D., Harada, K., Wanatabe, H., and Saitoh, M., 1977, Cryoprostatectomy under ultrasonic control. in "Ultrasound in medecine. Clinical aspects" Vol. 3A., Plenum Press, New York.

ACOUSTIC MEASUREMENT IN TISSUE MIMICKING PHANTOMS BY MINIPROBE TRANSDUCER

K. Soetanto, S. H. Wang, J. M. Reid., S. Ohtsuki[1] and M. Okujima[2]

Biomedical Engineering & Science Institute, Drexel University, PA19104
and Department of Radiology, Thomas Jefferson University, PA 19107
[1] Tokyo Institute of Technology, Yokohama, Japan
[2] Toin Yokohama University, Yokohama, Japan

INTRODUCTION

Sound velocity and attenuation coefficient are two acoustic important parameters in ultrasonic measurement. These two parameters will change the values which depend on the states of measured locations. Thus, this phenomenon reveals the possibility to apply sound velocity and attenuation to differentiate the structure of tissues. However, to be adopted in clinical diagnosis one need not only to find the feasible methods to measure these two parameters, but also to obtain the accurate information of the whole measured organ.

Many techniques[1,2,3] have been developed to measure the speed of sound as well as attenuation coefficient. These techniques apply either in the reflection or the transmission mode has tried to avoid the sources of error in order to obtain the reasonable data during the measurement. The measurement of the acoustic parameters can also be divided into the absolute and relative method. The absolute method directly measures the acoustic parameters without taking any other data as reference. On the other hand, the relative method takes another data, which has the same measuring condition as measured materials, to be reference.

During the measurement, some sources of error could greatly influence the accuracy of result. These possible sources of error include the refraction effect of other, the estimation of target tissues and the path length, the diffraction losses, the distortion of the received waveform, the phase-cancellation effect due to the path length variation across the beam of transducer, and so on. According to Hill[4], the source of errors resulting from the estimation of the thickness of soft tissues and the diffraction losses could lead to 10% and 30% errors, respectively. Modern signal processing techniques and statistical analysis could remove most of the coupling noise of the received signal, but the source of errors as mentioned above will not easy to be removed. To remove the sources of error, one needs to start to develop a feasible technique as well as to use precise equipment for the measurement.

A miniprobe transducer was constructed[5] by installing a pair of transducer at the tips of a caliper. The transducers on the miniprobe are made of PVDF material and have the resonance frequency of 3 MHz. The square shape of miniprobe transducer has the area of 3x3 mm^2. The caliper is an accurate tool to measure distance. The caliper can measure the length of 130 mm and the absolute accuracy is equal to 0.05 (mm). Since the combination of transducers with caliper, the estimation of path length becomes easy and accurate. Furthermore, small size of transducers on the miniprobe enables the measurement locally. By using miniprobe transducer, we may improve some sources of error, such as the estimation of distance, alignment of transducer. Throughout this work, the substitution method and the transmission mode has been used to measure both sound velocity and attenuation coefficient. By using the substitution method, the diffraction loss could be improved[4].

Throughout these works, a miniprobe transducer is applied to explore the measurement of sound velocity and attenuation coefficient. Before applying the measurement in living tissues, eight tissue mimicking phantoms, made by ATS Laboratories[7], with different scattering concentration were firstly measured. The same experiments were approached by using another commercial transducer, 2.25 MHz made by ATL Inc. The result reveals that applying miniprobe transducer to the measurement is more precise than the commercial probe.

MEASUREMENT OF SOUND VELOCITY

The basic principle of measuring sound velocity by ultrasonic system has the same idea as in physics. The measurement begins with measuring the path length of a measured material. The substitution method belongs to the relative method. This method begins with choosing a material of known sound velocity to be reference, water in our case. When using the transmission mode for the measurement, the transmitted and received pulse only travels single path length. Therefore, if the path length of a measured material as well as water has the same distance as the symbol of d and the temperature controlled as the same, the following relation can be obtained.

$$v \cdot t = v_w \cdot t_w \qquad (1)$$

where v, v_w mean sound velocity, t and t_w are the single-trip traveling time of a measured material and water, respectively. Besides to have the same path length for a measured material and water, the temperature and the energy of the transmitted pulse also need to be equal. Then the time difference can be measured as follows,

$$\Delta t = (t_w - t) \qquad (2)$$

Finally, sound velocity of a measured material can be calculated by the following equation.

$$v = \frac{v_w \cdot d}{d - \Delta t \cdot v_w} \qquad (3)$$

Throughout the measurement for sound velocity, water data published by Grosso[6] have been adopted as reference.

MEASUREMENT OF ATTENUATION COEFFICIENT

The attenuation coefficient is defined as attenuation over a path length, and usually apply dB/cm as the unit. In principle, attenuation is measured by calculating the energy loss from the transmitted and the received pulses. To measure the attenuation coefficient, we need to measure the path length very accurate, too. The estimation of attenuation coefficient can be approached either in the time or the frequency domain. In the time domain measurement. the transmitted pulse is assumed to contain a sufficient number of cycles of the wave such that the measurement is regarded as being made at a single frequency. However, the measurement in the time domain is possible to lead to large error than in the frequency domain because the shape of the time waveform is distorted by the dispersive attenuation and phase-cancellation effect[4]. The error of the measurement in the time domain could lead to 10% according to the results from Kak and Dines[7].

In ultrasonic system, the measured material can be considered analogously to the state variables in linear system, where the transmitted pulse is the input and the received pulse as the output signal. According to the theory of linear system, the relationship between input and output pulse may simply be written as following equation,

$$y(t,z)=u(t,z)*h(t,z), \qquad (4)$$

where $y(t,z)$ is the received time waveform, $u(t,z)$ is the transmitted time waveform, $h(t,z)$ symbolizes the states transfer function of a measured material, t is time in unit of 10^{-6} second, z is the path length, and $*$ symbol means the operation of convolution. If the attenuation coefficients of water and a measured material are assumed as α_w and α_t, the attenuation in the unit of (dB) can be described by the transmitted pulse and the received pulse as in the following,

$$\alpha_w \cdot d = -20 \cdot \log_{10}\left(\frac{P_w}{P_0}\right) \qquad (5)$$

where P_w is the received pulse after passed through a distance of d, and P_0 is the transmitted pulse. Similarly, an unknown attenuation of measured material also can be denoted as in the following,

$$\alpha_t \cdot d = -20 \cdot \log_{10}\left(\frac{P_t}{P_0}\right) \qquad (6)$$

where P_t is the received pulse from measured material. Due to the energy of transmitted pulses into either water or a measured material remain the same, therefore we may obtain the following equivalence by calculating the transmitted pulse from the above equations(5)&(6).

$$P_0 = P_w \cdot 10^{\frac{\alpha_w \cdot d}{20}} = P_t \cdot 10^{\frac{\alpha_t \cdot d}{20}} \qquad (7)$$

According to equation (7), the attenuation coefficients can be calculated by the pulses received from water and measured material. The relationship can be described as follows,

$$\frac{P_t}{P_w} = 10^{\frac{(\alpha_w - \alpha_t)d}{20}} \qquad (8)$$

Consequently, applying logarithm to equation (8), the attenuation coefficient of the measured material can be obtained by the following equation.

$$\alpha_t = \alpha_w - \frac{20}{d} \cdot \log_{10}(\frac{P_t}{P_w}) \qquad (9)$$

We may rewrite equation (9) as follows,

$$\alpha_t = \alpha_w + \alpha_{tw} \qquad (10)$$

where α_{tw} is equal to $(-\frac{20}{d} \cdot \log_{10}(\frac{P_t}{P_w}))$.

According to Hill[4], the attenuation coefficient of degassed, distilled water has a square-law frequency dependence, where α_w at 20^0C is equal to following,

$$\alpha_w = (25 \times 10^{-5}) \cdot f_c^2 \qquad (11)$$

where f_c is the center frequency of the applied transducer. Since the center frequency of miniprobe transducer is 3 (MHz), therefore α_w is equal to 0.00215 (dB/cm) in our application. According to equation (10), if α_{tw}. is significant larger than α_w, the estimation of α_t can be simplified as follows,

$$\alpha_t = \alpha_w. \qquad (12)$$

As applying the Fourier Transform to equation (4), the received spectrum will become as follows,

$$Y(f,z) = H(f,z) \cdot U(f,z), \qquad (13)$$

where Y and U are the received and transmitted spectrum, and f is frequency in units of MHz. Due to the frequency dependent attenuation of the measured material, the states transfer function of H(f,z) again can be assumed as follows,

$$H(f,z) = A_0 \cdot e^{-\alpha(f) \cdot z}, \qquad (14)$$

where $\alpha(f)$ is the frequency dependence of the attenuation coefficient in units of dB/cm, and A_0 is the magnitude of transfer function. To simplify the measurement, we can assume equation of the frequency dependent attenuation coefficient as follows,

$$\alpha(f)=\beta \cdot f^n , \qquad (15)$$

where n is the order of frequency dependence of the measured material. From equations (14) and (15), the attenuation coefficient in unit of dB/cm may be calculated as follows,

$$\alpha(f)=\frac{20}{z}\log_{10}(\frac{Y(f,z)}{A_0 \cdot U(f,z)})=\frac{20}{z}\log_{10}(\frac{A_0 \cdot U(f,z)}{Y(f,z)}) \qquad (16)$$

The bandwidth of a spectrum is defined as an interval of frequency within 3 dB of pressure waveform or 6 dB of power spectrum. When the range of attenuation coefficient within the bandwidth has a linear relationship, the order of n in equation (15) can be assumed as equal to one. Therefore, equation (15) can be simplified as follows,

$$\alpha(f)=\beta \cdot f \qquad (17)$$

Consequently, we can apply the homogeneous linear mean-square(MS) estimation[8] to find out the optimal value of β in equation (17). According to the estimation procedure, the MS error is as follows,

$$e=E\{(\alpha(f)-\beta \cdot f)^2\} \qquad (18)$$

The value of β can be estimated as follows,

$$\beta=\frac{E(\alpha \cdot f)}{E(f^2)} \qquad (19)$$

EXPERIMENTS

Both sound velocity and attenuation coefficient were experimented with a miniprobe transducer and a pair of 2.25 MHz transducer by the transmission mode. The miniprobe transducer has the feature introduced as former section. A pair of 2.25 MHz transducer have circular plane with the diameter of 13 mm, and the far field is approximately at 7 cm of the axial distance. These two kinds of transducer were applied to measure tissue-mimicking phantoms for the comparison of the results. The whole setup of experiment include a Lecroy 9450A digital oscilloscope, Panametrics 5052UA ultrasonic analyzer, and 386-PC with GPIB board inside the computer. The ultrasonic analyzer was used to be a pulsar for the transmitted pulse and an amplifier for the received signals. The continuous time waveforms from the received pulse were transmitted into digital oscilloscope, where the analog signals were digitized by a 400 MHz sampling rate, and 8-bits resolution. In order to eliminate the coupling noise during experiment, all collected signals have been averaged by a thousand sweeps of trigger. Finally digital data were stored in a 386 personal computer for further processing.

Consequently, eight tissues mimicking phantoms with thickness of 30 (mm) were measured. Totally 90 A-lines per phantom were collected. Before the experiments, all phantoms as well as beef tissues were stored in an incubator where temperature was set at 37^0C. Consequently, the experiments were done in a 37^0C water bath.

Table 1. Sound velocity of phantoms from A to H (mm/μsec)

phantoms	A	B	C	D	E	F	G	H
mean (miniprobe)	1.41077	1.41134	1.41450	1.41004	1.41410	1.41071	1.41084	1.42015
S. D.(%)	0.05861	0.08267	0.22571	0.04159	0.08895	0.07472	0.02776	0.06331
mean (2.25MHZ)	1.41162							1.42133
S. D.(%)	0.1701							0.1648

Table 2. Attenuation coefficient of phantoms A to H (the time and frequency domain)

phantoms	A	B	C	D	E	F	G	H
mean (time, miniprobe)	0.881	0.881	0.935	0.853	0.929	0.895	0.942	1.111
S. D.(%)	0.7	1.44	1.46	1.22	1.09	1.83	0.33	1.15
mean(time, 2.25MHz)	0.7014		0.7048					0.8215
S. D.(%)	0.7014		10.66					6.64
mean(freq.,mi niprobe)	0.89	0.89	0.942	0.864	0.951	0.906	0.964	1.182
S. D.(%)	0.99	1.3	1.71	1.02	1.39	1.7	0.77	1.22
mean(freq., 2.25MHz)	0.7391		0.7375					0.8570
S. D.(%)	10.89		1.71					7.17

RESULTS

To calibrate the measuring system for sound velocity, we have measured a 37^0C water by changing steps of 5 (mm) in the range of 5 to 100 (mm). The average of the result equals to 1.5239162 (mm/μsec), and which compared with Gross's data 1.523618 shows only 0.0852% in relative difference. Consequently, all sound velocity result by tissue mimicking phantoms were shown in Table 1. To measure the attenuation coefficient, the peak to peak pressure waveform was adopted to calculate the energy loss in the time domain. Consequently, 1024 sampling points of the time waveform were applied by the FFT algorithm to transform into the spectrum. Consequently, the spectra shift between water and phantoms were calculated. The results estimated by the time and frequency domain were shown in Table 2. The subtraction of attenuation coefficient from water data to phantom within a bandwidth of transducer between 1.95 (MHz) and 3.91 (MHz) shows linear relationship. The slope of all phantoms were estimated and shown in Table 3.

Table 3. Results of the slope of attenuation coefficient for all phantoms A to H

β	A	B	C	D	E	F	G	H
mean	0.3151	0.3117	0.3283	0.3000	0.3244	0.3231	0.3374	0.3683
S. D.(%)	0.25	0.64	0.49	0.25	0.23	0.58	0.12	0.35

DISCUSSION AND CONCLUSION

According to the results, the measurement by using miniprobe transducer shows more precisely than 2.25 MHz transducer in comparison with the standard deviation. The reasons include that miniprobe transducer is easier to estimate the path length than a pair of 2.25 MHz transducer. Furthermore, the far field of miniprobe transducer is 4.41 (mm) estimated in 37^0C water, which comparing to 70 (mm) by 2.25 MHz transducer. This far field and small size of miniprobe transducer reveal the feasibility to locally measuring a specific area of measured medium. On measuring sound velocity of water, miniprobe transducer shows highly accuracy. The results of the estimation of attenuation coefficient comparing to the manufacture's reports also shows reasonable. The wideband of miniprobe transducer suits for the estimation in the frequency domain, too.

Miniprobe transducer has proved its well function to measure both sound velocity and attenuation coefficient throughout this work. The measurements in tissue mimicking phantoms are preliminary for the further measurement in living tissues. We can apply miniprobe transducer for the measurement in living tissues during surgical operation, so that the refraction errors due to the influence of other tissues can be avoided.

ACKNOWLEDGEMENT

This work was supported by NIH Grant 1PO1CA52823-01.

REFERENCES

1. Wells, P. N. T., _Biomedical Ultrasonics_, Academic Press, London, 1977

2. Robinson, D. E., Ophir, J., Wilson, L. S., and C. F. Chen, " Pulse-echo ultrasound speed measurements: progress and prospects," Ultrasound Med. Biol., vol. 17, no. 6, pp. 633-646, 1991

3. Kuc, R., " Clinical application of an ultrasound attenuation coefficient technique for liver pathology characterization," IEEE Trans. on Biomed. Engi., BME-27, no. 6, pp. 312-319, 1980

4. Hill, C. R., <u>Physical Principles of Medical Ultrasonics</u>, John Wiley & Sons, INC., NY, 1986

5. Soetanto, K., Ohtsuki, S. and Okujima, M., " Miniprobe transducer for tissue characterization," IEEE Ultrasound Sympo., pp. 1417-1420, 1990

6. Grosso, V. A., and Mader, C. W.," Speed of sound in pure water," JASA, 52(5), pp. 1442-1446, 1972

7. Kak, A. and Dines, K. A., " Signal processing of broadband pulsed ultrasound: measurement of attenuation of soft biological tissue," IEEE Trans. Biomed. Engi., BME-25, no. 4, pp. 321-344,1978

8. Papoulis, A., Probability, <u>Random Variables, and Stochastic Processes</u>, 2nd ed., McGraw-Hill, NY 1984

COUNTING BIOLOGICAL CELLS WITH HIGH FREQUENCY FOCUSED ULTRASOUNDS

F Lefebvre, M Toubal, E Radziszewski, Y Deblock, B Nongaillard

I E M N
Institut d' Electronique et de Microélectronique du Nord
U M R   C N R S 9929
Département d' Opto - Acousto - Electronique
E N S I M E V
Université de Valenciennes et du Hainaut-Cambrésis
Le Mont Houy, B P 311
59304 Valenciennes Cedex France

INTRODUCTION

In biomedical laboratories, most of the work consists certainly on counting globules or sperm, bacteria monitoring under pharmaceutical factors, etc   At present, commonly used counters are based on either optical detection (light obscuration or scattering) or on the measurement of an electric field change However, these counters suffer from disadvantages such as the necessity of diluting or altering the media under test Our work consists in showing the feasibility of an acoustical particles counter (APC) which does not alter the biological media and wich provides direct counting without dilution In this paper, we will show that high frequency ultrasounds (more than 100 Mhz) produced by a focused acoustic sensor give a suitable solution for small particles detection (over a wide diameter range. 10 to 200 µm) The acoustic sensor performances are determinant for the APC For this reason, it is important to study the acoustic field produced by the focused system. We will present some theoretical approach which provides a complete description of the spatial characteristics of the sensor After basical considerations on the principle of the APC, we will present the system we realized in our laboratory which can extract cells concentration from acoustic reflexions on these particles Previously designed for biological applications, this apparatus can also be devoted to other kind of particles such solid or gazeous ones

BRIEF REVIEW OF SOME TECHNIQUES FOR COUNTING AND SIZING PARTICLES

In this part, we will present some methods curently used which we sometimes need to count or determine size of particle. All of these methods are based on the interaction of the particle with a radiation. Different kinds of radiation can be used such as optical, acoustical or electrical ones. They can be separated in two families: global techniques, which consider the interaction of the radiation with a set of particles, and local or statistical techniques which count particles individually.

Global Particles Measuring Systems (GPMS) measure physical properties of the solution by different maners like diffraction[1], diffusion[2], absorption[3] or attenuation[4]. Most of them need a complex modeling of the media and in general computer based off-line signal processing techniques to extract concentration or size information on the particles suspended in the solution. In particular techniques such as optical ones, dilution are necessary because of the opacity criterion. Some systems, like diffraction sizers, are not able to measure absolute concentration but only the size distribution which represents the proportion of objects related to a set of size classes.

Local techniques, also called Single Particle Counting Systems (SPCS), are based on the extinction of a radiation produced by an optical beam[5] or an electrical field[6] due to a particle passing through this beam. They need a calibrated flow of the solution through the detector cell in order to calculate the particle concentration. Problems of coincidence[7] due to several particles passing at the same time in the beam must be treated either by special processing or by dilution.

BASIC PRINCIPLE OF ACOUSTICAL COUNTING

The principle consists in using an ultrasonic field generated by a piezoelectric transducer mounted on sapphire rod, and focused by a spherical lens (figure1). When a particle passes through the focal zone, it reflects a part of the acoustic energy which is received by the same sensor. By counting electrical pulses due to acoustic reflections, the system is able to determine the number of objects passing through the beam per unit of time. A simple model show that the particle flux $\delta\phi$ is related to concentration n, the average speed of particles v and their size by:

$$\delta\phi = K\, n\, v\, \delta S$$

where δS is the active surface of detection and **K** a constant which is a function of the particle size. It is clear that the average speed must be known or measured for a given particle size. If those parameters are not kwown, the concentration can be calculated after a calibration with a reference solution.

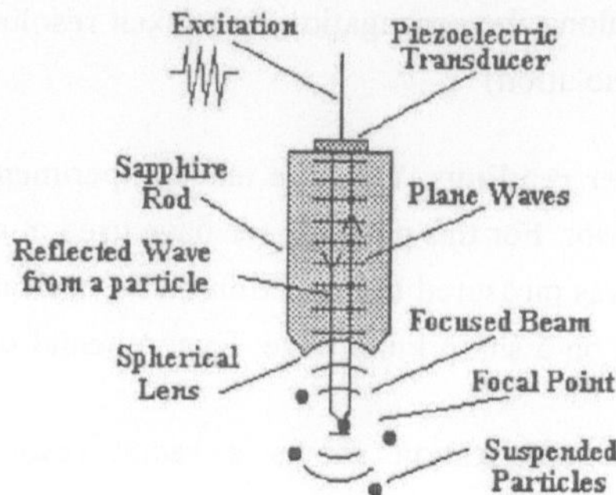

Figure 1. The focused acoustic sensor
acting in the particulate fluid

Maximal sensibility is obtained when the diameter of the focal zone is of the order of the particle diameter This criterion is achieved for high frequency ultrasounds (> 100Mhz) which provide a wide range of size detection (10 à 200µm) However, attenuation effects in the fluid surrounding the particles, limit the utilization of high frequency ultrasounds

The advanges of this system reside in several points Neither dilution nor sampling is necessary and the detection can be made directly in the solution containing biological cells Because of the focal zone size, only one particle can be detect at a time which supress coincidence problems An other important point is that fluid does not need to be set in motion because of the natural mobility of the living cells We must notice that this mobility is rather constant in wide range of concentration for a given pathology of the cells culture If the particles are static, mechanical agitation must be provide in order to homogenize the suspension In particular applications, it is not necessary to measure the concentration of particles in the fluid A simple counting is required just for monitoring the cells population kinetic

THEORETICAL AND EXPERIMENTAL STUDY OF FOCUSED ACOUSTIC FIELD

Acoustic field characteriscs are decisive for the performances of the APC It's the reason why, we gave a particular attention to the study of them both theoretically and experimentally. We used two different approaches for modeling acoustic focused beam. The fisrt technique is base on the solution of the wave equation for the acoustic potential at the boundary of the spherical lens This solution takes the form of the Rayleigh-Sommerfeld equation. This integral form has been processed by numerical methods The second approach consists on the decomposition into plane waves[8] which expressed that the acoustic field can be described in term of angular spectrum The later is obtained by the bidimentionnal Fourier transform of the spatial distribution of the field These methods are use to determine the size of the focal zone, both along the propagation axis (axial resolution) and around this axis at the focal plane (radial resolution)

Experimental field meter readings. We have made experimental readings of the acoustic field produced by our sensor For this purpose, we have used a mechanical scanning acoustic microscope Axial field was measured by reflection on an optical polished surface and radial field profile by reflection on a sharp knife edge Experimental curves are shown in figure 2 and 3

This experimental characterization shows a radial resolution about 7 µm and a longitudinal ones about 50 µm

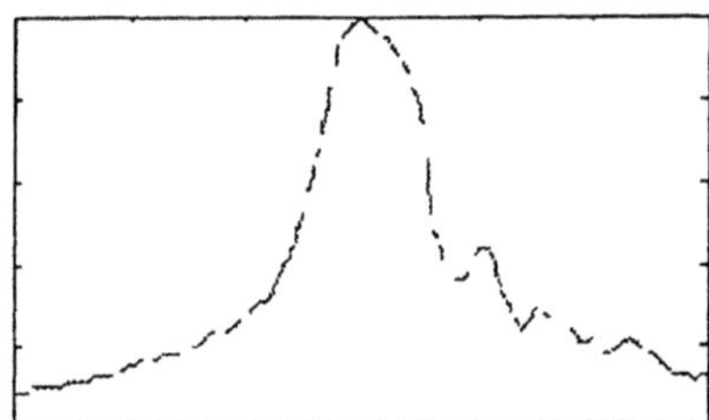

Figure 2. Experimental longitudinal field measurement

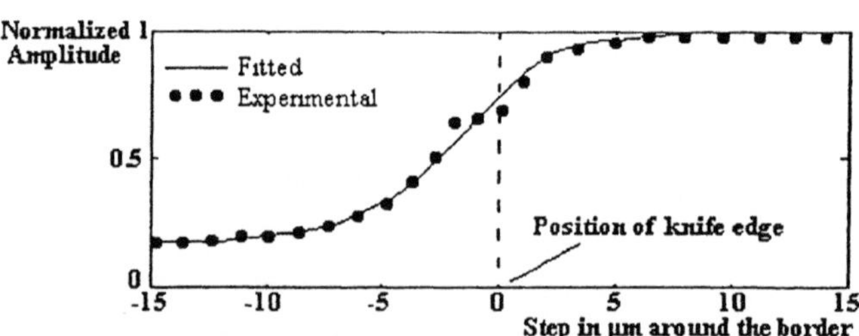

Figure 3. Radial acoustic field profile measured by the technique discribed in the text

ACOUSTIC PARTICLE COUNTER PROTOTYPE

We can see in figure 4 the block diagram of our counter A 200 Mhz focused acoustic sensor is driven by a high voltage pulser which produces narrow acoustic emission pulses in the media under test The reflected echoes arriving from the particles are amplified and detected by a hight speed detector The performances of the later allow us a straight detection of the high frequency echoes without intermediate demodulator Logical pulses

538

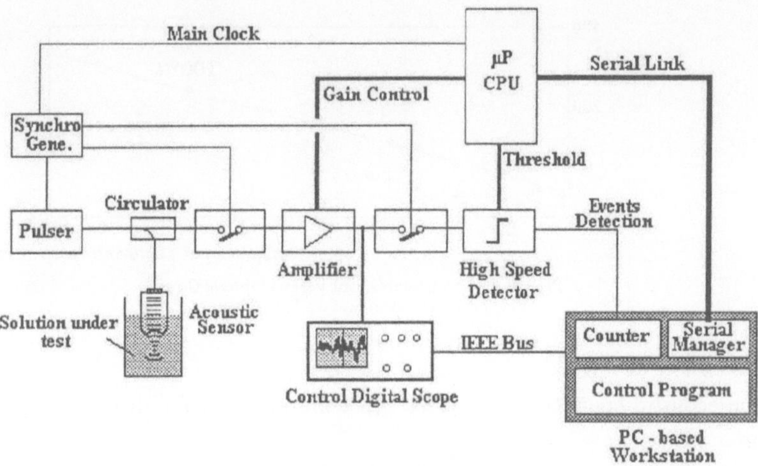

Figure 4. Principle of the propose APC

corresponding to the particle detection are counted by a specially developped PC-board after the treatment of multiple detections

All the parameters of the counter (gain, sampling threshold,) are controled by a microprocessor itself piloted by a PC based workstation by a serial link The computer collects the counts and provides statistical processes, data storage and display We developed a special oriented graphical environnment for this purpose The figure 5 shows a typical echogram obtained from a particle passing through the focal volume and its events detection signal Finally, a high frequency digital storage oscilloscope is used to control echograms and sampling conditions It can be driven by the computer by an IEEE bus

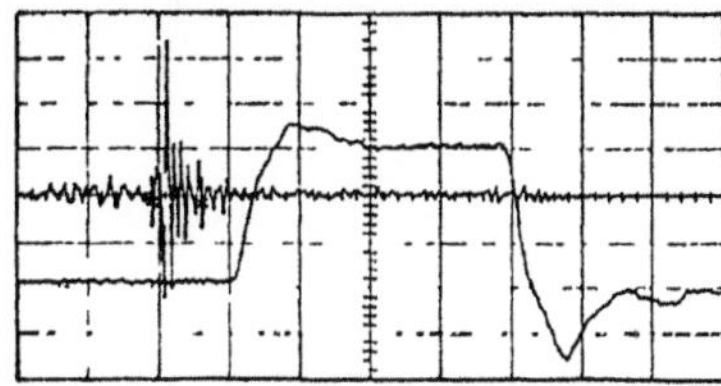

Figure 5. High frequency echogram from a particle and its logical detection signal

EXPERIMENTAL RESULTS

First experiments were conducted to verify the law between the flux of particles $\delta\phi$ and the concentration **n** For this purpose, we used several solutions of thetrahymenas (well known protozoa in biological world) the concentration of which was measured by an optical method using a classical Nageote cell We compared the results obtained with this technique

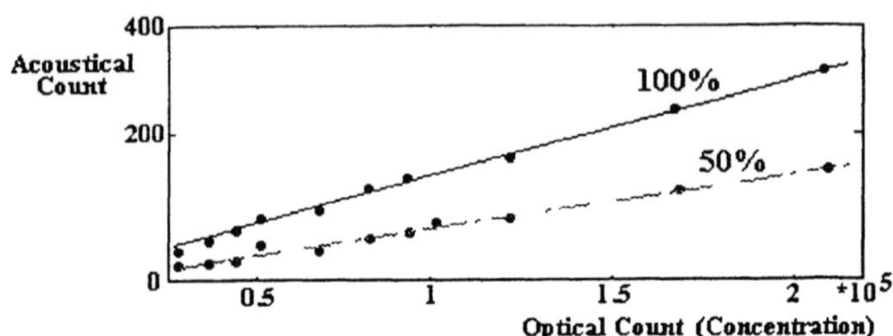

Figure 6. Acoustical Count versus Optical Count
for a solution and its 50% dilution

with the acoustic counting for different concentrations and for their corresponding 50%
dilution

Figure 6 shows the good agreement between optical measurement of concentration and
acoustical count We also evaluate the experimental accuracy of our system which is given
by by the relative dispersion $<\delta\phi>/\sigma$ of experimental data This precision is shown in figure
7 and point out the better ability of our system to work at higher concentrations rather than
lower ones

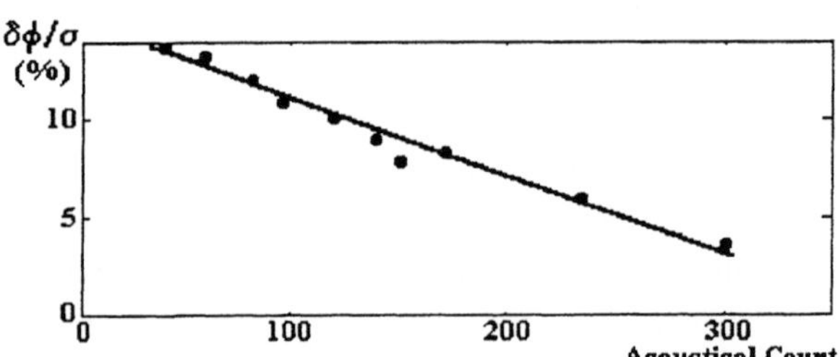

Figure 7. Relative dispersion in percent of experimental
data versus acoustical count

CONCLUSION AND FUTURE EXTENSIONS

This paper presents the first step of the developpement of an automatic acoustic particle
counter. We have shown that it was possible to measure the particle concentration by
counting them directly in the suspension The use of high frequency focused ultrasounds
provides a very high spatial resolution which is suitable for small particles detection In
addition, we must point out some advantages of ultrasounds compared to other radiations.
opacous media monitoring capability, harmless radiation for biological tissues at low imaging
power... At present, we are working on the interaction of the focused beam with the media
in order to qualify the perturbation of the particles motion in comparison to the natural
mobility of biological cells Those facts get us to add to our counter a time of flight based
velocimeter to take account of the average speed of the detected particles

REFERENCES

1. J Cornillault "Particle size analyser" App Optics, vol 11, n° 2, pp 265-268 (1972)

2. L Filipczinski "Diffusion ultrasonore dans le sang et détection de bulles de gaz" I T B M , vol 5, n° 1, pp 67-76 (1984)

3. G Gopala Krishna, A K W Anwar Ali, A Ahmad "Absorption of sound in human blood" Colloque de Physique, vol 51, n° 2, pp 311-414 (1990)

4. J R Allegra, S A Hawley "Attenuation of sound in suspensions and emulsions Theory and experiments" J A S A , vol 51, n °5, pp 1545-1564 (1972)

5. D J Holve, K D Annen "Optical particle counting, sizing, and velocimetry using intensity deconvolution" Opt Eng , vol 23, n° 5 (1984)

6. E C Gregg, K D Stedley "Electrical counting and sizing of mammalian cells in suspension" Bioph Journal, vol 5, pp 393-405 (1965)

7. J F Pisani, G H Thomson "Coincidence errors in automatic particle counters" Jour of Phys , vol 4, pp 359-361 (1971)

8. C F Quate , A Atalar , K K Wickramasinghe "Acoustic Microscopy with Mechanical Scanning-A review" Proc of the I E E E , vol 67, n° 8, pp 1092-1114 (1979)

ACOUSTICAL MONITORING OF THE PROCESS
OF FOCUSED ULTRASOUND
SURGICAL LESION FORMATION

C. R. Hill, R. L. Clarke, G. R. ter Haar and J. C. Bamber

Physics Department, Institute of Cancer Research, Royal Marsden
Hospital, Sutton, Surrey, SM2 5PT, UK

Although long established as a concept, focused ultrasound surgery has only become a qenerally practicable clinical technique with the recent development of high — resolution tomographic imaging methods. Ultrasonic imaging is of particular interest here, partly because of its physical compatibility with the surgical technique. It offers the impotant added feature, however, that acoustic property changes result from the lesion formation process, and are thus recordable as image data and provide potential non — invasive means for monitoring progress of the procedure.

The process of lesion formation is known to be predominantly thermal, but with possible involvement of cavitation mechanisms. With this in mind we are carrying out, and shall report on, three related sets of observations: (1) the time course of acoustic emission levels (sub — harmonic and anharmonic) from a lesion during the course of its formation; (2) the corresponding time courses of acoustic backscattering cross — section and attenuation coefficient of the lesion; and (3) the acoustic and corresponding optical microscopic characteristics of lesions at various stages of development.

Acoustical Imaging, Volume 20 Edited by Y Wei
and B Gu, Plenum Press, New York, 1993

ASSESSMENT OF BONE STRUCTURES BY ACOUSTICAL MICROSCOPY

H.-J. Hein, P. Czurratis[*], A. Bernstein[**]

Inst. of Applied Biophysics, Dep. of Medicine, MLU Halle/S
Str. der OdF 6, D-O-4020 Halle/S
[*] Leica GmbH, Lilienthalstr. 39-45, PF 1651, D-W- 6140 Bensheim 1
[**] Clinic of Orthopaedics, Dep. of Medicine, MLU Halle/S
Magdeburger Str. 22, D-O-4020 Halle/S
Germany

INTRODUCTION

The visual inspection and description of objects is continuously extended by analytical methods.

An example is the application of non optical information carriers. For their specifical interaction with the object follows new knowledge about the object structures.

One of the new methods and procedures is the acoustic scanning microscopy. There are two reasons for application of acoustical microscopy [1, 2, 3, 5].

1. The ability to penetrate tissues that are opaque to light.

2. The contrast in the picture depending on mechanical properties.

In biological structures the penetrating ultrasound wave is weak scattered by changing of elastical properties from tissues. Detaills are described in [4, 6, 7].

In the following we present first results for bone investigations.

METHODS

First of all we had to prepare the bone material. After the first attempts we found the application of methacrylate useful. The bone material is embedded in this material and has contact with the substrate. The surface is smoothed and the thickness of specimen was 50 μm. The roughness of the surface is lower than 1μm. The coupling medium is bidestilled water with room temperature. For frequencies of 1 GHz it is necessary to use temperatures from about 35 °C. We used two frequencies, 400 MHz and 1 GHz. From the 1 GHz pictures we produced V(z)-curves, from these the shear waves (Rayleigh) velocity is calculated in order to characterize the bone tissues.

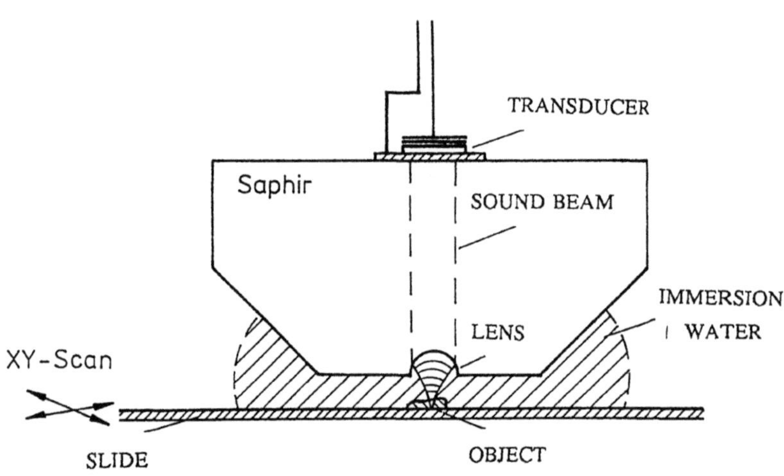

Fig.1: The construction of the acoustical lens with the position of object and the way of sound propagation

The measurements are carried out by acousto-microscope ELSAM. The principe of microscope lens is presented in fig.1.

An electrical high-frequency signal of about ten to twenty nanosecond pulse length is given on a transducer, where it is turned into a longitudinal ultrasonic wave. The sapphire lens on the bottom is covered by an antireflecting coating in order to reduce reflectivity losses for the mechanical wave.

The sound velocity in sapphire is 11000 m/s, the velocity in water is about 1500 m/s. This generates a refraction index of n=7.4 and a aberration-free image as well as a diffraction limited spot of focus.

The SAM is the sole microscope which is able to image structures situated below the surface.

For bone investigations the frequency of 400 MHz is more attractive because their penetration depth is higher. The resolution is about 4 μm. For the inspection of microstructures we used frequencies of 1 GHz. The disadvantage of such frequencies is the low penetration depth, the advantage the high resolution from about 1.5 μm.

For an effective tissue characterization it is necessary to produce the V(z)-curve. It is deduced from the following formula [7] :

$$v(z) = \int_0^{f\sin\alpha} U^2(r)\, P^2(r)\, R\left(\frac{r}{f}\right) \exp\left(-2jk_0 z\sqrt{1-\left(\frac{r}{f}\right)^2}\right) r\, dr \qquad \textbf{(1)}$$

where r/f = sine of the incidence angle for acoustic plane waves in the liquid
      α = opening half-angle of the lens
      k_0 = $2\pi/\lambda$ acoustic wave number in the liquid
      z = displacement of the object from the focal plane
      U = field generated by acoustic transducer
      P = lens pupil function
      R = acoustic reflectance function of the object
      r = radial position in the back focal plane of the lens
      f = focal length of the lens

The velocity of the Rayleigh waves is calculated from the V(z)-curves. The period of oscillation in V(z) is:

$$\Delta z = \frac{\Lambda_o}{2\,(1-\cos\theta_R)} \qquad \textbf{(2)}$$

and the velocity

$$v_r = v_0\sqrt{1-\left(1-\frac{v_0}{2f\Delta z}\right)^2} \qquad \textbf{(3)}$$

where Δz = period of the Rayleigh-oscillations
      v_0 = velocity in the coupling fluid
      λ_0 = wavelength in the coupling fluid
      f = frequency
      θ_R = diffraction angle.

That means by measuring the period of oscillations the Rayleigh velocity can be deduced directly [8].

RESULTS

We used specimen embedded in methacrylate as well as in water after the common histological procedure. Their thickness was 50 μm. The inhomogeneous structures of bones are clear to recognize in the pictures. Fig. 2 presents the bone (Pelvis) of an old patient. He had osteoporosis with malacide component. Fig. 2a shows a overview about a small region of interest. The light places are regions of high calcification inside of spongiosa structures. The dark holes correspond regions which are filled up with liquid lower attenuation.

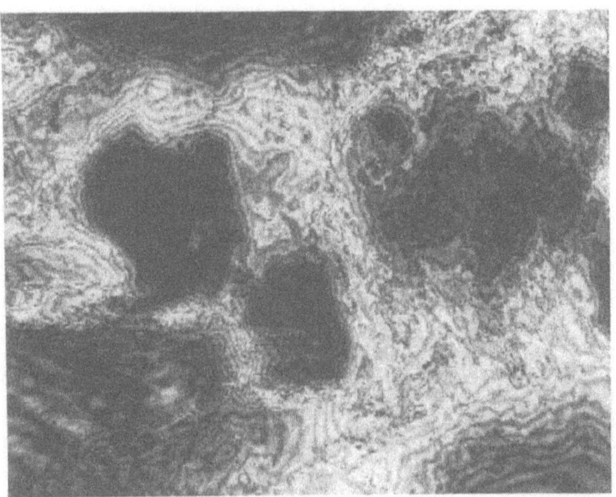

Fig. 2a.   The picture shows an overview about a part of spongiosa (Frequency 1 GHz, picture width 1 mm, z = - 12 μm)

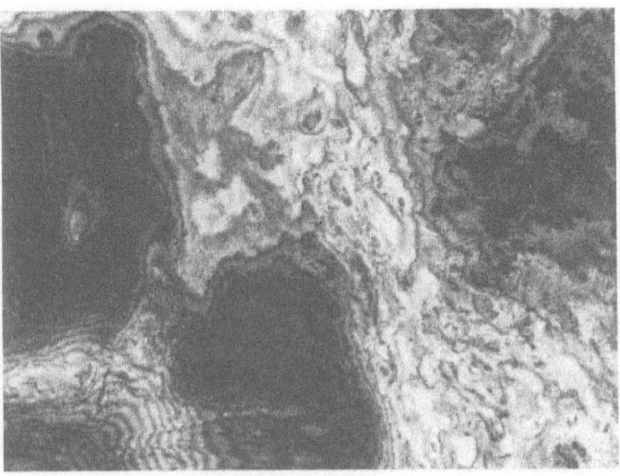

Fig. 2b.   The part of fig. 2a (Frequency 1 GHz, picture width 500 μm, z = -12 μm)

The Fig. 2b shows a two time magnification of the picture before. On the left side in the dark region a cell (like osteoblast) is to see which feeds the bone material; clear to see is the different increasing calcification from the margin to the inside.

This is demonstrated in a higher resolution by the fig. 2c in the bottom dark part.

Fig. 2c.   A further part of fig. 2a, but picture width is 200 μm, frequenzy 1 GHz, z = - 12 μm

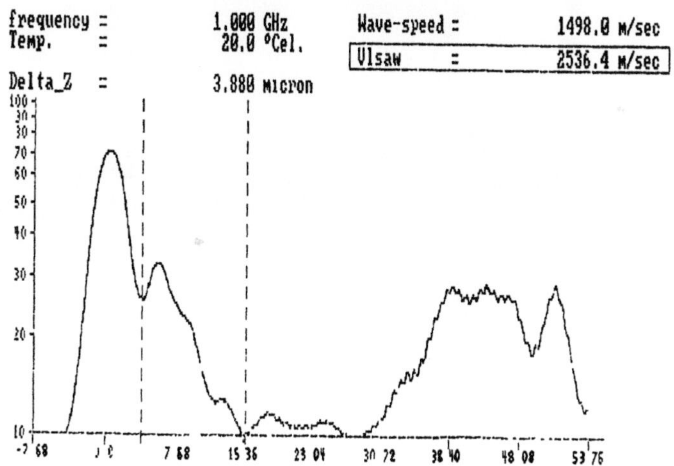

Fig. 3.   V(z)-curve for specimen imaged in figures 2. Sound velocity is very high, caused by the methacrylate

We determined the Rayleigh-velocity from V(z)-curves of this material. The velocity is very high because we had prepared in methacrylate. The propagating sound velocity in methacrylate higher is higher than in heterogeneous bone material, also depending on the poissons ratio. This is demonstrated in fig 3.

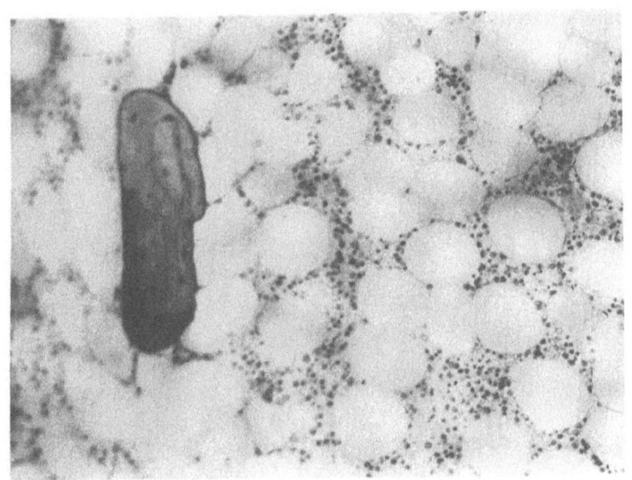

Fig. 4. Optical picture from bone marrow of the same specimen as before

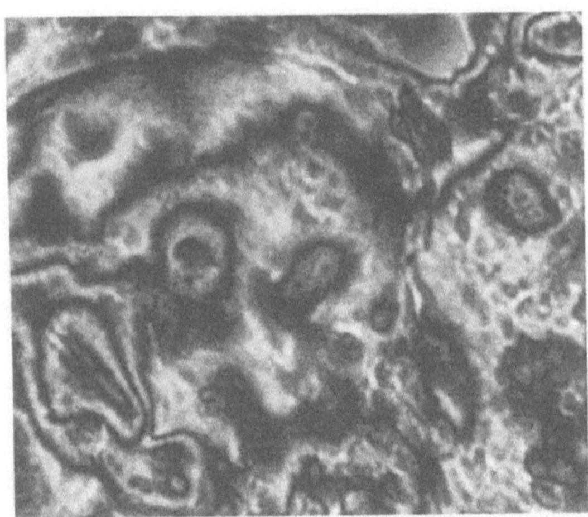

Fig 5. Acoustical image from bone marrow. The cells and vessels are clear to see. the frequency is 400 MHz, picture width 500 μm, z = -35μm

Fig. 4 shows an optical picture from the same specimen. It is noticed that the acoustic picture from the same region as presented in fig. 5 seems to be more informative and to have more contrast as the optical one.

The picture in fig. 5 is produced at a frequency of 400 MHz. The structures of the marrow are clear to differentiate in their acoustical properties for vessels and cell structures. The fig. 6 shows bone marrow from a healthy bone. The used frequency is 400 MHz.

Fig. 6.   This acoustical picture shows bone marrow from a young patient (Frequency 400 MHz, picture width 500 μm, z = - 58 μm)

In order to show the influence of material where the specimen are embedded we produced V(z)-curves and calculated the velocities. They are presented in fig. 7. The Rayleigh-velocity is significant lower as in the case described before in fig. 3. It is composed from velocities in water and bone.
In the fig. 8 a bifurcation of spongiosa in high resolution image is presented. That is a healthy bone and in a good calcification state.

Onother acoustic picture (fig.9) shows compact bone with "Haver"-channels on the top and one osteoblast cell in center of picture.

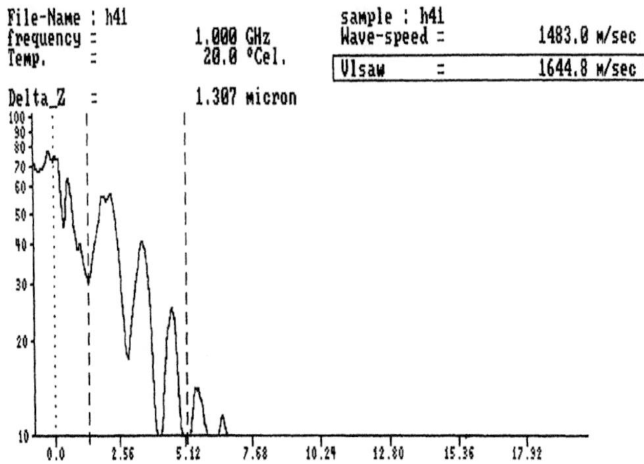

Fig. 7.   V(z)-curve from bone material embedded in water. Velocity is significant lower than in figure 3

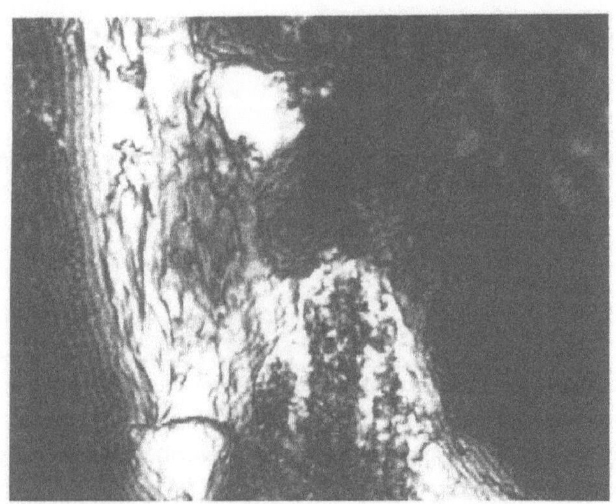

Fig. 8.   It shows the bifurcation of a spongiosa beam
(Frequency 1 GHz, picture width 200 μm, z = - 7μm)

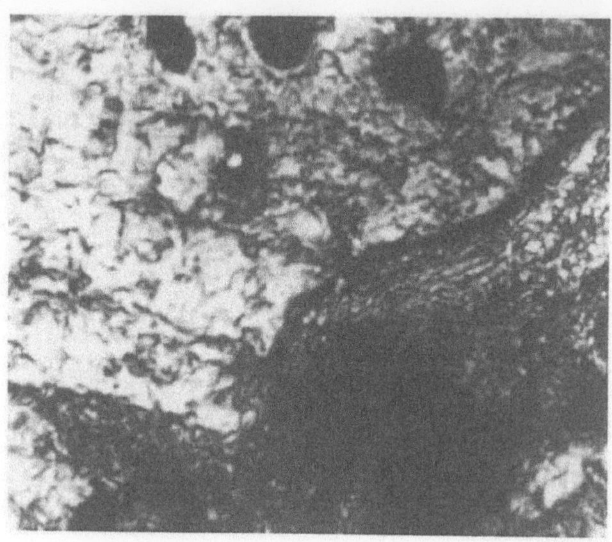

Fig. 9.   Here is imaged a compact bone where are to see on the top the "Haver" channels
(Frequency 400 MHz, picture width 500 μm, z = - 62.5μm)

CONCLUSION

The application of acoustic microscope seems to be attractive to assess mechanic properties of bone materials for tissue characterization.

However, the bone structure is strongly inhomogeneous, so that the result is depending on the preparation and the position inside from material, as well as the direction where specimen is cutted.

REFERENCES

1. A. Briggs. "An introduction to scanning acoustic microscopy", Oxford University Press, 1985
2. C.F. Quate, A Atalar and Wickramasinghe. "Acoustic microscopy with mechanical scanning- A review",
    Proc. of the IEEE, Vol. 67, No. 8 August 1979, p.1092
3. A. Thaer, M. Hoppe and W. J. Patzelt. "The ElSAM Acoustic Microscope",
    Leitz Mitteilungen für Wissenschaft und Technik, Vol. 2, No 3/4, 1982, S. 61-67
4. K. Itoh. "Studies of acoustic characterization of carcinoma with scanning acoustic microscope", *Asian Medical Journal*, Vol. 2, No 2 (1983)
5. L.W. Kessler, P.R. Palermo and A. Korpel. "Practical high resolution acoustic microscopy", Acoustical Holography, *Ed. by* G. Wade, Vol. 4, 1972
6. M. Hoppe and J. Bereiter-Hahn. "Applications of scanning acoustic microscopy-survey and new aspects",
    IEEE Transactions on Sonics and Ultrasonics, Vol. SU-32, No 2, March 1985, S. 289-301
7. J.A. Hildebrand and D. Rugar. "Measurement of cellular elastic properties by acoustic microscopy",
    J. Microscopy 134, 1984, P. 245-260
8. A. Briggs. "Acoustic Microscopy", Clarendon Press-Oxford, 1992

IN VIVO FLOW MEASUREMENT BY
ULTRASOUND TRANSVERSE DOPPLER

P. Tortoli[1], G. Guidi[1], C. Atzeni[1], and P. Pignoli[2]

[1]Electronic Eng. Dept., University of Florence, Florence, Italy
[2]Merate Hospital, Merate (Como), Italy

INTRODUCTION

As known, pulsed ultrasound systems allow the blood flow velocity to be detected, by measuring the shift between the frequency of the transmitted burst and the frequency of echoes reflected by moving erythrocytes. According to the Doppler principle, this shift is given by:

$$f_d = \frac{2}{\lambda}\, v \cos\theta \tag{1}$$

where λ is the transmitted wavelength, v the scatterers' velocity and θ the angle between the directions of ultrasound beam and flow.

The above equation implicitly predicts that one possible target moving with constant velocity leads to a single Doppler frequency component. However, this statement appears substantially incorrect when the shape of acoustical beams produced by practical ultrasound transducers is taken into consideration. Limited beamwidth in fact involves a limited scatterer transit time, which inherently leads to the broadening of the received signal spectrum[1]. Moreover, ultrasound beams are usually focussed at a depth F, i.e., they are formed by several rays oriented at different angles to each other. Therefore, each of these rays contributes with a corresponding frequency component to the overall Doppler spectrum.

These considerations have induced Newhouse et Al.[2] to develop a new theory which puts in relation the effective flow velocity with the bandwidth, B_d, of the received Doppler signal. It has finally been demonstrated that they are related through the relationship:

$$B_d = \frac{2}{\lambda}\, \frac{W}{F}\, v \sin\theta \tag{2}$$

where W and F are the aperture and focal length of the transducer, respectively.

Acoustical Imaging, Volume 20 Edited by Y. Wei
and B. Gu, Plenum Press, New York, 1993

It is convenient to point out that the well known Doppler equation is complemented, and not substituted by the new relationship. The former equation, in fact, relates the spectral mean frequency, f_d, to the axial velocity component, $v \cos\theta$, while the latter equation relates the spectral bandwidth, B_d, to the scatterers' "transverse" velocity component, $v \sin\theta$.

In particular, the two equations suggest that for transverse flow (i.e. for a 90° angle between the flow and the ultrasound beam axis), the "mean" Doppler shift is zero (according to eq.1), but it is surrounded by a full spectrum of Doppler frequencies, spanning a range B_d proportional to the velocity.

In this paper, the practical implications of eq.2 in "transverse" blood flow analysis are taken into consideration. A pulsed flowmeter designed for computerized measurements of both the Doppler mean frequency and bandwidth, has been used "in vitro" and "in vivo" to observe steady and pulsatile flow. A correct interpretation of the results emerging from these experiments, allows to detect significant velocity information even for a 90° Doppler angle, i.e. in a condition which is usually rejected in conventional Doppler ultrasound analysis. Based on these results, the capabilities of transverse Doppler approach are discussed.

EXPERIMENTS

In order to obtain reliable measurements of both the bandwidth and the mean frequency of the Doppler spectrum generated by blood flow, a specially designed real-time Fast Fourier Transform (FFT) unit has been used. The unit can analyze the In-phase and Quadrature signal components provided by any conventional Pulse Doppler instrument. The FFT output is directly interfaced to a Personal Computer, for spectral data display and post-processing[3]. Since we were interested to interpret the Doppler spectrum on the basis of the foregoing equations, it has been convenient to show the processed data in the classical form of spectrograms.

Examples are given in Fig.1, where the frequency vs. time grey-scale plot indicates the amplitude of the spectrum originated from the stationary motion of scatterers in the

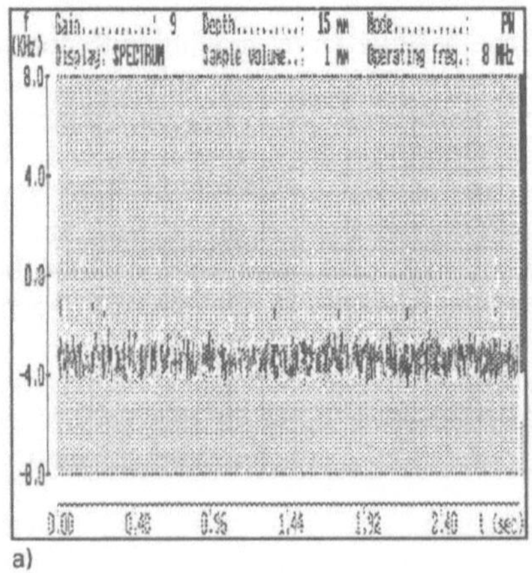

a)

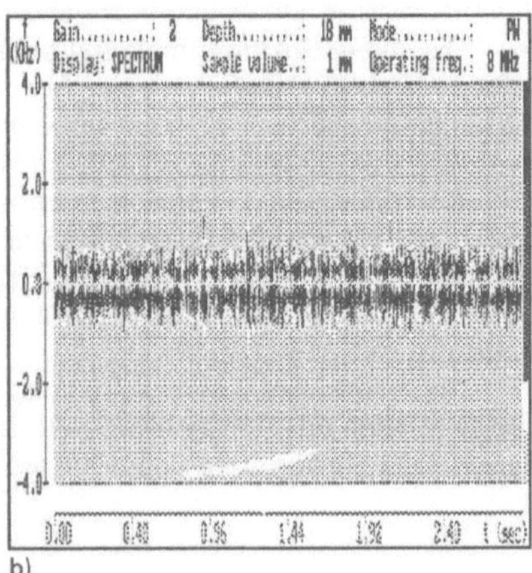

b)

Figure 1. Spectrograms obtained in real-time from analysis of steady flow in a tube-flow model. In a) $\theta \approx 45°$ while in b) $\theta = 90°$. In both cases, the pulsed Doppler system settings were such that only Doppler contributions from the center of the tube have been considered.

center of a tube-flow model. Fig. 1a was obtained by holding the ultrasound transducer at an angle θ of *about* 45° with respect to the tube, while in Fig. 1b θ was *exactly* 90°. In fact, while in the former case the angle could only be estimated qualitatively by observing the inclination between the probe and the tube, in the latter case the symmetry of the spectrogram around the time axis denotes the absence of axial velocity components, i.e. an accurate transverse orientation between flow and ultrasound beam.

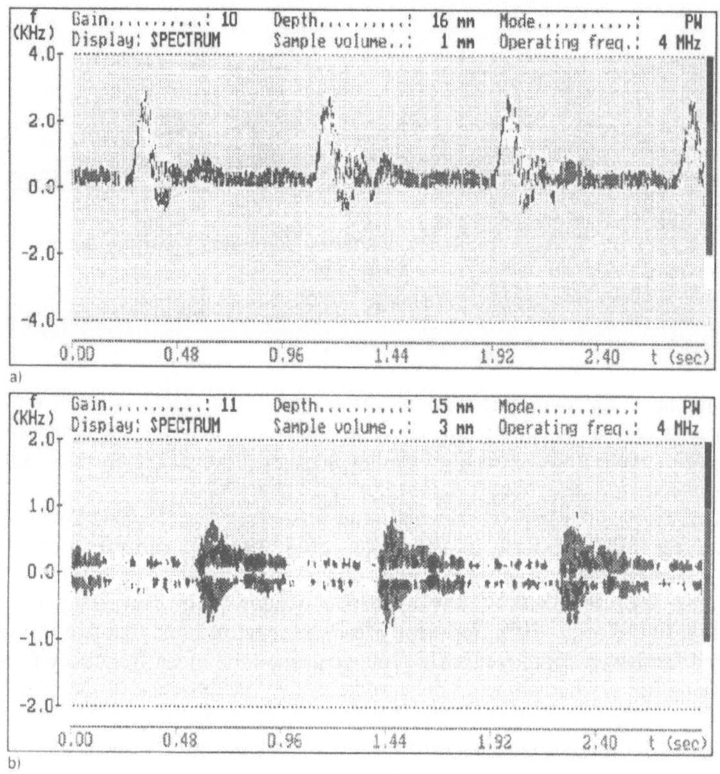

Figure 2. Spectrograms obtained in real-time from analysis of blood flow in a carotid artery. In a) $\theta \approx 70°$ while in b) $\theta = 90°$. In both cases, the pulsed Doppler system settings were such that only Doppler contributions from the center of the vessel have been considered.

This evaluation is strengthened by the consideration that for slight angle variations near 90°, the bandwidth is nearly invariant (according to eq. 2), while the spectrum position on the frequency axis (i.e., f_d), is much more sensitive (depending on the cosine of the angle θ: see eq. 1). In vitro flow experiments have confirmed that spectral symmetry is appreciably lost even for a deviation as low as 2° from 90°.

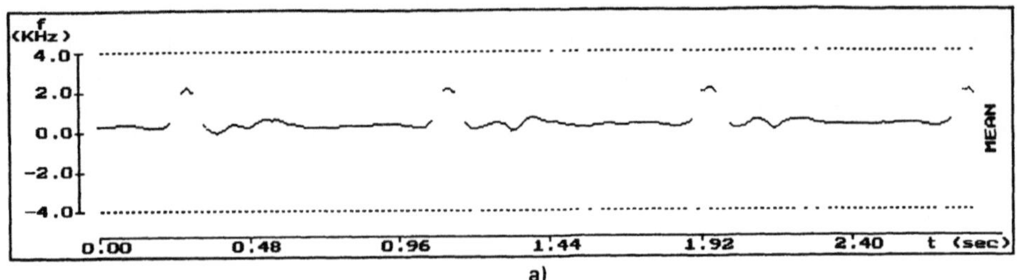

a)

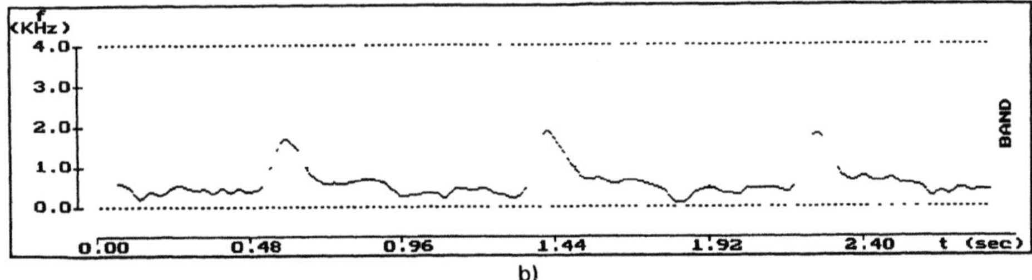

b)

Figure 3. Behavior of: a) Mean frequency sonogram computed from the spectra in Fig. 2a; b) Bandwidth sonogram computed from the spectra in Fig. 2b.

In summary, the two spectrograms differ mainly for their position on the vertical frequency axis, while the bandwidths (i.e., the difference between the upper and the lower edges of each spectrogram) are more similar (as it is predicted from eq.2, since $\sin(45°) \approx 0.7$).

Similar conclusions can be sketched after observing the spectrograms obtained from analysis of the pulsatile flow in the center of a common carotid artery (see Fig.2). However, this example gives greater evidence to a distinctive feature of transverse spectrograms, i.e. the lack of any spectral "window". In fact, Doppler analysis at conventional angles (e.g., 70°), frequently leads to spectrograms like that in Fig.2a, where the detected frequency components are grouped around the mean frequency f_d, which is far from 0 during the systolic phase of the cardiac cycle. The absence of Doppler components between 0 and the lower edge of the spectrum, thus constitutes a "window", which is justified by the temporary absence of low velocity components in the center of the vessel. On the other hand, it must be recalled that when the beam axis is at 90° to the flow, the artery is actually illuminated under an entire set of ultrasound rays, inclined at all angles included between 90° and the minimum angle connected to the degree of focussing. Each of these rays determines a corresponding Doppler shift frequency, in the range from zero up to the edge related to the most inclined rays. This is the reason why the spectrogram does not exhibit the aforementioned window even for instantaneous high velocities.

Apart from this difference, it can be shown that the two approaches yield comparable information. Ultrasound Doppler analysis based only on eq.1, in fact, aims, in general, at detecting the blood flow behavior through a single instantaneous frequency value. This can be obtained by substituting to each complete spectrum its corresponding mean frequency, so that a spectrogram like that in Fig.2a is modified in to the sonogram in Fig.3a, which gives the behavior of the axial velocity component vs time.

As said above, when the transducer is transversely oriented to the flow, the mean frequency is zero throughout all the cardiac cycle, while the bandwidth is proportional to the instantaneous velocity. Therefore, it is useful to compute the difference between the upper and the lower edges of the spectrogram in Fig.2b, to get the sonogram in Fig.3b, which gives the behavior of the true velocity (and not of only one of its components) vs time. It is immediate to recognize that the plots in Fig.3a and Fig.3b are equivalent. Similar results can be obtained for other human arteries.

DISCUSSION

First conclusion emerging from the above experiments is that it is not necessary to avoid a transverse beam-to-flow orientation, since this can lead to results equivalent to those obtained for different orientations. Furthermore, this condition involves a spectral symmetry which allows an accurate angle estimation. On the other hand, spectrograms obtained at 90° can appear less readable than those obtained at lower angles, because much of the received energy is filtered out by the high pass filters (HPFs') which are unavoidably present in Doppler electronics to reject contributions from fixed targets. In order to get good results, it is thus necessary to maintain the HPFs cutoff frequency at the lowest possible value, and to ensure a good signal-to-noise ratio in Doppler electronics and FFT. Moreover, a symmetrical spectrogram does not give any information about the flow direction, but possible solutions to this problem have already been proposed[4].

Finally, it can be pointed out that limiting Doppler analysis to the extraction of mean frequency, f_d, can lead to misleading results, because a mean frequency which is constantly zero, can be due either to an absence of flow or to a transverse flow. By considering the bandwidth of the received signal, this possible source of misunderstanding is avoided.

BIBLIOGRAPHY

1. Evans D.H., Mc Dicken W.N., Skidmore R., Woodcock J.P. (1989), *Doppler Ultrasound, Physics, Instrumentation, and Clinical Application*, John Wiley and Sons, pp. 131-133.

2. Censor D., Newhouse V.L. and Vontz T. (1988), "Theory of Ultrasound Doppler Spectra Velocimetry for arbitrary Beam and Flow Configurations", *IEEE Transactions on Biomedical Engineering*, BME-35, No. 9, pp. 740-751.

3. Tortoli P., Guidi G., Mariotti V., Newhouse V.L. (1992), "Experimental Proof of Doppler Bandwidth Invariance", *IEEE Transactions on Ultrasonics Ferroelectrics, and Frequency Control*, Vol. 39, No. 2, pp. 196-203.

4. Dickerson K.S., Newhouse V.L., Mariotti V. (1991), "Flow Direction Determination using the Transverse Doppler Spectrum", *Annual International Conference of the IEEE Engineering in Medicine and Biology Society*, Vol. 13, No. 1, pp. 193-194.

A NEW ALGORITHM OF ULTRASOUND DOPPLER
COLOR FLOW MAPPING SIGNAL PROCESSING

Jin Yi, Chen Siping, Yang Jianjun and Lu Jianliang

Ultrasound Department
Analogic Scientific Inc.
Shekou, Shenzhen 518067
P. R. China

INTRODUCTION

Color Doppler tomograph flow imaging utilizes Doppler effect between ultrasound beam and blood flow to estimate flow velocity in real-time, and maps the velocity with color code in two dimensions.

The processing of Color Flow Mapping (CFM), which differs from the conventional Doppler Methods, is using frequency analysis technique. Since a large number of two-dimensional space data should be estimated rapidly in real time, it is unable to use FFT. At present, auto-correlation method is widely applied to ultrasound diagnosis instruments. This method estimate average frequency shift of echo signal to calculate average velocity of flow by emitting ultrasound pulses repetitively in Pulses Repetition frequency (PRF). With series of echo sample points at specific space spots obtained, phase shift between two continuous sample points can be estimated in time-domain, frequency shift can be calculated and the velocity of flow at the space spot can also be obtained.

Auto-correlation method makes it possible to acquire accurate velocity in theoretical meaning in some volume and interval of samples. But it has been accepted in clinical application that color mapping is only of qualitative meaning. Average velocity can only be used to ease diagnosis, but other methods, such as Doppler's, must be used to get final conclusion. Furthermore, due to noise interference, the estimate precision will be reduced, thus it is difficult to get satisfactory result.

Based on the previous statements, we propose a new improved estimate algorithm of frequency called vector angle average method. This new method is also based on phase estimate in time-domain to obtain frequency shift and/or velocity. The new method has some differences with the auto-correlation method shown as follows:

(1) Use vector plane to analyze echo signal.

(2) Estimate instantaneous frequency shift of two continuous vectors.

(3) Average the instantaneous frequency shifts in equal weight to get average frequency shift and/or flow velocity.

To explain our method, we first describe the auto-correlation method briefly as shown by Kasai[1]. An ultrasound Doppler echo signal demodulated in quadrature can be expressed as $X(t)$:

$$X(t) = I(t) + j Q(t) \tag{1}$$

where I and Q denote the in phase and quadrature components, respectively. Its discrete expression is $X(n)$, and sample interval T is the reciprocal of PRF.

$$X(n) = I(n) + j Q(n) \tag{2}$$

For N points, auto-correlation function $R_{XX}(m)$ can be formed as:

$$R_{XX}(m) = \frac{1}{N} \sum_{n=0}^{N-1} X(n) X^*(n + m) \tag{3}$$

where $X^*(n)$ means conjugate. Auto-correlation method defined average shift of angle frequency $\bar{\omega}$ as follows:

$$\bar{\omega} = \frac{1}{T} \arctan \frac{\mathrm{Im}\left[R_{XX}(1)\right]}{\mathrm{Re}\left[R_{XX}(1)\right]} \tag{4}$$

where $\mathrm{Re}\left[R_{XX}(1)\right]$ and $\mathrm{Im}\left[R_{XX}(1)\right]$ are the real and imaginary parts of $R_{XX}(1)$, respectively, and the practical formula of $\bar{\omega}$ can be expressed as:

$$\bar{\omega} = \frac{1}{T} \arctan \frac{\displaystyle\sum_{n=0}^{N-1}\left[Q(n) I(n + 1) - I(n) Q(n + 1)\right]}{\displaystyle\sum_{n=0}^{N-1}\left[I(n) I(n + 1) + Q(n) Q(n + 1)\right]} \tag{5}$$

With the $\bar{\omega}$ above, flow velocity can be calculated. The velocity estimated by the present method denotes the amount of total flow divided by some interval of NT. The result may be affected significantly by high amplitude transient and random noise. To improve the estimate precision, a more direct way is to estimate instantaneous velocity at each interval instead of auto-correlation function, and average all the velocities to get an average velocity.

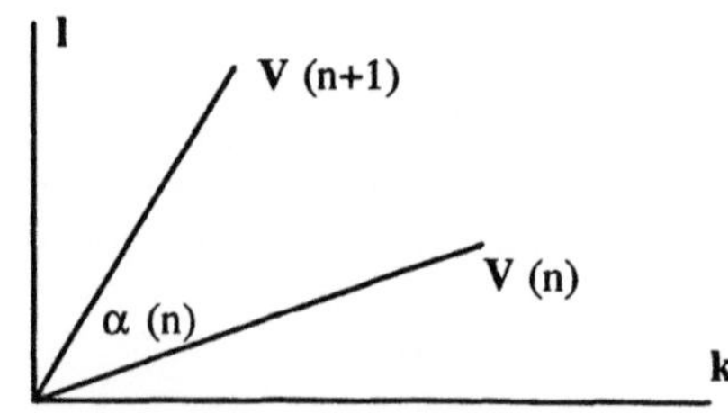

Figure 1.   Illustration of vectors $V(n)$, $V(n+1)$ and $\alpha(n)$

VECTOR ANGLE AVERAGE

This method is established on the basis of vector analysis. Sample series can be expressed as vectors. Suppose $V(n)$ is the vector numbered n, and $V(n+1)$ follows. Then $V(n) = kI(n) + lQ(n)$, $V(n+1) = kI(n+1) + lQ(n+1)$. As shown in Figure 1, $\alpha(n)$ is the angle between $V(n)$ and $V(n+1)$. In vector algebra, $\alpha(n)$ can be obtained through $V(n)$ and $V(n+1)$. Instantaneous angle velocity $\omega(n)$ can be defined as $\alpha(n)/T$. After all the angle velocities are got, average them to get a mean velocity $\bar{\omega}$. See Equations (6), (7) and (8).

$$\tan \alpha(n) = \frac{V(n) \times V(n+1)}{|V(n)| \cdot |V(n+1)|} \tag{6}$$

$$\omega = \frac{1}{T} \arctan \frac{Q(n)I(n+1) - I(n)Q(n+1)}{I(n)I(n+1) + Q(n)Q(n+1)} \tag{7}$$

$$\bar{\omega} = \frac{1}{N-1} \sum_{n=0}^{N-2} \arctan \frac{Q(n)I(n+1) - I(n)Q(n+1)}{I(n)I(n+1) + Q(n)Q(n+1)} \tag{8}$$

Compared Equation (8) to (5), the difference between the two methods can be considered as follows: In our method, estimating mean velocity is based on each instantaneous frequency shift, but in Kasai method which has no direct connection with continuous two sample data. We think the method is more reasonable in some applications because each instantaneous velocity is insensitive to transient signal amplitude. Eq. (7) shows the change of amplitude clearly. Due to insensitivity to transient, the algorithm becomes more sensitive to the variation of frequency. From clinical viewpoint, instantaneous velocities are more valuable than mean velocity. Furthermore, there is no need for the algorithm to accumulate real and imaginary parts of $R_{XX}(1)$, and thus simplifies hardware structure.

SIMULATION

The two methods are simulated individually with computer in the same condition below: single frequency sine wave, mixed with random noise; 12 bits full scale dynamic range; signal and noise have -20dB and -40dB, Respectively, maximum in amplitude of full scale; Inner data format use fix operation to emulate practical hardware process; Sample rate (PRF) was set at 4000 Hz with $N = 4$; signal frequency increase from -2000 to 2000Hz with step 1Hz. The estimated results are shown in Figure 2. It shows that the two methods have same phenomenon that the estimated results have more deviation at low frequency part, and has relatively fine precision at higher frequency part. The reason is that the quantities of samples are not enough. In Figure 2 (b), there are less total dispersion degree than in Figure 2 (a), it can show effects of the two methods. The simulation method may have satisfactory estimate even with only 4 samples. Anti-interference capacity and accuracy of the method have major effect on results.

CONCLUSION AND DISCUSSION

The method of vector angle average estimates the frequency shift more directly and concisely in either principle or practical hardware than auto-correlation method. Above experiments also prove that the method is superior to auto-correlation method in anti-interference and accuracy with short samples. In another word, its sensitivity is high. These improvements have important effect on clinical applications. Most velocities of blood flow are under-estimated due to low frequency of wall echo with high amplitude such as valves. These signals generally have 20—40dB high amplitude above blood echo. Thus many kinds of methods need special treatments to meet the request of low velocity flow. No matter how to decrease the effect by wall echo, inevitably real flows with low velocity are degraded by the same process. Hence, in the method proposed in this paper, the problems have been considered. It is therefore expected that the algorithm could be of significance in applications.

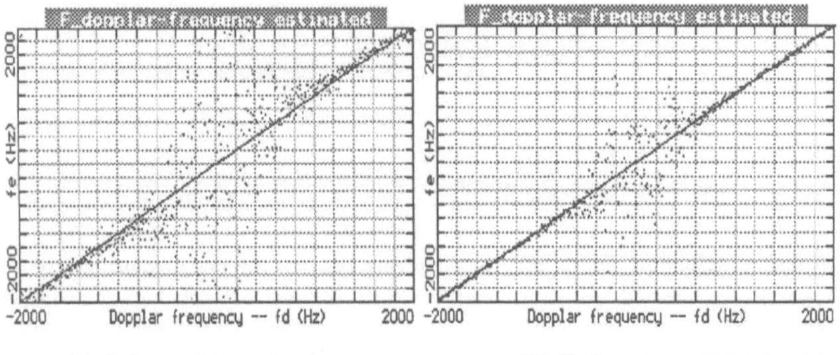

(a) Auto-correlation estimator   (b) Vector angle average estimator

Figure 2.   Output result of two kinds of frequency estimator. Solid lines express theoretical value

REFERENCES

1.   C. Kasai, K. Namekawa, A. Koyane and R. Omoto, Real-time two dimensional blood flow imagine using an auto correlation technique, *IEEE Trans. Sonics Ultrasonics*, SU—32, 458—463 (1985).

2.   Joseph Kisslo, B. David, et al, "Doppler Color Flow Imaging", Churchill Livingstone Inc, (1988).

PARAMETRIC ESTIMATION OF MEAN FREQUENCY AND SPECTRAL VARIANCE IN COLOR FLOW IMAGING

Sandro Dessì, Leonardo Masotti, Santina Rocchi

Department of Electronic Engineering
University of Florence
Via Santa Marta, 3 - 50139 Firenze
Italy

INTRODUCTION

Color Doppler imaging for the diagnosis of cardio-vascular diseases has become a widely applied technique in recent years. Fast and efficient techniques capable of evaluating blood flow characteristics using a few signal samples are required for real-time bi-dimensional mapping of blood velocity, flow direction and turbulence. In a Color Doppler system the above three parameters are color coded and superimposed on a morphologic image obtained using a standard echographic technique (B-mode).

A sector is scanned with a pulsed wave system and divided into a certain number of radial directions investigated by the transducer's sound beam. The scanning of the sector is repeated periodically with a particular frame rate. Given a certain frame rate value, each direction is investigated during a limited time interval (i.e. the sampling interval of the flow dynamics). The number of samples is limited by the finite velocity of ultrasound in biological tissues (about 1540 m/s). The samples obtained for a single gate of a particular investigated direction are elaborated by algorithms that estimate mean frequency and variance, corresponding respectively to the spatial average of the blood velocities and to turbulent flow.

The most commonly used techniques for power spectral density (PSD) estimation are based on the Discrete Fourier Transform applied to the signal samples. In the case of the Color Doppler imaging, these techniques cannot be applied because the small number of samples leads to insufficient frequency resolution and leakage in the spectral domain, due to the implicit data windowing. It must be noted, though, that the final aim of the Doppler signal processing is not estimation of the PSD in itself, but rather estimation of the two parameters which represent the biological situation of the blood flow, i.e. mean frequency and spectral variance. The solution to this problem is provided by a time-domain analysis of the Doppler signal based on the Wiener-Khinchine theorem.

Time-domain algorithms were analyzed: autocorrelation (AC) algorithms, based on the direct estimation of the autocorrelation function; PSD estimators, based on the autoregressive (AR) modelling technique and on Prony's model made of a sum of non-harmonically related dumped exponentials.

The aim of the present research is to individuate algorithms able to provide a consistent estimate of the two mentioned parameters from short data records. Therefore, only low order complex AR models, which are also the basis for the Prony model, can be used.

A first order complex model is able to provide estimates of the Doppler shift spectral parameters. The model and the characteristics of its power spectral density are univocally described by its single pole.

Simulated Doppler signals were applied to the above algorithms to assess their behaviour. Results show that a highly reliable mean frequency estimation is obtained by calculating the pole frequency. The efficiency of this estimator is equal to that of the best AC algorithm.

The AR technique and the Prony algorithm also provide a better spectral variance evaluation than the AC. The information contained in the pole modulus of a first order complex model yields a parametric estimation of the spectral variance, suitable of an easy hardware implementation.

ESTIMATION METHODS FOR POWER SPECTRAL DENSITY PARAMETERS

Different AC algorithms exist in the literature and some of them are actually implemented in Color Doppler echographs. In a previous research work, seven mean frequency estimators and four variance estimators were analyzed and tested to provide quantitative evaluation, [1]. The most efficient among them are hereafter expressed in terms of the autocorrelation function:

$$fm_{AC} = \frac{1}{2\pi\,T}\ \arctan\frac{Im\,[\,R(1)]}{Re\,[\,R(1)]} \qquad var_{AC} = \frac{2}{T^2}\left[\,1 - \frac{|\,R(1)\,|}{R(0)}\,\right] \qquad (1)$$

where T is the sampling interval.

Low order AR models are also used to detect mean frequency and variance, [2] [3]. If the Doppler demodulated complex signal is modelled as an AR process of order p the PSD of the Doppler shift is provided by the following expression, [4]:

$$P_{AR}(f) = \frac{\sigma^2\,T}{|\,1 + \displaystyle\sum_{k=1}^{p} a_k\,e^{(-j\,2\pi\,f\,k\,T)}\,|^2} \qquad (2)$$

where a_k are the model complex coefficients and σ^2 is the variance of the white-noise zero-mean driving process.

The AR coefficients are linked to the autocorrelation function by the Yule-Walker equations. A parallel between the AC algorithms and the spectral parameters of a 1st order AR model is provided by the solution of the Y-W equations for p = 1:

$$\begin{bmatrix} R(0) & R(-1) \\ R(1) & R(0) \end{bmatrix} \begin{bmatrix} 1 \\ a_1 \end{bmatrix} = \begin{bmatrix} \sigma^2 \\ 0 \end{bmatrix} \quad \Rightarrow \quad a_1 = -\frac{R(1)}{R(0)} \tag{3}$$

The AC algorithms yield:

$$fm_{AC} = \frac{1}{2\pi\,T}\arctan\frac{Im\,[\,R(1)]}{Re\,[\,R(1)]} = \frac{1}{2\pi\,T}\arctan\frac{Im\,[\,-a_1\,]}{Re\,[\,-a_1\,]} \tag{4}$$

$$var_{AC} = \frac{2}{T^2}\left[\,1 - \frac{|\,R(1)\,|}{R(0)}\,\right] = \frac{2}{T^2}\left[\,1 - |\,a_1\,|\,\right] \tag{5}$$

A connection between the above quantities and the PSD of a first order AR model can be individuated. The Z-plane representation of the AR model allows to rewrite the AR PSD as:

$$P_{AR}(f) = \frac{\sigma^2\,T}{1 + |\,p\,|^2 - 2\,|\,p\,|\cos[\,2\pi\,(\,f - fp\,)\,T\,]} \tag{6}$$

where $|p|$ is the modulus of the complex pole of the AR model, and fp is the pole frequency. The pole univocally defines the spectrum and its parameters.

The AR spectrum is symmetrical around the pole frequency. This frequency is also the mean frequency of the AR spectrum and the frequency at which the spectrum has its maximum. Therefore, the AC and AR mean frequency estimators are equivalent:

$$fm_{AC} = \frac{1}{2\pi\,T}\arctan\frac{Im\,[\,R(1)]}{Re\,[\,R(1)]} = \frac{1}{2\pi\,T}\arctan\frac{Im\,[\,p\,]}{Re\,[\,p\,]} = fm_{AR} \tag{7}$$

The same is not true for the AC variance estimator. In fact, the maximum value of the AR spectrum is:

$$MAX\,[\,P_{AR}(f)\,] = \frac{\sigma^2\,T}{(\,1 - |\,p\,|\,)^2} \tag{8}$$

The estimator var_{AC} is linked to the AR spectral variance according to the following expression:

$$var_{AC} = \frac{2}{T^2} \left[1 - \frac{|R(1)|}{R(0)} \right] = \frac{2}{T^2} \left[1 - |p| \right] \tag{9}$$

It can be observed that when the spectral variance increases var_{AC} increases, maintaining the same trend.

Despite this, var_{AC} is not the exact AR spectral variance, which is given by:

$$var_{AR} = \frac{\displaystyle\int_{-PRF/2}^{PRF/2} \frac{(f - fp)^2}{1 + |p|^2 - 2|p|\cos[2\pi(f - fp)T]} \, df}{\displaystyle\int_{-PRF/2}^{PRF/2} \frac{1}{1 + |p|^2 - 2|p|\cos[2\pi(f - fp)T]} \, df} \tag{10}$$

where $PRF = 1/T$ is the Pulse Repetition Frequency. The solution of this expression is proposed as a spectral variance estimator. Approximated numerical integration algorithms are applied to calculate its output. The complexity of the numerical solution does not limit the real time application, because a look-up table can provide an easy hardware implementation.

Many different methods exist which estimate the model single pole [4][5]. The following three were analyzed in the present research work: Yule-Walker via Levinson-Durbin algorithm (YW), Burg (BU), Least Square (LS).

The pole module calculated by applying these methods was used to calculate the output of equation 10. Results show that all these methods provide a good estimate of the Doppler spectral variance, but none of them can be considered the best in all the simulated situations.

Despite a clear improvement with respect to the AC algorithms, large oscillations around the theoretical variance value are still observed when only four samples are used. In the case of the AR model, though, these oscillations are mainly due to the inefficient evaluation of the pole modulus. A quantitative estimate of the spectral variance is obtained using 16 samples.

The results are reported in Figs. 1 and 2.

THE PRONY ALGORITHM

Another method for power spectrum parameter estimation was analyzed, which is based on a model consisting of a sum of non-harmonically related dumped exponentials.

This method, originally formulated by Prony, is characterized by a high frequency resolution similar to that obtained with the AR technique. In fact, unlike the DFT which is based on a sum of harmonically related sinusoids, the Prony exponentials are not constrained to be harmonically related, [4][5].

The frequency position of the exponentials is derived from the signal samples via an AR procedure. Therefore the computational complexity of the Prony method is equivalent to that of the AR technique.

The results, shown in Figs. 1 and 2, are similar to those obtained using the AR method.

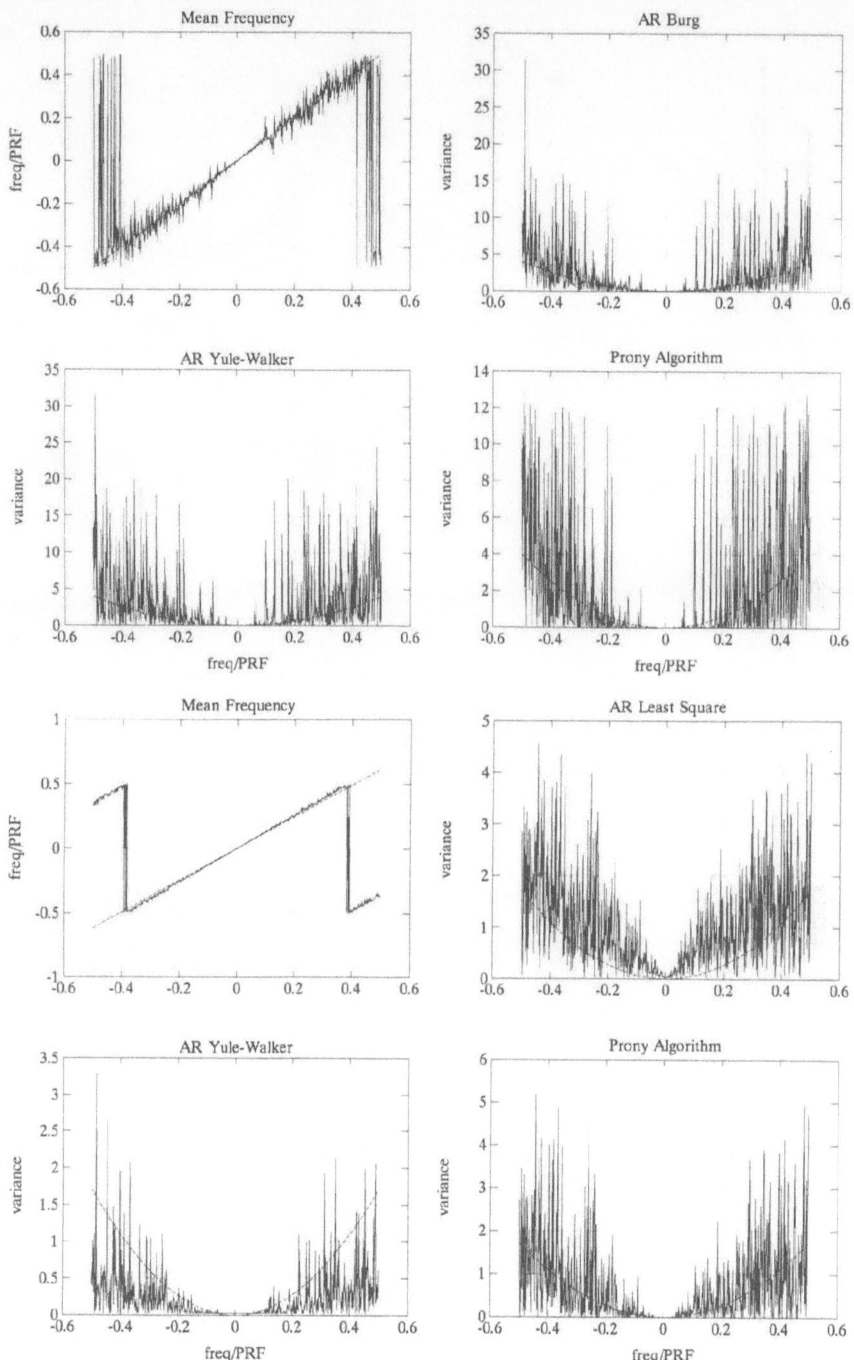

Figure 1. Top half: 4 samples, flat spectrum. Bottom half: 4 samples, asymmetrical spectrum.

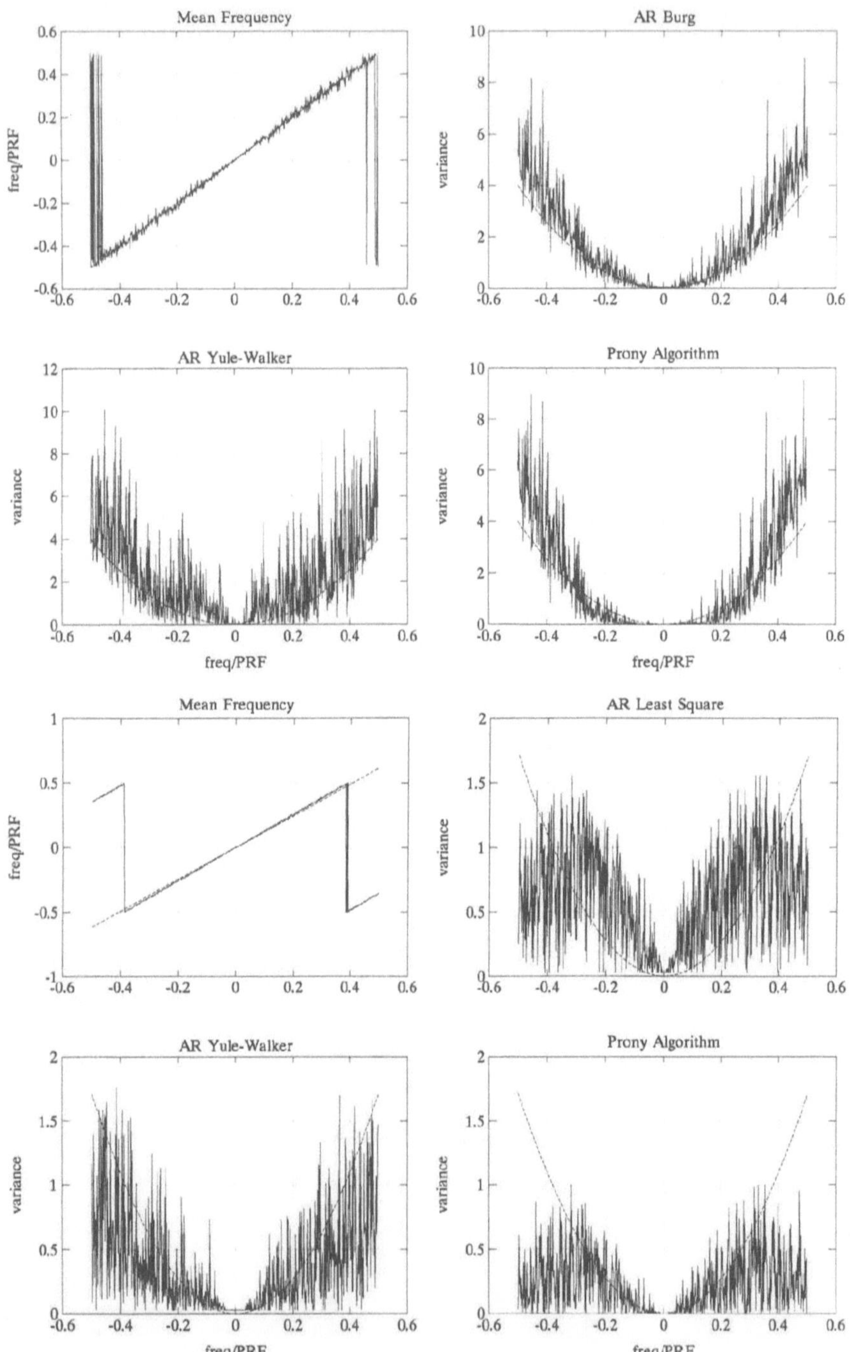

Figure 2. Top half 16 samples, flat spectrum   Bottom half: 16 samples, asymmetrical spectrum

CONCLUSIONS

Simulated Doppler signals were produced to evaluate the estimator behaviour under different conditions of: number of samples, Pulse Repetition Frequency, spectrum central frequency, bandwidth, S/N ratio.

The results demonstrate that the mean frequency estimation is very efficient using the AC, AR or Prony technique.

The spectral variance estimation using the AR or the Prony techniques is better than that obtained with the AC method, but it still provides only qualitative information when four samples are used. A quantitative estimate is obtained when at least ten samples are considered.

REFERENCES

1. K.Kristoffersen,B.A.J.Angelsen;"A Comparison Between Mean Frequency Estimators for Multigated Doppler Systems with Serial Signal Processing";IEEE Trans.Biom.Eng.,Vol.BME-32,No.9,1985.
2. Y.B.Ahn,S.B.Park,:"Estimation of Mean Frequency and Variance of Ultrasonic Doppler Signal by Using Second-Order Autoregressive Model";IEEE Trans.Ultras.Ferr.Freq.Ctrl.,Vol.38,No.3,1991.
3. T.Loupas,W.N.McDicken:"Low-Order Complex AR Models for Mean and Maximum Frequency Estimation in the Context of Doppler Color Flow Mapping";IEEE Trans.Ultras.Ferr.Freq.Ctrl.,Vol.37,No.6,1990.
4. S.M.Kay,S.L.Marple Jr.:"Spectrum Analysis-A Modern Perspective";IEEE Proc.,Vol.69,No.11,1981.
5. S.L.Marple Jr.:"Digital Spectral Analysis with applications";Prentice Hall Inc.,Englewood Cliffs, New Jersey, 1987.

ULTRASONIC COLOR DOPPLER TECHNOLOGY AND ASU-01C

Chen Siping, Zhou Yiyi

Ultrasound Department
Analogic Scientific Incorporation
Shekou, Shenzhen, P. R. China

INTRODUCTION

In the past decades, the development of Doppler technology initiates a new field in medical application. The basic principle of Doppler blood flow measurement has been recognized widely since World War Ⅱ, and has been improved continuously by physicians and engineers. It become possible to diagnosis cardiac diseases precisely by using Doppler ultrasound. But the real-time and two dimensional display of heart and peripheral blood vessels are dependent on the latest color flow mapping (CFM) technology.

CFM utilizes ultrasound Doppler effect to estimate human body's real-time inner blood parameters, encodes the parameters to colors and displays the information on two dimensional image. This new technology developed in 1980's is set up on the basis of ultrasonic scan, moving target detect (MTI), auto correlate estimate, image processing, computer technology, etc. CFM opens up a totally new field of vision in medical ultrasound diagnosis.

CFM IN CLINICAL APPLICATION

Color Doppler imaging system displys blood flow information in terms of direction and relative velocity by real-time color mapping within conventional B-mode gray scale images. This enables Doppler data to be easily understood because of the avoidance of complex spectral velocity displays. Doppler color imaging systems assign a given color to the direction flow. In general, red represents the flow towards the transducer while blue represents the flow away from the transducer. Progressively increasing velocities are encoded in various hues of either red or blue. The brighter the hue, the faster the relative velocity. The duller the hue, the slower the velocity. Color is also used to distinguish turbulent flow from laminar flow. It is easier, more directly to identify valvular, congenital, and other forms of heart disease by using CFM than conventional Doppler approaches. In addition to the heart, the peripheral vascular system has been examined using this non-invasive percutaneous method. [1]

COLOR DOPPLER FLOW MAPPING SYSTEM

CFM system is a separate processor that creats the color flow image based on the returning echo data and then integrates them with the two–dimensional anatomic image. So the final image has both the anatomic structure and color blood flow data. CFM is developed on the basis of pulsed Doppler, they both use Doppler theory to detect blood flow velocity. Pulsed Doppler detects one selected point in blood flow on the image while CFM can detect the whole selected area. CFM ultrasonic scan method is different from B–mode and conventional pulsed Doppler method, it combines the both. At a given angle, a number of pulses are emitted , then change the angle to the next, repeat the process through the entire CFM sector. The time between successive pulses determines the pulse repetition frequency (PRF) .

There are two sample processes at each emit angle, time sample and space sample. Time sample is a successive sampling process for detecting echo from quadrature demodulators, it includes time and space resolving information. If time sampling frequency is f_t , then the space resolution identified by the two adjacent sample points is:

$$d = c/2f_t$$

where d is the space resolution and c is the sound propagating speed in human tissue.

Space sample is repeatedly sampling points at each emit angle by using pulse repeating transmit. Space sampling rate is the same as pulse repetition frequency (PRF) . Actually, space sample is horizontal grouping for the vertical time sampling signals. [2]

Color flow processor is the corn subsystem in CFM. It estimates blood velocity, direction and variance, etc. This analysis cannot be accomplished by using conventional FFT method, since real-time display is required in clinical application. A special designed high–speed processor is used which can accomplish the multiplication and the addition 20 million times a second. Auto-correlation is a general mathematical approach in color Doppler data analysis.

AUS-01C COLOR DOPPLER IMAGING SYSTEM

In 1990, Analogic Scientific Incorporation developed the first color Doppler imaging system, ASU-01C, in China. It is mainly used in cardiovascular diagnosis, and also used in abdominal, obstetrical diagnosis. More than 60 customers reported that this color flow imaging system has clear two dimensional image and bright color Doppler image. It can easily diagnose heart shunt and regurgitant flow, distinguish turbulent from laminar flow. Its sensitive PE, CW waves, blood flow velocity, variance detecting are very helpful for cardiovascular study.

The most useful application of ASU-01C is in the detection of valvular regurgitation, such as mitral regurgitation, aortic regurgitation, and tricuspid regurgitation. They are best detected during the color two–dimensional echocardiographic examination. When we use ASU-01C's variance detection mode, green color will be added to the regurgitant lesion because abnormal flows within the heart are usually turbulent. Regurgitant quantification is based roughly upon the size and configuration of the regurgitant jet. Valvular stenoses (mitral stenosis, aortic stenosis, tricuspid stenosis and pulmonic stenosis) and congenital heart disorders (ventricular septal defect and atrial septal defect, etc.) are also relatively easily identified by using ASU-01C.

ASU-01C has a variety of calculation and measurement functions, which enable the system to be suitable for many different applications. New advances are easily encoded in software. The hardware is also easily updated at board level. In 1991, we add a new cineloop function on ASU-01C with large capacity DRAM. With this function, the system can store a maximum frame by frame or be reviewed repeatedly in a loop in order to detect and observe transient phenomena. The image memory can also store approximately 5 minutes M-mode and Doppler spectra.

In addition to the conventional 2. 5 MHz, 3. 5 MHz, and 5. 0 MHz eletronic phased array transducer, ASU-01C also has smart 5. 0 MHz transesophagus transducer.

CFM DEVELOPING TENDENCY

Color Doppler imaging is being developed very rapidly and becomes an essential non-invasive method in cardiovascular dignosis. The use of this new sonographic imaging technique in various medical fields has been expanding. It has been further improved in theory, engineering and application. There are some developing tendencies in CFM:

a. Improve flow detecting sensitivity and imaging quality.

b. increase CFM frame frequency.

c. Develop new flow mapping method.

d. Improve transesophagus cardiogram detecting method.

e. Develop new quantity algorithm on CFM image.

f. Use high frequency transducer.

Analogic Scientific Inc. is developing an advanced digital CFM system. In the near future, a totally new generation CFM will be produced.

ACKNOWLEDGEMENT

We thank all the members of CFM developing team of Analogic Scientific Incorporation for their kind help and perfect cooperation.

REFERENCES

1. Junji Machi, Bernard Sigel, et al, Operative color Doppler imaging for vascular surgery, *J. Ultrasound Med.* , p65, Feb. 1992.
2. Jin Yi, Signal Processing Method in Color Doppler Blood Flow Mapping, Master Degree Dissertation of Zhejiang University, China, 1989.

DOPPLER ULTRASOUND SPECTRAL ESTIMATION TECHNIQUES: PRACTICAL CONSIDERATION

Shangkai Gao, Guofu Xu

Electrical Engineering Department
Tsinghua University
Beijing 100084, China

INTRODUCTION

Ultrasonic Doppler spectra are of significance in clinical use, as from which quantitative diagnostic information can be extracted. There have already been some papers which studied or compared a variety of Doppler ultrasound spectrum estimation methods. [1][2][3] As a results, many of them are, in principle,effective for Doppler signal analysis. However, following problems must be in consideration when we put these techniques in real instruments.

The time of calculation

Some of the effective methods are complex in calculation. The more complex the estimation methods, the longer the data processing time. This will become a problem in real time processing.

The length of data segments

Different methods are often suitable for different length of data segments. This is a problem which must be noted when different methods are used.

Data length (the number of bits)

Long length data are useful to get a precise results of the estimation, but the costs of the hardware will then be increased. Hence, a compromise must be made between the precisions and costs.

A comparative study of FFT, averaging FFT and AR methods for Doppler signal analysis is made in this paper from practical application point of view. The results in the paper will be useful for spectral estimator designs.

THE SIMULATION OF DOPPLER BLOOD FLOW SIGNALS

In order to evaluate the effectiveness of different estimation methods, the computer simulation is made for simulation Doppler signal analysis. The signal simulation program was based on Van Leeuwen's[4] and Mo's[5] methods. With the assumptions of stationary, laminar flow in a circular tube and long enough entrance length, a velocity pattern with a parabolic profile is then formed. Its corresponding single sided power density spectrum is generated according to

$$S(f) = (1 - \frac{f}{f_{max}})^{\frac{2-n}{n}} \qquad \text{for} \quad 0 < f < f_{max}$$

$$S(f) = 0 \qquad \text{otherwise}$$

where fmax is the maximal Doppler frequency, and n a velocity profile constant. Fig.1(a) shows an example of n=8, fmax=6.4 kHz.It will be used in our computer simulation as a template.

S(f) can be considered as a combination of many narrow-band spectra. The Doppler signal x(t) with narrow-band Gaussian noise character can then be formed as follows

$$x(t) \cong \sum_{m=1}^{M} a_m \cos(2\pi f_m t + \varphi_m)$$

where $\quad f_m = (m - 1/2)\,\triangle f$
$$a_m = \sqrt{2\,S(f_m)\,\triangle f}\;y_m$$

in which f=fmax/M and the set {ym} are independent chi-square random variables with two degrees-of-freedom [5] . An example of simulation signal x(t) is shown in Fig.1(b).

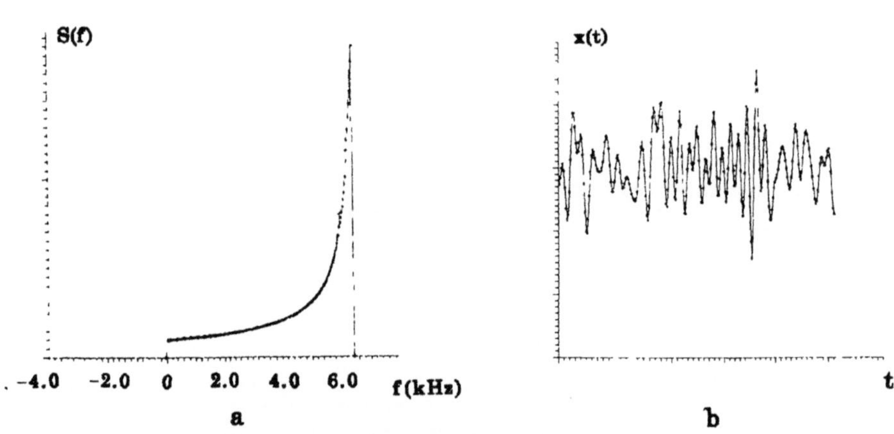

Fig.1 (a) Spectrum template of Doppler simulation signal
(b) An example of simulated Doppler signal

THE COMPARATIVE STUDY OF DOPPLER SPECTRAL ESTIMATION TECHNIQUES

The computer simulation and real Doppler signal analysis are made by FFT, averaging FFT and AR methods in this section. The resulted spectra are all compared with the spectrum template shown in Fig.1(a) to evaluate the effectiveness and the possibility of their applying to Doppler spectral estimation.

The influence of different signal to noise ratio (SNR)

Choosing fmax=6.4kHz, M=128, we calculated the simulated Doppler signal x(t) (see equa.(2)) first, and then added white noise to it leading different signal to noise ratio. Power spectra were estimated by 128 point FFT method and AR method in which Marple's algorithm with the model of seven orders was used.

The results are:
(1) AR method yields more statistical stability than FFT (Fig.2).
(2) AR method is quite sensitive to S/N ratio. The spectral peak position was shifted when S/N ratio is relative high. In the case of SNR=20dB, 800Hz shift appears in the AR spectrum (Fig.2(b)).

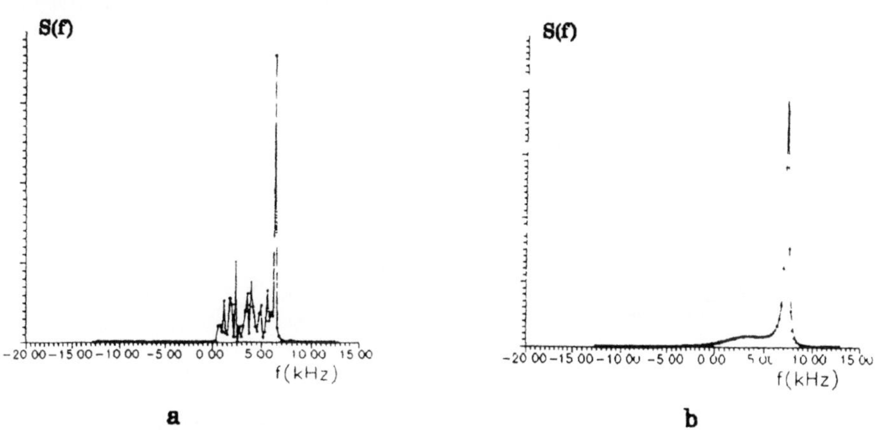

Fig.2 Spectral estimation with SNR-20dB. (a) 128 point FFT method. (b) 128 point AR method with a model of seven orders.

The effect of different length of data segments

Power spectra of the signals with data lengths of 256, 128, 64 points have been estimated respectively. It seems that FFT method are stable for the changes of data lengths. For the short data segments, if averaging FFT is used by averaging the spectra of several estimations from short data, the spectrum will become smooth as shown in Fig.3.

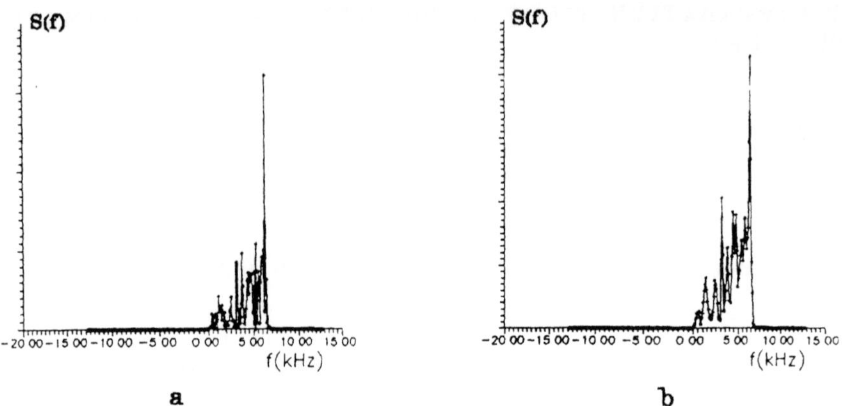

a   b

Fig.3 FFT method for different length of data segments.
(a) Single 256 point FFT. (b) Averaging FFT of two 128 point spectra.

A serious problem in AR method is the effect of different data lengths. In our computer simulation artificial spectral peaks were found (see Fig.4), which were also found in real Doppler signal spectral estimation.

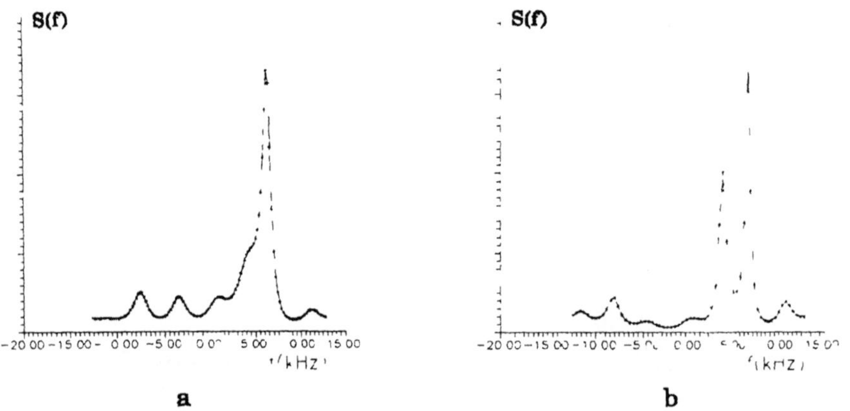

a   b

Fig.4 AR method for different length of data segments. (a) 128 point AR method. (b) 64 point AR method.

The results by choosing different number of data bit

The smaller the number of data bit, the lower the system costs. In our computer simulation there were no significant improvements in spectrum quality when input data are greater than 8 bit and output data are greater than 4 bit.

Fig.5 shows the dynamic power spectra of real Doppler blood flow signals during one cardiac cycle. The spectrum of 128 point two time's averaging FFT has correct trend of peak location. But the results of the 128 point AR method is relatively poor with some artificial peaks.

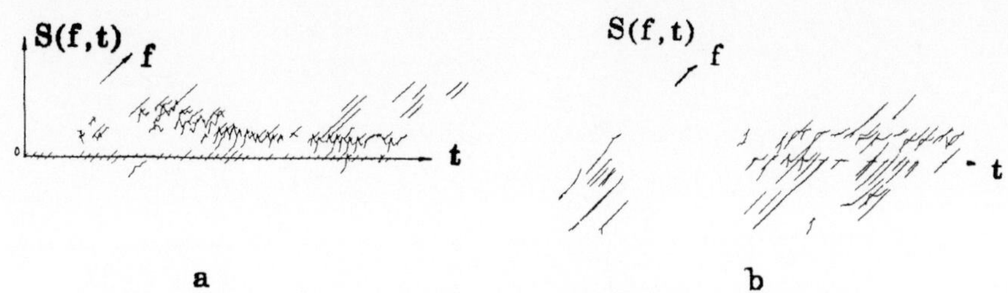

**Fig.5 The dynamic power spectra of real Doppler blood flow signal
(a) 128 point two time's averaging FFT. (b) 128 point AR method**

From practical considration, one should pay more attentions to length of data segment, model order and calculation time when using AR method.

CONCLUSIONS

(1) From calculation time point of view, FFT method has superiority. In general, averaging FFT method is better than either single FFT method or AR method.

(2) AR method is sensitive to the length of data segment, especially in real Doppler signal analysis. It is because that AR method is only suitable for stationary signals.

(3) The reality of estimated spectrum does not change if inputs are 8 bit data and outputs are constricted to 4 bit.

REFERENCES

1 P J Vaitkus and R S C Cobbold A comparati e study and assessment of Dopper ultrasound spectral estimation techniques Part I Estimation Methods Ultrasound in Med & Bio Vol 14, No 8 pp661 6⁻2 (1988)
2 P J Vaitkus, et al A comparative study and assessment of Dopper ultrasound spectral estimation techniques Part II Methods and Results Ultrasound in Med & Bio Vol 14, No 8, pp673 688(1988)
3 Jean-Yves David, et al Modern spectral analysis techniques for blood flow velocity and spectral measurements with pulsed Doppler ultrasound IEEE Trans on Biomedical Engineering Vol 38, No 6, pp589-596 (1991)
4 G H Van Leeuwen, et al Simulation of real-time frequency estimations for pulsed Doppler system Ultrasonic Imaging Vol 8, pp252-271 (1986)
5 Larry Y L Mo, et al 'Speckle in continuous wave Doppler ultrasound Spectra A Simulation Study IEEE Trans on Ultrasonics, Ferroelectrics, and Frequency Control Vol UFFC-33, No.6, pp747-753, November (1986)

DOPPLER ASSESSMENT OF THE EFFECTS OF PREMATURE VENTRICULAR DEPOLARIZATION UPON LEFT VENTRICLE FILLING AND EJECTION

Yanyan Hu,[1] Chungzeng Lu,[1] Chien-Sun Kuo [2]
Anthony N. DeMaria[2]

[1]Department of Cardiology
Shandong Medical University
Jinan 250012, P. R. China

[2]Department of Cardiology
University of Kentucky
Lexington, KY 40508, U. S. A.

INTRODUCTION

To investigate the influence of ventricular premature depolarization (VPD) on left ventricular ejection and filling, we examined 40 patients with pulsed Doppler echocardio-graphy. We hope that our investigation will provide evidence for the evaluation and treatment of VPD.

MATERIAL AND METHODS

The study group comprised 40 adult volunteers, 22 men and 18 women ranged in age from 13 to 65 years with a mean of 38, with normal LV function who were found free of heart disease by history, physical examination, ECG and echocardiogram.

Doppler Measurements. A commercially available phased- array echocardiographic Doppler system (SSH — 65A, TOSHIBA, JAPAN) was used. Pulsed-ware Doppler interrogation of mitral valve inflow velocities was performed using an apical four chamber, or five chamber, two-dimensional image to align the sample volume cursor perpendicular to and at the level of the mitral anulus or aortic anulus. The Doppler transmitral and transaortic velocity profile was digitized through the darkest gray scale and yielded the following parameters: the velocity integral of aortic flow (AVI) and mitral flow (MVI) in VPD, sinus beats (SB) before and after VPD. We calculated LV filling ratio (EMVI/NMVI) of any beat by {MVI of that beat/MVI of SB before VPD} and LV output ratio (EAVI/NAVI) by {AVI of that beat/AVI of SB before VPD}, prematurity index (PI=interval from the peak of T of SB to R of VPD/ RR interval of SB, also T-R/R-R).

RESULTS

I . AVI and MVI of VPD (EAVI and EMVI) were much lower than those of their sinus beats (NAVI and NMVI), ($p < 0.001$); the post beats of VPD (PAVI and PMVI) were higher than NAVI and NMVI ($p < 0.001$), (Figure 1).

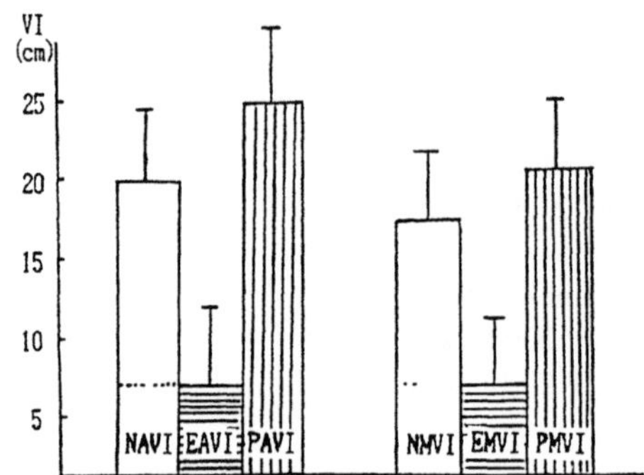

Figure 1.   Comparison between different velocity integrals

II . There were significant correlations both between PI and the output ratio (EAVI/ NAVI) of VPD and between PI and the filling ratio (EMVI/NMVI) of VPD ($r = 0.89$ and 0.84); when PI was less than 0.22, EMVI/NMVI was below 0.23 and EMVI/NMVI became zero (Figures 2 and 3).

DISCUSSION

Our study was conducted on 40 unmadicated subjects with normal LV function to compare AVI and MVI in sinus thythm with those in VPD and to evaluate the correlations both between PI and output ratio and between PI and filling ratio of VPD. In order to eliminate the influences of various factors on measurements, we evaluated the LV filling and the output with flow velocity integral produced by pulsed Doppler echocardiograph and used the method of comparison between indices in the same individual and the indices of output ratio and filling ratio. It was the first time these methods were used successfully for the quantitative study of left ventricular hemodynamics in VPD and the Doppler echocardiographic technique was suggested to be the noninvasive method of evaluating hemodynamics in arrhythmia.

The clinical significance of VPD mainly depends upon the influence of VPD on hemodynamics. We found that EAVI was much lower than NAVI [$p < 0.001$] and EAVI/NAVI was 0. 34, suggesting that the cardiac pumping function was impared in VPD, which is consistent.

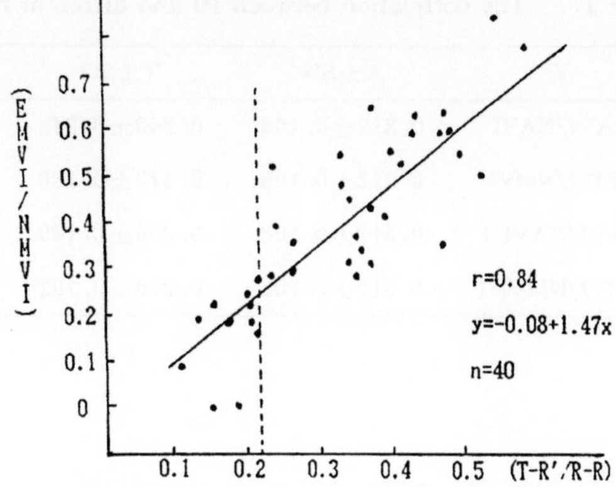

Figure 2.　　The correlation between PI and EMVI/NMVI

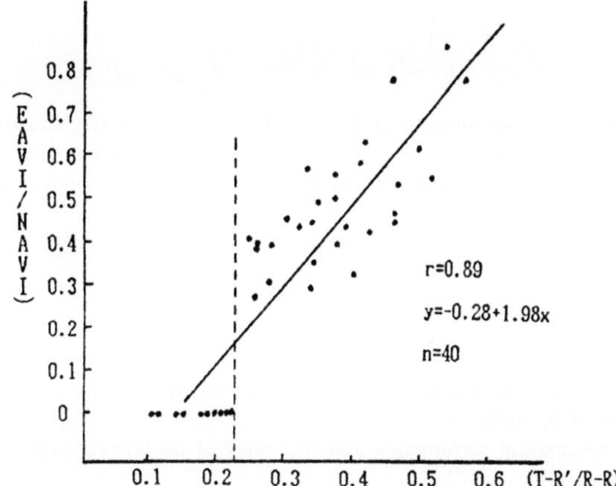

Figure 3.　　The correlation between PI and EAVI/NAVI

It was shown that PI (T-R'/R-R) was well correlated with EAVI/NAVI (r = 0. 89), and that the extent of the prematurity was an important factor for the decrease in LV output in VPD.

It was also shown that PI was well correlated with EMVI/NMVI, and that the extent of the prematurity was also an important factor for the decrease in LV filling in VPD (Figure 4).

We conclude that pulsed Doppler echocardiography is an accurate and useful method for the evaluation of hemodynamics in VPD. Stroke volume and filling volume in VPD decrease a great deal and those of post beat after VPD can not compensate enough for the decreases. The extent of the prematurity is an important factor which affects the LV hemodynamics adversely and the much early VPD should be taken in clinical practice.

Table 1. The correlation between PI and different ratios

X	Y	X±SD	Y±SD	r	p
T-R'/R-R	EAVI/NAVI	0.312±0.108	0.340±0.242	0.89	<0.001
T-R'/R-R	EMVI/NMVI	0.312±0.108	0.378±0.188	0.84	<0.001
T-R'/R-R	PAVI/NAVI-1	0.312±0.108	0.250±0.149	−0.09	>0.5
T-R'/R-R	PMVI/NMVI-1	0.312±0.108	0.276±0.182	−0.19	>0.05

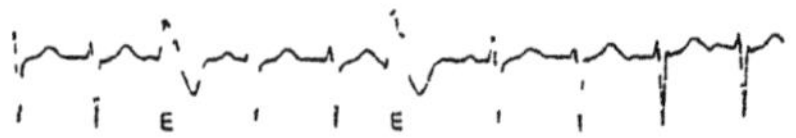

Figure 4. The simultaneously recording of pressures in LV and aortic artery showing lower LV systolic pressure in VPD. When VPD (the second VPD) occurred much earlier, the LV systolic pressure was lower than the aortic diastolic pressure and the aortic valves can not open

REFERENCES

1. H. Takada, et al, Experimental studies on myocardial contractility and hemodynamics in extrasystoles, *Japan J Circulation* 34: 419, 1970.
2. R. C. Boerth, et al, Mechanical performance and efficiency of the left ventricle during ventricular stimulation, *Am J Physiology* 221: 1686, 1971.
3. C. Layton, et al, Analog computer assisted beat—by—beat measurement of stroke volume and related variables in mans, *Cardiovasc Res* 10 (3) : 328, 1976.
4. K. Cohn, et al, The influences of ectopic beats and tachyarrhythmias on stroke volume and cardiac output, *J Electrocardiology* 14: 207, 1981.
5. M. F. Stoddard, et al, The effect of premature ventricular contraction on left ventricular relaxation, chamber stiffness and filling in human, *Am Heart J* 118 (4) : 725, 1989.
6. A. Benchimol, et al, Aortic flow velocity in man during cardiac arrhythmias measured with the Doppler catheter—flow meter systems, *Am J Heart* 78 (5) : 649, 1969.
7. L. Hatle, et al, "Doppler Ultrasound in Cardiology: Physiological Principle and Clinical Application", Second edition, Philadelphia, PA, Lea &. Febiger, 1985, pp306—318.
8. P. Calafiore, et al, Doppler echocardiography quantitation of volumetric flow rate, *Cardiol Clinic* 8 (2) : 191, 1990.
9. T. Tocche, Hemodynamic measurement by Doppler echocardiography, *Arch Mal Coeur* 83 (6) : 815, 1990.
10. J. Sanada, et al, Left ventricular blood flow dynamics by ventricular premature contraction: pulsed Doppler echocardiography study, *J Cardiol* 19 (1) : 263, 1989.

TWO DIMENSIONAL ECHOCARDIOGRAPHIC ASSESSMENT OF PULMONARY ATRESIA WITH INTACT VENTRICULAR SEPTUM

Shubao Chen, Ming Zhu

Xin Hua Children's Hospital
Shanghai Second Medical University
Shanghai 200092, P. R. China

The pulmonary atresia with intact ventricular septum (PA/IVS) has a high early and late mortality. It is accepted that the dismal outlook in PA/IVS may be related to the associated right ventricular hypoplasia and the presence of sinusoidal-coronary artery communications. [1,2] To clarify the value of 2DE in evaluating RV morphology, we report the echocardiographic features in 21 patients with PA/IVS, correlated with angiographic results.

METHODS

Patients. 5140 echocardiographic studies were performed from Mar. 1986 to Dec. 1989 at Xin Hua Children's Hospital. 21 patients with PA/IVS were identified. 11 were boys, 10 were girls. Ages ranged from 1 day to 11 years (mean 12 days), fifteen were less than 30 days.

Echocardiographic studies. 2DE study was performed with HP (77020) echoimaging system using a 5 MHz transducer. Diagnoses were made using all available tomographic planes, subxiphoid, apical, parasternal and suprasternal. According to Bull, et al[3] and Leung, et al[4], who defined a inlet portion incorporating the tricuspid valve apparatus, a trabecular portion lying above the insertion of the papillary muscles of the tricuspid valve towards the apex and an outlet (infundibulum) portion extending from the supraventricular crista to pulmonary valve, RV morphology was evaluated. RV inlet, trabecular and outlet portions were observed from the subxiphoid RV frontal view. Internal dimensions were measured including mitral and tricuspid valve annular dimensions in apical four chamber view, RV infundibular in subxiphoid frontal or parasternal short axis and main pulmonary artery in parasternal short axis.

Cardiac catheter studies. Selective RV, LV and retrograde aortic axial cineangiography was performed in 13 patients. RV morphology and sinusoidal coronary artery communications were evaluated. The internal dimensions of the tricuspid valve annulus, infundibulum and main pulmonary artery were measured.

Statistical analysis was performed using Student's test and linear regression.

RESULTS

In all the 21 patients, 2DE revealed atrioventricular concordance, intact ventricular septum, aortic–LV connected, a thickened and atretic pulmonary valve, patent foramen ovale or ASD with R–L shunting. PDA was seen in 19 patients, among them 3 patients have vertical ductus. Systemic-pulmonary arterial collateral vessles were seen in 2 patients.

Right ventricular morphology. 17 patients had a tripartite RV, 2 patients had a two component RV (inlet and infundibular portions), 2 patients had a single component RV with only an inlet portion. In 13 patients who had angiocardiography, the 2DE and angiocardiographic assessments of RV morphology were concordant. Vertical ductus were only seen in patients with single or two component RV.

Tricuspid valve. Tricuspid valve malformations was identified in 11 patients (52. 3%) : unguarded tricuspid valve 1, dysplastic tricuspid valve 3, tricuspid valve stenosis 4, tricuspid valve atresia 2 and Ebstein's anomaly 1. Two patients with tricuspid atresia had single component RV. One patient with dysplastic tricuspid valve and one patient with tricuspid valve stenosis had a two component RV. The 2DE and angiographic dimensions of the tricuspid annulus correlated well (r 0. 926). Angiographic measurements were 5—33% (mean 15%) higher than the 2DE measurement. In 15 neonates, the tricuspid annular diameter was larger in patients with a tripartite (range 7. 1—11 mm, mean 9 mm) than in those with one or two component RV (rande 6. 2—7. 6 mm, mean 7 mm). Significant differences in tricuspid annular diameter existed between the tripartite and one or two component RV (p$<$0. 01). The ratio of the tricuspid annular diameter to the mitral annular diameter was 0. 52—0. 76 (0. 64$\pm$0. 08) in the patients with single and two component RV, while it was 0. 89—1. 0 (0. 8$\pm$0. 13) in the patients with a tripartite RV. There was a significant difference in the ratio of the tricuspid annular diameter to the mitral annular diameter between the two groups (p$<$0. 01).

Right ventricular infundibular and main pulmonary dimensions. In 2 neonates who had single component RV, the infundibulum were atresia, the dimensions of the main pulmonary artery were 5 mm and 7 mm, respectively. Dimensions of the infundibulum and main pulmonary artery were 6 and 4 mm, and 8 and 6 mm, respectively, in 2 neonates with a two component RV. The infundibular diameter was 5—10 mm (mean 6. 8 mm), while the main pulmonary artery diameter was 5. 9—10 mm (mean7. 4 mm) in the remaining 11 neonates. There was no significant correlation between main pulmonary artery diameter and RV morphology. Angiocardiographic and echocardiographic dimensions of main pulmonary artery and infundibulum correlated well (r 0. 948, 0. 68). Right ventriculogram disclosed sinusoidal coronary artery communications in 2 patients. 2DE with color flow imaging demonstrated dilated sinusoidal in one patient, confirmed by right ventriculogram.

DISCUSSION

Greenwold, et al divided PA/IVS into two types according to right ventricular size. [5] Bull, et al[3] proposed a revised classification of PA/IVS, based on the tripartite approach to RV morphology which was developed by Goor, et al. Studies of blood flow velocity patterns within the RV suggest that the inlet portion contributes to the ventricular filling mechanism, while the trabecular and infundibulum portions contribute mainly to systolic pump function. A small right ventricular cavity may be due to ventricular wall hypertrophy with obliterated trabecular and infundibular cavities. After relief of RV obstruction, the hypertrophy may regress, so the trabecular cavities can dilate.

Tricuspid valve diameter is correlated well with RV morphology in angiography and pathology. Therefore it could be an indirect indicator of the potential developmental status of RV. The dimension of the tricuspid valve annulus is usually smaller than normal limits, and related to right ventricular cavity size in PA/IVS as confirmed in an autopsy study by Zuberbuhler, et al.[1] Autopsy data of Bull, et al[3] showed the larger the tricuspid valve annuli were in patients with tripartite RV, the smaller in those with only an inlet portion to the ventricular cavity. They had similar results in their angiographic study. Leung, et al.[4] found there was a significant difference in tricuspid annular diameter between tripartite and single component RV. The correlation between tricuspid annular diameter and RV morphology was also confirmed in this series. Previous studies along with ours showed that the results of angiographic, echocardiographic and autopsy measurements of tricuspid annular diameter correlated well. Angiographic estimates of tricuspid annular diameter were 43% greater than the autopsy measurements.[3] Echocardiographic estimates were 17% lower than the autopsy measurements.[8] In this series, angiographic estimates were 15% greater than the echocardiographic measurements. Tricuspid valve diameter may be so small that it could affect the right ventricular filling, which would adversely affect the postoperative results and right ventricular development.

An associated dysplastic tricuspid valve was seen in 52.3% of the patients with PA/IVS in this series. Zuberbuhler, et al[1] indicated that there was no correlation between the degree of the dysplasia of tricuspid valve and the right ventricular and tricuspid valve annular size. We found that all the patients with vertical ductus were in poor development of right ventricle. Although the pulmonary artery diameter is reduced, it is not a factor which may influence the surgical results. Severe hypoplasia of the pulmonary arteries, which is frequently seen in pulmonary atresia with VSD, is rarely found in PA/IVS. PA/IVS is frequently associated with infundibular stenosis. The smaller the right ventricular cavity, the more severe the infundibular stenosis to the point of atresia.

Myocardial ischemia and dysfunction may occur postoperatively in the patients with major sinusoidal coronary artery communications. Associated abnormal communication accounts for 8—55% of the patients with PA/IVS[6, 8], no less than 14% in this series. Leung, et al[4] indicated there was limitation of 2DE in detecting sinusoidal coronary artery communications. The clues indicate the existence of a dilated proximal coronary artery, abnormally large intramyocardial coronary arteries and irregular spaces. In the early stage of this study, not much attention was paid to the observation of sinusoidal coronary artery communications in 2DE study, Recently, a large intramyocardial sinusoid was detected with color flow imaging in a patient. We believe that color flow imaging is superior to 2DE in detecting sinusoidal coronary artery communications.

The results of this study suggest that the diagnosis and RV morphology assessment of PA/IVS can be correctly established by Doppler 2DE.

REFERENCES

1. J. R. Zuberbuhler, et al, Morphological variations in PA/IVS, *Br. Heart J.*, 1979; 41: 281.
2. M. De leval, et al, Pulmonary atresia and intact ventricular septum. Surgical management based on a revised classification, *Circulation*, 1982; 66: 272.
3. C. Bull, et al, Pulmonary atresia and intact ventricular septum. A revised classification, *Circulation*, 1982; 66: 266.
4. M. Leung, et al, Echocardiographic assessment of neonates with pulmonary atresia and intact ventricular septum, *J. Am. Coll. Cardiol.* 1988; 12: 719.
5. W. E. Greenwold, et al, Congenital PA/IVS: two types (abstr), *Circulation* 1956; 14: 945.
6. A. G. Patel, et al, Right ventricular volume determinations in 18 patients with PA/IVS, *Circulation* 1980; 61: 428.
7. H. P. Gutgesell, et al, Atrioventricular valve annular diameter: 2 dimensional echocardiographic-autopsy correlation, *Am. J. Cardiology* 1984; 53: 1652.
8. D. A. Fyfe, et al, Myocardial ischemia in patients with PA/IVS *J. Am. Coll. Cardiol.* 1986; 8: 402.

APPROACHES TO SOLVE NONLINEAR PROBLEMS
OF THE SEISMIC TOMOGRAPHY

Yang Wencai and Du Jianyuan

Beijing Computer Center of MGMR

Beijing 10083

P. R. China

INTRODUCTION

The problems of seismic tomography for the inhomogeneous earth are nonlinear, i. e., if the reference velocity and the velocity disturbance are considered as two parts of the medium velocity, the effect of the disturbance to wave propagation cannot be neglected. As the results, the conventional linearized procedures, such as the Born or Rytov approximations, cannot produce satisfied velocity images with high resolution.

The nonlinear optimization procedures of the least squares have been employed to solve the reflection seismic inverse problems (Torantola 1982; Fewcet, 1985; Norton, 1988), the success of these methods depends upon the goodness of the initial guess, and the computational cost of these procedures is very high.

In this paper I first introduce the successive linearization procedures for some nonlinear seismic tomographic problems. Then I compare the behaviors of the output sequence of nonlinear, iterations with the chaotic motion in studying forward problems. Understanding of the chaotic characteristics in nonlinear seismic inversions guides us to construct new approaches for better solutions. Numerical examples are given to illrustrate the approaches while their application to real data can be found in other paper given by the authors(Yang and Du,1992. a b).

SUCCESSIVE LINEARIZATION

For Traveltime Inversion

The traveltime is expressed by

$$T(s,r) = \int_{L(w(x),s,r)} W(x)dl + \delta T(s,r) \qquad (1)$$

where W is the slowness function, x is a point in the image domain, s and r are source and receiver points respectively, L is the raypath , δ T is the noise. The problem is to find $W(x)$ by given $T(s,r)$. It is nonlinear because the raypath depends upon the slowness.

Let

$$\lim_{k\to\infty} W_k(x) = W(x), \qquad\qquad k = 1,2,\cdots\infty \qquad (2)$$

$$W_k(x) = W_{k-1}(x) + \beta_k(x), \quad \beta_k(x) \leqslant W_{k-1}(x) \qquad (3)$$

we have

$$T(s,r) - T_{k-1}(s,r) = \int_{L_{k-1}} \beta_k(x)dl + \delta T(s,r) \qquad (4)$$

where k denotes iteration , based on $(k-1)$th slowness model L , one can calculate the traveltime T_{k-1}.

Giving an initial model and the traveltime data $T(s, r)$, we can calculate the theoretical traveltime and solve Equation(4) by minimizing the error term. Thus we have constructed an algorithm with (3) and (4) which produces a sequence of slowness estimates. The question is whether (2) can be satisfied when K grows, the traditional way to answer the question is finding the conditions for convergence. I shall discuss the problem later from a viewpoint of deterministic chaos.

FOR WAVE FIELD INVERSION

The acoustic wavefield $u(x,s;t)$ from a point pulse source satisfies

$$(\triangle - \frac{1}{c^2(x)} \frac{\partial^2}{\partial t^2})u(x,s,t) = - \delta(x - s) \qquad (5)$$

where $c(x)$ is the velocity at point x. The observation may be carried out at point r on the surface s, therefore observed data are represented by

$$u(r,s,t) = u(x,s,t), \qquad r \in s \qquad (6)$$

The tomographic inverse problem is finding c(x) by given u(r,s,t).

Let

$$\lim_{k \to \infty} C_k(x) = C(x) \qquad k = 1,2,\cdots,\infty \qquad (7)$$

$$C_k^{-2}(x) = C_{k-1}^{-2}(x)(1 + \alpha_k(x)), \ |\alpha_k(x)| \leqslant 1 \qquad (8)$$

and

$$u(x,s,t) = u_{k-1}(x,s,t) + v_k(x,s,t) \qquad (9)$$

where c is the velocity function, v_k is the velocity disturbance.

Putting (8) and (9) into equation (5) yields two equations, they are

$$(\Delta - \frac{1}{c_{k-1}^2(x)} \frac{\partial^2}{\partial t^2} u_{k-1}(x,s,t) = - \delta(x - s) \qquad (10)$$

and

$$(\Delta - \frac{1}{c_{k-1}^2(x)} \frac{\partial^2}{\partial t^2}), \ v_k(x,s,t) = \frac{\alpha_k(x)}{C_{k-1}^2(x)} u_k(x,s,t) \qquad (11)$$

The equation (11)is nonlinear because u contains the wavefield of the disturbance v.

After Born approximation and Fourier transform into the frequency domain, we have

$$(\Delta - \frac{\omega^2}{C_{k-1}^2(x)} v_k(x,s,\omega) \approx \frac{\alpha_k(x)}{c_{k-1}^2(x)} u_{k-1}(x,s,\omega) \qquad (12)$$

This equation has a solution

$$v_k(r,s,\omega) = \omega^2 \int G_{k-1}(x,s,\omega) G_{k-1}(x,r,\omega) C_{k-1}^{-2}(x) \alpha_k(x) dx \qquad (13)$$

where the Green function satisfies

$$(\Delta + \frac{\omega^2}{C_k^2(x)} G_k(x,y,\omega) = - \delta(x - y) \qquad (14)$$

where Y denotes either r or s so far we have the formulae for successive linearization of acoustic wavefield inversion. Giving initial guess of the velocity field c(x) and data u(r,s, ω), we can calculate the Green function G_{k-1} and the disturbance wavefield v_k from Eqs. (9),(10)and (11), solving equation (13) we can find a sequence of the updated velocity $C_k(x)$.

FOR LARGE DISTURBANCES

In equation (8), the condition of small disturbances is required for Born approximation for Equation (12). Actually the method can be further improved for large disturbance Earth models without direct employment of Born approximation. For the planar wave the equation is

$$(\triangle - \frac{1}{C^2(x)} \frac{\partial^2}{\partial t^2})u(x,t) = \alpha \tag{15}$$

Let

$$\lim_{k \to \infty} C_k(x) = C(x), \tag{16}$$

$$C_k^{-2}(x) = C_{k-1}^{-2}(x) + \gamma_k(x) \tag{17}$$

and

$$u(x,t) = u_{k-1}(x,t) + v_k(x,t) \tag{18}$$

where γ proportional to α in comparison with (8) denoting large slowness variation of the Earth, which corresponds to warefield $v_k(x,t)$ Putting (17) and (18) into (15) yields

$$(\triangle - \frac{1}{c_{k-1}^2(x)} \frac{\partial^2}{\partial t^2})\tilde{u}(x,t) = \gamma_k(x) \frac{\partial^2 u(\tilde{x},t)}{\partial t^2} \tag{19}$$

The solution of (19) becomes

$$u(x,t) = u_{k-1}(x,t) + \iint G_{k-1}(x,x',t,t')u(x',t')\gamma_k(x')dx'dt' \tag{20}$$

where the Green 's function satisfies

$$(\triangle - \frac{1}{c_{k-1}^2(x)} \frac{\partial^2}{\partial t^2}), G_{k-1}(x,x',t,t') = -\frac{\partial^2}{\partial t^2}\delta(t-t')\delta(x-x') \tag{21}$$

On geophone station $x=r$, equation (20) becomes after inserting (18). Where $u(r,t)$ is the seismic record. Since we do not know V_k in the

$$u(r,t) - u_{k-1}(r,t) = \iint G_{k-1}(x,r,t,t')(u_{k-1}(x,t') + v_k(x,t'))\gamma_k(x)dxdt' \quad (22)$$

right hand side of (22). If we set it equal to zero for a solution then Born approximation is employed as what we have done for equation (12). However, we can put

$$v_k = \mu \quad v_{k-1}$$

follow iterations to improve the approximation. As a result equation (22) turns to

$$u(x,t) - u_{k-1}(x,t) = \iint G_{k-1}(x,r,t,t')(u_{k-1}(x,t') + \mu v_{k-1}(x,t'))\gamma_K(x)dxdt'$$

$$(23)$$

where μ is a small number, $0<\mu<1$, Equations (23) and (17) can be used for successive linearization as follows. Given initial model and data, we can calculate initial G and u, then find γ_1 from (23) by setting $v_k=0$. Following iteration is to calculate $(K=1)$th u, G and then to solve (23) for γ_k. The iteration produces a sequence of velocity estimates $C_k(x)$, $k=1,2,\cdots$

GENERALIZED BUCKAS—GILBERT METHOD

Denote u as the seismic wavefield, then the forward equation can be expressed as

$$F \quad u(q(x);x,s,t) = 0 \quad (24)$$

where F is a pseudo—differential operator. At receiver points,

$$u(q(x);x,s,t) = d(r,s,t) \quad (25)$$

where d denotes seismic data, $q(x)$ is a set of functions describing the Earth model, x is a point in the volume under consideration, s the source point and r the receiver. The nonlinearity of equation (24) is represented by the fact that u is dependent upon $q(x)$ which appears in operator F as coefficient terms. If one separates $q(x)$ into two parts: the reference model $q_0(x)$ and the disturbance $w(x)$,

$$q(x) = q_0(x) + \mathrm{w}(x) \tag{26}$$

then one can find an operator G which maps w into u

$$G\mathrm{w}(x,s,t) = u(q;x,s,t) \tag{27}$$

where G is usually nonlinear. Applying small disturbance techniques or Newton's method for linearization, we can form iteration like

$$q_k = q_{k-1} + \mathrm{w}_k \tag{28}$$

and

$$H_k(\mathrm{w}_{k+1} - \mathrm{w}_k) = d(r,s,t) - u(q_k;r,s,t) \tag{29}$$

where H is a linear operator derived from G, such as the Frechet operator of G. Thus, by giving initial guess of q, one can calculate the synthetic seismic data $u(q_{k-1}(x);r,s,t)$ and find higher order approximation of $q(x)$.

CHAOTIC CHARACTERISTICS OF THE OUTPUT SEQUENCES

Whenever an iterative procedure is dealt with , one usually has to analyze convergence conditions and corresponding criteria, forcing termination of the iteration. We are so familiar with this numerical treatment as to think that convergence and divergence are only possible states dur ing the iteration. As a mate of fact, a nonlinear iteration sometimes could display chaotic behaviors, such as containing some phases with complicated phase transitions (Schuster, 1987). For forward problems, as we have known, nonlinear differential equations can lead to completely chaotic trajector ies in the phase space (e. g. Lovenz, 1962). For inverse problems of partial differential equations, therefore, we can expect that the successive linearization iteration would display some chaotic characteristics which cannot be explained only by the concepts of convergence and divergence. As a result, the features of whole output sequences of the iteration must be studied in comparison of corresponding deterministic chaos. In recent years I have been doing research on this field and published two papers, entitled "NONLINEAR CHAOTIC INVERSION OF SEISMIC TRACES: PART I AND II" showing the research results (Yang, 1992). In those papers I compare someideas in the deterministic chaos (for forward problems) with successive linearization iterations for nonlinear inversion of seismic traces (as a simplified 1D wave equation inversion) via carefully selected numerical experiments. The results shall not be discussed here in detail but conclusions are shown as follows.

During a nonlinear seismic inverse problem being solved viasuccessive linearization, the iteration produces an output sequence containing infinite possible corresponding estimates of the Earth parameters. When input data contain even small errors and some control parameters are varying during the iteration, this sequence will finally go to a state of disorder and irregularity if no criteria are applied to force the itera tion stopped according to the criteria. It is similar to a nonlinear system which leads to chaotic motion due to control parameter and inner entropy continuously increased.

For ill — posed linear inverse problems , continuously reducing regularization parameters , e. g. the damping parameter, etc. , may produce an output sequence of estimates with increasing variance. An estimate for the linear inverse problems does not belong to disorder and irregularity because the variance directly corresponds to errors in input data.

For nonlinear seismic inverse problems, the characteristics of the output sequence are different, i. e. the initial error in input data triggers the disorder and irregularity of the sequence, but owing to the nonlinearity of the inverse problems, it does not correspond to irregularity in the follow—up output. The nonlinearity is a necessary, but not a sufficient condition for generation of chaotic iterations similar to chaos theory forward problems, the Poincare map or system equation plays improtant role daring the iteration, but the quantitative measures of chaotic motion, such as Liapunov exponents, invariant measure, etc. , can not be used directly. I have redefined the exponents for nonlinear inversions and shown the corresponding phase space.

Traditionally in numerical analysis we use the concept of convergence (or divergence) which means fitting errors or variances decrease (or increase for divergence) during an iteration. These concepts are not sufficient for description of the nonlinear iteration of successive linearization which contains several phases of relative stable states together with relative sharp phase transition, something similar to a dynamic system of chaotic motion. However, the redefined Liapunov exponents and the trajectory in the corresponding phase space can serve as good indications of the iterative phases. For instance, in the first phase of many iterations the Liapunov exponents are often turn to a zero — line, correspondingly one can find a normal attractor often existing in the phase space which shows slow advanced trajectory. This phase used to be called"convergence", but it may be followed by more than one phase or states, not so simple as"divergence" means. As a result, simple criteria, such as least squares fitting, cannot display the essential features of the iteration for nonlin ear seismic inverse problems.

THE MULTI—PORT ITERATION

For a specified inverse problem, the goal of inversion should be finding an estimate of the problem having resolution as high as possible while keeping errors tolerable. This estimate should be the one corresponding to a special stable state (phase) during the iteration. Of course, this special state cannot occur autonomously unless some negative entropy flows being introduced into the nonlinear system. From the viewpoint of information theory, introducing negative entropy is equivalent to input new information to excite the special state. In order to input new information during iteration one should open new ports and configure a network to link different reconstruction modules with the information ports. I call this kind of iteration networks as the multiport system.

The multi—port system belongs to opening systems in which information exchanges with the outside. It likes to a dissipative system, for which volume elements shrink as time increases. But same differences exist between them: in a dissipative system phase transitions are usually caused by variation of external control parameters, while in a multi — port system new phase is reached by introducing new information. We are most interested in the so — called self — organized state in a dissipative system which is something like the special state we are seeking for the automatic control of the iteration which produces the best estimate of nonlinear inverse problems. But we are not sure that this special state can be excited in multi — port iteration due to the differences as mentioned before. Anyhow, we may perform some numerical experiments for the answer.

For seismic inversions, useful information are traveltime data, amplitude data, wavefield records and constraints obtained from well—logging and drilling. The different sorts of the information may be introduced through different ports which are connected in the network in the manner of cascade or parallel.

The cascade approach may consist of three stages: (a) modifed BPT algorithm producing an initial model of the inhomogeneous earth, (b) traveltime inversion by using LSQR or CGLS algorithms to produce new reference velocity fields for full waveform seismic tomography in the inverse scattering procedure which employs the inverse Radon transform. Because the cascade approach uses varying reference velocity fields which make velocity disturbance small, the assumptions of Born approximation are satisfied at each step of the successive linearization iteration. Sometimes the variation of amplitude of the first break is related to existence of interested objectives due to diffraction or scattering, we can use it to map the apparent attenuation factor which provides a supplemental image of the earth. Such a multi—port network is displayed in Fig. 1. (a).

The parallel approach of the successive linearization, shown in Fig. 2 (b), is an improvement of the successive Born approximations of the acoustic wavefield as stated before. The successive traveltime inversion and wavefield invertion construct two parallel lines which are linked after each iteration step to produce a new reference velocity agreeable to both the traveltime and wavefield data. Actually the parallel approach should not be classified as opening systems because no information is introduced after excitation. A typical multi — port network suggested as shown in Fig. 1 (c), where the self — organized constraints can be a special controller of artificial intelligence.

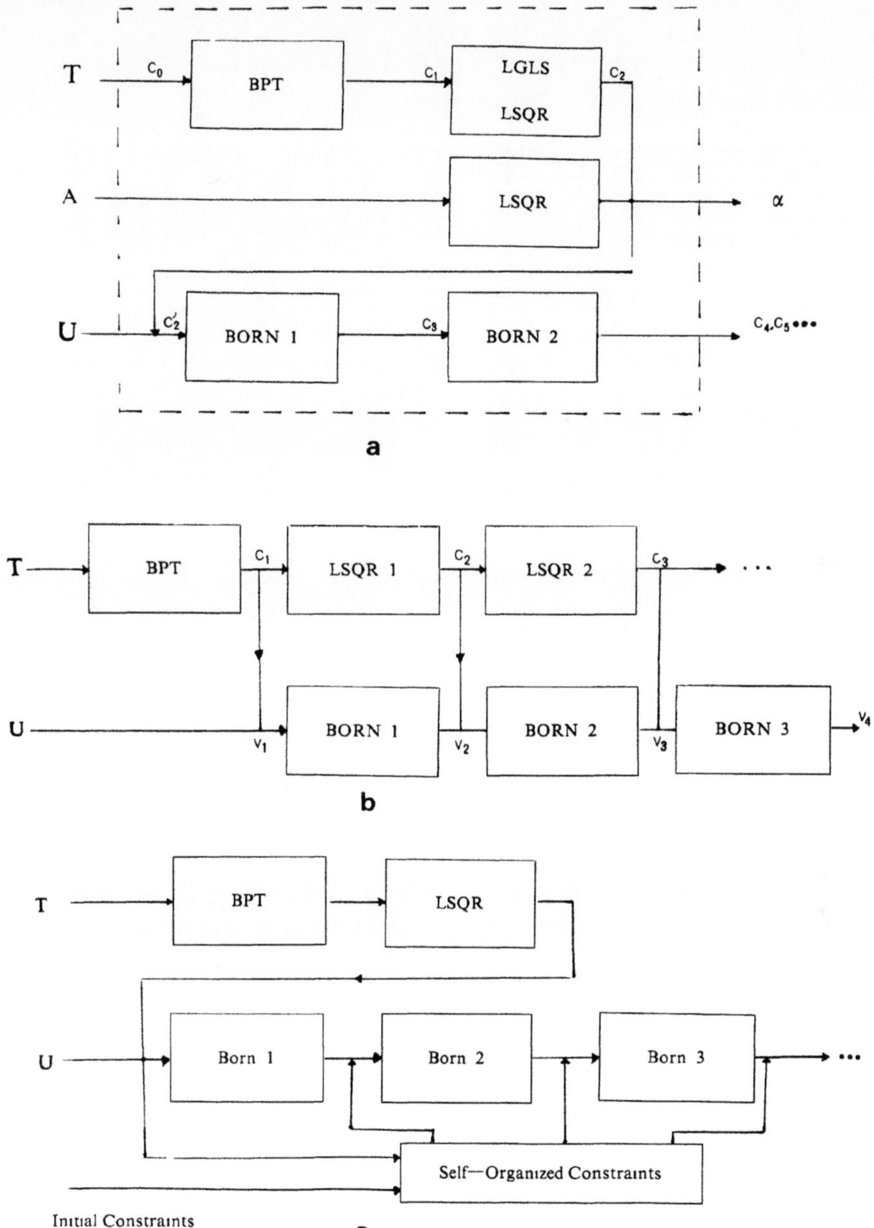

Figure 1. The multi-port Iteration networks for nonlinear seismic inversion. (a) The cascade approach, denotes are T traveltime data, A-amplitude data, U wavefield data; α the apparent attenuation factor, C-output sequence of Earth velocity. (b) The parallel approach, V-velocity disturbances. (c) A suggested hybrid approach, the block of self-Organized constraints means a special controller of artificial intelligence.

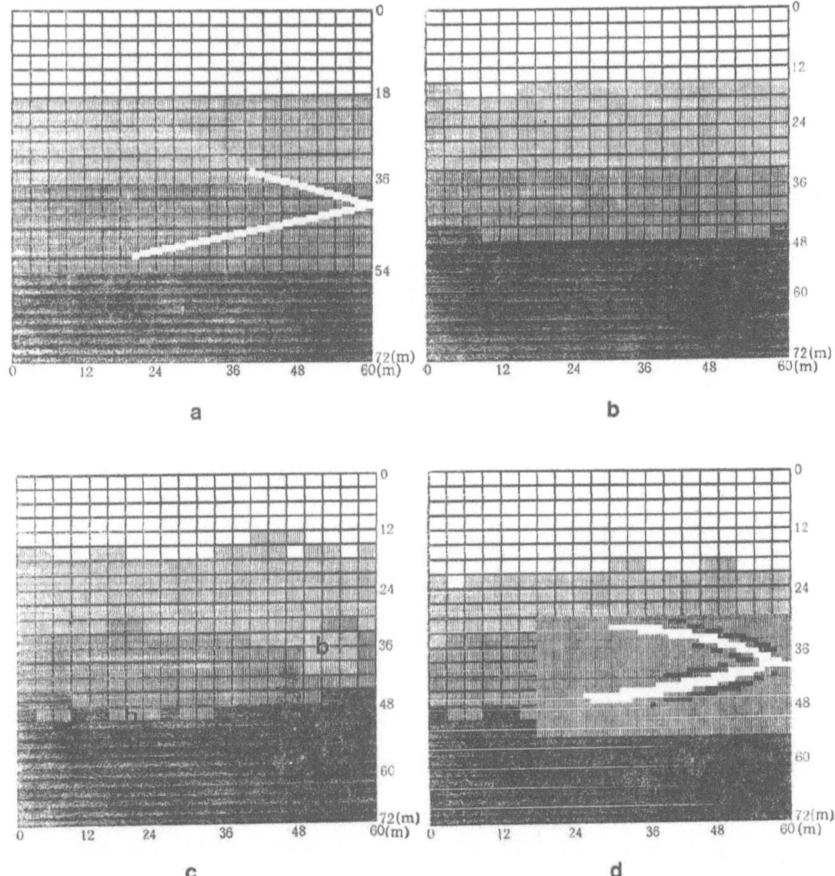

Figure 2. An Example for illustration of the multi-port iteration
shown in Fig.1 (a). (a) The earth model of two fault zones imbedded
in a vertically increased reference velocity. (b) The estimate from
modified BPT algorithm as output the first stage of the cascade
approach. (c) the estimate from LSQR as the second stage. (d) The
final estimate of the cascade approach after the wavefield
inversion.

In order to show how the multi — port iteration approaches produce improved tomographic images a numerical example is illustrasted in Figure 2. As shown in Fig. 2 (a), the earth model contains two fault zones existing in a vertically increased reference velocity field which is a linear function of the depth,increasing from 800m/s at top to 1100m/s at the bottom. The fault zones have a thickness of only 0. 5 meter and low velocity of 500m/s. A cross—hole survey is suggested to map the fault zones,separation of the holes are 60m with the hole of 72m. Both the shot spacing and the receiver spacing are equal to 3m,so the size of the pixels for traveltime inversion is selected as $3m \times 3m$. Based on the model the traveltime data and seismic records for 24 shots are calculated numerically, then 5% Gaussian noises are added into the seismic data for tomographic inversion by using the cascade approach as shown in Fig. 1(a). Fig. 2(b) shows the first output estimate produced by the BPT algorithm, it gives good estimate background velocity field but nothing can be found about the fault zones, because the resolution is very poor at the very first stage of the iterataion. As the successive linerization being continued,the resolution of the output sequence increases at the second stage of using the LSQR algorithm as shown in Fig. 2 (c). We can find in this image that two low velocity zones (marked by letter "b" in the figure) appear corresponding to the fault zones, but their shape and range are not clear. Using this estimate as the reference velocity field, the Born approximation at the third stage of the cascade iteration gives a clear image of the fault zones shown in the Fig. 2(d). The Born approximation procedure, using the inverse of the generalized Radon transform has been introduced by Yangand Du (1989,1991). With such a result, we can confirm the advantages of the multi—port iterations. The applicatrions of the approach to real data are described in references (Du and Yang, 1991, Yang and Du, 1992).

SUMMARY

In this paper I first show several procedures for successive linearization of nonlinear seismic inverse problems under the assumption that output sequence of the iteration would finally converge to the exact solutions of the problem when the step K approaches infinity. This assumption is based on the conventional viewpoint in numerical analysis. But for a practical inverse problem dealing with erroneous data, the output sequence finally becomes disorder when K approaches infinity. This fact seems controversial to the assumption, so we have to study the whole process of an nonlinear iteration and find several phases may exist in the output sequence. Each phase contains a set of estimates which are similar with little change in variance and resolution, while between the adjacent phases sharp changes can occur like the phase transition in a typical chaotic motion.

At this point I realized that some concepts and ideas in deterministic chaos might be useful for studying the characteristics of the nonlinear iteration because chaos also deals with nonlinear differential equations, though they usually belong to the so — called forward problems. Owing to the differences between forward and inverse problems, I had to make the analogy between the technical terms used in the chaos and the inversion, distinguish those related only to forward problems and redefine some parameters such as Lyapunov exponents. After numerical experiments, I believe that some common characteistics, caused by nonlinearity, appear in both nonlinear forward problems and the iteration of nonlinear inversion. These new undestanding may give us some idea to improve the nonlinear inversion approaches.

From the practical point of view, one may concern how to excite and identify the special phase of the iteration which steadily produce a set of estimates with high resolution and low variance in the output sequence. The multi — port approach is one of ways for excitng the special phase, while new Poincare maps with varying control parameter can also do the same, though we have not discuss them yet. As shown by numerical experiments, appropriately defined Lyapunov exponents can be a good measure for identifying the special phase. The other paramenters under considerations are fractal dimension, information dimension, etc. However as we are just beginning to study a new research field between the inversion and chaos, many problems are remained and much more researches should be continued.

ACKNOWLEDGMENTS

This word belongs to the project jointly suppored by the National Natural Science Foundation of China, Academia Sinica, Chinese National Petroleum Company and Daqin Petroleum Bureau, under project No. 49290200.

REFERENCE

Backus, G. E., and Gilbert, J. F., 1967, Numerical Application of a Formulism for Geophysical Inverse Problems. Geophys. J. R. astr. Soc., 13, 247—276.

Backus, G. E. and Gilbert, J. F., 1968, The Resolving Power of Grossearth Date. Geophys. J. R astr. Soc., 16, 169—205.

Backus, G. E., and Dilbert, J. F., 1970, Uniqueness in the Inversion of Inaccurate Gross Earth Data. Phil. Trans. Roy. Soc., London, Ser. A 266, 123—192.

Bishop, T. N., Bube, F. P., Cutler, R. T., Langan, R. T., Love, P. 1., Resnick, J. R. Shuey, R. T., Spindler, D. A., and Wyld, H. W., 1985, Tomographic Determination of Velocity and Depth in Laterally Varying Media; Geophysics, v. 50, p. 903—923.

Hao Bai—Lin 1984, Chaos. World Scientific, Singapore.

Hutson, V., and Pym, J. S., 1980, Applications of Functional Analysis and Operator Theory, Academic Press.

Lorenz, E. N. , Deterministic Nonperiodic Flow, J. Atmos. Sci. ,20 130—,1963.

Norton , S, J. ,1988, Iterative Seismic Inversion, Geophys. , J. , 94, No. 3, 475 — 468. Schuster, H. G. ,1987, Deterministic Chaos ,Combridge Press.

Tarantola, A. ,1984, Inversion of Seismic Reflection Data in the Acoustic Approximation, Geophys. , 49,1259—1266.

Tarantola, A. , 1987, Inverse Problem Theory: Methods for Data Filtering and model Parameter Estimation, Elsevier.

Yang Wencai, 1990, Seismic Velocity Imaging in an Inhomogeneous Layered Medium, Chinese J. Geophysics, 33, 265—283, Allerton Press Inc. New York.

Yang Wencai, 1992, Nonlinear Chaotic Inversion of Seismic Traces (I): Theory and Numerical Experiments, to be published in Chinese J. Geophysics Allerton Press Inc.

Yang Wencai, 1992, Nonlinear Choatic Inversion of Seismic Traces (II): Lyapunov Exponents and Attractors, to be published in Chinese J. Geophysics. Allerton Press Inc.

Yang Wencai and Du Jianyuan, 1992, Seismic Tomographic Imaging for Detecting Collapsing Pilars in Coal Mines, Geophis. & Geochem. Expl. ,16, 174—180 (in Chinese).

Du Jianyuan and Yang Wencai, 1992, Cascade Approach for Cross — Hole Seismic Tomography, Chinese J. Geophysics (in press), Allerton Press Inc.

APPLICATION OF NEURAL NETWORKS
TO SEISMIC IMAGING

Woon S.Gan and Cheong K.Gan

Acoustical Services Pte Ltd
29 Telok Ayer Street
Singapore 0104
Republic of Singapore

INTRODUCTION

Diffraction tomography has been applied to seismic imaging either with the assumption of small velocity variation in a homogeneous background[1] or using iterative method for nonlinear consideration[2]. In this paper, we propose the use of neutral networks to solve nonlinear seismic tomography problem. Neural networks are naturally suited to perform the inversion of tomography data due to its iterative and optimisation nature. Conventional inversion methods using the Fourier Diffraction Theorem assume weak scattering, resulting in a loss of accuracy. Frequency domain interpolation is prone to instability problems. Also there is a ill-posed problem of divergene when solving the nonlinear integral in diffraction tomography. Neutral networks approach is a better alternative without the above problems and gives a better resolution in seismic tomography when the medium is strongly inhomogeneous and the degree of nonlinearity is high.

SCOPE OF THE PROBLEM

The geophysical cross-holes imaging will be used. The diffraction tomography will consist of two parts: the forward problem and the reconstruction problem. Neural networks will be applied to both problems.

FORWARD PROBLEM

The forward problem concerns with the computation of the scattered field. To start with the inhomogeneous Helmholtz wave equation is used:

$$(\nabla^2 + k_0^2)u(\bar{r}) = -o(\bar{r})u(\bar{r}) \tag{1}$$

where

$$o(\bar{r}) = k_0^2\left[n^2(\bar{r}) - 1\right] \tag{2}$$
$$= \text{object function}$$

$u(\bar{r}) = u_0(\bar{r}) + u_s(\bar{r}) = $ total field, $u_0(\bar{r}) = $ incident field and $u_s(\bar{r}) = $ scattered field. It can be shown that the scattered field is[3] :

$$u_s(\bar{r}) = \int g(\bar{r} - \bar{r}')o(\bar{r}')u(\bar{r}')d\bar{r}' \tag{3}$$

where $\qquad g(\bar{r} - \bar{r}') = $ Green's function

$$= \frac{e^{jk_0(\bar{r}-\bar{r}')}}{4\pi(\bar{r} - \bar{r}')} \tag{4}$$

This is a Fredholm equation of the second kind and is nonlinear.

Eq.(3) can be written as

$$u_s(\bar{r}) = \int g(\bar{r} - \bar{r}')f(\bar{r}')d\bar{r}' \tag{5}$$

Let $f'(\bar{r}')$ be the estimated value for $f(\bar{r}')$. Then the corresponding scattered wave to $f'(\bar{r}')$ is expressed according to Eq.(5) as follows:

$$u_s'(\bar{r}) = \int g(\bar{r} - \bar{r}')f'(\bar{r}')d\bar{r}' \tag{6}$$

We propose the use of Hopfield neural network model to solve the problem.

Let Q be the square error of the estimated scattered wave to the measured scattered wave, then Q is expressed as follows:

$$Q = \int_{-\infty}^{\infty}\left|u_s(\bar{r}') - u_s'(\bar{r}')\right|^2 d\bar{r}' \tag{7}$$

If we can make the value of Q approximately zero by changing the value of the estimated $f'(\bar{r}')$, then $f'(\bar{r}')$ is approximately $f(\bar{r}')$. The reason we choose Hopfield neural network model in computing $u_s(\bar{r}')$ is that we can define an energy function corresponding to Q in a neural network of Hopfield model and this network has the characteristics to allow minimization of the energy function.

To apply Hopfield model to the scattered field $u_s(\bar{r}')$, we introduce the discrete representation of Eq.(7) as follows:

$$Q = \sum_{k=1}^{k_0}\left|u_s(\bar{r}_k') - u_s'(\bar{r}_k')\right|^2 \tag{8}$$

Substituting Eq.(6) into Eq.(8), we obtain the following equation:

$$Q = \sum_{k=1}^{k_0} \left| u_s(\bar{r}_k') \right|^2 + \sum_{i=1}^{l_0} \sum_{j=1}^{l_0} \sum_{k=1}^{l_0} R_e \{ g(\bar{r}_{kj}')$$

$$g^*(\bar{r}_{kj}') \} f'(\bar{r}_i') f'(\bar{r}_j') - \sum_{i=1}^{l_0} \sum_{k=1}^{l_0} R_e \{ u_s(\bar{r}_k')$$

$$g^*(\bar{r}_{ki}') \} f(\bar{r}_i') \tag{9}$$

where k_0 and l_0 are sampling numbers of the scattered field and object planes respectively. Re denotes the real part of complex numbers and asterisks * express the complex conjugates. Next step is to determine the correspondence between Q and the functional value in the neural network.

The fundamental architecture of the Hopfield network is shown in Fig.1 where each unit has a characteristic denoted by the function F.

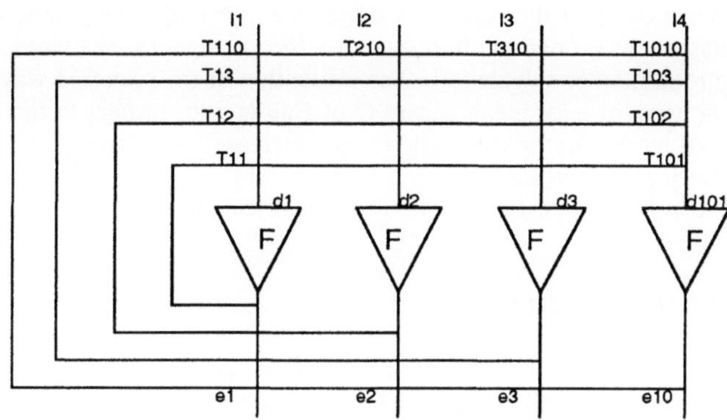

Fig. 1: Network architecture of
the Hopfield model

This function defines the relation between input d and output e as e=f(d) and each output is connected to all inputs of units with the coefficient T_{ij} without self-feedback, that is $T_{ij}=0$. In Fig.1, the relation between input and output of each unit can be written as follows:

$$d_i = \sum_{j=1}^{l_0} T_{ij} e_j + I_i \tag{10}$$

where I_i denotes an offset. Suppose that the function F is a lamp function, sigmoid function, step function etc, then output e_i of each unit changes the following energy function E to have the minimum value.

$$E = -\sum_{i=1}^{l_0} \sum_{j=1}^{l_0} T_{ij} e_j e_i - \sum_{i=1}^{l_0} I_i e_i \tag{11}$$

We suppose the evaluation function Q of Eq.(9) as energy function E of Eq.(11), the connection coefficient T_{ij} and offset I_i in Eq.(11) can be written as follows:

$$T_{ij} = -\sum_{k=1}^{k_0} R_e \{g(\bar{r}'_{kj})g^*(\bar{r}'_{ki})\} \tag{12}$$

$$I_i = \sum_{k=1}^{k_0} R_e \{u_s(\bar{r}'_k)g^*(\bar{r}'_{ki})\} \tag{13}$$

COMPUTER SIMULATION FOR SCATTERED FIELD

The simulation of scattered field using the Hopfield model network of Fig.1 is done by selecting one unit, randomly and calculating input into this unit from other outputs, resulting in changing the output of this unit. We continue this process until the output distribution shows the satisfactory focusing as a scattered wave. The improvement of the scattered field depends upon the repetition times of input-output calculations. Therefore computer simulation was done with respect to the change of scattered wave quality according to the repetition of calculation. The evaluation of the scattered wave can be done by calculating the term of least mean square Q of Eq.(8) with respect to the estimated and measured scattered wave. As the calculation times increase, the accuracy of scattered wave is improved, resulting in decrease of the value of Q. However, there is a saturation effect for decreasing the value of Q as the number of times of calculation increases.

RECONSTRUCTION PROBLEM

The reconstruction problem is more difficult than the forward problem because now both the object and the field inside the object are unknown. This means that it is necessary to design a procedure that simultaneously estimates both the object and the field inside the object. This procedure is made more difficult because the reconstruction is formed by illuminating the object by a number of different fields and the exact field inside the object must be calculated for each view. In geophysical imaging the receiver array is fixed in space (usually along the borehole axis) and the sources do not completely surround the formation being imaged. These changes in geometry from medical ultrasound tomography will have a pronounced effect on the resulting image quality of the reconstructions derived from the algorithm. Again the Hopfield neural network is proposed for the image reconstruction.

Let $\xi'(\bar{r}')$ be the vague reconstructed image using first order approximation and $\xi(\bar{r}')$ is the improved image by teaching the neural network.

Let Q' be the square error of the vague image to the improved image, then

$$Q' = \int_{-\infty}^{\infty} |\xi(\bar{r}') - \xi'(\bar{r}')|^2 \, d\bar{r}' \tag{14}$$

As before, we define an energy function corresponding to Q' in a neural network of Hopfield model and this network has the characteristics to allow minimization of the energy function.

Discrete representation of Eq.(14) gives

$$Q' = \sum_{k=1}^{k_0} \left| \xi(\bar{r}_k') - \xi'(\bar{r}_k') \right|^2 \tag{15}$$

The neural state update equation can be written as:

$$O^{(j+1)}(k) = O^{(j)}(k) + \Delta O(k) + \lambda I_k, 1 \leq k \leq M \tag{16}$$

This neural net processor works iteratively until a stable state of the net is reached to give a solution for the image reconstruction problem.

COMPUTER SIMULATION FOR THE IMAGE RECONSTRUCTION

Again the simulation of image reconstruction is done by selecting one unit, randomly and calculating input into this unit from other outputs, resulting in changing the output of this unit. This process is continued until the output distribution shows the satisfactory focusing as a reconstructed image.

Computer simulation is done with the following input parameters: 115ft wavelength, sampling number of the scattered field is 21 and sampling number of the image plane is 32. The improvement of the reconstructed image depends upon the repetition times of input-output calculations. Therefore computer simulation is done with respect to the change of image quality according to the repetition of calculation. The evaluation of the reconstructed image can be done by calculating the term of least mean square Q' of Eq.(15) with respect to the vague and improved image data. As the calculation times increase, the resolution of reconstructed image is improved, resulting in decrease of the value of Q' However, there is a saturation effect for decreasing the value of Q' as the number of times of calculation increase.

CONCLUSION

Neural networks is a useful tool for solving nonlinear problem as seismic diffraction tomography. The next step would be to look into a way of reducing the computation time.

REFERENCE

1   A Devaney,"Geophysical Diffraction Tomography", IEEE Trans. Geosci. Remote Sensing, Vol, GE-22, 3(1984).
2   B Gu, Y, Ji, J, Qiao, and H.Ji, "Several Problems of Geophysical Diffraction Tomography - Towards Practical Application", paper presented at the 19th International Symposium of Acoustical Imaging, Bochum, Germany, April 1991.
3   M. Slaney and A C Kak, "Imaging with Diffraction Tomography", Reprint No. TR-EE 85-5, Feb 1985.

A GENERALIZED RADON TRANSFORM METHOD FOR RECONSTRUCTING DENSITY AND LAME PARAMETERS OF ELASTIC MEDIUM USING FIELD DATA

Feng Yin and Yu Wei

Department of Physics
Southeast University
Nanjing 210018
P.R.China

INTRODUCTION

Recently more and more attentions are paid to the inversion of multi-dimensional wave equation inversion, for they can provide tomography picture with high resolution. In the past years, a large proportion of previous work on wave equation tomography has been based on a acoustical wave equation and great progress in this field has been made. But as we know, tomography methods using elastic wave equation not only can provide more elastic parameters of medium, but also are more rational than that based on the acoustical wave equation in geophysical exploration and non-destructive evaluation. However, in elastic inversion, people usually start from 1D elastic wave equation to inverse elastic parameters of medium[1] or use SH wave to inverse elastic parameters[2] because of the complexity of inversing 2D distribution of density and Lame parameters of elastic medium by 2D vector elastic wave equation. But many practical cases belong to two-dimensional inverse problems, thus, the methods using 1D elastic wave equation can not solve all problems. In addition, the method using SH wave can only inverse the density and shear modulus. Therefore, in order to inverse density and Lame parameters, we should use P wave

In this paper, at first, starting from the elastic wave equation, we derive an integral equation in frequency domain, in which the scattered data are connected with the density and Lame parameters perturbation. Then, corresponding integral equation in the time domain can be derived by Fourier transform and the relationship between scattered field and inversed parameters can be expressed by a generalized Radon transform. Under geometrical optic and Born approximation, the inverse generalized Radon transform operator (IGRT) can be derived approximately.

In this inverse Radon transform, inversed function in the left side of equation includes the density, Lame constants and scattering angle. Thus, after this inversed function is inversed by IGRT with different scattering angles for three times or more than three times, the density and Lame parameters can be determined by solving 3*3 linear algebraic equation system. At last, the computer numerical tests are given for Cross-holes exploration

ELASTIC SCATTERING THEORY

Consider below elastic wave equation in the isotropic inhomogeneous medium

$$\rho(\mathbf{r})\frac{\partial^2 s(\mathbf{r},t)}{\partial t^2} = \nabla \cdot \tau(\mathbf{r},t) \tag{1}$$

where the stress tensor of an incompressible solid is given by

$$\tau(\mathbf{r},t) = \lambda(\mathbf{r})I\nabla \bullet s(\mathbf{r},t) + \mu(\mathbf{r})(2\nabla s(\mathbf{r},t)$$

$$+I\times\nabla\times s(\mathbf{r},t)) \tag{2}$$

where the parameters $\rho(\mathbf{r})$, $\lambda(\mathbf{r})$ and $\mu(\mathbf{r})$ are the density and first and second Lame parameters distribution. I is the unit dyad, s($\mathbf{r}$) is the elastic displacement field. In order to facilitate the inversion algorithm, we neglect the scattering effects of shear waves and assume approximately

$$s(\mathbf{r},t) = \nabla\varphi(\mathbf{r},t) \tag{3}$$

where $\varphi(\mathbf{r},t)$ is the longitudinal elastic potential, substituting equation (3) into (1), we obtain

$$\rho(\mathbf{r})\frac{\partial^2\nabla\varphi(\mathbf{r},t)}{\partial t^2} = \nabla[\lambda(\mathbf{r})\nabla^2\varphi(\mathbf{r},t)] +$$

$$+ 2\nabla\bullet[\mu(\mathbf{r})\nabla\nabla\varphi(\mathbf{r},t)] \tag{4}$$

Adding $\rho_0\partial^2\nabla\varphi/\partial t^2 - \nabla(\lambda_0\nabla^2\varphi) - 2\nabla\bullet(\mu_0\nabla\nabla\varphi)$ to either sides of equation (4), here, the ρ_0, λ_0 and μ_0 being the homogeneous density and elastic constants respectively and performing the Fourier transform gives

$$(\nabla^2 + k^2)\nabla\varphi(\mathbf{r},\omega) = -k^2\gamma_\rho(\mathbf{r})\nabla\varphi(\mathbf{r},\omega) + \nabla[\gamma_\lambda(\mathbf{r})\nabla^2\varphi(\mathbf{r},\omega)]$$

$$2\nabla\bullet[(\gamma_\mu(\mathbf{r})\nabla\nabla\varphi(\mathbf{r},\omega)] \tag{5}$$

where

$$\gamma_\rho(\mathbf{r}) = \frac{\rho(\mathbf{r}) - \rho_0}{\rho_0}$$

$$\gamma_\lambda(\mathbf{r}) = \frac{\lambda(\mathbf{r}) - \lambda_0}{\lambda_0+2\mu_0}$$

$$\gamma_\mu(\mathbf{r}) = \frac{\mu(\mathbf{r}) - \mu_0}{\lambda_0+2\mu_0}$$

and $k = \omega/c_0$ is the free space wavenumber, c_0 is the velocity of longitudinal elastic waves, which equals to

$$c_0 = (\frac{\lambda_0 + 2\mu_0}{\rho_0})$$

Taking the divergence of both sides of equation (5) gives the scalar wave equation

$$(\nabla^2 + k^2)\Phi(\mathbf{r},\omega) = \overline{L}\varphi(\mathbf{r},\omega) \qquad (6)$$

where $\Phi(\mathbf{r},\omega) = \nabla^2\varphi(\mathbf{r},\omega)$ is known as the elastic dilatation and the differential operator L is

$$\overline{L} = -k^2\nabla\cdot\gamma_\rho(\mathbf{r})\nabla + \nabla^2\gamma_\lambda(\mathbf{r})\nabla^2 + 2\nabla\cdot\nabla\cdot\gamma_\mu(\mathbf{r})\nabla\nabla$$

let

$$\Phi(\mathbf{r},k) = \Phi_0(\mathbf{r},k) + \Phi_{sc}(\mathbf{r},k) \qquad (7)$$

where Φ_0 is the incident field, which equals to

$$\Phi_0(\mathbf{r},k) = G(\mathbf{r},\mathbf{r}_s,k) \qquad (8)$$

where $G(\mathbf{r},\mathbf{r}_s,k)$ is the 2D Green function, which satisfies

$$(\nabla^2 + k^2)G(\mathbf{r},\mathbf{r}_s,k) = -\delta(\mathbf{r}-\mathbf{r}_s) \qquad (9)$$

Within the Born approximation and geometrical optical approximation, the integral corresponding to equation (6) can be written as

$$\Phi_{sc}(\mathbf{r}_0,\mathbf{r}_s,k) = -\int_D G(\mathbf{r}_0,\mathbf{r},k)\overline{L}G(\mathbf{r},\mathbf{r}_s,k)\,d\mathbf{r} \qquad (10)$$

$$= -\int_D A(\mathbf{r}_0,\mathbf{r})e^{-i\omega T(\mathbf{r},\mathbf{r}_0)}\hat{L}A(\mathbf{r},\mathbf{r}_s)e^{-i\omega T(\mathbf{r},\mathbf{r}_s)}\,d\mathbf{r}$$

$$\hat{L} = \overline{L}/k2$$

where the A and T satisfy the transportation and eikonal equations respectively. On using Green's theorem in the plane and assume that $\gamma_\rho(\mathbf{r})$, $\gamma_\lambda(\mathbf{r})$ and $\gamma_\mu(\mathbf{r}) = 0$ on the boundary outside the object region, we obtain

$$\Phi_{sc}(\mathbf{r}_0,\mathbf{r}_s,k) = -k^2$$
$$* \int_D a(\mathbf{r}_0,\mathbf{r},\mathbf{r}_s)e^{(-i\omega\tau)}f(\mathbf{r},\vartheta)\,d\mathbf{r} \qquad (11)$$

where

$$a(\mathbf{r}_0,\mathbf{r},\mathbf{r}_s) = A(\mathbf{r}_0,\mathbf{r})A(\mathbf{r},\mathbf{r}_s)$$

$$\tau = T(\mathbf{r},\mathbf{r}_0)T(\mathbf{r},\mathbf{r}_s)$$

$$f(\mathbf{r},\vartheta) = \gamma_\lambda(\mathbf{r}) + \cos(\vartheta)\gamma_\rho + 2\cos^2(\vartheta)\gamma_\mu(\mathbf{r})$$

From the equation (11), we know that the scattered field is produced by three terms: first term involving density variation and $\cos(\vartheta)$, and second term involving compressibility fluctuation, and last term involving second elastic parameter fluctuation, and the directivity pattern of which depends upon $2\cos^2(\vartheta)$, ϑ is the angle between $\nabla T(\mathbf{r},\mathbf{r}_s)$ and $\nabla T(\mathbf{r},\mathbf{r}_0)$ as shown Fig.1.

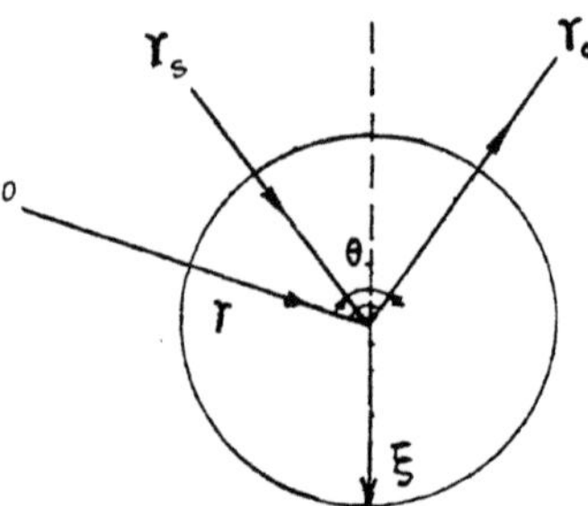

Fig.1 The angle between the incident ray and reflect ray

A GENERALIZED RADON TRANSFORM METHOD FOR INVERSING THREE PARAMETERS OF ELASTIC MEDIUM

Performing the inverse Fourier transform with respect to either sides of equation (11), integral equation in time domain corresponding to the equation (11) can be expressed as

$$\Phi_{sc}(\mathbf{r}_0,\mathbf{r}_s,t) = \int a(\mathbf{r}_0,\mathbf{r},\mathbf{r}_s)f1(\mathbf{r},\vartheta)\delta''(t-\tau)d\mathbf{r} \quad (12)$$

$$f1(\mathbf{r},\vartheta) = f(\mathbf{r},\vartheta)/c_0$$

the equation (12) is in the form of generalized Radon transform. According to the method of the paper (3), the equation (12) can be simplified as

$$\Phi_{sc}(\mathbf{r}_0,\mathbf{r},\mathbf{r}_s,\tau) = -\frac{a(\mathbf{r}_0,\mathbf{r},\mathbf{r}_s)c_0}{8\cos^3(\vartheta)} \times$$

$$\int f1(\mathbf{r}',\vartheta)\delta''(\xi*(\mathbf{r}-\mathbf{r}'))d\mathbf{r}' \quad (13)$$

where the definition of ξ is shown in fig.1. we known that the classic Radon transform is defined as

$$\Re f(p,\xi) = \int f(\mathbf{r}')\delta(p-\xi\cdot\mathbf{r}')d\mathbf{r}' \quad (14)$$

where the integral is calculated over the super-plane $p=\xi\cdot\mathbf{r}'$. therefore, the inverse Radon transform of $f(\mathbf{r}_0)$ can be expressed as

$$f(\mathbf{r}_0) = -\frac{1}{8\pi^2}\int d\xi\, \Re f''(\xi,p=\xi\cdot\mathbf{r}')$$

$$=-\frac{1}{8\pi^2}\int d\xi\, f(\mathbf{r}')\delta''[\xi\cdot(\mathbf{r}_0-\mathbf{r}')]d\mathbf{r}' \quad (15)$$

Comparing (14) with (16), the inversed function f1 for imaging can be derived as

$$f1(\mathbf{r},\vartheta) = \frac{\cos^3(\vartheta)}{\pi^2 c_0^3(\mathbf{r})} \int_{|\xi|=1} \frac{\Phi_{sc}(\mathbf{r}_0,\mathbf{r}_s,\tau(\mathbf{r}_0,\mathbf{r},\mathbf{r}_s))}{A(\mathbf{r},\mathbf{r}_s)A(\mathbf{r},\mathbf{r}_0)} d\xi \qquad (16)$$

the integral region is the surface of ball around a point **r** with unit length.

Similarly, for the two-dimensional case, using the 2D inverse Radon transform formula, we can derive the inversed function

$$f1(\mathbf{r},\vartheta) = \frac{\cos^2(\vartheta)}{\pi^2 c_0^2(\mathbf{r})} \int_{|\xi|=1} \frac{H[\Phi_{sc}(\mathbf{r}_0,\mathbf{r}_s,\tau(\mathbf{r}_0,\mathbf{r},\mathbf{r}_s))]}{A(\mathbf{r},\mathbf{r}_s)A(\mathbf{r},\mathbf{r}_0)} d\xi \qquad (17)$$

where $H[\cdots]$ represents the Hilbert transform, for homogeneous background, $A(\mathbf{x},\mathbf{y})= (C_0/(8\pi|\mathbf{x}-\mathbf{y}|))$.

COMPUTER SIMULATION RESULTS

Now, we consider a two-dimensional cross-holes imaging system as shown in fig.2, in which the longitudinal wave emitting from a point source is used to detect the object region

Fig.2   A model for imaging

Fig.3   Reconstucted results of ρ by our generalized Radon transform method.

Let us consider two models as shown in fig.3 for cross
-holes imaging, the hole deep is 160 m, the width of the
hole is 160 m. 48 source and 48 receivers are located
around the left hole, surface and right hole, the
distance between two sources and two receivers is 10 m.
The source function we used is four terms Blackman-Harris
window with duration 3.5 ms. The received waveforms are
calculated by moment method and the object region is
divided into 16*16 pixels.

When performing the inverse procedure, seven different
scattered angles ϑ are chosen respectively and the inversed
function $f(\mathbf{r},\vartheta_n)$ with a fixed angle ϑ_n is reconstructed by
equation (17) $(n=1,2,\cdots,7)$. Then we solve the 7*3 linear
algebraic equation system by LSQR algorithm to derive the
density and Lame parameters. These seven scattering
angles we used are $\vartheta_n = n\pi/14$ $(n = 0,1,2,\cdots,7)$

The fig.4, fig 5 and fig6 are the inversed results of
the density ρ, Lame parameters λ and μ of the model shown
in fig.2, respectively. In fig.4 to fig.6, a Laplace
operator is used to obtain the tomography pictures with
clear boundary.

**Fig.4 Reconstucted results of λ by our generalized
Radon transform method**

**Fig.5 Reconstucted results of μ by our generalized
Radon transform method**

CONCLUSION

A generalized Radon transform tomography algorithm for inversing the three parameters of elastic medium is given in this paper. The numerical results derived by us show that our method is effective to inverse the density and Lame parameters. But our method should satisfy below conditions:

1) the medium is non-dispensive
2) the medium is weak inhomogeneous so that the Born approximation could be applied;
3) the impulsed source is band limited, the maximum wave length should be smaller that the path between the scattered point and the receivers.
4) the waveform data for inversion are the P wave received in the far field.

Althought the Born approximation is limited in application. But it can give simple formula of the inversed results and estimated the property of the medium, so it still play an important role in the inverse problem field.

ACKNOWLEDGEMENT

We would like to thank J.S. Wang and Z.Z. Zhang in computational assistance when making program of the inverse Radon transform using acoustical wave field. The work is supported by the National Natural Science Foundation of China.

REFERENCES

[1] S.J. Norton and L.R. Testard, Reconstruction of one-dimensional inhomogeneities in elastic modulus and density using acoustical dimensional resonances, J. Acoust. Soc. Am., 79(4), 932-942, 1988
[2] M. A. Hooshyar, A. B. Weglein, Inversion of the two-dimensional SH elastic wave equation for the density and shear modulus, J. Acoust. Soc. Am., 78,1280,1986
[3] D. Miller, M.oristaglio, and G.Berlkin, A new slant on seismic imaging: Migration and integral geometry, Geophysics, 57,7 943-946,1984

A METHOD OF DIFFRACTION TOMOGRAPHY : COMPOUND SCANNING OF TRANSMITTER AND RECEIVER

Keinosuke Nagai[1] , Tomoki Yokoyama[1] and Wen Cao[2]

[1] Institute of Applied Physics, University of Tsukuba
Tsukuba, Ibaraki 305, Japan
[2] Daqing Petroleum Institute
Anda Heilongjiang 151400, P.R. China

INTRODUCTION

A new method to reconstruct ultrasonic cross-sectional images is dis-cussed. The most popular one to reconstruct them is a so-called pulseecho-method. Diffraction and scattering of waves severely degrade those images. The methods which adapt numerical processing to compensate diffraction and scattering effects and to reconstruct the clear image are generally called diffraction tomography. The method proposed in this paper is one of the diffraction tomography.

E.Wolf proposed diffraction tomography in optics[1]. In his method sinusoidal plane waves illuminate an object, which are scattered and observed at a point on a plane. Three dimensional image is reconstructed from observed data which have three dimensional parameters; the propagating direction of the illuminating wave and observation point on the two dimensional plane. R.Muller et al applied the method to ultrasonic imaging of two dimensions[2]. The data have two dimensional parameters; propagating direction of illuminating plane wave and observation point on a line. J.Devaney et al extended the method further to collect data on the arbitrary closed line[3]. It seems difficult to generate sinusoidal plane wave and to sweep its propagating direction over 360° . This may be the reason why no experiment has been reported on these methods, though many examination using numerical simulation have been done.   D.Nahamoo et al proposed the method which collected two dimensional data by using a usual transmitter and a receiver on each parallel lines, which were independently scanned[4]. They described the method to reconstruct two dimensional cross-sectional image from the data. But their experiment has not been reported neither, though the experiment on their method seems more practicable than those of the formers.

J.F.Greenleaf et al proposed the method which used pulse waves[5][6].

A transmitter faces to a receiver and an object is set between them. Time-of-flight is measured and recorded as a datum. The transmitter and the receiver simultaneously scans a pair of parallel lines. After the scanning the lines the direction of the pair of lines is rotated at step-angle and the data-acquisition is repeated. The angle of the rotation covers 180°. This combination of linear scanning and angular scanning is called compound scanning. So the data have two dimensional parameters. The method is different from their formers as it uses pulse waves and it assumes the propagation path is straight from the transmitter to the receiver. But they succeeded to reconstruct good images in their experiment.

A method of diffraction tomography using sinusoidal waves is proposed in this paper. Energy is distributed over long time range in sinusoidal wave, while it is concentrated in a short time range in pulse wave. Sinusoidal wave is often advantageous, therefore, over pulse wave when it propagates in dissipative medium as in human body. In this method the received wave is quadrature-detected and data are complex numbers. Applying the compound scanning of the transmitter and the receiver the data have two dimensional parameters. The procedure of data acquisition is described first. Then the theory and the procedure of image-reconstruction are told. The experimental result is displayed which demonstrates the feasibility of the method.

THEORY

Data-Acquisition

The geometry for data-acquisition is displayed in Figure 1. Sinusoidal wave (tone-burst) is shot from the transmitter T. It is scattered by an object and received by the receiver R. The received wave is then quadrature-detected and recorded as complex-number-datum. The data are collected changing the relative positions of T, R and the object to reconstruct image. The compound scanning is adapted in this paper; linear scanning of T and R simultaneously plus rotation of the pair of scanned lines.

Two dimensional problem is discussed here for the simplicity. The result is easily extended to three dimensional case. Let's consider Cartesian coordinates system (x,y) as in Figure 1. The scanned lines of T and R are $y=Y_t$ and $y=-Y_t$, respectively. The x-coordinates of T and R are commonly $x=x_t$. 'Scanning a pair of lines' means 'increasing x_t ' to obtain data. In other words the coordinates of T and R are (x_t ,Y_t) and $(x_t ,-Y_t)$, respectively and data are acquired by increasing x_t for a proper step. After the data-acquisition on a pair of lines the angle of the lines is rotated for a proper step and new data are acquired. The data-acquition are repeated, till the total angle becomes 180°.

Equation of Scattered Wave

An object is assumed to be embedded in homogeneous medium. It is also assumed to be small comparing to the distance from T to R and exists at the neighborhood of the origin of the coordinate system.

Let the coordinate of a point in the object (x,y) and it's position vector $\underline{r}$: $\underline{r} =(x,y)$. The observation-point, that is, the point of R, $(x_t ,-Y_t)$

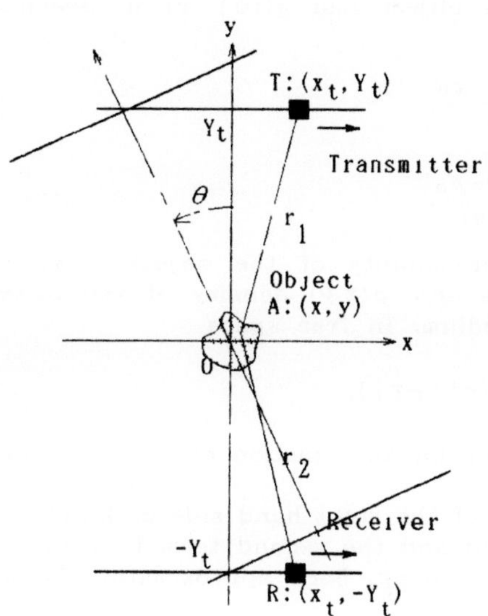

Figure 1. Geometry for data-acquisition of compound scanning diffraction tomography. Tone-burst wave is shot from the transmitter T. It is scattered by an object and received by the receiver R. The received wave is quadrature-detected and recorded as a complex-number-datum. T and R, facing each other, scan a pair of parallel lines; $y=Y_t$ and $y=-Y_t$, respectively, with same speed. The data are acquired with T and R at proper steps on the lines. Then the scanned lines are rotated for a proper angle from the x-axis and new data are acquired. The data-aquisitions are repeated, till the total angle becomes 180°.

as mentioned above. It's position vector $\underline{r}_0$; $\underline{r}_0 = (x_t , -Y_t)$. Incident sinusoidal pressure wave p_i. It's wavelength λ and the wave number k in the surrounding homogeneous medium.

Incident wave p_i is scattered by the object. Total pressure p at the observation-point are given as follows[7].

$$p(\underline{r}_0) = p_i(\underline{r}_0) + \int\int \{k^2 \gamma_c(\underline{r})p(\underline{r})g(\underline{r}_0 \mid \underline{r}) + \gamma_d(\underline{r})\nabla p(\underline{r}) \cdot \nabla g(\underline{r}_0 \mid \underline{r})\}dxdy \tag{1}$$

where γ_c and γ_d are quantities respectively depending on compressibility and density of the object and $g(r_0 \mid r)$ is Green's function written as followings.

$$\gamma_c(x,y) = \frac{\kappa(x,y) - \kappa_0}{\kappa_0}, \tag{2}$$

$$\gamma_d(x,y) = \frac{\rho(x,y) - \rho_0}{\rho(x,y)}, \tag{3}$$

where $\kappa(x,y)$ compressibility of the object , κ_0 compressibility of the surrounding medium and $\rho(x,y)$ density of the object and ρ_0 density of the surrounding medium. In free space

$$g(\underline{r}_0 \mid \underline{r}) = \frac{i}{4}H_0^{(2)}(k \mid \underline{r}_0 - \underline{r} \mid), \tag{4}$$

where $H_0^{(2)}$ 0th order Hankel function of the first kind.

The first term of the right hand side of Eq.(1) represents the incident wave p_i as mentioned and the second term indicates the scattered wave p_s . Assuming weak scattering, Born approximation is applied to obtain

$$p_s(\underline{r}_0) = \int\int \{k^2 \gamma_c(\underline{r})p_i(\underline{r})g(\underline{r}_0 \mid \underline{r}) + \gamma_d(\underline{r})\nabla p_i(\underline{r}) \cdot \nabla g(\underline{r}_0 \mid \underline{r})\}dxdy. \tag{5}$$

If the aperture of the transmitter is narrow, p_i could be regarded as the cylindrical wave. Normalizing the amplitude, it is represented by

$$p_i(x,y) = H_0^{(2)}(kr_1), \tag{6}$$

$$\text{where} \quad r_1 = \sqrt{(x-x_t)^2 + (y-Y_t)^2}.$$

Approximation

If an object is human body, considering the application of this imaging method to the medical diagnostics, $\rho(x,y) \doteqdot \rho_0$. Then $\gamma_d(x,y) \doteqdot 0$ from Eq.(3). If $2\pi r1, 2\pi r2 >> \lambda$, Hankel function could be approximated to

$$H_0^{(2)}(kr) = \frac{1+i}{\sqrt{\pi kr}}\exp\{-ikr\}. \tag{7}$$

Using these approximation Eq.(5) becomes

$$p_s(x_t,-Y_t) = - \frac{1}{\lambda} \iint \frac{1}{\sqrt{r_1 r_2}} \gamma_c(x,y) \exp\{-ik(r_1+r_2)\}dxdy. \qquad (8)$$

And further using the approximation if $|z| \ll 1$, $\sqrt{1+z} \doteq 1+\frac{1}{2}z$,

$$r_1 \doteq Y_t + \frac{1}{2Y_t}\{ -2Y_t y + y^2 + (x-x_t)^2 \}$$

$$r_2 \doteq Y_t + \frac{1}{2Y_t}\{ +2Y_t y + y^2 + (x-x_t)^2 \}$$

then

$$r_1 + r_2 \doteq 2Y_t + \frac{1}{Y_t}\{ y^2 + (x-x_t)^2 \}. \qquad (9)$$

In the calculation of Eq.(8), Eq.(9) is substituted in the argument of exponential function and Y_t is done in both r1 and r2 in the denominator. This yields

$$p_s(x_t,-Y_t) = - \frac{1}{\lambda Y_t} \iint \overline{\gamma}_c(x,y)\exp\{-i\frac{k}{Y_t}(x-x_t)^2\}dxdy, \qquad (10)$$

$$\text{where } \overline{\gamma}_c(x,y)=\gamma_c(x,y) \cdot \exp(-i\frac{k}{Y_t}y^2).$$

γ_c is considered to represent 'object'. If the amplitude of γ_c is used to display the 'object', $\overline{\gamma}_c$ also could be used. Because $|\overline{\gamma}_c| = |\gamma_c|$.

Reconstruction of Image

'Image-reconstruction' means to calculate $\gamma_c(x,y)$, that is $\overline{\gamma}_c(x,y)$, from acquired data $p_s(x_t,-Y_t)$ using the relationship of Eq.(10). It could be done by way of Fourier transform as described bellow.

Let two dimensional Fourier transform of $\overline{\gamma}_c(x,y)$ be $\overline{\Gamma}_c(u,v)$, that is,

$$\overline{\Gamma}(u,v)=\iint_{-\infty}^{\infty} \overline{\gamma}_c(x,y)\exp\{-i(ux+vy)\}dxdy. \qquad (11)$$

One dimensional Fourier transform of

$$h(x)=\exp\{-i\frac{k}{Y_t}x^2\} \qquad (12)$$

H(u), that is,

$$H(u)=\int_{-\infty}^{\infty} h(x)\{-iux\}dx.$$

$$=(i+1)\sqrt{\frac{Y_t\lambda}{2}}\exp\{\frac{Y_t}{4k}u^2\}. \qquad (13)$$

And further one dimensional Fourier transform of acquired data $p_s(x_t,-Y_t)$ $P_s(u)$,

$$P_S(u)=\int_{-\infty}^{\infty} p_S(x_t,-Y_t)\exp\{-iux_t\}dx_t.\tag{14}$$

Noting Eq.(10) is one dimensional convolution integral, Fourier transform of both side with regard to x_t yields

$$P_S(u)=-\frac{1}{\lambda Y_t}\overline{\Gamma}_c(u,0)\cdot H(u).\tag{15}$$

Equation (15) shows the two dimensional Fourier transform of the object $\overline{\Gamma}_c(u,v)$ is known on the u-axis, that is $\overline{\Gamma}_c(u,0)$ is known, from data acquired by the pair of scanned lines parallel to the x-axis. Two dimensional Fourier transform region is displayed in Figure 2. The u-axis corresponds to the x-axis, as mentioned in Eq.(11). One dimensional Fourier transform of data obtained by the pair of lines at angle θ from the x-axis, therefore, yields the value of $\overline{\Gamma}_c(u,v)$ on the line at the same angle θ from the u-axis. The relationships between the angle of the pair of scanned lines and the angle of the line on which $\overline{\Gamma}_c(u,v)$ could be obtained, resembles the relationship in the x-ray computerized tomography. Thus the value of whole $\overline{\Gamma}_c(u,v)$ could be obtained from data acquired by compound scanning.

Since $\overline{\Gamma}_c(u,v)$ is known in the whole (u,v), $\overline{\gamma}_c(x,y)$ is calculated by the inverse Fourier transform,

$$\overline{\gamma}_c(x,y) = \frac{1}{(2\pi)^2}\int\int_{-\infty}^{\infty}\overline{\Gamma}_c(u,v)\exp\{-i(ux+vy)\}dudv.\tag{16}$$

EXPERIMENT

Experiment on image-reconstruction of the object in water tank is made to examine the practicability of the theory discussed in the previous sections. The setup is shown in Figure 3. There is the water tank at the back, where the transmitter, the receiver and their scanning equipment are attached. In front of the tank, from left to right, oscilloscope to monitor the experiment, power-supply for scanning equipment, power-supply for signal processor, driving circuit for scanning equipment, signal processor including quadrature detector and the personal computer for the control of the experiment.

Circular sponge masks with the slit of 1 mm wide and 10 mm long at the center are bound to the surfaces of the transmitter and the receiver. Because the transmitter and the receiver have to be regarded as the line source and line sensor, respectively, in the two dimensional problem. Other parameters for data acquisitions are as follows.

distance from transmitter to receiver	250 mm, (Y_t =125 mm)
carrier frequency of the tone burst	2.0 MHz
wave length in the water	0.74 mm, (22° C)
sampling step on a scanned line	0.5 mm
number of samples on a scanned line	128
angular step for rotation of the lines	3.28°

Figure 4 shows the phantom tested in the experiment. It is Vinyl tube of which dimensions are as follows.

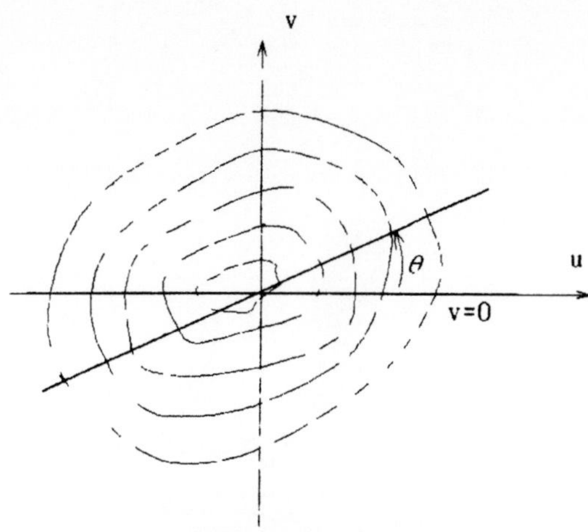

Figure 2. Two-dimensional Fourier transform of an object $\overline{\gamma}_c\ (x,y)$. It is denoted by $\overline{\Gamma}_c\ (u,v)$, where the spatial frequency u and v correspond to the real axis x and y, respectively. $\overline{\Gamma}_c\ (u,v)$ on the u-axis could be obtained from one-dimensional Fourier transform of data acquired on the pair of scanned lines parallel to the x-axis. $\overline{\Gamma}_c\ (u,v)$ on the line at an angle θ from the u-axis could be obtained, therefore, from data acquired on the pair of lines rotated for the same angle θ from the x-axis. Thus, $\overline{\Gamma}_c\ (u,v)$ in the whole (u,v) domain are obtained by rotating the pair of scanned lines successively at proper steps over 180° .

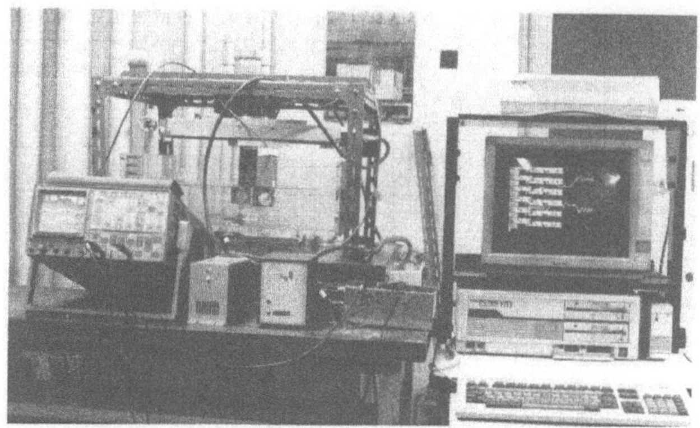

Figure 3. Experimental setup. The water-tank with scanning equipment at the back. The transmitter and the receiver are immersed in water in it. Oscilloscope for monitoring the experiment at the most left-hand side in front of the water-tank. To the right, power supply for the scanning equipment, power supply for signal processor, driving circuit for scanning equipment, signal processor including quadrature detector and the personal computer for the control of the experiment.

Figure 4. Phantom tested. Vinyl tube with 18.0 mm outer radius and 1.9 mm thick.

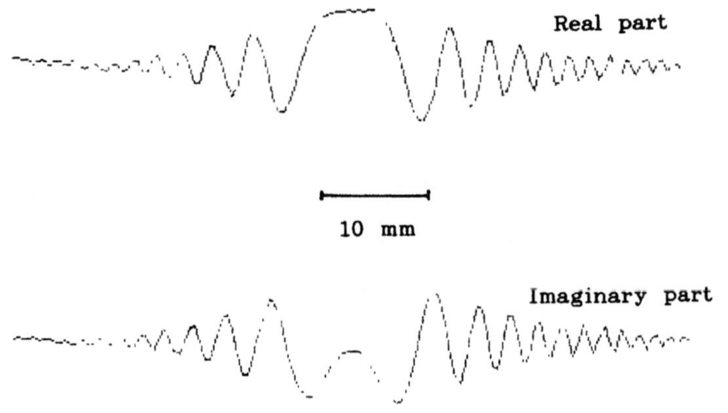

Figure 5. Data acquired on a pair of lines, which are complex numbers as the results of quadrature detection. Data acquisition area is 63.5 mm long at intervals of 0.5 mm. The abscissa represents the scanned line.

Figure 6. The reconstructed image. The image of the phantom displayed in Fig. 4 is reconstructed by the method discussed in this paper.

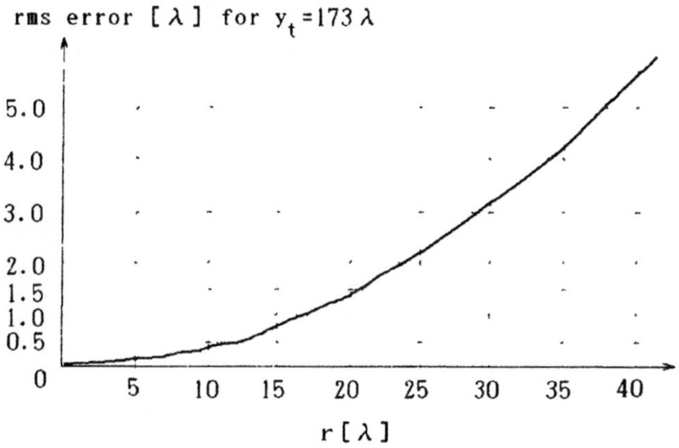

Figure 7. The root mean square error of Eq.(9) for the ultrasonic propagation-distance. The squared errors of the approximated equation (9) for the distance of transmitter-object-receiver are averaged for all positions of the transmitter and the receiver. The square root of the average is plotted with regard to the distance of the point-object from the origin of the coordinate system. The image-quality might be degraded if the error becomes more than one half of a wavelength.

outer radius	18.0 mm
thickness	1.9 mm

Data for the phantom acquired on a pair of scanned lines are displayed in Figure 5. Scattered wave received is quadrature detected. Real and imaginary parts of the complex data are obtained. The abscissa represents the scanned line. This is a kind of diffraction-pattern of the phantom.

Figure 6 shows the reconstructed image by the theory described in the previous sections from the experimentally acquired data. One can see the cross-section of the tube, especially the boundary between the tube and the surrounding medium (water). The clearness of the image demonstrates the feasibility of the theory.

CONCLUDING REMARKS

The experiment proves the main aspect of the theory is correct. But it is necessary to examine the details. The essential part of the imaging method proposed in this paper is that $h(x,y)$, as shown in Eq.(12), depends only on x and is independent of y. This is so if approximation of Eq.(9) holds.

The squared error of Eq.(9) is averaged for all sampling points in the geometry of the experiment. The squared root of the average is shown in Figure 7. For the symmetry of the geometry the error depends only on the distance r of the point object from the origin. As shown in this figure, if the distance is less than 10λ, the rms error is less than $\lambda/2$ and Eq.(9) is considered to hold. But if the distance exceeds 10λ, the error abruptly increases. It is necessary to suppress the error by any means if the object is large, for example, as increasing the distance from the transmitter to receiver. Other problems should be studied in the future.

REFERENCES

[1] E.Wolf,"Three-dimensional structure determination of semitransparent objects form holographic data", Optics Communications, 1, 4, pp.153-156(1969).

[2] R.Muller, M.Kaveh and R.Iverson,"A new approach to acoustic tomography using diffraction techniques", Acoustical Imaging 8, pp.615-628(1980).

[3] A.J.Devaney and G.Beylkin,"Diffraction tomography using arbitrary transmitter and receiver surface", Ultrasonic Imaging 6, pp.181-193(1984).

[4] D.Nahamoo, S.X.Pan and A.C.Kak,"Synthetic aperture diffraction tomography and its interpolation-free computer implementation", IEEE Tr. on Sonics and Ultrasonics, SU-31, 4, pp.218-229(1984).

[5] J.F.Greenleaf, S.A.Jhonson, W.F.Samayoa and F.A.Duck,"Algebraic reconstruction of spatial distributions of acoustic velocities in tissue from their time-of-flight profiles", Acoustical Holography 6, pp.71-90(1975).

[6] J.F.Greenleaf,"Multidimensional ultrasonic imaging and tissue characterization", Japanese Journal of Applied Physics, 30, Supplement.30-1, pp.17-22(1991).

[7] P.M.Morse and K.U.Ingard,"Theoretical Acoustics" p.140(1986). Princeton Unv. Press.

AN INVERSE SCATTERING METHOD FOR RECONSTRUCTING DISTRIBUTION OF THE VELOCITY PERTURBATION IN INHOMOGENEOUS BACKGROUND MEDIUM*

Feng Yin, J.S. Wang, B.L. Gu, Q.S. Li and Yu Wei

Department of Physics
Southeast University
Nanjing, 210018
P.R.China

INTRODUCTION

In wave equation tomography, there are two typical methods, one is analytical inverse methods, e.g., Geophysical diffraction tomography (GDT)[1], etc., Radon transform method[2], etc.; the other is iterative inverse scattering method in frequency domain (FD-IISM), e.g., Born IISM[3], etc In GDT, when the velocity variation is large, these methods fail, and the resolution of GDT is not high. In FD-IISM, because the inverse problem is ill-posed, the regularization method with smoothing condition is used to derive smoothing solution of inverse scattering problem, thus, the resolution of solution is reduced. In addition, its ability of antinoise is poor and it is very difficult to choose the damping factor.

In order to consider the strong scattering effect of wave in tomographic algorithm based on wave equation and reconstruct the velocity which varies largely and avoid difficulty to select the regularization factor used in FD-IISM, we adopt the maximum entropy as constrain condition in this paper and put forward an improved inverse scattering tomography algorithm of reconstructing the velocity in inhomogeneous background medium under basis of the paper[4]. By using maximum entropy Cambridge algorithm (MECA), we select a solution with maximum entropy as the present value of velocity modification from all solutions which fit the current scattered field in the present background and use it to modify the velocity distribution. At first, we discuss the scattering theory in inhomogeneous background medium, then, give maximum entropy Cambridge algorithm applied to inversing the velocity distribution in inhomogeneous medium and give its realization. At last, the numerical tests were done. The computational results show that this algorithm is valid to

---

*The work is Supported By The National Natural Science Foundation of China

inverse the velocity which varies largely when the incomplete
and the noisy field data are used.

SCATTERING THEORY IN THE VARIABLE BACKGROUND MEDIUM

Now consider below the acoustic equations with constant
density

$$\left(\nabla 2 + \frac{\omega^2}{C^2(r)}\right)U(r,\omega) = -\delta(r - r_s) \tag{1}$$

where $C(r)$ is the acoustic wave velocity of the medium, $U(r,\omega)$
scalar field (eg.,acoustic pressure),.angular frequency, ∇^2
Laplace operator. Define

$$a(r)=n^2(r) - n_b^2(r) \tag{2}$$

where C_0 is a constant velocity, n_b is the background refrac-
tivity index and $n(r)=C_0/v(r)$, let

$$U(r) = U^i(r) + U^s(r) \tag{3}$$

then, the integral equation that relates to formula (1) is

$$U^s(r) = k^2 \int_D \alpha(r')U(r')G(|r - r'|,n_b)dr' \tag{4}$$

where, $k= \omega/C_0$, D is object region, $G(|r-r'|)$ represents
Green function under present background. Suppose that there
are M scattered data measured U^s_j $(1\leq j\leq M)$ outside the object,
if the object is divided into N pixel, then $D = D_1 \cap D_2 \cap \cdots \cap D_N$
$(D_i \cup D_j=0,i=j)$
Assuming α_i is the value of $\alpha(r)$ in the ith pixel D_i,
which is image pixel to be inversed. By the moment method,
the equation (4) can be discretized as

$$U^s_j = \sum_{j=1}^{N}A_{ji}\alpha_i \qquad (1\leq j\leq M) \tag{5}$$

where the value of A_{ji}

$$A_{ji} = -G_{ji} \cdot k2 \cdot U_i \cdot \Delta S$$
$$G_{ji} = G(|r_j-r_i|,n_b)$$
$$U_i = U(r_i)$$

where ΔS is the area of ith pixel. On the other hand, from
the equation (5), the equation of the forward scattering
problem can be written as

$$H \cdot a = b, \tag{6}$$

$$H_{ji} = \delta_{ji} - \int\int_{Di} G(|r_j-r'|,n_b)dr'$$

where the a is an unknown column vector whose entries are the
the discretized value U_i, b is a vector whose entries is the
discretized value of the incident field in the object region.
For reconstructing the velocity distribution in the variable
background medium, the updated Green's function is obtained

by solving the forward scattering problem. However, the linear system of equations for calculating the Green's function is the same as equation (6) except that the entries in the column vector b is the incident field generated by a two-dimensional point source located at the corresponding receiver point.

MAXIMUM ENTROPY ALGORITHM FOR INVERSING THE VELOCITY PERTURBATION IN INHOMOGENEOUS BACKGROUND MEDIUM

Because A_{ji} in equation (5) includes the total field U_i and Green function G_{ji} in the object, so it's impossible to solve directly. In order to solve the equation (5) efficiently and overcome the disadvantages of the FD-IISM, we put forward an improved inverse scattering tomography algorithm for inversing the velocity perturbation in the variable background medium using the maximum entropy principle(MEIIST) under the basis of the paper [4] as following, define image entropy as

$$S(P_1, P_2, \cdots, P_N) = -\sum_{i=1}^{N} P_i \log P_i \tag{7}$$

$$P_i = \frac{\alpha_i}{\sum_{i=1}^{N} \alpha_i} \tag{8}$$

The solution of maximum entropy in this paper is to maximize (7) subject to the constrain (5). When the Lagrange method is used to solve this problem, the noise may be used as useful signal and its implementation is complicated to realize, therefore, we use the Cambridge algorithm instead of this method. When considering the effect of noise, equation (5) can be rewritten as

$$U_j = \sum_{i=1}^{N} A_{ji}\alpha_i + e_j \qquad (1 \le j \le M) \tag{9}$$

where, e_j is the noise in jth measured data, If $e_j \sim N(0, \sigma_j), \sigma_j$ the variance of the noise, then the square function

$$Q(\alpha) = \frac{1}{2}\sum_{j=1}^{M} 1/\sigma_j^2 (\sum_{i=1}^{N} A_{ji}\alpha_i - U_j)^2 \sim \chi^2(\alpha) \tag{10}$$

where

$$\alpha = (\alpha_1, \alpha_2, \cdots, \alpha_N)^T$$

Knowing from (9) that when $Q(\alpha)$ is too large, the data generated from α has great difference compared with the measured value, and when $Q(\alpha)$ is too small , we may take the noise as useful signal, this will lead to many artifacts in inversed image. From the statistical analysis, we should take such α which satisfies $M/2 \le \chi^2(\alpha) \le M$ as the confident inversed image. So the required maximum entropy image can be derived by maximizing

$$J = S - \lambda \cdot Q \tag{11}$$

subject to $M/2 \leq \chi^2(\alpha) \leq M$

THE REALIZATION OF MECA APPLIED TO INVERSING THE VELOCITY IN INHOMOGENEOUS BACKGROUND MEDIUM

ME Cambridge algorithm

The general non-linear optimum method can be used to select the maximum value of equation (11), but it is difficult to select the Lagrange factor. In order to select the maximum entropy effectively, the Cambridge algorithm was used to select the maximum entropy solution in the paper, that is

$$\alpha^{(q+1)} = \alpha^{(q)} + \delta\alpha \tag{12}$$

$$\delta\alpha = x_1 e_1 + x_2 e_2 + x_3 e_3$$

where the searching direction $e_k (1 \leq k \leq 3)$ are

$$(e_1)_i = \alpha_i \frac{\partial S}{\partial \alpha_i}$$

$$(e_2)_i = \alpha_i \frac{\partial Q}{\partial \alpha_i} \tag{13}$$

$$(e_3)_i = \sum_{k=1}^{N} \alpha_i \frac{\partial^2 S}{\partial \alpha_i \partial \alpha_k} \alpha_k \frac{\partial J}{\partial \alpha_k}$$

In order to determine suitable coefficients $x_k (1 \leq k \leq 3)$ for the search direction and eliminate linear dependence of the search direction and simplify the evaluation of metric tenser $-\nabla\nabla S$ and $-\nabla\nabla Q$, control of the algorithm pass into the subspace. The problem in the subspace become one of optimizing a quadratic function subject to quadratic constraints.

let $S(\mathbf{x}) = S(\alpha+\delta\alpha)$, $Q(\mathbf{x}) = Q(\alpha+\delta\alpha)$, $\mathbf{x} = (x1, x2, x3)$, $S_0 = S(\alpha)$, $Q_0 = Q(\alpha)$, then by three linear transform, the quadratic expressions of $S(\mathbf{x})$ and $Q(\mathbf{x})$ in $\mathbf{x}$ domain are transformed into $\mathbf{u}$ space, that is

$$S(\mathbf{u}) = S_0 + \sum_{K=1}^{3} S_k u_k + 0.5 \sum_{K=1}^{3} u_k^2 \tag{14}$$

$$Q(\mathbf{u}) = Q_0 + \sum_{K=1}^{3} Q_k u_k + 0.5 \sum_{K=1}^{3} \mu_k u_k^2 \tag{15}$$

where $S_k = (\nabla S, e_k)$, $Q_k = (\nabla Q, e_k)$, the relationship between $\mathbf{u}$ and $\mathbf{x}$ is given in paper [5]. When the $-\nabla\nabla S$ is used as metric tenser, the length-squared of the increment $\delta\alpha$ is

$$1 = \delta\alpha^T \cdot (-\nabla\nabla S) \cdot \delta\alpha = \sum_{k=1}^{3} u_k \tag{16}$$

In order to derive effective search step, the object function in the Cambridge algorithm is modified as

$$J(\mathbf{u}) = \beta \cdot S(\mathbf{u}) - Q(\mathbf{u}) - \gamma \cdot l^2$$

$$Q(\mathbf{u}) = M_1/2 \tag{17}$$

$$l^2 \leq l_0^2$$

where the step constraint $l^2 \leq l_0^2$ is added. By theoretical analysis and numerical tests, we find that the l_0^2 should be taken to be $\xi \cdot \Sigma \alpha_{0k}/q$, $0.1 \leq \xi \leq 0.5$, q is the iteration number. let

$$\frac{\partial J(\mathbf{u})}{\partial \alpha_n} = 0 \tag{18}$$

we obtain

$$u_k = \frac{\beta S_k - Q_k}{\beta + \mu_k + \upsilon} \tag{19}$$

putting equation (19) into equation(15), we have

$$Q + \sum_{k=1}^{3} \lambda_k \left(\frac{\beta S_k - Q_k}{\beta + \mu_k + \upsilon}\right) + \sum_{k=1}^{3} \mu_k \left(\frac{\beta S_k - Q_k}{\beta + \mu_k + \upsilon}\right)^2 = Q_{aim} \tag{20}$$

where

$$Q_{aim} = \max\left(0.5 * M_1, \ Q - 1/3 \sum_{k=1}^{3} Q_k^2/\mu\right)$$

in equation (20), at first, let $\upsilon = 0$, solve β and derive u_k ($1 \leq k \leq 3$) by equation and derive l^2 by equation (16), judge wether or not l^2 is less than l_0, if $l^2 > l_0^2$, then increase υ slowly, repeat the cycle, until $l^2 < l_0^2$, then inverse $\mathbf{u}$ into x_k by

$$\mathbf{x} = \mathbf{T}^T \mathrm{diag}(1/\sqrt{\lambda_1}, \ 1/\sqrt{\lambda_2}, \ 1/\sqrt{\lambda_3}) \mathbf{v}^T \cdot \mathbf{u} \tag{21}$$

where the definitions of all parameters in equation (20) are given in the paper [5], by equation (10), the model increment is derive, and use it to modify the model of velocity, repeat like these, until $|Q_{aim} - Q(\mathbf{u})| < \epsilon$, where ϵ is small number, stop iteration and use this α as the inversed results.

Realization of MECA

Knowing from the Cambridge algorithm mentioned above, the key of the algorithm is how to evaluate $\partial Q/\partial \alpha_i$, $\partial^2 Q/\partial \alpha_i \partial \alpha_k$ and $\partial J/\partial \alpha_i$ effectively and quickly. From the definitions of Q, S, J, we obtain

$$\frac{\partial Q}{\partial \alpha_i} = \sum_{j=1}^{M} \frac{1}{\sigma_j^2} \left[R(d^{(c)}{}_j - dj) \frac{\partial R(d_j^{(c)})}{\partial \alpha_i} + \right.$$

$$\left. I(d^{(C)} - d_J) \frac{\partial I(d_j^{(c)})}{\partial \alpha_i} \right. \tag{22}$$

where

$$\frac{\partial d_j^{(c)}}{\partial \alpha_i} = [\frac{\partial R(d_j^{(c)})}{\partial \alpha_i}, \frac{\partial I(d_j^{(c)})}{\partial \alpha_i}]$$

$$= \sum_{n=1}^{N} \frac{\partial A_{jn}}{\partial \alpha_i} \alpha_n + A_{ji}$$

where $d_j^s = U_j$ is the scattered field data under present background, $d^{(c)}_j$ evaluated data, and the here $\partial d_j^{(c)}/\partial \alpha_i$ includes $\partial A_{jn}/\partial \alpha_i$ which is omitted in the paper [4].

$$\frac{\partial A_{jn}}{\partial \alpha_i} = -k^2 \cdot G_{jn} \frac{\partial U_n}{\partial \alpha_i} - k^2 \cdot U_n \frac{\partial G_{jn}}{\partial \alpha_i} \quad (1 \leq j \leq M, 1 \leq i \leq N, 1 \leq n \leq M) \quad (23)$$

from (6), $\partial U_n/\partial a_i$ can be derived

$$\frac{\partial U}{\partial \alpha_i} = H^{-1} \cdot b \quad (1 \leq i \leq N) \tag{24}$$

where $\partial U/\partial \alpha_i = (\partial U_1/\partial \alpha_i, \cdots, \partial U_N/\partial \alpha_i)^T$, $b = (-I_{1i}U_i, \cdots, -I_{ni}U_i)^T$

$$\frac{\partial^2 Q}{\partial \alpha_i \partial \alpha_k} = \sum_{j=1}^{M} \frac{1}{\sigma_j^2} [R(\frac{\partial R(d_j^{(c)})}{\partial \alpha_k}) * R(\frac{\partial R(d_j^{(c)})}{\partial \alpha_i}) +$$

$$I(\frac{\partial R(d_j^{(c)})}{\partial \alpha_k}) * I(\frac{\partial R(d_j^{(c)})}{\partial \alpha_i})] + [R(d^{(c)}_j - d_j) \frac{\partial^2 R(d_j^{(c)})}{\partial \alpha_i \partial \alpha_k} +$$

$$I(d^{(c)} - d_J) \frac{\partial^2 I(d_j^{(c)})}{\partial \alpha_i \partial \alpha_k}) \tag{25}$$

where

$$\frac{\partial^2 (d_j^{(c)})}{\partial \alpha_i \partial \alpha_k} = \frac{\partial A_{jk}}{\partial \alpha_i} + \frac{\partial A_{ji}}{\partial \alpha_k}$$

and

$$\frac{\partial J}{\partial \alpha_i} = \frac{1}{\sum_{i=1}^{N} \alpha_i} \{ \sum_{j=1}^{N} P_j \log \alpha_j - \log \alpha_i] \} - \lambda \frac{\partial Q(\alpha)}{\partial \alpha_i} \tag{26}$$

λ the can be derived by setting $\nabla J = 0$,

$$\lambda = (\sum_{n=1}^{N} \alpha_n (\frac{\partial S}{\partial \alpha_n})^2 / \sum_{n=1}^{N} \alpha_n (\frac{\partial Q}{\partial \alpha_n})^2 \tag{27}$$

Up to now, we can obtain the procedures of our algorithm realization as follows:

1) Suppose there is a initial object distribution function ,eg.,homogeneous distribution etc.;
2) Solve the total field U_n and G_{mn} in object,then solve A_{mn};
3) When solving α in qth iteration, computing

   a) $J(\alpha)=S(\alpha)-\lambda\cdot Q(\alpha)$
   b) evalute

$$\frac{\partial U_n}{\partial\alpha_i},\frac{\partial G_{jn}}{\partial\alpha_i},\frac{\partial A_{jn}}{\partial\alpha_i},\frac{\partial d^{(c)}_j}{\partial\alpha_i},\frac{\partial Q}{\partial\alpha_i},\frac{\partial^2 Q}{\partial\alpha_i\partial\alpha_k}$$

$$(1\leq i,n\leq N\ ,\ 1\leq j\leq M)$$

4) modify model using $\delta\alpha$ the results derived by Cambridge alogo-rithm

$$\alpha^{(q+1)}=\alpha^{(q)}+\delta\alpha$$

5) evaluate $E_1^{(q)}=\frac{\|d-d^{(c)}\|}{\|d\|}\leq\epsilon_1,\ E_2^{(q)}=\frac{\|\alpha^{(q)}-\alpha^{(q-1)}\|}{\|\alpha^{(q-1)}\|}\leq\epsilon_2$

where ϵ_1 and ϵ_2 are small positive numbers. If above condition is satisfied,then stop iteration, if this condition is not satisfied, then,return to procedure (2), iteration again.

NUMERICAL TESTS

Fig.1 is a model for cross-holes, vertical seismic profiling (VSP) imaging. In this Figure, We take 9*9 data, and the size of pixels is 0.25 wave-length, then the object is divided into 8*8 pixels. At first, we let the disturbing level be 20%, forward Cross-holes data D1, VSP data D2 are generated from this model, then, 10% Gaussian Noise is added to D1 and D2 to derive the data D3 and D4 with noise. Fig.2 and Fig.3 are the tomographys by **MECA**, using D1 and D2 after twenty iterations. Fig.4 are the tomographys by (MECA), using Cross-holes D1 & VSP D2 after twenty iterations. Fig.5 and Fig.6 are the tomographys by **MECA**, using D1 and D2 after twenty iterations.

Fig.1 A model for imaging

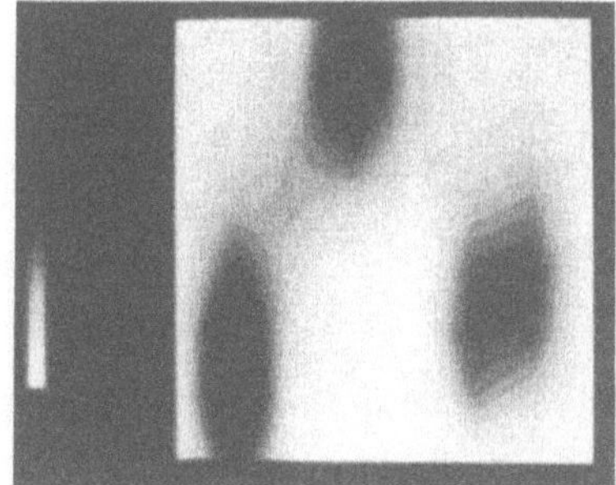

Fig.2 Reconstucted result by using CH data after twenty iterations.

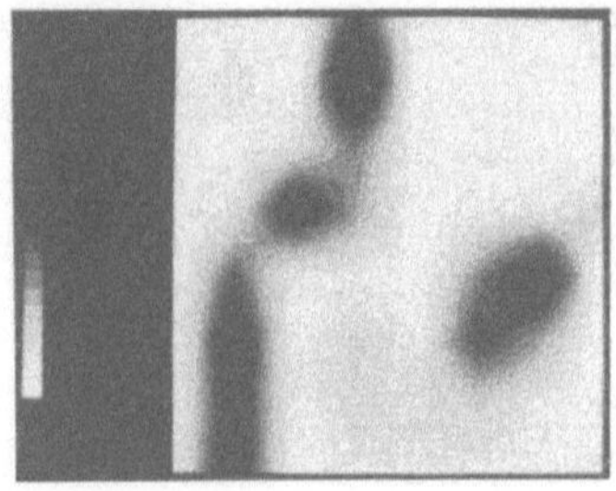

Fig.3 Reconstucted result
by using VSP data after
twenty iterations.

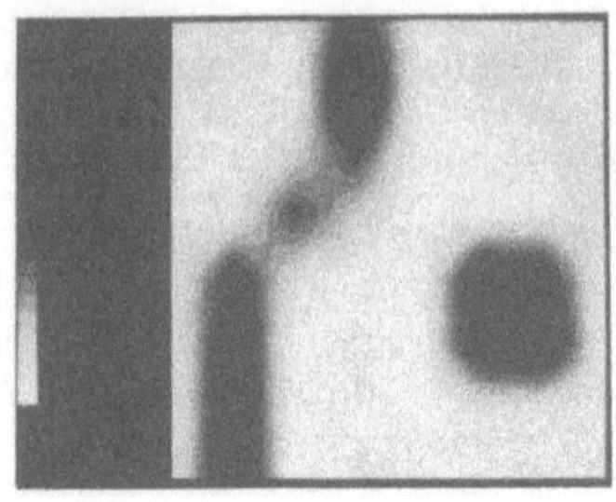

Fig.4 Reconstucted result
by using CH & VSP data
after twenty iterations.

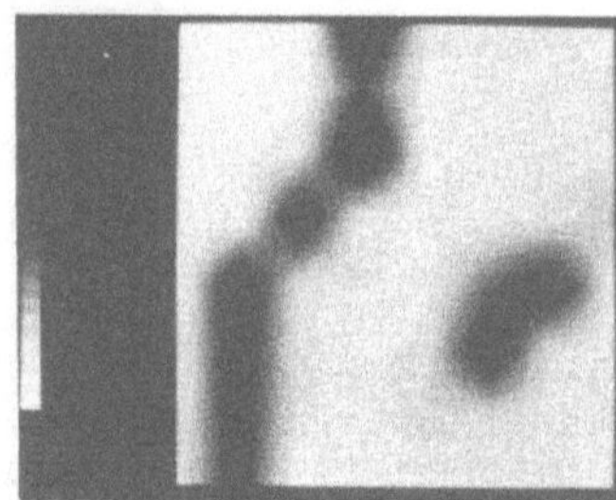

Fig.5 Reconstucted result
by using VSP data D3 after
twenty iterations.

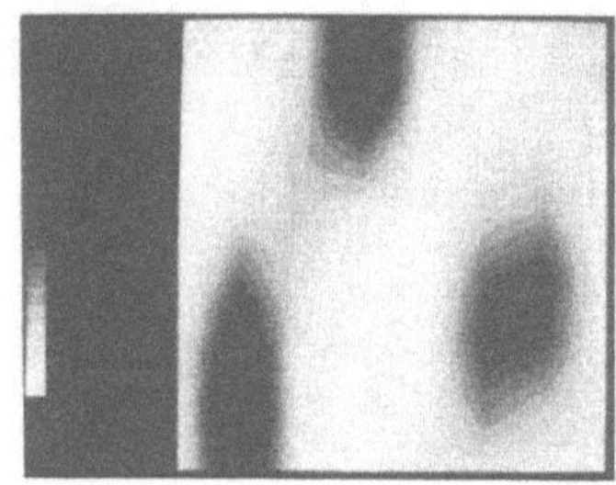

Fig.6 Reconstucted result
by using CH data D4 after
twenty iterations.

CONCLUSION

The numerical examples show that our algorithm is powerful tools inverse the incomplete and noisy field data for geophysical exploration and they are suitable in wide range which the born and Rytov fails and our algorithm avoid difficult to select the regularization factor in FD Born method.

REFERENCES

[1] Wu, R.S., and Toksoz, M.Nafi., Diffraction tomography tomography and multisource holography applied to seismic imaging, Geophysics, 51, 11-25, 1987.
[2] Miller, D., Oristaglio, M., and Beylkin, G. A new slant on seismic imaging: Migration and integral geometry, Geophysics, 52, 943-964, 1987.
[3] Chew, W.C. and Wang, Y.M., Recons-truction of Two-dimensional permitivity distribution using the distorted Born iterative method, IEEE trans. Med. imaging, 9, 218-225, 1990.
[4] F. Yin, J.S. Wang, Wave equation Tomography using Maximum entropy, Society of Geophysicists sixty-second International Annual meeting, New Oleans, U.S.A.
[5] Skilling, J., The Cambridge maximum entropy algorithm, Maximum entropy and Bayesian method in applied statistics, edited by Justice, J.H., Cambridge University press, 1986.

INVERSION FOR ρ, λ, μ IN LAYERED EARTH MODEL

Zheng Jia–Mao[1], Gu Ben–Li[1], Wei Yu[1]
and Li You–Ming[2]

[1]Southeast University, Nanjing, 210018
[2]Institute of Geophysics, Chinese Acadamy of Science
Beijing, 100101

ABSTRACT

The inverse method for density ρ, Lame's parameter λ and shear modulus μ in layered earth model is set up by means of 2–D one order elastic wave equation in vector form. Through transforming and decomposing the inverse problem for ρ,λ, μ in 2–D one order elastic wave equation in vector form into three independent inverse problems for single unknown coefficient in 1–D wave equation, we know that using P(or S) wave information can inverse impendance $k_p = \sqrt{\rho(\lambda + 2\mu)}$ (or $k_s = \sqrt{\rho\mu}$), using P–S coupling wave information can inverse density ρ, so ρ, λ, μ are all determined.

INTRODUCTION

It is well–known that there are many inverse methods for single parameter of medium characteristics in the layered earth model [1-6], but only few inverse methods for three elastic parameters because of the complexity. It is natural that the more we know the parameters of medium characteristics , the more we know the medium. So it is great important to determine more parameters of medium characteristics. In this paper, starting with 2–D one order elastic wave equation in vector form, we set up the inverse method for density ρ, Lame's parameter λ and shear modulus μ in the layered earth model. Through transforming and decomposing the inverse problem for ρ, λ, μ in 2–D one order elastic wave equation in vector form into three independent inverse problems for single unknown coefficient in 1–D wave equation, we know that using P(or S) wave information can inverse impedance $k_p = \sqrt{\rho(\lambda + 2\mu)}$ (or $k_s = \sqrt{\rho\mu}$),

Acoustical Imaging, Volume 20 Edited by Y. Wei
and B. Gu, Plenum Press, New York, 1993

using P–S coupling wave information can inverse density ρ, so ρ, λ, μ are all determined.

This paper is organized as follows: the transformation and decomposition from the inverse problem for ρ, λ, μ into three independent inverse problems for single unknown coefficient in 1–D wave equation are given out in section II. The theoretical analytic results and the numerical test results are presented in section III and section IV respectively. Section V concludes by summarizing the results of the paper.

TRANSFORMATION AND DECOMPOSITION OF THE BASIC MODEL

Supposing the earth is a linear isotropic elastic lossless plane stratified medium and has only small displacement and no hypocentre inside it, for convenience, we study the elastic wave problem in a plane which is perpendicular to the surface of the earth. In the state of plane strain, 2–D elastic wave equation which describes the process of elastic oscillation propagation in the plane can be written in the following one order vector form

$$\left(A\frac{\partial}{\partial t} - \sum_{i=1}^{2} B_i \frac{\partial}{\partial x_i} \right)\vec{U} = 0 \qquad x_1 > 0, t > 0, -\infty < x_2 < \infty, \tag{1}$$

where $\vec{U} = (u_1, u_2, \sigma_{11}, \sigma_{12}, \sigma_{22})'$, (t, x_1, x_2) is time–point coordinate system, $x_1 > 0$ denotes the depth into the ground, with $x_1 = 0$ the ground surface, $(u_1, u_2)'$ particle velocity vector, σ_{ij} stess tensor, $a = (2\mu)^{-1}$, $b = \lambda[4\mu(\lambda + \mu)]^{-1}$,

$$A = \begin{bmatrix} \rho & 0 & 0 & 0 & 0 \\ 0 & \rho & 0 & 0 & 0 \\ 0 & 0 & a-b & 0 & -b \\ 0 & 0 & 0 & 2a & 0 \\ 0 & 0 & -b & 0 & a-b \end{bmatrix}, \quad B_1 = \begin{bmatrix} 0 & 0 & 1 & 0 & 0 \\ 0 & 0 & 0 & 1 & 0 \\ 1 & 0 & 0 & 0 & 0 \\ 0 & 1 & 0 & 0 & 0 \\ 0 & 0 & 0 & 0 & 0 \end{bmatrix}, \quad B_2 = \begin{bmatrix} 0 & 0 & 0 & 1 & 0 \\ 0 & 0 & 0 & 0 & 1 \\ 0 & 0 & 0 & 0 & 0 \\ 1 & 0 & 0 & 0 & 0 \\ 0 & 1 & 0 & 0 & 0 \end{bmatrix},$$

the density ρ, Lame's parameter λ and shear modulus μ are fuctions of only x_1.

The initial condition is taken of the form

$$\vec{U}\big|_{t<0} = 0, \qquad x_1 > 0, \ -\infty < x_2 < \infty. \tag{2}$$

The boundary conditions

$$\sigma_{1j}(0, x_2, t) = \delta(x_2, t), \qquad t > 0, \ -\infty < x_2 < \infty, \ j = 1,2 \tag{3}$$

are called the coupling pulse wave source conditions and the boundary conditions

$$u_j(0, x_2, t) = g_j(x_2, t), \qquad t > 0, \ -\infty < x_2 < \infty, \ j = 1,2 \tag{4}$$

638

are called the additional conditions which are particle velocities measured on the line of $\{\,(x_1, x_2)\mid x_1 = 0\,\}$ in two directions of the coordinate axes x_1 and x_2.

Equations (1)–(4) form a nonlinear inverse problem for $\rho(x_1), \lambda(x_1)$ and $\mu(x_1)$. It is the basic model in the paper.

Let $\tilde{U}(x_1, v, t)$ be the Fourier transformation of $\vec{U}(x_1, x_2, t)$,

$$\tilde{U} = R\vec{V}, \qquad z = \int_0^{x_1} \rho(\xi)\,d\xi,$$

$$R = \begin{bmatrix}
0 & 1/\sqrt{2} & 0 & 1/\sqrt{2} & 0 \\[4pt]
1/\sqrt{2} & 0 & 0 & 0 & 1/\sqrt{2} \\[4pt]
0 & k_s/\sqrt{2} & 0 & -k_s/\sqrt{2} & 0 \\[4pt]
k_p/\sqrt{2} & 0 & 0 & 0 & -k_p/\sqrt{2} \\[8pt]
0 & \dfrac{bk_s}{\sqrt{2}\,(a-b)} & \sqrt{\dfrac{\rho}{a-b}} & \dfrac{-bk_s}{\sqrt{2}\,(a-b)} & 0
\end{bmatrix}$$

$$\vec{W} = \vec{V}\big|_{v=0} \equiv (W_1, W_2, W_3, W_4, W_5)',$$

$$\vec{Y} \equiv \frac{\partial \vec{V}}{\partial v}\bigg|_{v=0} \equiv (Y_1, Y_2, Y_3, Y_4, Y_5)',$$

We can separate problem (1)–(4) into the following inverse problems for single unknown coefficient in 1–D wave equation

$$\left(\frac{\partial}{\partial t} - k_s\frac{\partial}{\partial z}\right)W_1 - \frac{k'_s}{2}(W_1 - W_5) = 0$$

$$\left(\frac{\partial}{\partial t} + k_s\frac{\partial}{\partial z}\right)W_5 - \frac{k'_s}{2}(W_1 - W_5) = 0$$

$$W_1\big|_{t<0} = W_5\big|_{t<0} = 0 \tag{5}$$

$$(W_1 - W_5)\big|_{z=0} = \frac{\sqrt{2}}{k_s(z)}\,\tilde{\sigma}_{12}(z,v,t)\big|_{z=v=0} = \frac{\sqrt{2}}{k_s(0)}\delta(t)$$

$$(W_1 + W_5)\big|_{z=0} = \sqrt{2}\,\tilde{u}_2(z,v,t)\big|_{z=v=0} = \sqrt{2}\,\tilde{g}_2(0,t)$$

$$\left(\frac{\partial}{\partial t} - k_p\frac{\partial}{\partial z}\right)W_2 - \frac{k'_p}{2}(W_2 - W_4) = 0$$

$$\left(\frac{\partial}{\partial t} + k_p\frac{\partial}{\partial z}\right)W_4 - \frac{k'_p}{2}(W_2 - W_4) = 0$$

$$W_2\big|_{t<0} = W_4\big|_{t<0} = 0 \tag{6}$$

$$(W_2 - W_4)\big|_{z=0} = \frac{\sqrt{2}}{k_p(z)}\,\tilde{\sigma}_{11}(z,v,t)\big|_{z=v=0} = \frac{\sqrt{2}}{k_p(0)}\delta(t)$$

$$(W_2 + W_4)\big|_{z=0} = \sqrt{2}\,\tilde{u}_1(z,v,t)\big|_{z=v=0} = \sqrt{2}\,\tilde{g}_1(0,t)$$

$$\left(\frac{\partial}{\partial t} - k_s \frac{\partial}{\partial z}\right) Y_1 - \frac{k_s'}{2}(Y_1 - Y_s) + \frac{\theta_1}{\rho} W_2 - \frac{\theta_2}{\rho} W_4 = 0$$

$$\left(\frac{\partial}{\partial t} + k_s \frac{\partial}{\partial z}\right) Y_s - \frac{k_s'}{2}(Y_1 - Y_s) + \frac{\theta_2}{\rho} W_2 - \frac{\theta_1}{\rho} W_4 = 0$$

$$Y_1\big|_{t<0} \equiv Y_s\big|_{t<0} = 0$$

$$(Y_1 - Y_s)\big|_{z=0} = 0 \tag{7}$$

$$(Y_1 + Y_s)\big|_{z=0} = \sqrt{2}\frac{\partial}{\partial v} \tilde{g}_2(v,t)\big|_{v=0} \equiv 2\tilde{g}(t),$$

where $\theta_1 = \dfrac{i}{2k_p}(k_p - k_s)(2k_s + k_p)$, $\theta_2 = \dfrac{i}{2k_p}(k_p + k_s)(k_p - 2k_s)$, k_s' and k_p' denote the derivative functions of k_s and k_p respectively. We shall see that $k_s(z)$ and $k_p(z)$ can be determined by solving inverse problem (5) and (6) respectively. As $k_s(z)$ and $k_p(z)$ are known, $\theta_1(z)$ and $\theta_2(z)$ are also known. So we can determine $\rho(z)$ by solving inverse problem (7).

THEORETICAL ANALYSIS

First, we study the well–posed properties of inverse problems (5), (6) and (7) in theory. Through singularity analysis and strict mathematical proof by the theory of P.D.E., we know that as $\rho(z)$, $k_p(z)$, $k_s(z)$ and their derivatives are all piecewise continuous functions of z, $k_s(z)$ can be determined by solving inverse problem (5), $k_p(z)$ can be determined by inverse problem (6) and $\rho(z)$ can be determined by inverse problem (7). In this case, inverse problems (5),(6) and (7) are all well–posed. Having known $\rho(z)$, $k_s(z)$ and $k_p(z)$, we can easily determine $\rho(x_1)$, $\lambda(x_1)$ and $\mu(x_1)$ by means of the following formulas $x_1 = \int_0^z \frac{1}{\rho(z)} dz$, $\mu = \dfrac{k_s^2}{\rho(z)}$, $\lambda = \dfrac{k_p^2 - 2k_s^2}{\rho(z)}$.

NUMERICAL TEST RESULTS

Using the method of characteristic and the following constitution relationship equations

$$W_1\left(z, \int_0^z k_s^{-1}(\xi)d\xi\right) = \frac{\sqrt{2}}{4}\frac{k_s'(z)}{\sqrt{k_s(0)k_s(z)}}$$

$$W_2\left(z, \int_0^z k_p^{-1}(\xi)d\xi\right) = \frac{\sqrt{2}}{4}\frac{k_p'(z)}{\sqrt{k_p(0)k_p(z)}}$$

$$Y_1\left(z, \int_0^z k_p^{-1}(\xi)d\xi\right) = -\frac{i}{\rho}\sqrt{\frac{1}{2k_p(z)k_p(0)}}[k_p(z) - 2k_s(z)]$$

we set up the downward recursive algorithms for solving $k_s(z)$ and $k_p(z)$ in inverse problems (5) and (6) , the downward recursive algorithm for solving $\rho(z)$ in inverse problem (7) and some computerized numerical tests. Numerical results of these downward recursive algorithms are presented in table 1 and table 2 .

Table 1   Inversion for $k_s(z)(\text{or} k_p(z))$ by (5) (or (6))

$k_s(z)$	inversion interval [0,T]	mesh width $\triangle$	noisy parameter d,d_1	the maximum of absolute errors	the maximum of relative errors(%)
$1+z$	[0,2]	0.01	0	0.0001	0.0005
			$d=0.1$	0.001	0.5
			$d_1=0.4$	0.02	1.0
$1+0.5sin(10z)$	[0,2]	0.01	$d=0$	0.009	0.6
			$d=0.05$	0.009	0.5
			$d_1=0.3$	0.028	2

Table 2   Inversion for $\rho(z)$ by (7) (as $(k_s(z)=1+z,\ \ k_p(z)=1.5k_s(z))$

$\rho(z)$	inversion interval [0,T]	mesh width $\triangle$	noisy parameter d,d_1	the maximum of absolute errors	the maximum of relative errors(%)
$(1+0.5z)^{-1}$	[0,2]	0.01	0	0.0001	0.005
			$d=0.03$	0.019	1.2
			$d_1=0.3$	0.075	5
$1+z^2$	[0,2]	0.01	0	0.0004	0.04
			$d_1=0.3$	0.05	6
$(1+0.5sin(10z))^{-1}$	[0,2]	0.01	0	0.0055	0.4
			$d=0.03$	0.0214	1.5
			$d_1=0.3$	0.08	6

where d and d_1 are artificial noisy parameters which are used to generate artificial noises on the input data $W_j(0,t)$, j= 1, 2, 4, 5, $Y_j(0,t)$, j= 1, 5 in the following two ways

(1) $W_j(0,(2N-1)\triangle)=W_j(0,(2N-1)\triangle)[1+(-1)^{N+1}d]$,

(2) $W_j(0,(2N-1)\triangle)=W_j(0,(2N-1)\triangle)[1+(-1)^{N+1}d_1 R(2N-1)]$,

and $\{R(2N-1)\}$ is a random number sequence on the interval [0,1].

These numerical test results show that the accuracy and stability of the downward recursive algrithms are all good.

CONCLUSIONS

It has been shown in the paper that the density,Lame's parameter and shear modulus in the lossless layered earth model can be uniquely determined from the particle velocity fields on the line in its surface at all time due to the transverse and longitudinal pulse plane waves simultaneously at normal incidence. The theoretical analysis and numerical test results show that the inverse problems for k_p, k_s and ρ are all well–posed. It is noticed that the necessary and sufficient condition to uniquely determine ρ, λ, μ are $\tilde{g}_1(0,t)$, $\tilde{g}_2(0,t)$ and $\left.\frac{\partial}{\partial v}\tilde{g}_2(v,t)\right|_{v=0}$. These are only three moment information of the $u_1(0,x_2,t)$ and $u_2(0,x_2,t)$ with respect to x_2. Combining with the initial condition (2) and the boundary source condition (3), we also know that these moment information fully decide the ρ, λ, μ and the whole wave field in the lossless layered earth model.

REFERENCES

1 R.G.Newton, "Inversion of reflection data for layered media, a review of exact methods", Geophys J.R. Astr. Soc., vol. 65, pp191–215, 1981

2 G.Q. Xie, "A new iterative method for solving the coefficient inverse problem of the wave equation", Comm. Pure and Appl. Math., XXXIX(2), pp307–332,1986

3 P.B.Kenneth, "The one–dimensional inverse problem of reflection seismology", SIAM. Review, vol25(4), pp497–559, 1983

4 P.B.Kenneth, "Convergence of numerical inversion methods for discontinuous impedance profiles", SIAM. J. Numer. Anal., vol22(5), pp924–946, 1985

5 P.B.Kenneth, "Numerical methods for reflection inverse problems, convergence and non–impulsive sources", SIAM. J. Numer. Anal., vol23(2), pp227–258, 1986

6 G.Q.Zheng,"Inverse Problem in One Dimensional Wave Equation",Sci. Sinica (A)no7, pp701–721,1988

A LAYER-STRIPPING ALGORITHM
IN MULTI-DIMENSIONAL RAY TRAVEL-TIME INVERSION

Hong Liu, Youming Li, Fanling Meng and Xing Gao

Institute of Geophysics
Chinese Academy of Sciences
Beijing, 100101, Box 9701
P. R. China

INTRODUCTION

The purpose of this paper is giving a new method for the inversion of ray travel-time. The inversion of ray travel-time is a nonlinear problem which is conventionally solved by iteration method with step-by-step linearization [1]. Although the iteration method has many advantages, the convergence problem can not be avoided, which may lead to long computing time and multi-solutions. In this paper, we try another method for the nonlinear problem, that is recursion method. The problem is solved without iteration process, therefore, the computing time is saved and unique solution can be obtained.

In this paper, only complete-date reconstruction is considered, it is assumed that the travel-times of ray between any two points on the boundary of image region are known. During inversion process, by means of the finite difference formula of the eikonal equation, the slowness of the layer nearest to the outer boundary of the image region is evaluated firstly. Then the travel-times between the points on the outer and the inner boundaries of the layer are solved. After doing this, the travel-times between the points on the inner boundary of the layer are worked out. In this way, the problem return to otherwise original problem but with a contracted image region, so that the process above repeats again and the slowness in the new layer can be inversed. This procedure is continued until the slowness in all image region is worked out.

In next sections, the fundamental idea of the layer-stripping is introduced, then a practical algorithm is described, in the end the computation result is showed.

FUNDAMENTAL IDEA OF LAYER-STRIPPING

For the sake of understanding, we introduce the layer-stripping method on a rectangle grid.

We assume that the travel-times between any two points on the boundary of image region are known. The slowness in every rectangle cell is assumed homogeneous and unknown. By means of a finite-difference approximation of eikonal equation, the formulations of inversion and the travel-time extrapolation are given. The procedure of layer stripping is on the base of those formulations.

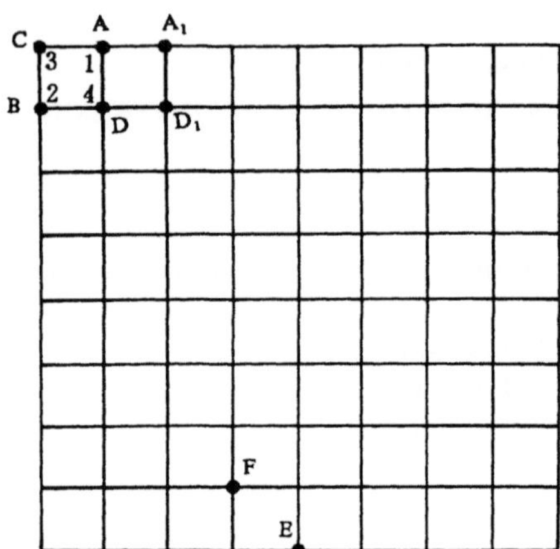

Figure 1.   Sketch map of inner-layer recursion and inter-layer recursion

The eikonal equation is a fundamental equation of ray theory which is high frequency approximation of wave equation[2] and can be written as

$$\left(\frac{\partial T}{\partial x}\right)^2 + \left(\frac{\partial T}{\partial y}\right)^2 = S^2 \tag{1}$$

or

$$(\text{grad } T)^2 = S^2$$

where T is the travel-time function, S is the slowness function, x and y are two rectangle coordinates. By replacing the partial differentiation with finite difference, the equation can be discretized. As in Figure 1, points 1, 2, 3 and 4 are vertexes of a square, in which points 1 and 2 are the vertexes of a diagonal of the square, and points 3 and 4 are the vertexes of the other diagonal of the square; d is the sidelength of the square, S is slowness in the square.

$$\left[\frac{T_4 - T_3}{\sqrt{2}\,d}\right]^2 + \left[\frac{T_2 - T_1}{\sqrt{2}\,d}\right]^2 = S^2 \tag{2}$$

From which we have

$$T_4 = T_3 \pm \sqrt{2d^2 S^2 - (T_2 - T_1)^2} \tag{3}$$

Because points 1, 2, 3 and 4 are arbitrary observation points, so Eq. (3) can be used wherever the source point is.

In Eq. (3), the sign before the square root can be determined according to that which one of points 3 and 4 is near the source point.

Eq. (3) is used to extrapolate travel-time function.

When the source point and the observation point are near enough, say, the source is at point 1 and the observation is at point 2 in Figure 1, we have $T_4 = T_3$, $T_1 = 0$, therefore from Eq. (2)

$$S = \frac{T_2}{\sqrt{2}\,d} \equiv \frac{T_{12}}{\sqrt{2}\,d} \tag{4}$$

where T_{12} indicates that the travel-time is observed at point 2 and the source is at point 1. We will use formula (4) to evaluate the slowness.

The fundamental idea of layer-stripping is extrapolating the travel-time continuously in order to get the travel-time between two neighbor points, then to inverse the slowness.

As in Figure 1, the large square is a image region. The travel-times between every two points on the side of the region are known. The layer to be firstly inversed is consisted of squares near the side of the region. Points A, A_1, C, B and E are on the outer boundary of the layer. D, D_1 and F on the inner boundary of the layer. A, C, B and D are four vertexes of square 1, and A_1, A, D and D_1 are four vertexes of square 2.

According to the assumption, both A and B are on the outer boundary of the layer, so the travel-time between A and B is known, by Eq. (4), we have

$$S_1 = \frac{T_{AB}}{\sqrt{2}\,d} \tag{5}$$

where S_1 is the slowness in square 1.

So, S_1 can be evaluated. Also, according to the assumption, the travel-times from an arbitrary point E on the outer boundary to points A, C and B are known by Eq. (3), we have

$$T_{ED} = T_{EC} \pm \sqrt{S_1^2 2d^2 - (T_{EA} - T_{EB})^2} \tag{6}$$

If point E passes through all the points on the outer boundary, then all the travel-times from the points on the outer boundary to point D can be obtained, So the travel-time from point A_1 to point D can be obtained, then we get

$$S_1 = \frac{T_{AB}}{\sqrt{2}\,d} \tag{7}$$

Because S_2 is known, we can obtain the travel-time from any point on the outer boundary to point D_1 by extrapolating from travel-time EA, EA_1 and ED. The recursion can be repeated among the squares of the layer. Then, the slowness of the layer can be solved out. In the same time, the travel-time from the points on the outer boundary of the layer to inner boundary are known.

According to reciprocity theorem, the travel-time from F to A is equal to the travel-time from A to F. So, from the above procedure, travel-times T_{FA}, T_{FC} and T_{FB} from any point F on the inner boundary of the layer to outer boundary A, C and B are known. Therefore T_{FD} can be worked out by Eq. (4).

Let point F pass through all the points on the inner boundary. The travel-time between any two points on the inner boundary can be worked out.

Now, the problem left return to the original one, but the image region is different. The first outer layer is stripped, the image region is contracted. The travel-time between any two points on the boundary of new image region is known. Then a new recursion can be carried out. When all the recursions are finished, the slowness of the whole image region will be solved out.

PRACTICAL ALGORITHM

There are two types of recursion in the above algorithm. One type is inner layer recursion. Another is inter-layer recursion. After testing the algorithm above, we found that the inner layer recursion causes errors accumulating quickly. In order to reduce the accumulated error, we use a more rough but more stable algorithm in which only inter- layer recursions are remained.

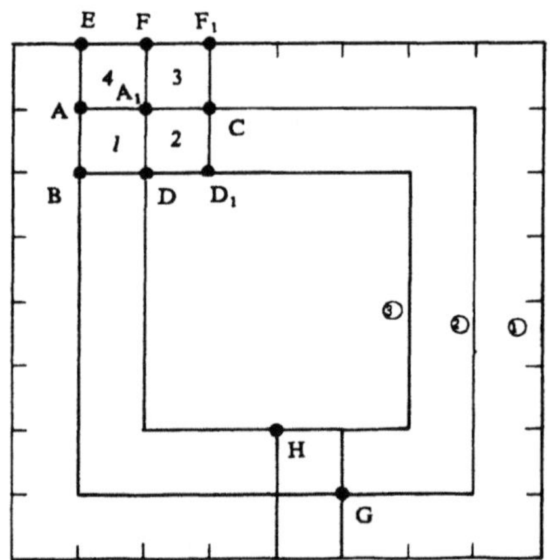

Figure 2.   Sketch map of only inter-layer recursion

Let us look at three boundaries in Figure 2. Boundaries ① and ② are the outer and inner boundaries of the first layer, respectively. Boundary ③ is the inner boundary of the second layer. Points E, F and F_1 are on boundary ①; C, A and A_1 are on boundary ②; B, D and D_1 are on boundary ③. We assume that the slowness of the first layer have been worked out and the travel-time between any two points on boundaries ① and ② are known.

We use the travel-time between points A and C and that between points A and A_1 to get the slowness in squares 1 and 2 of the second layer.

We can approximate that

$$S_1 = \frac{T_{AC}}{d} \, , \qquad\qquad S_2 = \frac{T_{AA_1}}{d} \tag{8}$$

Then, for any source point G on boundary ②, the travel-time from source point G to point D can be worked out by approximate formula

$$T_D = T_F \pm \sqrt{4d^2\left(\frac{S_1 + S_2 + S_3 + S_4}{4}\right)^2 + (T_{A_1} - T_C)^2} \tag{9}$$

646

After all the travel-times between the points on boundaries ② and ③ are obtained, the source point H is moved to boundary ③. Then, by Eq. (9) again, the travel-time between any points on boundary ③ will be known by extrapolation.

Our practice show that the algorithm given in this section is more stable than the algorithm given in last section, because the inner layer recursion is avoided.

FORWARD AND INVERSE EXAMPLE

We will give some examples to show the algorithm introduced in last section. The data for inversion are calculated by forward computation. We will describe the ray-tracing method we used in the forward computation firstly.

The forward travel-time algorithm is called interface grid minimum travel-time ray-tracing. The principle of the algorithm is based on Fermat's principle of minimum travel-time. The algorithm also considers the theorem that if a trajectory AB is the shortest path between points A and B, then any section CD of the trajectory are the shortest path between points C and D.

Our method is similar to Mosor's algorithm[3]. The difference is that we only discretize the interface between any two pixels, not discretize the pixel.

The model of slowness is divided into different square pixel. The slowness in every pixel is constant. The side of every square is divided into several sections, say, n sections. There are $n + 1$ discrete points on every side, as shown in Figure 3.

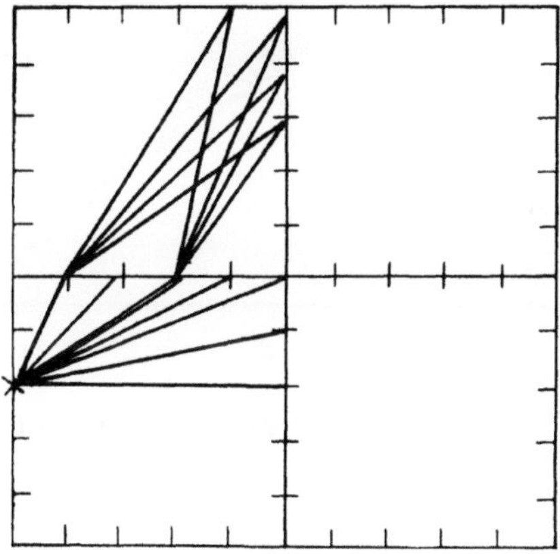

Figure 3.   Sketch map of the interface grid minimum travel-time ray-tracing ($n = 5$)

The algorithm can be divided into two parts, forward accumulation and backward tracing. During the forward accumulation, the information on the travel-time and secondary source co-ordinates is recorded from which ray comes to the observation point. During the backward tracing, secondary sources are traced to obtain the ray trajectory.

For a source put on any position, firstly, the travel-time from the source to the discrete point on the side of the square, on which the source is put, is worked out by simple arithmetic calculation. Then, the secondary source points are moved to the point which the travel-times have been worked out. The travel-time from unknown point to the source point passing through the secondary source point is given by the travel-time of the secondary source added with travel-time between secondary source and an unknown point. This travel-time is compared with the travel-time stored in the unknown point and the shorter travel-time and its secondary source coordinates are recorded. In this way, when every secondary source on the side of calculation square has been computed, the travel-time and the secondary source coordinates on every unknown point will be the shortest travel-time and true secondary source coordinates. When secondary source points have passed through all discrete points of image region, the forward accumulation procedure finished.

The true ray-trajectory is found by back tracing. From any discrete point, its secondary source point can be found, which is its last ray point that should be connected. From the point found, its secondary point also can be found. In this way, ray trajectory can be traced.

Figure 4 is a slowness model. It is divided into 32×32 squares. Figure 4 shows a ray-tracing result of a source.

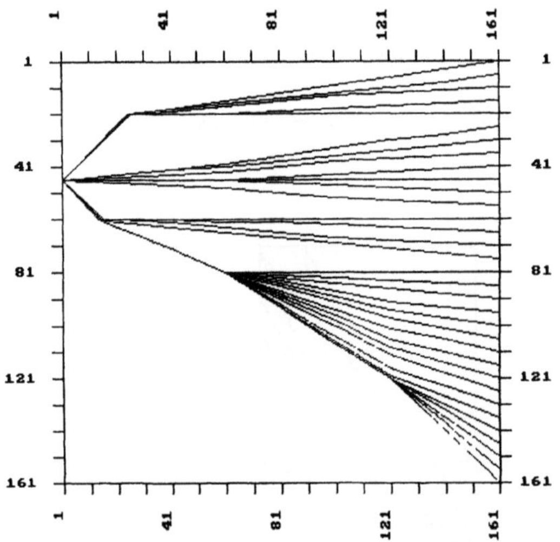

Figure 4.   Result of the interface grid minimum travel-time ray-tracing

Applying the ray-tracing algorithm above, we calculate the travel-time data to test the layer-stripping inversion. For a 32×32 pixel model, there are 128 boundary points.We get $128 \times (128-1)/2$ travel-time data.

We make inversion for slowness from those data. The calculating time of the layer-stripping is compared with that of conventional iteration method[1, 4]. On AST386 personal computer, the layer-stripping inversion for 32×32 slowness pixel need about 1 miniute, but the 10 times iteration inversions need about 1 hour.

We found that the algorithm of layer-stripping is very unstable. In this stage of algorithm, better results occur in the case that the number of discrete pixels is less.

Figure 5 is a model with 8×8 pixels. Figure 6 is the result inversed by the layer stripping algorithm. Because the number of discrete pixel is less, we get a good result.

Let us sum up this section. From the research now, we know that the main problem of the layer-stripping method is how to keep stable. When the recursion have more steps, the results of inversion will become worse. The computation of the algorithm is very fast. In order to make all

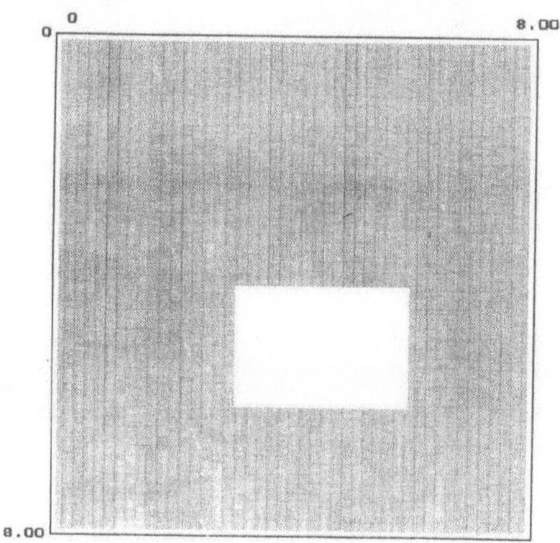

Figure 5.   Supposed slowness model

the algorithms stable, it is desired to control the part which had been inversed by the part which hadn't been inversed. It means that the inversion algorithm may become a implicit version other than the explicit version now.

CONCLUSION

This paper introduced a recursion algorithm for the ray travel-time inversion. The recursion algorithm based on the finite difference approximation of eikonal equation and a layer-stripping procedure. The algorithm is a nonlinear algorithm and in principle, the slowness is not necessary to be small perturbation. The forward date are calculated by the interface grid minimum travel-time ray tracing. The inversion of the data is carried out by the layer-stripping algorithm. The example show that the algorithm is very fast. Further work is needed to improve the stableness of the algorithm.

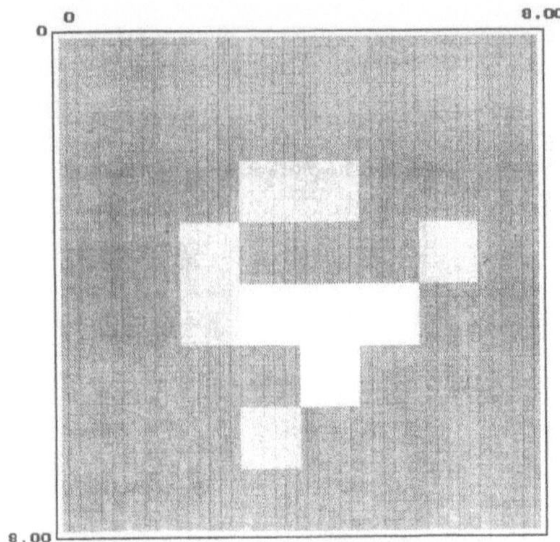

Figure 6.   Inversion result of layer-stripping.

REFERENCES

1.   N. D. Bregman, R. C. Bailey, C. H. Chapman, Crosshole seismic tomography, *Geophysics*, 54: 200—215 (1989).
2.   V. Cerveny, I. A. Molotkov, I. Psencik, "Ray Method in Seismdogy", Prague: Univerzita Karlova, (1977).
3.   T. J. Moser, Shortest path calculation of seismic rays, *Geophysics*, 56: 59—67 (1991).
4.   H. Liu, Y. Li, X. Gao, An improvement on the maximum or minimum cross entropy method in the cross-well ray tomography, in: "Acoustical Imaging", Plenum publishing corporation, New York (1993).

WELL–CONTROL INVERSION OF MECHANICAL PARAMETERS

Jiaqi Liu and Shoudong Wang

Dept. of Mathematics
Harbin Institute of Technology
P. R. China

SUMMARY

We present a new method for the inversion of the impedance between two wells and out–line the theoretical background. The main differences from other inverse techniques are that it not only uses the surface seismic data but also uses the impedance in two wells. It uses the $1-D$ inverse method to solve $2-D$ inverse problem. To solve the functional extreme value problem, the damped least–square method is used. The method has been tested on many numerical examples. The results show that the method has higher accuracy and stableness. The method will be applied to the data obtained from the practice.

INTRODUCTION

At present, the inversion of wave equation is placed in difficult circumstances to the practice,the main reasons are that
(1) The inversion is ill–posed.
(2) The wavelet can not be determined precisely.
(3) The given data do not fit the mathematical model and have the higher noise level.
(4) The computing cost to inverse the $2-D$ or $3-D$ wave equation is high. Where(1), (2) are caused by the lack of the known information.

The common inverse method for wave equation is to determine the parameters only using the surface seismic data, not usi ng the other information. In the practice, the surface seismic data have the higher noise level, so the inverse results is poor. It is necessary to use the other information in the inverse process. In this paper, we utilize the well log, which is available in the practice, in our inverse process.

THEORY

The $1-D$ wave equation model for the inversion can be written as

$$\frac{\partial}{\partial z}\left(k(z)\frac{\partial u}{\partial z}\right)=\rho(z)\frac{\partial^2 u}{\partial t^2} \qquad\qquad L>z>0\ ,\qquad t>0 \qquad\qquad (1)$$

$$u|_{t=0}=\frac{\partial u}{\partial t}\Big|_{t=0}=0 \qquad\qquad L\geqslant z\geqslant 0 \qquad\qquad (2)$$

$$k(z)\frac{\partial u}{\partial z}\Big|_{z=0}=g_1(t) \qquad\qquad t>0 \qquad\qquad (3)$$

$$u|_{z=L}=0 \qquad\qquad t>0 \qquad\qquad (4)$$

$$u|_{z=0}=f(t) \qquad\qquad t>0 \qquad\qquad (5)$$

where u is the displacement ; $g_1(t)$ is source function; $f(t)$ is the data received on the surface, the additional conditions for the inversion; z is the vertical coordinate; $k(z)$ is the elastic coefficient of the media; $\rho(z)$ is the density of the media. In the inversion, $k(z)$ and $\rho(z)$ are unknown and the objective of the inversion is to calculate $k(z)$ and $\rho(z)$.

Generally, the following transform is introduced

$$c(z)=\sqrt{k(z)/\rho(z)}$$

$$x(z)=\int_0^z \frac{1}{c(s)}\,ds$$

$$\sigma(z)=\sqrt{\rho(z)\times k(z)}$$

This can transform $(1)-(5)$ into

$$\frac{\partial}{\partial x}\left(\sigma(x)\frac{\partial u}{\partial x}\right)=\sigma(x)\frac{\partial^2 u}{\partial t^2} \qquad\qquad T>x>0\ ,\qquad t>0 \qquad\qquad (1)'$$

$$u|_{t=0}=\frac{\partial u}{\partial t}\Big|_{t=0}=0 \qquad\qquad T\geqslant x\geqslant 0 \qquad\qquad (2)'$$

$$\sigma(x)\frac{\partial u}{\partial x}\Big|_{x=0}=g(t) \qquad\qquad t>0 \qquad\qquad (3)'$$

$$u|_{x=T}=0 \qquad\qquad t>0 \qquad\qquad (4)'$$

$$u|_{x=0}=f(t) \qquad\qquad t>0 \qquad\qquad (5)'$$

where $c(z)$ is the wave speed; $x(z)$ is the traveltime coordinate; $\sigma(x)$ is the acoustic impedance of the media; T is the time that it takes for the wave to propagate to L. Thus there is only one unknown parameter $\sigma(x)$. We assume that the media underground are approximately horizontal and layered. Let $\sigma_i(x)$ denote the $i-$th trace impedance. The position of the well is at first trace, then $\sigma_i(x)$ can be written as the piecewise function

$$\sigma_1(x)=\begin{cases}\sigma^{(1)} & 0\leqslant x<s_1^{(1)}\\[4pt]\sigma^{(2)} & s_1^{(1)}\leqslant x\leqslant s_2^{(1)}\\[2pt]\ \vdots & \qquad\vdots\\[2pt]\sigma^{(n)} & s_{n-1}^{(1)}\leqslant x\leqslant T\end{cases} \qquad\qquad (6)$$

where $\sigma^{(i)}$ is the value of $\sigma_i(x)$ in the $i-$th layer; $s_i^{(1)}$ is the location of the interface between the $i-$th layer and the $(i+1)-$th layer, so the $\sigma_2(x)$ can be written as

$$\sigma_2(x) = \begin{cases} \sigma^{(1)} & 0 \leqslant x < s_1^{(2)} \\ \sigma^{(2)} & s_1^{(2)} \leqslant x \leqslant s_2^{(2)} \\ \vdots & \vdots \\ \sigma^{(n)} & s_{n-1}^{(2)} \leqslant x \leqslant T \end{cases} \qquad (7)$$

where the $\sigma^{(i)}$ is known, so the problem of inversing $\sigma_2(x)$ is transformed into the problem of inversing $s_i^{(2)}$, $s_i^{(2)}$ can be calculated through minimizing the function

$$\| N(s^{(2)}) - f(t) \| = min \qquad (8)$$

where $N(s^{(2)})$ represents a maping defined by $(1)'-(4)'$ and (6), $s^{(2)}$ is a vector consisting of $s_i^{(2)}$. (8) can be solved by damped least−square method.

Through the similar process, $s_i^{(3)}$, $s_i^{(4)}$, $\cdots$, $s_i^{(m)}$ can all be calculated. if $\sigma_m(x)$ is also known, by the similar method, $\sigma_{m-1}(x)$, $\sigma_{m-2}(x)$, $\cdots$, $\sigma_1(x)$ can be calculated from $\sigma_m(x)$. Thus, the layers that pass through the well at $m-$th trace and do not pass through the well at first trace can be found out.

In order to find out the layers that neither pass through the well at first trace nor the well at $m-$th trace, we can further inverse the impedance point by point under the binding condition−the impedance that has just been calculated. The binding inversion can be achieved by substituting (8) with

$$\| N'(\sigma_i(x)) - f(t) \| + \alpha_1 \|\sigma_i(x) - \sigma_c \| + \alpha_2 A(\sigma_i(x)) = min \qquad (9)$$

where $N'(\sigma_i(x))$ represents a maping defined by $(1)'-(4)'$; α_1, α_2 are changeable factors and are identified by the experience; σ_c is the impedance that has just been calculated; $A(\sigma_i(x))$ is a stable function.

NUMERICAL EXAMPLE

To demonstrate the effectiveness of the inversion method, we show one of the many numerical calculations we have performed. The true model is shown in figure 1. The parameters of every region are shown in table 1. Figure 2 is the synthetic seismic profile. It is made by $2-D$ wave equation and the incident wave is the planewave.

Table 1. The parameters of the media

Number of the region	Density $\rho(10^3 \text{kg/m}^3)$	velocity v(m/s)
1	2.18	2800
2	2.22	3000
3	2.18	2800
4	2.26	3200
5	2.22	3000
6	2.26	3200
7	2.31	3500
8	2.22	3000
9	2.29	3400
10	2.26	3200
11	2.31	3500

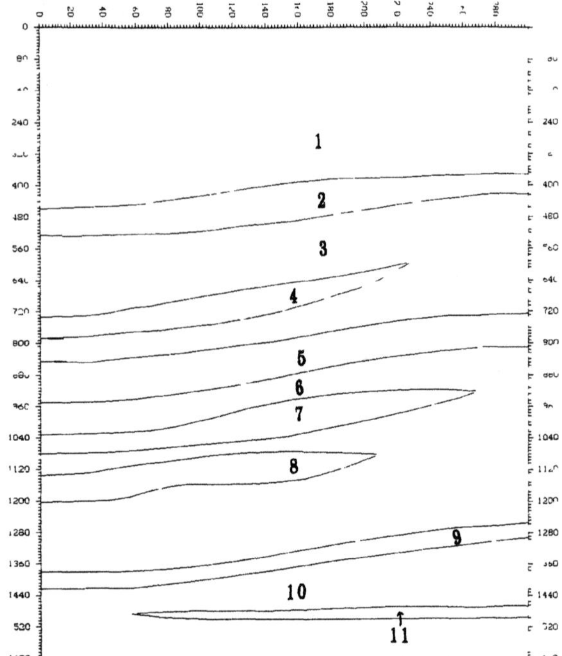

Figure 1. The true model

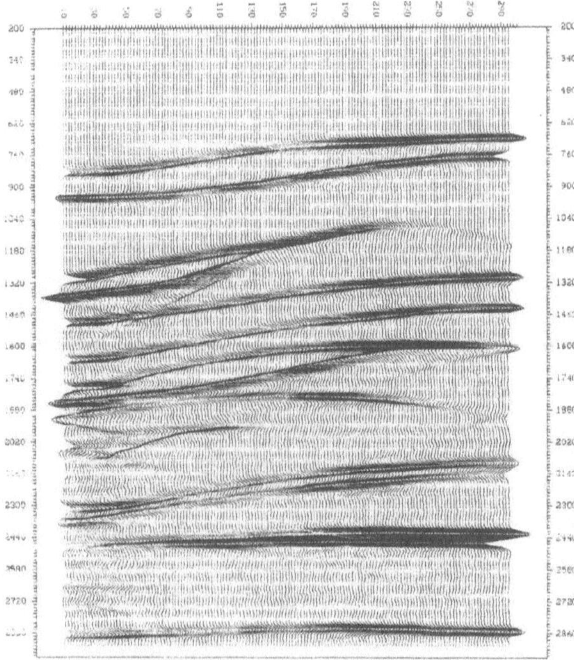

Figure 2. The synthetic seismic Profile

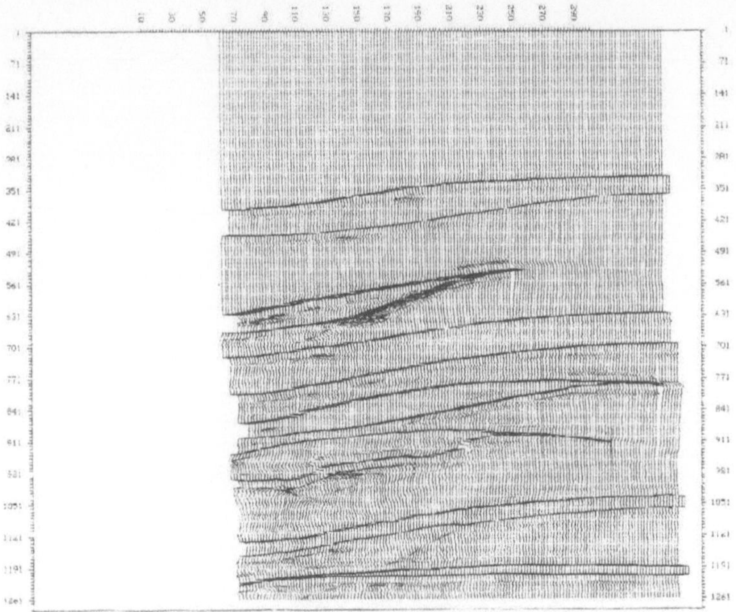

Figure 3. The results of the inversion of the
seismic data

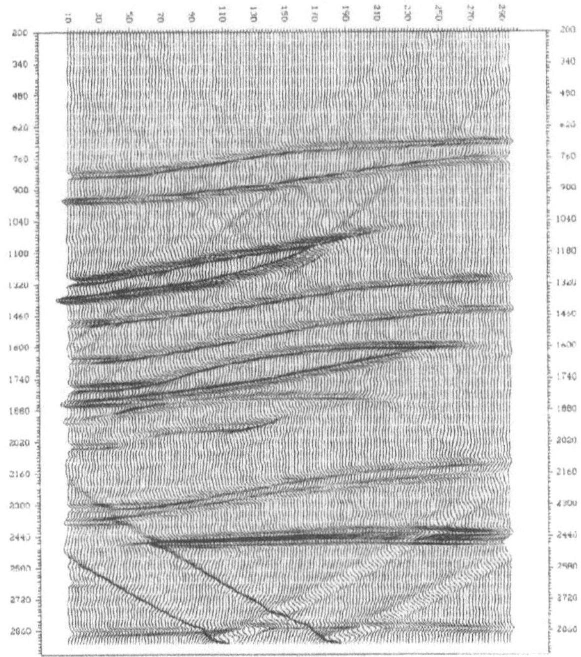

Figure 4. The migration profile of figure 2

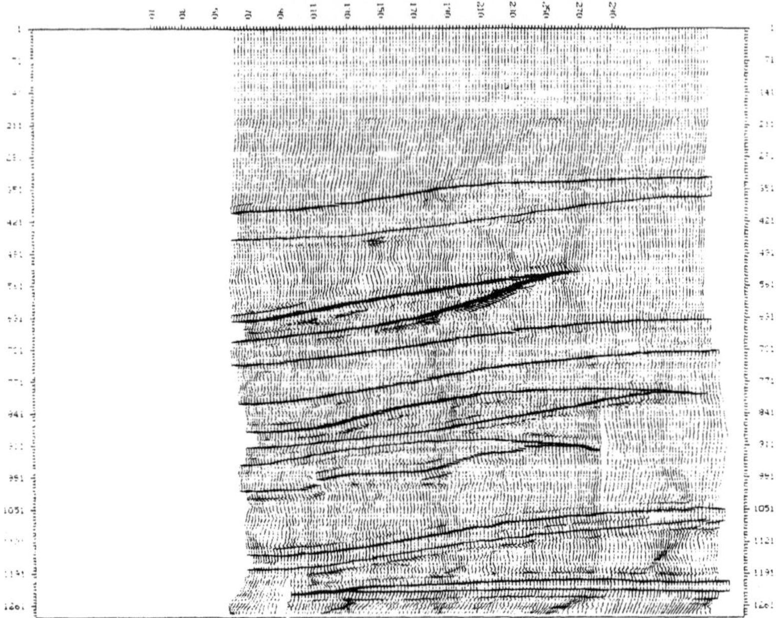

Figure 5. The results of the inversion of the migration data

The inverse procedure is as follows
1. Inverse the interfaces from the left to the right.
2. Inverse the interfaces from the right to the left .
3. Binding inverse point by point.

Figure 3 shows the results of the inversion of the seismic data. Figure 4 is the migration profile of figure 2. The results of the inversion of the migration data are shown in figure 5.

CONCLUSION

We have developed a new method for the inversion of the impedance between two wells. The numerical results show that our method have higher accuracy and stableness.

ACKNOWLEDGMENTS

We thank the GRI, Bureu of Oil Geophysical Prospecting for the use of its computing facilities. We also thank prof. Zhenxing Fan for very useful discussions.

REFERENCE

Cohen, J. K. and Bleistein, N., 1979, Velocity inversion procedures for acoustic wave, geophysics, 44, 1077 − 1085.

Tarantola, A., 1987, Inversion problem theory: method for data fitting and model parameter estimation, Elsevier science Pub. B. V.

AN IMPROVEMENT ON THE MAXIMUM OR MINIMUM CROSS ENTROPY METHOD IN THE CROSSWELL RAY TOMOGRAPHY

Hong Liu, Youming Li and Xing Gao

Institute of Geophysics
Chinese Academy of Science
Beijing 100101, Box 9701
P. R. China

INTRODUCTION

In the ray tomography and diffraction tomography, the methods of maximum entropy (ME) and minimum cross entropy (MICE) are widely used to improve the quality of the reconstruction image. [1,2] There are two kinds of ME or MICE methods: the gradient algorithm and dual algorithm. The gradient algorithm of ME or MICE method can be formulated into different forms according to different tomography types and inversion methods. [1] The dual algorithm[2] of ME or MICE method, however, has united form for ray tomography and diffraction tomography. This is because the only difference between different tomographies is the estimated region in the wave-number domain of image function. The dual algorithm proposed by Lo[2] has two shortcomings. Firstly, because the algorithm reconstructs the wave-number spectrum directly from observation data according to Fourier slice theorem, the interpolation in the spectrum domain are needed, which may reduce the computing efficiency. Secondly, when the estimated region in the wave-number domain is an angle region as in the straight ray tomography, the improvement of reconstruction image by the ME or MICE method is not significant. This paper will work on those two aspects.

This paper is divided into three sections. In first section, the fundamental idea of the ME or MICE is introduced. It is pointed out that when the dual algorithm is used, the known spectrum information can be obtained by 2-D Fourier transform from an image reconstructed by any inversion method without the ME or MICE method. The interpolation in the wave-number domain is avoided. In the second section, the analytic characters of dual algorithm of the ME or MICE method are discussed. Based on this, it is proved that the wave-number spectrum in the region estimated by the ME or MICE method will be zero, if the estimated region is an angle region. This is the reason why the improvement of image by the ME or MICE method is not significant. Considering on this, we give a new version of dual algorithm which can avoid the shortcoming. In the third section, the other parts of iteration inversion method of ray traveltimes are described briefly, the computing examples with the ME or MICE are showed

THE FUNDAMENTAL IDEA OF THE MAXIMUM OR MINIMUM CROSS ENTROPY

In this section, we review briefly the base idea that improves the quality of image by the ME or MICE method.

In the interwell tomography, due to incomplete information reconstructions, the computing image has distortion from true image. The incomplete information can be divided into two types, the first type is due to limited aperture and discrete sampling, which lead to ray overlay incomplete. The second type is incomplete angle overlay. The computation of Lo^2 shows that the influence of the first type on the image is small, the influence of second type is large. So we will pay attention to the second type. The Fourier slice theorem of straight ray tomography and diffraction tells us that due to the limited observation angles overlay, the wave-number spectrum in some region will lost. We hope that by the principle of ME or MICE, the information in

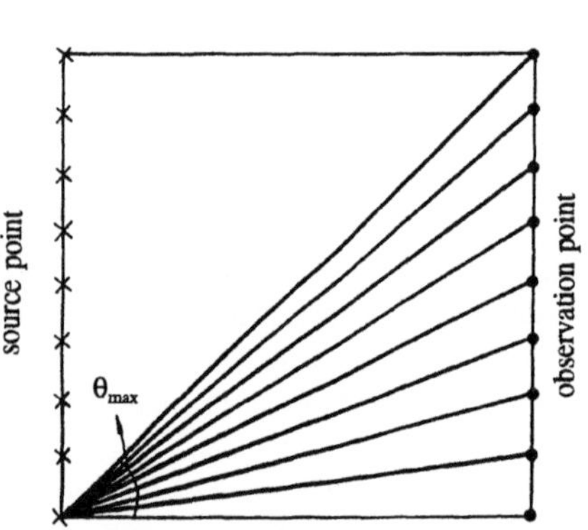

Figure 1. Observation angle limit
of cross-well ray tomography

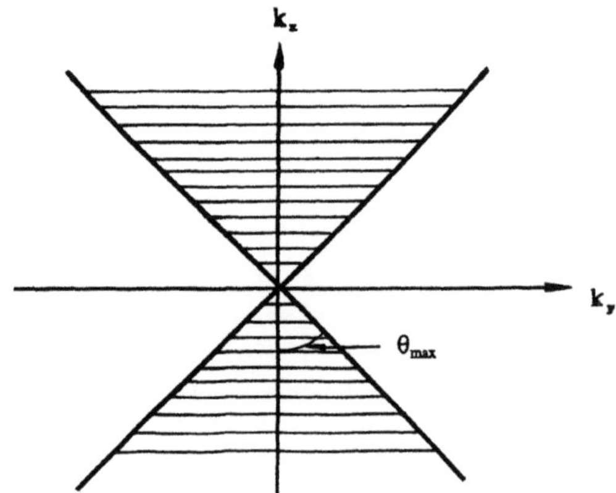

Figure 2. Incomplete overlay of wave-number
spectrum of slowness function

this region will be estimated. This is a constraint optimization problem. The constraint condition is observation data. The object function is entropy or cross entropy. Generally, this problem can be solved by gradient method[1]. The dual algorithm[2] gives another idea which inverts the problem to a dual problem. The dual problem is to find a function whose data in some region of dual domain are constrained to zero, and the data in the known region of wave-number domain remain unchange. This problem can be solved by the iteration between dual domain and wave-number domain.

We show the problem of incomplete observation angle overlay in the case of straight ray tomography. For the case of interwell, the number of sources and receivers is limited. So the aperture is limited. This makes the angle between measured ray and horizontal line less than θ_{max}. This means that the travel-time we get belongs to the rays with angle limited in $-\theta_{max} < \theta < \theta_{max}$. According to Fourier slice theorem, in the wave-number domain of slowness, by the cross-well measurement, only the information in the shade region of Figure 2 can be got, this is the region between two straight lines coming from the origin with an angle $2\theta_{max}$ included between them. We define the region out the shade region as region Ω. The information in region Ω

is unknown. Because of this, the reconstruction image of slowness will be distorted. The aim of ME or MICE is to estimate the wave-number spectrum in the region Ω. So that the image distortion may be reduced. The function which ME or MICE will find is that its wave-number spectrum in the known region is equal to the known wave-number spectrum and its wave-number spectrum in the unknown region makes entropy $\int \log O(x, y) \, dxdy$ maximum or cross entropy $\int O(x, y) \log O(x, y) \, dxdy$ minimum, where (x, y) are space coordinates. It can be proved that the requirement above is equivalent to that

$$R(k_x, k_y) \equiv \iint \frac{1}{O(x, y)} \exp\left[i(k_x x + k_y y)\right] dxdy \tag{1}$$

is zero in the region Ω, or

$$R_1(k_x, k_y) \equiv \iint \left[\log O(x, y) - 1\right] \exp\left[i(k_x x + k_y y)\right] dxdy \tag{2}$$

is zero in the region Ω.

If a function in a known region of the wave-number domain equals to the known spectrum of slowness and is also zero in the region Ω in the dual domain, that is the domain of $R(k_x, k_y)$, $R_1(k_x, k_y)$, then, this function satisfies the requirement of ME or MICE. Because of the entropy maximum or cross entropy minimum, the inhomogeneity in the new image may be reduced.

In the algorithm proposed by Lo[2], the wave-number spectrum is given directly by measured data, according to Fourier slice theorem. In order to invert the data of polar coordinates, as in the straight ray tomography, or data on the circle arcs, as in the diffraction tomography, into the data of rectangle coordinates, the interpolation in the wave-number domain is necessary. This version of algorithm is also not suitable for a constraint inversion which is often required.

We suggest that the wave-number spectrum in the known region can be got by 2-D Fourier transform, Because by whatever method such as ART, SIRT of CGLS, the reconstruction image $O(x, y)$ can always be got, which can be used to get spectrum in the known region.

In this way, the formulae of ME or MICE method do not depend on the inverse method or type of tomography. For every type of tomography, it only needs to know the region in which the information should be estimated by the ME or MICE method. For example, a bending ray tomography can be assumed to have unknown region similar to one of the straight rays, so we can use ME or MICE to estimate its information of an angle region in wave-number domain. This is the first improvement of our work.

UNDERSTANDING ME OR MICE BY 2-D WIENER-HOPF THEOREM

1-D Wiener-Hopf theorem[3] is related to a half space analytic function in complex plane with a half-axis zero function by 1-D Fourier transform. The theorem has been widely applied in wave equation theory and information theory.

2-D Wiener-Hopf theorem[4, 5] is related to a 2-D complex function analytic in a certain region of 2-D complex space with a 2-D complex function which is zero in an angle region by 2-D Fourier transform. In the straight line tomography, the function described by the ME or MICE method can be considered with 2-D Wiener-Hopf theorem. Because the estimated region in wave-number domain is an angle region, so the image function in dual domain is zero in the angle region according to dual algorithm of the ME or MICE method. By 2-D Wiener-Hopf theorem, it is analytic in a certain region of 2-D complex space in the image domain, if we make the analytic extension of the real image domain to complex domain. For an analytic function in a region, if it is not equal to zero, then its reciprocal fraction and its exponent function will also be analytic in the region. So, the image function is analytic in the certain region of 2-D complex plane. By means of the 2-D Wiener-Hopf technique, the image function in the wave-number domain will be zero in the estimating region.

So, by Wiener-Hopf theorem, if Ω is an angle region,

$$\int \frac{1}{O(x, y)} \exp\left[i(k_x x + k_y y)\right] dx dy \tag{3}$$

is zero in region Ω, or

$$\int \left[\ln O(x, y) - 1\right] \exp\left[i(k_x x + k_y y)\right] dx dy \tag{4}$$

is zero in region Ω, the

$$\int O(x, y) \exp\left[i(k_x x + k_y y)\right] dx dy \tag{5}$$

will be zero in region Ω.

So, if estimated region Ω is an angle region, the ME or MICE will not change the original image. In this case, the ME or MICE will not have significant effect.

The above discussion is for the case of straight ray. For the case of bending ray, if the ray is almost straight, the ME or MICE will not have significant effect, too.

From the above discussion, if we want to improve the image, we should not use ME or MICE alone. The improvement is suggested as follows:

In the reconstruction before ME or MICE, the hard constraint or soft constraint should be applied, so that the reconstruction gives some information in the Ω region which will not be changed by the ME or MICE. The estimated wave-number spectrum of slowness by the ME or MICE are the left part of the Ω region, because the estimated region is not an angle region, so ME or MICE will give some improvement on the quality of image.

COMPUTATION EXAMPLE

Now we briefly describe the computation procedure.

The ray travel-time inversion problem is solved by an iteration process. Iteration step length is worked out by the inversion of linear perturbation. During linear inversion, the ray tracing is carried out by the interface grid minimum travel-time algorithm. [6] The algorithm approximates the slowness by block uniform base function. The interface of every block is discretilized. The Fermat's minimum travel-time principle is applied to search the ray trajectory as shown in Figure 1. After the length of ray in every block is solved out, the ART or SIRT linear inversion is carried out. During linear inversion, the upper and lower bounds of the slowness function or the imaging range are applied. This will make the information in Ω region increased. Then, the dual algorithm of ME or MICE will be used during which, only no information part of Ω region in wave-number domain will be estimated

Figure 3 is a model, Figure 4 is the result without the ME, Figure 5 is the result of ME without hard constraint. Figure 6 is the result of ME with hard constraint.

It can be seen that ME method without hard constraint gives no improvement to the reconstruction image, the ME method with hard constraint gives an improvement to the reconstruction image.

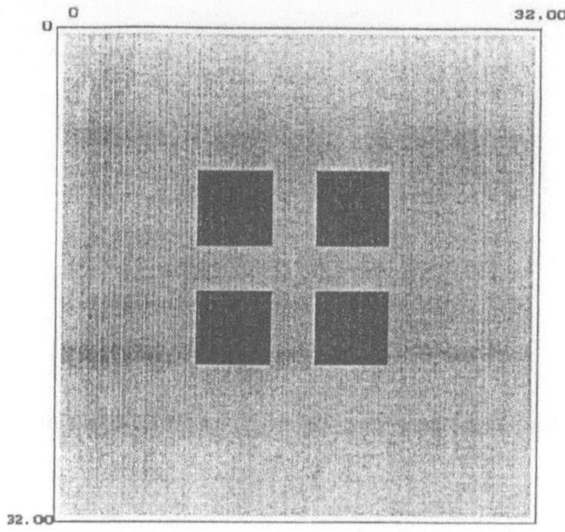

Figure 3.   Supposed slowness model

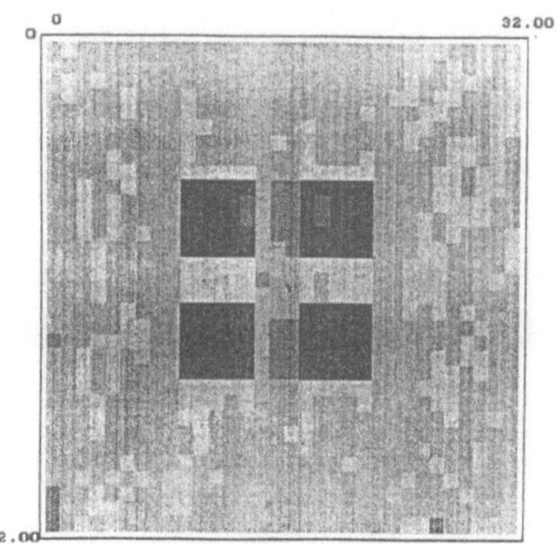

Figure 4.   Reconstruction image by ART with hard constraint

CONCLUSION

There are two improvements on the dual algorithm of ME and MICE. Firstly, the wave-number spectrum data are obtained by 2-D Fourier transform from reconstruction image. This extends the application of the dual algorithm. Secondly, the information in unknown region in wave-number domain is particularly given by hard constraint or soft constraint. Then the ME or MICE method is applied to estimate the left part of the information. Thus a good result can be obtained.

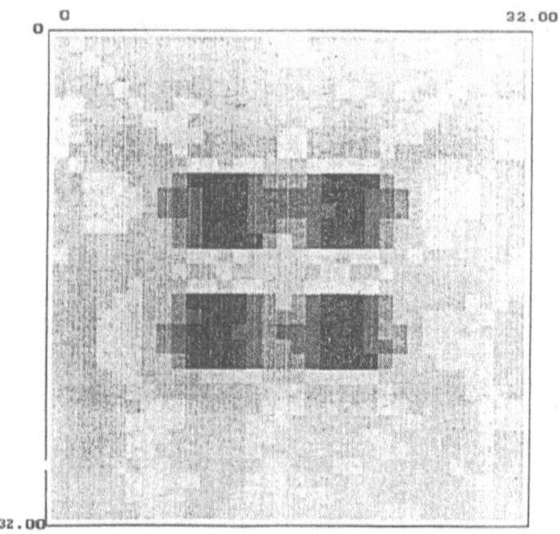

Figure 5.   Reconstruction image by ART and ME without hard constraint

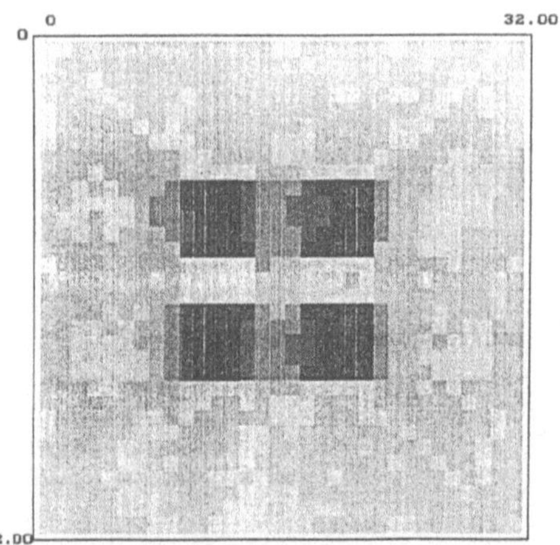

Figure 6   Reconstruction image by ART and ME with hard constraint

REFERENCES

1.   S. J. Wernecke, Maximum entropy image reconstruction, *IEEE Trans. on Computers*, C26 351—364 (1977).

2.   T. W. Lo   "Seismic Borehole Tomography", Ph. D Thesis, MIT, (1987).

3.   B. Noble, "The Wiener-Hopf Technique", Pergamon, New York (1958).

4.   H. Liu, On analytic method to scattering by a dielectric wedge, *ACTA Electronica Sinica*, 15 (5)   121—123 (1987).

5.   E. M. Stein, G. Weiss, "Introduction to Fourier Analysis on Euclidean Spaces", Princeton University Press, Princeton, New Jersey (1971).

6.   H. Liu, Y. Li, F. Meng, X. Gao, A layer-stripping algorithm in multi-dimensional ray travel-time inversion, in "Acoustical Imaging", Plenum Publishing Corporation, New York (1993).

7.   H. Liu, "Electromagnetic Borehole Image", Post-Doctoral Report, Institute of Geophysics, Chinese Academy of Sciences, Beijing (1991).

GEOTOMOGRAPHIC RECONSTRUCTION OF VELOCITY AND
RELATIVE ATTENUATION IN ACOUSTIC TRANSMISSION PROSPECTING

Guoliang Hu

Department of Territory Planning
Wuhan Municipal Planning Commission
Wuhan 430014, P. R. China

ABSTRACT

In this paper, the author describes a method of estimating velocities and relative attenuations from a cross borehole acoustic transmission survey using straight ray tomography to produce velocity and relative attenuation models which fit the traveltimes and amplitudes. Acoustic transmission data of model tank and real field are processed. Since the reconstructed images are consistent with known model or other geological results and the modeled traveltimes and relative attenuations closely match those traveltimes and relative attenuations from the actual data, the geotomographic reconstruction is proved to be a useful approach in acoustic transmission data interpretation.

The processing was carried out on a microcomputer and the program is readily adaptable to most field configurations. The field data processing techniques have been tested in a mine.

INTRODUCTION

Borehole acoustic transmission method (BATM), which was developed in recent years, is a new underground acoustic prospecting approach and is shown to be effective in mineral exploration[1]. Traveltimes and amplitudes of transmission waves are measured in BATM so as to find subsurface bodies such as ores, fractures, etc. , The most important measurement system is of crosshole type. The principle and data collection of BATM are similar to those in medical tomography[2,3] thus some approaches of CT are applied to data processing and interpretation in BATM.

In geophysics, tomographic images had been reconstructed from both seismic and electromagnetic measurements by Corlson et al.[4], Frank et al.[5] and Lizhen Liu[6]. Tienwhen Lo et al.[7] studied the tomographic inversion of velocity by the data of model experiments. Some theoretical problems about geotomography were discussed by Herman[8] and Nobel[9]. In the above applications, only one parameter (velocity or attenuation) is used to image.

In this paper, the author describes a method of geotomographic reconstruction of two parameters (velocity and relative attenuation) in BATM. The algebraic reconstruction techniques (ART) and the simultaneous iterative reconstruction techniques (SIRT) have been successfully applied to solving inverse equation composed of acoustic data. Model tank and real data are processed by this method. Tomograms of velocity and relative attenuation are clearly illustrated.

METHODS

There are many different borehole source-receiver geometries which can be used in BATM. This paper is concerned with the cross-borehole type where source and receiver are both positioned in boreholes. The experiment involved placing sources in one borehole and receivers in another borehole and subsequently recording the wave field which is transmitted between boreholes. The source-receiver geometry for cross-borehole survey in BATM is similar to one in seismic measurement[10]. The wave field mainly contains information on traveltime and amplitude, which can be used to reconstruct velocity and relative attenuation images.

RELATIVE ATTENUATION TOMOGRAPHY[6]

Based on the ray optics approximation, the acoustic wave transmission equation in subsurface lossy media can be simply expressed as follows:

$$A = I \cdot \exp\left(-\int_R X(r)\,dr\right) \cdot \frac{f}{R} \tag{1}$$

where I is the power factor of acoustic wave source and has a relation to transmitter, r is the propagation path of wave which is presumed to be straight, dr is the differential element of r, f is another factor related to radiation patterns of sources to the projection for calculations, $X(r)$ is the absolute attenuation constant, A is the measured field amplitude.

The relation of the background field strength B of surrounding rock with apparent attenuation constant of the surrounding rock X_0 is as follows:

$$B = I \cdot \exp\left(-\int_R X_0 dr\right) \cdot \frac{f}{R} \tag{2}$$

Eq. (2) divided by Eq. (1) gives

$$\frac{B}{A} = \exp\left(\int_R (X(r) - X_0)\,dr\right) \tag{3}$$

The region between boreholes is represented by a two dimensional area where the distributions of velocity and relative attenuation are calculated. The region to be imaged is discreted into rectangular pixels[11]. For a lot of measurements, the author can obtain a set of linear equations from Eq. (3):

$$[C_s] = [D] \cdot [\Delta X] \tag{4}$$

where $C_s = \ln(B/A)$, $\Delta X = X - X_0$, ΔX is referred to relative attenuation. $[\Delta X]$ is the relative attenuation vector, $[C_s]$ is the amplitude ratio vector, $[D]$ is called distance matrix in which the elements are the length of segments intercepted by the pixels. The reconstruction of ΔX overcomes difficulties in calculating I.

VELOCITY TOMOGRAPHY

Similarly, another acoustic wave transmission equation is

$$T = \int_R S\,(r)\,\mathrm{d}r \tag{5}$$

where T is the measured field traveltime, $S\,(r)$ is the slowness in subsurface media. $V = 1/S$, where V is the velocity.

After using the same discrete model in relative attenuation tomography, Eq. (5) yields

$$[Y] = [D] \cdot [Z] \tag{6}$$

where $[Y]$ is the vector composed of traveltimes, $[Z]$ is the unknown velocity vector.

Eqs. (4) and (6), which have the same form mathematically, can share the existed various reconstructive algorithms. Eqs. (4) and (6) are typically ill-posed systems and involve computing pseudoinverses of a rectangular matrix. Despite the ill-posed natures of such systems, ART[12], and SIRT[9] are adequate to solve Eqs. (4) and (6).

COMPUTATION PROGRAM

The inverse process will demand a large computational effort if the above approximation is not used. To simplify the computation, it was decided to try simple algorithms (ART and SIRT) that could be quickly programmed and run on a microcomputer. The velocity and relative attenuation to be imaged are shared in these algorithms. These methods have one notable advantage. It is easy to apply and is undemanding in terms of considerable computer resources.

The computation program requires the following input: a) the north south, east west pixel reference of each source/receiver location, b) the traveltimes and amplitudes of transmission wave between each location, amplitudes of transmission wave between the same distance as above in surrounding background media, c) the specification of the representative parameters pixel spacing and dimension, and d) some special data such as restricted values of parameters.

The computer output yields: a) representative velocities and relative attenuation of each pixel for ART and SIRT, b) a ranked scale mapping (color block section, contour section, 3-D pictures) of two parameters, and c) iterative numbers, etc.

RESULTS OF TWO MODELS AND SOME REAL DATA

Acoustic transmission experiment simulating geotomography was carried out in a modeling water tank. In order to demonstrate the effect of velocity lows and highs with respect to the background material, the author used two models: air and glycerine spheres. Water is regarded as background media. The velocity of the water is 1438m/s. The absolute attenuation of the water is $14.\,2 \times 10^{-3}1/\mathrm{cm}$. The velocities of air and glycerine sphere models are 343 m/s and 1750m/s, respectively. The absolute attenuations of air and glycerine sphere models are $675 \times 10^{-3}1/\mathrm{cm}$ and $137 \times 10^{-3}1/\mathrm{cm}$, respectively. In the model, two boreholes separated by 74cm are considered. We placed a set of transmitter in one borehole and 2 receivers, which had a separation of 5 or 10cm, in the other borehole.

The air sphere is considered as a low-velocity and high-attenuation contrast anomaly, and the glycerine sphere is regarded as a high-velocity and high-attenuation contrast anomaly. Figures 1 and 2 give the reconstructed images of air and glycerine sphere models, respectively. The shape, center and relative acoustic properties of the inverted targets are apparently pointed out in Figures 1 (b), 1 (c), 2 (b) and 2 (c).

A crosshole acoustic field measurement was performed at a Pb-Zn mine. The two holes used were 305m deep and 86m apart. About 10 explosion were set off in one of the boreholes and registrations were collected at 2 movable geophones in the opposite borehole. All the transmission wave traveltimes were measured for 400 rays. The objective of the acoustic measure-

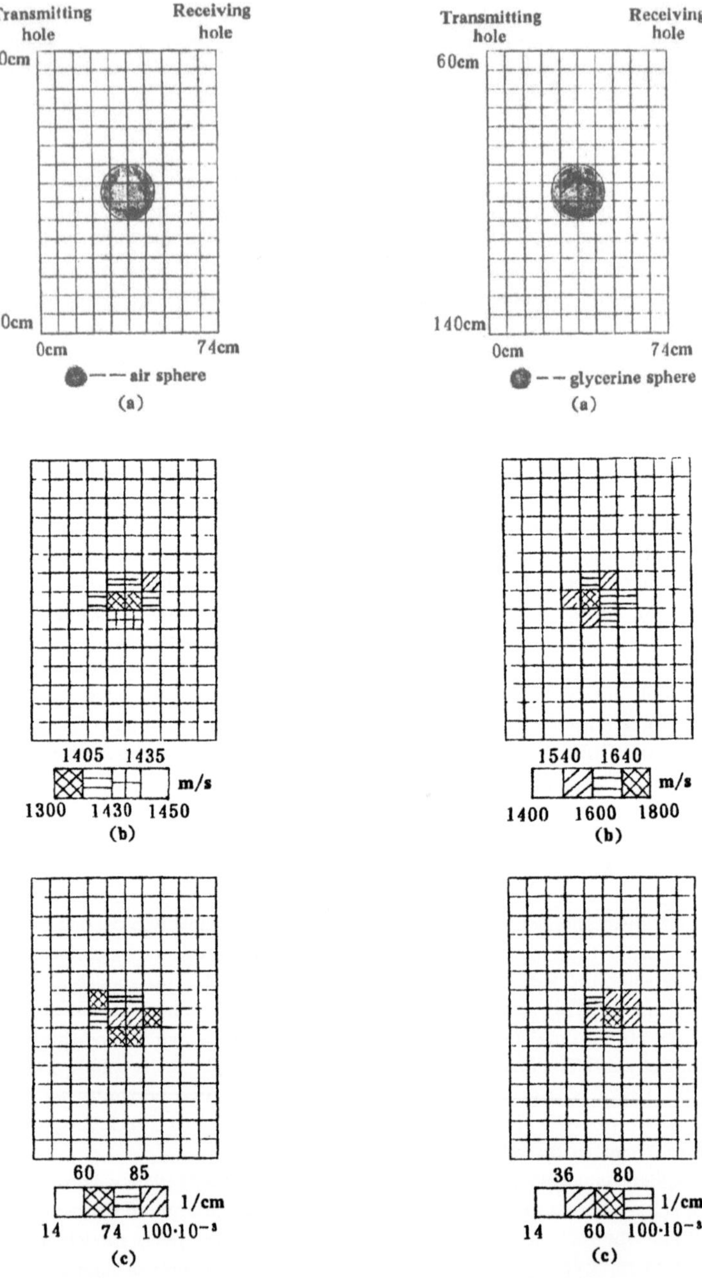

Figure 1. Tomograms of air sphere- water modeledexperimental data. (a) Geometry, (b) Color block section of reconstructed velocity by SIRT, (c) Color block section of reconstructed relative attenuation by SIRT.

Figure 2. Tomograms of glycerine sphere- water modeled experimental data. (a) Geometry, (b) Color blocksection of reconstructed velocity by ART, (c) Color block section of reconstructed relative attenuation by ART.

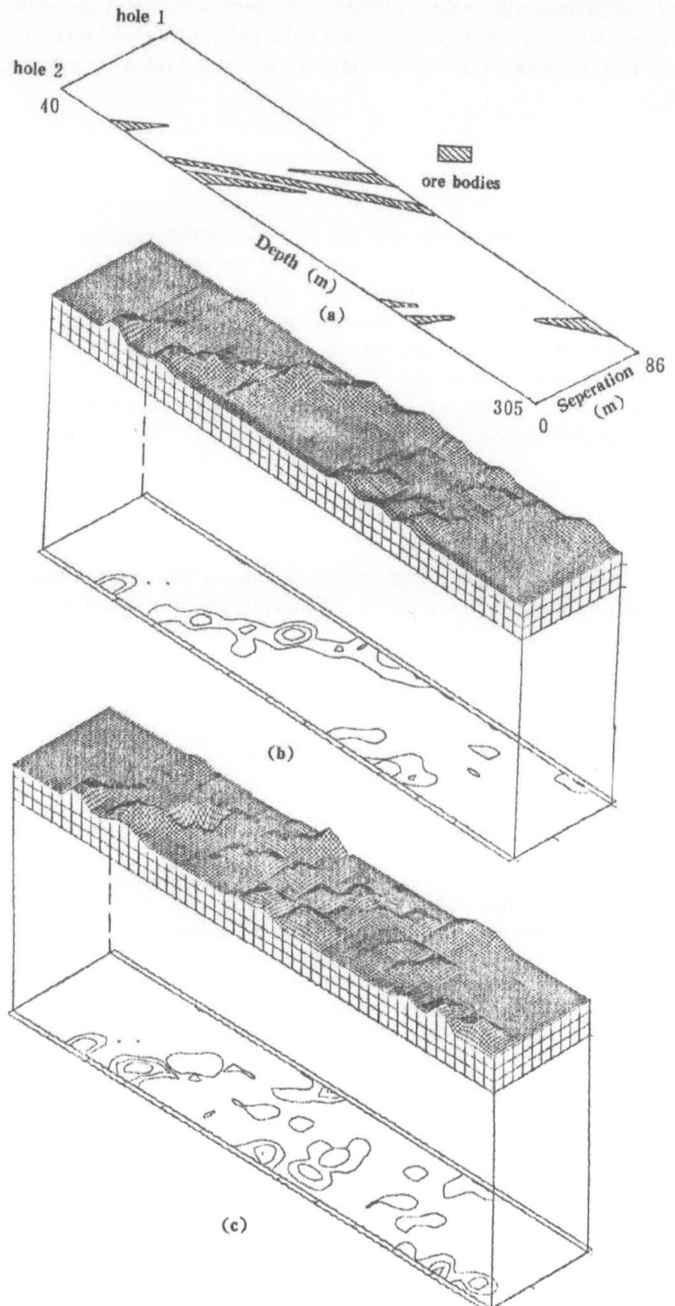

Figure 3　Tomograms of crosshole real acoustic data (hole 1—hole 2) in a Pb-Zn mine　(a) Geometry and ore bodies obtained by geological method　(b) 3 D map of reconstructed velocity by ART　(c) 3-D map of reconstructed velocity by SIRT

ments was to map special sections of the Pb-Zn ores in terms of transmission wave velocity and attenuation for comparison with other geotechnical properties. Figure 3 shows the reconstructed results of crosshole data. The distributions of ores to be obtained by geological method are given in Figure 3 (a). The inversion anomalies in Figures 3 (b) and 3 (c) are consistent with those in Figure 3 (a).

CONCLUSION

In this paper, computerized tomography has been successfully applied to acoustic transmission prospecting. The inversion model is founded under the condition of optics approximation. The distributions of velocity and relative attenuation are acquired in the model. Thus the information of the two parameters are highly desirable to narrow the range of possible interpretations. The results show that the reconstructed images are useful in identifying special regions on the section of acoustic prospect area.

Further trials will be carried out to determine the usefulness of other tomographic algorithms applied to borehole acoustic transmission data.

ACKNOWLEDGEMENTS

The author would like to thank Prof. Hulian Wang for many important instructions and suggestions and is grateful to senior engineer Zhengying Lei, Yan Zhao and Pengfei Gao for their valuable help.

This work was partially supported by Graduate School and Computer Center of China University of Geosciences, Wuhan, China.

REFERENCES

1. Zhengying Lei, Application of acoustic transmission method to mineral exploration, *Expanded Abstracts of 89'BISEG*, Beijing (1989).
2. R. Gordon et al, Algebraic reconstruction techniques for three dimensional electron microscopy and X-ray photography, *J. Theor. Biol.*, 29 (6), 471—481 (1970).
3. G. T. Gullberg et al, Three dimensional reconstruction in nuclear medicine emission imaging, *IEEE Trans. Nucl. Sci.*, 21 (3), 2—20 (1974).
4. B. L. Corlson and C. A. Balanis, Pulse time delay method for geophysical tomography, *IEEE Trans. Geosci. Remote Sensing*, 26 (5), 511—517 (1988).
5. M. S. Frank and C. A. Balanis, Methods for improving the stability of eletromagnetic geophysical inversion, *IEEE Trans. Geosci. Remote Sensing*, GE—25 (3), 18—27 (1989).
6. Lizhen, Liu, Reconstruction of the distribution of relative attenuations in subsurface media: a new tomographic technique, *ACTA Geophysical Sinica*, 33 (1), 98—105 (1990).
7. Tienwhen Lo et al., Ultrasonic laboratory tests of geophysical tomographic reconstruction, *Geophysics*, 53 (7), 231—239 (1988).
8. G. T. Herman, "Image Reconstruction from Projections", New York Academic Press Inc., New York (1980).
9. G. Nobel, "Seismic Tomography with Application in Global Seismology and Exploration Geophysics", D. Reidal Publishing Company, New York (1987).
10. S. Ivansson, Tomographic modeling used for cross data analysis, in: "Borehole Geophysics for Mining and Geotechnical Applications", P. G. Killeen, ed, Geological Survey of Canada, Canada (1986).
11. Guoliang, Hu, The tomography inversion of acoustic data in well, *Computerized Tomography Theory and Applications*, 1 (2), 1—7 (1992).
12. R. Gordon, A tutorial on ART, *IEEE Trans. Nucl. Sci.*, 21 (1), 78—93 (1974).

APPLICATION OF SEISMIC TOMOGRAPHY
TO THE HYDROELECTRIC ENGINEERING EXPLORATION

Ouyang Daijun,[1] Liu Jiaqi[2]
Huang Jinli,[1] Wu Jianchen,[2] Wei Mingguo[2]

[1]Geophysical Exploration Brigade
Changjiang River Water
Conservance Commission (YVPO), China
[2]Harbin Institute of Technology

INTRODUCTION

In the early 1980's, we used low velocity segments in average velocity fanned figure of seismic rays to judge the existence of geological anomalies and used the intersection method to locate their positions, but we couldn't determine the shapes of anomalies, and if the low velocity segments in fan figure were not obvious, it is difficult to distinguish the anomalies reliably. Recently, with CT technology, we carried out some experiments in two different rock areas, one is in the granite area in Three Gorges dam site; the other is in the limestone area in Goupitan dam site in the middle reach of Wujiang river, Guizhou Province. Compared with medical CT, geo-CT is more complicated because there are more unknown factors and rapidly changed parameters, limited observation data and directions, and greater medium velocity change gradient (more than 5%). With the practice of engineering geo-CT, we have studied some useful methods of inversion and got some good results.

CURVED RAY TRACING AND STRAIGHT RAY INVERSION

In hydroelectric engineering geology, close attention is paid to some broken rocks because there are great differences in velocity and energy absorption between broken and unbroken rocks. Curved ray tracing method can fit the rays quite well. Using traditional shooting method, the initial shoot angle should be adjusted continuously so as to get a suitable one that the ray can just propagate from the source to the receiver. Thus the method will take a lot of computation time. In order to reduce the computation cost, we evaluate the shortest time track between two points according to Fermat principle, the steps of the method are as follows:

Acoustical Imaging, Volume 20 Edited by Y. Wei
and B. Gu, Plenum Press, New York, 1993

Giving an initial angle $a(0) = a_0$,

$$x(t) = V \cos a(t)$$
$$y(t) = V \sin a(t)$$
$$a(t) = V_x \sin a(t) - V_y \sin a(t)$$

where t is the traveltime. For a fixed source, giving a suitable angle, we can solve the primary problem to make the ends of the tracks be just between some two receivers. Then giving an angle increment, we will get a new angle

$$a_1 = a_0 + \sum_{i=1}^{l} \Delta a_i$$

The above primary problem can be solved continuously and get many tracks, we can choose some of the tracks whose ends are well distributed between some two receivers, the traveltimes are evaluated by interpolation method. It can reduced the computation cost by more than 50 percent.

Because of using primary problem to evaluate boundary problem, this method has some drawbacks. If a boundary exists, there is a blind zoon which the scanning couldn't reach. So the traced ray track is not the least traveltime. In order to overcome these drawbacks, a fast search method was adopted. According to the principle that wave always inclines to the larger velocity side, we first consider the initial track a straight line, then take the middle point of this line, from this point, search forward according to the velocity field gradient in this point, then we can get a new line PR (Figure 1).

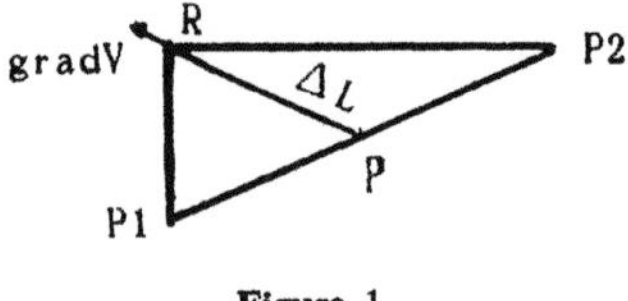

Figure 1

When wave propagates along this line (PR), the traveltime is the smallest. We then take the middle point of P_1R, RP_2 and repeat the above steps again and again until the traveltime difference between two adjacent tracks becomes very small. The effect will be better if the tracing combined with above two methods.

Besides the tracing method, it is also very important to choose the inversion method.

The inversion we used is the following iteration method:

1. Give a constant to an initial velocity, so all the tracks are straight lines;

2. Inverse the velocity distribution with stable algorithm such as ART;

3. Use tracing method, get new traveltimes, go to step 2, repeat the iteration until a better inversion result is gotten.

A mass storage computer must be used in scanning method. If collected data are up to 1000, the grid number is also 1000, and the matrix elements will be up to 1000K. This is beyond the IBM PC/XT capacity. In field, we used straight ray ART algorithm to get initial result. And a method of saving memory was taken and a virtual driver was adopted to make up inadequate memory, but the computer run time increased.

ARRANGEMENT OF OBSERVATION SYSTEM

It is difficult to observe in all directions in geo-CT just like in medical CT. Generally only two directions (the boreholes or two adits) can be observed. Therefore, it is very important to arrange observation direction reasonably and efficiently. For stratified medium (such as fault zone), if its apparent dipping tendency is consistent with the direction of borehole or adit, the low velocity segments which reflect the fault can't be discerned because the abnormal information of rays distributes homogeneously in every path. If oblique, some parts of rays must parallel with anomaly trend by means of increasing synchronous high difference or observation direction. Most rays should oblique to the strike of geology body as far as possible instead of perpendicular, but for single anomaly such as karst cavity the requirement for observation direction is not so strict.

GRID SIZE AND OBSERVATION DATA

Any kind of low velocity anomaly (such as shattered fault zones, karst cavities, etc.) is considered a dangerous rock mass if its size is only about 1—2 meters, and must be found out in preliminary design stage. As wave propagates in bad rocks, its frequency is about 500Hz and velocity less than 2000m/s, and the wavelength is within 4 meters. The distance between boreholes or adits may be some tens meters in geoexploration. In detailed survey, the interval between receivers is selected to be about 1 meter. With enough correct data, the inversion result will be unique. The more accurate the data, the better the inversion result is. Figure 2 and Figure 3 are the inversion results with almost the same grid size but different data (the data used in Figure 2 are twice as in Figure 3). There is no continued false anomaly shown in Figure 2. But a false anomaly in right side of F54 fault is appeared in Figure 3. The size of grid reflects the resolution in CT inversion, the larger the grid size, the lower the resolution is. Figure 3 and Figure 4 are the inversion results with same rays but different grid sizes. Figure 3 shows good continued anomaly although a false anomaly appeared. The relationship between the amount of observation data and the number of grids (overdetermination and underdetermination) affects severely the elements variance of inversion. The results on overdetermination condition are better than those on underdetermination condition when grid size is satisfactory in reflecting geology anomaly under the condition of larger amount observatioɪ. Jata. Figure 2 is the inversion image in which the amount of data (2200) is approximately equal to the number of grids. The stretch of low velocity zone, which is corresponded to F54 fault, is shown clearly and there is no false anomaly. If the collected data in field is of high precision but the number of the data is limited, the grid size must be determined first to reflect the geology anomaly effectively. And the inversion result is significant good in reflecting the low velocity anomaly even in underdetermination condition and can get useful information. False anomaly still exists but can be distinguished by other methods. Figure 3 is the inversion image with 968 data and 1. 14m × 1. 16m grid size. The low velocity zone also responded to F54 fault well but a false anomaly zone appeared in the right side of main anomaly zone. With some observation data, if the overdetermination condition is first considered, then the grid size and grid amount, no significant inversion image, can be obtained. Figure 4 shows this result with 968 data and 1. 5m × 1. 5m grid size. So the inversion image is a compromised result of variance and resolution.[1] On the conditions of high precision collected and limited data number, the little variance (appears as overdetermined condition) will reduce the resolution, and the little grid size (high resolution) will increase the variance. Consideration must be given to the above two conditions to get optimal result. Checking method is: inversion with different grid size, comparing the results with known geology conditions to get best inversion results which they fit best.

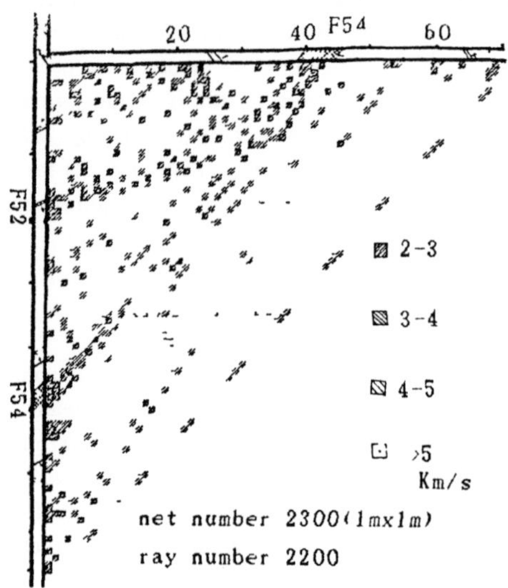

Figure 2.   Velocity grey graph between adits at Sandoupin dam site

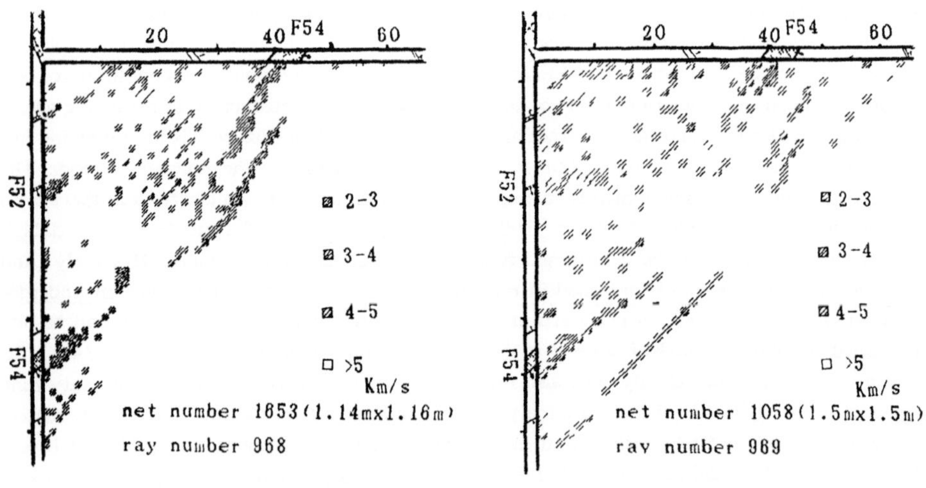

Figure 3.   Result of pretreatment   Figure 4.

DATA PRETREATMENT

Reliable, high precision observation data are the base of inversion. Although special attentions had been paid to exciting and receiving conditions, point location measurement, instrument concordance, etc. , random error still existed. Original data with error would damage the inversion image severely, especially if the velocity difference between anomaly body and around rock was small. So, the collected data must be checked in the following two ways: One is sampling check to interchanging channel, linked channel and repeated shot record in field, the sampling rate is 10—20%. Errors of sampling check data defer to normal state distribution. The accuracy standard of original measured data may be determined by sampling check. We take 0. 2ms as the accuracy standard in granite rock area in Three Gorges dam site because the mean square value of measured data is 0. 2ms. The other is total check in house included plotting time-distance curves and removing those were suddenly changed over standard but not caused by the wall of boreholes or adits. Besides above sampling check methods, the standard also can be determined by the minimum read accuracy according to the record scale of different instruments. This standard is changeable with different instruments and geology condition. For ES–1225 seismograph, the time standard is 0. 2ms and for ES–1210 the time standard is 0. 5ms. Figure 5 shows the inversion image with no rule because the data were not pretreated. Figure 3 is the inversion image with the same data after pretreatment, the fault location and extended direction are shown clearly.

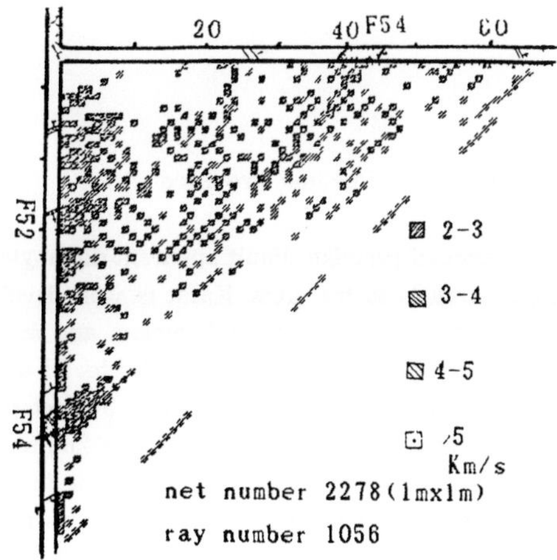

Figure 5.   Unpretreated result

EXAMPLES

1. Three Gorges Project in Chang Jiang River

The stratum is granite rock of pre-simian F54, fault across No. 8 adit in ship lock line. Its stretch is NE57°, the dip angle is 78°, the width of influence zone is 2. 5—6 meters, the wave velocity in it is 2km/s, and that in the rocks around is 4km/s; another fault F52, stretch E-W, dip angle 30 degree, influence zone width 2 meters, wave velocity in good rock is over 6km/s. The geophone array is arranged oblique and parallel to the faults, the geophone interval is 1. 25 meters. ES-1225 seismograph is used to record seismicwave. After pretreatment, the remained 2200 data were considered reliable. Grid size was selected $1m \times 1m$, the up and low velocity limits were 7km/s and 2km/s, respectively. It was iterated 80 times. Figure 2 shows the result that the two low velocity zones corresponding with F54 and F52, but a little more wide and not good continuous in middle parts.

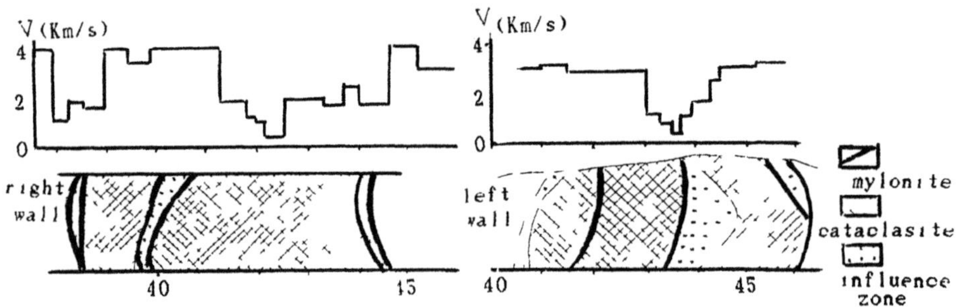

Figure 6.   F54 fault in No. 8 adit, Geological section and acoustic wave curve

2. Guopitan Project in Wujiang River, Guizhou Province

The stratum is thick limestone of permian, Fault apparent dip angle is about 70 —80°. The interlayer shear faulted zone parallels to the adits. Karst is well developed, most of the karst cavities are 0. 5 meter in diameter, some are about 5 meters and some are full of loess and mud. Data were collected with ES-1210 seismograph in adits No. 21 and No. 23, both of which were about 100 meters long. The distance between them was 34 meters. The geophone interval was 3 meters, 423 data were collected. The grid size was $2m \times 2m$, total 816 grids, the up and low velocity limits were 6500m/s and 500m/s, respectively. The inversion result showed low velocity zone concentrated near No. 21 adit and well responded to the fault and karst on the wall of adit. Approximate horizontal low velocity zone near No. 23 adit also responded to inter-layer compressional faulted zone, the low velocity zone near the entrance of the adit concerned with the interlayer tension zone. Late geology material indicated that large ω ($\omega < 0. 1L/m. m$) parts responded low velocity ($v < 3km/s$ and 4km/s) zone, and small ω ($\omega < 0. 001L/m. m$) corresponded to the good rock ($v > 5km/s$) zone in Z7 borehole. Karst cavity of 2. 3 diameters and mud fulled concerned with the velocity zone that is greater than 4km/s. It is unsatisfied that the direction of low velocity zone do not respond to the direction of fault on the wall of a-dit.

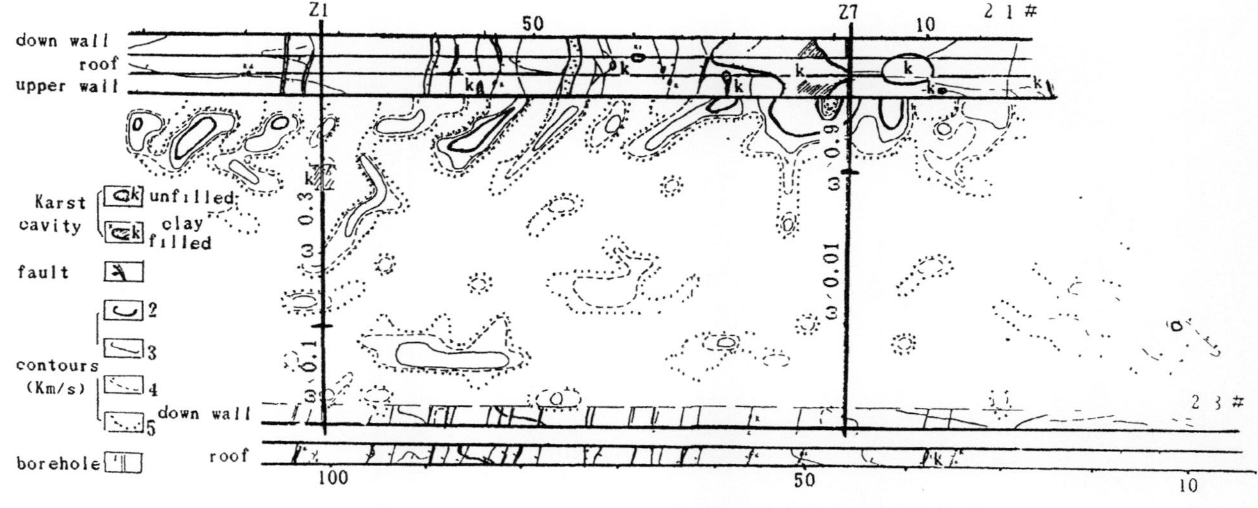

Figure 7.   Velocity contours of adits 21#—23# in Goupitan (Wujiang River) Project

REFERENCE

1. Yang Wencai, "Geophysical Inversion &. Seismic Tomography", Publishing House of Geology, Beijing China (1989).

CROSS-HOLE SEISMIC COMPUTER TOMOGRAPHIC TECHNIQUE
FOR UNDERGROUND POWERHOUSE OF XIAOLANGDI PROJECT
ON THE YELLOW RIVER

Guoxian Yang, Keni Zhang, Piwu Li

The Design Institute of YRCC
Zhengzhou 450003, P. R. China

ABSTRACT

In order to explore the structure of strata in the area of underground powerhouse of Xiaolangdi Project, two groups of cross-hole prospecting by means of seismic computer tomographic technique (hereinafter referred to CT) were conducted. Time domain data of 3324 seismic rays were obtained.

The findings of the work, according to geologists' analyses, are of high reliability: firstly, they are basically identical with the statistical values of the sound wave data, microseismic and borehole sound velocity data; secondly, the CT velocity stratification coincides with the positions of boring logs.

The final conclusion is that apparent low velocity zones exist, but no harmful faults.

DESCRIPTION

Xiaolangdi Multipurpose Dam Project is located on the main stem of the Yellow River. The underground powerhouse is in vicinity of the crossing point of the T-shaped ridge on the left bank of the proposed dam site. The location of the powerhouse has a least elevation of 103. 6 m, in which the distribution of T_1^{3-1} to T_1^{6-1} exists, it is comprised of sandstone, siltstone and mudstone.

In order to know the development of strata here by means of CT technique, three boreholes were set in line (T_{622}, T_{630} and T_{628}, see Figure 1) from northeast to southwest on the site. The hole depths are 153. 94 m, 184. 26 m and 182. 00 m, respectively. The hole spacing is 82 m with a diameter 76 mm and the hole top elevations are 243. 03 m, 273. 61 m and 272. 08 m, respectively.

Acoustical Imaging, Volume 20 Edited by Y. Wei
and B. Gu, Plenum Press, New York, 1993

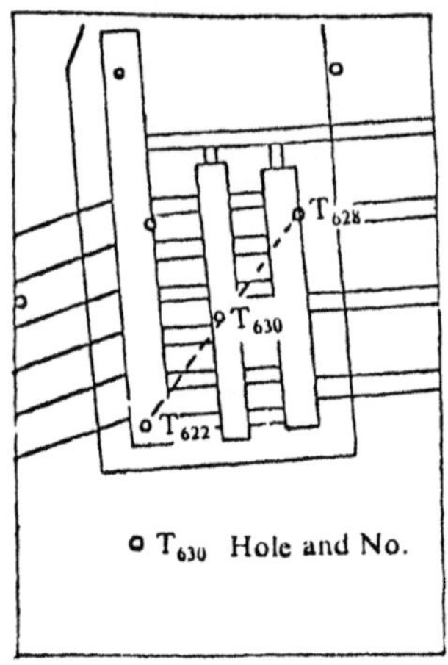

Figure 1.   Distribution of boreholes

INSTRUMENTATION SYSTEM AND EQUIPMENT

The 20kJ electric spark was used for ignition. The cascade-connected detecting cable was used for receiving signals. Observation on site was carried out by ES-1225 seismograph and data acquisition by a portable computer Zenith-171.

Two groups of tests were carried out. The middle hole (T_{630}) received signal (cascade-connected detecting cable set inside) and the end holes were for ignition. Both detecting points and receiving points were spaced 5 m apart. The instrumentation system is shown in Figure 2.

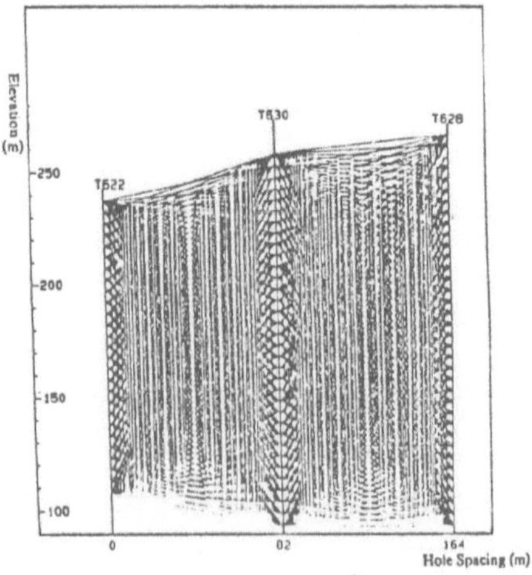

Figure 2.   CT instrumentation system

DATA PROCESSING

Self-adaptive filtering was done first during data processing in order to reduce noise and thus highlight the effective elements. The imaging principle is according to Reference[1]. The cascade approach[2] was used in calculation and the calculation methods of different resolution were incorporated to improve the resolutions of images.

The data processing flow chart is shown in Figure 3.

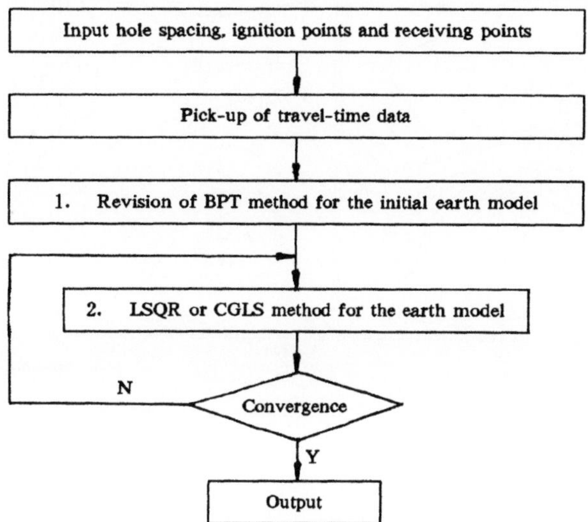

Figure 3.   Data processing flow chart

FINDINGS

The data obtained show that for deep and big spacing holes the approach of cross-hole seismic tomography with the electric spark and cascade-connected detecting cable is workable and very effective, and that the cascade calculations can improve the resolution of images.

The major findings from the tests are:

1) Two grades of low velocity zones are determined shown in Figure 4. The wave velocity of grade I is less than 1700 m/s and that of grade II is 1700—2200 m/s. This information provides a basis for the budget making and design of underground excavation.

2) The bending of velocity contours marked by the dotted line in Figure 5 indicates that local fissures may exist in the strata between T_{622} and T_{630}.

3) The velocities coincide well with the statistical values based on sound wave data, microseismic and sound velocity data measured from 1979 to 1991 (see Table 1.). Thus we can say the tomographic approach has brought about the real situation of the velocity field in the strata between the holes.

4) The rock strata shown by the tests are almost identical with the boring logs (Figure 5).

5) The images prove that no harmful faults exist in the underground power house area.

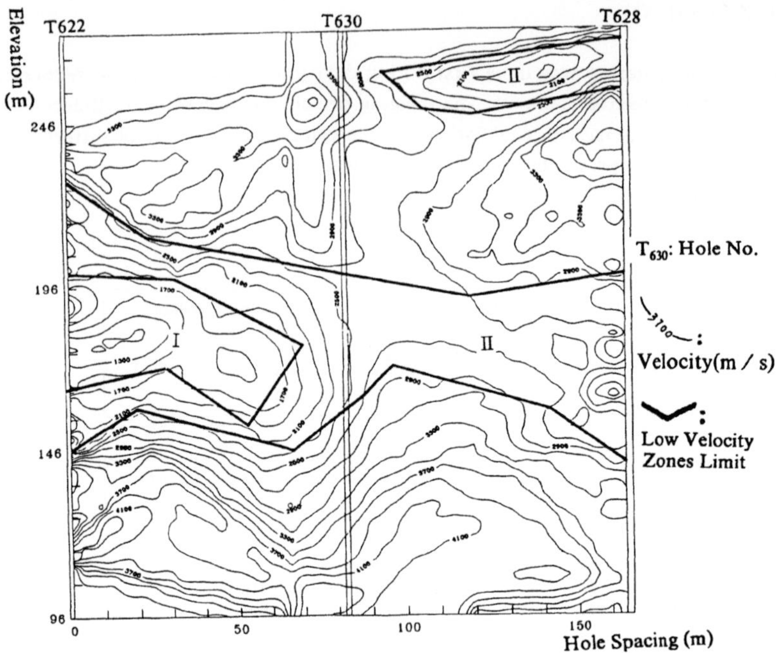

Figure 4. Velocity contours by tomography

Table 1. Comparison of wave velocities in m/s

Data source	Unit						
	T_1^{6-1}	T_1^{5-3}	T_1^{6-2}	T_1^{5-1}	T_1^4	T_1^{3-2}	T_1^{3-1}
CT	2100—3300	2700—3500	2700—3500	2500—2900	1300—3300	2000—4100	4000
Previous stat.	3800	3600—4000	2500—4200	2000—4300	480—3800 (frequent 1800)	4500	3900—5300

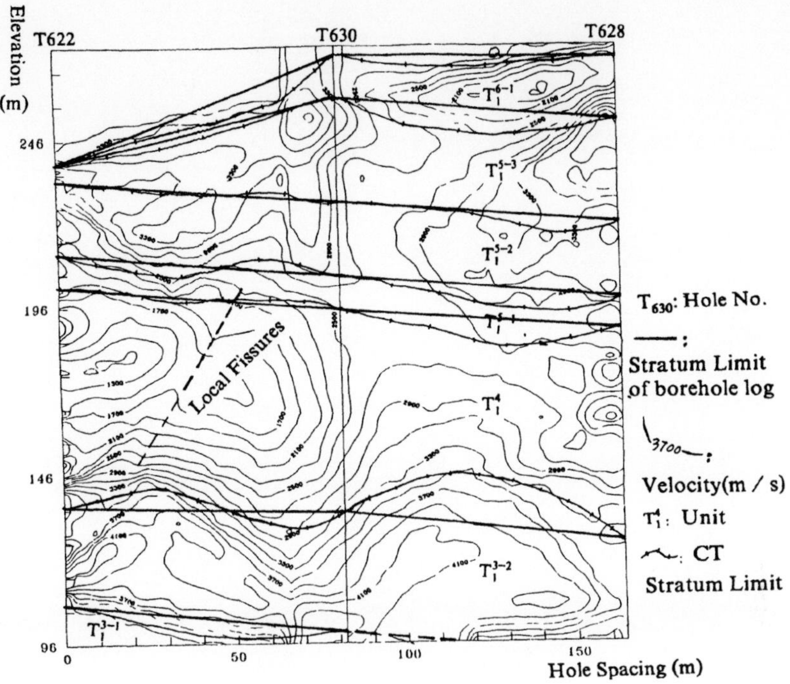

Figure 5.   Result of tomography

REFERENCES

1.   Yang Wencai, "Geophysical Inversion Seismic Tomography, " Geology Publishing, (1989).

2.   Du Jianyuan, Yang wencai, *The Journal of Geophysics*, 34, 771—780 (1991).

3.   Wang Zhendong, "Shallow Layer Seismic Prospecting Applied Technique, " Geology Publishing House, (1988).

TOMOGRAPHIC IMAGING FOR TWO-PHASE FLOW

BY ULTRASONIC MODELLING SYSTEM

Guo Chun Lu, Bjørn Ursin, Ole Bernt Lile and Jostein Sveen

Division of Petroleum Engineering and Applied Geophysics
The Norwegian Institute of Technology
The Trondheim University
N 7034 Trondheim, Norway

INTRODUCTION

In the last decade, ultra–sonic tomographic imaging techniques have been applied in many engineering investigations to determine inner constructions from external measurements, such as NDT (Non–Destructive Testing) and medical ultrasonic diagnosis, and so on. All applications consist of three procedures, measurement, data analysis as well as display.

For multi–phase flow research, it is desirable to obtain a real–time image about the phase patterns in the cross section of a tube. Motived by the efficiency of acoustic measurement and advanced tomographic reconstruction techniques, we will develop an ultra–sonic tomographic imaging system to obtain real–time image for the estimation of multi–phase flow patterns.

This paper reports the experiments using the tomography technique to image the two–phase flow patterns by the ultrasonic modelling system.

An acoustic measurement instrumentation was designed for our experiments, the fan geometrical scanning was performed by the relative rotation between the receiver array and the two–phase flow model (gas/water or oil/water). The first arrival traveltimes were picked up automatically from received transmissive signals by the microprocessor in an ultra–sonic instrument (EPOCH II). The straight–ray tomographic reconstruction software has been developed. The tube geometry was taken into account in the tomographic imaging algorithm. Our experiments show that the tomography imaging results had a good agreement with the two–phase flow models.

LABORATORY EXPERIMENT

The experiments were carried out in the laboratory of Continental Shelf and Petroleum Technology Research Institute (IKU). We have been assisted by laboratory personnel from IKU and mechanical engineers from the Petroleum Technology Center, NTH.

The stepping motor was controlled by IBM compatible PC as well as relevant integrated circuits. The pulse steps for one complete rotation was 2880.

The receiver array was composed of 7 transducers with angle interval of 15^0. This receiver array was rotated by the stepping motor. One transducer was fixed as the acoustic transmitter. The conventional fan scanning geometry was performed by the relative rotation between the receiver array and the two–phase flow model (the physical model). Figure 1 shows the sketch of the mechanical facilities.

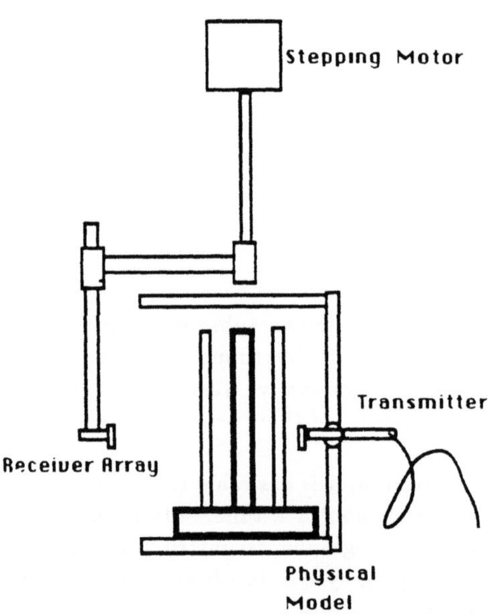

Figure 1. Sketch of the mechanical facilities for the acoustic measurement.

The transmitters and receivers used in our experiment were Panametrics, 12.7mm (0.5"), 1 MHz, P-wave transducer. In order to assure optimum acoustic measurements with high first arrival traveltime picking precision, we chose EPOCH II (Panametrics, Model 2100) as the ultrasonic instrument. This instrument is a completely digital and portable microprocessor–controlled ultrasonic flaw detector, designed for non–destructive testing purposes.

The EPOCH II consists of a transmitter, a receiver. The parameters of generated pulses and received signals are listed in Table 1.

Table 1.Parameters of generated pulses and received signals in EPOCH–II.

MHz Setting	Pulse Repetition Frequency	Display Update Rate	Sample Rate
0.5–2.25 MHz	60 Hz	60 Hz	20 MHz
0.5–6.00 MHz	120 Hz	60 Hz	40 MHz
0.5–15.0 MHz	240 Hz	60 Hz	80 Mhz

A significant feature of EPOCH II is that the traveltime which corresponds to the depth of the flaw in material can be automatically detected with the internal precise microprocessor. It provides sound path, surface distance and "edge" or "peak" depth to 0.025 mm resolution.

In our experiments, the "through transmission" mode, rather than the "echo" mode, was used for the first arrival traveltime picking. The traveltime picking resolution can be approximatively estimated from digitizing sample rate (Krajewski et al. 1989; Bregman et al. 1989).

Tektronix 468 oscilloscope with internal waveform digitizer was used for our experiments. The purpose of using Tektronix 468 was to display a clear and stable waveform and to digitize it for logging it into a computer. The storage system of Tektronix 468 provides an average signal storage of 2^n times received signals. This function assures a high signal/noise ratio.

In our experiments, two kinds of the two–phase flow were simulated by using the physical models:

1. Gas bubble in water (the gas/water flow),
2. Oil bubble in water (the oil/water flow).

For the gas/water flow, the gas phase was simulated by the perspex cylinder filled with air. These "gas" cylinders with different size were fixed in a frame which can be rotated in the center of the mechanical facility. Figure 2 shows the cross section of the gas/water flow model.

For the oil/water flow, the oil phase was simulated by the glass cylinder filled with oil. Figure 3 shows the cross section of the oil/water flow model.

The set–up of the ultra–sonic measurement system is shown in figure 4. The EPOCH II generated a series of high frequency pulses. These pulses were used as the source power for the transmitter, and a trigger signal to an oscilloscope. The received transmissive signal was amplified, filtered and rectified for micro–processing and display in EPOCH II. Simultaneously, the received signal as the RF mode was logged on the oscilloscope for extensional display and storage. The RF signal was then digitized and logged into a computer.

The ASYST software (A Scientific System) was used for receiver array scanning, waveform displaying and logging.

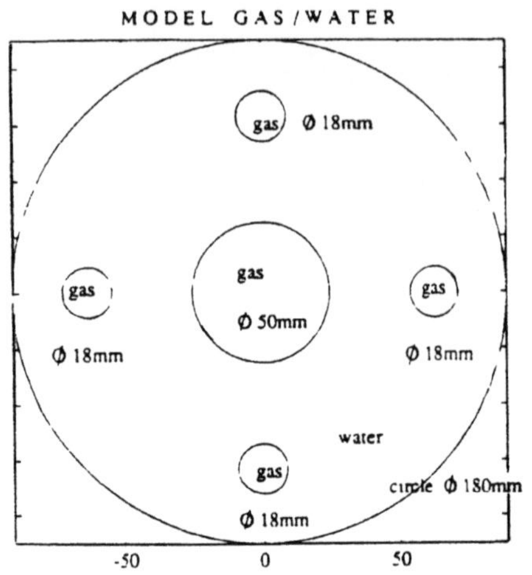

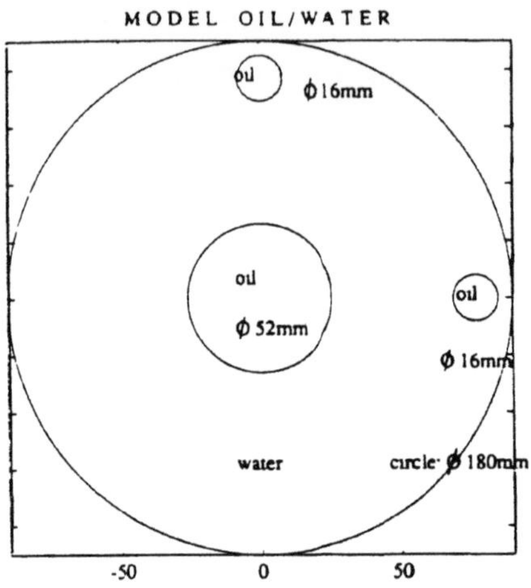

Figure 2. Cross section of the gas/water flow model.

Figure 3. Cross section of the oil/water flow model.

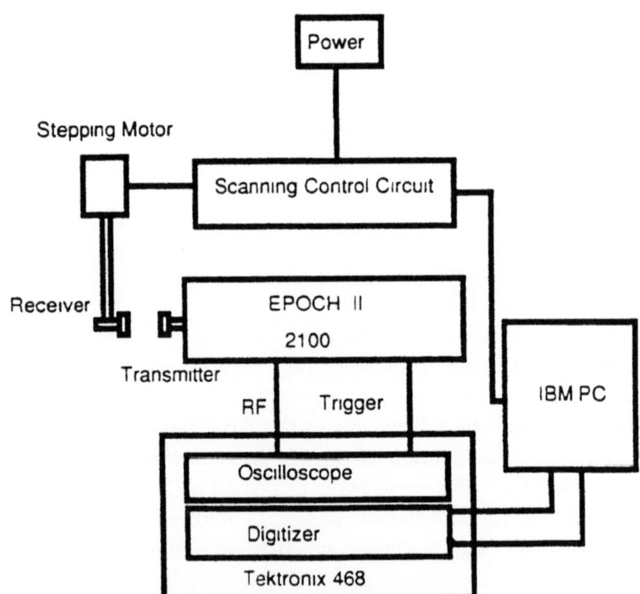

Figure 4. Set-up of the acoustic measurement system.

MEASUREMENT RESULTS AND TOMOGRAPHIC IMAGING

In the experiment, The angle spacing of the sources and the receivers was 15^0. The total number of rays is $\frac{n(m-1)}{2}$, where m, n are the number of sources and receivers, respectively. For our case, the maximum number of rays was 278. Most traces contained reliable first–break energy in the waveform records, and the traveltimes in these traces were picked directly with the help of EPOCH II without any difficulties. For some faint traces with very low signal levels, we chose different settings in EPOCH II, such as gate start, gate level, gate width, reference level and rejection, and so on. Then the first arrival picks were averaged to give a single value as the traveltime.

ART (Algebraic Reconstruction Techniques) algorithms were chosen from the various tomographic imaging methods. These algorithms pose the integral equation in the matrix form. Then the matrix equation is solved using appropriate relaxation, and taking advantage of the sparsity of the matrix. The ART algorithms are generally simple and can be easily modified to different data geometries and to restricted angular coverage as well as irregular sampling. The ART algorithms belong to a category of summation expansion methods originated in X-ray tomography in which the paths of energy are approximated by the straight rays between sources and receivers.

The traveltimes of the acoustic waves crossing a two–dimensional circle region can be represented as a line integral:

$$t = \int_{Ray} \frac{1}{V(x,y)} ds, \tag{1}$$

where $V(x,y)$ is the velocity distribution in the imaged region. The integration is along the particular raypath determined by this velocity distribution. The inversion problem is now to estimate $V(x,y)$ from the observed traveltimes and the raypaths which subject to $V(x,y)$. Based on the assumption of straight raypaths, a set of linear equations can be obtained after discretizating the imaged region. The final mathematical model for the traveltime tomography is written using the matrix form:

$$\mathbf{T} = \mathbf{B.S}, \tag{2}$$

Where $\mathbf{T}$ is a vector of the traveltimes, and $\mathbf{S}$ is a vector of the slownesses (the reciprocity of velocity). $\mathbf{B}$ is called the ray matrix in terms of the medical tomography, a simple form of $\mathbf{B}$ is a matrix with elements b_{ij}, corresponding to the length of the i–th ray which penetrates the j–th cell.

The tomographic imaging algorithms designed for tube geometry have been applied to the real data sets from the acoustic experiments for the two physical models.

The cell was square size with the gridnode spaced 5 mm apart in both horizontal and vertical directions, resulting in $36 \times 36 = 1296$ gridcells in total. The fine grid was used to avoid unnecessary loss of resolution. Stability problem did not arise because of the relaxation in the solutions.

Gas/water model

Figure 5 is the tomograms reconstructed from the data set for the gas/water model after 12 iteration steps. The initial velocity distribution was chosen as a homogeneous medium with a constant velocity of 1400 m/s. The relaxation coefficient was 0.8. Five velocity anomalies (the gas bubbles) with different sizes can be recognized and distinguished from each other.

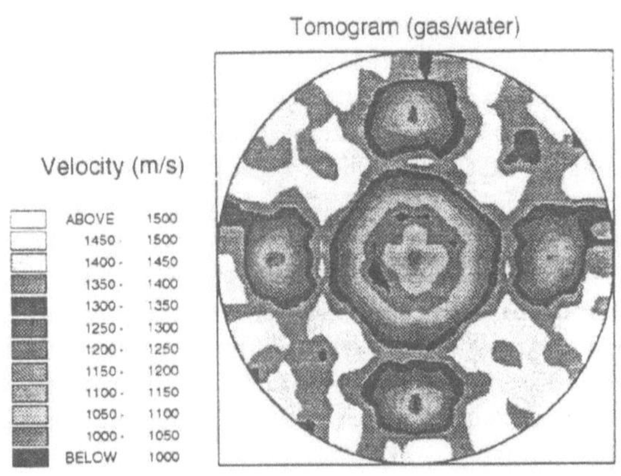

Figure 5. Tomogram for the gas/water flow model.

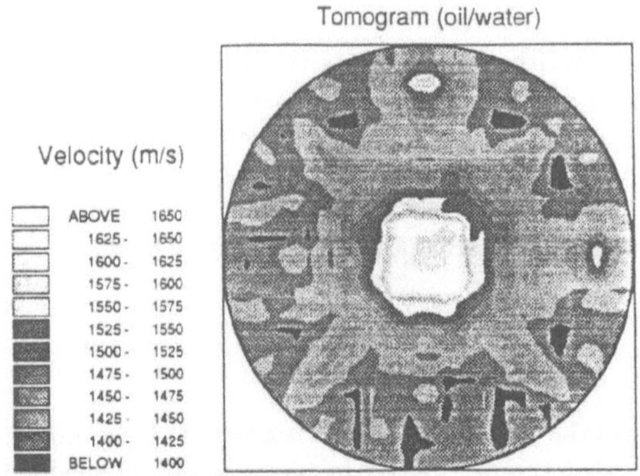

Figure 6. Tomogram for the oil/water flow model.

Oil/water model

Figure 6 is the tomograms reconstructed from the data set for the oil/water model after 15 iteration steps. The same initial velocity guess and relaxation coefficient as the gas/water model were chosen. A big oil bubble in the center of the imaged area can be clearly recognized from the tomograms. However, two small oil bubbles in the vicinity of the boundary of the imaged area were reconstructed with distortion.

CONCLUSION

We have outlined a methodology for tomographic imaging of the two–phase flow and applied it to a real ultrasonic data from physical models.

(1) The first arrival traveltimes can be picked up from most traces in the acoustic waveform records. The information of traveltime differences, existed in different source–receiver pairs, is available to image the two–phase flow patterns using tomographic reconstruction techniques.

(2) The traveltime tomographic imaging software designed for the tube geometry has been tested with real data to show a good convergence and reliability.

(3)The strong reflected or diffracted energy can be detected in the acoustic data, and in consequence the full waveform tomographic imaging techniques using reliable information about the reflection or diffraction are expected to provide higher qualified images than using traveltimes only. We will develop the diffraction tomography to image the two–phase flow patterns by using the acoustic measurements in the future.

ACKNOWLEDGMENT

G. Lu wishes to thank Amerada Hess, Saga Petroleum A/s, Norsk Hydro, and Statoil for financial support of this research, and thank Ø. Baltzersen in IKU for his helps with laboratory work.

REFERENCES

Bregman, N.D., Bailey, R.C., and Chapman, C.H., 1989, Crosshole seismic tomography: Geophysics, **54**, 200-215.

East, R.J.R., Worthington, M.H., and Goulty, N.R., 1988, Convolutional back–projection imaging of physical models with crosshole seismic data: Geophysical Prospecting, **36**, 139–148.

Lines, L.R., and Lafehr, E.D., 1989, Tomographic modeling of a cross–borehole data set: Geophysics, **54**, 1249–1257.

Krajewski, C., Dresen, L., Gelbke, C., and Rüter,H., 1989, Iterative tomographic methods to locate seismic low-velocity anomalies: a model study: Geophysical Prospecting, **37**, 717-751.

Miller, D., Oristaglio, M., and Beylkin, G., 1987, A new slant on seismic imaging: migration and integral geometry: Geophysics, **52**, 943-964.

Peterson, J.E., Paulsson, B.N.P., and Mcevilly, T.V., 1985, Application of algebraic reconstruction techniques to crosshole seismic data: Geophysics, **50**, 1566–1580.

Pratt, R.G., and Worthington, M.H., 1988, The application of diffraction tomography to cross-hole seismic data: Geophysics, **52**, 1284–1294.

Pratt, R.G., and Worthington, M.H., 1990a, Inverse Theory applied to multi-source cross-hole tomography. Part 1: acoustic wave-equation method: Geophysical Prospecting, **38**, 287–310.

Pratt, R.G., and Worthington, M.H., 1990b, Inverse Theory applied to multi-source cross-hole tomography. Part 2: Elastic Wave-equation method: Geophysical Prospecting, **38**, 311–329.

Tarantola, A., 1984, Inversion of seismic reflection data in acoustic approximation: Geophysics, **49**, 1259–1266.

FAR-FIELD RADIATION PATTERNS AND ENERGIES FROM
DIFFERENT SEISMIC SOURCES IN A FLUID-FILLED BOREHOLE

Yinbin Liu,[1] Youmin Li,[1] Rushan Wu,[1] Shoumian Yu[2]

[1]Institute of Geophysics
Chinese Academy of Sciences
Beijing 100101, P. R. China
[2]Department of Physics, Shandong University
Jinan 250100, P. R. China

ABSTRACT

The far-field radiation patterns and energies from different axissymmetrical seismic sources in a fluid-filled borehole are studied in detail under a low-frequency approximation. The range of low-frequency is discussed. The radiation patterns of the compressional and shear waves from different sources are numerically studied. The analytical expressions for the energies of the tube wave, the radiation compressional wave, and the radiation shear wave are derived. From the calculations we find that a point volume source emits almost all ($>99\%$) its energy as tube wave which travels along the borehole and not out into the formation, and the radiation shear energy is greater than the radiation compressional energy. The radiation energy of the compressional and shear waves is greater in soft formation than in hard formation. The tangential stress source radiates all its energy into the formation as compressional and shear waves, while it does not excite the tube wave in a fluid-filled borehole, in addition, the radiation shear energy is about ten times greater than the radiation compressional energy. The radiation energies from a point volume source and a radial stress source are directly proportional to ω^4, and the radiation energy from a tangential stress source is proportional to ω^2. The array source constituted by identical individual source can much enhance the radiation energy.

INTRODUCTION

For an empty borehole, Heelan (1953) and White (1960) discussed the far-field radiation patterns and energies of P-wave and S-wave from the radial and tangential forces acting on the borehole wall, Greenfield (1978) showed the radiation pattern from a point volume source acting on the wall of the borehole. White and Senbush (1963) formulated the effect of borehole fluid on the seismic radiation patterns combining Heelan's solution with the tube wave inside the borehole. Lee and Blach (1982), Lee, et al (1984), and Lee (1986) derived the far-field radiation patterns from axissymmetrical and nonaxissymmetrical sources in a fluid-filled borehole and discussed the effect of the borehole on the seismic radiation into the surrounding medium. Chen et al (1990) studied experimentally the basic physical properties of the various sources in open holes and cased holes. Winbow (1991a, b) analyzed the radiation energies of seismic source in open holes and cased holes. This paper briefly derived the formulas on the radiation properties and studied in detail the radiation patterns and energies, from different axissymmetrical downhole sources.

POINT VOLUME SOURCE

Consider a cylindrically circular, fluid filled borehole with radius a in a homogeneous elastic medium of density ρ_2, compressional velocity v_c, and shear velocity v_s. The fluid medium has compressional velocity v_1 and density ρ_1. In frequency domain the elastic wave field inside and around the borehole can be described by scalar functions φ' (inside), φ, and ψ (around), respectively. In a cylindrical coordination system, three scalar functions satisfy the wave equations.

Boundary condition requires that radial stress and displacement are continuous and the tangential stress is zero on the wall of the borehole, the radiation condition requires that the waves vanish at infinity. Then we have

$$\Phi' = \int_{-\infty}^{+\infty} \left[SK_0(\mu_1 r) + A(k, \omega) I_0(\mu_1 r) \right] e^{ikz} dk$$

$$\Phi = \int_{-\infty}^{+\infty} B(k, \omega) K_0(\mu_c r) e^{ikz} dk \tag{1}$$

$$\Psi = \int_{-\infty}^{+\infty} C(k, \omega) K_1(\mu_s r) e^{ikz} dk$$

$$\mu_1 = \sqrt{k^2 - \omega^2/v_1^2} \qquad \mu_c = \sqrt{k^2 - \omega^2/v_c^2}$$

$$\mu_s = \sqrt{k^2 - \omega^2/v_s^2} \qquad k = \omega/v$$

where S is the source coefficient, ω denotes the angle frequency, v denotes the phase velocity, k denotes the wave number, and I_i and K_i are modified Bessel functions of the ith ($i = 0, 1$) order.

Generally speaking, the frequencies using by the cross borehole geophysics and reverse VSP are several hundred Hz, and the receivers are far from the downhole sources, so we can take the first order approximation. The phase velocity of the refracted compressional wave is larger than the formation compressional velocity v_s. For $f \ll v_1/2\pi a$, i. e., the Bessel functions have approximation formulas

$$I_p(x) \approx \frac{x^p}{2^p \Gamma(1+p)}; \qquad K_p(x) \approx \frac{x^{p-1}\Gamma(p)}{x^p}; \qquad K_0(x) \approx \ln\frac{1}{x} \qquad (x \to 0) \tag{2}$$

Equation (2) can be well satisfied from $x < 0.1$, which is corresponding to the low frequency radiation condition of $f < 350$Hz for the borehole radius of $a = 10$cm, when softer the condition to $x < 0.2$, we can obtain the low frequency radiation of $f < 700$Hz, which is generally satisfied for the cross borehole and RVSP.

The refracted compressional and refracted shear waves are corresponding to the contributions of the compressional branch-cutand shear branch-cut, i. e. ,

$$\Phi = \int_{-k_c}^{k_c} B(k, \omega) K_0(\mu_c r) e^{ikz} dk, \qquad \Psi = \int_{-k_s}^{k_s} C(k, \omega) K_1(\mu_s r) e^{ikz} dk \tag{3}$$

where $k_c = \omega/v_c$, $k_s = \omega/v_s$.

Obviously, Integral (3) does not contain the pole points of the tube wave, in the far-field approximation, the directions of k and R are the same, we can move $B(k, \omega) = B(k_c \cos \varphi, \omega)$ from Integral (3), according to the stationary phase method we can derive the directional factors of the compressional and shear

$$D_P(\varphi) = \left| \frac{1 - v_s^2 \cos^2\varphi/v_c^2}{\rho_1/\rho_2 + v_s^2/v_1^2 - v_s^2\cos^2\varphi/v_c^2} \right| \tag{4a}$$

$$D_S(\varphi) = \left| \frac{2\sin\varphi\cos\varphi}{\rho_1/\rho_2 + v_s^2/v_1^2 - \cos^2\varphi} \right| \tag{4b}$$

Figures 1 and 2 show the radiation patterns of the compressional and shear waves in hard $(v_s > v_1)$ and soft $(v_s < v_1)$ formations. The radiation patterns in two conditions are similar.

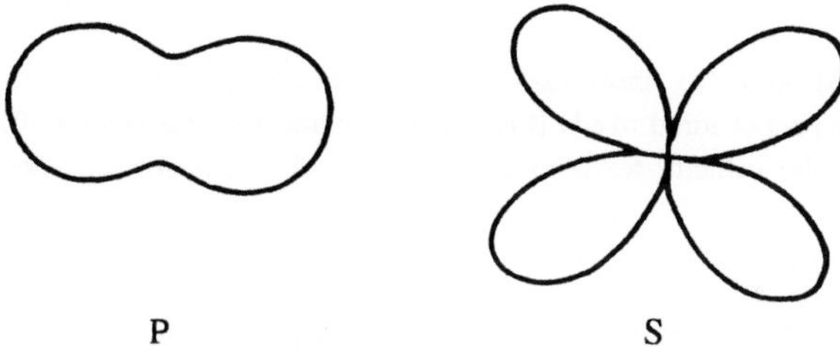

Figure 1. Radiation patterns in hard formation from a point volume source, acting on the axis of the borehole ($f = 200$Hz, $l = 0.1$m, $n = 4$, $\rho_2/\rho_1 = 2.2$, $v_1 = 1500$m/s, $v_s = 2400$m/s, $v_c = 4000$m/s)

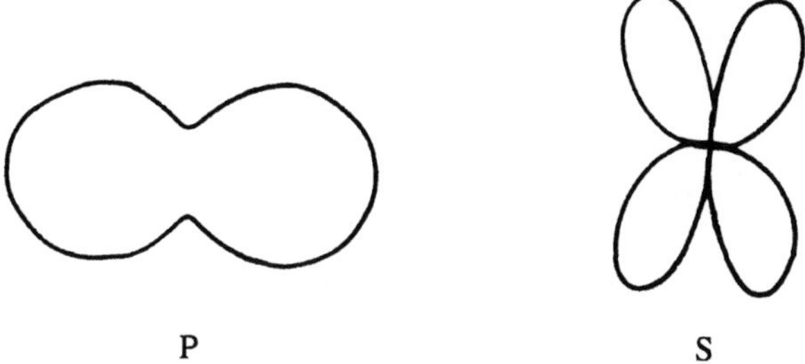

Figure 2. The same as Figure 1 for a soft formation ($\rho_2/\rho_1 = 2.0$, $v_s = 1300$, $v_c = 2000$m/s)

According to the displacement and stress, we can calculate the proportion of source energy that actually radiate into the formation. The intensity I_R of the radiation is

$$I_R = T_R \cdot V_R \tag{5}$$

where T_R is the purely radial component of stress tensor and V_R is the formation particle velocity in the radial direction. By integrating over all angles the total P-wave energy radiated at a given frequency across a spherical surface of large radius is

$$E_p = 4\pi\rho_2/v_c\left(\frac{\omega^2\rho_1 V_0}{4\pi\rho_2}\right)^2 I\,(P) \tag{6a}$$

$$I\,(P) = 4 - \frac{b_1^2\,(2b_2^2 - 1)^2}{2b_2^2\,(1 - b_1^2 b_2^2)} + \frac{b_1}{4b_2^3}\,(12b_2^4 - 4b_2^2 - 1)\ \ln\left|\frac{1 - b_1 b_2}{1 + b_1 b_2}\right| \tag{6b}$$

where $b_1 = v_c/v_s$, $\quad b_2^2 = \rho_1/\rho_2 + v_s^2/v_l^2$.

Similarly, we can derive the radiation energy of the S-wave.

$$E_s = 16\pi\rho_2/v_s\left(\frac{\omega^2\rho_1 V_0}{4\pi\rho_2}\right)^2 I\,(S)$$

$$I\,(S) = -\,1.5 + \frac{1}{4b_2}\,(1 - 3b_2^2)\ \ln\left|\frac{1 - b_2}{1 + b_2}\right| \tag{7}$$

Equations (6) and (7) show that the radiation energies of the compressional and shear waves are directly proportional to ω^4. It is worth showing that Equations (6) and (7) do not fit the case that the tube velocity tends to the shear velocity, this special case will be discussed in another paper.

The tube energy can be written as

$$E_T = 8\pi\rho_1 v_T\left(\frac{\omega V_0}{4\pi a}\right)^2 \tag{8}$$

Equation (8) shows that the radiation energy of tube wave is directly proportional to ω^2.

With frequency increasing, the energy increasing of the compressional and shear waves is faster than that of the tube wave. In other words, the high frequency signal is beneficial to radiation energy. In the soft formation, the radiation compressional energy is less than 0.1% of the tube energy and the radiation shear energy is less than 0.5% of the tube energy for $f = 200$Hz. In the hard formation, the radiation compressional and shear energies are less than 0.01% of the tube energy for $f = 200$Hz. Therefore, the radiation energy of the compressional and shear waves is greater in the soft formation than in the hard formation.

RADIAL STRESS SOURCE

Let us consider an axissymmetrical radiant force acting on the wall of a fluid-filled borehole, similar to the analysis of the point volume source mentioned above the radiation factors of the compressional and shear waves can be written as:

$$D_P^R(\varphi) = \left| \frac{(1 - v_1^2\cos^2\varphi/v_c^2)\,(1 - 2v_s^2\cos^2\varphi/v_c^2)}{(\rho_1/\rho_2 + v_s^2/v_1^2 - v_s^2\cos^2\varphi/v_c^2)} \right|$$

$$D_S^R(\varphi) = \left| \frac{2\,(1 - v_1^2\cos^2\varphi/v_s^2)\sin\varphi\cos\varphi}{\rho_1/\rho_2 + v_s^2/v_1^2 - \cos^2\varphi} \right| \tag{9}$$

Figure 3 shows the radiation patterns of the compressional and shear waves in the hard formation from a radial source. Similar to the point volume source, we can derive the radiation energy from a radial stress source as follows:

P-wave

$$E_P^R = 4\pi\rho_2/v_c\left(\frac{a^2\omega^2 E^r}{4\rho_2\mu v_1^2} \right)^2 I^R(P) \tag{10}$$

where

$$I^R(p) = 1 + \frac{4}{3b_1^2}\left(2\frac{\rho_1}{\rho_2} + \frac{3}{5b_1^2} - 1 \right) + 8\frac{\rho_1}{\rho_2}(b_2^2 - 1)$$

$$+ \frac{\rho_1^2}{\rho_2^2}I(P) - \frac{\rho_1}{\rho_2}\frac{b_1}{b_2}(2b_2 - 1)^2\ln\left| \frac{b_1 b_2 + 1}{b_1 b_2 - 1} \right| \tag{11}$$

S-wave

$$E_S^R = 16\pi\rho_2/v_s\left(\frac{a^2\omega^2 E^r}{4\rho_2\mu v_1^2} \right)^2 I^R(s) \tag{12}$$

where

$$I^R(s) = \frac{2}{15} + \frac{4}{3}\frac{\rho_1}{\rho_2} - 2\frac{\rho_1}{\rho_2}b_2^2 + \frac{\rho_1^2}{\rho_2^2}I(s) - \frac{\rho_1}{\rho_2}b_2(1 - b_2^2)\,\ln\left| \frac{b_2 + 1}{b_2 - 1} \right| \tag{13}$$

Tube wave

$$E_T^R = \frac{8\pi\rho_1 v_T v_1^4}{a^2\omega^2 v_s^4}\left(\frac{a^2\omega^2 E^r}{4\rho_2\mu v_1^2} \right)^2 \tag{14}$$

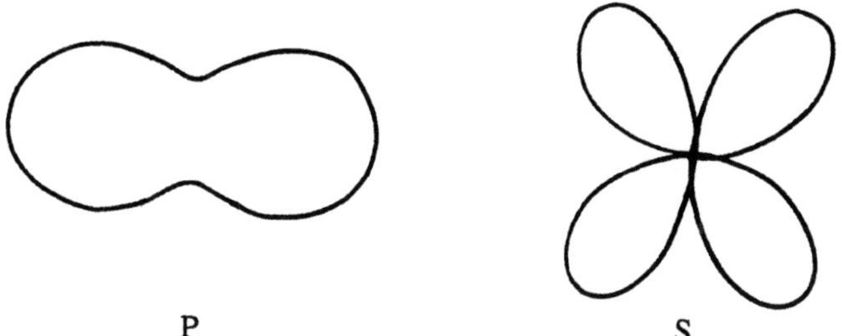

Figure 3. Radiation patterns from a radial stress source in the hard formation (P-wave with a scale factor of 1 : 2)

The radiation energies of the compressional and shear waves are proportional to ω^4, and the tube energy is proportional to ω^2. With frequency increasing, the energy increase of the compressional and shear waves is faster than that of the tube energy. In the soft formation, the radiation compressional and shear energies are less than 1% of the tube energy, and the radiation compressional energy is larger than radiation shear energy for $f = 200\text{Hz}$. In hard formation, the radiation energies of the compressional and shear waves are less than 6% of the tube energy, and the radiation shear energy is larger than radiation compressional energy for $f = 200\text{Hz}$. Therefore, the radiation energies of the compressional and shear waves are greater in the hard formation than in the soft formation.

TANGENTIAL STRESS SOURCE

Consider an axissymmetrical tangential stress source acting on the wall of a borehole, according to the boundary condition we can derive the directional factors of the compressional and shear waves

$$D_p^z(\varphi) = \left| \frac{\cos\varphi}{v_c} \right|; \qquad D_s^z(\varphi) = \left| \frac{\sin\varphi}{v_s} \right| \tag{15}$$

The radiation patterns from a tangential stress source and a tangential stress array source are shown in Figure 4.

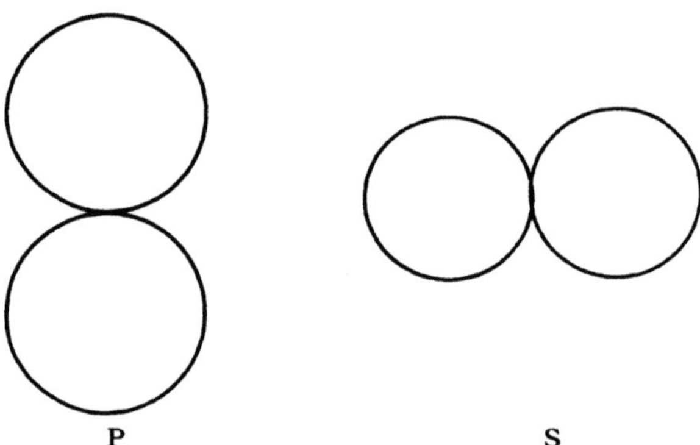

P   S

Figure 4. Radiation patterns in the hard formation from a tangential stress source acting on the wall of the borehole

We can derive the radiation energy of the compressional and shear waves from a tangential stress force as

$$E_p^z = \frac{a^2\omega^2 E^z}{3\pi\rho_2 v_c^3}, \qquad E_s^z = \frac{2a^2\omega^2 E^z}{3\pi\rho_2 v_s^3}, \qquad \frac{E_p^z}{E_s^z} = \frac{1}{2}\left(\frac{v_s}{v_c}\right)^3 \tag{16}$$

where E^z is the source coefficient. The radiation energies of the compressional and shear waves are proportional to ω^2. For most of the sedimentary rock, its Poisson's ratio is 0.25, $E_p/E_s = 10\%$. This shows that the radiation shear energy is 10 times larger than the radiation compressional energy.

Similarly, we can derive the radiation patterns and energies from a point n volume array source, radial stress array source, and a tangential stress array source constituted by identical individual source, respectively. The results show that the radiation energy from each array source is n times source.

CONCLUSIONS

The far-field radiation patterns from an axissymmetrical impulsive source acting on either the borehole axis or the borehole wall are derived under low-frequency approximation. The following conclusions can be made:

(1) The far-field low frequency approximation given in this paper fits the downhole source whose frequencies are less than 700Hz for $a = 10$cm of the borehole radius.

(2) A point volume source acting on the borehole axis and a radial stress source acting on the borehole wall generate the tube wave inside the borehole and the compressional and shear waves in the formation, and a tangential stress source acting on the borehole wall does not generate the tube wave inside the borehole but generates only the compressional and shear waves in the formation.

(3) A point volume source emits almost all ($>99\%$) its energy as tube wave and a tangential stress force emits 95% its energy as tube wave for $f = 200$Hz.

(4) The radiation energies of the compressional and shear waves from the point volume source and the radial stress source are proportional to ω^4 and the energy from tube wave is proportional to ω^2, the radiation energy of the compressional and shear waves from the tangential stress source is proportional to ω^2.

(5) The radiation energy from the point volume source is larger in the soft formation than in the hard formation, while the radiation energy from the radial stress force is larger in the hard formation than in the soft formation.

(6) The radiation shear energy is 10 times larger than the radiation compressional energy from a tangential stress source.

(7) The array source constituted by identical individual source can much enhance the radiation.

REFERENCES

Heelan, P. A, Radiation from a cylindrical sources of finite length, *Geophysics*, 1953, 18 (4) : 685—697.

Lee, M. W. and Balch, A. M, Theoretical seismic wave radiation from a fluid-filled borehole, *Geophysics*, 1982, 47 (9) : 1308—1314.

Lee, M. W, Low frequency radiation from point sources in a fluid-filled borehole, *Geophysics*, 1986, 51 (9): 1801—1807.

Winbow, G. A, Seismic sources in open and cased borehole, *Geophysics*, 1991, 56 (7): 1040—1050.

INFLUENCE OF MEASURING FIELDS, TRAFFIC LANES AND SECTIONS ON THE
CHARACTERISTICS OF THE LONGITUDINAL ULTRASONIC WAVE PROPAGATION

Romuald J.Sztukiewicz [1], Krystyna Łybacka [2]

[1] Institute of Civil Engineering
[2] Institute of Mathematics
Technical University
61-138 Poznan, Poland

INTRODUCTION

The condition of flexible pavement surface layer was described by means of the ultrasonic method. This method was also used to observe changes taking place in the condition of the surface layer in time. "In situ" ultrasonic tests of the surface layer have been carried out in an experimental section in Serbska St., Poznan. Between November 1984 and September 1990, the measurements were taken 18 times. In the article for the 19th International Symposium on Acoustical Imaging, the ultrasonic surface method and the technology of taking the measurements were described. Results of the measurements were presented in the form of an ultrasonic image of the surface layer.

The goal of this work is to demonstrate the influence of measuring fields, traffic lanes and sections on the characteristics of longitudinal ultrasonic wave propagation by means of a multifactor variance analysis.

THE EXPERIMENTAL SECTION

Ultrasonic measurements in the pavement of Serbska St., Poznan, have been being taken since 1984. It is a 2-roadway street. each roadway has 3 traffic lanes. In each roadway, 3 measuring sections were designated at the distance of 30 m from each other. In each section, 3 measuring fields were designated in each traffic lane[1]. The measuring fields (3.0 x 0.3m each) were designated both in places with the most frequent load and with sporadic load (traffic lane axis). Each measuring field was to be a separate, representative volume element (RVE). The surface layer of flexible pavement was made of moderate granular asphalt concrete of closed structure, 0.05 m thick, according to technological requirements. In each measuring field, 6 propagation times of longitudinal ultrasonic wave were registered at the constant distance between the ultrasonic transducers. Then, the distribution of the characteristics of longitudinal ultrasonic wave propagation was designated in the form of longitudinal ultrasonic wave propagation velocity and the coefficient of wave velocity variation in each of the 54 measuring fields[2].

VARIANCE ANALYSIS

In order to determine the influence of measuring fields, traffic lanes and sections on the result of ultrasonic measurements, multifactor variance analysis was applied[3]. Longitudinal ultrasonic wave velocity and wave velocity variation coefficient were assumed as statistic characteristic X with normal distribution in general population. Measuring sections (a=6 classes) were assumed as factor A, traffic lanes (b=3 classes) as factor B, measuring fields (c=3 classes) as factor C, and repetitions as k=1. Therefore, there were N=abck=54 classes. From each of the classes, a simple sample was taken, consisting of 3 observations. Each observation was understood as the mean result obtained in a longer time interval. The procedure of variance analysis for ultrasonic measurements will be presented separately for longitudinal ultrasonic wave velocity and for wave velocity variation coefficient.

Longitudinal Ultrasonic Wave Velocity

In order to write the table of variance analysis, in the first stage the sum of the main factor (Σx), its square $(\Sigma x)^2$ and the quotient $(\Sigma x^2/N)$ were calculated. In the second stage the sums of all the main factors – sections (ΣA), lanes (ΣB), and fields (ΣC) – as well as their squares, were calculated in accordance with the following formulas:

$$\text{factor A} = \sum \frac{(\Sigma A)^2}{bck} - \frac{(\Sigma x)^2}{N} = 100000 \tag{1}$$

$$\text{factor B} = \sum \frac{(\Sigma B)^2}{ack} - \frac{(\Sigma x)^2}{N} = 20000 \tag{2}$$

$$\text{factor C} = \sum \frac{(\Sigma C)^2}{abk} - \frac{(\Sigma x)^2}{N} = 0 \tag{3}$$

In the third stage all primary co-operations were calculated:
1. co-operation: sections and traffic lanes (AxB)

$$(AxB) = \sum \frac{(\Sigma AB)^2}{ck} - \frac{(\Sigma x)^2}{N} - A_{ss} - B_{ss} = 10000 \tag{4}$$

where the ss index stands for the sum of squares,
2. co-operation: sections and measuring fields (AxC)

$$(AxC) = \sum \frac{(\Sigma AC)^2}{bk} - \frac{(\Sigma x)^2}{N} - A_{ss} - C_{ss} = 10000 \tag{5}$$

3. co-operation: traffic lanes and measuring fields (BxC)

$$(BxC) = \sum \frac{(\Sigma BC)^2}{ak} - \frac{(\Sigma x)^2}{N} - B_{ss} - C_{ss} = 10000 \tag{6}$$

In the fourth stage the secondary co-operation of all the factors was calculated (AxBxC):

$$(AxBxC) = \sum \frac{(\Sigma ABC)^2}{k} - \frac{(\Sigma x)^2}{N} - A_{ss} - B_{ss} - C_{ss} - AB_{ss} - AC_{ss} - BC_{ss} = 10000 \quad (7)$$

in the fifth stage the total sum was calculated:

$$\Sigma x^2 - \frac{(\Sigma x)^2}{N} = 160000 \quad (8)$$

Results of the calculations are presented in table 1.

Table 1. Variance Analysis – Influence of Sections, Lanes and Fields on Ultrasonic Wave Velocity

Source of Variation	Sum of Squares	Variance	Variance Estimator	(Calculated) F Statistics	Snedesor's F Distribution Quantiles (p=0.95)
Sections (A)	100000	a-1=5	20000	40	$F_{0.05;5;20} = 2.71$
Lanes (B)	20000	b-1=2	10000	20	$F_{0.05;2;20} = 3.49$
Fields (C)	0	c-1=2	0	0	$F_{0.05;2;20} = 3.49$
AxB	10000	(a-1)(b-1)=10	1000	2	$F_{0.05;10;20} = 2.35$
AxC	10000	(a-1)(c-1)=10	1000	2	$F_{0.05;10;20} = 2.35$
BxC	10000	(b-1)(c-1)=4	2500	5	$F_{0.05;4;20} = 2.87$
AxBxC	10000	(a-1)(b-1)(c-1)=20	500		
Total	160000	N-1=53			

Analisis of the variance table demonstrated that wave velocity was different for different traffic lanes (B) and different measuring sections (A). However, it did not show clearly different velocities in different measuring fields (C). It also did not show that section influence depended on the lane (AxB) or on the field (AxC). On the other hand, lane influence depended on the field (BxC) to a large extent.

The mean ultrasonic wave velocities in the sections were calculated (regardless lanes or fields). They are the following:

$$A_1 = 3328.5351 \qquad A_2 = 3383.5187$$
$$A_3 = 3328.0026 \qquad A_4 = 3324.9192$$
$$A_5 = 3304.9192 \qquad A_6 = 3238.3646$$

had been calculated. the means were arranged according to their values: $B_2 < B_3 < B_1$. Then their differences were designated and compared with SSD:

$$|B_2 - B_3| = 30.6557 > SSD$$
$$|B_2 - B_1| = 52.4923 > SSD$$
$$|B_3 - B_1| = 21.8366 > SSD$$

This implies that velocity in lane 2 was clearly different from the others. The difference between the velocities in lanes 1 and 3 was also significant.

The measuring fields were also analyzed. After the means for the fields:

$$C_1 = 3312.6811$$
$$C_2 = 3333.7820$$
$$C_3 = 3307.6032$$

and the smallest significant diference:

$$SSD = 16.323$$

had been calculated, the means were arranged according to their values: $C_3 < C_1 < C_2$. and their differences were designated and compared with SSD:

$$|C_3 - C_1| = 5.0779 < SSD$$
$$|C_3 - C_2| = 26.1788 > SSD$$
$$|C_1 - C_2| = 21.1009 > SSD$$

This implies that there were no clear differences between velocities in fields 1 and 3. On the other hand, there was a clear depedence between the middle field (2) and the external fields (1 and 3).

In a similar way, the depedences of section on lane (AxB), section on field (AxC) and lane on field (BxC) were analyzed.

Longitudinal Ultrasonic Wave Velocity Variation Coefficient

Writing the table of variance analysis for velocity variation coeficient required carrying out calculations similar to those presented above. in accordance with formulas 1 through 8. The results are presented in table 2.

Analysis of variance table demonstrated that the influence of measuring sections and traffic lanes on variation coefficient was essential, while the influence of measuring fields was insignificant. It was found that co-operation of sections and lanes (AxB) and of lanes and fields (BxC) was essential, while that of sections and fields (AxC) was not. The mean coefficients of wave velocity variation were calculated for different sections (regardless lanes and fields):

$$A_1 = 1.97 \qquad A_2 = 2.18$$
$$A_3 = 1.96 \qquad A_4 = 1.89$$
$$A_5 = 1.95 \qquad A_6 = 1.70$$

According to formula 9, the standard error of the means difference was:

$$S_{D(A)} = 0.052915$$

Table 2. Variance Analysis - Influence of Sections, Lanes and Fields on Ultrasonic Wave Velocity Variation Coefficient

Source of Variation	Sum of Squares	Variance	Variance Estimator	(Calculated) F Statistics	Snedecor's F Distribution Quantiles (p=0.95)
Sections (A)	0.476930	a-1=5	0.095386	7.57	$F_{0.05;5;20}$ = 2.71
Lanes (B)	0.237423	b-1=2	0.118712	9.42	$F_{0.05;2;20}$ = 3.49
Fields (C)	0.054806	c-1=2	0.027403	2.17	$F_{0.05;2;20}$ = 3.49
AxB	0.333235	(a-1)(b-1)=10	0.033324	2.64	$F_{0.05;10;20}$ = 2.35
AxC	0.268552	(a-1)(c-1)=10	0.026855	2.13	$F_{0.05;10;20}$ = 2.35
BxC	0.168961	(b-1)(c-1)=4	0.042240	3.35	$F_{0.05;4;20}$ = 2.87
AxBxC	0.252001	(a-1)(b-1)(c-1)=20	0.012600		
Total	1.791908	N-1=53			

Then the quantile value was read from Duncan's test table:

$$t_{0.05; 6} = 2.34$$

Therefore, the smallest significant diffrence was:

$$SSD = 0.052915 \times 2.34 = 0.123821$$

After the means had been arranged:

$$A_2 > A_1 > A_3 > A_5 > A_4 > A_6$$

their differences were designated and compared with the value of the SSD:

$$|A_2 - A_1| = 0.21 > SSD$$
$$|A_2 - A_3| = 0.22 > SSD$$
$$|A_2 - A_5| = 0.23 > SSD$$
$$|A_2 - A_4| = 0.29 > SSD$$
$$|A_2 - A_6| = 0.48 > SSD$$

This implies that section 2 differs clearly from all the other sections. Differences were also designated for section 6. It was found that section 6 also differed from all the other sections. However , no essential differences were found between sections 1, 3, 4 and 5.

After the means had been calculated, the standard error of the means difference was designated:

$$S_{D(A)} = \sqrt{\frac{2V_E}{bc}} = \sqrt{\frac{2 \times 500}{9}} = 10.5409 \tag{9}$$

Then Duncan's test was carried out and limiting ranges for the sections (24.6658), lanes (23.0846) and fields (23.0846) were designated.

After the means had been arranged according to their values:

$$A_6 < A_5 < A_4 < A_3 < A_1 < A_2$$

their differences were designated and compared with the value of the limiting range, e.g.:

$$
\begin{aligned}
|A_6 - A_2| &= 145.1541 > \text{than the limting range,} \\
|A_6 - A_5| &= 66.4278 > \text{than the limting range,} \\
|A_5 - A_4| &= 20.1268 < \text{than the limting range,} \\
|A_5 - A_3| &= 23.2102 < \text{than the limting range,} \\
|A_5 - A_1| &= 23.7427 < \text{than the limting range,} \\
|A_5 - A_2| &= 78.7263 > \text{than the limting range.}
\end{aligned}
$$

On the same basis, the following conclusions for the measuring sections were formulated:
- velocity in sections 6 differs clearly from all the others
- velocities in sections 1, 3, 4 and 5 differ clearly from those in sections 2 and 6,
- velocity in section 2 differs clearly from all the others.

Standard errors for different levels of the factors were calculated:

$$S_{D(B)} = \sqrt{\frac{2V_E}{ac}} = 7.4535 \tag{10}$$

$$S_{D(C)} = \sqrt{\frac{2V_E}{ab}} = 7.4535 \tag{11}$$

$$S_{D(AB)} = \sqrt{\frac{2V_E}{c}} = 18.2574 \tag{12}$$

$$S_{D(AC)} = \sqrt{\frac{2V_E}{b}} = 18.2574 \tag{13}$$

$$S_{D(BC)} = \sqrt{\frac{2V_E}{a}} = 12.9099 \tag{14}$$

After the means for the lanes:

$$
\begin{aligned}
B_1 &= 3342.7984 \\
B_2 &= 3290.3061 \\
B_3 &= 3320.9618
\end{aligned}
$$

and the Smallest Significant Difference (SSD):

$$SSD = 16.323$$

The depedence of the coefficient of longitudinal ultrasonic wave velocity variation on the traffic lanes (factor B) was analyzed. For that purpose the mean values of the variation coefficient were calculated (regardless sections and fields):

$$B_1 = 1.29$$
$$B_2 = 1.38$$
$$B_3 = 1.22$$

According to formula 10, the standard error of the means difference was:

$$S_{D(B)} = 0.0374165$$

Duncan's test value was:

$$t_{0.05;\ 3} = 2.19$$

The smallest significant difference was:

$$SSD = 2.19 \times 0.0374165 = 0.0819421$$

After the means had been arranged:

$$B_2 > B_1 > B_3$$

their differences were designated and compared with the value of the SSD:

$$B_2 - B_1 = 0.09 > SSD$$
$$B_2 - B_3 = 0.16 > SSD$$
$$B_1 - B_3 = 0.07 < SSD$$

It was found that differences between lane 2 and the other lanes were essential, while the difference between lanes 1 and 3 was not.

The last analysis connected with the variation coefficient of longitudinal ultrasonic wave velocity was carried out in relation to measuring fields (factor C). The mean values of the variation coefficient were calculated (regardless sections and lanes):

$$C_1 = 1.3398$$
$$C_2 = 1.2770$$
$$C_3 = 1.2684$$

According to formula 11, the standard error of the means differences was:

$$S_{D(C)} = 0.0374165$$

Duncan's test value was:

$$t_{0.05;\ 3} = 2.19$$

The smallest significant difference was:

$$SSD = 2.19 \times 0.0374165 = 0.0819421$$

After the means had been arranged:

$$C_1 > C_2 > C_3$$

their differences were designated and compared with the SSD:

$$|C_1 - C_2| = 0.06 < SSD$$
$$|C_1 - C_3| = 0.07 < SSD$$
$$|C_2 - C_3| = 0.01 < SSD$$

It was found that the measuring fields did not differ significantly from each other, therefore, there was no essential influence of the field on the value of the variation coefficient of longitudinal ultrasonic wave velocity.

In a similar way, the depedence of the influence of section and lane (AxB) and of lane and field (BxC) on variation coefficient was analyzed.

SUMMARY

The results of variance analysis confirm the possibility to register the changes taking place in measuring fields, traffic lanes and sections by means of the ultrasonic method. The detailed conclusions connected with the significance of the influence of longitudinal ultrasonic wave velocity and wave velocity variation coefficient can be explained by different external reasons and reactions. Explanation of the significance of some factors was attempted. It is easy to notice that the results of variance analysis imply that section 6 is different from the other sections, both in reference to velocity and variation coefficient. The explanation is obvious: section 6 was made from a mineral-asphaltic mix of a different composition and brought from a different bitumen-mixing plant than the mix in the other sections. Significant differences were also found between lane 2 and the other lanes (1 and 3), both for velocity and variation coefficient. They result from different traffic load on each traffic lane.

Analysis of variance table for measuring fields showed that ultrasonic wave velocity did not differ significantly in different fields. Further calculations, however, showed that this was only true in case of fields 1 and 3, while there were significant differences between field 2 and the other fields (1 and 3). The differences resulted from wheel track over the traffic lanes. At the same time, analysis of variation coefficient showed no essential differences between the measuring fields. Provided that wave velocity variation coefficient might be homogeneity measure, homogeneity would actually be the same in all measuring fields within the traffic lane.

Similar conclusions can be formulated on the basis of (BxC) products for longitudinal ultrasonic wave velocity. For wave velocity variation coefficient it was found that the influence of section and lane (AxB) and of lane and field (BxC) products was sigificant, which can be explained by the changes in homogeneity in different sections and traffic lanes.

REFERENCES

1. R.J.Sztukiewicz. "Ultrasonic Imaging of the Asphaltic Concrete Surface Layer", 19th International Symposium on Acoustical Imaging, Conf. Proc., Plenum Press, New York, vol.19, 601:605 (1991).
2. R.J.Sztukiewicz. "Applications of Ultrasonic Methods in Asphalt Concrete Testing", Ultrasonics, vol.29, 5:12 (1991).
3. D.Bobrowski, K.Łybacka. "Wybrane metody wnioskowania statystycznego", Wyd. Polit. Pozn., (1988).

A NEW GENERATION SIDE SCAN SONAR

Pierre Alais, Pascal Challande, François Ollivier, Nicolas Cesbron

Laboratoire de Mécanique Physique
Université Pierre et Marie Curie (Paris 6)
78210 Saint-Cyr-l'Ecole, France

INTRODUCTION

While it is now possible, with multibeam echosounders, to obtain true bathymetric images of the sea floor up to ranges of 10 km [1,2], the best way of imaging the underwater relief with high resolution at lower ranges remains the side scan sonar [3]. This 30 years old technique is however limited in several aspects that will be discussed later. Recent studies concern the development of a synthetic aperture side scan sonar [4] analog to the already classical synthetic aperture radars operating from planes or satellites. One difficulty for this technique is to obtain an accurate knowledge of the trajectory of the fish, mostly perturbed by navigation problems. Waiting for the non obvious development of this revolutionary technique, it remains very interesting to achieve now better performances with more classical systems.

I. THE SIDE SCAN SONAR

Imaging the sea floor with a side scan sonar is now a well known and simple technique : a fish is towed by the ship and two lateral transducers, working as well as emitters and receivers, deliver lateral beams, open in the vertical plane orthogonal to the fish and narrow in the axial direction, so that only a narrow strip of the sea floor is illuminated and delivers a reverberation information which is decoded in the horizontal lateral distance y, taking into account the distance obtained from the time of flight and the height h of navigation above the floor which is assumed nearly horizontal. The axial x coordinate is swept by the mechanical motion of the fish and covered by recurrent shots in accordance with the speed of navigation and the covered maximum range (Fig.1) [5]. Information is delivered mostly through shadows due to local relief and a corresponding lack of reverberation (Fig.2). The rusticity of the system, easiness of operation and ability to sweep a large width of the sea floor, even for low depths, are obvious advantages. However, several features limit the possibilities of this technique. If the y resolution of the image is easy to achieve by using short emitting pulses or compression of frequency modulated pulses, the x resolution orthogonal to the beam is related to the x aperture of the transducers, which imposes a constant diverging angle $\Delta\theta$ and a variable resolution $\Delta x = r\Delta\theta$, so that the image is not homogeneous in the y direction

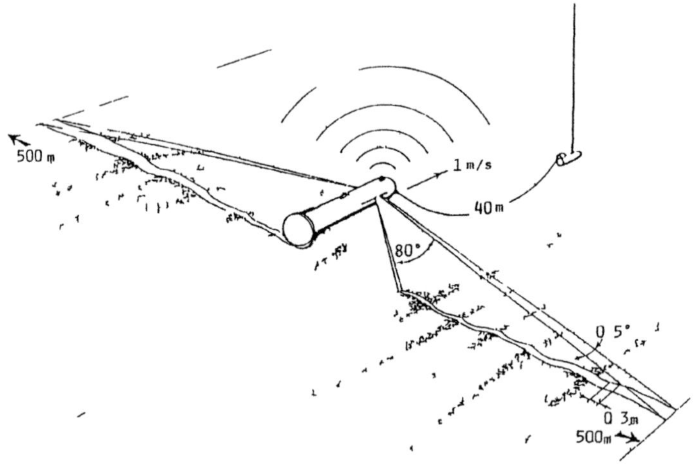

Figure 1 The acoustical towed system[5] developed by IFREMER in France
frequency 180 kHz immersion 0 5000m

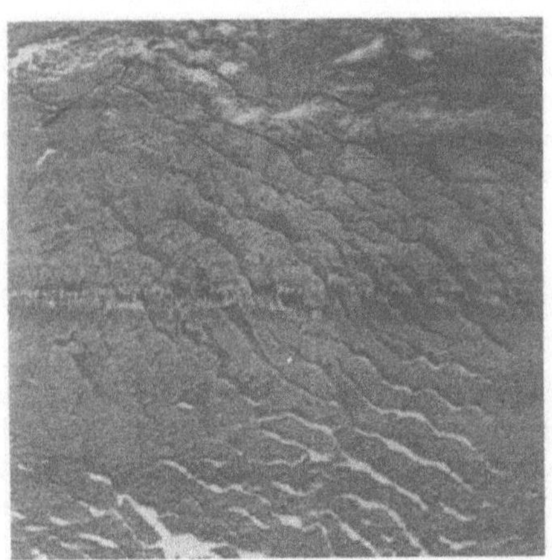

Figure 2 Field of Asymetric Dunes
(3 m high 80 m wavelength white shadows)

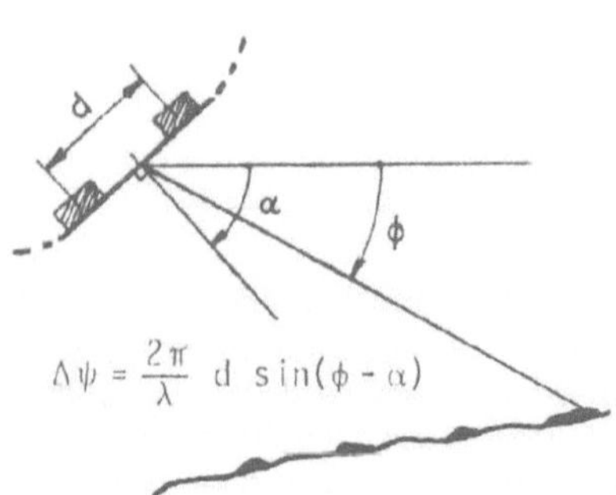

$$\Delta\psi = \frac{2\pi}{\lambda} \, d \, \sin(\phi - \alpha)$$

Figure 3 Interferometric Geometry
to obtain the grazing angle

Moreover, the speed of navigation is drastically limited by the necessity of not sweeping between two shots more than the minimal resolution obtained for the lower ranges and an elementary computation gives

$$V < 1/2 \, C \, \Delta\theta \, \sin \phi_{min}$$

which imposes, for example for $\phi_{min} = 10°$ and $\Delta\theta = .25°$, V < .6 m/s, a very severe condition, which limits either the speed of exploration or the angular resolution $\Delta\theta$. This last one is also obviously limited by the parasitic motion of the fish, essentially in terms of yaw and pitch, which should remain less than the angular resolution, a condition difficult to achieve practically.

With the classical side scan sonar, there is also a total lack of information on the main relief of the floor, which is assumed to be horizontal in the image formation. For this reason, bathymetric versions have been developed using at least two receivers, permitting through a phase measurement to appreciate the grazing angle of incoming echoes in the vertical plane of illumination (Fig. 3) [6,7,8], so that a bathymetric information may be obtained to build a relief map of the floor and a corrected image. We propose in the following a preliminary study for developing a new generation side scan sonar able to overcome the limitations encountered by the classical systems and to obtain more accurate information with higher speeds of exploration. The idea, which has already been used in military systems, is to achieve a dynamical focusing with acoustical arrays and an electronically controlled deflection of the beam to compensate the motions of the fish. The study should be done up to sea tests for a prototype achieving performances fitting many civil applications.

II. THEORETICAL ASPECTS

We have retained a relatively low frequency of operation, i.e. 100 kHz, to obtain a practical range of 1000 m suitable for exploring the continental shelf, and a maximum aperture of 4 m, which seems a maximum acceptable length for a fish and permits a lateral angular resolution of nearly 1/250 radian. To achieve a deflection of $\pm 4°$ of the beam, a value which may authorize the use of a fish operating in bad conditions, the receiving arrays have been designed with elementary transducers at the pitch of 12 cm with an angular acceptance at 3 dB of 1/8 radian, i.e. nearly 8°. Both receiving arrays have 32 such transducers with a total aperture length of 384 cm. The emitting array has been designed with a smaller pitch of 8 cm to achieve a greater angular acceptance and, in the same time, to reduce the grating lobes obtained in the echographic operation. Figure 4.c shows the optimal echographic focusing obtainable at 250 m with a deflection of 0, 2 and 4°, which shows a lateral resolution of 1 m at 3 dB and 2 m at 20 dB. Figure 4.b exhibits the focusing performances obtained from the receiving array. The first grating lobes, rejected at ± 7.5 degrees from the main lobes, appear on the left of this figure for the imposed deflection of 4° and are completely swept in the echographic focusing due to the emitting directivity represented on figure 4.a. Grating lobes, associated with the emitting array, are rejected at $\pm 12°$, due to the smaller pitch of this array, and do not overlap with the receiving grating lobes except in higher ranks, levels of which are not annoying. For lower ranges, a nearly constant value of the lateral resolution may be achieved by reducing the apodized aperture of the array and figure 5 shows what can be obtained with a choice of 16 focusing apertures differing both in focal lengths and aperture sizes. The apodized apertures have been determined according to a gaussian law with different widths. The retained constant 3 dB resolution of 1 m corresponds to a practical speed of 1 m/s. The emitted beam obviously may not be focused, but on the contrary has a divergence electronically controlled of .5, 1, 2 or 4° according to the sea state and the motion of the fish. It is a necessary condition to be sure to illuminate the intersection of the sea floor and the vertical plane orthogonal to the mean path imposed to the fish, and the angular correction electronically controlled may just compensate the yaw motion of the vehicle. During the receiving operation, the electronically tilted angle is a corrective combination of the yaw and pitch motion depending on the

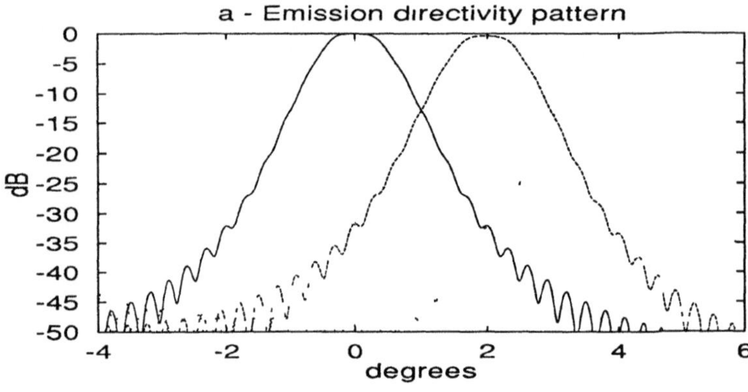

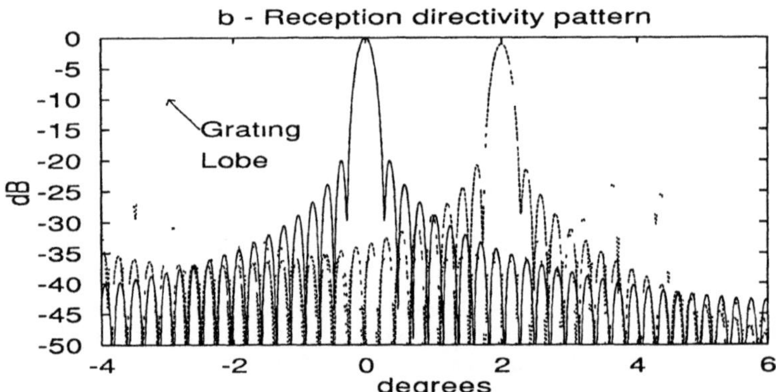

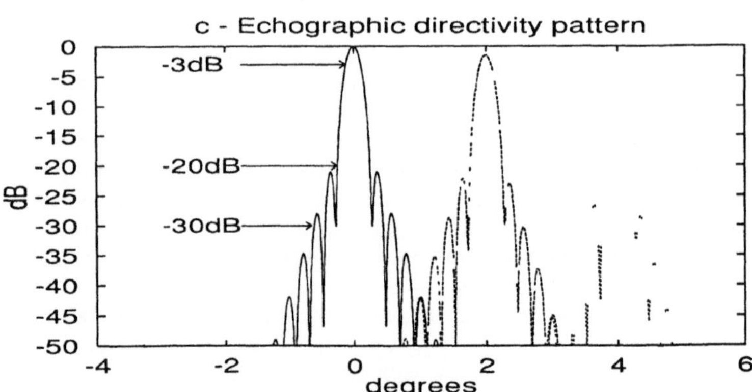

Figure 4. Directivity curves obtained at a range of 250 m for a deflection
of 0° ———— , 2° - - - - - - , 4°...........
a. Emission directivity for a divergence angle of 2° ,
b. Reception focusing with a gaussian apodizing aperture ;
c. Echographic focusing

grazing angle, i.e. of range and time to get a focusing direction remaining exactly in the vertical plane orthogonal to the mean path. The different divergences of the emitted beam have practically no effect on the echographic response, except for a difference in level which has to be taken into account by the computer while building the image.

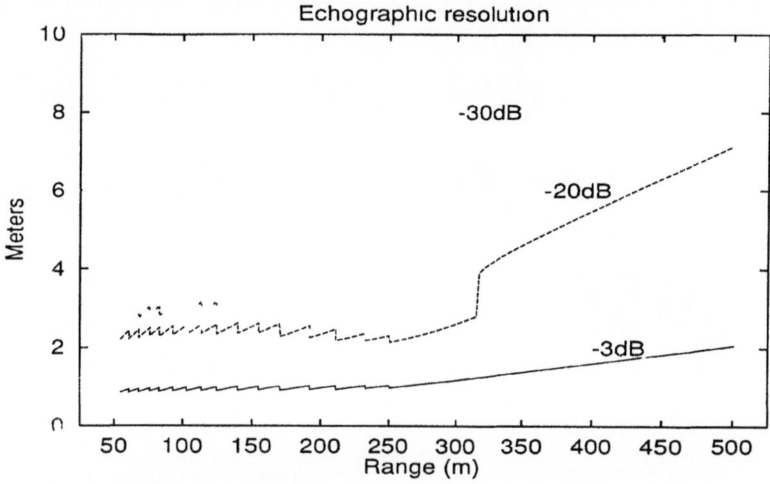

Figure 5. Optimal lateral resolution curves at -3, -20 and -30 dB
for a chosen resolution of 1m at 3 dB (16 focal zones).

III. TECHNOLOGICAL ASPECTS

The transducers have been built with composite ceramics. Orthogonally to the array, they have an angular acceptance of 60° at 6 dB (echographically), which should be sufficient by choosing an adequate grazing orientation around 40° and accepting lower sensitivity for the directions near the vertical. They have a bandwidth of 20 kHz around 100 kHz and guarantee a good phase response. Figure 6 shows the geometry of one elementary unit containing 4 receivers and 3 emitters. The whole array is made of 16 units with 2 on each end without emitters. The emitted signal is a 2 kHz bandwidth chirp of 8 ms duration, so that the electronic control of the diverging beam and the tilted angle may be operated by a control of the phase and relative amplitude, which may be numerically obtained by programmable memories on each emitting channel, as shown on figure 7.a.

The receiving channels (Fig. 7.b) are simple heterodyning circuits with a fixed frequency reference near 96 kHz, also numerically synthetized with a phase determined to compensate the delay associated to the focal length and the tilted angle which must be insured dynamically during the reception. As shown earlier, a choice of 16 focal lengths is sufficient and the angle is selected within a choice of 32 values with a pitch of .25° with a total variation of 8 degrees. The focusing operation results from the summation of the different intensities obtained by the receiving channels and the signal obtained from a current amplifier corresponds, for a single target, to a demodulated 2 kHz bandwidth chirp of 8 ms and 4 kHz central frequency (Fig.8). This signal is digitized and may be numerically correlated in line through a numerical dedicated circuitry with the signature of the elementary target, so that this linear operation gives, for a point target, the autocorrelation result, which is a 4 kHz moderately compressed signal of less than 1 ms duration, suitable for both amplitude and phase measurements. It should be noted that as the receiving operation is entirely linear, the phase of the results in the low frequency compressed signal represents the phase of the high frequency focused signal and that the difference of phase observed between results coming from the two receiving arrays may be used to extract bathymetric information. Amplitude and phase are numerically extracted in line by a dedicated numerical circuitry.

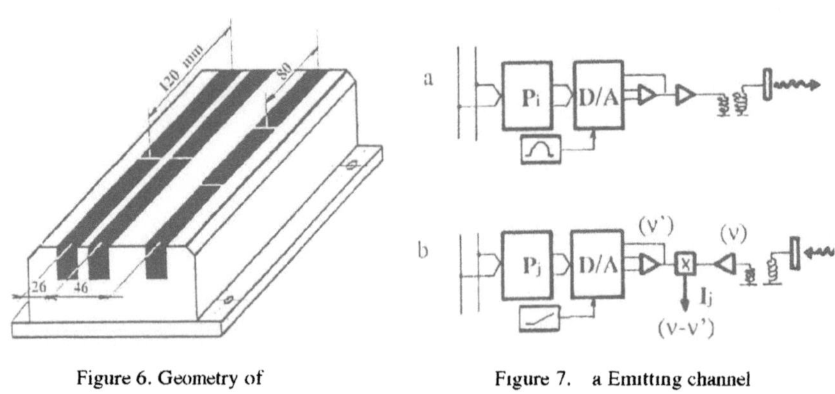

Figure 6. Geometry of
the elementary unit

Figure 7.   a Emitting channel
b.Receiving channel

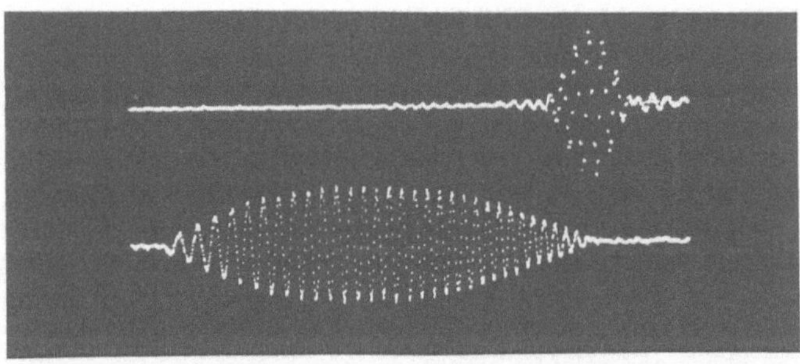

Figure 8. The 2 kHz demodulated chirp of 4 kHz initial frequency and the correlation product

A computer may achieve in real time the image formation, taking into account the correction in distance due to the height h of navigation above the floor. In the same time, it may give a bathymetric profile taking into account the bathymetric angle coming from the phase difference of the received signal and the correction which may be afforded, taking into account the roll motion of the vehicle. The data representing the roll, yaw and pitch motions of the vehicle are obtained from gyroscopic systems and must be transferred to the computer which evaluates in real time the correcting deflection of the beam both at emission and during reception.

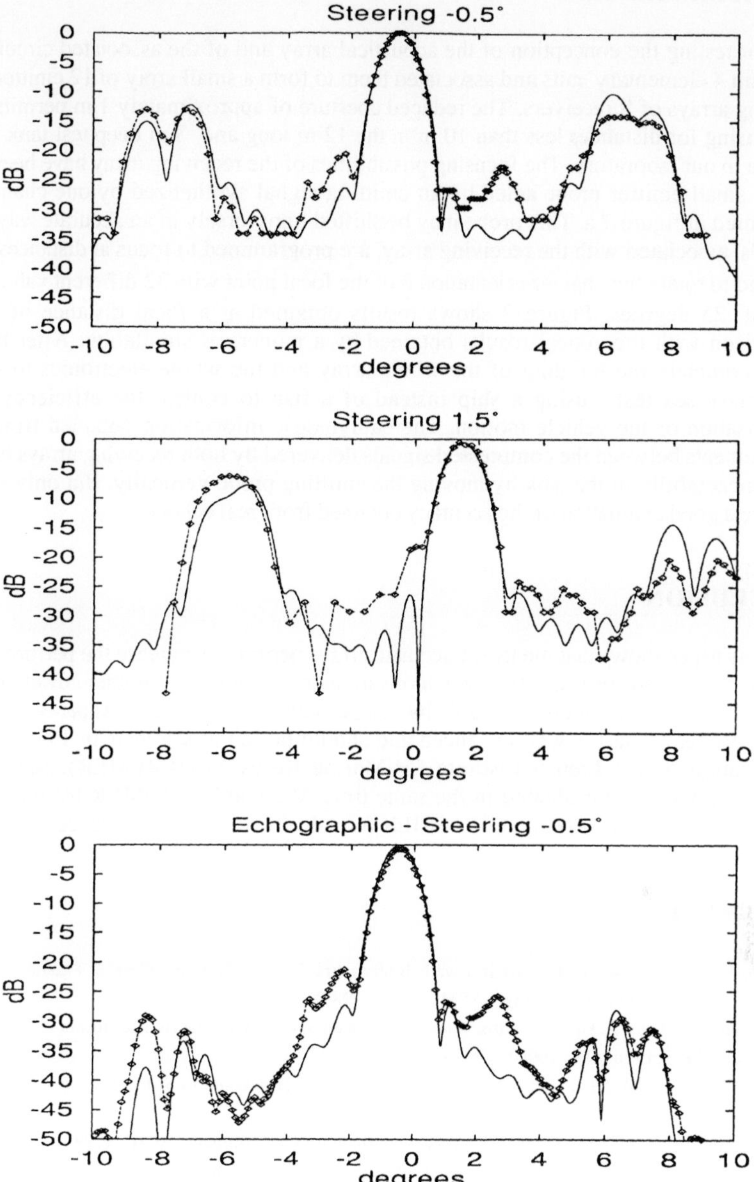

Figure 9. Comparison between theory (———) and experiment (-o--o-)
with the small array (1 m) and a probe set at 6 m.

a. Focusing performances with a deflection of - 0.5°.
    The probe emits a 500 µs signal.

b. Focusing performances with a deflection of 1.5°

c. Echographic performances with a deflection of - 0.5°

The experimental curve has been obtained from separate measurements of the emission
with the probe used as a receiver (negative focal length of 55 m).

IV. EXPERIMENTS

For testing the conception of the acoustical array and of the associated circuitry, we have built 4 elementary units and associated them to form a small array of 12 emitters and 2 receiving arrays of 8 receivers. The reduced aperture of approximately 1 m permits to test the focusing for distances less than 10 m in the 12 m long and 3 m deep test tank that we may use in our laboratory. The focusing possibilities of the receiving array have been tested using a small emitter probe acted by an emitting signal synthetized by one channel , as represented in figure 7.a. This probe may be shifted transversely in an accurate way. The 8 channels, associated with the receiving array, are programmed to focus at distances of 5 to 10 m and to rotate the angular orientation θ of the focal point with 32 different values at the pitch of .25 degrees. Figure 9 shows results obtained at a focal distance of 6 m in comparison with theoretical results obtained by a numerical simulation. After that, we should complete the building of the whole array and the whole electronics to achieve preliminary sea tests, using a ship instead of a fish to control the efficiency of the compensation of the vehicle motion. The bathymetric information obtained from phase measurements between the compressed signals delivered by both receiving arrays has been tested successfully in the tank by moving the emitting probe vertically. But only sea tests will give a good evaluation of the accuracy obtained from real echoes..

CONCLUSION

This paper shows that the use of acoustic arrays permits to increase the performance of a classical side scan sonar drastically in terms of lateral resolution. Another advantage is the possibility of using a vehicle less stable by compensating electronically its parasitic motion. The experimental part is well advanced and should be led to sea tests in 1993, with the collaboration of the French Institute for Marine Research (IFREMER). Bathymetric performances will be evaluated in the same time. We thank IFREMER for its financial support, and R. LALIMAN and C. SASSIER for valuable technical assistance.

REFERENCES

1. C. de Moustier, "State of the Art in Swath Bathymetry Survey Systems, Current Practices and New Technology in Ocean Engineering", ASME 29-38 (1988).

2. E. Hammerstad and F. Pohner, "Ultra Wide Swath Deep Sea Interferometric Multibeam Echo Sounder", Journées ISP Toulon (Déc. 1990).

3. B.W. Flemming, M. Klein and P.N. Denbigh, "Recent Developments in Side Scan Sonar Techniques", W.G.A. Russell-Cargill Edit.

4. European Community MAST Project, Acoustical Imaging Development (ACID) (1990).

5. A. Farcy, M. Voisset, J.M. Augustin and P. Arzelies, "Acoustic Imagery of the Sea Bed", Acoustical Imaging 16, 589-599 (1988).

6. P.N. Denbigh, "Swath Bathymemtry : Principles of Operation and an Analysis of Errors", IEEE Journal of Oceanic Eng. 14, 4 (1989).

7. R.L. Cloet and C.R. Edwards, "The Bathyscan Precision Swathe Sounder", Proc. MTS-IEEE Oceans 86 Conf., 1, 153-162 (1986).

8. J.G. Blackinton, D.M. Hussong and J.G. Kosalos, "First Results from a Combination Side Scan Sonar and Sea Floor Mapping System (Sea Marc II)", Proc. Offshore Technol. Conf. OTC 1, 307-314 (1983).

A REAL TIME IMAGING SONAR WITH SINGLE BEAM
USING ELECTRONIC SCANNING

Tian Tan, Fan Shibin and Guan Hao

Department of Underwater Acoustics
Harbin Shipbuilding Engineering Institute
Harbin 150001, P. R. China

INTRODUCTION

It is an effective method of detecting and imaging targets in a certain sector in the area of underwater acoustics by using a small size array of high frequency with narrow beam and mechanical scanning mode. Although the mechanical scanning sonar has an advantage of simplicity in circuit, its data rate and imaging speed are low, so it can not meet some application requirements of underwater moving vehicles whose mobility is increasingly improved. The reason is that the array has to stay at a direction and can not be rotated until the echo from targets in the longest range to be detected reaches the receiving point. A within pulse electronic beam scanning sonar is different from the mechanical scanning one. It radiates a short pulse to insonify a sector and its receiver samples at all angles in a range unit (resolution cell) related to the pulse duration. Hence the information about the distance and bearing of all the targets in the sector is obtained in a transmitting-receiving period.

By using time-variable phase compensation to each element of an equally spaced line array, a receiving beam can scan in a certain sector periodically. The modulated scanning method is a good scheme to get the time-variable phase shift with less complicated hardware and is suitable for use in a small underwater vehicle. The required synchronous oscillators for modulated scanning can be produced with digital technique, so some complexity in using analog phase-locked oscillators in the early time can be avoided. Consider that the output data from the scanned beam are stored in rows, to avoid the distortion of displayed picture, the calculation of transformation from rectangular coordinate to polar coordinate must be performed. This calculation can be completed in real time by using digital signal processor (DSP) and when the state angles of the array is aquired, the electronic stabilization of the beam can also be realized. A complete test system has been built in laboratory, in which the update rate of picture on the screen reaches 10 frames per second in a range of 40 m and a sector of 23° (double propagation ways are included). Some application software, including sector, rectangular, magnified and panoramic display have been programmed. Some experiments have been done in the test tank and the results are satisfactory.

BASIC PRINCIPLE OF BEAM SCANNING USING VARIABLE PHASE SHIFT

For a narrow bandwidth system, it is convenient to obtain beams in different directions by using phase compensation. In an equally spaced line array, the receiving signal of the kth element of the array, which has been phased, is

$$x_k (t) = A \cos (2\pi ft + k\varphi - k\beta) \tag{1}$$

where

$$\varphi = \frac{2\pi d}{\lambda} \sin\theta \tag{2}$$

and d is the spacing between elements, λ the sound wave length in water, θ the incident angle of plane wave, β the compensated phase between adjacent elements. After summing all the signals along elements, the normalized output of the array is obtained as

$$R(t, \theta) = \frac{\sin[N(\varphi - \beta)/2]}{N\sin[(\varphi - \beta)/2]} \cos \left[2\pi ft - \frac{N-1}{2}(\varphi - \beta)\right] \tag{3}$$

When the compensated phase $\beta = \varphi$, all the signals will be summed in the phase at the incident direction of sound wave, thus a beam at the direction of θ is produced. If β is changed with time linearly, i. e.

$$\beta = 2\pi f_s t - \pi, \qquad t \in [0, T_s] \tag{4}$$

the signal beam will sweep a sector as the time changes, and the scanning period is

$$T_s = 1/f_s \tag{5}$$

and the range resolution also depends on T_s. In this case, all the angles are sampled in a range resolution unit related to the pulse duration, so that all the information about distance and direction of all the targets in the sector can be obtained in a transmitting-receiving period.

Several methods of obtaining time-variable phase shift have been developed.[1,2] The modulated scanning method is the one having less hardware. If all the signals from elements are modulated by oscillators in turn with their frequency difference of f_s and a component of sum frequency is taken after summing all the modulated outputs, the output of scanned beam is obtained (see Figure 1). Assume that the modulated signals of the elements (the amplitude is omitted) are

$$x'_k (t) = \cos (2\pi ft + k\varphi) \cos [k(2\pi f_s t - \pi) + \psi_k] \tag{6}$$

if the initial phase of each oscillator is zero, i. e. $\psi_k = 0$, then after summing all the sum of frequency components, the normalized output is

$$R(t, \theta) = \frac{\sin\left[(\varphi + 2\pi f_s - \pi)\, N/2\right]}{N\sin\left[(\varphi + 2\pi f_s t - \pi)/2\right]}$$

$$\times \cos\left[2\pi\left(f - \frac{N-1}{2}f_s\right)t - \frac{N-1}{2}(\varphi - \pi)\right] \tag{7}$$

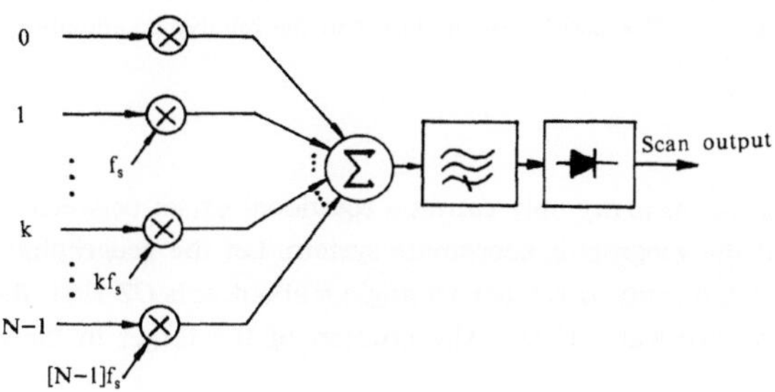

Figure 1. The principle of modulation scanning

The envelope of $R(t, \theta)$ is a slowly variable signal about time t, and is just the output of the scanned beam. Hence, it is seen that the direction of main lobe is changed as time t, so the beam scanning in the space is realized, and the scanning rate is equal to the frequency difference of oscillators used in each element of the array.

SCANNING CONVERSION AND BEAM STABILIZATION[3]

In general, the displayed picture on screen is rectangular, but the real detected area is a sector, so some distortion will occur. To present the target picture in a sector with reality, the required coordinate transformation from rectangular display to sector display must be performed. Besides, because of the roll, pitch and sway of the vehicle, the direction of the beam may deflect the desired direction even lose the target. Some measures for beam stabilization must be taken.

Assume that the line array is located at axis Y' in the relative coordinate system. In general, the measured original sonar data are distance r', azimuth α' and the elevation φ' of the target, as shown in Figure 2. The position of target M in the coordinate system is

$$x' = r' \cos\varphi' \cos\alpha' \tag{8a}$$

$$y' = r' \cos\varphi' \sin\alpha' \tag{8b}$$

$$z' = r' \sin\alpha' \tag{8c}$$

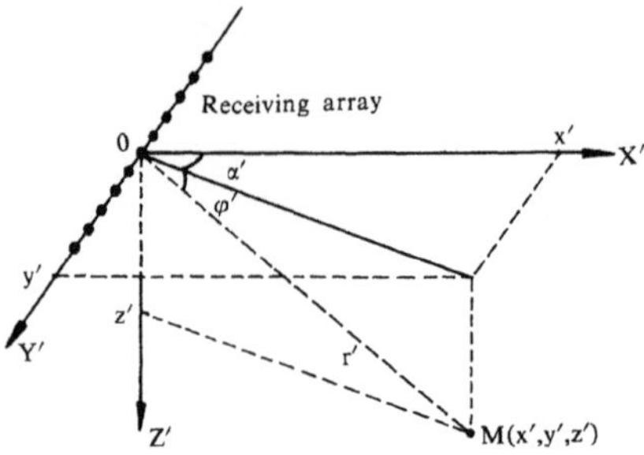

Figure 2. The position of the target in the relative coordinate system

Note that the motion of the array only causes a rotational offset between the array's relative co-ordinate system and the geographic coordinate system. Let the geographic coordinate system be $OXYZ$. Assume that the array is rotated an angle α about axis OZ first, then rotated an angle β about OY, finally rotated θ about OX. The position of the target in the geographic coordinate system is

$$\begin{bmatrix} x \\ y \\ z \end{bmatrix} = \begin{bmatrix} \cos\alpha & -\sin\alpha & 0 \\ \sin\alpha & \cos\alpha & 0 \\ 0 & 0 & 1 \end{bmatrix} \begin{bmatrix} \cos\beta & 0 & \sin\beta \\ 0 & 1 & 0 \\ -\sin\beta & 0 & \cos\beta \end{bmatrix} \begin{bmatrix} 1 & 0 & 0 \\ 0 & \cos\theta & -\sin\theta \\ 0 & \sin\theta & \cos\theta \end{bmatrix} \begin{bmatrix} x' \\ y' \\ z' \end{bmatrix} \tag{9}$$

In fact, the sonar operator can rotate the array to some direction α_t. Hence α in Eq. (9) can be viewed as an algebraical sum of α_y (the measured heading) and α_d, i. e. $\alpha = \alpha_y + \alpha_t$. For a PPI display, the real position is given as

$$\alpha_r = \mathrm{tg}^{-1}(1/x) \tag{10a}$$

$$r_r = (x^2 + y^2 + z^2)^{1/2} \tag{10b}$$

Eqs. (8), (9) and (10) are the basic operations required for displaying the target at the correct position. However, for a high resolution scanning sonar with a line array, φ' may not be measured, so we have $\varphi' = 0$. When the direction of the array, the heading of the ship and the relative azimuth are only considered (i. e. θ, $\beta = 0$), Eq. (9) becomes

$$x = r' \cos\ (\alpha_t + \alpha_y + \alpha') \tag{11a}$$

$$y = r' \sin\ (\alpha_t + \alpha_y + \alpha') \tag{11b}$$

Eq. (11) is the operation needed for scanning conversion and beam stabilization in bearing. From the obtained angle parameters, each sonar datum can be calculated by Eq. (11), so that the address of each picture cell in the picture memories can be obtained and hence the correct position display of the target on the screen will be realized.

HARDWARE DESCRIPTION

The system has been built in laboratory, which is composed of an arc transmitting array with a beamwidth of 23°, a receiving line array of 16 elements with beamwidth of 1. 5°, a transmitter, a modulation scanning receiver, the data acquisition system, a scanning converter, the picture memory and monitor. The schematic diagram of the system is shown in Figure 3, in which on the left is the underwater part. The operating frequency of the system is 333 kHz and the beam scanning rate is $f_s = 10$ kHz. MC1595Ls are employed to act as double-balanced modulators. Synchronous coherent oscillating are produced by reading EPROMs. If the read rate is f_r, and the number of written points in one period is N for frequency f_r, when

$$\frac{N}{f_r} = \frac{1}{f_s} \tag{12}$$

the sinusoidal waves with the frequency of $if_s (i = 0, 1, ... , N - 1)$ are all in phase just after $T = 1/f_s$. In the system, $N = 512$, the read rate is $f_r = 5. 12$ MHz. The single stage modulation is employed to avoid the complexity in designing filters. All the outputs of modulators are added and passed through a high-pass filter to take the sum of frequency components. Because the beam output is a vary narrow signal with only a few periods and the general detector is hard to meet the requirement, we use an absolute detector followed by a high order low pass filter to abstract the output envelope.

For the full range of 40 m and resolution of 7. 5 cm, the picture has about 512 lines. If the number of samples in each range cell is 64, there will be 32 K picture points (each point is 4

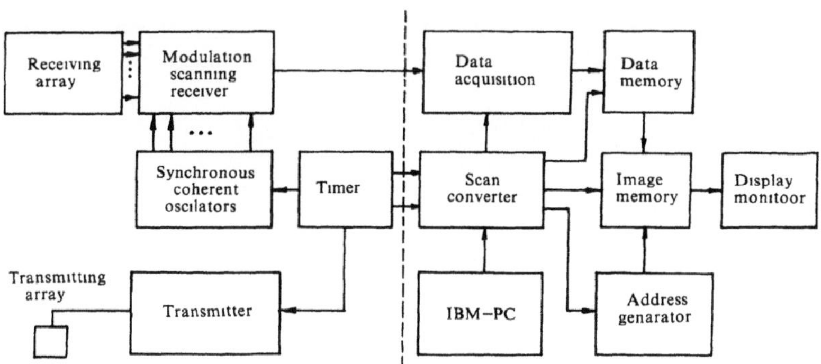

Figure 3.   The schematic diagram of imaging sonar with single beam using electronic scanning

bits) in one frame of sonar picture. To update the picture with a high rate, a DSP chip TMS320C25[4] is employed to perform the calculation in Eq. (11). The heading data measured by the bearing sensor is transferred to a signal with the corresponding frequency, which is countered and sent to the data memory. This data is acquired after the sonar data are acquired. The frame and line synchronous signal from underwater part are provided to TMS320C25, which is used for controlling the data memory to start reading data in frame and line in order to assure the displayed data to correspond the correct direction of beam in space. There are two sets of memories composed of RAMs, each of which has a capacity of $32K \times 8$ bit. They work alternately under the control of TMS320C25. When TMS320C25 is in state "hold", the contents in the current memory is displayed, which enables the picture to be frozen. It should be pointed out that the two memories with each capacity of $32K \times 8$ bit are necessary for beam heading stabilization. That is because when there is a sway, the direction of the beam will be changed and hence to display the target at the correct position, the display sector must be extended. In this system the display sector is extended to two times of the practical scanning sector of the sonar. The update rate of the picture mainly depends on the longest range to be detected and the calculating time for coordinate transformation. The update rate of the picture reaches 10 frames per second in the range of 40 m in this system.

SOME EXPERIMENTAL RESULTS

Some experiments were done in test tank with the size of $12 \text{ m} \times 6 \text{ m} \times 4 \text{ m}$. The transmitting and receiving array of the system were mounted together with their vertical beamwidth of $10°$. The transmitting pulse width was 0. 1 ms. The targets were two steel plates with their size of $20 \text{ cm} \times 20 \text{ cm}$. The spacing between them was 20 cm both in range and in direction. The array and the targets were all placed 2. 17 m below water surface. Figure 4 is a typical displayed picture when the targets are 5. 5 m far from the array. The full range displayed in this picture is 20 m in both the sector and rectangular formats. In fact, when two targets are separated for 10 cm in range, the system can still display them clearly.

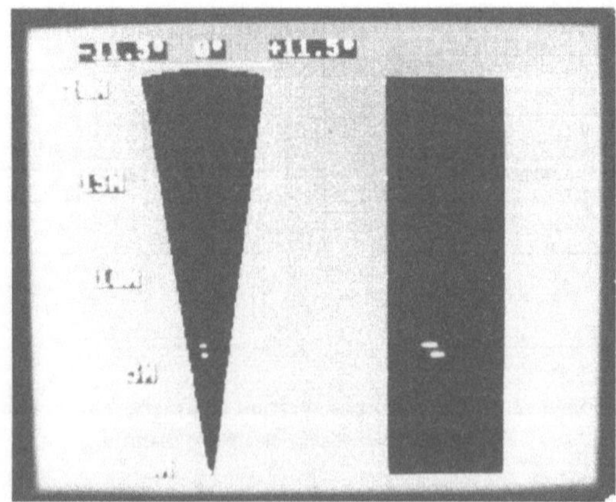

Figure 4.   The picture of two targets on screen taken from test tank

SUMMARY

Since within pulse scanning technique is employed, the data rate of the sonar is higher than that of the mechanical scanning sonar. The use of modulation scanning method makes the sonar system simpler in hardware. Furthermore, by using the high speed DSP, the update rate of the displayed picture has reached 10 frames per second in the range of 40 m and in a sector of 23°.

REFERENCES

1. R. B. Mitson, A sector scanning sonar system: Ten years of use, in: "Proc. Acoustic Conf. ", Lowestoft, 1979.

2. A. D. Goodson, et al, Seavision——A new sector scanning sonar, in: "Proc. IEE colloq. on Underwater Navigation", March, 1984.

3. Tian Tan, Liu Wenshuai and Guan Hao, A scan converter for high resolution in-within pulse Electronic scanning sonar, *Journal of Harbin Shipbuilding Engineering Institute*, Vol. 12, No. 3, Sept. , 1991.

4. "Digital Signal Processing Application with the TMS320 Family", Texas Instruments Incorporated, 1987.

3D ACOUSTIC CAMERA FOR UNDERWATER IMAGING

Rolf Kahrs Hansen, Poul Arndt Andersen

OmniTech as
Nedre Åstveit 12
N-5083 Øvre Ervik
Bergen, Norway

INTRODUCTION

Underwater work operations are often restricted by soft bottom materials stirred up by current, moving vehicles, thrusters, tools etc. The turbidity might be so severe that the use of video systems is prevented. Subsea exploration and production of hydrocarbons call for a wide variety of underwater work and inspection activities. As water depths increase as well as the emphasis on diver safety, the remotely operated vehicle (the ROV) plays an increasing role as a "subsea work horse". ROVs are normally equipped with sonars and video systems as well as navigation sensors, but there is no real-time imaging device covering the work area in case of turbid water.

OmniTech is in the process of building a real-time acoustic camera addressing the requirements of the ROVs. The project is carried out in close cooperation with Statoil, which is a major norwegian oil company with large activities in the North sea and elsewhere in the world. Since the camera is meant to do service during work and inspection operations, real-time performance is of vital importance and the primary range to be covered is between about 0.5 meters and 10 meters.

A small scale prototype has been constructed and tested in the sea. Further testing will take place and a full scale prototype is being constructed with scheduled offshore testing late 1993. This paper addresses the operational requirements for an acoustic camera to be applied as specified above, the technical solutions are discussed and some experimental results are presented.

TECHNICAL AND OPERATIONAL REQUIREMENTS

Since the objective is to build a sensor providing support during work and inspection near the sea bottom, the range of the instrument has to be up to 10 meters. Real time operation is a definite requirement. Not only is it important that inspection is fast and cost effective, but any work operation is impractical if there is no type of real time imaging sensor available. In view of the actual speed of operations under water, several images per second should be generated.

In some cases the need for a large field of view and fast updating is more important than the need for resolution and detailed information. In other cases more detailed images are important and a limited field of view can be tolerated. Thus a multifrequency approach could be used in order to provide the operator with several options with respect to resolution and field of view

Resolution tends to increase the size of the acoustic array and thus increase the number of detectors. This often leads to costly and heavy systems with large computational requirements. The objective has been to achieve the best possible resolution - with the complexity and cost within reasonable limits.

In order to provide the operator with maximum information on the surroundings, 3D information is required An acoustic pulse is emitted and reflected from objects at various distances from the camera Generation of lateral images at various distances allows a 3D representation of the environment - information that provides not only visual information, but also measures on distances and the scale of objects An artist's impression of the operation of the camera is shown in figure 1 and illustrates how detection at several distances provides 3D information

Figure 1 Operation of the acoustic camera from an ROV   artists impression

TECHNICAL SOLUTIONS

Image Generating Algorithm

In order to ensure maximum flexibility and low complexity, it is important that the algorithm does not place restrictions on placement and number of detectors in the detector array and with respect to the operating distance and the distance to the axis perpendicular to the camera face Spectral decomposition of the aperture field[1] offers this kind of flexibility Image generating matrices have to be preprocessed, however, for a set of distances and fields of view, but this is not considered a major drawback since the cost, size and power consumption of modern electronic storage circuits tend to decrease

Let g represent the acoustic field in the object plane as a complex vector and A represent the transformation of the field from object plane to aperture plane (where the detectors are located) Then the aperture field f is a complex vector and

$$f = A\,g \tag{1}$$

A is constructed using Green's function[2] assuming omnidirectional receivers By letting A be non-square, we can have more image pixel elements than we have detector elements The best estimate of the object is the image vector b

$$b = A^H (A\,A^H)^+ (f + n) \tag{2}$$

which is the minimum norm solution of (1)[3] $^+$ denotes the pseudoinverse, ^{H} denotes the complex conjugate and transpose and we have added a noise term n

Since AA^H is square and Hermitian, it can be decomposed using the spectral theorem[4].

$$\underline{b} = A^H \left(\sum_{i=1}^{k} \lambda_i^{-1} \, \underline{q}_i \, \underline{q}_i^H \, \underline{f} + \sum_{i=1}^{k} \lambda_i^{-1} \, \underline{q}_i \, \underline{q}_i^H \, \underline{n} \right) \tag{3}$$

λ_i represent the eigenvalues of AA^H and $\underline{q}_i$ represent the corresponding eigenvectors so that the order of eigensets or spectral components is according to the size of the eigenvalues. All the $\underline{q}_i$'s constitute a complete set of orthonormal eigenvectors and hence decompose the aperture vector $\underline{f}$ as well as the noise vector $\underline{n}$ into a set of orthogonal spectral components. The number of spectral components to be included is governed by the spatial distribution of the vectors (the included vectors must span the whole field of view) as well as the signal to noise ratio.

Equation (3) can be represented as one simple complex matrix multiplication with parameters stored in RAM. It can be shown[1] that an aperiodic linear array with only 11 elements may be more efficient than a 24 element strictly periodic array. Efficient in this context means resolving ability in the presence of noise relative to hardware complexity. This property is utilized in the design of the full scale prototype detector array.

Hardware Implementation

The technical solutions aim at utilizing modern electronics and microprocessor technology. The camera is realized using transputer technology and parallel processing The overall processing requirement is about 160 MFLOP, somewhat dependant upon the kind of man-machine-interface that is being used. Use of transputers allows placement of a transputer node under water at the vehicle itself with communication at full speed - 20 Mbits per second - via optical fibre to surface processors. The surface processors are transputers hosting 1860 64 bit processors via shared RAM. The number of parallel processors can be adjusted according to the computational burden. The small scale prototype in operation to day uses only one 1860 processor but can easily be upgraded to full processing power.

The full scale prototype will produce lateral images at 10 different distances per transmitted pulse and will operate at 150 kHz, 300 kHz or 600 kHz. The frequency can be selected by the operator and allows the use of three alternatives with respect to field of view and resolution. The size of the camera face will be 200 mm diameter and the number of detectors will be limited to about 400. As pulse rate will be 5 per second, 50 images will be generated per second. The operating range will be between 0.5 m and 10 m. The field of view at 150 kHz will be nearly 180 degrees, 60 degrees at 300 kHz and 30 degrees at 600 kHz.

The transmitters will be separate units connected to the camera by cable, making it possible to generate pulses from different angles and with several transmitters at the same time.

A small scale prototype has been built with 225 detectors arranged as a 7.5 cm $\cdot$ 7.5 cm array with the detectors arranged as a strictly periodic array. The resolution at 150 kHz is 8 degrees, at 300 kHz 4 degrees and at 150 kHz 2 degrees. As the electronics do not include parallel AD converters, only one image frame per pulse is captured. The camera is presently running at a frame rate of 3 per second. Hence the small scale prototype is a reduced version with respect to aperture size, processing power, resolution and 3D visualization.

The step from the small scale prototype to the full scale one will involve improved transmitter power, parallel AD converters and the grabbing of 10 frames per pulse, more processing power, improved resolution, better signal to noise ratio and reduced sidelobe levels.

EXPERIMENTAL RESULTS

Experiments confirm the assumptions with respect to resolution and imaging performance. Figure 2 shows the small scale prototype housing with the 225 element array covered by elastomer in the front. The full scale prototype will have about the same size as the small scale one - cylindrical with diameter 300 mm and length 350 mm.

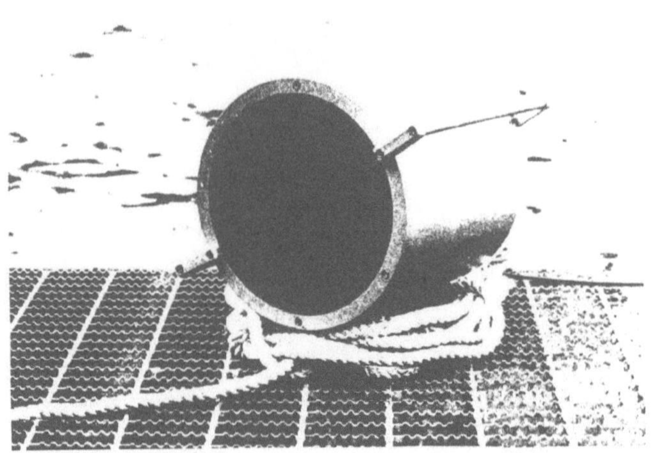

Figure 2. Housing of the small scale prototype. The elastomer in the front covers a 15·15 element array in the centre.

Experiments in laboratory and in the sea have been carried out in order to evaluate the camera performance. Imaging in the sea has been done at 300 kHz and a distance of 3 m with various simple objects. Figure 3 shows the experimental image of two air filled litre bottles. The frontal image represents a 4 m · 4 m area. Processing is done at a rate of 3 per second. Both frontal view and side and top projections are shown. Thus all the information necessary to place the image in a 3D visualization is available. Both the distance to the acquisition plane and the length of the transmitted pulse can be selected by the operator or he may choose to let the camera select the distance based on the received acoustic power.

Experiments with the small scale prototype will be continued parallel with the realization of the full scale one.

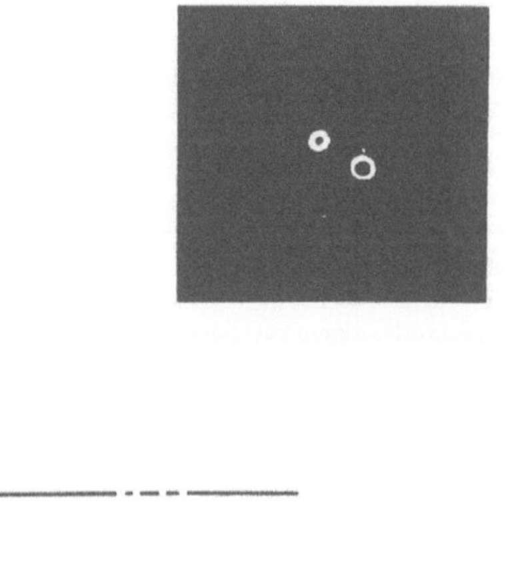

Figure 3. Experimental image produced by the small scale prototype using 300 kHz at 3 m distance. Frontal image and side and top projections.

ACKNOWLEDGEMENTS

The authors would like to thank Statoil for supporting the development of the acoustic camera and the Royal Norwegian Council for Scientific and Industrial Research for supporting the initial research work. We would also like to thank Gvt. Auth. Transl. Ada B. Kahrs for linguistic support.

REFERENCES

1. R.K.Hansen, Acoustical imaging using spectral decomposition of the aperture field, *In* :
   Acoustical Imaging, Volume 19, H.Ermert and H.-P. Harjes,ed., New York (1992), pp.103-107.

2. J.W. Goodman. "Introduction to Fourier Optics", McGraw Hill, New York (1968).

3. D.G. Luenberger. "Optimization by Vector Space Methods", Wiley, New York (1969), p.161.

4. G. Strang, "Linear Algebra and its Applications", Harcourt Brace Jovanovich, Orlando (1980), p.224.

THE IMAGE RECONSTRUCTION OF UNDERWATER NOISE SOURCES BY USING THE FRESNEL INTEGRAL

Yang Desen, Xie Changsheng, and Piao Shengchu

Department of Underwater Acoustical Engineering
Harbin Shipbuilding Engineering Institute (HSEI)
Harbin 150001, P. R. China

INTRODUCTION

The image reconstruction has been proven to be a useful tool for studying underwater noise sources. From the sound images we can get characterizations such as exact location and intensity distribution of every source. In this paper we make the sound field transformation from measurement plane to sources plane by using Fresnel integral and get the mutual intensity of sound waves in sources plane, and then obtain the radiation sound images. From this principle, the paper gives out two experiments for fixed sources and moving sources in the water, respectively, and shows the images of these sources. It is clear in the images that where is most strong and where is weakest in mutual intensity on sources plane. Therefore, this method is applicable to studying the location or mapping of radiation sound sources underwater.

THEORY OF ACOUSTICAL IMAGE

We consider the noise sources distributing on the surface of source plane S, see Figure 1. The measurement plane Σ is parallel with S. The distance between S and Σ is R, $A(x, y, t)$ is the pressure connected with the sources intensity in element dS on the sources plane. For the convenience of mathematics and measurements we consider a reference point $M_0(\xi_0, \eta_0, R)$ on the measurement plane. The mutual intensity of sound wave between any point and $M_0(\xi_0, \eta_0, R)$ on the measurement plane is defined as

$$J(\xi, \eta, R) = <P(\xi, \eta, R, t) \cdot P^*(\xi_0, \eta_0, R, t)> \tag{1}$$

where $< \cdot >$ is the time average, $*$ is the complex conjugate. $P(\xi, \eta, R, t)$ and $P(\xi_0, \eta_0, R, t)$ can be written as

Acoustical Imaging, Volume 20 Edited by Y. Wei
and B. Gu, Plenum Press, New York, 1993

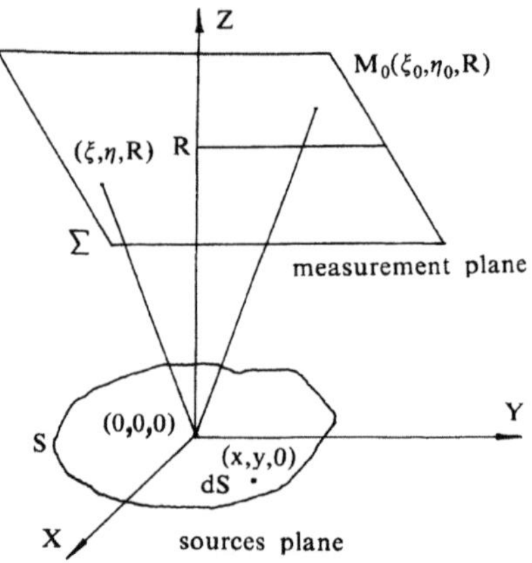

Figure 1.   The typical mode of sources plane and measurement plane

$$P(\xi, \eta, R, t) = \iint_S \frac{1}{R_1} A\left(x, y, t - \frac{R_1}{C}\right) dxdy \tag{2}$$

$$P(\xi_0, \eta_0, R, t) = \iint_S \frac{1}{R_2} A\left(x, y, t - \frac{R_2}{C}\right) dxdy \tag{3}$$

$R_1^2 = (x - \xi)^2 + (y - \eta)^2 + R^2$, $R_2^2 = (x - \xi_0)^2 + (y - \eta_0)^2 + R^2$. As shown by Ghatak[1], if Fresnel nearfield limit is R_F, $R_F < \dfrac{(2d)^2}{\lambda}$, λ is the wave length. ($2d$) is the largest dimension of sources, the signal is narrow band wave, and the measurement plane is infinite. The relations of mutual intensity between the measurement plane and the sources plane can be shown by Eqs. (4) and (5)

$$
\begin{aligned}
J(\xi, \eta, R) &= -\frac{1}{i\lambda R} \exp(-iKR) \iint_{-\infty}^{+\infty} I(x, y, 0) \\
&\quad \cdot \exp\left\{-\frac{iK}{2R}\left[(\xi - x)^2 + (\eta - y)^2\right]\right\} dxdy \\
&= -\frac{1}{i\lambda R} \exp(-iKR) \iint_S I(x, y, 0) \\
&\quad \cdot \exp\left\{-\frac{iK}{2R}\left[(\xi - x)^2 + (\eta - y)^2\right]\right\} dxdy
\end{aligned} \tag{4}
$$

$$I(x, y, 0) = -\frac{1}{i\lambda R} \exp(iKR) \iint_{\Sigma} J(\xi, \eta, R) \exp\left\{\frac{iK}{2R}[(x - \xi)^2 + (y - \eta)^2]\right\} d\xi d\eta \quad (5)$$

where $K = 2\pi/\lambda$, $J(\xi, \eta, R)$ and $I(x, y, 0)$ are mutual intensities in the measurement and sources planes, respectively. Eqs. (4) and (5) are called a pair of Fresnel transformation. As shown by Ghatak[1] and Ueha[2], it is complex to calculate (4) or (5) directly. But we can calculate them by Fourier transformation. We rewrite Eq. (4) into another form:

$$J(\xi, \eta, R) = -\frac{1}{i\lambda R} \exp(-iKR) \iint_{S} I(x, y, 0) \cdot \exp\left[\frac{iK}{2R}(x^2 + y^2)\right]$$

$$\cdot \exp\left[-\frac{iK}{2R}(\xi^2 + \eta^2)\right] \cdot \exp\left[\frac{iK}{R}(\xi x + \eta y)\right] dx dy \quad (6)$$

or

$$J(\xi, \eta, R) \cdot \exp\left[\frac{iK}{2R}(\xi^2 + \eta^2)\right] = -\frac{1}{i\lambda R} \exp(-iKR) \iint_{S} I(x, y, 0)$$

$$\cdot \exp\left[-\frac{iK}{2R}(x^2 + y^2)\right] \cdot \exp\left[\frac{iK}{R}(\xi x + \eta y)\right] dx dy \quad (7)$$

Similarly, Eq. (5) can be rewritten as:

$$I(x, y, 0) \cdot \exp\left[-\frac{iK}{2R}(x^2 + y^2)\right] = \frac{1}{i\lambda R} \exp(iKR) \iint_{\Sigma} J(\xi, \eta, R)$$

$$\cdot \exp\left[\frac{iK}{2R}(\xi^2 + \eta^2)\right] \cdot \exp\left[-\frac{iK}{R}(\xi x + \eta y)\right] d\xi d\eta \quad (8)$$

We define

$$W(\xi, \eta) = J(\xi, \eta, R) \cdot \exp\left[\frac{iK}{2R}(\xi^2 + \eta^2)\right] \quad (9)$$

$$\widetilde{W}\left(\frac{Kx}{R}, \frac{Ky}{R}\right) = I(x, y, 0) \cdot \exp\left[-\frac{iK}{2R}(x^2 + y^2)\right] \quad (10)$$

Eqs. (7) and (8) can thus represented as

$$W(\xi, \eta) = -\frac{1}{i\lambda R} \exp(-iKR) \iint_{S} \widetilde{W}\left(\frac{Kx}{R}, \frac{Ky}{R}\right) \cdot \exp\left[\frac{iK}{R}(\xi x + \eta y)\right] dx dy \quad (11)$$

$$\widetilde{W}\left(\frac{Kx}{R}, \frac{Ky}{R}\right) = \frac{1}{i\lambda R} \exp(iKR) \iint_{\Sigma} W(\xi, \eta) \cdot \exp\left[-\frac{iK}{R}(x\xi + y\eta)\right] d\xi d\eta \quad (12)$$

These two integrals are of the type of Fourier transformation. So we can calculate Fresnel integral by the aid of Fourier transformation and this is convenient.

The resolution of image reconstruction is an important problem. From Eqs. (11) and (12) we can get the measurement intervals in measurement plane both in X and Y directions:

$$\Delta\xi < \lambda R/a \tag{13}$$

$$\Delta\eta < \lambda R/b \tag{14}$$

Where a and b are the lengthes of sources plane in X and Y directions, respectively.

EXPERIMENTAL RESULTS

As mentioned above it should be satisfied to get the reconstructed image of underwater sound sources that $R > (2d)$, $R < R_F$, $\Sigma \to \infty$ and Eqs.(13) and (14). From this theory, two experiments are accomplished in a tank whose three dimensions are $11\ \text{m} \times 5\ \text{m} \times 4\ \text{m}$ in HSEI. For the reason of actual application we select two kinds of sources, one is fixed five elements cross array, and the other is a little moving ship. The calculated intensities of the reconstructed image are distributed from 0 to 35 in an arbitrary scale. The images are obtained with full scale intensities but if the levels are higher than 30, they shall be changed to 35 and if the levels are lower than 10, they shall be changed to 0. The images are drawn with a thermal sensitive plotter. To locate sound sources, the shape and the position are drawn in white in reconstructed images. i. e. white position denotes higher intensity and vice versa.

The Image of Fixed Sound Sources

The sound sources system is shown in Fig. 2. There are five point sources in the system and source 5 is not a real sound source but a diffraction source. The signal is narrow band white noise and its centre frequency is $f_0 = 50\ \text{kHz}$. The array is 2 m deep in the water. Other parameters are $R = 1\ \text{m}$, $\Delta\xi = \Delta\eta = 4\ \text{cm}$, and $\Sigma = 60\ \text{cm} \times 60\ \text{cm}$. The measurement plane and block diagram of the experiment are shown in Figure 3. We use a fixed hydrophone to measure reference signal and two hydrophones to do scanning measurement in plane. The reconstructed image of sound sources underwater is shown in Figure 4. Compared with the sources in Figure 2, we think the image shows clearly the face of sound sources. Although the fifth source is not radiation sound, the image shows the lower grey scale too because of its diffraction.

Image of Moving Ship

The shape of the ship is shown in Figure 5. The bottom area is about $40\ \text{cm} \times 20\ \text{cm}$ and it is excited by an exciter. The measurement plane is formed by the measurement of a line array when the ship is moving. The line array is 31 cm deep in the water and consisted of 13 hydrophones, the interval of every two elements is 6 cm. The reference hydrophone moves with the ship, and their speed is 16. 4 cm/s. The signal is single frequency sound wave whose f_0 is 6. 1kHz. When the ship is moving the continue signals are measured by line array. And then the signals are divided into discrete points signals in measurement plane. The area of measurement plane is $72\ \text{cm} \times 60\ \text{cm}$. The block diagram is shown in Figure 6. Figure 7 is the image reconstructed.

CONCLUSION

We have discussed the image reconstruction of underwater noise sources by using Fresnel integral as a tool of sound field transformation and completed two experiments. We think this may be an applicable method for getting clear images of sound sources underwater if only the parameters chosen satisfy the rquirement of Fresnel integral. And the calculation of Fresnel integral by the aid of Fourier transformation is convenient to some extent.

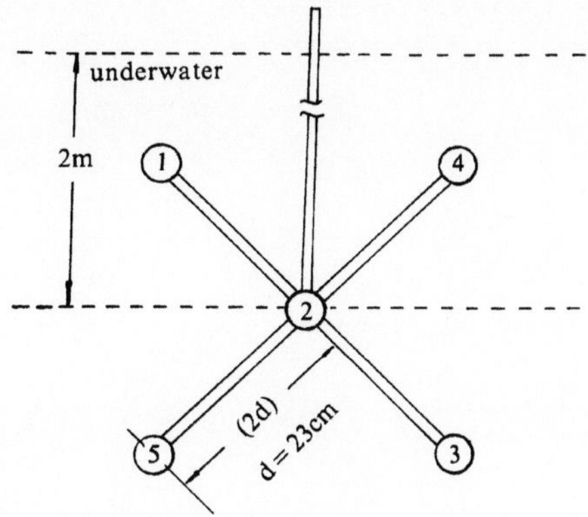

Figure 2. Five elements cross array sound sources system

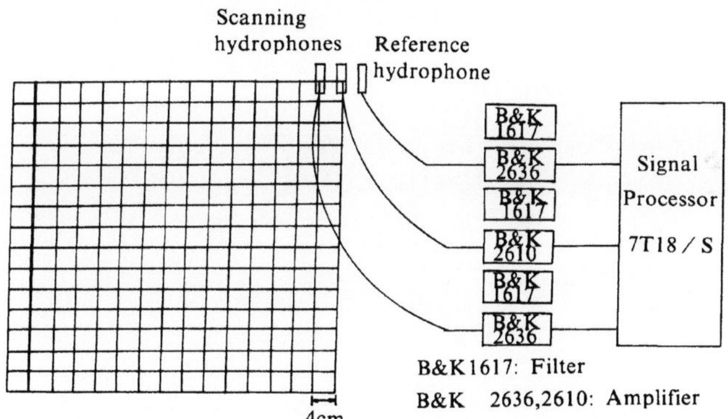

Figure 3. Measurement plane and block diagram of measurement setup for fixed sources image experiment

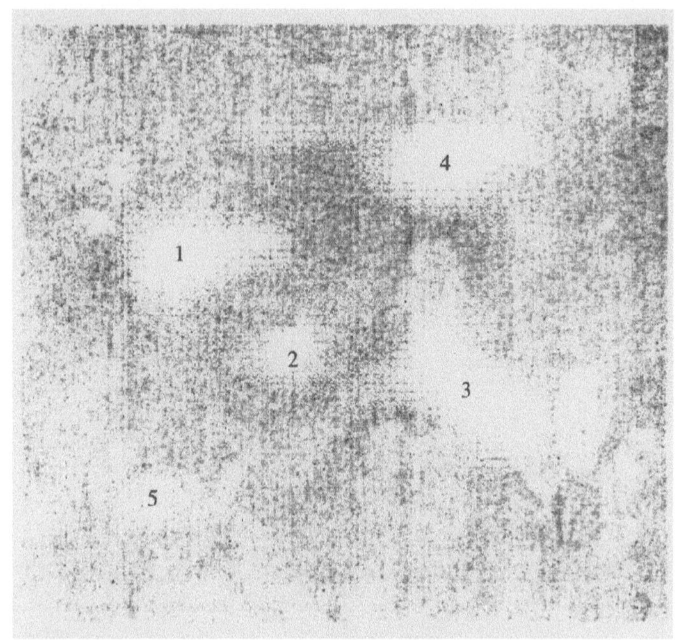

Figure 4. The sound image of five elements cross array underwater

Figure 5. Little ship with an exciter

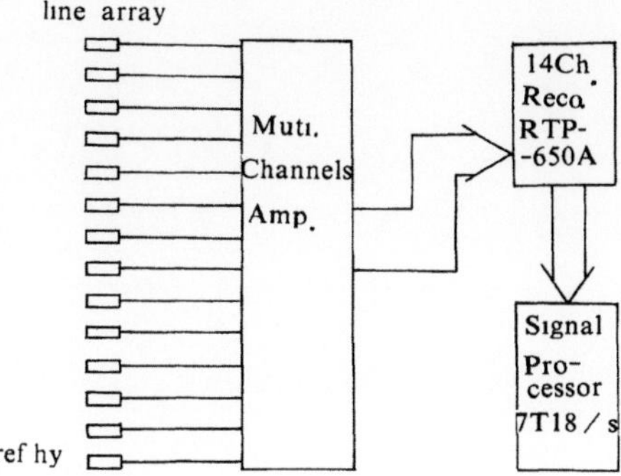

Figure 6.　Block diagram of measurement setup for little moving ship's image experiment

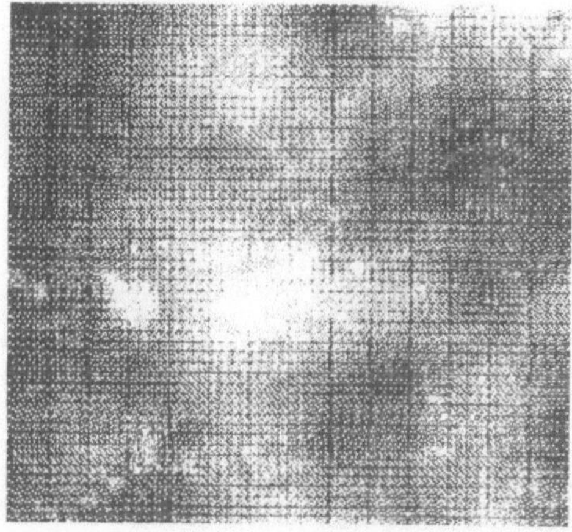

Figure 7.　The sound image of little moving ship in water

REFERENCES

1.　A. K. Ghatak and K. Thyagarajan, "Contemporary Optics", Plenum Press, New York (1978).

2.　S. Ueha, Imaging of Acoustic Radiation Sources with Acoustical Holography, *Optica ACTA*, Vol. 23, No. 2, pp107—104, (1976).

3.　W. A. Veronesi and J. D. Maynard, Digital Holohraphic Reconstruction of sources with arbitrarily shaped surfaces, *J. Acoust. Soc. Am.*, Vol. 85, No. 2, February (1989).

A MULTIPLE SHOTS 3-D HOLOGRAPHIC SONAR
USING A SET OF ORTHOGONALIZED MODULATING SIGNALS

Yasutaka Tamura,[1] Yoshihiko Aochi,[1] Shinya Terasaki,[1] Osamu Takano,[1] Takao Akatsuka,[1] Chiaki Ishihara[2] and Norio Ishii[2]

[1]Faculty of Engineering, Yamagata University, Jyonan 4-3-16 Yonezawa 992
[2]Akishima Laboratories (Mitsui Engineering and Shipbuilding) Inc.
1-50 Tsutsujigaoka 1-chome Akishima Tokyo 196

Introduction

Holographic sonar, which incorporates a transducer array, is capable of high speed data acquisition. We have proposed a high speed imaging system using transmitting pulses modulated with Walsh functions [1]. Walsh functions have a number of merits with respect to modulating the functions of acoustical imaging systems; 1 they are easy to generate and process because of their binary properties, ② the duration time of pulses becomes short, and ⦿3 an image can be reconstructed by means of a single transmitting and receiving process (single shot imaging).

However, the reconstructed image with a single shot process produces many artifacts. These are thought to arise as a result of nonorthonormal correlations of the received wavefronts corresponding to distinct target locations.

In this paper, we propose a set of modulating functions to reduce the artifacts. The modulating functions have zero cross-correlations following the precepts of Hadamard matrix's characteristics [2]. Here we introduce the orthogonality of the functions. The experimental results of 3-D holographic sonar using the proposed functions clearly demonstrate the feasibility of this method.

Data acquisition system

Fig.1 is a schematic drawing of the data acquisition system. Ultrasonic waves are simultaneously transmitted from n transmitters. We used sinusoidal waves of frequency f_0

modulated by a system of Walsh functions synchronized to a clock signal. The period of the clock signals, Δt, is equal to an integral multiple of the sine wave's period, $1/f_0$. The transmitting and receiving process is repeated n times. With the w-th transmitting and receiving cycle, the transmitting signal $u_p^{(w)}$ corresponds to the p-th ($p = 0, 1, \ldots, n-1$) transmitter, and the waveform detected by i-th receiver is denoted by $u_p^{(w)}(t)$ and $r_i^{(w)}(t)$, respectively. The origins ($t = 0$), these functions are set as at the start position of each transmitting pulse.

The transmitting waveform functions are given by

$$u_p^{(w)}(t) = \sum_{i=1}^{n} W_{\zeta(p,w)\,i}\; f(t - (i-1)\Delta t) \quad for\ i = 1,2,\cdots,n \tag{1}$$

where,

$$f(t) = \begin{cases} e^{j\,2\pi f_0 t} & 0 \le t \le \Delta t \\ 0 & \Delta t < t \end{cases} \tag{2}$$

is a sinusoidal pulse-burst of width, Δt, W_{ij} denotes a i-j component of the $n \times n$ Hadamard matrix, and the column number, $\zeta(p, w)$, has a one to one correspondence with the transmitter number p, that is

$$\begin{cases} \zeta(p, w) = 1, 2, \ldots, n \\ \zeta(p, w) \neq \zeta(q, w) & if\ p \neq q \end{cases} \tag{3}$$

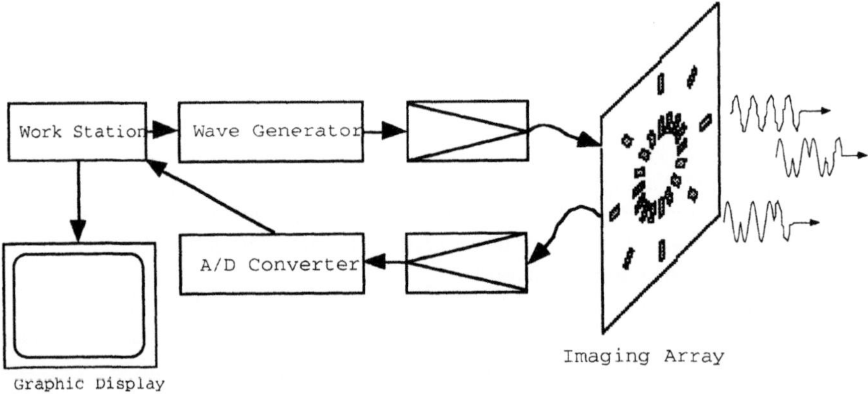

Fig.1 Holographic Sonar System

Reconstruction of Image

The reconstructed object function $S(x)$ corresponding to a position x calculated by the sum of the correlations between the received echo functions and the expected received signal for a single point target located at x.
Accordingly, $S(x)$ is given by

$$S(x) = \left| \sum_{l=0}^{n-1} \sum_{i=0}^{m-1} \sum_{j=0}^{n-1} \phi_{ij}^{(l)}(\tau(\mu_i, x) + \tau(\nu_j, x)) \right|^2 \qquad (4)$$

where

$$\phi_{ij}^{(w)}(\tau) = \int r_i^{(w)}(t) u_j^{(w)}(t-\tau)^* dt \qquad (5)$$

is a cross-correlation function between the transmitted signal $u_j^{(w)}$ at the j-th transmitter and the received echo $r_i^{(w)}$ at the i-th receiver at the w-th cycle. The suffix " * " denotes a complex conjugate operator. $\tau(\gamma, x)$ is a delay (time of flight) between the target and the transducer at the position x and γ respectively.

Mod 2 repetition

We proposed a method of transmission for multiple shots holography. We called this method "Mod 2 repetition". We also considered a method, "Cyclic repetition", for comparison. The column number function $\zeta(p, w)$ for the "Mod 2 repetition" is given by

$$\zeta(p, w) = p \oplus w + 1 \qquad (p, w = 0, 1, 2, \cdots, n-1) \qquad (6)$$

and for the "Cyclic repetition",

$$\zeta(p, w) = (p + w) \bmod n + 1 \qquad (p, w = 0, 1, 2, \cdots, n-1) \qquad (7)$$

where $\oplus$ expresses modulo 2 addition for every corresponding bits of binary represented numbers.

Orthogonal property of the transmitting pulses

In this section, the cross-correlations of the transmitting pulses are introduced and the orthogonal properties of the pulses demonstrated. The cross-correlation function for i-th transmitting signal and j-th one is expressed by

$$\varphi_{ij}(\tau) = \sum_{l=0}^{n-1} \int u_i^{(l)}(t) u_j^{(l)}(t-\tau)^* dt = \begin{cases} n^2 \int f(t) f(t-\tau)^* dt & i = j \\ 0 & i \neq j \end{cases} \qquad (8)$$

Transmitting signal for Mod 2 is

$$u_p^{(l)}(t) = \sum_{i=0}^{n-1} W_{(p \oplus l)i} f(t - i\Delta t) \qquad (9)$$

So, Substituting (9) for (8) and rearranging, we have :

$$\varphi_{ij}(\tau) = \sum_{l=0}^{n-1} \int \left\{ \sum_{i=0}^{n-1} W_{(p \oplus l)i} f(t - i\,\Delta t) \right\} \left\{ \sum_{j=0}^{n-1} W_{(p \oplus l)j} f((t - j\,\Delta t) - \tau) \right\}^{*} dt$$

$$= \sum_{l=0}^{n-1} \sum_{i=0}^{n-1} \sum_{j=0}^{n-1} W_{pi} W_{li} W_{qj} W_{lj}\, \varphi((j-i)\,\Delta t + \tau) \tag{10}$$

Hadamard matrix's characteristic:

$$W_{(p \oplus l)i} = W_{pi} W_{li} \tag{11}$$

Here, applying other characteristic:

$$\sum_{l=0}^{n-1} W_{li} W_{lj} = n\,\delta_{ij} \tag{12}$$

We can rearrange (10) moreover:

$$= \sum_{i=0}^{n-1} \sum_{j=0}^{n-1} W_{pi} W_{qj} \left\{ \sum_{l=0}^{n-1} W_{li} W_{lj} \right\} \varphi((j-i)\,\Delta t + \tau)$$

$$= n \sum_{i=0}^{n-1} W_{pi} W_{qj} \delta_{ij}\, \varphi(\tau)$$

$$= n^2 \delta_{pq}\, \varphi(\tau) \tag{13}$$

Hence, we have :

$$\varphi_{pq}(\tau) = \begin{cases} n^2 \varphi(\tau) & \text{for } p = q \\ 0 & \text{other} \end{cases} \tag{14}$$

for $\tau < T_p$, where T_p is the period of repetition. Though the *Mod 2* repetition is simultaneous transmitting method, it is equivalent to a time-divided transmitting method in which an array transmits a single-pulse from a single transmitter in turn.

EXPERIMENTS

The appearance of the transducer array using the experimental system is shown in Fig. 2. The circular array consists of 16 transmitters and 8 receivers. The frequency of the carrier wave, f_0, was 50kHz (Wavelength λ =6.9mm, and clock period Δt was $2 \times 1/f_0 = 40\mu sec$). Fig. 3 shows a block diagram of the experimental system. Transmitting pulses were generated by a wave-generator, and the received waves reflected from objects are digitized by an A/D converter, and transferred to a personal computer. Then the data was send to an HP9000 workstation and the image reconstructed.

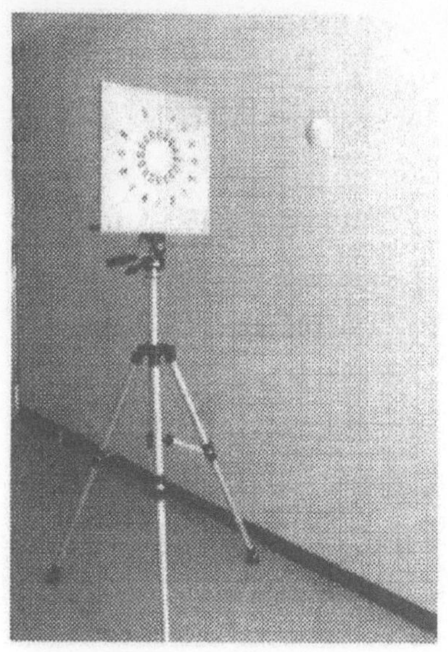

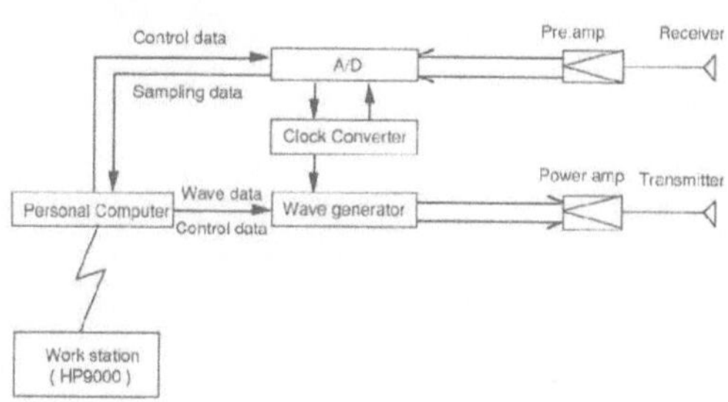

Fig 2 Appearance of the transducer array
(and a sample target)

Fig 3 Block diagram of the experimental
system

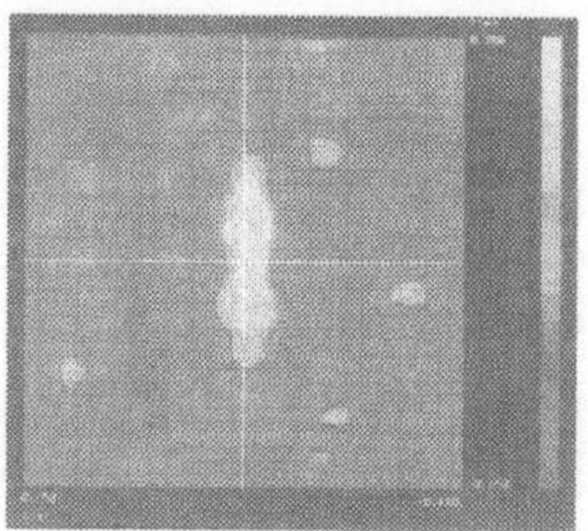

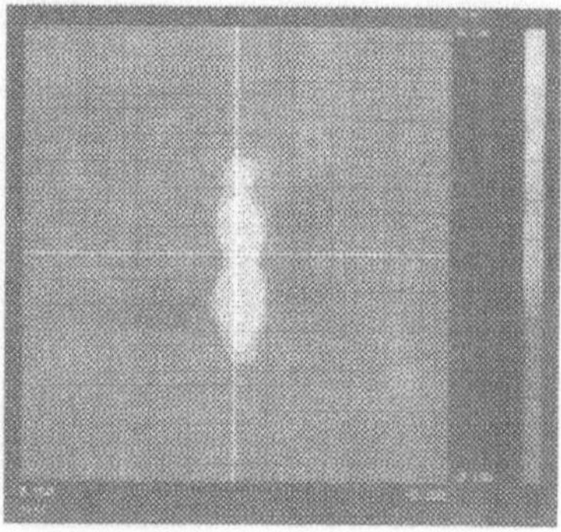

(a) one shot transmitting

(b) simple cyclic transmitting

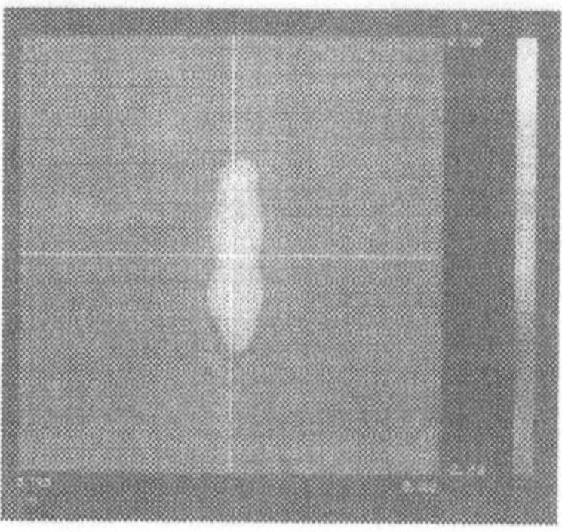

(c) *Mod 2* transmitting

Fig 4 The reconstructed focal-plane images

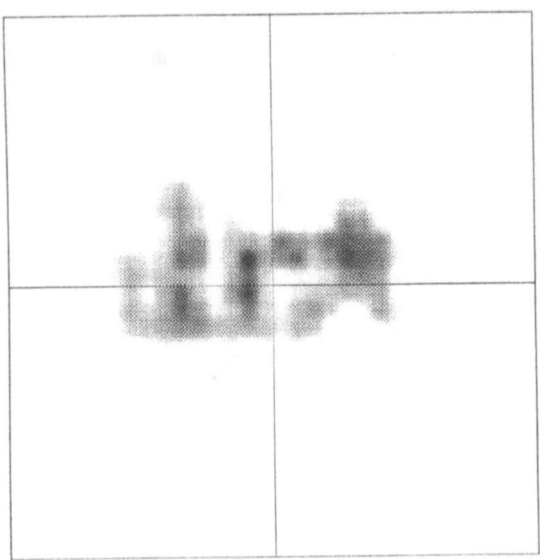

(a) focal-plane imagie

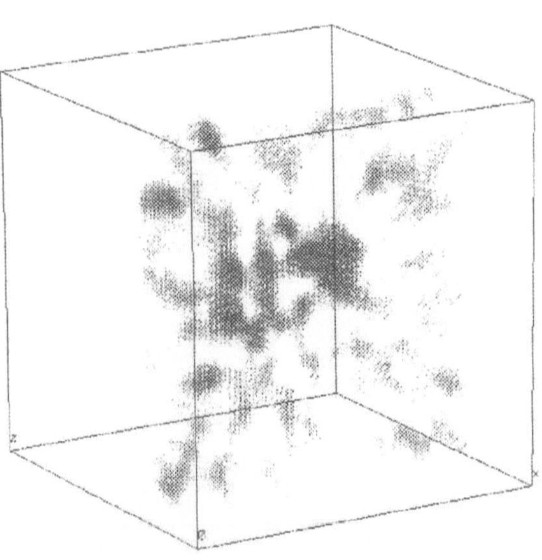

(b) 3-D imagie

Fig 5   The reconstructed image for Mod 2

The reconstructed focal-plane images are shown in Fig. 4. The size of the images is 64 λ x64λ square, and the intensities were increased 4 times so that the artifacts would be easy to see. The target was a acrylic rod of 15mm in diameter set 1m away from the array. Figs. 4 (a) is a reconstruction following only one transmitting and receiving process (single shot imaging). The reconstructed image shows many artifacts. Figs. 4 (c) shows the result following a *Mod 2* repetition method. Compared with the simple cyclic repetition (Figs. 4 (b)) , the reconstructed image has fewer artifacts. Fig. 5 shows the 3-D image for the *Mod 2* repetition method. The target consisted of acrylic plates formed into the Chinese characters " 山大 " which is the abbreviated form of " Yamagata University ".

By simulation, for 3-D voxel data for a single target set 1m away from the array, the ratios of the main peak to maximum sub peak were about 9dB for the one shot method, and 16dB for the *Mod 2* method.

CONCLUSIONS

Using multi-transmitting methods, artiifacts can be reduced in acoustical holographic imaging systems which uses transmitting pulses modulated by Walsh functions. It would be natural and effective to introduce these methods, considering that real-system repeats transmitting and receiving.

REFERENCES

1. Y.Tamura and T.Akatsuka : Trans. of SICE, 24-3, 213/219 (1988)
2. Y.Aochi, S.Terasaki, O.Takano, Y.Tamura, T.Akatsuka, T.Ishihara and N.Ishii: Proc. of SICE, 30, 227/228 (1991)

HIGH-RESOLUTION ARRAY PROCESSING EXPERIMENTS WITH MULTI-BEAM ECHO SOUNDER II: ANALYSIS AND FURTHER RESULTS

Timo-Pekka Jäntti

Hollming Ltd Electronics
26100 Rauma
Finland

INTRODUCTION

Acoustic bathymetry is acoustic remote sensing to measure the depths of water in oceans, seas, and lakes. Multibeam echo sounders are acoustic equipment used in acoustic bathymetry. High-resolution array processing methods are signal processing methods, whose resolution capabilities can exceed the conventional $\theta_3 \sim \lambda/L$ - formula by several orders. These concepts with a brief description of four high-resolution methods were mentioned in[1].

The use of the described high-resolution array processing methods in our experiments was justified by a theoretical model, also briefly presented in[1]. We then presented some results and at the end discussed the validity of the theoretical model.

In this paper we continue that work, presenting some new results with analysis.

MEASUREMENTS

The data was measured on April 26th, 1988 on the Mediterranean with 16 elements of the 57 element receiving array of the multibeam echo sounder. The elements were spaced in $\lambda/2$ intervals in the array. A 10 ms pulse of 15kHz CW was transmitted, and after an appropriate delay the scattered signals were sampled at 5 µs intervals simultaneously on all 16 elements. Only the inphase component was recorded. The delay was adjusted so, that first samples consist of noise only. Before sampling the signals were amplified and filtered in narrow-band filters, having a passband of 2 kHz at 15 kHz center frequency. The depth was approximately 2300 m. More details of the measurement arrangement are found in[2].

The above described dataset is used here instead of the one used in[1], as the sampling interval of that other dataset, i.e. 61 µs, is too low for a detailed time-domain analysis.

Acoustical Imaging, Volume 20 Edited by Y. Wei
and B. Gu, Plenum Press, New York, 1993

Containing both inphase and quadrature phase components, that data is, however, appropriate for angle-of-arrival estimation, as used in[1].

THEORETICAL MODEL

A general model to describe the signal $x_i(t)$ measured at element i of an array of N elements at time t was presented in[1]

$$x_i(t) = \int d\theta \; s(t-\tau_i(\theta),\theta) + n_i(t) \qquad i=1,...,N \tag{1}$$

where $s(t,\theta)$ is the signal density impinging the array from angle θ at time t, $\tau_i(\theta)$ is the delay between transmission and reception at element i from angle θ, and $n_i(t)$ is the noise at element i at time t. In[1] we made some restrictions on $s(t,\theta)$ and $n_i(t)$, so as to satisfy the assumptions usually made when analyzing the theoretical performance of the high-resolution array processing methods.

The model (1) is quite general. In this paper we relax the assumptions made in[1], i.e. the assumption of ergodicity and even gaussianity of the signal density $s(t,\theta)$, and also the whiteness of the noise $n_i(t)$. Also the point source assumption is relaxed. From now on we merely try to show the validity of such a presentation and to find the properties of $s(t,\theta)$, $\tau_i(\theta)$, and $n_i(t)$ based on measured data, i.e. to invert the measured data. This relaxation means, that we have to be careful, when drawing conclusions on the results, when using high-resolution methods, for example.

We still are willing to keep the far-field assumption, i.e.

$$\tau_i(\theta) = \tau(\theta) + i\cdot\frac{d}{c}\cdot\sin(\theta) \qquad i=1,...N \tag{2}$$

where $\tau(\theta)$ is the common delay for all elements in the array, for signals arriving from angle θ, d is the spacing between elements and c the velocity of sound propagation in the sea. This is justified by the assumption, that the signal density $s(t,\theta)$ represents scattering at the seabed at the depth of 2300 m, approximately.

We also assume, that the scattering is originated as the interaction of the transmitted signal and the seabed, the transmitted signal known to consist of a single pulse of a CW sinusoid. The model remains valid even if the signals penetrate the seabed and therefore are scattered from lower layers.

RESULTS AND ANALYSIS

Figure 1 shows two 80 ms samples of the received signals, that of element 1 in (a), and that of element 2 in (b). The angle of arrival is around 28 degrees. The large and rapid fluctuations in both samples is noticeable. The amplitude of the envelope varies some 26 dB, from 1000 to about 50, the latter being almost the level pure noise, whose rms value is 42 in all elements, ±3 variation between elements. It is of interest to remember, that the transmitted pulse length was 10 ms, and to notice that some of the distinguishable pulses are much shorter than 10 ms. Also worth mentioning is the difference in appearance of both signal samples. Despite of general similarity, they have noticeable differences, although the elements are separated by only one half of the wavelength.

The notions mentioned above are true for all 16 elements and for longer samples.

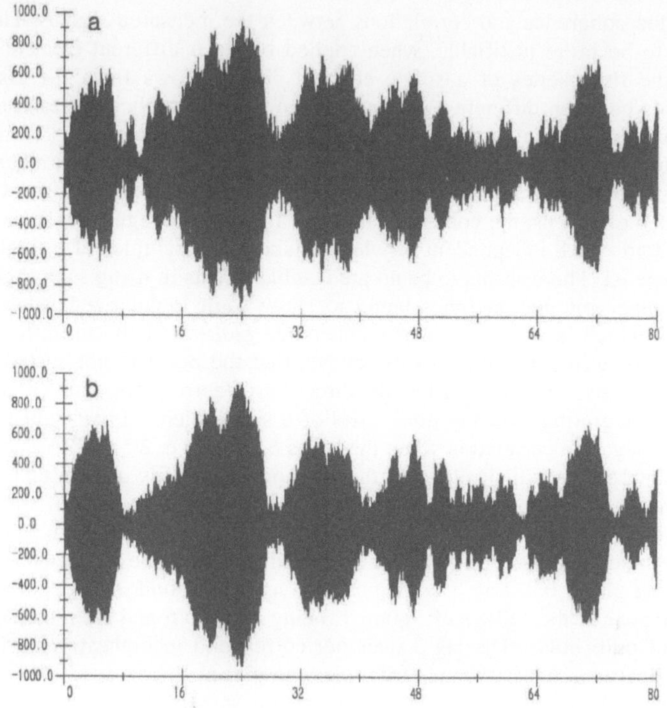

Figure 1. Two 80 ms samples. Element 1 in (a) above, and element 2 in (b) below. Time in ms.

Trying be more precise we define the unity-normalized coherence function $\gamma_{ij}(\tau)$ as[3]

$$\gamma_{ij}(\tau) = \frac{\langle x_i(t)x_j^*(t+\tau)\rangle}{\sqrt{I_iI_j}} \qquad i,j=1,...,N \qquad (3)$$

where the operation $\langle\ \rangle$ means either expectation over the ensemble, or time average whichever is appropriate, the $x_i(t)$ is the wave or signal at spatial point i, corresponding here the element i at time t, $x_i^*(t)$ is the complex conjugate of $x_i(t)$, τ is the delay between the two signals, and I_i is the intensity, defined as

$$(4)$$

$$I_i = \langle x_i(t)x_i^*(t)\rangle \qquad i=1,...,N$$

As we have no access to the ensemble, we must be satisfied with time averaging. Strictly this requires infinite integration time and ergodicity, which is questionable here. As shown by Middleton et al. both theoretically[4], and experimentally[5], the reverberation itself is a non-stationary process and ensemble expectations should be practiced. On the other hand, our application relies on the model, where the pulse travels constantly, which by no means makes the situation easier.

Since we here are interested in very short time intervals when analyzing the fine details of the scattered signals, we use time averaging to get at least a crude quantitative measure of the coherence and correlations between the measured signals. This procedure is expected to be more justifiable, when applied between different elements, than when applied in the time series of a single element. This follows from the assumption the measurements between different elements contain delayed replicas of each other, i.e. the non-stationarity affects more the time domain than the space domain.

We start with noise, a short sample consisting of pure noise was recorded on each element before the actual scattered signal begins. When integrated over short interval, say 20 snapshots, i.e. 100 μs, the coherence function fluctuates irregularly, having any values between -0.8 and +0.8 independent of whether taken over samples of a single element or between elements. There seems to be no predictable effects in using such short integration intervals. Longer integration times begin to show some regularized patterns, the most noticeable of which is a nearly constant coherence $\gamma_{ij}(0) = 0.5$ between different elements, when the lag is zero. This leads us to believe, that the noise is not spatially white, but more coherent noise is arriving from the direction of zero angle, than from other directions. When integrating over the time series of a single element, we notice, that there is rather strong negative correlation when the lag is 6 and 7, i.e. 30 and 35 μs corresponding 180 degrees at 15 kHz, which suggests, that the noise is heavily influenced by the narrow-band filtering.

The short time coherence between different elements, when they in addition to noise contain the scattered signals, varies from time to time. The highest coherence appear when lag is 3, being about 0.9. Lag 3 corresponds to angles around 26.7 degrees. The lowest coherence appear at the valleys of Figure 1, being about 0.6 and even below, resembling the values of pure noise. The lag 3 does not correspond to highest value in these low-coherence situations, but it changes from place to place.

In Figures 2 and 3 two special situations are shown. The Figure 2 is a zoom of Figure 1 at about 5 to 6 ms, and Figure 3 is a zoom of Figure 1 at about 61 to 62 ms.

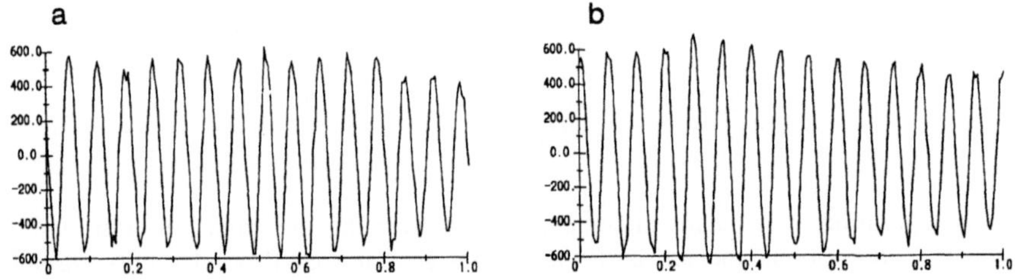

Figure 2. A zoom of Figure 1 at 5 to 6 ms. Element 1 in (a) left, element 2 in (b) right. Time in ms.

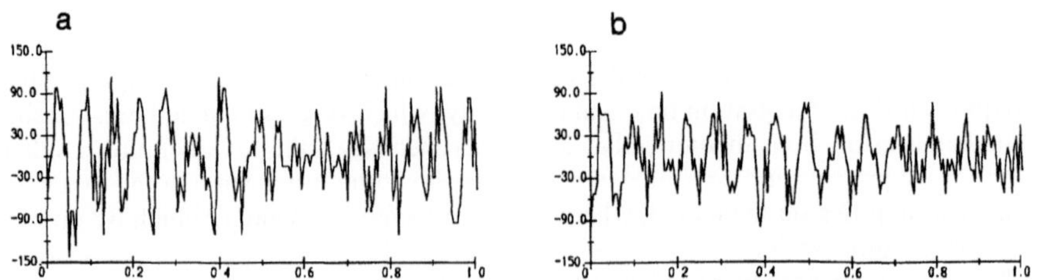

Figure 3. A zoom of Figure 1 at 61 to 62 ms. Element 1 in (a) left, element 2 in (b) right. Time in ms.

The samples in Figure 2 are nearly perfectly coherent, at lag 3 they have a coherence of 0.98, and it increases at other appropriate lags, e.g. -10 lag has a coherence of 0.997. We have not interpolated new points between existing samples, but it is obvious, that this procedure would still increase the coherence values, as the appropriate lag corresponding the delay caused by angle-of-arrival would be found. In the Figure 4 (a) the MUSIC[1] estimate of the angle-of-arrival corresponding the samples of Figure 2.

The samples of Figure 3 resemble noise. They appear about 55 ms later than those in Figure 2, and behave completely different. Their maximum coherence appear at zero lag, that being about 0.6. There is no same kind of relation between coherences calculated between different lags as there should be if one sample consists of delayed replicas of the other sample, but merely irregularity resembling that of noise. It is difficult to determine what is the angle-of-arrival in this case. In the Figure 4 (b) below is the MUSIC estimate for the angle-of-arrival corresponding the samples of Figure 3.

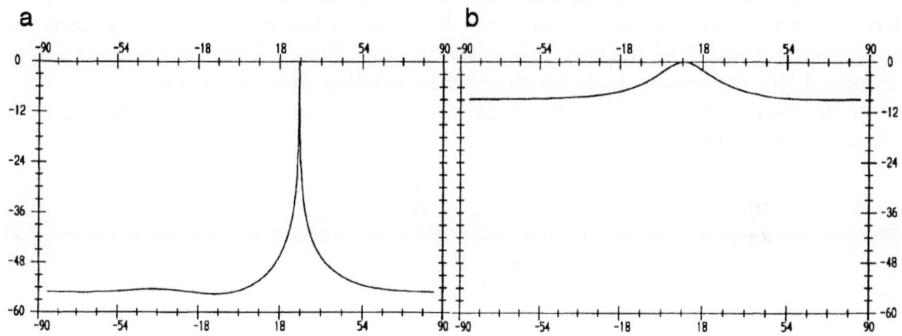

Figure 4. The MUSIC estimates of angle-of-arrival. Plot (a) on the left corresponds to the samples of the Figure 2, and plot (b) on the right corresponds to the samples of the Figure 3. In both cases 20 snapshots and 16 elements was used. Ordinate in dB, abscissa in degrees.

The MUSIC estimate of angle-of-arrival in Figure 4 (a) is 26.58 degrees, that of the ESPRIT[1] is 28.51 degrees, and that of the CBF[1] is 26.44 degrees. The real data was made analytic by interpolating the quadrature component with moving a 11 point spline interpolation method with an accuracy of 50 ns. The analyticity can be seen in the Figure 4 as the plots are only one sided, i.e. contain only one side of the spatial spectrum. In the MUSIC and the ESPRIT methods spatial processing[1] was used, so that the effective length of the array was only 3 elements. Although longer arrays give similar results, this short array was the best. The CBF was implemented as zero padded 1024 point complex Fast Fourier Transformation.

The result of Figure 4 (b) is in doubt. The estimate of MUSIC is 9.52 degrees, that of ESPRIT is -7.72 degrees, and that of CBF is 24.83 degrees. These results are compatible with those mentioned earlier, when calculating the coherence of this sample. The implementations in this case are equal to those mentioned above for the case of Figure 4 (a).

In the MUSIC and ESPRIT methods the power σ^2 of spatially white noise is found as the minimum eigenvalues of the covariance matrix[6]

$$(R - \lambda_i I)\hat{e}_i = 0 \qquad i = 1,..M \tag{5}$$

where R is the covariance matrix, I is the identity matrix, λ_i are the eigenvalues, and $\hat{e}_i$

the corresponding eigenvectors, and M is the effective array length after spatial processing. If the number of source *is* $K < M$, then theoretically the *M-K* smallest eigenvalues are all equal to σ^2. Based on this, the noise contribution can be subtracted from the covariance matrix.

The theoretical result of noise subtraction in the MUSIC and ESPRIT methods is not applicable here. The spread of eigenvalues is so high, that the value of the treshold below which eigenvalues could be identified as noise, varies arbitrarily. Never have we seen a smooth set of equal eigenvalues, which could be termed as noise, simulations being the exceptions, however. This can also be concluded from the results, that show no change whether noise is subtracted or not. This is compatible with the results mentioned above, which state that the coherence of noise is irregular when integrated over short intervals. This is true for the whole 80 ms sample.

We also calculated the MUSIC, ESPRIT, and CBF estimates, based on different lags between the elements in both cases. Only small lags, from -5 to 5 were used. The performance of all three methods was independent of the lags in the data of Figure 4 (a). What the lags do, is that they change peak location on the angular axis, e.g. three lags, corresponding about 26.7 degrees, moves the peak of Figure 4 (a) near zero. In the case of Figure 4 (b), the results made no more sense whether delayed or not

In the Figure 5, we present the angle-of-arrival estimates of the MUSIC and CBF methods for the 80 ms sample.

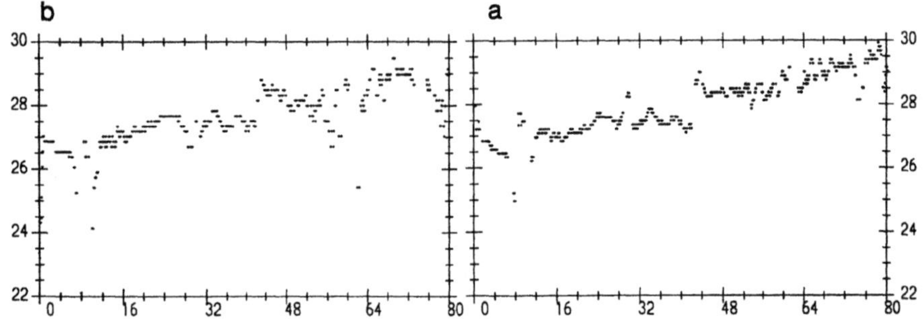

Figure 5. The angle-of-arrival estimates of the MUSIC on the right (a), and the CBF on the left (b) methods for the 80 ms sample. Angles in degrees, time in ms

Both plots in the Figure 5 show a lot of deviation, the trend and structure being the same in both, however. The greatest deviations, which we believe represent totally wrong estimates, occur at the valleys in the Figure 1, e.g. around 8 ms, 10 ms, 29 ms, 61 ms, and 75 ms. It is, however, interesting to observe, that also these wrong estimates have similar structure in both plots. The ESPRIT estimates, not shown, have even greater deviation, filling the whole framed area.

Finally we mention the attempts to correlate the measured signals with a simulated transmission pulse, i.e finite length sinusoid with a single frequency. We used pulse lengths between 100 μs and 200 ms at 15 kHz. The absolute value of short term coherence varied between nearly zero to 1, i.e.nearly perfect correlation. The absolute value of the long term coherence was surprisingly low being around 0.65 at maximum, with slight deviation between different elements. This value occured when the pulse length varied between 3 and 200 ms We also varied the frequency in 10 Hz steps between 12 and 18 kHz. The absolute value of short term coherence was occassionally at maximum at a frequency different than 15 kHz In these cases the deviations could be very large. The absolute value of long term coherence was always at maximum at exactly 15 kHz.

DISCUSSION

The performance of angle-of-arrival estimation methods change continuously with time as the pulse travels along the sea floor. Although difficult to measure quantitatively and based partly on qualitative reasoning, this change in performance can be predicted. The performance is good when the coherence function calculated between different elements is high, and bad when the coherence function is low, and/or fluctuates irregularly. These correspond to the peaks or values above a threshold, and to the valleys or values below the threshold of the envelope of the received signal respectively, see Figure 1. There are two notions to be made immediately. The performance is vaguely defined here, as we do not know the true bottom profile. It is merely based on the quality of the plots like those in Figure 4, the differences in the estimates of different methods, the scatter plots of the estimates, like those in Figure 5, and for example the eigenvalue spectrum of the covariance matrix. We have some knowledge about the bottom profile, so that the greatest deviations in the estimates can be identified. The estimate of coherence is poor, as it is based on short term integration. We have used this measure only as a relative measure, to give us the direction in change of the true coherence function.

The best way to proceed would be to use true ensemble averages, as then we could calculate both coherence function and covariance matrix. If the process is non-stationary, even this is not a justification to the use of the high-resolution array methods, at least if the performances are to be compared with the theoretical values, which so far have based in stationarity assumption. One of the goals of this research however is to test the methods in real environments leading to development better and more robust methods if necessary.

The performance of the methods studied here and those studied earlier[1,2] is satisfactory most of the time, around 90 ± 5 % of the time, and this performance can be predicted as mentioned above. In exception the ESPRIT method, whose performance has been poorer so far. We have not found any cure to improve the estimates of the poor coherence case, the best we can suggest is to predict their occurrence and to reject them. This kind of procedure is satisfactory when these methods are used in the sea floor mapping equipment. This kind of treatment of data may be even better than that of long integration together with resulting smoothing. When judging the performance of these methods, we must remind that the Figure 5, for example, has a lot of information, not usually presented. The real time mapping equipment currently in use, have a resolution of one to two degrees, so there would be only one to three estimates in the Figure 5.

As we have special interest to study the sea floor with very high-resolution, we are not satisfied with the performance, however. It is clear, that the changes in the amplitude of the received waveforms and those of coherence are reactions of something, containing a lot of information. A simple model would be, that the low coherence regions correspond to complete reflections at the sea bed, whereas those of high coherence correspond to nearly perpendicular surfaces, perpendicular to the direction of propagation, on the sea floor to cause the backscattering to be more coherent at the bottom interaction. This is by no means unique, and a measurement in different directions would serve us with more information on this.

The building of better models and the inversion of them is what is needed. In real time applications we cannot measure over all space, but it is not clear how big the measurement platform must be, if enough a priori information is available.

The benefits of using wide band signals would be twofold. Whereas the full inversion requires all frequencies, we could be satisfied with a few separate frequencies, selected so that their scattering properties fits a wide enough range of materials and objects of the sea bed. The pulse compression techniques found useful in radar could provide us with additional information. The low coherence of the received signals with the simulated transmitted pulse shows that also this needs more research.

Finally we would like to comment on the point source versus extended source modelling. It is clear that when the finite length pulse hits the sea bed it always extends over some region. We have simulated a few extended source models[7], and these preliminary results show, that in the perfect coherence case, the high-resolution methods model internally the extended source to consist of several point sources, and try to estimate the locations of these. In our experiments with real data we have never seen this to occur, be it the influence of non-perfect coherence or not. The plot in Figure 4 (a) is so clear, that we expect the scattering at the sea bed to have a lot of resemblance to a point source model.

What is to be remembered, when we speak of the use of high-resolution methods, their benefits come from their resolving properties, not the plots like Figure 4 (a) or (b) themselves. That is, even the CBF method can be made to look like Figure 4 (a), when a proper functional relation is used. This does not, however, increase the resolution of the CBF to resolve two or more closely spaced targets. So far the sea floor reliefs have been too smooth to cause the pulse to stretch in several simultaneous directions with comparable amplitude. In the case of a single source, a phase locked loop, to determine the phase of the signal of each element, would perform the task of estimation easily. This was actually tested when we correlated the simulated transmission pulse with the measured signals. The short integration time results were comparable to those achieved with the high-resolution methods. It would be very interesting to have data, where the bottom relief changes fast enough to cause several angles-of-arrival to see the real benefits of the high-resolution methods.

REFERENCES

1. T-P. Jäntti,High-resolution array processing experiments with multibeam echo sounder, *in:* "Acoustical Imaging 19," H.Ermert, and H-P.Harjes,ed., Plenum Press, New York (1992).
2. T-P. Jäntti,High-resolution beamformer for multibeam echsounder, *in:* "Proceedings of the Techno-Ocean'88, vol. II," Kobe, Japan (1988).
3. K. Rohlfs,"Tools of Radio Astronomy," Springer Verlag, Heidelberg (1986).
4. D. Middelton, A statistical theory of reverberation and similiar first-order scattered fields part I, *IEEE Trans. on Inf. Theory* IT-13:372 (1967).
5. T.D.Plemmons and J.A.Shooter, D.Middelton, Underwater acoustic scattering from lake surfaces part I, *The J. Acoust. Soc. of Amer.* 52:1487 (1972).
6. S.U.Pillai,"Array signal processing," Springer Verlag, Berlin (1989).
7. T-P.Jäntti,The influence of extended sources on the theoretical performance of the MUSIC and ESPRIT methods: narrow-band sources, *in:* "Proceedings of ICASSP-92, vol 2," San Francisco, USA (1992).

A STUDY ON THE TRANSFORM CODING OF UNDERWATER SONIC IMAGES

Zhu Weiqing, Xu Wen and Zhang Hailan

Institute of Acoustics, Academia Sinica
Beijing, 100080, P.R.China

INTRODUCTION

In recent years,the Unmanned Untethered Vehicle(UUV) has developed quickly especially in the area of underwater objects searching. The application of ultra high resolution imaging sonar to searching results in a reduced search swath. In order to store digital sonar images and transmit them to the experiment ship by acoustic waves,image data must be compressed significantly.

Transform coding is a type of source coding that is used to compress the image data without substantial loss of fidelity. It has been studied for TV images[1] and SAR images[2] and the results have shown favorable compression ratios of the order of 10 to 1 and 6 to 1, respectively, without noticeable loss of picture quality.This is due to the strong correlation between pixels of these images.

The study on the statistics of sonar images shows that the space correlation of sonar images is weaker than that of TV images[3]. However,some correlation does exist that can be exploited in order to reduce the average number of bits per pixel needed to code the image within a given fidelity criterion, which may be laxer than that for TV images.

In this paper, five types of transform including Fast Fourier Transform(FFT),Walsh-Hadamard Transform(WHT), Haar Transform(HAT), Slant Transform(SLT) and Discrete Cosine Transform(DCT) are tested to sonar images, the spectral properties are studied and the results are given with coding rate 0.5 bits/pixel.

METHOD AND PROCEDURE

For an $M * N$ rectangular image $U = \{u(m,n)\}$ the transform pair is

$$V = Am\,U\,An$$

$$U = Am^{-1}\,V\,An^{-T}$$

where Am and An are $M*M$ and $N*N$ unitary matrices, respectively. The elements $v(k,l)$ are called the transform coefficients and $V = \{v(k,l)\}$ is called the transformed image. Most unitary transforms have a tendency to pack a large fraction of the average energy of the image into a relatively few components of the transform coefficients. Since the total energy is preserved, this means that many of the transform coefficients will contain very little energy. While the input image elements are highly correlated, the transform coefficients tend to be uncorrelated. So the redundancy or predictability should no longer be present in the transform domain. Transform coding is based on this principle as shown by Jain[4].

A practical transform coding algorithm for images is shown in Figure 1.

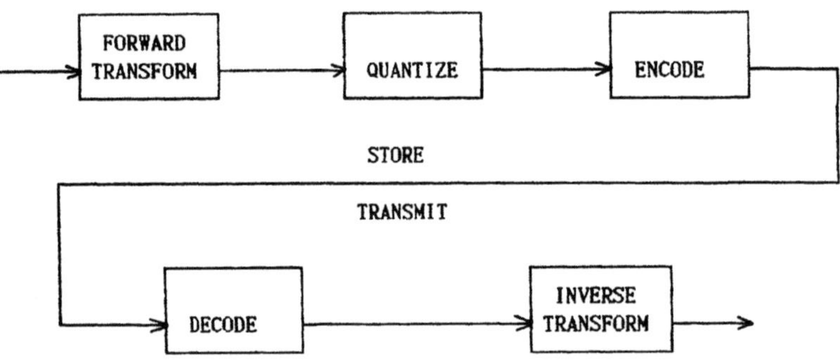

Figure 1. Diagram of a practical transform coding algorithm.

Sonar images in Figure 2 were produced by an M851 imaging sonar made in our institute and were obtained from experiments in Haizi Reservoir near Beijing and in the sea near Dalian. They are digitized at 7 bits/pixel. In Figure 2(a) the bottom is smooth and the echo is weak. Looking from center to right, a shipwreck can be seen lying on the sea bottom. In Figure 2(b) the echo is stronger and a very strong line at the lower left shows the echoes from a dam.

RESULTS

To make transform coding practical , a given image is divided into small rectangular blocks and each block is transform coded, independently. Five types of transform are applied to several sonar images. They are compared on the basis of the fraction of the total energy occurring in the 16 lowest order coefficients for the $16*16$ transform. Numerical results are listed in Table 1.

The effect of underwater sonar imaging depends on many factors, determined and undetermined, such as target position, target size, bottom or surface scattering, multipath propagation, propagation loss etc. Random noises including impulsive noise and Gaussian noise can be found in sonar images. Thus some preprocessing is needed

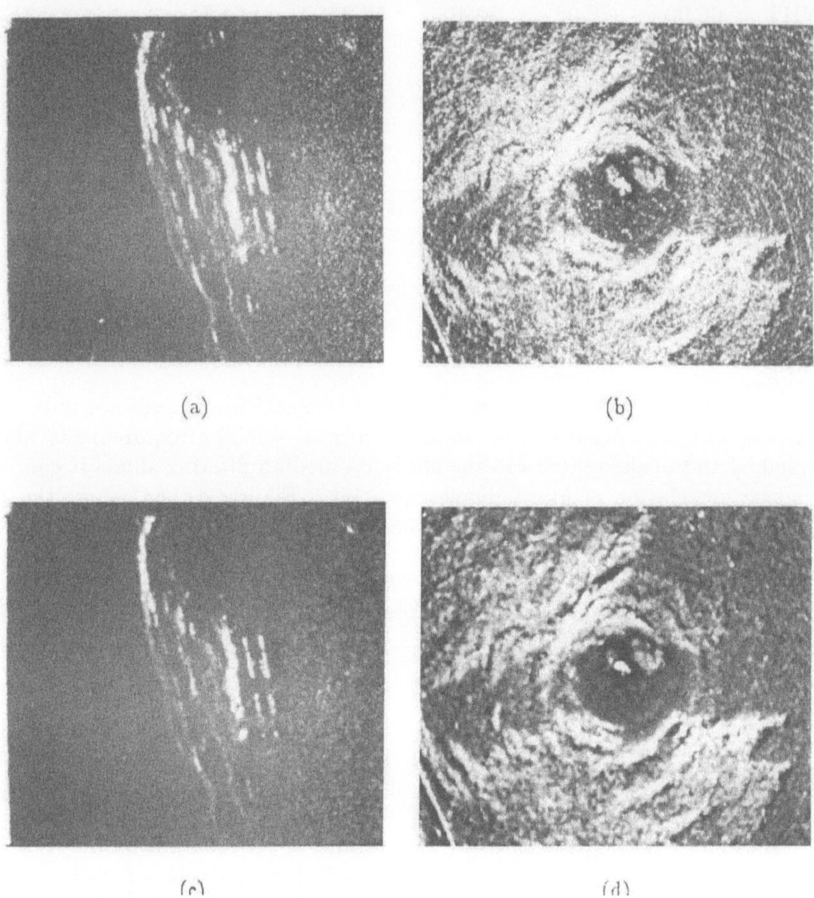

Figure 2. Sonar images (a)side scan, in the sea near Dalian. (b)polar scan, in Haizi Reservoir near Beijing. (c)median filtered image of (a). (d)median filtered image of (b).

Table 1. Energy compression results for Figures 2(a),(b), their median filtered images and 2 TV images. The transform block size is $16 * 16$.

	Fig.2(a)	Fig.2(c)	Fig.2(b)	Fig.2(d)	TV image(a)	TV image(b)**
FFT	46.78*					
WHT	60.09	79.08	29.40	59.93	64.71	96.81
HAT	60.09	79.05	29.40	59.11		
SLT	64.00	81.71	30.47	62.16		
DCT	66.27	85.08	31.93	65.40	71.59	98.23

*Data listed in the table are percentages of the total energy occurring in the 16 lowest order transform coefficients.

**The results of TV images are cited from Clarke[1]

to decrease the effects of noises.Such preprocess algorithm should not only suppress the noises but also preserve the edges of targets well.The requirements above are satisfied by the median filter and the study on median filtering shows it can improve the space correlation of input images greatly[5]. Figures 2(c),(d) show the median filtered images of Figures 2(a),(b) and Table 1 lists the transformed results of Figures 2(c),(d) also.

For an $N * M$ image divided into $M * N/p * q$ blocks, each of size $p * q$, the main storage requirement for implementing the transform is reduced by a factor of $M * N/p * q$. The computational load is reduced by a factor of $log_2 M * N/log_2 p * q$ for a fast transform requiring a $N * log_2 N$ operation to transform an $N * 1$ vector.

Table 2. Comparison of the performance of DCT with different block sizes for Fig.2(c).

	a*	b**	storage reduction factor	computation reduction factor
$8 * 8$	49.34	83.33	1600	2.77
$16 * 16$	59.38	85.08	400	2.08
$32 * 32$	64.34	85.91	100	1.66

*Data listed in this column are percentages of the total energy occurring in the first lowest order transform coefficient in $8 * 8$ block, 4 lowest in $16 * 16$ block and 16 lowest in $32 * 32$ block.

**Data listed in this column are percentages of the total energy occurring in the 4 lowest order transform coefficients in $8 * 8$ block , 16 lowest in $16 * 16$ block and 64 lowest in $32 * 32$ block.

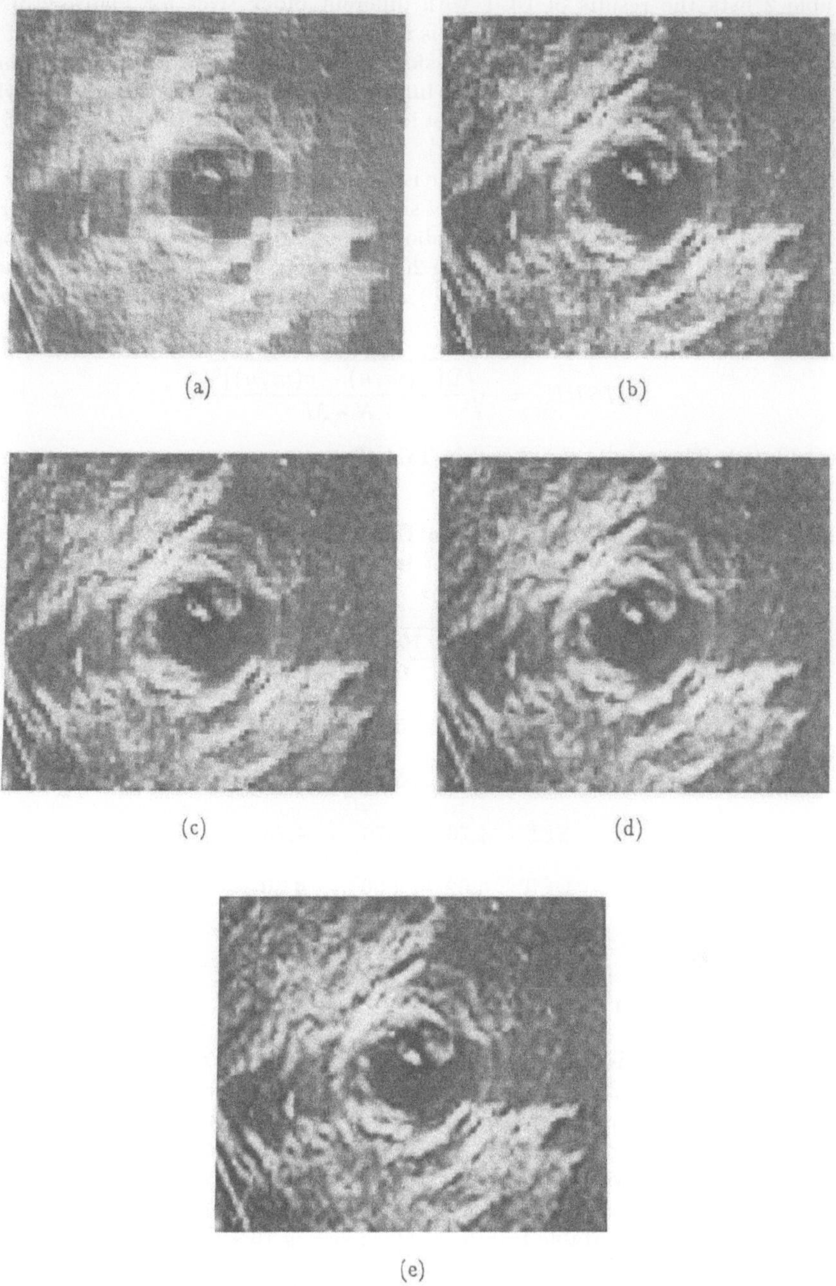

Figure 3. Reconstructed images with coding rate 0.5 bits/pixel for Figure 2(d). The transform type is (a)FFT,(b)WHT,(c)HAT,(d)SLT,(e)DCT.

Table 2 lists the results of DCT with different block sizes for Figure 2(c). It shows that although the operation count is not greatly reduced,the complexity of the hardware for implementing small size transform is reduced significantly . Furthermore the performance of $16*16$ is much better improved than that of $8*8$ transform while the performance of $32*32$ transform is not improved greatly.So, the $16*16$ transform is most suitable.

It is worth noting that the sonar image is "rough" in some way. Usually , only the sea bottom line, target position and target size etc. are interested.Laxer fidelity criterion is allowed as long as the information above is not lost. The reconstructed images with coding rate 0.5 bits/pixel for Figure 2(d) are shown in Figure 3. Mean Square-Root Error (MSRE) between Figure 2(c) and its reconstructed image $\{u'(m,n)\}$, i.e.

$$MSRE \;=\; \sqrt{\frac{\Sigma\{u'(m,n)-u(m,n)\}^2}{N*M}}$$

are calculated. The results are shown in Table 3.

Table 3. Comparison of MSRE between Figure 2(c) and its reconstructed images with 0.5 bits/pixel coding rate. Different transform block sizes and different transform types are used.

	$8*8$	$16*16$	$32*32$
FFT	5.66	6.34	
WHT	5.19	5.19	5.19
HAT	5.19	5.19	5.19
SLT	4.76	4.76	4.76
DCT	4.62	4.30	4.20

DISCUSSION AND CONCLUSION

Transform coding is used to compress sonar images. From the analysis of the spectral properties of sonar images and from the measurement of MSRE between the original image and its reconstructed image, it is derived that : 1. The performance of the transform coding of sonar images is worse than that of TV images. The result is in good agreement with that of the statistics study. 2. According to the effects of transforms, DCT is the best, SLT is better .It is the same as the situation for TV images. 3. The most suitable transform block size is $16*16$.In this way ,the speed of implementing algorithm is high enough and the result is almost the best and the hardware system for processing is easily realized.

When median filter has been used to preprocess the sonar images ,the performance of transform coding of sonar images is similar to that of TV images. The reconstructions obtained at 0.5 bits per pixel are quite satisfactory due to less of details in sonar images. Higher compression ratio will be obtained if some adapted algorithm are adopted.

REFERENCES

1. R.J.Clarke,"Transform Coding of Images",Academic Press,London(1985).

2. Tony Gioutsos and Susan Werness,"Transform coding of synthetic aperture radar(SAR) images",ICASSP, vol.2(1987).

3. Zhang Hailan ,Xu Wen and Zhu Weiqing,"A study on the statistics of sonic images",Chinese Journal of Acoustics,Vol.10 No.2(1991).

4. A.K.Jain, "Fundamentals of Digital Image Processing", Prentice Hall Inc., New Jersey(1989).

5. G.R.Arce,N.C.Gallagher,and T.A.Nodes,Median filters:Theory for one- and two-dimensional filters,in "Advances in Computer Vision and Image Processing " T.S.Huang, ed.,JAI Press Inc.,Connecticut(1986).

6. Huang Zhi,"Practical Program Library for Image Processing and Recognition", Tianjin Science and Technology Press,Tianjin(1989,in chinese).

A UNIVERSAL EXPERIMENTAL SYSTEM
FOR UNDERWATER ACOUSTICAL IMAGING

Enfang Sang, Xiaoyu Qiao, Lifu You, and Jingyi Zhao

Department of Underwater Acoustical Engineering
Harbin Shipbuilding Engineering Institute
Harbin 150001, P. R. China

INTRODUCTION

Since human activities of ocean exploration and development increase day by day, two-dimension (C-mode scanning) and three-dimension high resolution acoustical imaging are more and more needed. People's interests are not only in detecting the presence, bearing and distance of undersea objects by the traditional sonar, but also in getting their shape or geometry information with enough details. The undersea video camera can get images with much higher resolution. However, in undersea case, the visible distance of video camera is very limited, while the distance information of the object is more easily obtained in acoustical imaging process. Therefore, especially in middle distance (5—30 metres), acoustical imaging is a preferable choice.

For developing the sight of undersea robots, the researches of C-mode scan imaging and acoustical holography with high resolution have often been demonstrated [1,2,3]. However, for these kinds of imaging there is no real time engineering system available until now. The difficulties exist in imaging acoustics, fast imaging algorithms and their real time hardware realization. To study these problems and develop new technologies, a universal experimental system for underwater acoustical imaging has been developed, being presented in this paper. There are several features in this system: the hardware is programmable, including the selection of working mode, sampling frequency, recording block length and so forth, and the multi-functional hardware and very rich software sources are furnished. Therefore, many kinds of acoustical imaging experiments can be performed with this system.

THE SYSTEM STRUCTURE

The block diagram of the universal experimental system is shown in Figure 1.

In this system, the parallel data acquisition set is composed of thirty-three parallel A/D converters (extendible). The sampling mode, sampling rate, sampling data block length and starting delay time are all programmable. The maximum sampling rate per channel is 15 M words/s and is divided into 128 steps. The selections are preset by IBM microcomputer. Therefore it could cover very broad frequency range. To match such a high sampling speed, parallel storage has to be adopted. In sampling mode, external address generator addresses simultaneously the thirty-three parallel memory with 4K bytes size each. The memories are divided into three pages and mapped directly into the internal memory of IBM microcomputer.

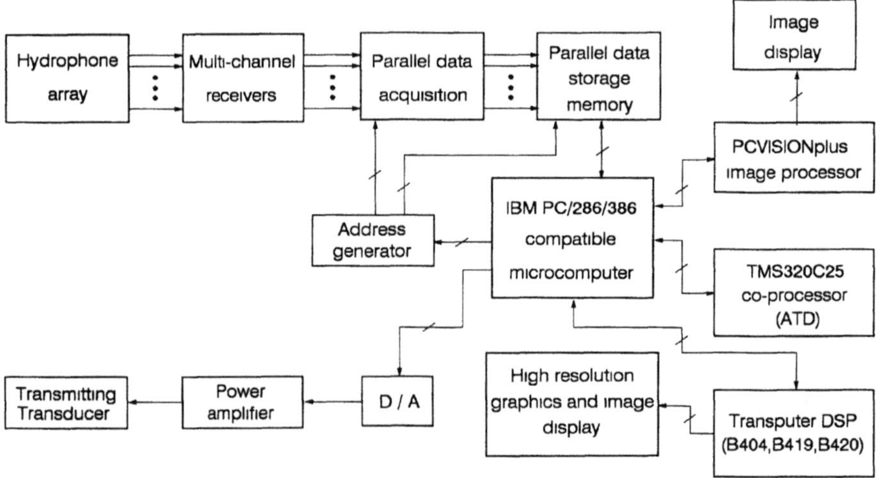

Figure 1. The block diagram of the universal experimental system for underwater acoustical imaging.

Since 2D and 3D imaging consume quite a lot of time, the TMS320C25 processor (ATD board) and INMOS Transputer processor boards (4 IMS-B404, 1 IMS-B420, 1 IMS-B419) are incorporated with microcomputer as co-processors. There is a digital image processing board in the system, which possesses the image grob, frame saving, real-time image display and the fundamental image processing functions.

The analog parts contained in the system are the wideband receivers and filters which are changeable to match different frequency signals.

In order to obtain both horizontal and vertical resolutions, two-dimension acoustical hydrophone arrays are built up. At present, a cross-like receiving array composed of 33 hydrophones and a single transmitting transducer are available.

THE BASIC MATHEMATICAL MODE AND THE SOFTWARE

The basic mathematical mode of the acoustical imaging is

$$G\left(x_i, y_k\right) = \left| \sum_i \sum_j \tilde{V}\left(a_i, b_j\right) \exp\left[\frac{2\pi}{\lambda} r\left(a_i, b_j, x_i, y_k\right)\right] \right|$$

where $G(x_l, y_k)$ is the backscattering strength from the coordinate (x_l, y_k) of the objective plane, $r(a_i, b_j, x_l, y_k)$ is the compensation distance from the point (a_i, b_j) of the receiver plane to the point (x_l, y_k) of the objective plane, $\widetilde{V}(a_i, b_j)$ is the receiving signal at the point (a_i, b_j) of the receiver plane.

The above geometry is shown in Figure 2.

For focussing in the near field or beamforming in the far field, the complex receiving signal $\widetilde{V}(a_i, b_j)$ must be formed and compensated with a reletive phase $\exp\left[\dfrac{2\pi}{\lambda} r(a_i, b_j, x_l, y_k)\right]$. For this purpose, it is necessary to sample orthogonally the receiving signal.

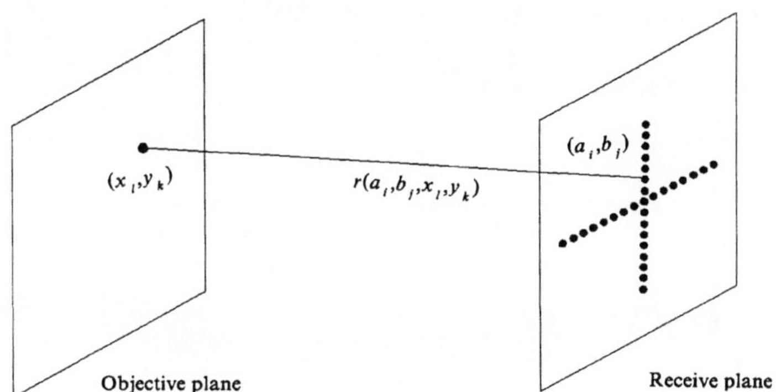

Figure 2. The geometry of acoustical imaging

The software of this system comprises four parts: initialization, image data acquisition, imaging and post-processing. The QUICK BASIC, MSC 5. 0, ASSEMBLE, OCCAM, PARALLEL C languages have been used individually or jointly for programming. The initialization software has the functions of the mode selection, setting sampling frequency, delay time and working parameters. The data acquisition software implements real-time sampling and storage. Imaging software includes various beamforming algorithms: the ordinary time and frequency domain beamforming FFT and Chirp-Z transform beamforming algorithms, focussing and its fast algorithms. The image post-processing software contains the image enhancement, restoration, feature detection and neural network classification. All the classical digital image processing programs and the new developed programs especially for the acoustical image are organized into a layer-built menu forming a software package.

EXPERIMENTAL RESULTS

By using this system, a series of experimental studies of C-mode acoustical imaging have been practised in a water pool. The receiving array with 33 hydrophones is of equal space and cross-like form array. The beamwidth of transmission is about 22 degrees. The signal frequency is 330 kHz and the source level is 180 dB.

In Figures 3 (a), 3 (b) and 3 (c), the acoustical images of one, two and three point sources
are shown, respectively. The distances between the objects and the receiving plane are about 8
meters. In Figure 4, an image of a square board (320mm×280mm) with 7. 5 meters distance
from the receiving plane are shown. The smoothing, linear and nonlinear enhancement algorithms have been used in post-processing process.

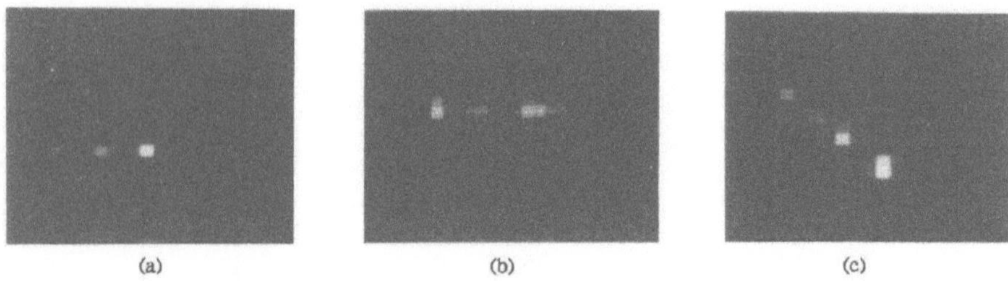

(a)　　　　　　　　　(b)　　　　　　　　　(c)

Figure 3.   (a) The acoustical image of one point source
             (b) The acoustical image of two point sources
             (c) The acoustical image of three point sources

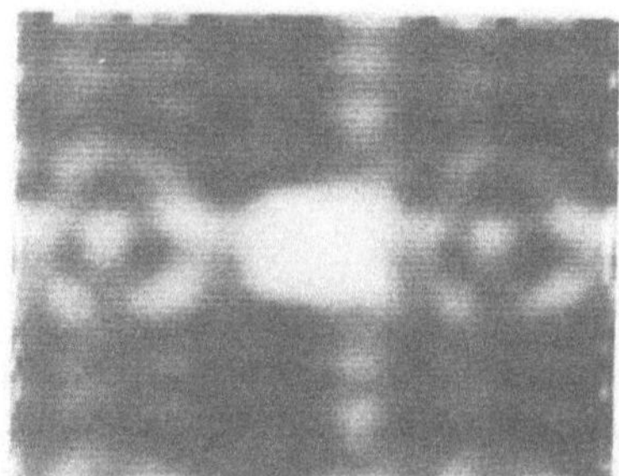

Figure 4. The acoustical image of a square board

CONCLUSION

Since the system described in this paper has the complete, advanced and programmable
hardware, and the rich software, it is a universal experimental system that is suitable to various
acoustical imaging studies and developments.

REFERENCES

1. Jules S. Jaffe and Philippe M. Cassereau, Multimode imaging for underwater robots, SPIE, Vol. 768, International Symposium on Pattern Recognition andAcoustical Imaging, 1987.
2. M. El-Banna, O. Abdel Alim, and M. Ezz-El-Arad, Focusing of linear arrays, Ultrasonic Symposium,
   1982.
3. Y, Tamura, et al, Ultrasound holographic imaging system using high speed scanning with pulses modulated by Walsh functions, Trans. of SICE, 24—3, 213—219, 1988.

UNIVERSAL HIGH RESOLUTION IMAGING SONAR
SIMULATION TOOLKIT

O. George and R. Bahl

Centre for Applied Research in Electronics
Indian Institute of Technology, Delhi
New Delhi-110016 , India

INTRODUCTION

We have been working towards the development of meaningful computer models for building imaging sonar systems and the external environment in which they operate[1-5]. This paper presents a Universal High Resolution Imaging Sonar Simulation Toolkit which is a software development tool for building imaging sonar simulators operating in a variety of underwater scenarios. They can be used in several unprecedented simulation applications like designing and testing the performance of sectorscan and sidescan sonar systems, tracking of underwater vehicles, underwater vehicle navigation, object search and location on the sea bed, obstacle avoidance etc. The toolkit could also be used to simulate an animated sequence of images during a particular "run" along a user selectable path/route in a 3D bounded ocean volume. The Toolkit is kept flexible and modular and to enable a user to build a wide variety of applications simply and quickly, the programs have been interfaced to X11 Window System. The user interface has been written using low level Xlib routines so that it can run on many types of workstations without changes. The application program has two major modules. In the 3D ocean scenario module, the user can create a number of moving and stationary objects and position them anywhere in the ocean volume with different types of sea bottom (flat, sloping and a few complex 3d surfaces). Once the 3D scene is generated, the module computes the visible coordinates using a new algorithm for the removal of hidden/occluded surfaces. The second module is the acoustic imaging sonar module which can model sectorscan/sidescan sonar, acoustic shadows, terrain shadows in case of non-flat surface, specular reflection, receiver array geometry, number of receiver elements, receiver beam pattern, overlapping beams, absorption and transmission loss, bottom reverberation, ambient noise, target strength etc. The toolkit provides 3d acoustic sectorscan sonar images (bearing, elevation and range) and 2d acoustic sidescan images (bearing and range).

USER INTERFACE

The user interface have been written solely with Xlib to have the entire simulation toolkit flexible and user-friendly. Higher level subroutine libraries known as toolkit widgets have been avoided so that the program can run on many workstations without changes. The program implements a set of user interface features such as menus, command bottom and keyboard and allows applications to manipulate these features. We could setup all the parameters of the ocean and the sonar module and run the simulation toolkit from this menu-driven program. The toolkit can also be run directly from command line.

3-D OCEAN SCENARIO MODULE

3-D Object And Terrain Modelling

We have developed a general simulation toolkit to model the ocean by creating multiple 3-d moving/stationary objects anywhere in the ocean volume, sea bottom and sea surface as densely packed surfaces of point scatterers. We superimpose a number of basic shapes like sphere, cylinder and cone to generate complex objects (refer[3,4] for more detail). Removal of overlapping surfaces in case of complex or multiple objects (with possibility of occlusion) has been achieved by modified ray tracing algorithm[4,5] which removes all those coordinates which fall in the shadow volume behind the object.

The sea bottom surface is generated using grid-based terrain modelling. The sea bottom is considered as a very high density data points (depending on the average sea bottom grain size). The terrain surfaces may be modelled by polynomial equations. The basic general polynomial equation for surface representation is

$$z = a_0 + a_1x + a_2y + a_3x^2 + a_4y^2 + a_5xy + a_6x^3 + \ldots \tag{1}$$

where z_i is the height of an individual point i, x_i and y_i are the rectangular coordinates of the point i and a_0, a_1, a_2... are the coefficients of the polynomial. We have considered the polynomial $z = a_0$ and $z = a_1x$ or $z = a_5xy$ respectively for flat and slopping bottom.
We could also generate complex surfaces using a set of mathematical functions[6]. A simple egg-tray shape can be generated using an equation of the form

$$z = a*\sin^2(bx)*\sin^2(by) \tag{2}$$

where a and b are relief amplitude and scaling factor respectively.

In case of non-flat terrain, in addition to the generation of surface coordinates we compute the surface normal which is used to generate visible coordinates of the terrain surface.

Accurate 3-D Visible Coordinate Generation

The next step is to set up the imaging scenario based around one or more objects generated by the ocean module. The local coordinates and the surface normals of the composite object, sea bottom and the sea surface scatterers are first transformed to the world coordinates depending upon the scenario set up w.r.t the sonar (fig.1). We generate the visible coordinates after computing the transformed coordinates of the entire imaging scenario by selecting those coordinates whose surface normals have a component towards the sonar view point. If the scenario has submerged/surfaced objects, then the object coordinate below/above the bottom/surface are removed.

Composite Object Shadow on Sea bottom/surface

Depending upon the geometrical scenario set up, the target may cast shadows on the sea bottom or on the sea surface or both. The coordinates of the scatterers which are lying in the shadow region are computed from the visible coordinates by mapping each visible coordinate either to the sea bottom or to the sea surface, depending upon the position of the visible coordinates. The area to be shadowed depends upon the position of the visible coordinates

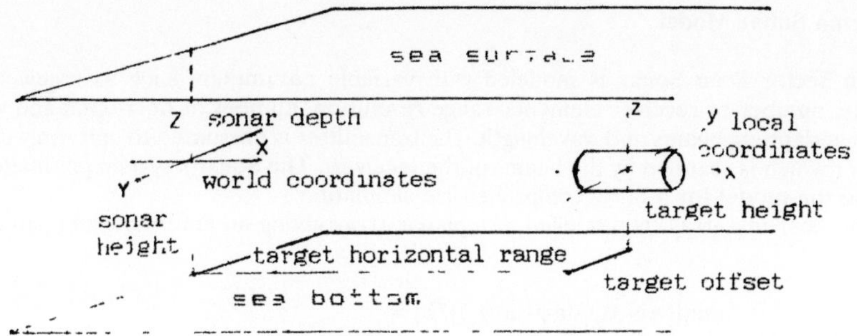

Figure 1. Target and the world coordinate system.

above/below the sea bottom/surface. Once the shadow region is defined, all the scatterers within this region are removed from the visible coordinates of the imaging scenario.

DYNAMIC SCENARIO MODELLING

The program can operate in two different modes. In the "Static" mode it generates one image of the static scenario and in the "Animation" mode it generates a sequence of images to animate a dynamic scenario and hence be used in several unprecedented simulation applications. These include tracking of underwater vehicles, navigation of undersea vehicles, object search and location on the sea bed, mission planning and control for Autonomous underwater vehicles and obstacle avoidance.

The present version of the program can simulate platform speed and direction, platform dynamics (roll, pitch and yaw), object speed and direction, object dynamics (roll, pitch and yaw), scenario with multiple stationary/moving objects.

In the animation mode the program accepts the surface coordinates of the composite object, sea bottom and the sea surface and first converts them to the world coordinates by taking into account the geometrical scenario set up. Then, for each snap-shot of the changing scenario, depending on the relative speed of the target and the rate of turn of the sonar platform, we compute the visible coordinates of the scenario including the object shadow on sea bottom/surface for visual display and acoustic imaging program. Once the initial parameters are selected, the program is fully autonomous and generates a sequence of perspective and acoustic image snap shots. These images are stored and later converted to bit maps for viewing the composite image sequence rapidly on a high resolution monitor.

ACOUSTIC IMAGING SONAR MODULE

We have modelled two types of sonars in the acoustic imaging sonar module. These are sectorscan and sidescan sonars. The visible coordinates of the scenario are given as input to this module. For large objects (dimensions above say 1 meter), the total number of visible coordinates to be generated would require huge storage space and program run time. In order to reduce the storage requirement and the enormous run time, we have modelled large objects to be consisting of a number of square plates whose dimensions can be chosen depending on the size of the object. After having decided the size of this plate we generate a data file for storing the magnitude (Target Signal Level) of the plate for incremental angles from 0° to 90° orientation using the standard Target Strength formula for a rectangular plate[7].

Sectorcsan Sonar Model

The Sector Scan Sonar is modeled with variable parameters such as receiver array geometry, number of receiver elements, range resolution, number of horizontal and vertical beams, overlapping beams and wavelength. The transmitter is presumed to uniformly insonify the sector which is scanned by the beams of the receiver . The Sonar's system parameters are built into the model for a fairly comprehensive simulation.

The receiving array is modelled as a planar array giving an antenna beam pattern[8,9]:

$$b_N(\psi,\theta,\psi_s,\theta_s) = \frac{\sin[(\pi N_z d_z(\sin\psi-\sin\psi_s))/\lambda]}{N_z\sin[(\pi d_z(\sin\psi-\sin\psi_s))/\lambda]} * \frac{\sin[(\pi N_y d_y(\sin\theta\cos\psi-\sin\theta_s\cos\psi_s))/\lambda]}{N_y\sin[(\pi d_y(\sin\theta\cos\psi-\sin\theta_s\cos\psi_s))/\lambda]} \tag{3}$$

where $b_N(\psi,\theta,\psi_s,\theta_s)$ is the 3-d beam pattern steered to direction (ψ_s,θ_s), N_z and N_y are the number of vertical and horizontal receiver array elements respectivily, d_z and d_y are the vertical and horizontal array element spacing respectivily and λ is the wavelength.

As N_z, N_y, d_z and d_y are user selectable, we could simulate any 3-D beam pattern and steer it horizontally and vertically by introducing appropriate phase shifts to the array elements. The scanned sector can also be decided depending upon the number of horizontal and vertical beams and their beam separation.

The return signal from the imaging scenario is spatially sampled and beamforming simulated. The signal output from the beamformer is used to generate a (3D) C-scan image which paints bearing, elevation and range as rectangular coordinates. The C-scan format adopted in this model consists of 20 bearing cells plotted horizontally and 10 vertical cells plotted vertically and 250 range cells plotted in depth. The cell sizes are defined in terms of bearing, elevation and range resolution of the Sonar. The image intensity is stored in 255 levels after the necessary normalization.

Scenario Imaging. The program calculates the complex signal return from each visible voxel/plate of the scenario by performing a running sum of individual signals within the range-bearing-elevation resolution cell of the Sonar. The return signal from the imaging scenario is assumed to be processed in a receiver that compensates for the spreading and absorption losses occurring in the water. For every range cell, the program steers the beam to all the bearing and elevation cells and for each steered direction the signal contribution of all the voxels/plates (lying in the range cell under consideration) depending on the beam pattern are coherently added and the resultant is stored in the relavent resolution cell of the sonar. After computing the bearing and elevation signal return from all the range cells, the program generates the acoustic image of the scenario by computing the (scalar) magnitude of the summed signals for each range-bearing-elevation cell. For ease of display the image is normalised to maximum level of 255.

Example of composite acoustic image generated form sectorscan sonar is shown in fig.2. The acoustic image is shown as multiple B-scan images for clarity.

Sidescan Sonar Model

The simulation of sidescan sonar was basically taken up to study and develop a model for high frequency backscatter from different types of seabottom. The sonar is modelled similar to sectorscan sonar except that only one beam is transmitted and the spatial coverage is obtained by moving the sonar physically through the water perpendicular to the direction

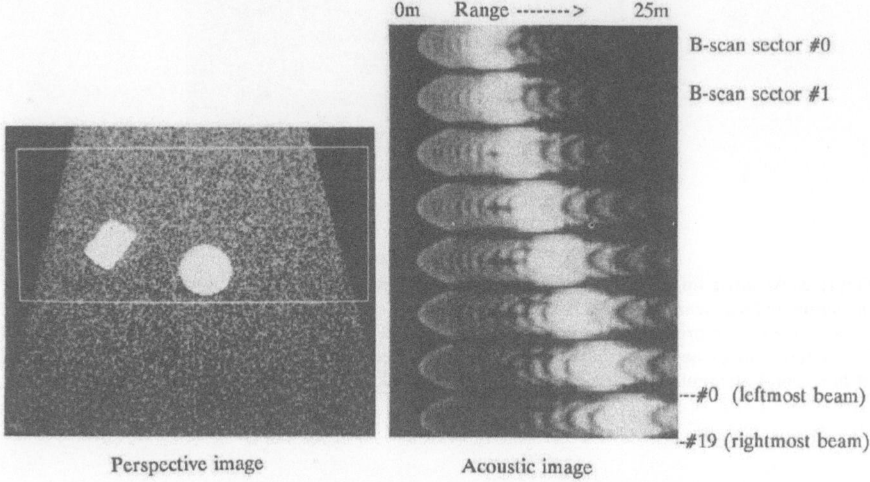

Figure 2. Perspective and composite acoustic image of a scenario having multiple objects. Environment Geometry: Horizontal distance and offset of the cylinder from the sonar 10m and -3m respectivily and height above the seabottom 2m, horizontal distance and offset of the sphere from the sonar 10m and 0.5m respectively and height above the seabottom 0.2m, channel height 35m. Sonar parameters: Sonar height above the seabottom 15m, frequency 180kHz, range resolution 0.1m, number of horizontal and vertical elements and element spacing 40 and 0.5λ respectivily, shading rectangular and number of horizontal and vertical beams 19 and 8 respectivily.

of sound transmission. The receiving array is modelled as a linear array giving an antenna beam pattern[8,9]:

$$b_N(\psi,\theta) = \frac{\sin[(\pi N_y d_y(\sin\psi)/\lambda]}{N_y \sin[(\pi d_y(\sin\psi)/\lambda]} \qquad (4)$$

where $b_N(\psi,\theta)$ is the 3-d beam pattern, N_y is the number of horizontal receiver array elements, d_y is the spacing between the elements and λ is the wavelength.

During each echo-ranging cycle, the program samples the received signal at the receiver array and after beamformation it generates one line at a time in a rectangular display format. The basic image format is the standard B-scan image which paints range and bearing as rectangular coordinates.

Example of composite acoustic image generated from sidescan sonar is shown in fig.3.

SEA-BOTTOM BACKSCATTERING MODEL

We have developed a model for bottom backscattering for frequencies above 100 Khz for sandy sea bottom. Very little modelling work has been reported in this frequency range. The most widely used is the composite roughness model[10] developed by D.R.Jackson to predict bottom backscatter as a function of grazing angle and environmental conditions for the frequencies 10 - 100 kHz.

In the model developed the sediment is assumed to be homogenous. This assumption is reasonable because a small amount of very high frequency energy may pass through the

Figure 3. Acoustic image of the seabottom dominated by anisotropic sand waves or ripples generated by the simulated sidescan sonar. Seabottom Parameters : Sediment mean grain size 0.166mm, porosity 39%, density 1.966, compressional wave velocity ratio 1.133 and Max. height of the ripple 0.07m. Sonar Parameters: Sonar height above the seabottom 2.5m, frequency 180kHz, range resolution 0.05m, number of horizontal elements and element spacing 20 and 0.5λ respectivily and shading triangular.

water-sediment interface and interact with the subbottom (attenuation coefficient is about 100dBm^{-1} at 100Khz and this factor increases with frequency[11]).This makes it possible to ignore subbottom inhomogeneities. For sandy bottoms where there is little biological activities we could assume very little volume scattering and more homogeneous sediments.

For millimetre length acoustic waves ($>$100Khz.), it is likely that the most important inhomogeneity is the graininess of the sediment. When the wavelength approaches the actual grain sizes, the very high frequency may begin to show a frequency dependence. We intend to do low level modelling to cater for most of the sand grain properties like mean grain size, porosity, density and compressibility etc.

Since the topography of the seabed is an important factor, we have considered the following:
* large features that block propagation and create "terrain self shadows",
* intermediate sized features that represent large scale roughness and
* small scale features that represent most of the sand grain properties.

We have considered the grid spacing of the bottom terrain to be equal to the mean sand grain size to model the small scale features. This will lead to a very high density of points and would require huge disk space for storing the surface coordinates. Our approach here involves the use of facet models. We can approximate a rough surface through a series of small planar "facets" each tangential to the actual surface. Facet models treat the scattering from the assemblage of such facets by taking into account both their reradiation patterns and distribution of their slopes.

The reradiation pattern of the facet, which represents the actual seabottom patch, is determined by placing sand grains at each of the grid points of the facet and from this well-packed grain structure we remove sand grains at random depending upon the porosity. We assume the sand grains to be spherical, not fixed and rigid, but possess a compressibility. Since the ratio of the circumference to wavelength of the sand grain is much less than unity, we could use the Rayleigh backscattering formula for small spheres[7] to find the signal contribution from individual sand grains. By performing a running coherent sum of these signals from 0° to 90° facet orientation, we generate the re-radiation pattern of the facet.

Next, we generate the terrain surface using these facets. The terrain module generates the facet location and orientation depending upon the type of surface selected by the user. The intermediate scale features can be modelled by perturbing the facet orientation randomly depending upon the roughness of the seabottom.

After the terrain surface is generated, the simulated sidescan sonar is used to find the average backscattering strength for various grazing angles. The backscattering strength for various grazing angles is calculated by

$$BS(\theta_i) = 10\log(P_i) - 10\log A_i \tag{5}$$

where
θ_i = grazing angle,
P_i = rel. power level from range-bearing cell i
   = (rel. voltage level V_i)^2 and
A_i = area exposed to the acoustic beam in the ith range-bearing cell.

Relative voltage level V_i is generated by the sidescan sonar by computing the magnitude of the summed signals from all the facets falling in the range cell R_i. The $BS(\theta_i)$ is then averaged over about 10 echo ranging cycle.

We have compared the results of our high frequency backscattering prediction model to expremental data reported by Stanic[12]. The environment and the sonar parameters measured at the site were fed to the model and it was found that the predicted scatterring strength at 180kHz (fig.4) for grazing angle 5°-30° was a reasonable fit to the experimental points.

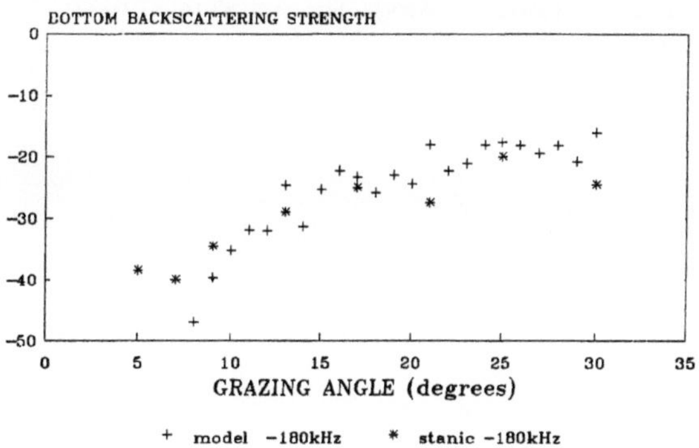

Figure 4. Comparison of high frequency backscattering model and experimental data reported by Stanic[12]

SUMMARY AND CONCLUSION

This paper has outlined the methodology of Acoustic Imaging Sonar Simulation Toolkit. Facilities available for simulation are the construction of complex 3-d object shapes, specification of Sonar parameters, Sonar platform dynamics, setting up of imaging scenario, dynamic scenario for animation, and provision of visual perspective images for reference. The acoustic images simulated have included specular reflections from the object, acoustic shadows, complex terrain modelling, bottom reverberation, array shading and grating lobe effects. The acoustic images may be displayed after normalisation to 0-255 range in a chosen range window. The flexibility of the program allows one to generate realistic images for an unlimited number of scenarios and object shapes. This feature is of paramount importance in studying and predicting the performance of Sectorscan and Sidescan Sonars and is expected to be the mainstay for the design of signal and image processing techniques for such systems.

REFERENCES

1. R.Bahl and J.P.Powers, "Computer Model of a High-Resolution Imaging Sonar", Technical Report No:NPS 62-90-011 Naval Postgraduate School, Monterey, CA 93943, (1990).
2. O.George, N.Rajpal, R.Bahl, T.B.Rao, V.Natrajan, "3-D High Resolution Sonar Imaging", 19th

International Symp. on Acoustical Imaging, Bochum, Germany (1991).

3. O.George and R.Bahl, "THIRISM-2 : 3-D High Resolution Imaging Sonar Model", National Symposium on Ocean Electronics, Cochin University of Science and Technology, Kochi-6820022, India (1991).

4. O.George and R.Bahl, Technical Report CARE/91-3 Indian Institute of Technology Delhi, New Delhi-110016, India.

5. O.George and R.Bahl, "High Resolution Imaging Sonar Simulation Toolkit", International Symposium on Ocean Technology, NIO, Goa, India (1992).

6. G.Petrie and T.J.M.Kennie, TERRAIN MODELLING IN SURVEYING AND CIVIL ENGINEERING, Whittles Publishing (1990).

7. R.J.Urick, PRINCIPLES OF UNDERWATER SOUND, McGraw-Hill, New York (1975).

8. L.J.Ziomek, UNDERWATER ACOUSTICS - A Linear Systems Theory Approach, ACADEMIC PRESS, INC, (1985).

9. W.S.Burdic, UNDERWATER ACOUSTIC SYSTEM ANALYSIS, Prentice Hall, Englewood Cliffs, New Jersey 07632 (1991).

10. D.R.Jackson, D P. Winebrenner and Akira Lshimaru, "Composite roughness model of backscattering," J.Acoust. Soc. Am. 79 (5) (1986).

11. E.L.Hamilton, "Compressional wave attenuation in marine sediments," Geophysics 37, 620-646 (1972).

12. S.Stanic, K.B.Briggs, P.Fleischer, R.I.Ray, and W.B.Sewyer, "Shallow-water high-frequency bottom scattering off Panama City, Florida," J. Acoust. Soc. Am. 83 (6) (1988).

AUTOMATIC OBJECT IDENTIFICATION FROM

SONAR IMAGE SHADOW

N. Rajpal[1], S. Banerjee[2], and R. Bahl[1]

[1]Centre for Applied Research in Electronics
[2]Department of Computer Science & Engineering
Indian Institute of Technology Delhi
New Delhi-110016, India

INTRODUCTION

High resolution sector scan and side scan sonars are useful in underwater acoustic imaging for obstacle avoidance search of mines, submarines, lost wrecks etc. The manual analysis of images is difficult because of huge information volume and unfavorable environmental conditions. Autonomous underwater vehicles equipped with imaging sonars and automatic object recognition systems are currently being designed.

In this paper, we first present the highlights of a model for simulating 2-d high resolution sonar images[1] and then an automatic object identification algorithm from acoustic shadow in these images. The process of object identification is broadly divided in two parts. Firstly, pre-processing steps for extraction of the features of object from its shadow. Secondly, the shape matching task, which matches the object features against the models' features to find out the presence of models in the scene.

FEATURES OF SONAR IMAGE SIMULATOR

There is a need to have an acoustic image acquisition system for studying algorithms for object identification. Such a system should have the capability to provide repeatable performance in a number of controlled experiments. Experience has shown that undersea data collection is a very expensive affair in terms of facilities, resources and time. There is therefore, sufficient motivation to develop a simulator for acoustic images. However, such a simulator should provide "realistic" data in a variety of scenarios.

The Sonar simulator developed by us[1] is being constantly upgraded to include more facilities and user-friendly features. It presently has the following set-up facilities:
1.Creation of various object shapes and orientations
2.Selection of scanning Sonar parameters
3.Scene description with reference to sea surface and bottom

The simulator generates the acoustic image data with the following capabilities:
1.Bottom reverberation dependent on grazing angle
2.Object echos including specular effect
3.Acoustic shadow of the object on the sea bottom

Figure 1 Simulated Sonar Image in B-Scan Format.

A typical acoustic image simulated for a sector-scanning Sonar is shown in figure 1. The existence of a "shadow" on the sea bottom is a key feature of such images. It is apparent that the object echo is lost within reverberation and is therefore not very useful in identifying the object. On the other hand, the acoustic shadow is more easily extracted from the background. It has been shown[1] that the shadow gets converted to the silhouette when the acoustic image is processed by a non-linear transformation of the range axis. Such an image is extremely useful for identification of the object.

OBJECT IDENTIFICATION

As apparent from the figure 1, the acoustic shadow corresponds with the shape of the object. The 2-d silhouette can easily be extracted from the shadow. For object identification we use a silhouette based object recognition technique using random associative search[2]. The process of object identification is broadly divided into two parts.

Pre-Processing

The pre-processing broadly involves: (i) segmentation of the image into shadow and background, (ii) contour tracing and (iii) feature extraction for matching algorithm. The features required for object algorithm discussed in this paper are the points of maximum curvature on the silhouette.

Segmentation. Segmentation involves thresholding and region growing. As we are using sonar image shadow for object recognition, the purpose of segmentation is to separate out shadow region from the rest of the image. The shadow region being dark, intensity of pixels in this region is very close to zero. The image is first converted into binary image by using threshold value slightly higher than zero, the pixels below this threshold are converted to one and above this to zero. The regions of interest (image shadow) from the binary image are extracted using connectivity algorithm (Algorithm I). The smaller regions which may arise due to noise can be dropped in this process.

Algorithm I Connectivity Algorithm
Procedure CONNECTIVITY(A , L)
Notation: A is the starting point of the region set (the points of value one in the binary image belong to the region set) and L is the label for the region being grown. The starting point can be searched by scanning the image in horizontal or vertical direction. C is the current point

being examined. The steps are:

    0. Set intensity value of pixel A to L

    1. For all the 8 neighbours of A perform step 2.

    2. If C (one of the 8 neighbors of A) belongs to the region set

    Then call procedure connectivity recursively as

        CONNECTIVITY (C , L).

    3. End of Algorithm.

Contour Tracing. This step gives the boundaries of the shadow regions obtained from the connectivity algorithm given above. The Contour tracing algorithm is applied for tracing the outer boundaries of the region. The algorithm starts with a pixel on the boundary, which can be found by scanning the image in horizontal or vertical direction, and traverses the boundary in counterclockwise direction and stores the coordinates of the boundary points. The details of the algorithms are given in[3].

Feature Extraction. The object silhouette from the shadow boundary is extracted by applying a non-linear range transformation based on the sonar position with respect to the bottom. The points of maximum curvature are obtained by polygon approximation of the silhouette. The boundary obtained in the previous step is in the form of a closed curve. Our aim is to find a polygon that approximates the boundary while having as small a number of vertices as possible. Fitting a polygon is helpful in determining the features that are important for the description of the shape. The main task is to find the lines and the break points of the polygon. The algorithm is based on the split-and-merge strategy. The basic idea is to examine groups of points and check whether they are approximately collinear. If they are not, then the group is split until collinearity condition is satisfied. On the other hand, groups are merged if the result will be a group with approximately collinear pixels. In this way the split-and-merge process gives the polygon approximation to the boundaries. The details of algorithm are given in[3].

Shape Matching

The next step after the pre-processing is matching the features of the observed object against the model features already computed and stored. The shape matching method discussed here is based on a random associative search technique. The algorithm matches the critical point set (maximum curvature) of the scene against the point sets of the models simultaneously. The method takes care of the affine transformation and partial occlusion. The class of affine transformation includes the similarity transforms, and in addition allows "skew". Such distortions typically arise when an object is observed under different viewing directions. The invariance of the coordinates in transformed affine bases is used as the guiding principle for the matching process.

Problem Definition. The matching problem can be defined as follows. A function, which maps the set of interest points in the scene to the set of points in the knowledge base, is an interpretation for the scene. All the points mapped to the points of one model in the knowledge base represent the corresponding part of that model in the scene. Our problem is to search in the space of all such possible functions to find the correct interpretation using the affine transformation property explained as follows.

A 2-d silhouette and its transformation can be approximated as an affine 2-d correspondence, namely, there is a non-singular 2X2 matrix A and a 2-d translation vector b, such that each point x in the 2-d silhouette is translated to the corresponding point Ax + b in the transformed silhouette. Our problem is to recognize the objects in the scene and for each recognized object to find the affine transformation that gives the best match of the model of the object and its transformed image in the scene. As there are six parameters (2X2 matrix A and a 2-d vector b) in the affine transformation (T), it can be uniquely defined by the transformation of three non-collinear points in the scene. Further, any affine transformation applied to the set of points in the scene will not change the set of coordinates based on the ordered basis triplet obtained from the three non-collinear points in the scene. This fact can be explained as follows. Consider the set of M points in the scene and pick any ordered subset of three non-collinear points. The two linearly independent vectors based on these vectors are a 2-d linear basis. One can express the coordinates of all model points in this basis. (The basis points will have the coordinates (0,0), (1,0), (0,1), respectively.) Let e_{00}, e_{10}, e_{01}, be an ordered affine basis triplet in the plane. The affine coordinates (x,y) of a point v are

$$v = x(e_{10} - e_{00}) + y(e_{01} - e_{00}) + e_{00} .$$

Application of an affine transformation T will transform the point v to

$$Tv = x(Te_{10} - Te_{00}) + y(Te_{01} - Te_{00}) + Te_{00} .$$

Hence Tv has the same affine coordinates (x,y) in the basis triplet Te_{00}, Te_{10}, Te_{01} .

Using the above invariant property of the affine coordinates we formulate an error function over the space of interpretations. Then the point matching problem reduces to one of finding the interpretation which reduces the error functional and can be solved using the random associative search technique. We have defined the point matching problem in the framework of Relaxation Labelling[4] such that all consistent and unambiguous labellings correspond to the minima of error functional. The Labelling problem is then solved using a cooperating team of Learning Automata[5,6].

Representation Of the Scene and Models. The scene and models are represented by the ordered sequence of the interest points (maximum curvature) on the 2-d silhouette traced in the counterclockwise direction. The scene is represented as the object set, O = {O[1],O[2],...O[N]} and models as set L = {L[1],L[2]...L[M]}. Here O[i] are the interest points on the 2-d silhouette of the scene. The model set L consists of p disjoint models say model[1] , model[2] ..... model[p]. Each model[i] in L represents the interest points on 2-d silhouette of that model exactly in the same way as the object O above. Let msize[model[i]] represents the number of points in the model[i] then Σ_i msize[model[i]] = M.

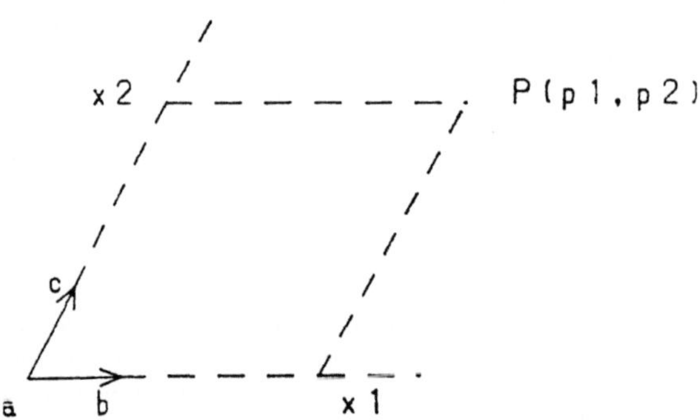

Figure 2 Coordinates of the point P in the affine basis triplet (a, b, c).

Problem Formulation. Using the invariance of the affine coordinates of the point under affine transformation as explained earlier we formulate an error function E(f) whose minimum gives the correct interpretation of the scene. Before defining the error function we need to explain the method of computation of interest point coordinates in the ordered basis triplet.

The computation of the coordinates $x(x_1,x_2)$ of the interest point $p(p_1,p_2)$ in the affine basis (a, b, c) with 'a' as origin can be formulated as a solution of the linear system of the two equations in two unknowns Ax = d, where A is 2X2 matrix and its two columns are the difference vectors of the basis interest points b-a, and c-a, respectively, and the vector d is p-a. These vectors a, b, c and p are represented in the image coordinates, whereas the solution vector x gives the representation of the point p in the affine basis (a, b, c) coordinates as shown in figure 2.

Consider the interpretation f = (f[1],f[2],...f[N]) which maps a point in the scene O[i] to a point in the model set f[i], where f[i] belongs to set L. The error functional E(f) over the space of possible interpretations f, the minimum of which will give the correct interpretation f* is defined as follows. First, we define error function $E_i(f)$ (definition 2) for every interest point O[i] under the interpretation f. The total error E(f) under interpretation f is given by the sum of errors at each object interest point O[i]. Before defining the error function we need the following definitions.

Definition 1: The preceding and succeeding neighbours of a point in the scene are given by,

   PNGH(O[i]) = O[N] if i = 1
              = O[i-1] otherwise.
   SNGH(O[i]) = O[1] if i = N
              = O[i+1] otherwise.

The preceding and succeeding neighbours of a model point L[i] which belongs to model[k] are given as,

if r = msize[model[1]] + msize[model[2]].....+ msize[model[k-1]].

   PNGH(L[i]) = L[r + msize[model[k]]]   if i = r+1
              = L[i-1] otherwise.
   SNGH(L[i]) = L[r+1] if i = r + msize[model[k]]
              = L[i+1] otherwise.

Definition 2: The error $E_i(f)$ incurred at O[i] under the interpretation function f is given by the average of $e_{ij}(f[i],f[j])$ over all those j such that (j <> i) and (f[i] and f[j] belongs to the same model) . The $e_{ij}(f[i],f[j])$ over all j given under the above conditions is defined as:

Let $O_j(x,y)$ be the coordinates of the point O[j] in affine basis (O[i], O[s], O[p]) computed as explained above. Here O[s] and O[p] are the succeeding and the preceding neighbours of the point O[i].

Let L[s] and L[p] be the succeeding and preceding neighbours of the point f[i] in the model to which f[i] belongs and $f_j(\alpha,\beta)$ be the coordinates of the point f[j] in affine basis (f[i], L[s], L[p]).

Then,

$e_{ij}(f[i],f[j])$ = 0   if the distance between the points $O_j(x,y)$ and $f_j(\alpha,\beta)$ is less than the allowed tolerance factor.

              = 1   if the above distance is more than the tolerance factor.

The tolerance factor depends upon the minimum distance between the points in the model to which f[i] belongs. In the ideal case the tolerance factor should be zero.

The total error function E(f) under interpretation f is given by the sum of $E_i(f)$ over all the object points O[i]. The correct interpretation f* can be found by minimizing the E(f).

Minimization of E(f) Through Relaxation Labelling. In this section we describe the problem of finding minimum of E(f) and thus the correct interpretation of the scene in the framework of Relaxation Labelling. The method was suggested by Banerjee.S[5]. The details of Relaxation Labelling technique are given in [6]. We present a continuous labelling problem where all consistent and unambiguous labellings are local minima of E(f) and vice-versa. Thus, by solving the labelling problem, we find a local minimum of E(f), closest to the initial assignment, but the interpretation f obtained may not be the global minimum. The local minima in E(f) arise when a portion of the scene closely resembles several portions of the model. In most cases the local minimum by proper choice of initial assignments gives a near correct interpretation of the scene with a small percentage of wrong assignments. This will be clear from the results of the algorithm given later.

The sets O = {O[1], O[2], ... O[N])} and L = {L[1], L[2], ... L[M]} described earlier in this paper form, respectively, the object and the label sets in the Relaxation Labelling paradigm. The neighbour relation over the object set is defined through the concept of the preceding and the succeeding neighbours, given in Definition 1. The compatibility expression $r_i(f)$ is specified through the shape matching constraints in the following way:

$$r_i(f) = 1 - E_i(f).$$

As is clear from the definition of the $E_i(f)$ that its value lies between 0 and 1, hence the value of $r_i(f)$ lies between 1 and 0 (as is clear from the above equation). The following theorem establishes the correspondence between the local minima of $E(f)$ and the consistent and unambiguous labelling for the Relaxation Labelling problem.

Theorem 1: The labelling problem posed above with the $r_i(f[i],f[j])$ as defined above is such that a point $f \in K^*$ is consistent if and only if $E(f)$ assumes a local minimum at f.
For the proof of the above theorem see[5].

In this method, by proper choice of the initial distribution of the label probabilities, and then using Learning Automata algorithm given in[6], we can obtain the local minima of the error functional which is close to the initial probability distribution. The initial probabilities to the object label pairs are assigned on the basis of some local measurements explained below.

For each object point O[i], compute the following:
$$D(O[i],L[p]) = 1/ (1 + e_{ik}(L[p],L[q]) + e_{ij}(L[p],L[r]))$$
$$\text{for } p = 1, 2, \dots\dots M$$
where O[k] = PNGH(PNGH(O[i])) , O[j] = SNGH(SNGH(O[i]))
     L[q] = PNGH(PNGH(L[p])) , L[r] = SNGH(SNGH(L[p]))
We assign the label probabilities as
     P(i,p) = D(O[i],L[p]) / Z
          for p = 1, 2, ... M
Where the normalizing constant Z is given as
     Z = D(O[i],L[1]) + D(O[i],L[2]) + ... D(O[i],L[M]).

RESULTS

Figure 3 shows the boundary of the shadow of an object extracted from the image in figure 1. Figure 4 shows the polygon approximation of the silhouette obtained after applying the range transformation on the shadow boundary shown in figure 3. The lower portion of the silhouette is distorted because of occlusion caused by range overlap of the target echo. Table 1 shows the result of matching the shape shown in figure 4 against the model shape shown in figure 5.

Figure 3 Boundary of the Shadow extracted from the image shown in figure 1.

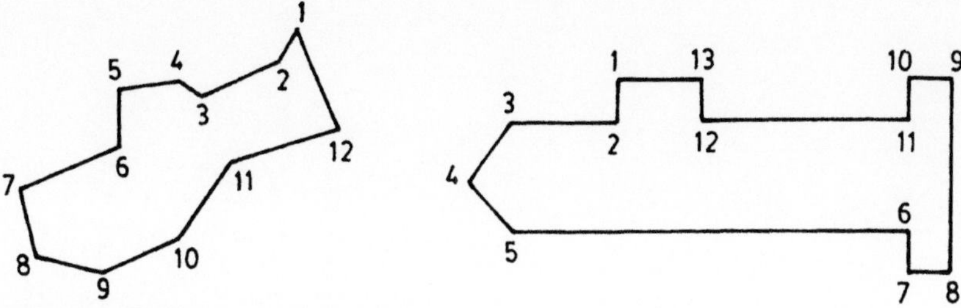

Figure 4 The Scene.
(Polygon approximation of the silhouette
obtained after applying range transformation
on the shadow boundary shown in figure 3.)

Figure 5 The Model.

TABLE 1. Final Label assignments obtained after applying the matching algorithm on the
Scene and the Model shown in figure 3 and 4 respectively (* indicates no match)

OBJECTS:	1	2	3	4	5	6	7	8	9	10	11	12
LABELS:	10	11	12	13	1	2	3	4	5	*	*	*

CONCLUSION

The 2-d silhouette based object identification method discussed in this paper is a novel
application of "Relaxation Labelling", and compares very favourably with the techniques found
in the literature; both in terms of accuracy and speed[7]. Further, the algorithm is completely
parallel and consequently, can be easily implemented on a parallel architecture for real time
applications such as in autonomous underwater vehicles. The shape matching technique
discussed here is a general 2-d silhouette based matching method and can be used for 2-d
intensity and 3-d range images, from which the silhouette of the objects can be extracted. The
high resolution sonar simulator discussed in here 2 is also useful for optimising sonar
parameters for specific applications.

REFERENCES

1. Bahl,R. and Powers,J.P. (1990), 'Computer Model of a High-Resolution Imaging Sonar',
   Technical Report, NPS62-90-011, Naval Postgraduate School, Monterey, California.
2. Rajpal,N., Banerjee,S. and Bahl,R. (1991), 'Silhouette Based Object Recognition Using
   Random Associative Search', Technical Report, CARE/91-2, Indian Institute of
   Technology, Delhi, New Delhi.
3. Pavlidis,T (1982) 'Algorithms for Graphics and Image Processing' Springer Verlag, Berlin-
   Heidelberg.
4. Hummel,R.A and Zucker,S.W. (1983),'On the Foundations of Relaxation Labelling
   Processes', IEEE Trans. PAMI-5, pp. 267-286.
5. Banerjee,S. and Sastry,P.S. (1989), Technical Report, Dept. of Elec. Engg., Indian Institute
   of Science, Bangalore.
6. Thathachar,M.A.L and Sastry,P.S.(1986),'Relaxation Labeling with Learning Automata',
   IEEE Trans. PAMI-8, pp. 256-268.
7. Zerr,B. (1990) 'Automatic Classification of Objects in Sonar Images', Conference
   Proceedings, Undersea Technology, at Novotel, London, England.

PARTICIPANTS

Dr. P. Alais
Laboratoire de Mecanique Physique
Universite Paris 6
2, Place de la Gare de Ceinture
F-78210 ?
France

Prof. Forrest Anderson
Impulse Imaging Corp.
P.O. Box 1400
Bernalillo, NM 87004

Mr. Yoshinao Aoki
Dept. of Information Engineering
Hokkaido University
Sapporo, Japan

Prof. Virgil Bardan
Geophysical Department of
  Bucharest University and
  Computer Center of IPGG
Str. Coralilor 20
Sect. 1, Bucharest 78449
Romania

Prof. D.D. Bennink
Department of Materials
University of Oxford
Parks Road, Oxford OX1 3PH
United Kingdom

Dr. E. Biagi
Dipartimento di Ingegneria Universita di
  Firenze
Via s.Marta 3,50319 Firenze
Italy

Mr. A. Benkhelifa
Laboratoire d'Imagerie Parametrique-
  URA CNRS 1458
15 rue de l'Ecole de Medecine 75006
Paris, France

Prof. V.A. Burov
Dept. of Acoustics
Faculty of Physics
Moscow M.V. Lomonosov State University
Moscow 119899
Russian Federation

Mr. L. Capineri
Lab. di Ultrasuoni e Controlli Non Distruttivi
Dipt. di Ingegneria Elettronica
Univ. degli Studi di Fireze
Via S. Marta 3
50139 Firenze
Italy

Dr. Gan-Changmin
Moscow State University
119899, Moscow
Russian Federation

Mr. L.A. Chernozatonskii
Center of Acoustic Microscopy
Institute of Chimical Physics
Russian Academy of Sciences
Moscow 117977, Kosygin Str, 4
Russian

Mr. H. Dabirikhah
Dept. of Electronic and Electrical Engineering
King's College
University of London Strand
London WC2R 2LS
United Kingdom

Dr. Zhan Dejun
Wuhan Institute of Physics
Chinese Academy of Sciences
Wuhan, Hubei, 430071
P.R. China

Dr. Yang Desen
Institute of Acoustics
Harbin Shipbuilding
  Engineering Institute(HSEI)
Harbin, P.R. China

Mr. S. Dessi
Department of Electronic Engineering-
  University of Florence
Via Santa Marta
3-50139 Firenze
Italy

Dr. Sang Enfang
Dept. of Underwater
  Acoustical Engineering
Harbin Shipbuilding Eng. Institute
Harbin 150001
China

Mr. A.G. Every
Physics Department
University of The Witwataerstrand
P.O. Wits 2050
South Africa

Zhen Xiang Fan
GRI, Bureau of Oil
Geophysical Prospecting
P.R. China

Dr. Yin Feng
Department of Physics
Southeast University
Nanjing 210018
P.R. China

Mr. K.W. Ferrara
Center for Image Processing and
  Integrated Computing
University of California
Davis, CA 95616

Mr. M. Fink
Laboratoire Ondes el Acoustique
University Paris VII
E.S.P.C.I.
10 rue Vanquelin
75231 Paris Cedex 05
France

Prof. Louis Fishman
Department of Mathematical and
  Computer Sciences
Colorado School of Mines
Golden, CO 80401

Pro. A. Fort
Laboratorio di Ultrasuoni e Controlli
  Non Distruttivi
Dipartimento di Ingegneria Elettronica
Universita degli Studi di Firenze
Via S. Marta 3, 50139 Firenze
Italy

Mr. O. George
Centre for Applied Research in Electronics
Indian Institute of Technology, Delhi
New Delhi-110016
India

Dr. W.S. Gan
Acoustical Services Pte Ltd.
29 Telok Ayer Street
Singapore 0104
Republic of Singapore

Dr. A.V. Glazkov
Department of Acoustics
Faculty of Physics
Moscow M.V. Lomonosov State University
Moscow 119899
Russian Federation

Dr. X.F. Gong
Institute of Acoustics
Nanjing University
Nanjing 210008
P.R. China

Mr. A.A. Gorjunov
P.P. Shirnov Institute of Oceanology
Russian Academy of Science
Russia, 117218
Moscow B-218 Krasikov Street, 23
Russian Federation

Prof. James F. Greenleaf
Biodynamics Research Unit
Department of Physiology
  and Biophysics
Mayo Clinic/Foundation
Rochester, MN 55905

Mr. Wei Han
Norwegian Institute of
  Wood Technology
N-0314 Oslo 3
Norway

Dr. Benli Gu
Department of Biomedical Engineering
Southeast University
Nanjing, China

Mr. Rolf Kahrs Hansen
N-5083 Ovre Ervik
Norway

Dr. H.J. Hein
Inst. of Applied Biophysics
Clinics of Orthopedic
Department of Medicine
MLU Halle
Germany

Dr. C.R. Hill
Physics Department
Institute of Cancer Research
Royal Marsden
Hospital, Sutton, Surrey
SM2 5PT
United Kingdom

Prof. G. Hofelmann
Dept. of Communication Engineering
Duisburg University
Germany

Dr. Liu Hong
Institute of Geophysics
Chinese Academy of Sciences
Beijing, 100101
P.R. China

Dr. Gushing Hu
Dept. of Mach. Engineering
Zhejiang University
Hangzhou, Zhejiang 310027
P.R. China

Dr. Yanyan Hu
Sandong Medical University
Jian, P.R. China

Dr. Vladimir Ivanov
VNII Geoinformsystem
Varshavskoe shosse, 8
Moscow, 113105
Russia

Dr. X. Jia
Groupe de Physique des Solides
Universite Paris 7
Tour 23, 2 Place Jussieu
75251 Paris Cedex 05
France

Mr. Li Jiaou
Beijing Aeronautical Manufacturing
  Technology Research Institute
P.R. China

Dr. Timo-Pekka Jantti
Hollmling Ltd. Electronics
Rauma - Finland

Mr. Leon K. Jodlowski
Military Technical Academy
01-489 Warszawa
Poland

Dr. Jerzy Kapelewski
Inst. of Technical Physics
WAT S. Kaliskiego Str. 00-908 Warsaw
Poland

Mr. Michael Y. Kiernan
GE Nuclear Energy
P.O. Box 780
Wilmingtibm NC 28402

Mr. Kiyohito Koyama
Faculty of Engineering
Yamagata University
Jonan 1-3-16
Yonezawa 992, Japan

Prof. A. Kulik
Ecole Polytechnique Federale de Lausanne
Institute de Genie Atomique
CH-1015 Laussanne
Switzerland

Mr. C.Q. Lan
Wuhan Institute of Physics
The Chinese Academy of Sciences
Wuhan 430071
P.R. China

Prof. K.J. Langenber
Dept. of Electrical Engineering
University of Karssel
3500 Kassel
Germany

Mr. P. Laugier
1URA CNRS 1458 Imagerie Parametriqye
15 rue de l'Ecole de Medecine
75006 Paris, France

Dr. Vladim Levin
Center of Acoustic Microscopy
Institute of Chemical Physics
Russian Academy of Sciences
Moscow, 117977 Kosygin Str.4
Russia

Mr. G.R. Li
Dept. of Materials Sciences
   and Engineering
Yamagata University
Yonezawa 992
Japan

Mr. W. Li
Erasmus University Rotterdam
Department of Diagnostic Radiology
University Hospital Leiden
The Netherlands

Dr. Jing Liu
Shandong Medical University
Jian, P.R. China

Dr. Tang Lonji
Human Computing Center
Changsha, 410012
China

Mr. S.L. Lopatnikov
Vniigeoinformsystem
Varshavskoe Shosse 8
Moscow, 113105
Russian Federation

Mr. Guochun Lu
Division of Petroleum Engineering and
   Applied Geophysics
Norwegian Institute of Technology
University of Trondheim
N-7034 Trondheim
Norway

Mr. Jian-yu Lu
Biodynamics Research Unit
Department of Physiology and Biophysics
Mayo Clinic/Foundation
Rochester, MN 55905

Dr. Jiaqi Lu
Dept. of Mathematics
Harbin Institute of Technology
P.R. China

Dr. Zhen-Qiu Lu
Department of Physics
Nankai University
Tianjin 300071
P.R. China

Mr. Konstantin Maslov
Inst. of Chemical Physics
Russian Academy of Sciences
Kosygin Str. 4
Moscow, 117334
Russian Federation

Prof. U. Moser
Institute of Biomedical Engineering and
  Medical Informatics
University and ETH Zurich
Moussonstsr. 18
CH-8044 Zurich

Prof. K. Nagai
University of Tsukuba
Tsukuba Ibaraki 305
Japan

Mr. J. Nakayama
Department of Electronics and
  Information Science
Kyoto Institute of Technology
Matsugasaki, Kyoto 606
Japan

Mr. U. Netzelmann
Frauhofer-Institute for
  Nondestructive Testing
D-6600 Saarbrucken
Fed. Rep. of Germany

Mr. William D. O'Brien
University of Illinois
Urbana, IL

Mr. Jong Cheul Park
Dept. of Electrical Engineering
Korea Advanced Institute of
  Science and Technology
P.O. Box 150
Chongyangni, Seoul
Korea

Dr. Chengbin Peng
Earth Resources Laboratory
Department of Earth
Atmospheric and Planetary Sciences
Massachusetts Institute of Technology
Cambridge, MA 02139

Mr. Riccardo Pini
Institute of Gerontology and
  Geriatronics
University of Florence
Florence, Italy

Mr. Stanislaw J. Pogorzelski
Environmental Acoustics Laboratory
University of Gdansk
Wita Stwosza 57, 80-952 Gdansk
Poland

Dr. M. Pollakowski
Ruhr University
Bechum Institut fur
  Hochrequenz-technik University
150, D-4630 Bochum
Germany

Dr. Donghai Qiao
Inst. of Acoustics
Academia Sinica
Beijing 10008
P.R. China

Dr. Zhen-Qiulu
Department of Physics
Nankai University
Tianjin, 300071
P.R. China

Dr. Liu Quangyin
Jiangsu Cancer Research Institute
P.R. China

Dr. N. Rajpal
C.A.R.E.
IIT Delhi
New Delhi-11016
India

Mr. M. Ravez
Lab. d'Opto-Acousto-Electronique
URA 832 CNRS ENSIMEV
Univ. De Valenciennes
BP311, 59304 Valenciennes Cedex
France

Prof. A.P. Sarvazyan
Institute of Theoretical and
  Experimental Biophysics
Acad. Sci.
Pushchino 142292
Russia

Dr. S. Sathish
Analytical Services and Materials Inc.
107 Research Drive
Hampton, VA 23666

Mr. Takuso Sato
The Graduate School at Nagatsuta
Tokyo Institute of Technology
4259 Nagatsuta, Midori-ku
Yokohama-shi
Japan

Dr. Chen Shubao
Second Medical University
Shanghai 200092
P.R. China

Dr. Chen Siping
Analogic Scientific Incorporation
Shekou, ShenZhen
P.R. China

Mr. K. Soetanto
Biomedical Engineering
Drexel University
Philadelphia, PA 19104

Dr. Yao Sun
Department of Automatic Control
Harbin Shipbuilding
Engineering Institute
Harbin, China

Dr. R.J. Sztukiewics
Institute of Civil Engineering
Technical University
Poznan, Poland

Mr. E. Ya. Tagunov
Department of Acoustics
  Faculty of Physics
Moscow M.V. Lomonosov State University
Moscow 119899
Russian Federation

Mr. Yasutaka Tamura
Faculty of Engineering
Yamagata University
Jonan 4-3-16 Yonezawa 992
Japan

Dr. Tian Tan
Dept. of Underwater Acoustics
Harbin Shipbuilding Engineering Institute
Harbin 150001, Heilongjiang
P.R. China

Mr. J.M. Thijssen
Biophysics Laboratory
Department of Opthalmology
University Hospital
6500 HB Nijmegen
The Netherlands

Dr. T. Tommasi
Department of Biophysical and
  Electronic Engineering
University of Genova
Via Opera Pia 11a, 16145
Genova, Italy

Dr. P. Tortoli
Electronic Engineering Department
University of Florence
Florence, Italy

Prof. M. Toubal
I.E.M.N. Department of O.A.E.
U.M.R.-C.N.R.S.-N 9929
Universite de Valenciennes
Le Mont Houy-BP 311
59304 Valenciennes Cedex
France

Dr. Madvallev U.
Physical Technical Institute of
  Tajik Academy of Sciences
734063, Dushambe
Ainy Str. 299/1
Russian Federation

Mr. M. Urchulutegui
Departamento Fisica de Materiales
Facultad Ciencias Fisicas
Universidad Complutense
28040 Madrid, Spain

Prof. G. Wade
Dept. of Electrical and
  Computer Engineering
University of California
Santa Barbara, CA 93106

Dr. X.L. Wang
Wuhan Institute of Physics
The Chinese Academy of Sciences
Wuhan, 430071
P.R. China

Dr. Zhao-Qiu Wang
Lab. of Photoacoustic Sciences
Institute of Acoustics
Nanjing University
Nanjing 210008
China

Dr. Sumio Watanable
Research and Development Center
RICON Co., Ltd
16-1, Shin-ei-cho, Kohoku-ku
Yokohama, 223
Japan

Dr. Zhu Weiqing
Institute of Acoustics
Academia Sinica
Beijing, 100080
P.R. China

Prof. Yang Wenoai
Beijing Computer Center of Geoloy,
  100083
P.R. China

Dr. Xia Xianhua
Wuhan Institute of Physics
The Chinese Academy of Sciences
Wuhan, Hubei, 430071
P.R. China

Dr. Huan Xiu-Chang
81st Hospital
Nanjing, China

Dr. Katsuyuki Yamamoto
Division of Biomedical Engineering
Hokkaido University
Sapporo, Japan

Dr. Nuan-Min Yang
Department of Physics
Nanjing University
Nanjing, 210008
P.R. China

Prof. Jin Yi
Member of Chinese Society of
  Biomedical Engineering
Analogic Scientific Inc.
She Kou, Shen Zhen 518067
China

Dr. Liu Yinbin
Institute of Geophysics
Chinese Academy of Sciences
Beijing, 100101
P.R. China

Mr. T. Zhai
Dept. of Materials
University of Oxford
Parks Road, Oxford OX1 3PH
United Kingdom

Mr. D. Zhang
National Microelectronics Research Center
Lee Maltings
Prospect Row
Cork, Ireland

Dr. Mei Zhang
Department of Cardiology
Affiliated Hospital
Shandong Medical University
Jinan, Shandong
P.R. China

Mr. Dong Zheng
Institute of Acoustics
Nanjing University
Nanjing 210008
P.R. China

Dr. Jiamao Zheng
Southeast University
210018 Nanjing
P.R. China

Dr. Zhao Zheying
Inst. of Acoustics
Academia Sinica
P.O. Box 2712
Beijing 10008
P.R. China

Dr. Kangyuan Zhou
Univ. of Science and
  Technology of China
230026 Hefei, Anhwi
China

Dr. Xiaojun Zhou
Dept. of Mach. Eng.
Zhejiang University, Hangzhou
Zhejiang 310027
P.R. China

Prof. Jerzy K. Zieniuk
Inst. of Fundamental Technology Research
Polish Academy of Sciences
00-049 Warszawa
Poland

Dr. Xian-Hua Zou
Peking Union Medical College Hospital
Chinese Academy of Medical Sciences
Beijing, China

INDEX

MIX
Papier aus verantwortungsvollen Quellen
Paper from responsible sources
FSC® C105338

If you have any concerns about our products,
you can contact us on
ProductSafety@springernature.com

In case Publisher is established outside the EU,
the EU authorized representative is:
Springer Nature Customer Service Center GmbH
Europaplatz 3, 69115 Heidelberg, Germany

Printed by Libri Plureos GmbH
in Hamburg, Germany